VOLUME FIVE HUNDRED AND THIRTY ONE

METHODS IN ENZYMOLOGY

Microbial Metagenomics, Metatranscriptomics, and Metaproteomics

METHODS IN ENZYMOLOGY

Editors-in-Chief

JOHN N. ABELSON and MELVIN I. SIMON
Division of Biology
California Institute of Technology
Pasadena, California

Founding Editors

SIDNEY P. COLOWICK and NATHAN O. KAPLAN

VOLUME FIVE HUNDRED AND THIRTY ONE

METHODS IN ENZYMOLOGY

Microbial Metagenomics, Metatranscriptomics, and Metaproteomics

Edited by

EDWARD F. DeLONG

Morton and Claire Goulder Family
Professor in Environmental Systems
Massachusetts Institute of Technology
Dept. of Civil and Environmental Engineering
Dept. of Biological Engineering
Cambridge, MA, USA
and
Center for Microbial Ecology: Research and Education
University of Hawaii, Manoa Honolulu, HI

AMSTERDAM • BOSTON • HEIDELBERG • LONDON
NEW YORK • OXFORD • PARIS • SAN DIEGO
SAN FRANCISCO • SINGAPORE • SYDNEY • TOKYO
Academic Press is an imprint of Elsevier

Academic Press is an imprint of Elsevier
525 B Street, Suite 1800, San Diego, CA 92101-4495, USA
225 Wyman Street, Waltham, MA 02451, USA
Radarweg 29, PO Box 211, 1000 AE Amsterdam, The Netherlands
The Boulevard, Langford Lane, Kidlington, Oxford, OX5 1GB, UK
32 Jamestown Road, London NW1 7BY, UK

First edition 2013

For information on all Academic Press publications
visit our website at store.elsevier.com

ISBN: 978-0-12-407863-5
ISSN: 0076-6879

Printed and bound in United States of America
13 14 15 16 11 10 9 8 7 6 5 4 3 2 1

CONTENTS

Contributors *xiii*
Preface *xxi*
Volumes in Series *xxiii*

Section I
Preparation and Genomic Analyses of Single Cells from Natural Populations

1. In-Solution Fluorescence *In Situ* Hybridization and Fluorescence-Activated Cell Sorting for Single Cell and Population Genome Recovery **3**

Mohamed F. Haroon, Connor T. Skennerton, Jason A. Steen, Nancy Lachner, Philip Hugenholtz, and Gene W. Tyson

1. Introduction 4
2. Technical Protocols and Considerations 6
3. Case Study: Targeted Isolation of Anaerobic Methanotrophic Archaea for Single Cell and Population Genomics 15
4. Summary 17

Acknowledgments 17
References 18

2. Whole Cell Immunomagnetic Enrichment of Environmental Microbial Consortia Using rRNA-Targeted Magneto-FISH **21**

Elizabeth Trembath-Reichert, Abigail Green-Saxena, and Victoria J. Orphan

1. Introduction 22
2. Methods 24
3. Results and Discussion 33
4. Summary 41

Acknowledgments 41
References 42

3. Coupling FACS and Genomic Methods for the Characterization of Uncultivated Symbionts **45**

Anne Thompson, Shellie Bench, Brandon Carter, and Jonathan Zehr

1. Introduction 46
2. Fluorescence-Activated Cell Sorting 48

3. Metagenomic Sequencing of Uncultured Symbiont Populations 51
4. Determining Host Identity and Metabolism 55
References 58

4. Optofluidic Cell Selection from Complex Microbial Communities for Single-Genome Analysis **61**

Zachary C. Landry, Stephen J. Giovanonni, Stephen R. Quake, and Paul C. Blainey

1. Introduction 62
2. Optical Tweezing for Cell Isolation 64
3. Optical Hardware Setup 70
4. Microfluidics Configuration and Setup 78
5. Procedure for Sorting and Amplifying Single Cells 80
6. Software 83
7. Performance and Limitations 85
Acknowledgments 88
References 88

5. Quantifying and Identifying the Active and Damaged Subsets of Indigenous Microbial Communities **91**

Corinne Ferrier Maurice and Peter James Turnbaugh

1. Introduction 92
2. Flow Cytometry and Fluorescent Dyes 93
3. Experimental Procedures 95
4. Experimental Validation 99
5. Troubleshooting 103
Acknowledgment 106
References 106

Section II
Microbial Community Genomics—Sampling, Metagenomic Library Preparation, and Analysis

6. Preparation of BAC Libraries from Marine Microbial Populations **111**

Gazalah Sabehi and Oded Béjà

1. Introduction 111
2. BAC Library Construction Procedures 112
References 121

7. Preparation of Fosmid Libraries and Functional Metagenomic Analysis of Microbial Community DNA **123**

Asunción Martínez and Marcia S. Osburne

1. Introduction 124
2. General Considerations: Genomic Library Vector and Host 125
3. Construction of Fosmid Environmental Libraries 128
4. General Considerations for Working with Fosmids 131
5. Functional Screens 133
6. Conclusions 139

References 139

8. Preparation of Metagenomic Libraries from Naturally Occurring Marine Viruses **143**

Sergei A. Solonenko and Matthew B. Sullivan

1. On the Importance of Environmental Viruses and Viral Metagenomics 144
2. The DNA Viral Metagenomic Sample-to-Sequence Pipeline 147
3. The Library Preparation Process 149
4. Conclusions 159

Acknowledgments 160

References 160

Section III

Microbial Community Transcriptomics—Sample Preparation, cDNA Library Construction, and Analysis

9. Preparation and Metatranscriptomic Analyses of Host–Microbe Systems **169**

Jarrad T. Hampton-Marcell, Stephanie M. Moormann, Sarah M. Owens, and Jack A. Gilbert

1. Introduction 170
2. Materials 171
3. Components 172
4. Experiments 175
5. Sequencing and Data Analysis 184

Acknowledgments 185

References 185

10. Preparation of Microbial Community cDNA for Metatranscriptomic Analysis in Marine Plankton **187**

Frank J. Stewart

1. Introduction 188

2. Sample Collection and RNA Preservation 192
3. RNA Extraction and Standardization 198
4. rRNA Depletion 201
5. RNA Amplification and cDNA Synthesis 209
6. Summary 212
Acknowledgments 213
References 213

11. Sequential Isolation of Metabolites, RNA, DNA, and Proteins from the Same Unique Sample 219

Hugo Roume, Anna Heintz-Buschart, Emilie EL Muller, and Paul Wilmes

1. Introduction 220
2. Sampling and Sample Preprocessing 222
3. Sample Processing and Small Molecule Extraction 226
4. Sequential Biomacromolecular Isolation 228
5. Quality Control 230
6. Outlook 235
References 235

12. Use of Internal Standards for Quantitative Metatranscriptome and Metagenome Analysis 237

Brandon M. Satinsky, Scott M. Gifford, Byron C. Crump, and Mary Ann Moran

1. Introduction 238
2. Method Overview 241
3. DNA Template and Vector Design for Internal RNA Standards 242
4. mRNA Standard Preparation 244
5. DNA Standard Preparation 247
6. Internal Standard Addition 247
7. Internal Standard Recoveries and Quantification 248
8. Dataset Normalization Using Internal Standards 248
Acknowledgments 249
References 250

13. Sample Processing and cDNA Preparation for Microbial Metatranscriptomics in Complex Soil Communities 251

Lilia C. Carvalhais and Peer M. Schenk

1. Introduction 252
2. Working with RNA 253
3. Sampling 254

4. RNA Extraction of Complex Soil Communities 255
5. Assessing RNA Quality and Quantity 256
6. Soil Samples with Low Microbial Biomass 258
7. Removing Inhibitors 258
8. rRNA Subtraction/mRNA Enrichment 259
9. mRNA Amplification 261
10. cDNA Synthesis 262
11. Current State and Future Prospects 263
References 263

Section IV
Microbial Community Proteomics—Sampling, Sample Preparation, Spectral Analysis, and Interpretation

14. Sample Preparation and Processing for Planktonic Microbial Community Proteomics 271

Robert M. Morris and Brook L. Nunn

1. Introduction 272
2. Sample Collection 273
3. Sample Preparation 278
4. Extracting and Digesting Proteins 281
5. Mass Spectrometry 283
References 286

15. Sample Handling and Mass Spectrometry for Microbial Metaproteomic Analyses 289

Ryan S. Mueller and Chongle Pan

1. Introduction 290
2. Experimental Considerations and Options 291
3. Generalized Procedure 295
Acknowledgment 299
References 300

16. Molecular Tools for Investigating Microbial Community Structure and Function in Oxygen-Deficient Marine Waters 305

Alyse K. Hawley, Sam Kheirandish, Andreas Mueller, Hilary T.C. Leung, Angela D. Norbeck, Heather M. Brewer, Ljiljana Pasa-Tolic, and Steven J. Hallam

1. Introduction 306
2. Exploring Oxygen Minimum Zone Community Structure 307

3. Detecting Oxygen Minimum Zone Proteins 314
Acknowledgments 325
References 325

Section V
Microbial Community Informatics: Computational Platforms, Comparative Analyses, and Statistics

17. Assembling Full-Length rRNA Genes from Short-Read Metagenomic Sequence Datasets Using EMIRGE **333**

Christopher S. Miller

1. Introduction: The Utilitarian Small Subunit rRNA Gene in Microbial Ecology 334
2. Overview of the EMIRGE algorithm 337
3. EMIRGE Outputs: Reconstructed Sequences with Estimated Abundances 342
4. Practical Considerations When Running EMIRGE 343
5. Choosing a Candidate Database of rRNA Genes 347
6. Conclusions and Future Outlook 348

References 348

18. Computational Methods for High-Throughput Comparative Analyses of Natural Microbial Communities **353**

Sarah P. Preheim, Allison R. Perrotta, Jonathan Friedman, Chris Smilie, Ilana Brito, Mark B. Smith, and Eric Alm

1. Introduction 354
2. Sequencing Terminology 355
3. Sequence Processing 356
4. Community Analysis 362
5. Summary 368

References 368

19. Advancing Our Understanding of the Human Microbiome Using QIIME **371**

José A. Navas-Molina, Juan M. Peralta-Sánchez, Antonio González, Paul J. McMurdie, Yoshiki Vázquez-Baeza, Zhenjiang Xu, Luke K. Ursell, Christian Lauber, Hongwei Zhou, Se Jin Song, James Huntley, Gail L. Ackermann, Donna Berg-Lyons, Susan Holmes, J. Gregory Caporaso, and Rob Knight

1. Introduction 372
2. QIIME as Integrated Pipeline of Third-Party Tools 373

3. PCR and Sequencing on Illumina MiSeq 375
4. QIIME Workflow for Conducting Microbial Community Analysis 377
5. Other Features 427
6. Recommendations 438
7. Conclusions 438
Acknowledgments 439
References 439

20. Disentangling Associated Genomes **445**

Daniel B. Sloan, Gordon M. Bennett, Philipp Engel, David Williams, and Howard Ochman

1. Introduction 446
2. Sequencing 446
3. Data Analysis and Genome Assembly 451
4. Postassembly Analysis 457
5. Summary 460
Acknowledgments 461
References 461

21. Microbial Community Analysis Using MEGAN **465**

Daniel H. Huson and Nico Weber

1. Introduction 465
2. Taxonomic Analysis 468
3. Functional Analysis 470
4. Sequence Alignment 474
5. Comparison of Samples 477
6. Conclusion 483
References 484

22. A Metagenomics Portal for a Democratized Sequencing World **487**

Andreas Wilke, Elizabeth M. Glass, Daniela Bartels, Jared Bischof, Daniel Braithwaite, Mark D'Souza, Wolfgang Gerlach, Travis Harrison, Kevin Keegan, Hunter Matthews, Renzo Kottmann, Tobias Paczian, Wei Tang, William L. Trimble, Pelin Yilmaz, Jared Wilkening, Narayan Desai, and Folker Meyer

1. Introduction 488
2. Pipeline and Technology Platform 489
3. Web Interface 495
4. How to Drill Down Using the Workbench 510

5. MG-RAST Downloads 515
6. Discussion 519
7. Future Work 521
Acknowledgments 521
References 521

23. A User's Guide to Quantitative and Comparative Analysis of Metagenomic Datasets **525**

Chengwei Luo, Luis M. Rodriguez-R, and Konstantinos T. Konstantinidis

1. Introduction 526
2. How to Assemble a Metagenomic Dataset 528
3. How to Determine the Fraction of the Community Captured in a Metagenome 533
4. How to Identify the Taxonomic Identity of a Metagenomic Sequence 535
5. How to Determine Differentially Abundant Genes, Pathways, and Species 538
6. Limitations and Perspectives for the Future 540
Acknowledgments 542
References 542

Author Index *549*
Subject Index *581*

CONTRIBUTORS

Gail L. Ackermann
Biofrontiers Institute, University of Colorado, Boulder, Colorado, USA

Eric Alm
Biological Engineering, Massachusetts Institute of Technology, Cambridge, Massachusetts, USA

Mary Ann Moran
Department of Marine Sciences, University of Georgia, Athens, Georgia, USA

Oded Béjà
Faculty of Biology, Technion—Israel Institute of Technology, Haifa, Israel

Daniela Bartels
Argonne National Laboratory, Lemont, and University of Chicago, Chicago, Illinois, USA

Shellie Bench
Stanford University, Stanford, California, USA

Gordon M. Bennett
Department of Ecology and Evolutionary Biology, Yale University, New Haven, Connecticut, USA

Donna Berg-Lyons
Biofrontiers Institute, University of Colorado, Boulder, Colorado, USA

Jared Bischof
Argonne National Laboratory, Lemont, and University of Chicago, Chicago, Illinois, USA

Paul C. Blainey
Department of Biological Engineering, Broad Institute and Massachusetts Institute of Technology, Cambridge, Massachusetts, USA

Daniel Braithwaite
Argonne National Laboratory, Lemont, and University of Chicago, Chicago, Illinois, USA

Heather M. Brewer
Environmental and Molecular Sciences Laboratory at PNNL, Richland, Washington, USA

Ilana Brito
Biological Engineering, Massachusetts Institute of Technology, Cambridge, Massachusetts, USA

J. Gregory Caporaso
Department of Biological Sciences, North Arizona University, Flagstaff, Arizona, and Institute for Genomics and Systems Biology, Argonne National Laboratory, Argonne, Illinois, USA

Brandon Carter
University of California, Santa Cruz, Santa Cruz, California, USA

Lilia C. Carvalhais
School of Agriculture and Food Sciences, The University of Queensland, Brisbane, Queensland, Australia

Byron C. Crump
College of Earth, Ocean and Atmospheric Sciences, Oregon State University, Corvallis, Oregon, USA

Narayan Desai
Argonne National Laboratory, Lemont, and University of Chicago, Chicago, Illinois, USA

Mark D'Souza
Argonne National Laboratory, Lemont, and University of Chicago, Chicago, Illinois, USA

Philipp Engel
Department of Ecology and Evolutionary Biology, Yale University, New Haven, Connecticut, USA

Corinne Ferrier Maurice
FAS Center for Systems Biology, Harvard University, Cambridge, Massachusetts, USA

Jonathan Friedman
Physics, and Computational and Systems Biology, Massachusetts Institute of Technology, Cambridge, Massachusetts, USA

Wolfgang Gerlach
Argonne National Laboratory, Lemont, and University of Chicago, Chicago, Illinois, USA

Scott M. Gifford
Department of Civil and Environmental Engineering, Massachusetts Institute of Technology, Cambridge, Massachusetts, USA

Jack A. Gilbert
Argonne National Laboratory, Argonne, and Department of Ecology and Evolution, University of Chicago, Chicago, Illinois, USA

Stephen J. Giovanonni
Department of Microbiology, Oregon State University, Corvallis, Oregon, USA

Elizabeth M. Glass
Argonne National Laboratory, Lemont, and University of Chicago, Chicago, Illinois, USA

Antonio González
Biofrontiers Institute, University of Colorado, Boulder, Colorado, USA

Abigail Green-Saxena
Division of Biological Sciences, California Institute of Technology, Pasadena, California, USA

Steven J. Hallam
Department of Microbiology & Immunology, and Graduate Program in Bioinformatics, University of British Columbia, Vancouver, British Columbia, Canada

Jarrad T. Hampton-Marcell
Argonne National Laboratory, Argonne, and Department of Ecology and Evolution, University of Chicago, Chicago, Illinois, USA

Mohamed F. Haroon
Australian Centre for Ecogenomics, School of Chemistry and Molecular Biosciences, The University of Queensland, St. Lucia, Queensland, Australia

Travis Harrison
Argonne National Laboratory, Lemont, and University of Chicago, Chicago, Illinois, USA

Alyse K. Hawley
Department of Microbiology & Immunology, University of British Columbia, Vancouver, British Columbia, Canada

Anna Heintz-Buschart
Luxembourg Centre for Systems Biomedicine, University of Luxembourg, Esch-sur-Alzette, Luxembourg

Susan Holmes
Department of Statistics, Stanford University, Stanford, California, USA

Philip Hugenholtz
Australian Centre for Ecogenomics, School of Chemistry and Molecular Biosciences, and Institute for Molecular Bioscience, The University of Queensland, St. Lucia, Queensland, Australia

James Huntley
Biofrontiers Institute, University of Colorado, Boulder, Colorado, USA

Daniel H. Huson
Center for Bioinformatics, University of Tübingen, Tübingen, Germany

Kevin Keegan
Argonne National Laboratory, Lemont, and University of Chicago, Chicago, Illinois, USA

Sam Kheirandish
Department of Microbiology & Immunology, University of British Columbia, Vancouver, British Columbia, Canada

Rob Knight
Biofrontiers Institute, and Howard Hughes Medical Institute, University of Colorado, Boulder, Colorado, USA

Konstantinos T. Konstantinidis
Center for Bioinformatics and Computational Genomics; School of Biology, and School of Civil and Environmental Engineering, Georgia Institute of Technology, Atlanta, Georgia, USA

Renzo Kottmann
Max-Planck Institute für Marine Biologie, Bremen, Germany

Nancy Lachner
Australian Centre for Ecogenomics, School of Chemistry and Molecular Biosciences, The University of Queensland, St. Lucia, Queensland, Australia

Zachary C. Landry
Department of Microbiology, Oregon State University, Corvallis, Oregon, USA

Christian Lauber
Cooperative Institute for Research in Environmental Sciences, Boulder, Colorado, USA

Hilary T.C. Leung
Department of Microbiology & Immunology, University of British Columbia, Vancouver, British Columbia, Canada

Chengwei Luo
Center for Bioinformatics and Computational Genomics; School of Biology, and School of Civil and Environmental Engineering, Georgia Institute of Technology, Atlanta, Georgia, USA

Asunción Martínez
Department of Civil and Environmental Engineering, Massachusetts Institute of Technology, Cambridge, Massachusetts, USA

Hunter Matthews
Argonne National Laboratory, Lemont, Illinois, USA

Paul J. McMurdie
Department of Statistics, Stanford University, Stanford, California, USA

Folker Meyer
Argonne National Laboratory, Lemont, and University of Chicago, Chicago, Illinois, USA

Christopher S. Miller
Department of Integrative Biology, University of Colorado Denver, Campus Box 171, P.O. Box 173364, Denver, Colorado, USA

Stephanie M. Moormann
Argonne National Laboratory, Argonne, Illinois, USA

Robert M. Morris
School of Oceanography, University of Washington, Seattle, Washington, USA

Andreas Mueller
Department of Microbiology & Immunology, University of British Columbia, Vancouver, British Columbia, Canada

Ryan S. Mueller
Department of Microbiology, Oregon State University, Corvallis, Oregon, USA

Emilie EL Muller
Luxembourg Centre for Systems Biomedicine, University of Luxembourg, Esch-sur-Alzette, Luxembourg

José A. Navas-Molina
Department of Computer Science, University of Colorado, Boulder, Colorado, USA

Angela D. Norbeck
Pacific Northwest National Laboratory, Richland, Washington, USA

Brook L. Nunn
Department of Genome Sciences, and Department of Medicinal Chemistry, University of Washington, Seattle, Washington, USA

Howard Ochman
Department of Ecology and Evolutionary Biology, Yale University, New Haven, Connecticut, USA

Victoria J. Orphan
Division of Geological and Planetary Sciences, California Institute of Technology, Pasadena, California, USA

Marcia S. Osburne
Department of Molecular Biology and Microbiology, Tufts University School of Medicine, Boston, Massachusetts, USA

Sarah M. Owens
Argonne National Laboratory, Argonne, and Computation Institute, University of Chicago, Chicago, Illinois, USA

Tobias Paczian
Argonne National Laboratory, Lemont, and University of Chicago, Chicago, Illinois, USA

Chongle Pan
Oak Ridge National Laboratory, Oak Ridge, Tennessee, USA

Ljiljana Pasa-Tolic
Environmental and Molecular Sciences Laboratory at PNNL, Richland, Washington, USA

Juan M. Peralta-Sánchez
Biofrontiers Institute, University of Colorado, Boulder, Colorado, USA

Allison R. Perrotta
Civil and Environmental Engineering, Massachusetts Institute of Technology, Cambridge, Massachusetts, USA

Sarah P. Preheim
Biological Engineering, Massachusetts Institute of Technology, Cambridge, Massachusetts, USA

Stephen R. Quake
Department of Bioengineering, and Department of Applied Physics, Stanford University, Stanford, California, USA

Luis M. Rodriguez-R
Center for Bioinformatics and Computational Genomics, and School of Biology, Georgia Institute of Technology, Atlanta, Georgia, USA

Hugo Roume
Luxembourg Centre for Systems Biomedicine, University of Luxembourg, Esch-sur-Alzette, Luxembourg

Gazalah Sabehi
Faculty of Biology, Technion—Israel Institute of Technology, Haifa, Israel

Brandon M. Satinsky
Department of Microbiology, University of Georgia, Athens, Georgia, USA

Peer M. Schenk
School of Agriculture and Food Sciences, The University of Queensland, Brisbane, Queensland, Australia

Connor T. Skennerton
Australian Centre for Ecogenomics, School of Chemistry and Molecular Biosciences, and Advanced Water Management Centre, The University of Queensland, St. Lucia, Queensland, Australia

Daniel B. Sloan
Department of Ecology and Evolutionary Biology, Yale University, New Haven, Connecticut, USA

Chris Smilie
Computational and Systems Biology, Massachusetts Institute of Technology, Cambridge, Massachusetts, USA

Mark B. Smith
Microbiology, Massachusetts Institute of Technology, Cambridge, Massachusetts, USA

Sergei A. Solonenko
Department of Ecology and Evolutionary Biology, University of Arizona, Tucson, Arizona, USA

Se Jin Song
Department of Ecology and Evolutionary Biology, University of Colorado, Boulder, Colorado, USA

Jason A. Steen
Australian Centre for Ecogenomics, School of Chemistry and Molecular Biosciences, The University of Queensland, St. Lucia, Queensland, Australia

Frank J. Stewart
School of Biology, Georgia Institute of Technology, Atlanta, Georgia, USA

Matthew B. Sullivan
Department of Ecology and Evolutionary Biology, and Department of Molecular and Cellular Biology, University of Arizona, Tucson, Arizona, USA

Wei Tang
Argonne National Laboratory, Lemont, and University of Chicago, Chicago, Illinois, USA

Anne Thompson
University of California, Santa Cruz, Santa Cruz, California, USA

Elizabeth Trembath-Reichert
Division of Geological and Planetary Sciences, California Institute of Technology, Pasadena, California, USA

William L. Trimble
Argonne National Laboratory, Lemont, and University of Chicago, Chicago, Illinois, USA

Peter James Turnbaugh
FAS Center for Systems Biology, Harvard University, Cambridge, Massachusetts, USA

Gene W. Tyson
Australian Centre for Ecogenomics, School of Chemistry and Molecular Biosciences, and Advanced Water Management Centre, The University of Queensland, St. Lucia, Queensland, Australia

Luke K. Ursell
Biofrontiers Institute, University of Colorado, Boulder, Colorado, USA

Yoshiki Vázquez-Baeza
Biofrontiers Institute, University of Colorado, Boulder, Colorado, USA

Nico Weber
Center for Bioinformatics, University of Tübingen, Tübingen, Germany

Andreas Wilke
Argonne National Laboratory, Lemont, and University of Chicago, Chicago, Illinois, USA

Jared Wilkening
Argonne National Laboratory, Lemont, and University of Chicago, Chicago, Illinois, USA

David Williams
Department of Ecology and Evolutionary Biology, Yale University, New Haven, Connecticut, USA

Paul Wilmes
Luxembourg Centre for Systems Biomedicine, University of Luxembourg, Esch-sur-Alzette, Luxembourg

Zhenjiang Xu
Biofrontiers Institute, University of Colorado, Boulder, Colorado, USA

Pelin Yilmaz
Max-Planck Institute für Marine Biologie, Bremen, Germany

Jonathan Zehr
University of California, Santa Cruz, Santa Cruz, California, USA

Hongwei Zhou
Department of Environmental Health, School of Public Health and Tropical Medicine, Southern Medical University, Guangzhou, China

PREFACE

Since its inception, microbial science has frequently experienced Kuhnian leaps forward in understanding, often precipitated by technical and methodological advances. Leeuwenhoek's simple spherical lenses, the electron microscope, pure culture technique, and recombinant DNA technology all exemplify such paradigm-shifting tools and methods that have transformed our understandings of the microbial world. Likewise, over the past 20 years, cultivation-independent DNA sequence-based microbial surveys have transformed our understanding of the natural microbial world in remarkable ways, from the deep sea to the human microbiome. New perspectives of microbial community structure, new gene families, patterns of gene distributions in the environment, and the nature of host–symbiont interactions and evolution have all been revealed by applying new "omics" approaches to microbial communities. Most recently, single-cell characterization, high-throughput "next-generation" sequencing methods, and new bioinformatic approaches are all advancing our understanding of microbial ecology in a variety of contexts.

This volume of *Methods in Enzymology* attempts to encapsulate many late-breaking and transformative methods and approaches that are now accelerating advances in microbial biology and ecology. The chapters are loosely organized into five sections: single-cell population genomics, microbial community genomics, microbial community transcriptomics, microbial community proteomics, and microbial community informatics. Each chapter encapsulates a specific developing area and enabling technology, under the general framework of microbial community "omics."

The authors have well captured the technical details, theory, application, and excitement of these newly developing disciplines. I am deeply appreciative to all the authors for their outstanding efforts and to the Editors in Chief as well for the push to initiate this new volume. We all hope that the information and summaries provided will support and enable new research and discovery, for workers new to the field as well as seasoned veterans.

Edward F. DeLong

METHODS IN ENZYMOLOGY

VOLUME I. Preparation and Assay of Enzymes
Edited by SIDNEY P. COLOWICK AND NATHAN O. KAPLAN

VOLUME II. Preparation and Assay of Enzymes
Edited by SIDNEY P. COLOWICK AND NATHAN O. KAPLAN

VOLUME III. Preparation and Assay of Substrates
Edited by SIDNEY P. COLOWICK AND NATHAN O. KAPLAN

VOLUME IV. Special Techniques for the Enzymologist
Edited by SIDNEY P. COLOWICK AND NATHAN O. KAPLAN

VOLUME V. Preparation and Assay of Enzymes
Edited by SIDNEY P. COLOWICK AND NATHAN O. KAPLAN

VOLUME VI. Preparation and Assay of Enzymes (*Continued*)
Preparation and Assay of Substrates
Special Techniques
Edited by SIDNEY P. COLOWICK AND NATHAN O. KAPLAN

VOLUME VII. Cumulative Subject Index
Edited by SIDNEY P. COLOWICK AND NATHAN O. KAPLAN

VOLUME VIII. Complex Carbohydrates
Edited by ELIZABETH F. NEUFELD AND VICTOR GINSBURG

VOLUME IX. Carbohydrate Metabolism
Edited by WILLIS A. WOOD

VOLUME X. Oxidation and Phosphorylation
Edited by RONALD W. ESTABROOK AND MAYNARD E. PULLMAN

VOLUME XI. Enzyme Structure
Edited by C. H. W. HIRS

VOLUME XII. Nucleic Acids (Parts A and B)
Edited by LAWRENCE GROSSMAN AND KIVIE MOLDAVE

VOLUME XIII. Citric Acid Cycle
Edited by J. M. LOWENSTEIN

VOLUME XIV. Lipids
Edited by J. M. LOWENSTEIN

Volume XV. Steroids and Terpenoids
Edited by Raymond B. Clayton

Volume XVI. Fast Reactions
Edited by Kenneth Kustin

Volume XVII. Metabolism of Amino Acids and Amines (Parts A and B)
Edited by Herbert Tabor and Celia White Tabor

Volume XVIII. Vitamins and Coenzymes (Parts A, B, and C)
Edited by Donald B. McCormick and Lemuel D. Wright

Volume XIX. Proteolytic Enzymes
Edited by Gertrude E. Perlmann and Laszlo Lorand

Volume XX. Nucleic Acids and Protein Synthesis (Part C)
Edited by Kivie Moldave and Lawrence Grossman

Volume XXI. Nucleic Acids (Part D)
Edited by Lawrence Grossman and Kivie Moldave

Volume XXII. Enzyme Purification and Related Techniques
Edited by William B. Jakoby

Volume XXIII. Photosynthesis (Part A)
Edited by Anthony San Pietro

Volume XXIV. Photosynthesis and Nitrogen Fixation (Part B)
Edited by Anthony San Pietro

Volume XXV. Enzyme Structure (Part B)
Edited by C. H. W. Hirs and Serge N. Timasheff

Volume XXVI. Enzyme Structure (Part C)
Edited by C. H. W. Hirs and Serge N. Timasheff

Volume XXVII. Enzyme Structure (Part D)
Edited by C. H. W. Hirs and Serge N. Timasheff

Volume XXVIII. Complex Carbohydrates (Part B)
Edited by Victor Ginsburg

Volume XXIX. Nucleic Acids and Protein Synthesis (Part E)
Edited by Lawrence Grossman and Kivie Moldave

Volume XXX. Nucleic Acids and Protein Synthesis (Part F)
Edited by Kivie Moldave and Lawrence Grossman

Volume XXXI. Biomembranes (Part A)
Edited by Sidney Fleischer and Lester Packer

Volume XXXII. Biomembranes (Part B)
Edited by Sidney Fleischer and Lester Packer

Volume XXXIII. Cumulative Subject Index Volumes I-XXX
Edited by Martha G. Dennis and Edward A. Dennis

Volume XXXIV. Affinity Techniques (Enzyme Purification: Part B)
Edited by William B. Jakoby and Meir Wilchek

Volume XXXV. Lipids (Part B)
Edited by John M. Lowenstein

Volume XXXVI. Hormone Action (Part A: Steroid Hormones)
Edited by Bert W. O'Malley and Joel G. Hardman

Volume XXXVII. Hormone Action (Part B: Peptide Hormones)
Edited by Bert W. O'Malley and Joel G. Hardman

Volume XXXVIII. Hormone Action (Part C: Cyclic Nucleotides)
Edited by Joel G. Hardman and Bert W. O'Malley

Volume XXXIX. Hormone Action (Part D: Isolated Cells, Tissues, and Organ Systems)
Edited by Joel G. Hardman and Bert W. O'Malley

Volume XL. Hormone Action (Part E: Nuclear Structure and Function)
Edited by Bert W. O'Malley and Joel G. Hardman

Volume XLI. Carbohydrate Metabolism (Part B)
Edited by W. A. Wood

Volume XLII. Carbohydrate Metabolism (Part C)
Edited by W. A. Wood

Volume XLIII. Antibiotics
Edited by John H. Hash

Volume XLIV. Immobilized Enzymes
Edited by Klaus Mosbach

Volume XLV. Proteolytic Enzymes (Part B)
Edited by Laszlo Lorand

Volume XLVI. Affinity Labeling
Edited by William B. Jakoby and Meir Wilchek

Volume XLVII. Enzyme Structure (Part E)
Edited by C. H. W. Hirs and Serge N. Timasheff

Volume XLVIII. Enzyme Structure (Part F)
Edited by C. H. W. Hirs and Serge N. Timasheff

Volume XLIX. Enzyme Structure (Part G)
Edited by C. H. W. Hirs and Serge N. Timasheff

Volume L. Complex Carbohydrates (Part C)
Edited by Victor Ginsburg

Volume LI. Purine and Pyrimidine Nucleotide Metabolism
Edited by Patricia A. Hoffee and Mary Ellen Jones

Volume LII. Biomembranes (Part C: Biological Oxidations)
Edited by Sidney Fleischer and Lester Packer

Volume LIII. Biomembranes (Part D: Biological Oxidations)
Edited by Sidney Fleischer and Lester Packer

Volume LIV. Biomembranes (Part E: Biological Oxidations)
Edited by Sidney Fleischer and Lester Packer

Volume LV. Biomembranes (Part F: Bioenergetics)
Edited by Sidney Fleischer and Lester Packer

Volume LVI. Biomembranes (Part G: Bioenergetics)
Edited by Sidney Fleischer and Lester Packer

Volume LVII. Bioluminescence and Chemiluminescence
Edited by Marlene A. DeLuca

Volume LVIII. Cell Culture
Edited by William B. Jakoby and Ira Pastan

Volume LIX. Nucleic Acids and Protein Synthesis (Part G)
Edited by Kivie Moldave and Lawrence Grossman

Volume LX. Nucleic Acids and Protein Synthesis (Part H)
Edited by Kivie Moldave and Lawrence Grossman

Volume 61. Enzyme Structure (Part H)
Edited by C. H. W. Hirs and Serge N. Timasheff

Volume 62. Vitamins and Coenzymes (Part D)
Edited by Donald B. McCormick and Lemuel D. Wright

Volume 63. Enzyme Kinetics and Mechanism (Part A: Initial Rate and Inhibitor Methods)
Edited by Daniel L. Purich

Volume 93. Immunochemical Techniques (Part F: Conventional Antibodies, Fc Receptors, and Cytotoxicity)
Edited by John J. Langone and Helen Van Vunakis

Volume 94. Polyamines
Edited by Herbert Tabor and Celia White Tabor

Volume 95. Cumulative Subject Index Volumes 61–74, 76–80
Edited by Edward A. Dennis and Martha G. Dennis

Volume 96. Biomembranes [Part J: Membrane Biogenesis: Assembly and Targeting (General Methods; Eukaryotes)]
Edited by Sidney Fleischer and Becca Fleischer

Volume 97. Biomembranes [Part K: Membrane Biogenesis: Assembly and Targeting (Prokaryotes, Mitochondria, and Chloroplasts)]
Edited by Sidney Fleischer and Becca Fleischer

Volume 98. Biomembranes (Part L: Membrane Biogenesis: Processing and Recycling)
Edited by Sidney Fleischer and Becca Fleischer

Volume 99. Hormone Action (Part F: Protein Kinases)
Edited by Jackie D. Corbin and Joel G. Hardman

Volume 100. Recombinant DNA (Part B)
Edited by Ray Wu, Lawrence Grossman, and Kivie Moldave

Volume 101. Recombinant DNA (Part C)
Edited by Ray Wu, Lawrence Grossman, and Kivie Moldave

Volume 102. Hormone Action (Part G: Calmodulin and Calcium-Binding Proteins)
Edited by Anthony R. Means and Bert W. O'Malley

Volume 103. Hormone Action (Part H: Neuroendocrine Peptides)
Edited by P. Michael Conn

Volume 104. Enzyme Purification and Related Techniques (Part C)
Edited by William B. Jakoby

Volume 105. Oxygen Radicals in Biological Systems
Edited by Lester Packer

Volume 106. Posttranslational Modifications (Part A)
Edited by Finn Wold and Kivie Moldave

Volume 107. Posttranslational Modifications (Part B)
Edited by Finn Wold and Kivie Moldave

Volume 108. Immunochemical Techniques (Part G: Separation and Characterization of Lymphoid Cells)
Edited by Giovanni Di Sabato, John J. Langone, and Helen Van Vunakis

Volume 109. Hormone Action (Part I: Peptide Hormones)
Edited by Lutz Birnbaumer and Bert W. O'Malley

Volume 110. Steroids and Isoprenoids (Part A)
Edited by John H. Law and Hans C. Rilling

Volume 111. Steroids and Isoprenoids (Part B)
Edited by John H. Law and Hans C. Rilling

Volume 112. Drug and Enzyme Targeting (Part A)
Edited by Kenneth J. Widder and Ralph Green

Volume 113. Glutamate, Glutamine, Glutathione, and Related Compounds
Edited by Alton Meister

Volume 114. Diffraction Methods for Biological Macromolecules (Part A)
Edited by Harold W. Wyckoff, C. H. W. Hirs, and Serge N. Timasheff

Volume 115. Diffraction Methods for Biological Macromolecules (Part B)
Edited by Harold W. Wyckoff, C. H. W. Hirs, and Serge N. Timasheff

Volume 116. Immunochemical Techniques
(Part H: Effectors and Mediators of Lymphoid Cell Functions)
Edited by Giovanni Di Sabato, John J. Langone, and Helen Van Vunakis

Volume 117. Enzyme Structure (Part J)
Edited by C. H. W. Hirs and Serge N. Timasheff

Volume 118. Plant Molecular Biology
Edited by Arthur Weissbach and Herbert Weissbach

Volume 119. Interferons (Part C)
Edited by Sidney Pestka

Volume 120. Cumulative Subject Index Volumes 81–94, 96–101

Volume 121. Immunochemical Techniques (Part I: Hybridoma Technology and Monoclonal Antibodies)
Edited by John J. Langone and Helen Van Vunakis

Volume 122. Vitamins and Coenzymes (Part G)
Edited by Frank Chytil and Donald B. McCormick

Volume 123. Vitamins and Coenzymes (Part H)
Edited by Frank Chytil and Donald B. McCormick

Volume 124. Hormone Action (Part J: Neuroendocrine Peptides)
Edited by P. Michael Conn

Volume 125. Biomembranes (Part M: Transport in Bacteria, Mitochondria, and Chloroplasts: General Approaches and Transport Systems)
Edited by Sidney Fleischer and Becca Fleischer

Volume 126. Biomembranes (Part N: Transport in Bacteria, Mitochondria, and Chloroplasts: Protonmotive Force)
Edited by Sidney Fleischer and Becca Fleischer

Volume 127. Biomembranes (Part O: Protons and Water: Structure and Translocation)
Edited by Lester Packer

Volume 128. Plasma Lipoproteins (Part A: Preparation, Structure, and Molecular Biology)
Edited by Jere P. Segrest and John J. Albers

Volume 129. Plasma Lipoproteins (Part B: Characterization, Cell Biology, and Metabolism)
Edited by John J. Albers and Jere P. Segrest

Volume 130. Enzyme Structure (Part K)
Edited by C. H. W. Hirs and Serge N. Timasheff

Volume 131. Enzyme Structure (Part L)
Edited by C. H. W. Hirs and Serge N. Timasheff

Volume 132. Immunochemical Techniques (Part J: Phagocytosis and Cell-Mediated Cytotoxicity)
Edited by Giovanni Di Sabato and Johannes Everse

Volume 133. Bioluminescence and Chemiluminescence (Part B)
Edited by Marlene DeLuca and William D. McElroy

Volume 134. Structural and Contractile Proteins (Part C: The Contractile Apparatus and the Cytoskeleton)
Edited by Richard B. Vallee

Volume 135. Immobilized Enzymes and Cells (Part B)
Edited by Klaus Mosbach

Volume 136. Immobilized Enzymes and Cells (Part C)
Edited by Klaus Mosbach

Volume 137. Immobilized Enzymes and Cells (Part D)
Edited by Klaus Mosbach

Volume 138. Complex Carbohydrates (Part E)
Edited by Victor Ginsburg

Volume 139. Cellular Regulators (Part A: Calcium- and Calmodulin-Binding Proteins)
Edited by Anthony R. Means and P. Michael Conn

Volume 140. Cumulative Subject Index Volumes 102–119, 121–134

Volume 141. Cellular Regulators (Part B: Calcium and Lipids)
Edited by P. Michael Conn and Anthony R. Means

Volume 142. Metabolism of Aromatic Amino Acids and Amines
Edited by Seymour Kaufman

Volume 143. Sulfur and Sulfur Amino Acids
Edited by William B. Jakoby and Owen Griffith

Volume 144. Structural and Contractile Proteins (Part D: Extracellular Matrix)
Edited by Leon W. Cunningham

Volume 145. Structural and Contractile Proteins (Part E: Extracellular Matrix)
Edited by Leon W. Cunningham

Volume 146. Peptide Growth Factors (Part A)
Edited by David Barnes and David A. Sirbasku

Volume 147. Peptide Growth Factors (Part B)
Edited by David Barnes and David A. Sirbasku

Volume 148. Plant Cell Membranes
Edited by Lester Packer and Roland Douce

Volume 149. Drug and Enzyme Targeting (Part B)
Edited by Ralph Green and Kenneth J. Widder

Volume 150. Immunochemical Techniques (Part K: *In Vitro* Models of B and T Cell Functions and Lymphoid Cell Receptors)
Edited by Giovanni Di Sabato

Volume 151. Molecular Genetics of Mammalian Cells
Edited by Michael M. Gottesman

Volume 152. Guide to Molecular Cloning Techniques
Edited by Shelby L. Berger and Alan R. Kimmel

Volume 153. Recombinant DNA (Part D)
Edited by Ray Wu and Lawrence Grossman

Volume 154. Recombinant DNA (Part E)
Edited by Ray Wu and Lawrence Grossman

Volume 155. Recombinant DNA (Part F)
Edited by Ray Wu

Volume 156. Biomembranes (Part P: ATP-Driven Pumps and Related Transport: The Na, K-Pump)
Edited by Sidney Fleischer and Becca Fleischer

Volume 157. Biomembranes (Part Q: ATP-Driven Pumps and Related Transport: Calcium, Proton, and Potassium Pumps)
Edited by Sidney Fleischer and Becca Fleischer

Volume 158. Metalloproteins (Part A)
Edited by James F. Riordan and Bert L. Vallee

Volume 159. Initiation and Termination of Cyclic Nucleotide Action
Edited by Jackie D. Corbin and Roger A. Johnson

Volume 160. Biomass (Part A: Cellulose and Hemicellulose)
Edited by Willis A. Wood and Scott T. Kellogg

Volume 161. Biomass (Part B: Lignin, Pectin, and Chitin)
Edited by Willis A. Wood and Scott T. Kellogg

Volume 162. Immunochemical Techniques (Part L: Chemotaxis and Inflammation)
Edited by Giovanni Di Sabato

Volume 163. Immunochemical Techniques (Part M: Chemotaxis and Inflammation)
Edited by Giovanni Di Sabato

Volume 164. Ribosomes
Edited by Harry F. Noller, Jr., and Kivie Moldave

Volume 165. Microbial Toxins: Tools for Enzymology
Edited by Sidney Harshman

Volume 180. RNA Processing (Part A: General Methods)
Edited by James E. Dahlberg and John N. Abelson

Volume 181. RNA Processing (Part B: Specific Methods)
Edited by James E. Dahlberg and John N. Abelson

Volume 182. Guide to Protein Purification
Edited by Murray P. Deutscher

Volume 183. Molecular Evolution: Computer Analysis of Protein and Nucleic Acid Sequences
Edited by Russell F. Doolittle

Volume 184. Avidin-Biotin Technology
Edited by Meir Wilchek and Edward A. Bayer

Volume 185. Gene Expression Technology
Edited by David V. Goeddel

Volume 186. Oxygen Radicals in Biological Systems (Part B: Oxygen Radicals and Antioxidants)
Edited by Lester Packer and Alexander N. Glazer

Volume 187. Arachidonate Related Lipid Mediators
Edited by Robert C. Murphy and Frank A. Fitzpatrick

Volume 188. Hydrocarbons and Methylotrophy
Edited by Mary E. Lidstrom

Volume 189. Retinoids (Part A: Molecular and Metabolic Aspects)
Edited by Lester Packer

Volume 190. Retinoids (Part B: Cell Differentiation and Clinical Applications)
Edited by Lester Packer

Volume 191. Biomembranes (Part V: Cellular and Subcellular Transport: Epithelial Cells)
Edited by Sidney Fleischer and Becca Fleischer

Volume 192. Biomembranes (Part W: Cellular and Subcellular Transport: Epithelial Cells)
Edited by Sidney Fleischer and Becca Fleischer

Volume 193. Mass Spectrometry
Edited by James A. McCloskey

Volume 194. Guide to Yeast Genetics and Molecular Biology
Edited by Christine Guthrie and Gerald R. Fink

Volume 195. Adenylyl Cyclase, G Proteins, and Guanylyl Cyclase
Edited by Roger A. Johnson and Jackie D. Corbin

Volume 196. Molecular Motors and the Cytoskeleton
Edited by Richard B. Vallee

Volume 197. Phospholipases
Edited by Edward A. Dennis

Volume 198. Peptide Growth Factors (Part C)
Edited by David Barnes, J. P. Mather, and Gordon H. Sato

Volume 199. Cumulative Subject Index Volumes 168–174, 176–194

Volume 200. Protein Phosphorylation (Part A: Protein Kinases: Assays, Purification, Antibodies, Functional Analysis, Cloning, and Expression)
Edited by Tony Hunter and Bartholomew M. Sefton

Volume 201. Protein Phosphorylation (Part B: Analysis of Protein Phosphorylation, Protein Kinase Inhibitors, and Protein Phosphatases)
Edited by Tony Hunter and Bartholomew M. Sefton

Volume 202. Molecular Design and Modeling: Concepts and Applications (Part A: Proteins, Peptides, and Enzymes)
Edited by John J. Langone

Volume 203. Molecular Design and Modeling: Concepts and Applications (Part B: Antibodies and Antigens, Nucleic Acids, Polysaccharides, and Drugs)
Edited by John J. Langone

Volume 204. Bacterial Genetic Systems
Edited by Jeffrey H. Miller

Volume 205. Metallobiochemistry (Part B: Metallothionein and Related Molecules)
Edited by James F. Riordan and Bert L. Vallee

Volume 206. Cytochrome P450
Edited by Michael R. Waterman and Eric F. Johnson

Volume 207. Ion Channels
Edited by Bernardo Rudy and Linda E. Iverson

VOLUME 208. Protein–DNA Interactions
Edited by ROBERT T. SAUER

VOLUME 209. Phospholipid Biosynthesis
Edited by EDWARD A. DENNIS AND DENNIS E. VANCE

VOLUME 210. Numerical Computer Methods
Edited by LUDWIG BRAND AND MICHAEL L. JOHNSON

VOLUME 211. DNA Structures (Part A: Synthesis and Physical Analysis of DNA)
Edited by DAVID M. J. LILLEY AND JAMES E. DAHLBERG

VOLUME 212. DNA Structures (Part B: Chemical and Electrophoretic Analysis of DNA)
Edited by DAVID M. J. LILLEY AND JAMES E. DAHLBERG

VOLUME 213. Carotenoids (Part A: Chemistry, Separation, Quantitation, and Antioxidation)
Edited by LESTER PACKER

VOLUME 214. Carotenoids (Part B: Metabolism, Genetics, and Biosynthesis)
Edited by LESTER PACKER

VOLUME 215. Platelets: Receptors, Adhesion, Secretion (Part B)
Edited by JACEK J. HAWIGER

VOLUME 216. Recombinant DNA (Part G)
Edited by RAY WU

VOLUME 217. Recombinant DNA (Part H)
Edited by RAY WU

VOLUME 218. Recombinant DNA (Part I)
Edited by RAY WU

VOLUME 219. Reconstitution of Intracellular Transport
Edited by JAMES E. ROTHMAN

VOLUME 220. Membrane Fusion Techniques (Part A)
Edited by NEJAT DÜZGÜNEŞ

VOLUME 221. Membrane Fusion Techniques (Part B)
Edited by NEJAT DÜZGÜNEŞ

VOLUME 222. Proteolytic Enzymes in Coagulation, Fibrinolysis, and Complement Activation (Part A: Mammalian Blood Coagulation

Factors and Inhibitors)
Edited by Laszlo Lorand and Kenneth G. Mann

Volume 223. Proteolytic Enzymes in Coagulation, Fibrinolysis, and Complement Activation (Part B: Complement Activation, Fibrinolysis, and Nonmammalian Blood Coagulation Factors)
Edited by Laszlo Lorand and Kenneth G. Mann

Volume 224. Molecular Evolution: Producing the Biochemical Data
Edited by Elizabeth Anne Zimmer, Thomas J. White, Rebecca L. Cann, and Allan C. Wilson

Volume 225. Guide to Techniques in Mouse Development
Edited by Paul M. Wassarman and Melvin L. DePamphilis

Volume 226. Metallobiochemistry (Part C: Spectroscopic and Physical Methods for Probing Metal Ion Environments in Metalloenzymes and Metalloproteins)
Edited by James F. Riordan and Bert L. Vallee

Volume 227. Metallobiochemistry (Part D: Physical and Spectroscopic Methods for Probing Metal Ion Environments in Metalloproteins)
Edited by James F. Riordan and Bert L. Vallee

Volume 228. Aqueous Two-Phase Systems
Edited by Harry Walter and Göte Johansson

Volume 229. Cumulative Subject Index Volumes 195–198, 200–227

Volume 230. Guide to Techniques in Glycobiology
Edited by William J. Lennarz and Gerald W. Hart

Volume 231. Hemoglobins (Part B: Biochemical and Analytical Methods)
Edited by Johannes Everse, Kim D. Vandegriff, and Robert M. Winslow

Volume 232. Hemoglobins (Part C: Biophysical Methods)
Edited by Johannes Everse, Kim D. Vandegriff, and Robert M. Winslow

Volume 233. Oxygen Radicals in Biological Systems (Part C)
Edited by Lester Packer

Volume 234. Oxygen Radicals in Biological Systems (Part D)
Edited by Lester Packer

Volume 235. Bacterial Pathogenesis (Part A: Identification and Regulation of Virulence Factors)
Edited by Virginia L. Clark and Patrik M. Bavoil

Volume 236. Bacterial Pathogenesis (Part B: Integration of Pathogenic Bacteria with Host Cells)
Edited by Virginia L. Clark and Patrik M. Bavoil

Volume 237. Heterotrimeric G Proteins
Edited by Ravi Iyengar

Volume 238. Heterotrimeric G-Protein Effectors
Edited by Ravi Iyengar

Volume 239. Nuclear Magnetic Resonance (Part C)
Edited by Thomas L. James and Norman J. Oppenheimer

Volume 240. Numerical Computer Methods (Part B)
Edited by Michael L. Johnson and Ludwig Brand

Volume 241. Retroviral Proteases
Edited by Lawrence C. Kuo and Jules A. Shafer

Volume 242. Neoglycoconjugates (Part A)
Edited by Y. C. Lee and Reiko T. Lee

Volume 243. Inorganic Microbial Sulfur Metabolism
Edited by Harry D. Peck, Jr., and Jean LeGall

Volume 244. Proteolytic Enzymes: Serine and Cysteine Peptidases
Edited by Alan J. Barrett

Volume 245. Extracellular Matrix Components
Edited by E. Ruoslahti and E. Engvall

Volume 246. Biochemical Spectroscopy
Edited by Kenneth Sauer

Volume 247. Neoglycoconjugates (Part B: Biomedical Applications)
Edited by Y. C. Lee and Reiko T. Lee

Volume 248. Proteolytic Enzymes: Aspartic and Metallo Peptidases
Edited by Alan J. Barrett

Volume 249. Enzyme Kinetics and Mechanism (Part D: Developments in Enzyme Dynamics)
Edited by Daniel L. Purich

Volume 250. Lipid Modifications of Proteins
Edited by Patrick J. Casey and Janice E. Buss

VOLUME 251. Biothiols (Part A: Monothiols and Dithiols, Protein Thiols, and Thiyl Radicals)
Edited by LESTER PACKER

VOLUME 252. Biothiols (Part B: Glutathione and Thioredoxin; Thiols in Signal Transduction and Gene Regulation)
Edited by LESTER PACKER

VOLUME 253. Adhesion of Microbial Pathogens
Edited by RON J. DOYLE AND ITZHAK OFEK

VOLUME 254. Oncogene Techniques
Edited by PETER K. VOGT AND INDER M. VERMA

VOLUME 255. Small GTPases and Their Regulators (Part A: Ras Family)
Edited by W. E. BALCH, CHANNING J. DER, AND ALAN HALL

VOLUME 256. Small GTPases and Their Regulators (Part B: Rho Family)
Edited by W. E. BALCH, CHANNING J. DER, AND ALAN HALL

VOLUME 257. Small GTPases and Their Regulators (Part C: Proteins Involved in Transport)
Edited by W. E. BALCH, CHANNING J. DER, AND ALAN HALL

VOLUME 258. Redox-Active Amino Acids in Biology
Edited by JUDITH P. KLINMAN

VOLUME 259. Energetics of Biological Macromolecules
Edited by MICHAEL L. JOHNSON AND GARY K. ACKERS

VOLUME 260. Mitochondrial Biogenesis and Genetics (Part A)
Edited by GIUSEPPE M. ATTARDI AND ANNE CHOMYN

VOLUME 261. Nuclear Magnetic Resonance and Nucleic Acids
Edited by THOMAS L. JAMES

VOLUME 262. DNA Replication
Edited by JUDITH L. CAMPBELL

VOLUME 263. Plasma Lipoproteins (Part C: Quantitation)
Edited by WILLIAM A. BRADLEY, SANDRA H. GIANTURCO, AND JERE P. SEGREST

VOLUME 264. Mitochondrial Biogenesis and Genetics (Part B)
Edited by GIUSEPPE M. ATTARDI AND ANNE CHOMYN

VOLUME 265. Cumulative Subject Index Volumes 228, 230–262

VOLUME 281. Vitamins and Coenzymes (Part K)
Edited by DONALD B. MCCORMICK, JOHN W. SUTTIE, AND CONRAD WAGNER

VOLUME 282. Vitamins and Coenzymes (Part L)
Edited by DONALD B. MCCORMICK, JOHN W. SUTTIE, AND CONRAD WAGNER

VOLUME 283. Cell Cycle Control
Edited by WILLIAM G. DUNPHY

VOLUME 284. Lipases (Part A: Biotechnology)
Edited by BYRON RUBIN AND EDWARD A. DENNIS

VOLUME 285. Cumulative Subject Index Volumes 263, 264, 266–284, 286–289

VOLUME 286. Lipases (Part B: Enzyme Characterization and Utilization)
Edited by BYRON RUBIN AND EDWARD A. DENNIS

VOLUME 287. Chemokines
Edited by RICHARD HORUK

VOLUME 288. Chemokine Receptors
Edited by RICHARD HORUK

VOLUME 289. Solid Phase Peptide Synthesis
Edited by GREGG B. FIELDS

VOLUME 290. Molecular Chaperones
Edited by GEORGE H. LORIMER AND THOMAS BALDWIN

VOLUME 291. Caged Compounds
Edited by GERARD MARRIOTT

VOLUME 292. ABC Transporters: Biochemical, Cellular, and Molecular Aspects
Edited by SURESH V. AMBUDKAR AND MICHAEL M. GOTTESMAN

VOLUME 293. Ion Channels (Part B)
Edited by P. MICHAEL CONN

VOLUME 294. Ion Channels (Part C)
Edited by P. MICHAEL CONN

VOLUME 295. Energetics of Biological Macromolecules (Part B)
Edited by GARY K. ACKERS AND MICHAEL L. JOHNSON

VOLUME 296. Neurotransmitter Transporters
Edited by SUSAN G. AMARA

VOLUME 313. Antisense Technology
(Part A: General Methods, Methods of Delivery, and RNA Studies)
Edited by M. IAN PHILLIPS

VOLUME 314. Antisense Technology (Part B: Applications)
Edited by M. IAN PHILLIPS

VOLUME 315. Vertebrate Phototransduction and the Visual Cycle
(Part A)
Edited by KRZYSZTOF PALCZEWSKI

VOLUME 316. Vertebrate Phototransduction and the Visual Cycle (Part B)
Edited by KRZYSZTOF PALCZEWSKI

VOLUME 317. RNA–Ligand Interactions (Part A: Structural Biology
Methods)
Edited by DANIEL W. CELANDER AND JOHN N. ABELSON

VOLUME 318. RNA–Ligand Interactions (Part B: Molecular Biology
Methods)
Edited by DANIEL W. CELANDER AND JOHN N. ABELSON

VOLUME 319. Singlet Oxygen, UV-A, and Ozone
Edited by LESTER PACKER AND HELMUT SIES

VOLUME 320. Cumulative Subject Index Volumes 290–319

VOLUME 321. Numerical Computer Methods (Part C)
Edited by MICHAEL L. JOHNSON AND LUDWIG BRAND

VOLUME 322. Apoptosis
Edited by JOHN C. REED

VOLUME 323. Energetics of Biological Macromolecules (Part C)
Edited by MICHAEL L. JOHNSON AND GARY K. ACKERS

VOLUME 324. Branched-Chain Amino Acids (Part B)
Edited by ROBERT A. HARRIS AND JOHN R. SOKATCH

VOLUME 325. Regulators and Effectors of Small GTPases
(Part D: Rho Family)
Edited by W. E. BALCH, CHANNING J. DER, AND ALAN HALL

VOLUME 326. Applications of Chimeric Genes and Hybrid Proteins
(Part A: Gene Expression and Protein Purification)
Edited by JEREMY THORNER, SCOTT D. EMR, AND JOHN N. ABELSON

VOLUME 340. Drug–Nucleic Acid Interactions
Edited by JONATHAN B. CHAIRES AND MICHAEL J. WARING

VOLUME 341. Ribonucleases (Part A)
Edited by ALLEN W. NICHOLSON

VOLUME 342. Ribonucleases (Part B)
Edited by ALLEN W. NICHOLSON

VOLUME 343. G Protein Pathways (Part A: Receptors)
Edited by RAVI IYENGAR AND JOHN D. HILDEBRANDT

VOLUME 344. G Protein Pathways (Part B: G Proteins and Their Regulators)
Edited by RAVI IYENGAR AND JOHN D. HILDEBRANDT

VOLUME 345. G Protein Pathways (Part C: Effector Mechanisms)
Edited by RAVI IYENGAR AND JOHN D. HILDEBRANDT

VOLUME 346. Gene Therapy Methods
Edited by M. IAN PHILLIPS

VOLUME 347. Protein Sensors and Reactive Oxygen Species (Part A: Selenoproteins and Thioredoxin)
Edited by HELMUT SIES AND LESTER PACKER

VOLUME 348. Protein Sensors and Reactive Oxygen Species (Part B: Thiol Enzymes and Proteins)
Edited by HELMUT SIES AND LESTER PACKER

VOLUME 349. Superoxide Dismutase
Edited by LESTER PACKER

VOLUME 350. Guide to Yeast Genetics and Molecular and Cell Biology (Part B)
Edited by CHRISTINE GUTHRIE AND GERALD R. FINK

VOLUME 351. Guide to Yeast Genetics and Molecular and Cell Biology (Part C)
Edited by CHRISTINE GUTHRIE AND GERALD R. FINK

VOLUME 352. Redox Cell Biology and Genetics (Part A)
Edited by CHANDAN K. SEN AND LESTER PACKER

VOLUME 353. Redox Cell Biology and Genetics (Part B)
Edited by CHANDAN K. SEN AND LESTER PACKER

VOLUME 370. RNA Polymerases and Associated Factors (Part C)
Edited by SANKAR L. ADHYA AND SUSAN GARGES

VOLUME 371. RNA Polymerases and Associated Factors (Part D)
Edited by SANKAR L. ADHYA AND SUSAN GARGES

VOLUME 372. Liposomes (Part B)
Edited by NEJAT DÜZGÜNEŞ

VOLUME 373. Liposomes (Part C)
Edited by NEJAT DÜZGÜNEŞ

VOLUME 374. Macromolecular Crystallography (Part D)
Edited by CHARLES W. CARTER, JR., AND ROBERT W. SWEET

VOLUME 375. Chromatin and Chromatin Remodeling Enzymes (Part A)
Edited by C. DAVID ALLIS AND CARL WU

VOLUME 376. Chromatin and Chromatin Remodeling Enzymes (Part B)
Edited by C. DAVID ALLIS AND CARL WU

VOLUME 377. Chromatin and Chromatin Remodeling Enzymes (Part C)
Edited by C. DAVID ALLIS AND CARL WU

VOLUME 378. Quinones and Quinone Enzymes (Part A)
Edited by HELMUT SIES AND LESTER PACKER

VOLUME 379. Energetics of Biological Macromolecules (Part D)
Edited by JO M. HOLT, MICHAEL L. JOHNSON, AND GARY K. ACKERS

VOLUME 380. Energetics of Biological Macromolecules (Part E)
Edited by JO M. HOLT, MICHAEL L. JOHNSON, AND GARY K. ACKERS

VOLUME 381. Oxygen Sensing
Edited by CHANDAN K. SEN AND GREGG L. SEMENZA

VOLUME 382. Quinones and Quinone Enzymes (Part B)
Edited by HELMUT SIES AND LESTER PACKER

VOLUME 383. Numerical Computer Methods (Part D)
Edited by LUDWIG BRAND AND MICHAEL L. JOHNSON

VOLUME 384. Numerical Computer Methods (Part E)
Edited by LUDWIG BRAND AND MICHAEL L. JOHNSON

VOLUME 385. Imaging in Biological Research (Part A)
Edited by P. MICHAEL CONN

VOLUME 386. Imaging in Biological Research (Part B)
Edited by P. MICHAEL CONN

VOLUME 387. Liposomes (Part D)
Edited by NEJAT DÜZGÜNEŞ

VOLUME 388. Protein Engineering
Edited by DAN E. ROBERTSON AND JOSEPH P. NOEL

VOLUME 389. Regulators of G-Protein Signaling (Part A)
Edited by DAVID P. SIDEROVSKI

VOLUME 390. Regulators of G-Protein Signaling (Part B)
Edited by DAVID P. SIDEROVSKI

VOLUME 391. Liposomes (Part E)
Edited by NEJAT DÜZGÜNEŞ

VOLUME 392. RNA Interference
Edited by ENGELKE ROSSI

VOLUME 393. Circadian Rhythms
Edited by MICHAEL W. YOUNG

VOLUME 394. Nuclear Magnetic Resonance of Biological Macromolecules (Part C)
Edited by THOMAS L. JAMES

VOLUME 395. Producing the Biochemical Data (Part B)
Edited by ELIZABETH A. ZIMMER AND ERIC H. ROALSON

VOLUME 396. Nitric Oxide (Part E)
Edited by LESTER PACKER AND ENRIQUE CADENAS

VOLUME 397. Environmental Microbiology
Edited by JARED R. LEADBETTER

VOLUME 398. Ubiquitin and Protein Degradation (Part A)
Edited by RAYMOND J. DESHAIES

VOLUME 399. Ubiquitin and Protein Degradation (Part B)
Edited by RAYMOND J. DESHAIES

VOLUME 400. Phase II Conjugation Enzymes and Transport Systems
Edited by HELMUT SIES AND LESTER PACKER

VOLUME 401. Glutathione Transferases and Gamma Glutamyl Transpeptidases
Edited by HELMUT SIES AND LESTER PACKER

VOLUME 402. Biological Mass Spectrometry
Edited by A. L. BURLINGAME

Volume 403. GTPases Regulating Membrane Targeting and Fusion
Edited by William E. Balch, Channing J. Der, and Alan Hall

Volume 404. GTPases Regulating Membrane Dynamics
Edited by William E. Balch, Channing J. Der, and Alan Hall

Volume 405. Mass Spectrometry: Modified Proteins and Glycoconjugates
Edited by A. L. Burlingame

Volume 406. Regulators and Effectors of Small GTPases: Rho Family
Edited by William E. Balch, Channing J. Der, and Alan Hall

Volume 407. Regulators and Effectors of Small GTPases: Ras Family
Edited by William E. Balch, Channing J. Der, and Alan Hall

Volume 408. DNA Repair (Part A)
Edited by Judith L. Campbell and Paul Modrich

Volume 409. DNA Repair (Part B)
Edited by Judith L. Campbell and Paul Modrich

Volume 410. DNA Microarrays (Part A: Array Platforms and Web-Bench Protocols)
Edited by Alan Kimmel and Brian Oliver

Volume 411. DNA Microarrays (Part B: Databases and Statistics)
Edited by Alan Kimmel and Brian Oliver

Volume 412. Amyloid, Prions, and Other Protein Aggregates (Part B)
Edited by Indu Kheterpal and Ronald Wetzel

Volume 413. Amyloid, Prions, and Other Protein Aggregates (Part C)
Edited by Indu Kheterpal and Ronald Wetzel

Volume 414. Measuring Biological Responses with Automated Microscopy
Edited by James Inglese

Volume 415. Glycobiology
Edited by Minoru Fukuda

Volume 416. Glycomics
Edited by Minoru Fukuda

Volume 417. Functional Glycomics
Edited by Minoru Fukuda

Volume 418. Embryonic Stem Cells
Edited by Irina Klimanskaya and Robert Lanza

VOLUME 419. Adult Stem Cells
Edited by IRINA KLIMANSKAYA AND ROBERT LANZA

VOLUME 420. Stem Cell Tools and Other Experimental Protocols
Edited by IRINA KLIMANSKAYA AND ROBERT LANZA

VOLUME 421. Advanced Bacterial Genetics: Use of Transposons and Phage for Genomic Engineering
Edited by KELLY T. HUGHES

VOLUME 422. Two-Component Signaling Systems, Part A
Edited by MELVIN I. SIMON, BRIAN R. CRANE, AND ALEXANDRINE CRANE

VOLUME 423. Two-Component Signaling Systems, Part B
Edited by MELVIN I. SIMON, BRIAN R. CRANE, AND ALEXANDRINE CRANE

VOLUME 424. RNA Editing
Edited by JONATHA M. GOTT

VOLUME 425. RNA Modification
Edited by JONATHA M. GOTT

VOLUME 426. Integrins
Edited by DAVID CHERESH

VOLUME 427. MicroRNA Methods
Edited by JOHN J. ROSSI

VOLUME 428. Osmosensing and Osmosignaling
Edited by HELMUT SIES AND DIETER HAUSSINGER

VOLUME 429. Translation Initiation: Extract Systems and Molecular Genetics
Edited by JON LORSCH

VOLUME 430. Translation Initiation: Reconstituted Systems and Biophysical Methods
Edited by JON LORSCH

VOLUME 431. Translation Initiation: Cell Biology, High-Throughput and Chemical-Based Approaches
Edited by JON LORSCH

VOLUME 432. Lipidomics and Bioactive Lipids: Mass-Spectrometry–Based Lipid Analysis
Edited by H. ALEX BROWN

Volume 433. Lipidomics and Bioactive Lipids: Specialized Analytical Methods and Lipids in Disease
Edited by H. Alex Brown

Volume 434. Lipidomics and Bioactive Lipids: Lipids and Cell Signaling
Edited by H. Alex Brown

Volume 435. Oxygen Biology and Hypoxia
Edited by Helmut Sies and Bernhard Brüne

Volume 436. Globins and Other Nitric Oxide-Reactive Protiens (Part A)
Edited by Robert K. Poole

Volume 437. Globins and Other Nitric Oxide-Reactive Protiens (Part B)
Edited by Robert K. Poole

Volume 438. Small GTPases in Disease (Part A)
Edited by William E. Balch, Channing J. Der, and Alan Hall

Volume 439. Small GTPases in Disease (Part B)
Edited by William E. Balch, Channing J. Der, and Alan Hall

Volume 440. Nitric Oxide, Part F Oxidative and Nitrosative Stress in Redox Regulation of Cell Signaling
Edited by Enrique Cadenas and Lester Packer

Volume 441. Nitric Oxide, Part G Oxidative and Nitrosative Stress in Redox Regulation of Cell Signaling
Edited by Enrique Cadenas and Lester Packer

Volume 442. Programmed Cell Death, General Principles for Studying Cell Death (Part A)
Edited by Roya Khosravi-Far, Zahra Zakeri, Richard A. Lockshin, and Mauro Piacentini

Volume 443. Angiogenesis: *In Vitro* Systems
Edited by David A. Cheresh

Volume 444. Angiogenesis: *In Vivo* Systems (Part A)
Edited by David A. Cheresh

Volume 445. Angiogenesis: *In Vivo* Systems (Part B)
Edited by David A. Cheresh

Volume 446. Programmed Cell Death, The Biology and Therapeutic Implications of Cell Death (Part B)
Edited by Roya Khosravi-Far, Zahra Zakeri, Richard A. Lockshin, and Mauro Piacentini

VOLUME 447. RNA Turnover in Bacteria, Archaea and Organelles
Edited by LYNNE E. MAQUAT AND CECILIA M. ARRAIANO

VOLUME 448. RNA Turnover in Eukaryotes: Nucleases, Pathways and Analysis of mRNA Decay
Edited by LYNNE E. MAQUAT AND MEGERDITCH KILEDJIAN

VOLUME 449. RNA Turnover in Eukaryotes: Analysis of Specialized and Quality Control RNA Decay Pathways
Edited by LYNNE E. MAQUAT AND MEGERDITCH KILEDJIAN

VOLUME 450. Fluorescence Spectroscopy
Edited by LUDWIG BRAND AND MICHAEL L. JOHNSON

VOLUME 451. Autophagy: Lower Eukaryotes and Non-Mammalian Systems (Part A)
Edited by DANIEL J. KLIONSKY

VOLUME 452. Autophagy in Mammalian Systems (Part B)
Edited by DANIEL J. KLIONSKY

VOLUME 453. Autophagy in Disease and Clinical Applications (Part C)
Edited by DANIEL J. KLIONSKY

VOLUME 454. Computer Methods (Part A)
Edited by MICHAEL L. JOHNSON AND LUDWIG BRAND

VOLUME 455. Biothermodynamics (Part A)
Edited by MICHAEL L. JOHNSON, JO M. HOLT, AND GARY K. ACKERS (RETIRED)

VOLUME 456. Mitochondrial Function, Part A: Mitochondrial Electron Transport Complexes and Reactive Oxygen Species
Edited by WILLIAM S. ALLISON AND IMMO E. SCHEFFLER

VOLUME 457. Mitochondrial Function, Part B: Mitochondrial Protein Kinases, Protein Phosphatases and Mitochondrial Diseases
Edited by WILLIAM S. ALLISON AND ANNE N. MURPHY

VOLUME 458. Complex Enzymes in Microbial Natural Product Biosynthesis, Part A: Overview Articles and Peptides
Edited by DAVID A. HOPWOOD

VOLUME 459. Complex Enzymes in Microbial Natural Product Biosynthesis, Part B: Polyketides, Aminocoumarins and Carbohydrates
Edited by DAVID A. HOPWOOD

VOLUME 460. Chemokines, Part A
Edited by TRACY M. HANDEL AND DAMON J. HAMEL

VOLUME 461. Chemokines, Part B
Edited by TRACY M. HANDEL AND DAMON J. HAMEL

VOLUME 462. Non-Natural Amino Acids
Edited by TOM W. MUIR AND JOHN N. ABELSON

VOLUME 463. Guide to Protein Purification, 2nd Edition
Edited by RICHARD R. BURGESS AND MURRAY P. DEUTSCHER

VOLUME 464. Liposomes, Part F
Edited by NEJAT DÜZGÜNEŞ

VOLUME 465. Liposomes, Part G
Edited by NEJAT DÜZGÜNEŞ

VOLUME 466. Biothermodynamics, Part B
Edited by MICHAEL L. JOHNSON, GARY K. ACKERS, AND JO M. HOLT

VOLUME 467. Computer Methods Part B
Edited by MICHAEL L. JOHNSON AND LUDWIG BRAND

VOLUME 468. Biophysical, Chemical, and Functional Probes of RNA Structure, Interactions and Folding: Part A
Edited by DANIEL HERSCHLAG

VOLUME 469. Biophysical, Chemical, and Functional Probes of RNA Structure, Interactions and Folding: Part B
Edited by DANIEL HERSCHLAG

VOLUME 470. Guide to Yeast Genetics: Functional Genomics, Proteomics, and Other Systems Analysis, 2nd Edition
Edited by GERALD FINK, JONATHAN WEISSMAN, AND CHRISTINE GUTHRIE

VOLUME 471. Two-Component Signaling Systems, Part C
Edited by MELVIN I. SIMON, BRIAN R. CRANE, AND ALEXANDRINE CRANE

VOLUME 472. Single Molecule Tools, Part A: Fluorescence Based Approaches
Edited by NILS G. WALTER

VOLUME 473. Thiol Redox Transitions in Cell Signaling, Part A Chemistry and Biochemistry of Low Molecular Weight and Protein Thiols
Edited by ENRIQUE CADENAS AND LESTER PACKER

VOLUME 489. The Unfolded Protein Response and Cellular Stress, Part A
Edited by P. MICHAEL CONN

VOLUME 490. The Unfolded Protein Response and Cellular Stress, Part B
Edited by P. MICHAEL CONN

VOLUME 491. The Unfolded Protein Response and Cellular Stress, Part C
Edited by P. MICHAEL CONN

VOLUME 492. Biothermodynamics, Part D
Edited by MICHAEL L. JOHNSON, JO M. HOLT, AND GARY K. ACKERS

VOLUME 493. Fragment-Based Drug Design Tools, Practical Approaches, and Examples
Edited by LAWRENCE C. KUO

VOLUME 494. Methods in Methane Metabolism, Part A
Methanogenesis
Edited by AMY C. ROSENZWEIG AND STEPHEN W. RAGSDALE

VOLUME 495. Methods in Methane Metabolism, Part B
Methanotrophy
Edited by AMY C. ROSENZWEIG AND STEPHEN W. RAGSDALE

VOLUME 496. Research on Nitrification and Related Processes, Part B
Edited by MARTIN G. KLOTZ AND LISA Y. STEIN

VOLUME 497. Synthetic Biology, Part A
Methods for Part/Device Characterization and Chassis Engineering
Edited by CHRISTOPHER VOIGT

VOLUME 498. Synthetic Biology, Part B
Computer Aided Design and DNA Assembly
Edited by CHRISTOPHER VOIGT

VOLUME 499. Biology of Serpins
Edited by JAMES C. WHISSTOCK AND PHILLIP I. BIRD

VOLUME 500. Methods in Systems Biology
Edited by DANIEL JAMESON, MALKHEY VERMA, AND HANS V. WESTERHOFF

VOLUME 501. Serpin Structure and Evolution
Edited by JAMES C. WHISSTOCK AND PHILLIP I. BIRD

VOLUME 502. Protein Engineering for Therapeutics, Part A
Edited by K. DANE WITTRUP AND GREGORY L. VERDINE

VOLUME 503. Protein Engineering for Therapeutics, Part B
Edited by K. DANE WITTRUP AND GREGORY L. VERDINE

VOLUME 504. Imaging and Spectroscopic Analysis of Living Cells
Optical and Spectroscopic Techniques
Edited by P. MICHAEL CONN

VOLUME 505. Imaging and Spectroscopic Analysis of Living Cells
Live Cell Imaging of Cellular Elements and Functions
Edited by P. MICHAEL CONN

VOLUME 506. Imaging and Spectroscopic Analysis of Living Cells
Imaging Live Cells in Health and Disease
Edited by P. MICHAEL CONN

VOLUME 507. Gene Transfer Vectors for Clinical Application
Edited by THEODORE FRIEDMANN

VOLUME 508. Nanomedicine
Cancer, Diabetes, and Cardiovascular, Central Nervous System, Pulmonary and Inflammatory Diseases
Edited by NEJAT DÜZGÜNEŞ

VOLUME 509. Nanomedicine
Infectious Diseases, Immunotherapy, Diagnostics, Antifibrotics, Toxicology and Gene Medicine
Edited by NEJAT DÜZGÜNEŞ

VOLUME 510. Cellulases
Edited by HARRY J. GILBERT

VOLUME 511. RNA Helicases
Edited by ECKHARD JANKOWSKY

VOLUME 512. Nucleosomes, Histones & Chromatin, Part A
Edited by CARL WU AND C. DAVID ALLIS

VOLUME 513. Nucleosomes, Histones & Chromatin, Part B
Edited by CARL WU AND C. DAVID ALLIS

VOLUME 514. Ghrelin
Edited by MASAYASU KOJIMA AND KENJI KANGAWA

VOLUME 515. Natural Product Biosynthesis by Microorganisms and Plants, Part A
Edited by DAVID A. HOPWOOD

VOLUME 516. Natural Product Biosynthesis by Microorganisms and Plants, Part B
Edited by DAVID A. HOPWOOD

VOLUME 517. Natural Product Biosynthesis by Microorganisms and Plants, Part C
Edited by DAVID A. HOPWOOD

VOLUME 518. Fluorescence Fluctuation Spectroscopy (FFS), Part A
Edited by SERGEY Y. TETIN

VOLUME 519. Fluorescence Fluctuation Spectroscopy (FFS), Part B
Edited by SERGEY Y. TETIN

VOLUME 520. G Protein Couple Receptors Structure
Edited by P. MICHAEL CONN

VOLUME 521. G Protein Couple Receptors
Trafficking and Oligomerization
Edited by P. MICHAEL CONN

VOLUME 522. G Protein Couple Receptors
Modeling, Activation, Interactions and Virtual Screening
Edited by P. MICHAEL CONN

VOLUME 523. Methods in Protein Design
Edited by AMY E. KEATING

VOLUME 524. Cilia, Part A
Edited by WALLACE F. MARSHALL

VOLUME 525. Cilia, Part B
Edited by WALLACE F. MARSHALL

VOLUME 526. Hydrogen Peroxide and Cell Signaling, Part A
Edited by ENRIQUE CADENAS AND LESTER PACKER

VOLUME 527. Hydrogen Peroxide and Cell Signaling, Part B
Edited by ENRIQUE CADENAS AND LESTER PACKER

VOLUME 528. Hydrogen Peroxide and Cell Signaling, Part C
Edited by ENRIQUE CADENAS AND LESTER PACKER

VOLUME 529. Laboratory Methods in Enzymology: DNA
Edited by JON LORSCH

SECTION I

Preparation and Genomic Analyses of Single Cells from Natural Populations

CHAPTER ONE

In-Solution Fluorescence *In Situ* Hybridization and Fluorescence-Activated Cell Sorting for Single Cell and Population Genome Recovery

Mohamed F. Haroon[*], Connor T. Skennerton[*,†], Jason A. Steen[*], Nancy Lachner[*], Philip Hugenholtz[*,‡], Gene W. Tyson[*,†,1]

[*]Australian Centre for Ecogenomics, School of Chemistry and Molecular Biosciences, The University of Queensland, St. Lucia, Queensland, Australia
[†]Advanced Water Management Centre, The University of Queensland, St. Lucia, Queensland, Australia
[‡]Institute for Molecular Bioscience, The University of Queensland, St. Lucia, Queensland, Australia
[1]Corresponding author: e-mail address: g.tyson@uq.edu.au

Contents

1. Introduction 4
   1.1 Fluorescence *in situ* hybridization 4
   1.2 Combining FISH and FACS 5
2. Technical Protocols and Considerations 6
   2.1 Sample preparation 6
   2.2 FISH probe selection 6
   2.3 In-solution FISH 8
   2.4 Fluorescence-activated cell sorting 13
   2.5 Post-sorting 14
3. Case Study: Targeted Isolation of Anaerobic Methanotrophic Archaea for Single Cell and Population Genomics 15
4. Summary 17
Acknowledgments 17
References 18

Abstract

Over the past decade, technological advances in whole genome amplification, microfluidics, flow sorting, and high-throughput sequencing have led to the development of single-cell genomics. Single-cell genomic approaches are typically applied to anonymous microbial cells with only morphology providing clues to their identity. However, targeted separation of microorganisms based on phylogenetic markers, such as the 16S rRNA gene, is beginning to emerge in the single-cell genomics field. Here, we describe

Methods in Enzymology, Volume 531
ISSN 0076-6879
http://dx.doi.org/10.1016/B978-0-12-407863-5.00001-0

an in-solution fluorescence *in situ* hybridization (FISH) protocol which can be combined with fluorescence-activated cell sorting (FACS) for separation of single cells or populations of interest from environmental samples. Sequencing of DNA obtained from sorted cells can be used for the recovery of draft quality genomes, and when performed in parallel with deep metagenomics, can be used to validate and further scaffold metagenomic assemblies. We illustrate in this chapter the feasibility of this FISH–FACS approach by describing the targeted recovery of a novel anaerobic methanotrophic archaeon.

1. INTRODUCTION

Genome sequencing has become a routine method for characterizing microorganisms. However, the great majority of microbial species cannot be easily obtained in pure culture, and therefore culture-independent approaches are required to access the genomes of uncultivated species. Shotgun sequencing of environmental DNA (metagenomics) has emerged as a front-runner technology for this purpose (Hugenholtz & Tyson, 2008); however, methods that fractionate communities into simpler subsets (down to individual cells) are valuable complementary approaches to metagenomics. Fluorescence-activated cell sorting (FACS) is a powerful, high-throughput approach of fractionating microbial communities based on light scattering properties of cells (Fuchs, Zubkov, Sahm, Burkill, & Amann, 2000). Fluorescence *in situ* hybridization (FISH) using ribosomal RNA (rRNA)-targeted probes is a widely used approach in microbial ecology to visualize groups of related cells (Amann, Krumholz, & Stahl, 1990; DeLong, Wickham, & Pace, 1989; Wagner, Horn, & Daims, 2003). Here, we describe an approach that combines FISH and FACS to obtain phylogenetically identified single cells and populations for the purpose of genome sequencing. Related population and single-cell genomic approaches are presented in Chapters 2–4.

1.1. Fluorescence *in situ* hybridization

FISH in combination with epifluorescence microscopy has been used for over two decades to visualize morphology and spatial arrangement of targeted microbial cells in environmental samples (Amann, Binder, et al., 1990; Amann, Krumholz, et al., 1990; DeLong et al., 1989). Fluorescently labeled oligonucleotides (typically 15–20 bp long) that specifically target rRNAs, the major structural components of the ribosome, are hybridized to whole cells, thereby causing the cell to fluoresce (DeLong et al., 1989). The method takes advantage of singular properties of rRNAs, including natural amplification of the target site by the presence of hundreds to thousands of copies of ribosomes per cell, and variable sequence conservation, which allows phylogenetically broad (e.g., phylum) to narrow (e.g., species) groups to be targeted. Multiple probes with different fluorophores can be used in combination to highlight multiple target groups simultaneously. Traditionally, cells are preserved by chemical fixation, which is also thought to assist in probe permeability (Hugenholtz, Tyson, & Blackall, 2002). A number of adaptations of the method have been published, including amplification of the fluorescence signal via CARD (catalyzed reporter deposition)–FISH (Pernthaler, Pernthaler, & Amann, 2002), dual-labeling of probes (Stoecker, Dorninger, Daims, & Wagner, 2010), and removal of the fixation steps to facilitate downstream applications (Yilmaz, Haroon, Rabkin, Tyson, & Hugenholtz, 2010).

1.2. Combining FISH and FACS

FISH and flow cytometry were used in combination shortly after the development of FISH, as the potential of cytometry to quantify targeted populations was immediately apparent (Amann, Binder, et al., 1990; Wallner, Amann, & Beisker, 1993; Wallner, Fuchs, Spring, Beisker, & Amann, 1997). Combining FISH with FACS was a natural progression to sort labeled cells for downstream molecular analyses (Kalyuzhnaya et al., 2006). Initially, 16S rRNA genes were amplified and sequenced from sorted populations (Fuchs et al., 2000), but more recently, FISH–FACS has been used to link metabolic pathways or genes to specific phylogenetic groups (Kalyuzhnaya et al., 2006; Miyauchi, Oki, Aoi, & Tsuneda, 2007), and for single-cell genomics (Podar et al., 2007; Woyke et al., 2010). FISH labeling of cells also enables populations rather than just single cells to be sorted, which reduces the need and biases of whole genome amplification (see Section 2.5). However, the conservation of rRNA genes may result in the sorting of populations that are genomically heterogeneous due to inadequate phylogenetic resolution. FISH–FACS has a number of advantages over other methods for obtaining single cell and population genomes, including higher throughput and/or greater specificity than dilution (Zhang et al., 2006), micromanipulation (Ishoey, Woyke, Stepanauskas, Novotny, & Lasken, 2008; Woyke et al., 2010), and microfluidic (Marcy et al., 2007) approaches.

In this chapter, we describe the procedures and technical considerations required for FISH–FACS, which include (i) sample preparation, (ii) probe selection, (iii) in-solution hybridization, (iv) preparation for FACS, and (briefly) (v) post-sort molecular analyses (Fig. 1.1).

2. TECHNICAL PROTOCOLS AND CONSIDERATIONS

2.1. Sample preparation

Sample fixation is commonly thought to be necessary for preservation of cell structure and to facilitate probe hybridization. Typically, samples for FISH are preserved using chemical fixatives, primarily paraformaldehyde (PFA) that can be stored successfully at −20 °C for months to years.

PFA fixation protocol

1. Add 3 volumes of sample to 1 volume of 4% PFA in 1× PBS (final concentration 1% PFA) and incubate at 4 °C for 2–4 h.
2. Centrifuge the sample at 10,000 × *g* for 2 min to pellet biomass and discard the supernatant.
3. Wash the cells (≥2 times) by resuspending the biomass in 1× phosphate-buffered saline (PBS) and repeat centrifugation.
4. Resuspend the fixed biomass in 1:1 ratio of 100% ethanol and 1× PBS. Store at − 20 °C for long-term preservation.

While fixation preserves cell structure by cross-linking of proteins and DNA molecules (Hayat, 2000), it can make homogenization of samples for FACS more difficult and has been shown to decrease DNA yield from sorted populations (Yilmaz et al., 2010). It may therefore be preferable to avoid fixation if genome sequencing is the primary goal. For some cell and sample types, fixation appears to be unnecessary (Yilmaz et al., 2010). A drawback of this approach is the need for fresh samples, although we have found that some unfixed samples can be stored in 10–20% glycerol at −20 °C for several days to weeks prior to hybridization. In this chapter, we focus on fixation-free sample preparation methods for FISH–FACS and demonstrate the effects of using fixed and unfixed samples in control experiments with a typical Gram-negative organism, *Escherichia coli*, and a typical Gram-positive organism, *Bacillus megaterium*.

2.2. FISH probe selection

Oligonucleotide probe choice depends on the community composition and the desired specificity of cell sorting. For example, if the community

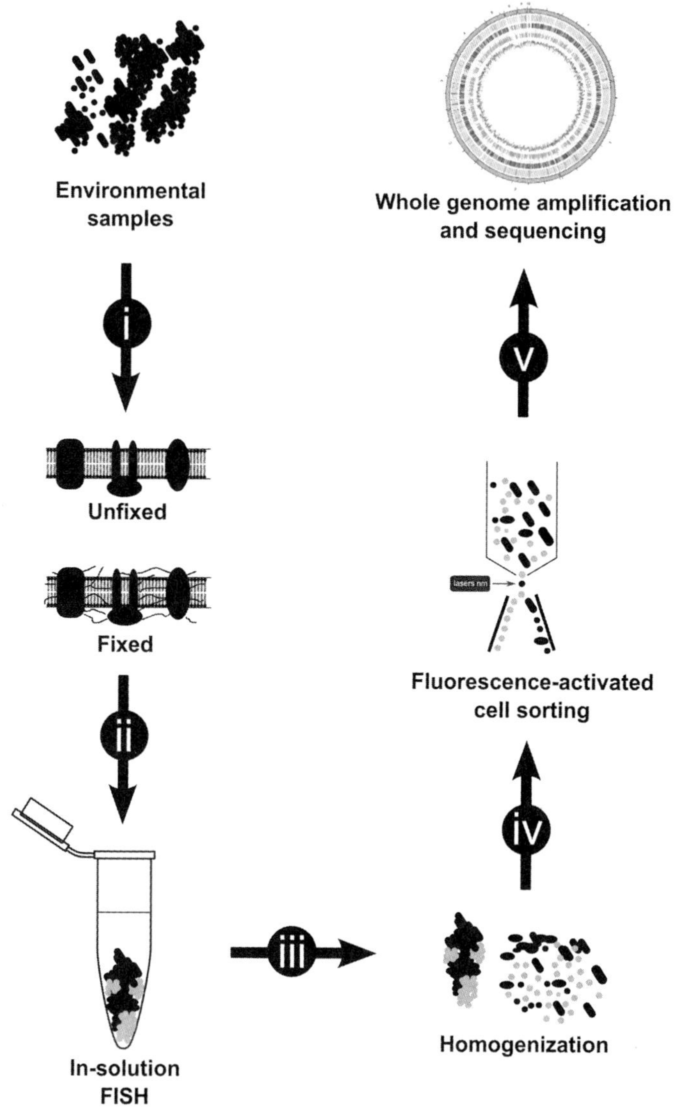

Figure 1.1 Flow diagram showing the key steps (i–v) in the in-solution FISH–FACS approach from environmental sample through to the generation of single cell or population genome sequences. (i) Cells can be preserved using PFA or in some instances processed without fixation (unfixed) and used in subsequent steps. (ii) In-solution FISH is then performed using a fluorescently labeled probe targeting the population of interest. (iii) After hybridization, cell aggregates are dispersed via mild sonication and filtration in preparation for FACS. (iv) Finally, the target population can be separated from the rest of the community by FACS based on their fluorescent signal and (v) used as the input for genome sequencing. (For color version of this figure, the reader is referred to the online version of this chapter.)

contains a single archaeal species, then a domain-level archaeal probe (e.g., Arc915; Stahl & Amann, 1991) would be sufficient to specifically enrich for that population (see Section 3). If multiple archaeal species are present, however, then one or more higher specificity probes would be required. If such probes do not exist for a population of interest, design and evaluation of new probes can be a major undertaking (Hugenholtz et al., 2002). Fortunately, hundreds of FISH probes that target a wide range of bacteria and archaea have been experimentally verified and cataloged (Loy, Maixner, Wagner, & Horn, 2007).

The choice of fluorophore for the probe is dependent on both the properties of the sample and FACS configuration. Some samples will have high autofluorescence under certain wavelengths (e.g., sediments and herbivore feces), and therefore the emission spectrum of the fluorophore should be outside of the autofluorescent range, so that labeled cells can be distinguished from background fluorescence with FACS. It is also important to ensure that fluorophores are compatible with the available FACS lasers.

2.3. In-solution FISH

The standard FISH protocol used for visualization of cells by microscopy involves the following steps: (i) attachment of fixed cells to slides, (ii) sample dehydration with ethanol to facilitate hybridization, (iii) hybridization with probes, and (iv) washing to remove unincorporated probe. For FISH–FACS, the protocol has been modified such that cells are not attached to a slide and steps (ii)–(iv) are performed in-solution so that cells can be subsequently sorted (Kalyuzhnaya et al., 2006; Wallner et al., 1993). It is advisable to verify cell labeling via microscopy after hybridization and to both sonicate and filter samples prior to FACS to prevent clogging of FACS fluidic system. To optimize labeling efficiency of in-solution FISH with fixed and unfixed cells, we tested two key parameters using *E. coli* and *B. megaterium*, namely, ethanol dehydration and hybridization incubation time. For these experiments, cells were grown overnight in Luria broth at 37 °C to a density of $\sim 10^9$ cells mL^{-1} before fixation with PFA as described above. Unfixed samples were prepared by centrifugation of the cells and resuspension in 10% glycerol. Prior to hybridization, samples were washed three times with sterile $1\times$ PBS to remove any small particulates that may cause autofluorescence.

2.3.1 *Ethanol dehydration*

Ethanol dehydration of cells is thought to improve hybridization efficiency by drawing more probe into cells via a greater osmotic gradient produced by the dehydration. Note that this step was omitted in fixation-free FISH to streamline the method and reduce cell loss (Yilmaz et al., 2010), but the effect of this omission on labeling efficiency was not evaluated.

Ethanol dehydration protocol

1. Prepare 50%, 80%, and 98% ethanol solutions in 50-mL falcon tubes. These solutions can be reused for multiple experiments and stored on the bench at room temperature.
2. Add 100 μL of fixed or unfixed cells to a 1.5-mL microcentrifuge tube and pellet the biomass.
3. Resuspend with 500 μL of 1 × PBS, centrifuge at 10,000 × *g* for 2 min to pellet cells, and discard the supernatant.
4. Resuspend the pellet in 500 μL of the 50% ethanol solution and incubate at room temperature for 3 min.
5. Pellet by centrifugation at 10,000 × *g* for 2 min and discard the supernatant carefully with a 200-μL pipette. Make sure the pellet is visible at all times.
6. Repeat steps 4 and 5 with 80% ethanol solution.
7. Repeat steps 4 and 5 with 98% ethanol solution.
8. Remove any residual ethanol by drying in a heating block set to 40 °C for up to 15 min (depending on volume of residual ethanol left in tube).

Using both microscopy and FACS, we found that ethanol dehydration increased the number of detectable cells substantially (up to 2.7-fold increase) in unfixed samples for the control organisms, but had no appreciable effect on PFA-fixed cells (Table 1.1). We therefore recommend using ethanol dehydration with unfixed but not fixed samples.

Table 1.1 Hybridization efficiency of *Escherichia coli* and *Bacillus megaterium* bacteria under four different conditions

		PFA fixed	Unfixed
E. coli	No ethanol dehydration	96%	37%
	Ethanol dehydration	92%	88%
B. megaterium	No ethanol dehydration	78%	71%
	Ethanol dehydration	86%	91%

Cells were hybridized using the in-solution FISH protocol outlined in Section 2.3 using a EUBmix338-FITC probe. Prior to hybridization, cells were either fixed in 1% PFA and stored (as described in Section 2.1) or stored in 10% (v/v) glycerol (unfixed treatment). The ethanol dehydration series (Section 2.3.1) steps were removed to determine their overall effect on the hybridization efficiency. The hybridization efficiency was calculated by staining cells post-hybridization with DAPI and calculating the percentage of cells that were excited with a 488 nm laser (FITC labeled) versus cells that were excited by the UV laser (DAPI stained).

Table 1.2 Volumes of formamide and water in hybridization buffer, and NaCl and EDTA in the wash buffer for varying concentrations of formamide

Hybridization buffer			Wash buffer	
Formamide (%)	Amount of formamide (μL) *X*	Amount of water (μL) *Y*	Amount of NaCl (μL)	Amount of EDTA (μL)
0	0	1598	2250	0
5	100	1498	1575	0
10	200	1398	1125	0
15	300	1298	795	0
20	400	1198	537.5	125
25	500	1098	372.5	125
30	600	998	255	125
35	700	898	175	125
40	800	798	115	125
45	900	698	75	125
50	1000	598	45	125

2.3.2 Hybridization

Hybridization relies on DNA–RNA binding of fluorescent probes to 16S rRNA molecules. Due to the secondary and tertiary structure of the ribosome, different probes will have varying degrees of access to their binding sites (Behrens et al., 2003). Probe access can be improved using formamide, which relaxes secondary structure of the ribosome by partial denaturation. Formamide also controls the stringency of probe–target binding. Other parameters that can affect hybridization efficiency and stringency include incubation time, temperature, salt concentration, and sample mixing.

Hybridization protocol

1. Prepare the hybridization buffer by adding 360 μL of 5 *M* NaCl, 40 μL of 1 *M* Tris–HCl, *X* μL of formamide; and add *Y* μL milli-Q water to 2-mL microfuge tube (see Table 1.2 for formamide and milli-Q water volumes).
2. Add 2 μL of 10% w/v sodium dodecyl sulfate (SDS) into the lid of the microfuge tube, close the lid, and mix the hybridization buffer by inverting the tube two or three times.
3. Thaw an aliquot of probe on ice in a covered container to protect the fluorophore from photobleaching.

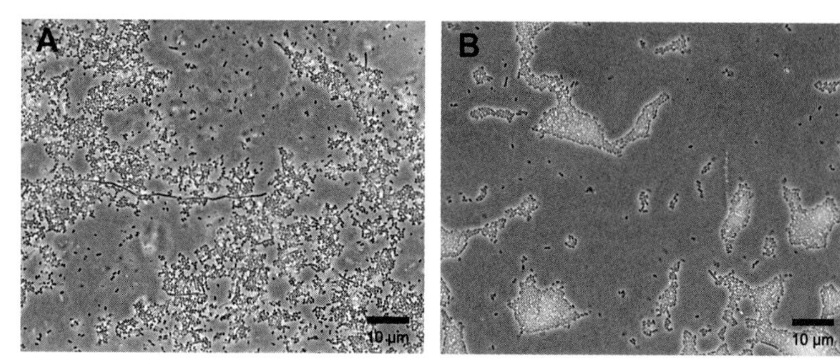

Figure 1.2 In-solution FISH performed on unfixed *E. coli* cells using a (A) 1.5 h and (B) 2.5 h incubation times. Cells were hybridized using EUB338mix (Daims, Brühl, Amann, Schleifer, & Wagner, 1999) with a FITC fluorophore before a subsample was dried onto a glass slide and visualized by epifluorescence microscopy (630× magnification). The shorter hybridization time resulted in only ~20% of cells fluorescencing, as opposed to the 2.5 h incubation where nearly all cells were detectably fluorescent. (See color plate.)

4. Add 45 μL of hybridization buffer to the sample and resuspend. A sonicating water bath can help in the resuspension of cells.
5. Add the probe/s to a final concentration of 2.5 ng μL^{-1}. *Note*: it is important to include a no probe control in all experiments for each sample type to access sample autofluorescence.
6. Carefully mix using a pipette or via vortexing for 2–3 s.
7. Incubate the tubes on a rotator at 46 °C for a minimum of 2.5 h (see below) in dark conditions.

Increasing the incubation time for unfixed cells from 1.5 to 2.5 h improved the labeling efficiency (number of detectably labeled cells) by up to fivefold for both control organisms (Fig. 1.2). We therefore recommend testing incubation times with samples to optimize labeling efficiency for fixation-free in-solution FISH.

2.3.3 Washing conditions

After hybridization, remove any nonspecifically bound or unbound probes from the cells using a series of washing steps.

Washing protocol

1. Add 250 μL of 1 *M* Tris–HCl, *X* μL 5 *M* of NaCl, *Y* μL of 0.5 *M* EDTA (see Table 1.2 for volumes of NaCl and EDTA); add milli-Q water such that the final volume of the solution is 12.5 mL, and finally add 12.5 μL of 10% w/v SDS.

2. Place the wash buffer either in a water bath or in a heating block set to 48 °C. It is important that this solution has had time to equilibrate to the wash temperature before washing.
3. Centrifuge the samples at 10,000 × g for 2 min to pellet cells and discard supernatant.
4. Add 500 μL of the preheated wash buffer, vortex to mix, pellet by centrifugation at 10,000 × g for 2 min and discard supernatant.
5. Resuspend the samples in 500 μL of wash buffer and incubate the samples at 48 °C in a water bath for 20 min.
6. Pellet cells via centrifugation, discard the supernatant, and resuspend the sample in 500 μL of ice-cold 1 × PBS. Repeat this step twice.

2.3.4 Labeling validation and preparation for FACS

Prior to FACS (which is generally an expensive undertaking), it is advisable to confirm that cells are labeled and dispersed. Cell aggregation presents a challenge to effective single-cell recovery. The small capillary tubes and nozzle sizes of cell sorters limit the size of cell aggregates that can be effectively sorted without clogging or damaging the instrument. If a sample is prone to aggregation or originates from a biofilm, various procedures can be used to homogenize the sample. In general, however, we recommend mild sonication to remove cell aggregates.

Sonication conditions must be optimized for each sample type with the goal of breaking up aggregates without lysing cells. This balance is a greater issue for unfixed cells as they are more prone to lysis than fixed cells. Optimal conditions can be determined by varying sonication time on aliquots of the hybridized cells, typically somewhere between 15 s and 2 min (e.g., see Section 3). Biofilms may require longer sonication times or different techniques to obtain a homogenous suspension of single cells such as probe sonication, mild chemical, or enzymatic disruption (Chen & Stewart, 2000; Dusane et al., 2008), or passage of the sample through a fine gauge syringe to disrupt the biofilm.

Sonication and visualization protocol

1. Perform mild water-bath sonication for 15 s to 2 min on the labeled sample to break up the cell aggregates.
2. Apply 5–10 μL of sonicated sample onto a microscope slide.
3. Stain cells with DAPI (0.5 μg mL^{-1}).
4. Add DABCO (or any other mounting media suitable for fluorescence microscopy) to the slide and place a coverslip over the sample. View the sample on an epifluorescence or laser scanning confocal microscope under 630 × or 1000 × magnification.

2.4. Fluorescence-activated cell sorting

After confirming that the cells are positively labeled, FACS is used to obtain cells of interest typically in microfuge tubes or microtiter plates. There are many different models of flow sorters, each with their own setup procedures, sorting accuracies, and laser configurations. For the control experiments and case study, we used a BD FACSAria III flow sorter with ultraviolet, 488 and 530 nm lasers for detection of cells labeled with DAPI (UV), and detection of probes labeled with FITC (488 nm) and Cy3 (530 nm). It is recommended that the manufacturer's instructions for setup, maintenance, and cleaning are followed, but also consider an additional cleaning of the capillary lines of the flow sorter with 1% sodium hypochlorite (bleach) solution prior to every sample to remove any residual biological material in the capillary lines. When performing single-cell sorting experiments, FISH labeling was assessed using scattergrams of forward scatter (FSC) versus side scatter (SSC), FSC versus FITC/DAPI, SSC versus FITC/DAPI, and FITC versus DAPI (Fig. 1.3).

FACS protocol

1. Run a negative (no probe) control of each sample setup and adjust the voltages of the flow sorter such that the cells appear in the lower left quadrant of the FSC versus SSC scattergram.
2. Run the sample hybridized with a probe targeting the cells of interest. There should be a noticeable shift in the fluorescence intensity of cells in the channel corresponding to the fluorophore of the probe used. Previous results suggest that scattergrams of FITC versus Cy3/DAPI should give clear separation of the positively labeled cells from the flanking populations (see Section 3). *Note*: We recommend performing multistage sorts to obtain pure single cells from the targeted population.
3. Select (gate) the regions of the scattergram representing the population of interest for sorting.
4. For multistage sorts, perform the first sort at a higher speed (20,000 cells s^{-1}) with less stringency on the purity, to enrich the targeted population. Sort 10^6 cells into a 5-mL tube, which is then used as the input for the second sort.
5. For subsequent sorts of the enriched target population, use a slower sorting speed and more accurate settings (the FACSAria III has a "single cell" mode) to produce higher purity enrichments of a single population or for single-cell sorting.

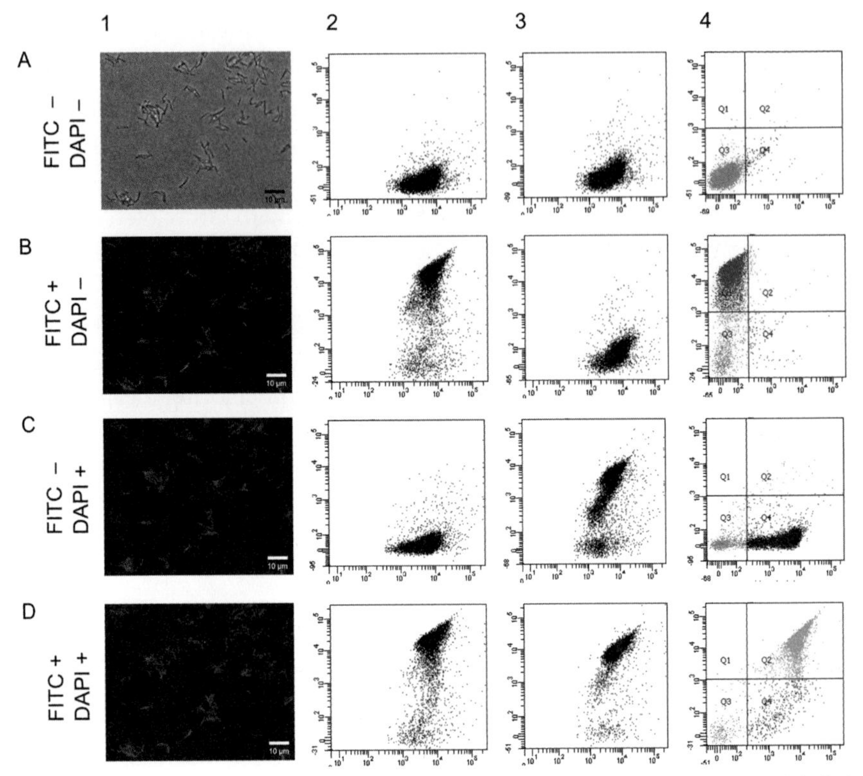

Figure 1.3 FISH–FACS outputs for unfixed *B. megaterium* cells (A) unlabeled, (B) labeled with EUBmix338-FITC, (C) stained with DAPI, and (D) labeled with EUBmix338-FITC and stained with DAPI. From left to right, (1) cells imaged using epifluorescence microscopy, flow cytometry scattergrams of (2) SSC versus FITC, (3) SSC versus DAPI, and (4) DAPI versus FITC. In column 4, scattergrams are partitioned into four quadrants. The Q1 quadrant gated cells (in red) were labeled with EUBmix388-FITC only, Q2 quadrant gated cells (in purple) were labeled with both EUBmix338-FITC and DAPI, Q3 quadrant (in yellow) is background, while Q4 quadrant gated cells (in green) are stained with DAPI only. (See color plate.)

6. Multiple subsets of target cells can be sorted into microfuge tubes or 96-well microtiter plates containing sterile sheath fluid (for storage) or alkaline lysis buffer (for subsequent DNA extraction).

2.5. Post-sorting

For single cell and population genome sequencing, we recommend processing sorted cells immediately; however, they can be stored in sterile sheath fluid at −20 °C for a period of days. Sorted populations of cells (but not single cells) can be confirmed by visualization of a sample aliquot using epifluorescence microscopy. Processing of genomic DNA typically involves alkaline lysis, neutralization, and whole genome amplification by multiple displacement amplification (MDA) (Lasken & Stockwell, 2007). MDA introduces coverage bias and chimeras (Ishoey et al., 2008; Lasken & Stockwell, 2007) which can be minimized by reducing MDA incubation time or volume (Marcy et al., 2007) and pooling of MDA products from multiple cells (Woyke et al., 2010). *Note*: MDA is highly susceptible to contamination especially when working with single cells, and extra precautions should be taken preparing reagents and the work area (Woyke et al., 2011). Whole genome amplification is covered in more detail in Chapter 3. After MDA, we recommend confirming the identities of the single or population amplified genomes using 16S rRNA gene amplicon sequencing (see Section 3). Amplified genomic DNAs can then be shotgun sequenced on a range of high-throughput platforms and bioinformatically processed. Computational analyses of sequence data is described in Section V of this volume.

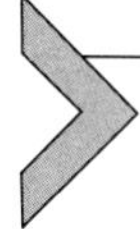

3. CASE STUDY: TARGETED ISOLATION OF ANAEROBIC METHANOTROPHIC ARCHAEA FOR SINGLE CELL AND POPULATION GENOMICS

Anaerobic methanotrophic archaea (ANME) have been implicated in anaerobic oxidation of methane (AOM), thereby playing an important role in global methane cycling (Boetius et al., 2000; Hinrichs, Hayes, Sylva, Brewer, & DeLong, 1999). ANME can use a variety of electron acceptors to perform AOM with (Haroon et al., 2013; Milucka et al., 2012) or without syntrophic bacterial associations (Beal, House, & Orphan, 2009; Boetius et al., 2000). Recently, we demonstrated that a representative of the ANME-2d clade, "*Candidatus Methanoperedens nitroreducens*", performs AOM concomitant with the reduction of nitrate to nitrite in a bioreactor fed with nitrate, ammonium, and methane (Haroon et al., 2013). Metabolic reconstruction from a metagenomic assembly of the *M. nitroreducens* genome revealed a complete reverse methanogenesis pathway and genes for nitrate reduction. However, it was necessary to confirm that nitrate reduction originated from *M. nitroreducens* and not from another member of the microbial community. To achieve this, we employed fixation-free FISH–FACS (as described in the sections above) to obtain single cells and populations (sets of ~10,000 cells) of *M. nitroreducens* for genome sequencing.

Fresh samples from a bioreactor performing AOM coupled to denitrification were collected and washed twice with 1 × PBS prior to in-solution fixation-free FISH. Cells were resuspended in FISH hybridization buffer (Section 2.3.2) with 40% formamide and labeled with the general archaeal (Arc915-FITC; Stahl & Amann, 1991) and ANME-2-specific probes (Darch872-Cy3; Raghoebarsing et al., 2006) at 46 °C for 2.5 h. After hybridization, cells were pelleted and washed (recipe in Section 2.3.2) at 48 °C for 20 min. After washing, cells were centrifuged and resuspended in ice-cold 1 × PBS. Prior to FACS, aggregates of *M. nitroreducens* cells (Fig. 1.4) were homogenized using a Branson Sonifier 250 for 2 × 15 s sonication cycles.

FACS scattergram plots of FITC versus Cy3 showed a distinct population shift on the Cy3 axis compared to the negative control (Fig. 1.5). This population of Darch872-labeled cells (*M. nitroreducens*) was initially sorted into a 5-mL tube at high speed (20,000 cells s^{-1}). The cells from this first sort were reanalyzed at a slower speed using the "single cell mode" of the BD FACSAria III. Multiple subsets (single cells up to 10^5 cells) of this higher purity enrichment were then sorted into a 96-well microtiter plate. The sorted cells were alkaline lysed and amplified using MDA. To validate the sorting specificity, the 16S rRNA genes of the single amplified cells and populations were PCR amplified, with only archaeal primers producing products. One amplified single-cell genome and one population genome

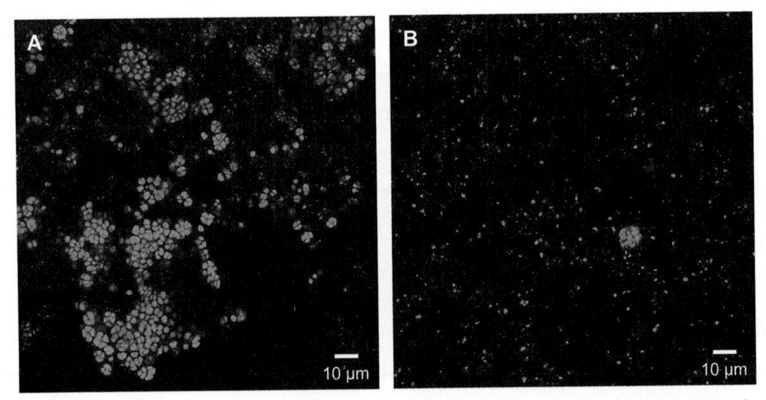

Figure 1.4 *M. nitroreducens* cells labeled with Arc915-FITC and Darch872-Cy3 (A) before sonication and (B) after 2 × 15 s probe sonication. (For color version of this figure, the reader is referred to the online version of this chapter.)

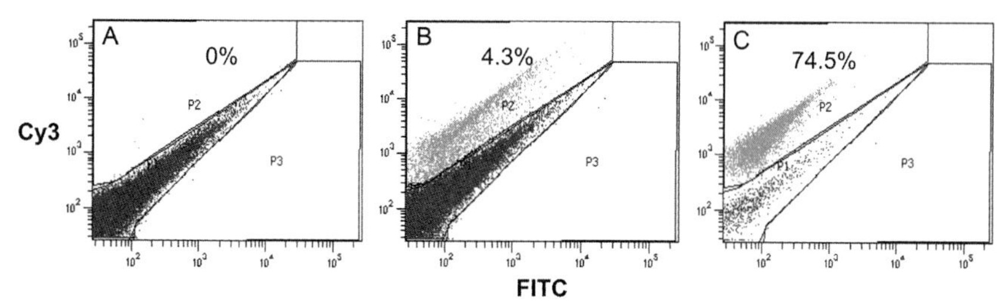

Figure 1.5 FACS scattergrams showing FITC versus Cy3 for bioreactor samples. (A) Unlabeled sample, (B) first rapid sort with specific Darch872-Cy3 probe which targets *M. nitroreducens*, and (C) second sort to ensure high purity of target population single cells. Each plot contains 50,000 events. The P2 region of the plot indicates the events gated for sorting. (See color plate.)

were then selected for shotgun sequencing on the Illumina platform. The sequence data were assembled and mapped to the metagenome-derived of the *M. nitroreducens* genome confirming the fidelity of the original assembly. We also note that a near-complete genome of the targeted archaeon could be assembled *de novo* from the single cell and population data.

4. SUMMARY

In-solution fixation-free FISH is a promising modification of conventional FISH for use with FACS to obtain genome sequences of phylogenetically identified target cells and populations. The approach has a number of caveats including the need for probes specific to the target population of interest, samples amenable to homogenization for FACS, and in most cases, cells that can be labeled without fixation to avoid complications with DNA extraction. We recommend using ethanol dehydration on unfixed samples and a minimum hybridization time of 2.5 h based on tests with model Gram-negative and Gram-positive organisms. Using an optimized approach for a bioreactor community, we demonstrated the feasibility of this FISH–FACS for the targeted separation and genome sequencing of an uncultivated, phylogenetically novel archaeal population.

ACKNOWLEDGMENTS

We thank Joshua Daly and Yun Kit Yeoh for providing the isolates used for validation experiments and Michael Nefedov for assistance with FACS.

REFERENCES

Amann, R. I., Binder, B. J., Olson, R. J., Chisholm, S. W., Devereux, R., & Stahl, D. A. (1990). Combination of 16S rRNA-targeted oligonucleotide probes with flow cytometry for analyzing mixed microbial populations. *Applied and Environmental Microbiology, 56*(6), 1919–1925.

Amann, R. I., Krumholz, L., & Stahl, D. A. (1990). Fluorescent-oligonucleotide probing of whole cells for determinative, phylogenetic, and environmental studies in microbiology. *Journal of Bacteriology, 172*(2), 762–770.

Beal, E. J., House, C. H., & Orphan, V. J. (2009). Manganese- and iron-dependent marine methane oxidation. *Science, 325*(5937), 184–187.

Behrens, S., Rühland, C., Inácio, J., Huber, H., Fonseca, Á., Spencer-Martins, I., et al. (2003). In situ accessibility of small-subunit rRNA of members of the domains bacteria, archaea, and eucarya to Cy3-labeled oligonucleotide probes. *Applied and Environmental Microbiology, 69*(3), 1748–1758.

Boetius, A., Ravenschlag, K., Schubert, C. J., Rickert, D., Widdel, F., Gieseke, A., et al. (2000). A marine microbial consortium apparently mediating anaerobic oxidation of methane. *Nature, 407*(6804), 623–626.

Chen, X., & Stewart, P. S. (2000). Biofilm removal caused by chemical treatments. *Water Research, 34*(17), 4229–4233.

Daims, H., Brühl, A., Amann, R., Schleifer, K., & Wagner, M. (1999). The domain-specific probe EUB338 is insufficient for the detection of all bacteria: Development and evaluation of a more comprehensive probe set. *Systematic and Applied Microbiology, 22*(3), 434–444.

DeLong, E. F., Wickham, G. S., & Pace, N. R. (1989). Phylogenetic stains: Ribosomal RNA-based probes for the identification of single cells. *Science, 243*(4896), 1360–1363.

Dusane, D. H., Rajput, J. K., Kumar, A. R., Nancharaiah, Y. V., Venugopalan, V. P., & Zinjarde, S. S. (2008). Disruption of fungal and bacterial biofilms by lauroyl glucose. *Letters in Applied Microbiology, 47*(5), 374–379.

Fuchs, B. M., Zubkov, M. V., Sahm, K., Burkill, P. H., & Amann, R. (2000). Changes in community composition during dilution cultures of marine bacterioplankton as assessed by flow cytometric and molecular biological techniques. *Environmental Microbiology, 2*(2), 191–201.

Haroon, M. F., Hu, S., Shi, Y., Imelfort, M., Keller, J., Hugenholtz, P., et al. (2013). Anaerobic oxidation of methane coupled to nitrate reduction in a novel archaeal lineage. *Nature*, Advance online publication.

Hayat, M. A. (2000). *Principles and techniques of electron microscopy: Biological applications.* Cambridge: Cambridge University Press.

Hinrichs, K. U., Hayes, J. M., Sylva, S. P., Brewer, P. G., & DeLong, E. F. (1999). Methane-consuming archaebacteria in marine sediments. *Nature, 398*(6730), 802–805.

Hugenholtz, P., & Tyson, G. W. (2008). Microbiology: Metagenomics. *Nature, 455*(7212), 481–483.

Hugenholtz, P., Tyson, G. W., & Blackall, L. L. (2002). Design and evaluation of 16S rRNA-targeted oligonucleotide probes for fluorescence in situ hybridization. In M. A. Muro & R. Rapley (Eds.), *Gene probes*, Vol. 179, (pp. 29–42). London: Humana Press.

Ishoey, T., Woyke, T., Stepanauskas, R., Novotny, M., & Lasken, R. S. (2008). Genomic sequencing of single microbial cells from environmental samples. *Current Opinion in Microbiology, 11*(3), 198–204.

Kalyuzhnaya, M. G., Zabinsky, R., Bowerman, S., Baker, D. R., Lidstrom, M. E., & Chistoserdova, L. (2006). Fluorescence in situ hybridization-flow cytometry-cell sorting-based method for separation and enrichment of type I and type II methanotroph populations. *Applied and Environmental Microbiology, 72*(6), 4293–4301.

Lasken, R., & Stockwell, T. (2007). Mechanism of chimera formation during the multiple displacement amplification reaction. *BMC Biotechnology, 7*(1), 19.

Loy, A., Maixner, F., Wagner, M., & Horn, M. (2007). probeBase—An online resource for rRNA-targeted oligonucleotide probes: New features 2007. *Nucleic Acids Research, 35*, D800–D804.

Marcy, Y., Ishoey, T., Lasken, R. S., Stockwell, T. B., Walenz, B. P., Halpern, A. L., et al. (2007). Nanoliter reactors improve multiple displacement amplification of genomes from single cells. *PLoS Genetics, 3*(9), e155.

Marcy, Y., Ouverney, C., Bik, E. M., Lösekann, T., Ivanova, N., Martin, H. G., et al. (2007). Dissecting biological "dark matter" with single-cell genetic analysis of rare and uncultivated TM7 microbes from the human mouth. *Proceedings of the National Academy of Sciences of the United States of America, 104*(29), 11889–11894.

Milucka, J., Ferdelman, T. G., Polerecky, L., Franzke, D., Wegener, G., Schmid, M., et al. (2012). Zero-valent sulphur is a key intermediate in marine methane oxidation. *Nature, 491*(7425), 541–546.

Miyauchi, R., Oki, K., Aoi, Y., & Tsuneda, S. (2007). Diversity of nitrite reductase genes in "Candidatus Accumulibacter phosphatis"—Dominated cultures enriched by flow-cytometric sorting. *Applied and Environmental Microbiology, 73*(16), 5331–5337.

Pernthaler, A., Pernthaler, J., & Amann, R. (2002). Fluorescence in situ hybridization and catalyzed reporter deposition for the identification of marine bacteria. *Applied and Environmental Microbiology, 68*(6), 3094–3101.

Podar, M., Abulencia, C. B., Walcher, M., Hutchison, D., Zengler, K., Garcia, J. A., et al. (2007). Targeted access to the genomes of low-abundance organisms in complex microbial communities. *Applied and Environmental Microbiology, 73*, 3205–3214.

Raghoebarsing, A. A., Pol, A., van de Pas-Schoonen, K. T., Smolders, A. J. P., Ettwig, K. F., Rijpstra, W. I. C., et al. (2006). A microbial consortium couples anaerobic methane oxidation to denitrification. *Nature, 440*(7086), 918–921.

Stahl, D. A., & Amann, R. (1991). Development and application of nucleic acid probes. In E. Stackebrandt & M. Goodfellow (Eds.), *Nucleic acid techniques in bacterial systematics*. Chichester, England: John Wiley and Sons Ltd.

Stoecker, K., Dorninger, C., Daims, H., & Wagner, M. (2010). Double labeling of oligonucleotide probes for fluorescence in situ hybridization (DOPE-FISH) improves signal intensity and increases rRNA accessibility. *Applied and Environmental Microbiology, 76*(3), 922–926.

Wagner, M., Horn, M., & Daims, H. (2003). Fluorescence in situ hybridisation for the identification and characterisation of prokaryotes. *Current Opinion in Microbiology, 6*(3), 302–309.

Wallner, G., Amann, R., & Beisker, W. (1993). Optimizing fluorescent in situ hybridization with rRNA-targeted oligonucleotide probes for flow cytometric identification of microorganisms. *Cytometry, 14*(2), 136–143.

Wallner, G., Fuchs, B., Spring, S., Beisker, W., & Amann, R. (1997). Flow sorting of microorganisms for molecular analysis. *Applied and Environmental Microbiology, 63*(11), 4223–4231.

Woyke, T., Sczyrba, A., Lee, J., Rinke, C., Tighe, D., Clingenpeel, S., et al. (2011). Decontamination of MDA reagents for single cell whole genome amplification. *PLoS One, 6*(10), e26161.

Woyke, T., Tighe, D., Mavromatis, K., Clum, A., Copeland, A., Schackwitz, W., et al. (2010). One bacterial cell, one complete genome. *PLoS One, 5*(4), e10314.

Yilmaz, S., Haroon, M. F., Rabkin, B. A., Tyson, G. W., & Hugenholtz, P. (2010). Fixation-free fluorescence in situ hybridization for targeted enrichment of microbial populations. *The ISME Journal, 4*(10), 1352–1356.

Zhang, K., Martiny, A. C., Reppas, N. B., Barry, K. W., Malek, J., Chisholm, S. W., et al. (2006). Sequencing genomes from single cells by polymerase cloning. *Nature Biotechnology, 24*(6), 680–686.

CHAPTER TWO

Whole Cell Immunomagnetic Enrichment of Environmental Microbial Consortia Using rRNA-Targeted Magneto-FISH

Elizabeth Trembath-Reichert*, Abigail Green-Saxena†, Victoria J. Orphan*[,1]

*Division of Geological and Planetary Sciences, California Institute of Technology, Pasadena, California, USA
†Division of Biological Sciences, California Institute of Technology, Pasadena, California, USA
[1]Corresponding author: e-mail address: vorphan@gps.caltech.edu

Contents

1. Introduction 22
2. Methods 24
   2.1 Samples and controls used in Magneto-FISH capture experiments 24
   2.2 Magneto-FISH 25
   2.3 DNA processing 31
   2.4 Quantification 31
   2.5 PCR and cloning 31
   2.6 Phylogenetic analysis of 16S rRNA and metabolic genes (*mcrA*) 32
3. Results and Discussion 33
   3.1 Evaluating the quantification and specificity of captured targets using general and species-specific FISH probes 33
   3.2 Optimization for metagenomics 37
   3.3 Optimization for other environmental systems 40
4. Summary 41
Acknowledgments 41
References 42

Abstract

Magneto-FISH, in combination with metagenomic techniques, explores the middle ground between single-cell analysis and complex community characterization in bulk samples to better understand microbial partnerships and their roles in ecosystems. The Magneto-FISH method combines the selectivity of catalyzed reporter deposition–fluorescence *in situ* hybridization (CARD–FISH) with immunomagnetic capture to provide targeted molecular and metagenomic analysis of co-associated microorganisms in the environment. This method was originally developed by Pernthaler et al. (Pernthaler

Methods in Enzymology, Volume 531
ISSN 0076-6879
http://dx.doi.org/10.1016/B978-0-12-407863-5.00002-2

et al., 2008; Pernthaler & Orphan, 2010). It led to the discovery of new bacterial groups associated with anaerobic methane-oxidizing (ANME-2) archaea in methane seeps, as well as provided insight into their physiological potential using metagenomics. Here, we demonstrate the utility of this method for capturing aggregated consortia using a series of nested oligonucleotide probes of differing specificity designed to target either the ANME archaea or their *Deltaproteobacteria* partner, combined with 16S rRNA and *mcrA* analysis. This chapter outlines a modified Magneto-FISH protocol for large- and small-volume samples and evaluates the strengths and limitations of this method predominantly focusing on (1) the relationship between FISH probe specificity and sample selectivity, (2) means of improving DNA yield from paraformaldehyde-fixed samples, and (3) suggestions for adapting the Magneto-FISH method for other microbial systems, including potential for single-cell recovery.

1. INTRODUCTION

As advancements in high-throughput sequencing technology allow deeper and more cost-effective means of sequencing complex microbial assemblages, we are left with more data, but not necessarily more means to understand them. The development of microbiological techniques to isolate and visualize environmental microorganisms *a priori* can be used to meaningfully parse environmental samples before metagenomic processing, and thereby provide additional context for downstream bioinformatic data interpretation. There is also increasing awareness that microbe–environment and microbe–microbe interactions are important factors in assessing microbial systems, their metabolic potential, and how these relationships affect larger scale processes such as ecosystem nutrient cycling.

A range of *in situ* techniques are currently available for physical separation of microorganisms of interest from environmental samples. Methods involving selection from a complex microbial sample often involve a stage of phylogenetic identification, such as 16S rRNA-based fluorescence *in situ* hybridization (FISH), coupled to a means of physical separation such as flow sorting (Amann, Binder, Olson, Chisholm, Devereux, and Stahl, 1990; Yilmaz, Haroon, Rabkin, Tyson, & Hugenholtz, 2010) (also see Chapters 1 and 3), optical trapping (Ashkin, 1997), microfluidics (Melin & Quake, 2007), or immunomagnetic beads (Safarık & Safarıkova, 1999). This is in contrast to separation methods where selection is based on a property other than identity, such as metabolic activity (Kalyuzhnaya, Lidstrom, & Chistoserdova, 2008), followed by downstream identification of the population exhibiting the property of interest. The majority of these methods have focused on single-cell analysis, rather than examining intact multi-species microbial associations, with the exception of intracellular microbial interactions (Yoon et al., 2011).

The Magneto-FISH method was originally developed by Pernthaler et al. (2008) to enrich for and characterize multi-species microbial associations in the environment. This technique was specifically developed for studying interspecies partnerships between anaerobic methane-oxidizing (ANME) archaea and sulfate-reducing *Deltaproteobacteria* (SRB) in anoxic marine sediments (Boetius et al., 2000; Orphan, House, Hinrichs, McKeegan, & DeLong, 2002). This method is based on 16S rRNA catalyzed reporter deposition (CARD)–FISH (identity) (Pernthaler, Pernthaler, & Amann, 2002) and immunomagnetic bead capture (separation) (Pernthaler & Orphan, 2010; Pernthaler et al., 2008). The Magneto-FISH method was shown to successfully concentrate the population of interest and aid in microbial association hypothesis development that could be further supported with metagenomics, microscopy, and isotope-labeling techniques. This technique provides a means to study metabolic potential at a level that is not defined in separate units of species identity, but operational groups of organisms that have evolved to serve a function, such as the symbiotic consortia mediating methane oxidation coupled to sulfate reduction. Magneto-FISH is also compatible with the physical challenges of sediment-associated ANME–SRB aggregates, namely, their heterogeneous morphology, wide size range (~3–100 μm diameter), and frequent association with mineral and sediment particles.

In evaluating the application of Magneto-FISH to other environmental populations, it is important to consider sample input constraints such as microbe size and morphology, sample output requirements such as yield and purity, and, of course, time and expense. Autofluorescent sediment particles and diverse ANME/SRB consortia size complicated the successful application of flow sorting approaches to the AOM system. In other environments, FAC sorting has been shown an effective means of cell separation, but often requires DNA amplification (Rodrigue et al., 2009; Woyke et al., 2011). Yield and purity are also often opposing constraints. For example, FAC sorting can provide high sample purity but may require significant instrument time for collecting sufficient material without including a post-amplification step (Woyke et al., 2011). Sample yield remains an issue with Magneto-FISH, as well. Initial application of Magneto-FISH required multiple displacement amplification (MDA) before construction of metagenomic libraries for 454 pyrosequencing (Pernthaler et al., 2008). However, advances in library preparation (e.g., Nextera XT) have significantly lowered the minimum DNA concentrations required. Magneto-FISH can be completed in a day and does not require the use of any specialized equipment beyond an epifluorescent microscope. The main expense is reagents, which scales with amount of sample processed and diversity of FISH probes needed. Another advantage is the versatility of this method. It is compatible with a broad range of oligonucleotide probes incorporated into the same, basic protocol; no instrument adjustment or recalibration is required between runs or with different microbial targets.

This chapter introduces three modifications to the Magneto-FISH protocol of Pernthaler et al. (2008) to improve DNA recovery and labor efficiency: (1) immuno-based attachment of magnetic beads for single-cell capture, (2) magnetic separation in a standard magnetic holder, and (3) DNA cross-link reversal incubation during extraction. Using this modified method, we evaluate the DNA recovery and microbial target specificity using a nested set of oligonucleotide probes and discuss (1) increasing target DNA yield for current pyrosequencing and metagenomics template requirements without amplification, (2) the relationship between sample purity and FISH probe specificity, (3) controls for association selectivity, and (4) DNA quality for metagenomic techniques.

2. METHODS

2.1. Samples and controls used in Magneto-FISH capture experiments

Sediment samples were collected in September 2011 from methane seeps within the S. Hydrate Ridge area off the coast of Oregon at a depth of 775 m using the ROV *JASON* and the R/V *Atlantis*. Marine sediment was collected in a push core (PC-47) associated with a sulfide-oxidizing microbial mat adjacent to an actively bubbling methane vent. A sediment slurry from the upper 0–12 cm depth horizon was prepared with 1 volume N_2 sparged artificial seawater to 1 volume sediment, overpressured with methane (3 bar) and incubated at 8 °C in a 1-l Pyrex bottle sealed with a butyl rubber stopper. A 4-ml sample from the incubation was collected on 19 November 2012. Samples were immediately fixed in 0.5 ml sediment aliquots in 1.5% paraformaldehyde (PFA) for 1 h at room temperature (fixation can alternatively be performed at 4 °C overnight). Samples were washed in 50% 1 × phosphate-buffered saline (PBS): 50% EtOH, then 75% EtOH: 25% DI water, and resuspended in 2 volumes (1 ml) 100% ethanol. Samples were centrifuged at 1000 × *g* for 1 min between wash steps. Samples were stored at −20 °C until usage.

As a control to test association specificity, 0.5 ml of sediment slurry was spiked with 10 μl of turbid *Paracoccus denitrificans*, strain ATCC 19367 extrinsic to hydrate ridge sediments. After addition, the sample was quickly vortexed, fixed, and washed as described earlier. 16S rRNA diversity surveys of the original sediment incubation sample supported the absence of *P. denitrificans* in the bulk sediment.

2.2. Magneto-FISH

A detailed protocol is provided in Table 2.1 and additional information and explanation of the major steps are provided below. When using Magneto-FISH with marine sediment samples, 100 μl of fixed sediment slurry (resuspended in 100% ethanol) is the recommended starting volume for the recovery of PCR-amplifiable DNA (technical support recommendations are generally based on cells per ml of sample as a starting point for optimizing a new system). The method has been tested with sediment volumes ranging from 75 to 3000 μl. Smaller sample sizes have higher target purity, but lower DNA yield. For the purposes of this chapter, all reagent amounts are given for the 100 μl starting sample size (small-scale prep) but can be scaled up as indicated for larger samples. There are two means to scale up these reactions: (1) using more of the starting sample with the same oligonucleotide probe or (2) using more of starting sample, but with different probes. With option 1, all sample aliquots can be combined during wash steps as indicated. For option 2, sample aliquots can be combined during the initial permeabilization stages but can no longer be combined after probes have been applied. All reagents should be sterilized by filtration (0.22 μm) prior to use, and sterile sample containers should be used in subsequent steps. Additionally, after fluor addition, samples should be treated as light-sensitive.

2.2.1 Permeabilization and inhibition of endogenous peroxidases

The TE pH 9 heating step serves to permeabilize cells and loosen sediment particles. The hydrogen peroxide addition inhibits endogenous peroxidases prior to the CARD reaction. To remove ethanol, spin sediment-ethanol slurry at 16,000 × *g* for 1 min, remove supernatant, and resuspend in Tris–EDTA (TE) (pH 9). When performing multiple reactions with the same sediment, they can be combined during these steps (i.e., for six captures, add 600 μl (original volume) of sediment slurry to 100 ml TE, pH 9) after removing ethanol. Sonication duration should be optimized for each system since it is a balance between loosening cell material and destroying cell material.

Table 2.1 Step-by-step Magneto-FISH method

Magneto-FISH
1. Permeabilization and inhibition of endogenous peroxidase **a.** Add 100 μl sediment slurry to 100 ml TE (pH 9) in a sterile 250-ml glass beaker (or other flat-bottomed vessel to maximize surface area). **b.** Microwave 2 min at 65 °C in a hybridization microwave (100% power) (BP-111-RS-IR, Microwave Research & Applications). **c.** Transfer to two 50-ml falcon tubes and spin at 5000 × *g* for 5 min at 4 °C (all spin steps should be performed in this manner unless otherwise indicated). **d.** Decant supernatant taking care to retain the sediment pellet by pouring slowly and all in one motion. **e.** Resuspend in 50 ml 1 × phosphate-buffered saline (PBS), 0.01 *M* sodium pyrophosphate (PPi), 0.1% H_2O_2 and incubate at room temperature for 10 min, inverting tubes occasionally to keep sediment in suspension. **f.** Sonicate for three 10-s pulses on setting 3 (~6V(rms) output power) (Branson Sonifier W-150 ultrasonic cell disruptor) at room temperature with sterile remote-tapered microtip probe (Branson) inserted into the liquid. **g.** Spin and decant.
2. Liquid CARD–FISH **a.** Resuspend sediment in 2 ml CARD buffer (0.9 M NaCl, 20 m*M* Tris–HCl, pH 7.5, 10% (w/v) dextran sulfate, 1% blocking reagent (in pH 7.5 maleic acid buffer), 0.02% (w/v) SDS) and transfer to a 2-ml Eppendorf tube. **b.** Add 20 μl of 50 ng μl^{-1} CARD probe and vortex (Vortex-Genie 2, MO BIO) briefly to mix. **c.** Wrap tubes in parafilm and tape to the sides of a beaker filled with DI water such that tubes float in an approximately horizontal orientation. **d.** Microwave for 30 min at 46 °C, power setting of 100%. **e.** Remove samples from the water bath and remove parafilm. **f.** Spin tubes at 10,000 × *g* for 2 min. **g.** Decant supernatant into formamide waste and resuspend hybridized sediment in 50 ml 1 × PBS. **h.** Incubate at room temperature for 10 min, shaking occasionally. **i.** Centrifuge, decant supernatant, resuspend in fresh 1 × PBS, centrifuge and decant again, leaving pellet. **j.** Resuspend in 2 ml amplification buffer (1 × PBS, 1% blocking reagent, 10% (w/v) dextran sulfate, 2 *M* NaCl) in 2-ml Eppendorf tube. **k.** Add 2 μl fluor-labeled tyramide (0.5 μg ml^{-1}), 2 μl biotin tyramide (0.5 μg ml^{-1}), and 5 μl 0.0015% H_2O_2. **l.** Wrap tube(s) in foil to protect from light and incubate with gentle shaking or rotating at 37 °C for 1.5 h. **m.** Spin at 10,000 × *g* for 2 min. **n.** Decant supernatant and resuspend in 50 ml 1 × PBS in 50-ml centrifuge tube.

Continued

Table 2.1 Step-by-step Magneto-FISH method—cont'd

- **o.** Incubate for 10 min at room temperature in the dark, shaking occasionally.
- **p.** Spin, resuspend in 50 ml 1 × PBS, and spin again.
- **q.** Resuspend in 49.5 ml 1 × PBS and 0.5 ml 10% blocking reagent in a 50-ml falcon tube.
- **r.** Microwave (BP-111-RS-IR, Microwave Research & Applications) in a vessel large enough to submerge 50-ml tubes for 20 min at 40 °C in DI water.
- **s.** Centrifuge, decant, and resuspend in 50 ml 1 × PBS, then centrifuge and decant again.
- **t.** Resuspend each sample in 1 ml 1 × PBS, 0.01 *M* PPi in a 1.5-ml Eppendorf tube.
- **u.** Counterstain a sample aliquot with DAPI and verify hybridization by microscopy.

3. Magnetic bead preparation and magnetic cell capture

- **a.** Sonicate sample in 1.5-ml tube for 5 s, setting 3, at room temperature to resuspend cells.
- **b.** Add 5 μl antifluor mouse monoclonal IgG antibody (Life Technologies) per 1 ml reaction volume and incubate at 4 °C for 20 min rotating to keep sediment in suspension (Hybridization Oven, VWR).
- **c.** While the sample is incubating, prepare beads:
  - **i.** Add 25 μl of Dynabeads Pan Mouse IgG (Life Technologies) per reaction to 1 ml of Buffer1 (1 × PBS, 0.1% BSA) and place in magnetic holder (Dynal MPC-1.5 ml).
  - **ii.** Invert holder and tube(s) multiple times to wash all beads down to magnet. Remove liquid with pipette and treat as azide waste. Remove tube from holder and resuspend washed beads in 30 μl of Buffer1.
- **d.** After 20 min incubation, spin sample at 300 × *g* for 8 min at 4 °C.
- **e.** Decant supernatant, resuspend sediment pellet in Buffer1, and spin again as in step 3d. Decant supernatant.
- **f.** Add 30 μl of washed beads and 1 ml Buffer1 per sample volume.
- **g.** Incubate 1.5 h at 4 °C in dark while rotating to keep sediment in suspension.
- **h.** Place sample(s) into magnetic holder slots. Invert multiple times and let sit 1 min until sediment has settled to the bottom of the tube. Remove liquid including all sediment while trying not to disturb magnetic beads.
- **i.** To wash beads and target cells, remove tube from magnetic holder and add 1 ml Buffer1 while aiming pipette tip at magnetic beads to resuspend them. If all beads are not resuspended when adding 1 ml, pipette up and down slowly to resuspend remaining beads from side of the tube. After a few washes, counterstain a sample aliquot with DAPI and verify bead attachment by microscopy. Repeat wash step at least nine more times (10 in total).
- **j.** Save any sample necessary for further microscopy before proceeding to DNA extraction.
- **k.** After final wash, resuspend washed beads and cells in 400 μl of TE buffer (pH 8).

Continued

Table 2.1 Step-by-step Magneto-FISH method—cont'd
DNA processing

1. Cell lysis and reversing cross-links in DNA
   - **a.** Add lysis reagents (10 μl 5 *M* NaCl and 25 μl 20% SDS) to 400 μl TE with beads from step 3k.
   - **b.** Remove liquid from screw cap 2 ml bead-beating tube with garnet sand (PowerSoil DNA Kit PowerBead Tubes, MO BIO).
   - **c.** Add total volume of sample and lysis reagents (435 μl) to bead-beating tube.
   - **d.** Bead beat at setting 5.5 for 45 s (FastPrep FP120, Thermo Electron Corp.).
   - **e.** Three rounds of alternating freeze/thaw (−80 °C and 65 °C were used in this study).
   - **f.** Incubate samples for at least 2 h, up to 48 h, in a 65 °C water bath.
2. DNA extraction
   - **a.** Add 0.5 ml phenol (pH 8, 0.1% hydroxyquinoline) to bead-beating tube.
   - **b.** Vortex to mix, and spin for 2 min at 16,000 × *g*.
   - **c.** Transfer supernatant to a new tube while avoiding particulates at TE/phenol interface.
   - **d.** Add 250 μl phenol and 250 μl chloroform:IAA (24:1).
   - **e.** Vortex to mix, spin 1 min at 16,000 × *g*, and transfer supernatant to new tube.
   - **f.** Add 500 μl chloroform:IAA, vortex briefly, and spin 2 min at 16,000 × *g*.
   - **g.** Add 200 μl TE to cellulose spin column (Microcon, Millipore) and then add DNA supernatant.
   - **h.** Spin 8 min, 14,000 × *g*. Wash DNA on spin column 3 with 500 μl TE.
   - **i.** Elute into new tube at 1,000 × *g* for 3 min, as per manufacturer's directions.
3. Concentration
   - **a.** Transfer DNA from elution tube to 1.5-ml maximum recovery centrifuge tube (Flex-Tubes 1.5-ml, Eppendorf) and bring volume up to 37.5 μl with TE.
   - **b.** Add 12.5 μl 10 *M* ammonium acetate (2.5 *M* final concentration), 0.2 μl linear acrylamide, and 125 μl cold EtOH (2.5 volumes).
   - **c.** Precipitate DNA overnight in wet ice (0 °C).
   - **d.** Spin 18,000 × *g* in a microfuge for 30 min at 4 °C to pellet DNA.
   - **e.** Decant supernatant, careful to retain pellet.
   - **f.** Lay tube on its side with cap open on a heat block at 65 °C to evaporate remaining liquid. Resuspend in 10 μl Tris–HCl (pH 8).

Step-by-step detailed instructions for Magneto-FISH protocol. Additional information and suggestions are included in the text for each section. Recommended equipment list: hybridization microwave (BP-111-RS-IR, Microwave Research & Applications), centrifuge (microtubes and 50-ml tubes), sonicator with tapered microtip probe (Branson Sonifier W-150 ultrasonic cell disruptor), rotating or shaking incubator/hybridization oven, magnetic holder (Dynal MPC-1.5 ml), water bath, bead-beating tubes with garnet sand (PowerSoil DNA Kit PowerBead Tubes, MO BIO) and bead beater (FastPrep FP120, Thermo Electron Corp.), cellulose spin columns (Microcon, Millipore), vortex (Vortex-Genie 2, M-O BIO), 1.5-ml maximum recovery centrifuge tubes (Flex-Tubes 1.5 ml, Eppendorf). Special reagents: linear acrylamide, dextran sulfate, blocking reagent, HRP probes, fluor-labeled tyramide(s), biotin tyramide, antifluor mouse monoclonal IgG antibody (Life Technologies), Dynabeads Pan Mouse IgG (Life Technologies).

Table 2.2 CARD–FISH probes, target organisms, and DNA recovery during Magneto-FISH

Sample	Target organism(s)	FA %	Total DNA (ng)	Bulk yield (%)
ANME-2c_760	Ar, ANME subgroup 2c	50	0.4	3
Seep-1a_1441	Ba, Desulfobacteraceae subgroup Seep-SRB1a	40	BD	–
Eel-MSMX_932	Ar, general ANME	35	0.9	8
Delta_495a	Ba, general δ-proteobacteria SRB	25	0.8	7
Arc_915	Ar, general domain-level archaea	25	1.2	11
Bulk sediment	–	–	11.0	–

CARD–FISH probes, microbial target organisms (Ar, archaea; Ba, bacteria), and associated formamide concentrations (FA %) used in this study with corresponding total DNA yield in nanograms quantified on a Qubit fluorometer from small-scale (100 μl) Magneto-FISH captures. The percent bulk yield is calculated by dividing the total DNA recovered for each Magneto-FISH capture by the total bulk DNA recovered from a same volume of paraformaldehyde-fixed sediment. Both loss during processing and selectivity of FISH probes used in Magneto-FISH contribute to the estimated percent bulk yield. Seep-1a_1441 DNA concentration was below detection (BD), but only 10% of sample was analyzed due to sample volume constraints. Probe references: Seep-1a_1441 (Schreiber et al., 2010), ANME-2c_760 (Knittel et al., 2005), Eel-MSMX_932 (Boetius et al., 2000), Arc_915 (Stahl and Amann, 1991), Delta_495a (Loy et al., 2002).

2.2.2 Liquid CARD–FISH

All oligonucleotide probes and corresponding formamide concentrations used are summarized in Table 2.2. When using a histological microwave for hybridization, formamide concentrations were lowered by 10% below the concentrations optimized for a conventional hybridization oven (Fike, Gammon, Ziebis, & Orphan, 2008). *Note*: A hybridization oven can also be used for liquid CARD–FISH, but probe incubation time should be increased to 2 h or more.

If doing multiple reactions, evenly divide sediment pellet among all samples in each CARD hybridization buffer with the appropriate formamide concentration. For histological microwave use, orient the beaker and samples such that only water, and no samples, is in the path of the temperature probe. Inverting or vortexing samples a few times during this incubation can improve mixing of probe and sample, since sediment will tend to settle out of suspension during the incubation. All samples with the same probe can be combined during wash steps, but different probe samples must be kept separate.

For the amplification reaction, samples must be evenly divided into their initial starting proportions (if started with 600 μl of slurry, then separate into six aliquots) for proper target to probe ratios, but like-samples can be recombined during subsequent wash steps. For a larger combined wash, samples can remain in a 50-ml tube with the appropriate amount of PBS and PPi after blocking reagent and washing steps. 2–3 μl of sample dried on a teflon slide is sufficient for verification of hybridization and CARD reaction success. Hybridized samples can also be stored overnight at 4 °C before proceeding with magnetic capture.

2.2.3 Magnetic capture

The magnetic capture consists of three main steps: (1) Incubation of antifluor antibodies with fluor-labeled cells, followed by two centrifugation wash steps to remove any unassociated antifluor. Increasing the centrifugation speed/force does not appear detrimental and could be optimal for other systems in order to retain more material (e.g., nonsediment-associated microbes, single cells, and smaller aggregates). According to technical support representatives, the fear is more in deforming cells with higher centrifugation speeds than losing the antibody attachment. We recommended saving supernatants from the washing steps until satisfied with magneto capture, in the event that steps need to be repeated or reoptimized during bead attachment. (2) Incubation of magnetic beads with antifluor attached cells. (3) Removing remaining sediment and cells that did not attach to beads using a magnetic tube holder. It is also recommended to retain the first two sediment washes until satisfied with magnetic capture and to evaluate efficiency (number of captured cells/cells remaining in wash). Bead resuspension between washes should be done as gently as possible to reduce the strain on bead–cell association. When performing larger reactions, multiply number of reactions by 1 ml Buffer1 to calculate volume of wash to use. Larger magnetic holders for 15- or 50-ml tubes may also be necessary (DynaMag, Life Technologies). To reduce larger volumes down to 400 μl for extraction, adding additional washes in increasingly smaller volumes before final suspension in TE may be helpful. After the final resuspension, it is easier to work with low retention tips as beads can stick to tips and tubes when in TE.

In Pernthaler et al. (2008), the magnetic beads and antifluor antibodies were incubated together before application to the sediment. Here, antifluor and magnetic beads are added in separate, successive reactions. We have found that addition of the antifluor antibodies independently, followed by subsequent addition of magnetic beads, resulted in higher recoveries, likely a result of improved antibody–cell hybridization, which may avoid steric hindrance caused by bulky magnetic beads during the attachment stage (R. S. Poretsky and V. J. Orphan, unpublished data). Pernthaler et al. (2008) also developed a separatory funnel apparatus outfitted with a neodymium ring magnet to allow large volumes of buffer to continually wash the magnetic beads and attached cells (Pernthaler & Orphan, 2010). To simplify this procedure, and increase the recovery of cells after magnetic capture, a conventional magnetic tube holder for 1.5- and 50-ml falcon tubes (Dynal) was used in combination with multiple washes to remove residual sediment particles and collect the bead-attached cells. We found that these modifications achieved a similar level of target cell enrichment with small samples.

2.3. DNA processing

2.3.1 Lysis and reversing PFA cross-links

Higher DNA yields have been reported after 48 h cross-link reversal incubation with no degradation of sample (Gilbert et al., 2007), but may not be necessary if fixation duration and time since fixation are short, or a different fixative is used. Gilbert et al. (2007) also provide a review of other published amendments to DNA extraction methods for PFA-fixed DNA that may provide further insight for optimizing this method for different sample types or downstream goals.

2.3.2 Extraction and concentration

DNA extraction and concentration methods are based on Sambrook and Russell (2001) and Crouse and Amorese (1987). Bead beating can be replaced by vortexing at maximum speed for 10 min. Freeze/thaw cycles can be performed at a range of freezing and thawing temperatures: −80 °C and 65 °C were chosen based on equipment available and for rapid cycling between states.

2.4. Quantification

The extremely low DNA concentrations from Magneto-FISH samples require the highest possible sensitivity for detection, reduction of sample loss during quantification, and minimization of contamination during processing or from reagents (Woyke et al., 2011). For DNA quantification prior to metagenomic library construction, the use of a Qubit fluorometer and HS dsDNA Assay kit (Life Technologies) is recommended, though it may require as much as half of the final DNA extract for the small-scale preparation (5 μl) to obtain a reading above detection.

2.5. PCR and cloning

Archaeal *16S rRNA* Primers, annealing 54 °C

- Arc23F (DeLong, 1992; Waldron, Petsch, Martini, & Nüsslein, 2007)—TCC GGT TGA TCC YGC C

- U1492R (Lane, 1991)—GGY TAC CTT GTT ACG ACT T

mcrA Primers, annealing 52 °C

- ME1 (Hales et al., 1996)—GCM ATG CAR ATH GGW ATG TC
- ME2 (Hales et al., 1996)—TCA TKG CRT AGT TDG GRT AGT

P. denitrificans, annealing 50 °C

- Bac27F (Lane, 1991)—AGA GTT TGA TYM TGG CTC
- PAR1244R (Neef et al., 1996)—GGA TTA ACC CAC TGT CAC

Hot start Taq DNA polymerases, such as HotMaster (5 PRIME), are recommended for PFA-fixed samples, especially when trying to amplify larger (>1000 bp) fragments such as full-length 16S rRNA (Imyanitov et al., 2006). All Magneto-FISH PCRs were 12.5 μl of total volume containing 1 μl DNA template. The following thermocycler conditions were used: 95 °C initial denaturation of 2 min, followed by 40 cycles of 94 °C for 20 s, annealing for 20 s at temperatures listed above for primers, 1–1.5 min extension at 72°C, and a final extension of 10 min at 72 °C. PCR reagents were used at the following concentrations: 1× HotMaster buffer with 25 m*M* Mg^{2+}, 0.22 m*M* dNTPs, 0.2 μ*M* forward and reverse primer, 0.2 U HotMaster Taq per μl reaction. Prior to cloning, an additional reconditioning PCR step of 5–8 cycles was performed in 25 μl, using 5 μl of template from the original PCR (Thompson, Marcelino, & Polz, 2002). Reconditioned PCRs were quantified by gel electrophoresis (1% gel, SYBR safe stain), filtered (MultiScreen PCR Filter Plate #MSNU03010, Millipore) to remove primers, and concentrated in 10 μl Tris–HCl (pH 8). Approximately 4 μl of PCR product was used per reaction according to guidelines for TOPO TA Cloning Kit for Sequencing with pCR4-TOPO Vector and One Shot Top 10 chemically competent *Escherichia coli* (Life Technologies). An ABI Prism 3730 DNA sequencer was used for all sequencing.

2.6. Phylogenetic analysis of 16S rRNA and metabolic genes (*mcrA*)

Translated methyl-coenzyme reductase alpha subunit (*mcrA*) nucleotide sequences were added to an *mcrA* database and aligned in ARB utilizing the ARB alignment features (Ludwig et al., 2004). 16S rRNA sequences were aligned using Silva online aligner (Quast et al., 2013) and then imported into ARB to verify alignment. Representative sequences were selected from the alignments and cropped to a common region containing no primers: 451 nucleotide containing positions for *mcrA* and 901 nucleotide containing positions for 16S rRNA. Sequences were then exported from ARB and phylogenies were computed using MrBayes (Ronquist et al.,

2012). Convergence was determined by an average standard deviation of split frequencies <0.01. Both phylogenies were computed by nucleotide. Inverse gamma rates and default recommendations from Hall (2004) were used for all other MrBayes parameters.

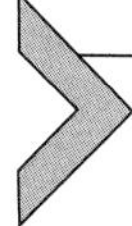

3. RESULTS AND DISCUSSION

3.1. Evaluating the quantification and specificity of captured targets using general and species-specific FISH probes

In the initial Magneto-FISH publication by Pernthaler et al. (2008), a clade-specific probe targeting the archaeal subgroup ANME-2c (probe ANME-2c_760; Knittel, Lösekann, Boetius, Kort, & Amann, 2005) was used to successfully enrich this group and physically associated bacteria from Eel River Basin methane seep sediments, increasing the percentage of recovered ANME-2c from 26% in the original sediments to 92% of the Magneto-FISH-captured archaeal diversity. Here, we expand upon this work, specifically evaluating how FISH probe selectivity affects Magneto-FISH microbial target selectivity. Five different CARD–FISH probes, including domain-level and group-specific probes targeting major methane seep archaeal and sulfate-reducing bacterial groups, were evaluated (Fig. 2.1 and Table 2.2). The three archaeal probes used were ANME-2c_760, Eel-MSMX_932 (general ANME; Boetius et al., 2000), and Arc_915 (general archaea; Stahl & Amman, 1991). Two bacterial probes, Seep-1a_1441 (Schreiber, Holler, Knittel, Meyerdierks, & Amann, 2010) and Delta_495a (Loy et al., 2002), were also used to target *Deltaproteobacteria* that commonly associate with ANME archaea. Seep-1a_1441 is a probe designed to hit a specific subgroup of the *Desulfococcus/Desulfosarcina* (DSS), shown to be a dominant partner of ANME-2c archaea in methane seeps (Schreiber et al., 2010). However, greater diversity of SRB and other bacteria exist in association with ANME in seeps (Holler et al., 2011; Knittel et al., 2003; Lösekann et al., 2007; Niemann et al., 2006; Orphan et al., 2002). Delta_495a targets a broader range of SRB and is expected to recover additional diversity if present in the sample. This allows investigation of the effectiveness of target species enrichment, as well as providing information on the breadth of associated ANME partners.

Total DNA recoveries from each Magneto-FISH capture ranged from below detection to 1.2 ng, depending on the specificity of the FISH capture probe (Table 2.2). The total DNA extracted for each sample was consistent

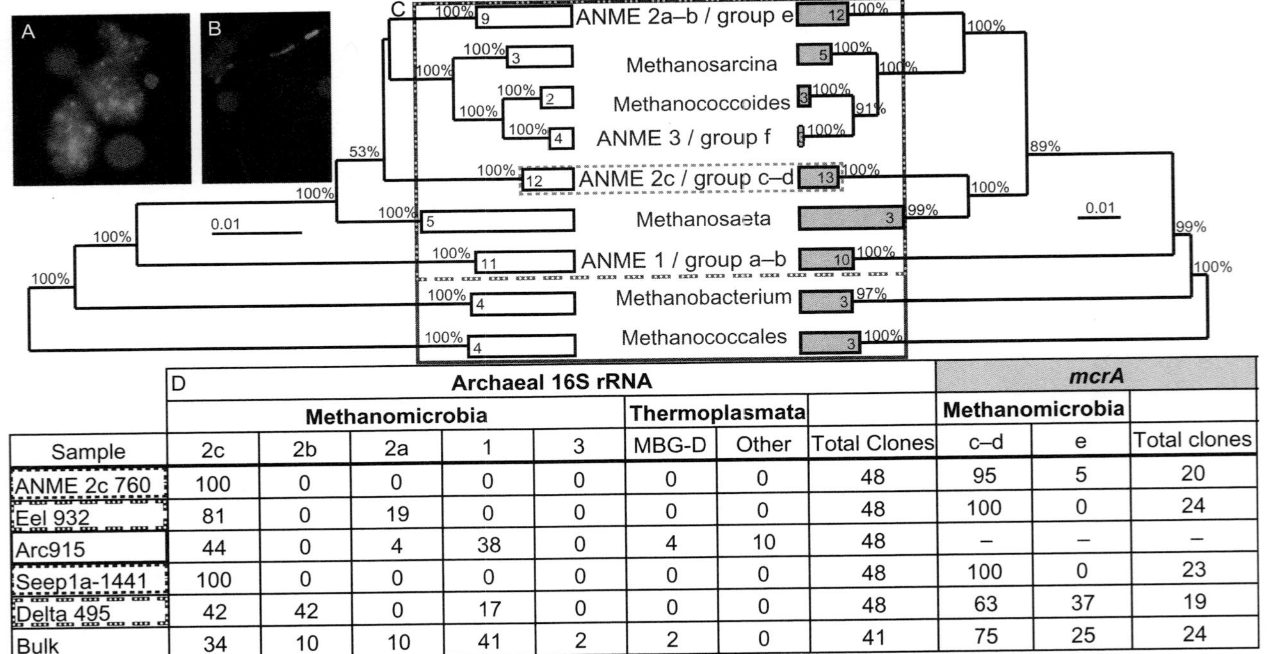

D	Archaeal 16S rRNA								mcrA		
	Methanomicrobia					Thermoplasmata			Methanomicrobia		
Sample	2c	2b	2a	1	3	MBG-D	Other	Total Clones	c–d	e	Total clones
ANME 2c 760	100	0	0	0	0	0	0	48	95	5	20
Eel 932	81	0	19	0	0	0	0	48	100	0	24
Arc915	44	0	4	38	0	4	10	48	–	–	–
Seep1a-1441	100	0	0	0	0	0	0	48	100	0	23
Delta 495	42	42	0	17	0	0	0	48	63	37	19
Bulk	34	10	10	41	2	2	0	41	75	25	24

Figure 2.1 Magneto-FISH probe specificity and related phylogenetic relationships. (A) CARD–FISH epifluorescent image of an ANME-2c (FITC) aggregate counterstained with DAPI. (B) CARD–FISH epifluorescent image of an ANME-1 (FITC) rod chain counterstained with DAPI. In both images, cy3 was overexposed to show beads (beads are 5 μm for scale). (C) Consensus trees of Archaeal 16S rRNA (white boxes) and *mcrA* (gray boxes) genes with ANME clade (16S rRNA) and Group (*mcrA*) names separated by a slash. All other group names apply to both trees. The target range of CARD–FISH probes ANME-2c_760 and Seep1a_1441 (dashed), Eel_932 and Delta_495 (dashed with dots), and Arc_915 (solid) are indicated in the trees and table. (D) Table includes percent of total archaeal clones from each Magneto-FISH capture for each archaeal group. *Thermoplasmata* was not included in the trees but is shown in the table to demonstrate the full diversity recovered. No *mcrA* group a–b or f was recovered from Magneto-FISH or Bulk samples, and is not included in the table. *mcrA* clone libraries were not constructed for the Arc_915 Magneto-FISH capture. (For interpretation of the references to color in this figure legend, the reader is referred to the online version of this chapter.)

with the predicted yield based on oligonucleotide probe specificity, where clade-specific probes (ANME-2c_760 and Seep-1a_1441) yielded lower DNA recoveries relative to Magneto-FISH captures with more general probes (Eel-MSMX_932, Delta_495a, and Arc_915). The DNA recovered from the group-specific Seep-1a_1441 probe is reported as not detected in the table; however, only 1 μl (10%) of the total DNA extract was quantified to preserve sample material. Typically, 5 μl (50%) was necessary for detection of other Magneto-FISH captures. Based on PCR amplification, the Seep-1a_1441 Magneto-FISH capture most likely recovered a DNA concentration similar to that observed with ANME-2c_760.

In an attempt to quantify the level of confidence in Magneto-FISH microbial associations, we spiked a bulk sediment sample with a known volume of an alien cultured organism, *P. denitrificans*. This pure culture has a diagnostic morphology and was not detected in any of our bulk sediment analyses. After confirming with FISH and microscopy that the introduced *P. denitrificans* cells were present in the sediment sample after fixation and at an environmentally relevant concentration (visible in each field of view, but not a dominant species), this spiked sample was used for Magneto-FISH with the Eel-MSMX_932 probe. Using a primer specific to *P. denitrificans* (Neef et al., 1996), DNA recovered from the capture did not reveal *P. denitrificans* contamination after 40 cycles of PCR. There was also positive PCR amplification of *P. denitrificans* from the spiked bulk sediment DNA extraction. Universal bacterial *16S rRNA* primers were used to confirm that the Eel-MSMX_932 + *P. denitrificans* sample did not have amplification inhibition. This suggests that microorganisms associated with target Magneto-FISH samples are unlikely present due to nonspecific attachment during the Magneto-FISH or sample preservation protocol.

To evaluate Magneto-FISH enrichment of target species, clone libraries for both archaeal 16S rRNA and *mcrA* were constructed (Fig. 2.1). *mcrA* encodes for an enzymatic step common to methanogenic and methanotrophic archaea (Hallam, Girguis, Preston, Richardson, & DeLong, 2003; Luton, Wayne, Sharp, & Riley, 2002). Conserved regions can be used as a measure of archaeal diversity in methane seeps, with similar tree topology to archaeal 16S rRNA (Hallam et al., 2003, 2011; Luton et al., 2002). Parallel analysis of 16S rRNA and metabolic gene diversity in Magneto-FISH capture experiments using complementary (or nested; Amann, Ludwig, & Schleifer, 1995) suites of FISH probes with differing specificities can assist in evaluating the affiliation of specific metabolic genes with a 16S rRNA phylotype. Results from five independent Magneto-FISH capture experiments using different probes on the same starting material recovered the predicted level of archaeal diversity, based on the specificity of the capture probe. For example, archaeal 16S rRNA diversity recovered from ANME-2c_760 and Seep-1a_1441 Magneto-FISH experiments was 100% affiliated with the ANME-2c group, with parallel *mcrA* analysis recovering 95% and 100% of *mcrA* groups c/d, respectively. The DSS-affiliated Seep-SRB1a group has been shown in environmental FISH surveys to predominately pair with ANME-2c (Schreiber et al., 2010). The abundance of ANME-2c in both the archaeal 16S rRNA and *mcrA* gene surveys from the SRB-targeted Seep-1a_1441 capture is consistent with these findings. These experiments support the results from Pernthaler et al. (2008), demonstrating that high specificity can be achieved with Magneto-FISH. These data also demonstrate the ability to corroborate a microbial association hypothesis, such as ANME-2c/Seep-SRB1a (Schreiber et al., 2010), with complementary Magneto-FISH experiments.

The interesting pattern of ANME-2 diversity in the more general Eel-MSMX_932 and Delta_495a Magneto-FISH samples is another example of the more nuanced information that can be recovered by this technique. Note that the simplified trees in Fig. 2.1 do not resolve the distinction between ANME-2a and ANME-2b; ANME-2b sequences form a coherent clade related to the ANME-2a group (see Figure 2 in Orphan et al., 2001). While ANME-2a and ANME-2b were equally represented in the bulk sediment diversity survey, these closely related archaeal groups showed differential distribution in the Eel-MSMX_932 and Delta_495a Magneto-FISH captures. While these two probes are expected to have similar levels of target group specificity in this system, ANME-2a was not detected in Eel-MSMX_932 samples and conversely ANME-2b was absent in the Delta_495a capture. The Eel-MSMX_932 probe was designed to target all Eel River Basin clones affiliated with the order *Methanosarcinales* (Boetius et al., 2000), but archaeal 16S rRNA ANME-2 diversity contained ANME-2c (81%) and ANME-2a (19%) sequences and no ANME-2b. The Delta_495a Magneto-FISH capture, selecting for general *Deltaproteobacteria*, recovered an equal number of ANME-2c and ANME-2b clones (42%), as well as 17% affiliated with ANME-1, but no ANME-2a. This would suggest that, in this sample, ANME-2c and ANME-2b might be more likely to form associations with *Deltaproteobacteria* than ANME-2a. These hypotheses can be tested with independent FISH hybridization experiments with the original sediment sample (see, e.g., Pernthaler et al., 2008).

Magneto-FISH can also aid in correlating diagnostic metabolic genes (e.g., *mcrA*, *dsrAB*, *aprA*, *nifH*, etc.) to 16S rRNA identity. Since 42% of the clones in the Delta_495a capture were ANME-2b, *mcrA* sequences that are distinct from the previously described ANME-1 group a–b or ANME-2c affiliated group c–d may be associated with ANME-2b, a currently undefined *mcrA* group designated here as e′. The bulk sediment distribution within ANME-2 archaeal 16S rRNA sequences alone is 64%—2c, 18%—2b, 18%—2a. The Delta_495a ANME-2 archaeal 16S rRNA distribution is 50%—2c and 50%—2b. The bulk sediment distribution of ANME-2 *mcrA* sequences is 75%—c–d (2c), 4%—e′ (2b), 21%—e (2a). The Delta_495a distribution of ANME-2 *mcrA* sequences is 63%—c–d (2c), 37%—e′ (2b), 0%—e (2a). Since all three ANME-2 groups are found in both archaeal 16S rRNA and *mcrA* clone libraries for bulk sediment, but only 2c (c–d) and 2b (e′) are found in Delta_495a, there are multiple lines of evidence to support the hypothesis of group e′ *mcrA*. It should also be noted that the *mcrA* primers are not complementary to the majority of ANME-1 sequences, so investigation of ANME-1 correlations between archaeal 16S rRNA and *mcrA* was not possible.

3.2. Optimization for metagenomics

Advances in library preparation and high-throughput sequencing protocols have significantly lowered the required amount of DNA for metagenomics (as low as 1 ng DNA with the Nextera XT). However, our small-scale Magneto-FISH captures yield DNA in amounts that are still below current thresholds without including a post-DNA amplification (e.g., MDA), similar to that used in single-cell genomics (Woyke et al., 2011 and other chapters in this volume) and in the Magneto-FISH ANME-2c metagenome (Pernthaler et al., 2008).

To determine where the protocol could be optimized to increase recovery and DNA yield, we evaluated the losses associated with the different steps of the Magneto-FISH protocol. The Magneto-FISH cell retention efficiency was assessed by extracting DNA from wash step supernatants during a large-scale Magneto-FISH ANME-2c_760 capture (Table 2.3). The DNA concentration of the supernatants was then compared to the amount of DNA extracted from PFA-fixed bulk sediment of the same initial volume (3 ml slurry). We estimate that ~6% total DNA is lost during the initial liquid CARD–FISH hybridization. An additional 28% is lost after the antibody (IgG) incubation, which can be improved by increasing the speed during centrifugation (discussed in Section 2). The sample remaining in the post-capture wash is due to both intended (selectivity from magnetic

Table 2.3 Sample retention efficiency by DNA recovery during major steps of the Magneto-FISH protocol

Sample	Total DNA (ng)	Bulk yield (%)
ANME-2c_760, post-capture sample	144.6	14
ANME-2c_760, post-liquid CARD–FISH	63.4	6
ANME-2c_760, post-IgG	276.3	28
ANME-2c_760, post-capture wash	430.5	43
Bulk	1000.0	–

DNA recovered from different stages of a large-scale ANME 2c-760 Magneto-FISH sample to examine losses and selectivity. Initial sample was from 3 ml of PFA-fixed slurry in EtOH. The percent bulk yield is calculated by dividing the total DNA recovered at each Magneto-FISH step (accounting wash volume differences) by the total bulk DNA recovered from a same volume of paraformaldehyde-fixed sediment. *ANME2c-760, post-capture sample* is target cells attached to beads at the end of the protocol. *ANME2c-760, post-liquid CARD–FISH* is a 50 ml 1 × PBS wash supernatant. *ANME-2c_760, post-IgG* is the supernatant after 300 × *g* spin to remove remaining antibody. *ANME-2c_760, post-capture wash* is the sediment and Buffer1 removed after the first wash when the sample is in the magnetic holder (remaining nontarget cells).

capture) and unintended (poor hybridization and/or unsuccessful magnetic capture) losses. The DNA yield from Magneto-FISH before the magnetic capture step can be estimated by adding the DNA recovered from the post-Magneto-FISH wash (430 ng) to the yield from Magneto-FISH sample (145 ng) for a total of 575 ng. Using the specific ANME-2c_760 capture probe, the DNA yield from ANME2c-760 Magneto-FISH is 25% of this estimated total yield. As 33% of the recovered bulk sediment clones are ANME-2c, this is close to the expected level of selectivity. Assuming ANME archaea are the dominant archaea and about one-third of the total microbial assemblage based on the ANME:SRB ratio of 1:3 from other Hydrate Ridge studies (Nauhaus, Albrecht, Elvert, Boetius, & Widdel, 2007; Orphan & House, 2009), and one-third of those archaea are ANME-2c (bulk clone library results, Fig. 2.1D), then one-sixth of the bulk sediment-extracted DNA would result in a theoretical yield of 166 ng. The experimental ANME-2c DNA yield (144.6 ng) is 87% of this theoretical yield.

We also examined DNA extraction efficiency by testing a range of methods to improve cell lysis, removal of formalin cross-links, and losses during DNA precipitation. As discussed in the methods, implementation of an extended heating step was found to reduce PFA cross-linking issues and yielded the greatest improvement to DNA extraction efficiency. The use of conventional organic extraction with phenol:chloroform resulted in higher yields than tested kit protocols (PowerSoil DNA Isolation Kit, MO BIO). Recovery of DNA after ethanol precipitation was enhanced by the use of ammonium acetate and linear acrylamide at 0 °C (Crouse & Amorese, 1987) and maximum recovery tubes. The theoretical yield of bulk sediment DNA per milliliter slurry is 10^{-6} g ml^{-1}. This is based on 10^7 aggregates per milliliter sediment slurry (calculated for this study) and estimates of 10^2 cells per aggregate (Nauhaus et al., 2007) and 10^{-15} DNA per cell (Button & Robertson, 2001; Simon & Azam, 1989). Although this calculation does not account for single cells that also contribute to the bulk DNA, single cells are estimated to be 10% or less of the total biomass at Hydrate Ridge (Nauhaus et al., 2007). This theoretical yield is the same order of magnitude as the bulk sediment DNA (experimental) yield of 1000 ng ml^{-1} sediment slurry, indicating efficient DNA extraction.

The age of the fixed sample (time since fixation) and length of fixation time can also impact the success of the Magneto-FISH capture and DNA recovery. Freshly fixed samples are recommended, when possible. We also evaluated ethanol as an alternative fixative to reduce cross-linking issues during DNA recovery. While CARD–FISH signals were not as bright, bead association was successful and expected clone diversity was recovered. Fixative choice and strength are recommended optimization areas for application of Magneto-FISH to other systems.

We also evaluated the ability of the Magneto-FISH procedure to meet metagenomic library preparation DNA concentration requirements without MDA amplification, by scaling up starting sample volume (large-scale Magneto-FISH prep). This large-scale prep is similar to the procedure originally reported in Pernthaler et al. (2008) and outlined in Schattenhofer and Wendeberg (2011) with a few modifications to the magnetic capture and washing steps (described in Section 2). In the large-scale Magneto-FISH prep, 3 ml of sediment slurry was used instead of 0.1 ml. From this volume of slurry, 48.2 ng DNA ml^{-1} slurry was obtained using the ANME-2c_760-specific probe. This is almost 14 times more DNA than a small-scale ANME-2c_760 capture and enough DNA for library preparation using the Nextera XT kit (minimum 1 ng) for Illumina miseq or highseq sequencing. However, the gain in total DNA yield also corresponded with a decrease in specificity. Only 53% of the 16S rRNA phylotypes associated with ANME-2c, compared with 100% in small-scale Magneto-FISH captures. The scaled-up protocol is still useful for enrichment of the target population, with 33% of the archaeal diversity associated with the ANME-2c target relative to 20% in the bulk sediment. For the larger volume

Magneto-FISH protocols, the incorporation of more extensive washing procedures using a separatory funnel apparatus (described in Pernthaler et al., 2008; Schattenhofer & Wendeberg, 2011) may aid in the removal of contaminating particles and enhance enrichment of the microbial target.

Sample specificity and DNA yield should therefore be optimized for downstream needs; if high specificity is required, then pooling many small-scale reactions is recommended; otherwise, one large-scale reaction may be sufficient. It is also recommended that any samples that need be compared are run together with the same conditions and reagents to reduce any methodological variation.

3.3. Optimization for other environmental systems

This Magneto-FISH protocol was developed and optimized for sediment-associated aggregated microorganisms, so optimal application to other systems likely requires adjustments to the liquid CARD–FISH protocol and washing steps for optimal cell recovery. Schattenhofer and Wendeberg (2011) reported enrichment of single SRB cells from hydrocarbon-contaminated sediment using a Magneto-FISH protocol similar to Pernthaler et al. (2008). Schattenhofer and Wendeberg incubated cells with magnetic beads already labeled with antibodies, which may reduce single cell loss during antibody wash steps in the method described here.

To evaluate the method presented here for single-cell Magneto-FISH, we focused on ANME-1. ANME-1 are found predominately as single cells or chains of single cells, rather than in association with SRB, at Hydrate Ridge and have a distinctive rod-shaped morphology (observation for this study and Knittel et al., 2005). When using the general Arc_915 probe to target all archaea in small-scale Magneto-FISH experiments, we were able to recover ANME-1 phylotypes at bulk sediment clone abundance. We also observed single cells and chains attached to beads, indicating the potential to enrich for nonaggregated cell types using this Magneto-FISH method. We then tried Magneto-FISH with an ANME-1-specific probe to select for a single-cell population. We used ANME-1_350 (Boetius et al., 2000) with 30% formamide. We confirmed single cells and chains attached to beads by microscopy (Fig. 2.1B). However, we did not recover quantifiable amounts of DNA and clone abundances were below bulk sediment ratios of ANME-1. Since ANME-1 represented 44% of the recovered archaeal bulk sediment diversity, this should not be due to issues with targeting too small a population.

A possible explanation is that more specific probes are more successful if they work at a higher stringency. When testing Magneto-FISH without adding probe or adding nonsense probes at 5–10% formamide, it is possible to collect nonspecifically bound aggregates. Nonspecific capture was confirmed by microscopy (beads attached to aggregates without any CARD signal) and DNA extraction yields. DNA yields from these samples were below or near the limit of detection, but similar to the DNA concentration of ANME-2c_760 and Seep-1a_1441 samples. However, ANME-2c and Seep-1a captures return only the expected single species and do not show signs of nonspecific binding. ANME-2c_760 (50%) and Seep-1a_1441 (40%) probes had higher formamide concentrations than ANME-1_350 (30%). Only less specific probes such as Arc_915 and Delta_495a (25% formamide) returned the expected population at lower formamide concentrations. Optimization of Magneto-FISH for other systems and/or nonaggregate forming populations may be more successful when utilizing probes with targeted, high specificities.

4. SUMMARY

Magneto-FISH provides a method to target microbial associations from environmental samples for metagenomic and other molecular analyses with high specificity. It is adaptable to a range of target populations within a system, working from the vast array of already vetted FISH probes or developing new ones. It is also an affordable technique since it does not require any special training or equipment beyond the contents of a normal microbiology laboratory. While the method was designed for ANME-2 aggregates and associated bacteria, it can be applied to, and optimized for, a range of microbial systems utilizing the recommendations described herein. By enriching for associations prior to metagenomic analysis, the genetic information obtained is for a working partnership that may otherwise be lost in a bulk environmental analysis. This middle ground will be invaluable in the effort to better understand all levels at which microbes function in an environment and, in particular, in understanding how microbial associations on small scales reflect larger scale chemical and nutrient cycling.

ACKNOWLEDGMENTS

We acknowledge Annelie (Pernthaler) Wendeberg, Rachel Poretsky, Joshua Steele for their contributions towards the development and optimization of Magneto-FISH. Stephanie Cannon, Jen Glass, Kat Dawson, Hiroyuki Imachi, and Caltech Genomics Center are also acknowledged for their assistance with various aspects of this project. Funding for this work was provided by the Gordon and Betty Moore foundation and a DOE early career grant (to V. J. O.) and NIH/NRSA training grant 5 T32 GM07616 (to E. T. R.).

REFERENCES

Amann, R. I., Binder, B. J., Olson, R. J., Chisholm, S. W., Devereux, R., & Stahl, D. A. (1990). Combination of 16s rrna-targeted oligonucleotide probes with flow cytometry for analyzing mixed microbial populations. *Applied and environmental microbiology, 6*, 1919–1925.

Amann, R. I., Ludwig, W., & Schleifer, K. H. (1995). Phylogenetic identification and in situ detection of individual microbial cells without cultivation. *Microbiological Reviews, 59*(1), 143–169.

Ashkin, A. (1997). Optical trapping and manipulation of neutral particles using lasers. *Proceedings of the National Academy of Sciences, 10*, 4853–4860.

Boetius, A., Ravenschlag, K., Schubert, C. J., Rickert, D., Widdel, F., Gieseke, A., et al. (2000). A marine microbial consortium apparently mediating anaerobic oxidation of methane. *Nature, 407*(6804), 623–626.

Button, D., & Robertson, B. R. (2001). Determination of DNA content of aquatic bacteria by flow cytometry. *Applied and Environmental Microbiology, 67*(4), 1636–1645.

Crouse, J., & Amorese, D. (1987). Ethanol precipitation: Ammonium acetate as an alternative to sodium acetate. *Focus, 9*(2), 3–5.

DeLong, E. F. (1992). Archaea in coastal marine environments. *Proceedings of the National Academy of Sciences of the United States of America, 89*(12), 5685–5689.

Fike, D. A., Gammon, C. L., Ziebis, W., & Orphan, V. J. (2008). Micron-scale mapping of sulfur cycling across the oxycline of a cyanobacterial mat: A paired nanoSIMS and CARD-FISH approach. *The ISME Journal, 2*(7), 749–759.

Gilbert, M. T., Haselkorn, P. T., Bunce, M., Sanchez, J. J., Lucas, S. B., Jewell, L. D., et al. (2007). The isolation of nucleic acids from fixed, paraffin-embedded tissues, which methods are useful when? *PLoS One, 2*(6), e537.

Hales, B. A., Edwards, C., Ritchie, D. A., Hall, G., Pickup, R. W., & Saunders, J. R. (1996). Isolation and identification of methanogen-specific DNA from blanket bog peat by PCR amplification and sequence analysis. *Applied and Environmental Microbiology, 62*(2), 668–675.

Hall, B. G. (2004). *Phylogenetic trees made easy: A how-to manual.* Sunderland: Sinauer Associates.

Hallam, S. J., Girguis, P. R., Preston, C. M., Richardson, P. M., & DeLong, E. F. (2003). Identification of methyl coenzyme M reductase A (mcrA) genes associated with methane-oxidizing archaea. *Applied and Environmental Microbiology, 69*(9), 5483–5491.

Hallam, S. J., Page, A. P., Constan, L., Song, Y. C., Norbeck, A. D., Brewer, H., et al. (2011). Molecular tools for investigating ANME community structure and function. *Methods in Enzymology, 494*, 75.

Holler, T., Widdel, F., Knittel, K., Amann, R., Kellermann, M. Y., Hinrichs, K. U., et al. (2011). Thermophilic anaerobic oxidation of methane by marine microbial consortia. *The ISME Journal, 5*(12), 1946–1956.

Imyanitov, E. N., Suspitsin, E. N., Buslov, K. G., Kuligina, E. Sh, Belogubova, E. V., Togo, A. V., et al. (2006). Isolation of nucleic acids from paraffin-embedded archival tissues and other difficult sources. In *The DNA book: Protocols and procedures for the modern molecular biology laboratory* (pp. 85–97). Sudbury, MA: Jones and Bartlett Publishers.

Kalyuzhnaya, M., Lidstrom, M., & Chistoserdova, L. (2008). Real-time detection of actively metabolizing microbes by redox sensing as applied to methylotroph populations in lake washington. *The ISME journal, 7*, 696–706.

Knittel, K., Boetius, A., Lemke, A., Eilers, H., Lochte, K., Pfannkuche, O., et al. (2003). Activity, distribution, and diversity of sulfate reducers and other bacteria in sediments above gas hydrate (Cascadia Margin, Oregon). *Geomicrobiology Journal, 20*(4), 269–294.

Knittel, K., Lösekann, T., Boetius, A., Kort, R., & Amann, R. (2005). Diversity and distribution of methanotrophic archaea at cold seeps. *Applied and Environmental Microbiology*, *71*(1), 467–479.

Lane, D. J. (1991). *16S/23S rRNA sequencing*. Chichester, England: John Wiley & Sons.

Lösekann, T., Knittel, K., Nadalig, T., Fuchs, B., Niemann, H., Boetius, A., et al. (2007). Diversity and abundance of aerobic and anaerobic methane oxidizers at the Haakon Mosby Mud Volcano, Barents Sea. *Applied and Environmental Microbiology*, *73*(10), 3348–3362.

Loy, A., Lehner, A., Lee, N., Adamczyk, J., Meier, H., Ernst, J., et al. (2002). Oligonucleotide microarray for 16S rRNA gene-based detection of all recognized lineages of sulfate-reducing prokaryotes in the environment. *Applied and Environmental Microbiology*, *68*(10), 5064–5081.

Ludwig, W., Strunk, O., Westram, R., Richter, L., Meier, H., Yadhukumar Buchner, A., et al. (2004). ARB: A software environment for sequence data. *Nucleic Acids Research*, *32*(4), 1363–1371.

Luton, P. E., Wayne, J. M., Sharp, R. J., & Riley, P. W. (2002). The mcrA gene as an alternative to 16S rRNA in the phylogenetic analysis of methanogen populations in landfill. *Microbiology*, *148*(11), 3521–3530.

Melin, J., & Quake, S. R. (2007). Microfluidic large-scale integration: The evolution of design rules for biological automation. *Annual Review of Biophysics and Biomolecular Structure*, *36*, 213–231.

Nauhaus, K., Albrecht, M., Elvert, M., Boetius, A., & Widdel, F. (2007). In vitro cell growth of marine archaeal-bacterial consortia during anaerobic oxidation of methane with sulfate. *Environmental Microbiology*, *9*(1), 187–196.

Neef, A., Zaglauer, A., Meier, H., Amann, R., Lemmer, H., & Schleifer, K. H. (1996). Population analysis in a denitrifying sand filter: Conventional and in situ identification of Paracoccus spp. in methanol-fed biofilms. *Applied and Environmental Microbiology*, *62*(12), 4329–4339.

Niemann, H., Lösekann, T., de Beer, D., Elvert, M., Nadalig, T., Knittel, K., et al. (2006). Novel microbial communities of the Haakon Mosby mud volcano and their role as a methane sink. *Nature*, *443*(7113), 854–858.

Orphan, V. J., Hinrichs, K. U., Ussler, W., Paull, C. K., Taylor, L. T., Sylva, S. P., et al. (2001). Comparative analysis of methane-oxidizing archaea and sulfate-reducing bacteria in anoxic marine sediments. *Applied and Environmental Microbiology*, *67*(4), 1922–1934.

Orphan, V. J., & House, C. H. (2009). Geobiological investigations using secondary ion mass spectrometry: Microanalysis of extant and paleo-microbial processes. *Geobiology*, 7(3), 360–372.

Orphan, V. J., House, C. H., Hinrichs, K. U., McKeegan, K. D., & DeLong, E. F. (2002). Multiple archaeal groups mediate methane oxidation in anoxic cold seep sediments. *Proceedings of the National Academy of Sciences of the United States of America*, *99*(11), 7663.

Pernthaler, A., Dekas, A. E., Brown, C. T., Goffredi, S. K., Embaye, T., & Orphan, V. J. (2008). Diverse syntrophic partnerships from deep-sea methane vents revealed by direct cell capture and metagenomics. *Proceedings of the National Academy of Sciences of the United States of America*, *105*(19), 7052–7057.

Pernthaler, A., & Orphan, V. J. (2010). *U.S. Patent No. 7736855. Process for separating microorganisms*. Washington, DC: U.S. Patent and Trademark Office.

Pernthaler, A., Pernthaler, J., & Amann, R. (2002). Fluorescence in situ hybridization and catalyzed reporter deposition for the identification of marine bacteria. *Applied and Environmental Microbiology*, *68*(6), 3094–3101.

Quast, C., Pruesse, E., Yilmaz, P., Gerken, J., Schweer, T., Yarza, P., et al. (2013). The SILVA ribosomal RNA gene database project: Improved data processing and web-based tools. *Nucleic Acids Research*, *41*(D1), D590–D596.

Rodrigue, S., Malmstrom, R. R., Berlin, A. M., Birren, B. W., Henn, M. R., & Chisholm, S. W. (2009). Whole genome amplification and de novo assembly of single bacterial cells. *PLoS One*, *4*(9), e6864.

Ronquist, F., Teslenko, M., van der Mark, P., Ayres, D. L., Darling, A., Höhna, S., et al. (2012). MrBayes 3.2: Efficient Bayesian phylogenetic inference and model choice across a large model space. *Systematic Biology*, *61*, 539–542.

Safarık, I., & Safarıkova, M. (1999). Use of magnetic techniques for the isolation of cells. *Journal of Chromatography B*, *33–53*.

Sambrook, J., & Russell, D. W. (2001). *Molecular cloning: A laboratory manual* (Vol. 1). Cold Spring Harbor, New York: Cold Spring Harbor Laboratory Press.

Schattenhofer, M., & Wendeberg, A. (2011). Capturing microbial populations for environmental genomics. In *Handbook of molecular microbial ecology I: Metagenomics and complementary approaches* (pp. 735–740). John Wiley & Sons, Inc.

Schreiber, L., Holler, T., Knittel, K., Meyerdierks, A., & Amann, R. (2010). Identification of the dominant sulfate-reducing bacterial partner of anaerobic methanotrophs of the ANME-2 clade. *Environmental Microbiology*, *12*(8), 2327–2340.

Simon, M., & Azam, F. (1989). Protein content and protein synthesis rates of planktonic marine bacteria. *Marine Ecology Progress Series, Oldendorf*, *51*(3), 201–213.

Stahl, D. A., & Amann, R. I. (1991). Development and application of nucleic acid probes in bacterial systematics. In E. Stackebrandt & M. Goodfellow (Eds.), *Sequencing and Hybridization Techniques in Bacterial Systematics* (pp. 205–248). Chichester, England: John Wiley and Sons.

Thompson, J. R., Marcelino, L. A., & Polz, M. F. (2002). Heteroduplexes in mixed-template amplifications: Formation, consequence and elimination by 'reconditioning PCR'. *Nucleic Acids Research*, *30*(9), 2083–2088.

Waldron, P. J., Petsch, S. T., Martini, A. M., & Nüsslein, K. (2007). Salinity constraints on subsurface archaeal diversity and methanogenesis in sedimentary rock rich in organic matter. *Applied and Environmental Microbiology*, *73*(13), 4171–4179.

Woyke, T., Sczyrba, A., Lee, J., Rinke, C., Tighe, D., Clingenpeel, S., et al. (2011). Decontamination of MDA reagents for single cell whole genome amplification. *PLoS One*, *6*(10), e26161.

Yilmaz, S., Haroon, M. F., Rabkin, B. A., Tyson, G. W., & Hugenholtz, P. (2010). Fixation-free fluorescence in situ hybridization for targeted enrichment of microbial populations. *ISME J*, *10*, 1352–1356.

Yoon, H. S., Price, D. C., Stepanauskas, R., Rajah, V. D., Sieracki, M. E., Wilson, W. H., et al. (2011). Single-cell genomics reveals organismal interactions in uncultivated marine protists. *Science*, *332*(6030), 714–717.

CHAPTER THREE

Coupling FACS and Genomic Methods for the Characterization of Uncultivated Symbionts

Anne Thompson*, Shellie Bench†, Brandon Carter*, Jonathan Zehr*,1

*University of California, Santa Cruz, Santa Cruz, California, USA
†Stanford University, Stanford, California, USA
[1]Corresponding author: e-mail address: jpzehr@gmail.com

Contents

1. Introduction 46
2. Fluorescence-Activated Cell Sorting 48
   2.1 Flow cytometer features 48
   2.2 Preparation for sorting 49
   2.3 Sorting 49
   2.4 Locating the target symbiont population 50
3. Metagenomic Sequencing of Uncultured Symbiont Populations 51
   3.1 Sampling to separate symbiotic partners 52
   3.2 Performing MDA on sorted cells 53
   3.3 Sequencing and bioinformatic analysis of symbiont genomes 53
4. Determining Host Identity and Metabolism 55
   4.1 Sorting single cells and screening by qPCR 56
   4.2 Nested PCR of 16S and 18S rRNA genes from positive qPCRs using universal primers 56
   4.3 Sequencing rRNA genes from nested PCR and phylogenetic analysis 58
References 58

Abstract

Symbioses between microbes are likely widespread and functionally relevant in diverse biological systems; however, they are difficult to discover. Most microbes remain uncultivated, symbioses can be relatively rare or dynamic, and intercellular connections can be delicate. Thus, traditional methods such as microscopy are inadequate for efficient discovery and precise characterization of novel interactions, their metabolic basis, and the species involved. High-throughput metagenomic sequencing of entire microbial communities has revolutionized the field of microbial ecology; however, genomic signals from symbionts can get buried in sequences from abundant organisms and evidence for direct links between microbial species cannot be gained from bulk samples.

Methods in Enzymology, Volume 531
ISSN 0076-6879
http://dx.doi.org/10.1016/B978-0-12-407863-5.00003-4

Thus, a specialized approach to the characterization of symbioses between naturally occurring microbes is required. This chapter presents methods for combining fluorescence-activated cell sorting to isolate and separate uncultivated symbionts with molecular biology techniques for DNA amplification in order to characterize uncultivated symbionts through genomic and metagenomic techniques.

1. INTRODUCTION

Symbioses occur between microorganisms in diverse biological systems. These associations range from delicate connections at cell surfaces to integration of one symbiont into the cell of another, similar to an organelle. The metabolic basis for the symbioses, the types of cellular structures that link cells, their dynamics, and the relative autonomy of the symbiotic partners, are likely as diverse and ecologically relevant as microbes themselves.

The majority of microbes remain uncultivated, so study of complex natural microbial communities is required to discover and characterize novel symbioses. Coupling of microscopy to fluorescence *in situ* hybridization (FISH) has been a successful approach to studying uncultivated symbioses (Orphan, House, Hinrichs, McKeegan, & DeLong, 2001). However, these traditional methods are limited for several reasons. Morphological features of the symbioses can be damaged during sample handling. This is relevant to epiphytic symbioses or symbioses between delicate cells. Use of untreated samples could reduce artifacts and preserve intercellular connections, but garnering enough precise measurements of the symbionts' cellular parameters for statistical power is tedious. To characterize metabolisms of symbionts, studies have applied stable isotope analysis along with FISH (Orphan et al., 2001). However, assessing the full metabolisms and evolutionary histories of symbionts is limited without genomic or metagenomic analysis and it would be difficult to recover intact cells from FISH preparations for whole-genome sequencing. An ideal approach to characterizing uncultivated symbionts will combine high-throughput methods that yield data on cellular parameters, phylogeny, and metabolism.

Fluorescence-activated cell sorting (FACS) is a widely chosen method for separating and concentrating cells from complex samples prior to genomic analysis. Other methods such as serial dilution (Zhang et al., 2006), micromanipulation (Hongoh et al., 2008; Kvist, Ahring, Lasken, & Westermann, 2007), laser-capture microdissection (Navin et al., 2011),

Raman tweezers (Brehm-Stecher & Johnson, 2004; Huang, Ward, & Whiteley, 2009), and microfluidics (Blainey, Mosier, Potanina, Francis, & Quake, 2011; Marcy et al., 2007) have also been used effectively to separate cells prior to genetic or genomic analysis. However, for characterizing uncultivated symbionts, these methods fall short of FACS in a number of ways.

FACS machines deposit single events (individual cells or multiple cells in association) into a variety of vessels quickly and accurately (Ibrahim & van den Engh, 2003, 2007). Precise assessment of pigment, cell size, membrane features, and genetic composition (cell-cycle analysis) can be obtained in minutes for thousands of events (Ibrahim & van den Engh, 2007). Contaminating DNA is limited because only a very small droplet of liquid from the source sample and sheath fluid accompanies each sorted event (Stepanauskas & Sieracki, 2007). Finally, FACS is a relatively gentle. The laminar flow fluidics of FACS prevents disruption of cells during sorting and this is critical in the study of potentially delicate symbioses (Ibrahim & van den Engh, 2007).

Genomic analysis can dramatically increase understanding of the basis for cellular interactions and the ecosystem function of symbionts beyond measuring cellular parameters (Tripp et al., 2010). However, current next-generation sequencing applications require more nucleic acids than can be obtained from thousands of sorted microbes. Thus, coupling FACS to nucleic acid amplification methods is essential.

MDA is a major advance over using PCR for whole-genome amplification (WGA) (Cheung & Nelson, 1996; Dean, Nelson, Giesler, & Lasken, 2001; Telenius et al., 1992; Zhang et al., 1992). MDA uses the Phi29 DNA polymerase to generate high quantities of double-stranded DNA that can be longer than 10 kb, making it an excellent starting material for sequencing applications and genomic analyses (Fig. 3.1).

Besides MDA, other nucleic acid amplification techniques can also be useful in characterizing microbial symbioses (Fig. 3.1). Especially in delicate associations, one symbiotic partner may remain unknown. In these cases, FACS can be coupled to nested PCR, which is applied to sorted single events (host–symbiont complexes), to amplify phylogenetically relevant genes for sequencing. This is a relatively low-cost method of identifying symbiotic partners before further sequencing efforts (Thompson et al., 2012) (Fig. 3.1).

Together FACS, MDA, and nested PCR can provide researchers with information on the species specificity of symbiotic associations, metabolic

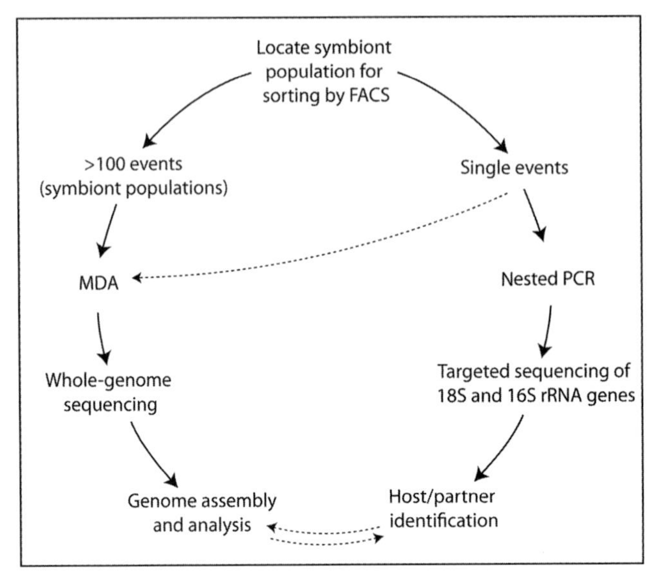

Figure 3.1 Strategy to characterize uncultivated microbial symbionts using FACS coupled to MDA for genomic analysis and to nested PCR for identification of unknown partner cells to suspected symbionts. Dashed line represents strategies for symbiont characterization that are effective following identification of all symbiotic partners.

potential, ecological function, and evolutionary history of uncultivated microbial symbionts. This chapter focuses on how FACS can be coupled with genomic analysis to identify and characterize uncultivated symbionts (Fig. 3.1).

2. FLUORESCENCE-ACTIVATED CELL SORTING

2.1. Flow cytometer features

While several high-throughput flow sorters are available, the BD Biosciences Influx™ Cell Sorter has been the FACS of choice for several recent studies on uncultivated symbionts (Cuvelier et al., 2010; Tripp et al., 2010; Vaulot et al., 2012; Zehr et al., 2008). The Influx™ can be equipped with 10 laser paths supporting collection of 24 data parameters, with the choice of lasers dependent on target cell types. A 488 nm laser (Sapphire Coherent) is commonly used to study phytoplankton as it excites chlorophyll *a* (measured with a band-pass optical filter of 692–40 nm) and other autofluorescent pigments such as phycoerythrin (572–27 nm). Unpigmented cells can be stained with SYTO-13 (Invitrogen, Carlsbad, CA) (Del Giorgio, Bird,

Prairie, & Planas, 1996; Stepanauskas & Sieracki, 2007) and 4′,6-diamidino-2-phenylindole and detected using the 488 nm or UV laser (355 nm) (JDS Uniphase Xcyte Laser), respectively. The multiple lasers of the Influx™ are especially useful when cells are hybridized to synthetic or antibody-based probes. Functional fluorophores are available across the light spectrum and are another means of distinguishing between cell populations via FACS (Orcutt et al., 2008).

2.2. Preparation for sorting

To minimize contamination, ultraclean conditions are required for cell sorting prior to DNA amplification. Sample collection tubes and the sheath fluid reservoir are treated with UV for 2 h before filling (Rodrigue et al., 2009). Sheath fluid may be as simple as a 1% NaCl solution prepared with UV-treated ddH_2O and heat combusted NaCl (450 °C for 4 h), or a commercially available formula such as BioSure (BioSure, Grass Valley, CA, USA). Sheath fluid and sample lines are cleaned by running warm water followed by bleach solution (5–10%) and extensive flushing with UV-treated, DNA-free ddH_2O (Rodrigue et al., 2009; Stepanauskas & Sieracki, 2007).

Nozzle diameter sizes from 70 to 200 μm are available for the Influx™. A 70-μm nozzle is commonly used and reliably sorts cells less than 10 μm in width. Larger nozzles provide more reliable sorting of large particles (up to 25 μm for the 140-μm nozzle) and blockages are less likely when processing samples containing large particles. However, sort droplet size increases with nozzle size, which increases chances of contamination from the sheath fluid. Lower sorting efficiency due to coincident drop occupancy is also a disadvantage of larger nozzle diameters. Choosing the smallest diameter nozzle that still allows sorting of the largest target cells is recommended.

2.3. Sorting

The Influx™ default for triggering data collection and sorting is light scatter in the forward direction (FSC). After preparing the fluidics, the laser alignment and focus are optimized using standard fluorescent beads (Spherotech, Lake Forest, IL). The Influx™ is equipped with FACS™ Software (BD Biosciences) providing users a flexible interface to optimize analysis settings, create sort gates, and collect data. Graphical interpretations and data analysis are conducted in FlowJo (Tree Star, Inc., Ashland, MA, USA).

Environmental samples may contain organisms or particles that are larger than the nozzle diameter. Prefiltration of samples with a pore size smaller than the nozzle diameter (Partec Celltrics, Swedesboro, NJ, USA) will reduce risk of clogging the sample line or nozzle opening. The maximum sort rate of the Influx™ is 40,000 events per second (Ibrahim & van den Engh, 2007). For environmental samples, diluting the sample to an event rate of less than 1000 events per second will reduce coincident drop occupancy and enhance sorting efficiency. When sorting single cells, additional dilutions or sorting target cell populations twice can help to guarantee pure single-event sorts (Rodrigue et al., 2009). Setting the sort mode drop count (1.0 Drop, 2.0 Drop, or 1.5 Drop) and sort mode attributes (Enrich, Single, Pure, Yield, or Recovery) allow choices in tradeoffs between sorting efficiency and purity (refer to the Influx™ user manual).

2.4. Locating the target symbiont population

Most environmental samples contain numerous populations of microbes in addition to the target symbiotic organisms. In order to proceed efficiently with DNA amplification and sequencing after sorting, it is necessary to identify a cell population that is enriched in the target symbionts. Identification of the best sort region is accomplished by screening sorted populations of unknown cells with a quantitative assay specific for one or both of the symbionts (Zehr et al., 2008). The sort region determined to contain the target organism(s) is a source of template for MDA or nested PCR.

Sort gates are created as a matrix of overlapping rows and columns that encompass all cell populations in the sample (Fig. 3.2). Cell populations known not to contain the target symbionts may be excluded from the sort gates to enrich sorts of target symbionts. For example, the strong phycoerythrin signal of the marine cyanobacterium *Synechococcus* can be used to exclude these cells from sorts.

Efficient and quantitative screening of row and column sorts is best carried out by a phylogenetically selective assay such as quantitative real-time PCR (qPCR) targeting one or both symbionts. The qPCR assay is applied to whole cells so sample losses that accompany DNA extraction are prevented. Preparation of all reagents, tubes, and plates is conducted in a UV-treated PCR workstation to prevent contamination. Before sorting, fill wells of qPCR tubes or plates with 10 μl of nuclease-free water. Initially, sort populations of 500 cells per gate into each well, in replicate, and screen alongside standards and no template controls (NTC) in a real-time thermal cycler. The area of overlap between positive (amplification of the target gene by qPCR) rows and columns pinpoints the sort region that contains the most target symbiont (Fig. 3.2).

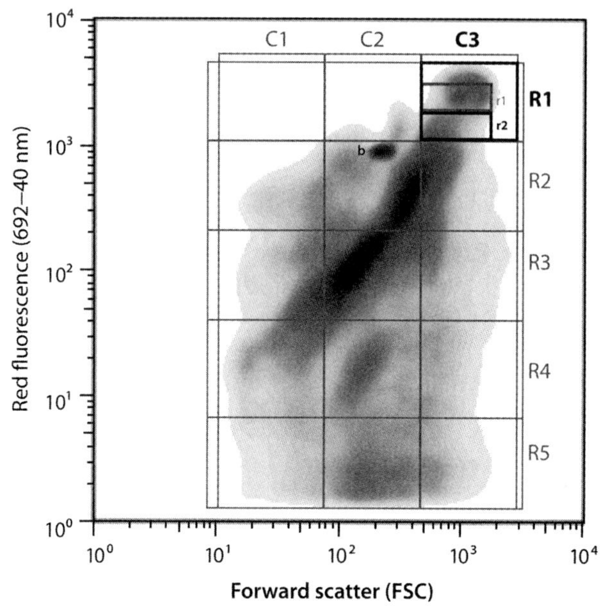

Figure 3.2 Flow cytometry density plot of 47,000 events from a coastal seawater sample showing multiple populations of unknown phytoplankton. Darker shading indicates higher numbers of cells. Sort gates of rows (R) and columns (C) are positioned to cover all phytoplankton populations. The sorts from each row and column are screened with a qPCR assay specific to one symbiont. In this example, the qPCR screen was positive (bold type) for cells from gates C3 and R1, meaning the target population is present where these gates overlap (large black box). With this information, additional sorts of cells from two smaller gates (r1 and r2) are conducted and the target symbiont population is pinpointed in r2. Standard 3 μm diameter beads (b) are included for internal reference.

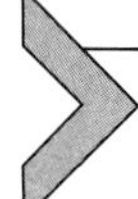

3. METAGENOMIC SEQUENCING OF UNCULTURED SYMBIONT POPULATIONS

Genomic analysis can be instrumental in determining the metabolism of uncultivated symbionts (Tripp et al., 2010). However, genome sizes of symbiotic partners can be vastly different. Especially in symbioses between bacteria (smaller genomes) and microbial eukaryotes, whose genomes can be much larger or present in multiple copy numbers, it is advantageous to isolate the one partner in an enriched sample to ensure adequate genome coverage following sequencing. WGA via MDA is effective at amplifying even very small amounts of DNA from sorted cells for next-generation sequencing.

The enzyme Phi29 and kits for MDA are commercially available through several companies. Previous work for sequencing symbiont populations used the GenomiPhi V2 DNA amplification kit (GE Healthcare Life Sciences) (Zehr et al., 2008), and more recent studies performing MDA on single cells or cell populations have used the Qiagen RepliG Mini or Midi kit (Bench et al., 2013; Reyes-Prieto et al., 2010; Yoon et al., 2011) or RepliPhi Phi29 DNA polymerase (Epicentre Biotechnologies) (Rodrigue et al., 2009) with success. Studies comparing a variety of WGA methods found that RepliG produced the highest yields of DNA with the least amplification bias across the genomes (Pan et al., 2008; Pinard et al., 2006). In MDA reactions with very little template DNA, a significant concern is the production of template-independent products (TIPs), which can contaminate sequencing processes and confound bioinformatic analyses. To avoid TIP contamination, start with larger numbers of cells (e.g., >5000 cells per reaction) when possible and limit the MDA reaction incubation time (Pan et al., 2008). The number of sorted cells will depend on their abundance in the sample, the expected diversity of the target population, difficulty of cell lysis, or numbers of nontarget contaminating cells in the sort gate. Amplification of genomic DNA from very few (1–100) *Prochlorococcus* cells reached a maximum product yield within 5 h of incubation, with evidence of TIP production after 6 h (Rodrigue et al., 2009). Based on those results, the protocol below keeps MDA incubation times at 5.75 h.

3.1. Sampling to separate symbiotic partners

Separation of symbiotic partners depends on the physical strength of their attachment. For delicate marine associations, symbiotic partners are separated by filtration to concentrate the sample, vortex mixing, and a freeze/thaw cycle. 100-fold concentrated samples are created by filtering 1 l of seawater onto Type GV Durapore filters (pore size depends on expected size of target symbionts) with 10 psi vacuum pressure. The damp filter is immediately placed in a 15 ml polypropylene tube with 10 ml of sterile (0.22 μm filtered) seawater with the cell-covered side submerged and facing the center of the tube. Vortexing resuspends the cells from the filter. The concentrated sample is frozen in liquid nitrogen and stored at −80 °C.

Prior to FACS analysis, the cell concentrate is thawed on ice then mixed again by vortexing. The freeze–thaw cycle should disrupt any delicate intercellular connections and leave the one symbiont intact as in a study on symbiotic cyanobacteria (Thompson et al., 2012; Zehr et al., 2008). Once thawed, the concentrated sample is used as soon as possible and certainly within a few hours.

3.2. Performing MDA on sorted cells

From the concentrated sample, sorts of the target cell population are directed into 100 μl aliquots of 1 × Tris–EDTA buffer (10 m*M* Tris, 1 m*M* EDTA, pH 8.0), in replicate, in sterile microfuge tubes. Briefly, spin the tubes to collect any cells that may have landed on the tube walls and store at −80 °C until MDA.

A modified version of a user-developed protocol for MDA using the Qiagen RepliG Midi DNA amplification kit is appropriate for starting with whole cells (Bench et al., 2013) and is as follows:

The 1.5-ml tubes used to receive sorted cells are marked where the pellet will collect, then centrifuged at 14,000 rpm (20,800 × *g*) for 40 min. Following centrifugation, use a fine pipette tip to remove supernatant, while avoiding the pellet, which may not be visible. Vortex or flick tubes vigorously to resuspend the cell pellet in any remaining liquid.

To each sample, add 2.5 μl of PBS and mix well. Then add 3.5 μl of Qiagen Buffer D2 (2.5 μl 1 m*M* DTT and 27.5 μl Buffer DLB) and mix well. Incubate the samples at 65 °C for 5 min, then place the samples on ice and add 3.5 μl of Qiagen Stop buffer. Next, prepare the reaction master mix in sterile UV-treated tubes. For one reaction, mix 10 μl RT-PCR grade H_2O (Ambion, Life Technologies), 1 μl RepliG Midi DNA Polymerase, and 29 μl RepliG Midi Reaction Buffer. Combine one aliquot of the master mix (40 μl) with the cell lysis mix in a sterile PCR tube and incubate at 30 °C for 5–16 h (5.75 h is recommended, see above) in a thermal cycler with a heated lid. Following isothermal amplification, heat samples to 65 °C for 5 min, cool to 4 °C, then store at −20 °C. Positive and negative control reactions are amplified alongside sorted cells. Prior to submission for next-generation sequencing, amplified genomic DNA should be quantified using Pico Green (Invitrogen Corporation, Carlsbad, CA) and checked for quality and fragment-size distribution using the Agilent 2100 Bioanalyzer (Agilent Technologies, Inc., Santa Clara, CA) DNA 7500 or 12000 chip.

3.3. Sequencing and bioinformatic analysis of symbiont genomes

The choice of platform for sequencing the genome of a newly isolated or identified symbiont will depend on a variety of factors, particularly genome size and available financial and computational resources. Nucleic acid sequencing technology has advanced rapidly in the past two decades, from capillary electrophoresis of dye-terminator Sanger sequencing to pyrosequencing and more recently to very high-throughput platforms which produce enormous amounts of sequence data (for a recent review, see Shokralla, Spall, Gibson, & Hajibabaei, 2012). Despite the advancing technologies, the older platforms remain available to researchers because each method has some advantages over the others. For example, Sanger sequencing is the most expensive (cost-per-base) technology, but none of the newer technologies can yet match its standard 1000-base read-length. So, for small-scale projects that require long continuous reads but do not require very deep coverage, Sanger sequencing is optimal. At the other end of the spectrum are the ultra-high-throughput technologies (e.g., SOLiD and Illumina) that have a very low cost-per-base and produce millions of shorter (50–250 bp) reads in one instrument run, making them a good choice for large genome or metagenome projects. One disadvantage of shorter reads is a reduced ability to assemble genomes without a reference (i.e., *de novo* assembly). However, the tremendous depth of coverage and especially the availability of mate-paired or paired-end sequencing have enabled successful *de novo* genome assembly of microbes (Iverson et al., 2012). Furthermore, a direct comparison of pyrosequencing to Illumina sequencing of a freshwater microbial community showed that, despite its shorter read lengths, paired-end Illumina data produced comparable assemblies in terms of contig length and the portion of the community represented (Luo, Tsementzi, Kyrpides, Read, & Konstantinidis, 2012). Of course, processing and analyzing the enormous amount of sequence data produced by the newest technologies also requires team members with bioinformatics expertise and significant computer power (typically, multiprocessor, multi-disk servers). Pyrosequencing technology falls between the two extremes in both cost and read-length and is a good option for small- to mid-sized genomes (or metagenomes) or for researchers without access to high-end computing capabilities. Paired-end libraries can also compensate for the lower depth of coverage provided by pyrosequencing to enable *de novo* genome assembly as evidenced by the study of a symbiont whose symbiotic lifestyle was only postulated at that time (Tripp et al., 2010).

A wide variety of bioinformatic tools are available for assembly and annotation of genomic data generated using next-generation sequence data. As with sequencing technologies, these tools are continually evolving and the choice of tool or pipeline is often determined by the user's experience and preferences. For sequence data from MDA product, it is important to screen out TIP or contaminant sequences. However, analyzing all the individual reads produced by current technologies requires significant computational time and processor power. Assembly can condense millions of reads into a manageable number of contigs. Pyrosequencing (i.e., 454) data can be assembled using the GS Data Analysis Software package (http://www.454.com/products/analysis-software/), but the best program for *de novo* assembly of Illumina data is debated (for detailed discussion of assemblers, see Paszkiewicz & Studholme, 2010; Zhang et al., 2011). The contigs generated by assembly can be screened using algorithms that identify open reading frames (ORFs) and by sequence similarity searches (e.g., BLAST). TIP data will be apparent as random sequences that do not have ORFs and will usually have no similarity to sequences in public databases. If DNA was amplified from a sample that had contaminating cells (e.g., if other species are present in the sort gate containing the target symbionts), contigs can be further screened by comparing them to genomes of closely related species. Automated identification of ORFs and corresponding functional and taxonomic annotation for microbial genomes (and metagenomes) can be obtained by submitting contigs to online services. These include RAST (Aziz et al., 2008) and the JGI IMG system (Markowitz et al., 2012), both of which were useful for annotating the genome of a cyanobacterial symbiont from FACS and MDA (Tripp et al., 2010; Zehr et al., 2008). Once a genome has been assembled and annotated, the subsequent analyses will often include detailed phylogenetic comparisons and determination of metabolic capabilities. Pathways that are novel or unexpectedly lacking can be particularly meaningful for microbes suspected of existing in symbioses.

4. DETERMINING HOST IDENTITY AND METABOLISM

Analysis of individual symbiotic associations is critical to unequivocally link the identity of an uncultivated symbiont to its host or symbiotic partner. Furthermore, relatively high-throughput analysis of single associations is essential to explore the species specificity of the association and this can be accomplished with nested PCR using universal primers for phylogenetically informative genes.

The physical connection between cells engaged in symbiosis can vary in its tolerance to sample handling. Preventing disruption of intercellular connections by turbulence, freezing, or by chemical treatment is essential to exploring intact associations. For open-ocean studies, the Influx™ can be equipped for the field (Thompson et al., 2012; Worden et al., 2012), which is required for analyzing untreated open-ocean samples. The target symbiont can be located among cell populations from the untreated samples with the same sort strategy described in Section 3.

4.1. Sorting single cells and screening by qPCR

Important considerations in applying nested PCR with universal primers to single cells include (1) one event is sorted into each well, (2) contaminating DNA is minimized, (3) the sample matrix (i.e., seawater) does not inhibit the PCR, and (4) target cells readily lyse following freeze/thaw cycles or during the first few cycles of PCR.

Aliquots of 10 μl (sorted events and NTCs) or 8 μl (standards) of nuclease-free water are added to wells as in Fig. 3.3. Single events, either single cells or multiple cells in association (i.e., the symbiont–host complex), are sorted individually into 72 wells of a 96-well PCR plate. No events are sorted into wells for standards or NTCs (Fig. 3.3). After sorting, cover plates with sterile sealing foil, briefly centrifuge, and then store at −80 or −20 °C until screening by qPCR. For qPCR, thaw plates and briefly centrifuge. Prepare master mix for qPCR and aliquot 15 μl to all wells. Carefully add 2 μl of standards, in replicate, to designated wells. Perform 45 cycles of amplification and analyze the results according to standard qPCR procedures. Move the entire volume (25 μl) of positive (amplification of approximately one gene copy) reactions to new sterile PCR tubes and store at −20 °C before nested PCR.

4.2. Nested PCR of 16S and 18S rRNA genes from positive qPCRs using universal primers

Nested PCR using universal primers for 18S and 16S rRNA genes is applied to the positive reactions from the qPCR assay to determine the phylogeny of the symbiotic partners. The nested PCR is useful for amplifying genes present in low abundance. Product from one round of PCR using "outer primers" to amplify a large fragment of the rRNA gene is used as template in a second round of PCR that targets a smaller region of the amplicon using "inner primers."

For nested PCR, use a high-performance polymerase mixture such as TaKaRa Ex Taq (Takara Bio, Inc.) to ensure amplification if targets are difficult to amplify. For the first round of nested PCR, use the outer primers EukA/B (Medlin, Elwood, Stickel, & Sogin, 1988) and Eub27F/Eub1492R

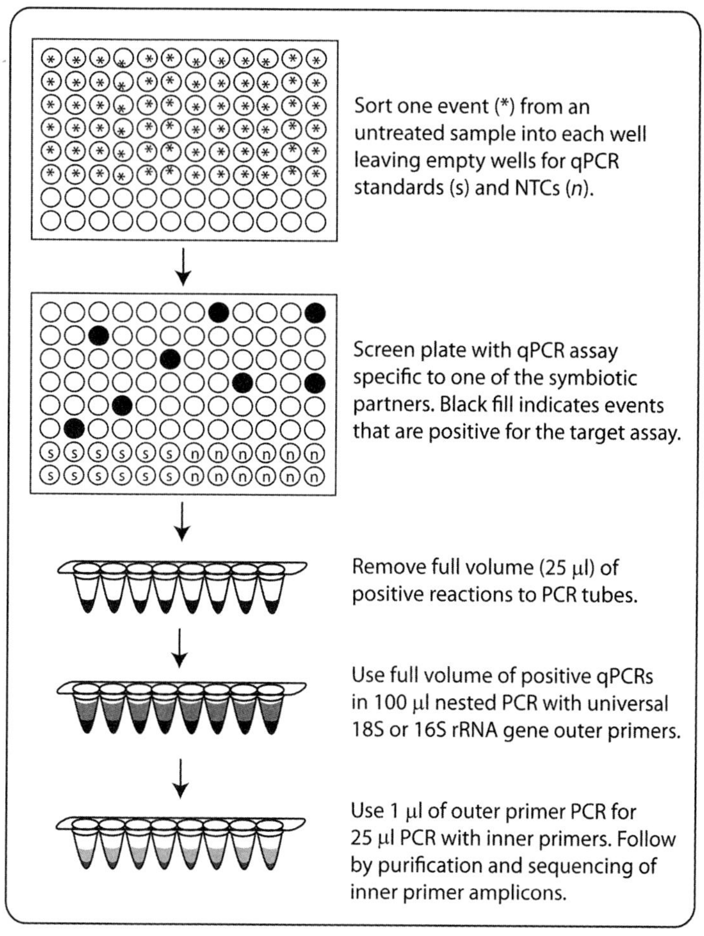

Figure 3.3 Strategy for application of nested PCR, with phylogenetically informative universal primers, to single-event sorts from untreated samples to determine the identity of uncultivated symbionts.

(Weisburg, Barns, Pelletier, & Lane, 1991) for amplification of 18S and 16S rRNA genes, respectively. 75 μl of PCR master mix should be added directly to the saved reaction from the qPCR assay (25 μl) and amplified for 35 cycles alongside positive and negative controls. Store completed outer primer reactions at −20 °C or immediately use 1 μl as template in 25 μl reactions for the second round of nested PCR with inner primers. Use the internal primers Euk18S-555F/1269R (López-García, Philippe, Gail, & Moreira, 2003) and 358F/907R (Lane, 1991) for the 18S and 16S rRNA reactions, respectively.

4.3. Sequencing rRNA genes from nested PCR and phylogenetic analysis

PCR products from the inner primers are analyzed by gel electrophoresis; any bands are cut from the gel, extracted (QIAquick Gel Extraction Kit), and cloned using kits such as pGEM-T (Promega, Madison, WI) or TOPO-TA (Life Technologies). The clones can be prepared by plasmid mini-prep and sequenced. Quality control and generation of contigs for identical or very similar sequences (>99%) can be generated in Sequencher (Gene Codes Corporation, Ann Arbor, MI, USA) or similar analysis software. Phylogenetic analysis and alignment of the sequences can be conducted through BLASTn to GenBank (Benson et al., 2013) or through other tools such as the Ribosomal Database Project (Wang, Garrity, Tiedje, & Cole, 2007), SILVA (Quast et al., 2013), or SINA (Pruesse, Peplies, & Glöckner, 2012). Negative controls often contain amplification products after the qPCR and nested PCR procedures, so it is important to work under as clean conditions as possible for highest efficiency of sequencing nested PCR products (Thompson et al., 2012).

REFERENCES

Aziz, R., Bartels, D., Best, A., DeJongh, M., Disz, T., Edwards, R., et al. (2008). The RAST Server: Rapid annotations using subsystems technology. *BMC Genomics*, *9*(1), 75.

Bench, S. R., Heller, P., Frank, I., Arciniega, M., Shilova, I. N., & Zehr, J. P. (2013). Whole genome comparison of six *Crocosphaera watsonii* strains with differing phenotypes. *Journal of Phycology*, *49*(4), 786–801.

Benson, D., Cavanaugh, M., Clark, K., Karsch-Mizrachi, I., Lipman, D., Ostell, J., et al. (2013). GenBank. *Nucleic Acids Research*, *41*(Database issue), D26–D42.

Blainey, P. C., Mosier, A. C., Potanina, A., Francis, C. A., & Quake, S. R. (2011). Genome of a low-salinity ammonia-oxidizing Archaeon determined by single-cell and metagenomic analysis. *PLoS ONE*, *6*(2), e16626. http://dx.doi.org/10.1371/journal.pone.0016626.

Brehm-Stecher, B. F., & Johnson, E. A. (2004). Single-cell microbiology: Tools, technologies, and applications. *Microbiology and Molecular Biology Reviews*, *68*(3), 538–559.

Cheung, V. G., & Nelson, S. F. (1996). Whole genome amplification using a degenerate oligonucleotide primer allows hundreds of genotypes to be performed on less than one nanogram of genomic DNA. *Proceedings of the National Academy of Sciences of the United States of America*, *93*(25), 14676–14679.

Cuvelier, M. L., Allen, A. E., Monier, A., McCrow, J. P., Messié, M., Tringe, S. G., et al. (2010). Targeted metagenomics and ecology of globally important uncultured eukaryotic phytoplankton. *Proceedings of the National Academy of Sciences of the United States of America*, *107*(33), 14679–14684.

Dean, F., Nelson, J., Giesler, T., & Lasken, R. (2001). Rapid amplification of plasmid and phage DNA using Phi29 polymerase and a multiply-primed rolling circle amplification. *Genome Research*, *11*, 1095–1099.

Del Giorgio, P. A., Bird, D. F., Prairie, Y. T., & Planas, D. (1996). Flow cytometric determination of bacterial abundance in lake plankton with the green nucleic acid stain SYTO 13. *American Society of Limnology and Oceanography*, *41*, 783–789, Waco, TX, ETATS-UNIS.

Hongoh, Y., Sharma, V. K., Prakash, T., Noda, S., Taylor, T. D., Kudo, T., et al. (2008). Complete genome of the uncultured Termite Group 1 bacteria in a single host protist cell. *Proceedings of the National Academy of Sciences of the United States of America*, *105*(14), 5555–5560.

Huang, W. E., Ward, A. D., & Whiteley, A. S. (2009). Raman tweezers sorting of single microbial cells. *Environmental Microbiology Reports*, *1*(1), 44–49.

Ibrahim, S. F., & van den Engh, G. (2003). High-speed cell sorting: Fundamentals and recent advances. *Current Opinion in Biotechnology*, *14*(1), 5–12.

Ibrahim, S. F., & van den Engh, G. (2007). Flow cytometry and cell sorting. *Advances in Biochemical Engineering/Biotechnology*, *106*, 19–39.

Iverson, V., Morris, R. M., Frazar, C. D., Berthiaume, C. T., Morales, R. L., & Armbrust, E. V. (2012). Untangling genomes from metagenomes: Revealing an uncultured class of marine Euryarchaeota. *Science*, *335*(6068), 587–590.

Kvist, T., Ahring, B., Lasken, R., & Westermann, P. (2007). Specific single-cell isolation and genomic amplification of uncultured microorganisms. *Applied Microbiology and Biotechnology*, *74*(4), 926–935.

Lane, D. J. (1991). 16S/23S rRNA sequencing. In M. Goodfellow & E. Stackebrandt (Eds.), *Nucleic acid techniques in bacterial systematics* (pp. 115–175). New York: John Wiley and Sons.

López-García, P., Philippe, H., Gail, F., & Moreira, D. (2003). Autochthonous eukaryotic diversity in hydrothermal sediment and experimental microcolonizers at the Mid-Atlantic Ridge. *Proceedings of the National Academy of Sciences of the United States of America*, *100*(2), 697–702.

Luo, C., Tsementzi, D., Kyrpides, N., Read, T., & Konstantinidis, K. T. (2012). Direct comparisons of Illumina vs. Roche 454 sequencing technologies on the same microbial community DNA sample. *PLoS ONE*, 7(2), e30087. http://dx.doi.org/10.1371/journal.pone.0030087.

Marcy, Y., Ouverney, C., Bik, E. M., Lösekann, T., Ivanova, N., Martin, H. G., et al. (2007). Dissecting biological "dark matter" with single-cell genetic analysis of rare and uncultivated TM7 microbes from the human mouth. *Proceedings of the National Academy of Sciences of the United States of America*, *104*(29), 11889–11894.

Markowitz, V. M., Chen, I.-M. A., Palaniappan, K., Chu, K., Szeto, E., Grechkin, Y., et al. (2012). IMG: The integrated microbial genomes database and comparative analysis system. *Nucleic Acids Research*, *40*(D1), D115–D122.

Medlin, L., Elwood, H. J., Stickel, S., & Sogin, M. L. (1988). The characterization of enzymatically amplified eukaryotic 16S-like rRNA-coding regions. *Gene*, *71*(2), 491–499.

Navin, N., Kendall, J., Troge, J., Andrews, P., Rodgers, L., McIndoo, J., et al. (2011). Tumour evolution inferred by single-cell sequencing. *Nature*, *472*(7341), 90–94. http://dx.doi.org/10.1038/nature09807.

Orcutt, K. M., Gundersen, K., Wells, M. L., Poulton, N. J., Sieracki, M. E., & Smith, G. J. (2008). Lighting up phytoplankton cells with quantum dots. *Limnology and Oceanography: Methods*, *6*, 653–658.

Orphan, V. J., House, C. H., Hinrichs, K.-U., McKeegan, K. D., & DeLong, E. F. (2001). Methane-consuming archaea revealed by directly coupled isotopic and phylogenetic analysis. *Science*, *293*(5529), 484–487.

Pan, X., Urban, A. E., Palejev, D., Schulz, V., Grubert, F., Hu, Y., et al. (2008). A procedure for highly specific, sensitive, and unbiased whole-genome amplification. *Proceedings of the National Academy of Sciences of United States of America*, *105*(40), 15499–15504.

Paszkiewicz, K., & Studholme, D. J. (2010). De novo assembly of short sequence reads. *Briefings in Bioinformatics*, *11*(5), 457–472.

Pinard, R., de Winter, A., Sarkis, G., Gerstein, M., Tartaro, K., Plant, R., et al. (2006). Assessment of whole genome amplification-induced bias through high-throughput, massively parallel whole genome sequencing. *BMC Genomics*, 7(1), 216.

Pruesse, E., Peplies, J., & Glöckner, F. O. (2012). SINA: Accurate high-throughput multiple sequence alignment of ribosomal RNA genes. *Bioinformatics*, *28*(14), 1823–1829.

Quast, C., Pruesse, E., Yilmaz, P., Gerken, J., Schweer, T., Yarza, P., et al. (2013). The SILVA ribosomal RNA gene database project: Improved data processing and web-based tools. *Nucleic Acids Research*, *41*(D1), D590–D596.

Reyes-Prieto, A., Yoon, H. S., Moustafa, A., Yang, E. C., Andersen, R. A., Boo, S. M., et al. (2010). Differential gene retention in plastids of common recent origin. *Molecular Biology and Evolution*, *27*(7), 1530–1537.

Rodrigue, S., Malmstrom, R. R., Berlin, A. M., Birren, B. W., Henn, M. R., & Chisholm, S. W. (2009). Whole genome amplification and de novo assembly of single bacterial cells. *PLoS ONE*, *4*(9), e6864. http://dx.doi.org/10.1371/journal.pone.0006864.

Shokralla, S., Spall, J. L., Gibson, J. F., & Hajibabaei, M. (2012). Next-generation sequencing technologies for environmental DNA research. *Molecular Ecology*, *21*(8), 1794–1805.

Stepanauskas, R., & Sieracki, M. E. (2007). Matching phylogeny and metabolism in the uncultured marine bacteria, one cell at a time. *Proceedings of the National Academy of Sciences of the United States of America*, *104*(21), 9052–9057.

Telenius, H., Carter, N. P., Bebb, C. E., Nordenskjöd, M., Ponder, B. A. J., & Tunnacliffe, A. (1992). Degenerate oligonucleotide-primed PCR: General amplification of target DNA by a single degenerate primer. *Genomics*, *13*(3), 718–725.

Thompson, A. W., Foster, R. A., Krupke, A., Carter, B. J., Musat, N., Vaulot, D., et al. (2012). Unicellular cyanobacterium symbiotic with a single-celled eukaryotic alga. *Science*, *337*(6101), 1546–1550.

Tripp, H. J., Bench, S. R., Turk, K. A., Foster, R. A., Desany, B. A., Niazi, F., et al. (2010). Metabolic streamlining in an open-ocean nitrogen-fixing cyanobacterium. *Nature*, *464*(7285), 90–94. http://dx.doi.org/10.1038/nature08786.

Vaulot, D., Lepère, C., Toulza, E., De la Iglesia, R., Poulain, J., Gaboyer, F., et al. (2012). Metagenomes of the picoalga Bathycoccus from the Chile coastal upwelling. *PLoS ONE*, *7*(6), e39648.

Wang, Q., Garrity, G., Tiedje, J., & Cole, J. (2007). Naive Bayesian classifier for rapid assignment of rRNA sequences into the new. *Applied and Environmental Microbiology*, *73*(16), 5261–5267.

Weisburg, W. G., Barns, S. M., Pelletier, D. A., & Lane, D. J. (1991). 16S ribosomal DNA amplification for phylogenetic study. *Journal of Bacteriology*, *173*(2), 697–703.

Worden, A. Z., Janouskovec, J., McRose, D., Engman, A., Welsh, R. M., Malfatti, S., et al. (2012). Global distribution of a wild alga revealed by targeted metagenomics. *Current Biology*, *22*(17), R675–R677.

Yoon, H. S., Price, D. C., Stepanauskas, R., Rajah, V. D., Sieracki, M. E., Wilson, W. H., et al. (2011). Single-cell genomics reveals organismal interactions in uncultivated marine protists. *Science*, *332*(6030), 714–717.

Zehr, J. P., Bench, S. R., Carter, B. J., Hewson, I., Niazi, F., Shi, T., et al. (2008). Globally distributed uncultivated oceanic N_2-fixing cyanobacteria lack oxygenic Photosystem II. *Science*, *322*(5904), 1110–1112.

Zhang, W., Chen, J., Yang, Y., Tang, Y., Shang, J., & Shen, B. (2011). A practical comparison of de novo genome assembly software tools for next-generation sequencing technologies. *PLoS ONE*, *6*(3), e17915. http://dx.doi.org/10.1371/journal.pone.0017915.

Zhang, L., Cui, X., Schmitt, K., Hubert, R., Navidi, W., & Arnheim, N. (1992). Whole genome amplification from a single cell: Implications for genetic analysis. *Proceedings of the National Academy of Sciences of the United States of America*, *89*(13), 5847–5851.

Zhang, K., Martiny, A. C., Reppas, N. B., Barry, K. W., Malek, J., Chisholm, S. W., et al. (2006). Sequencing genomes from single cells by polymerase cloning. *Nature Biotechnology*, *24*(6), 680–686. http://dx.doi.org/10.1038/nbt1214.

CHAPTER FOUR

Optofluidic Cell Selection from Complex Microbial Communities for Single-Genome Analysis

Zachary C. Landry*, Stephen J. Giovanonni*, Stephen R. Quake†,‡, Paul C. Blainey§,1

*Department of Microbiology, Oregon State University, Corvallis, Oregon, USA
†Department of Bioengineering, Stanford University, Stanford, California, USA
‡Department of Applied Physics, Stanford University, Stanford, California, USA
§Department of Biological Engineering, Broad Institute and Massachusetts Institute of Technology, Cambridge, Massachusetts, USA
[1]Corresponding author: e-mail address: pblainey@broadinstitute.org

Contents

1. Introduction 62
2. Optical Tweezing for Cell Isolation 64
   2.1 Applying forces to objects in solution with light 65
   2.2 Action at a distance 67
   2.3 Practical approach to SC-WGS with lower resources 67
   2.4 Avoiding contamination 68
   2.5 Integrating microscopy and cell sorting 68
   2.6 Low required sample volume 69
   2.7 Technical considerations for sorting using an optical trap 69
3. Optical Hardware Setup 70
   3.1 Specifying the laser and microscope 70
   3.2 Choosing an objective lens 74
   3.3 Alignment procedure and safety 75
4. Microfluidics Configuration and Setup 78
   4.1 Microdevice design 78
5. Procedure for Sorting and Amplifying Single Cells 80
6. Software 83
7. Performance and Limitations 85
Acknowledgments 88
References 88

Abstract

Genetic analysis of single cells is emerging as a powerful approach for studies of heterogeneous cell populations. Indeed, the notion of homogeneous cell populations is receding as approaches to resolve genetic and phenotypic variation between single

Methods in Enzymology, Volume 531
ISSN 0076-6879
http://dx.doi.org/10.1016/B978-0-12-407863-5.00004-6

cells are applied throughout the life sciences. A key step in single-cell genomic analysis today is the physical isolation of individual cells from heterogeneous populations, particularly microbial populations, which often exhibit high diversity. Here, we detail the construction and use of instrumentation for optical trapping inside microfluidic devices to select individual cells for analysis by methods including nucleic acid sequencing. This approach has unique advantages for analyses of rare community members, cells with irregular morphologies, small quantity samples, and studies that employ advanced optical microscopy.

1. INTRODUCTION

The ability to amplify very small amounts of DNA, even single molecules, has been a reality for more than a decade (Dean, 2002; Dean, Nelson, Giesler, & Lasken, 2001; Sykes et al., 1992; Vogelstein & Kinzler, 1999). This approach was eventually applied to individual microbial genomes (Kvist, Ahring, Lasken, & Westermann, 2006) and ultimately coupled with high-throughput sequencing technologies to analyze genomes of both laboratory-cultured organisms (Zhang et al., 2006) and uncultured environmental microbes (Marcy, Ouverney, et al., 2007). From this early work, single-cell whole-genome sequencing (SC-WGS) expanded from proof-of-concept to an established and popular technique that is applied widely in biology to supplement culture-based isolation/genomic approaches (Fig. 4.1). SC-WGS has been used extensively to access the genomes of yet-uncultivated organisms in environmental microbiology (Blainey, Mosier, Potanina, Francis, & Quake, 2011; Malmstrom et al., 2013; Marcy, Ouverney, et al., 2007; Stepanauskas & Sieracki, 2007; Swan et al., 2011; Woyke et al., 2009; Youssef, Blainey, Quake, & Elshahed, 2011) and problems in medicine, for example, studies of pathogens and components of the human microbiome (Marcy, Ouverney, et al., 2007; Pamp, Harrington, Quake, Relman, & Blainey, 2012). Applications of single-cell genomics to human genetics are also becoming more widespread (Fan, Wang, Potanina, & Quake, 2010; Wang, Fan, Behr, & Quake, 2012; Zong, Lu, Chapman, & Xie, 2012), and single-cell genetics has also been used for lineage analysis (Frumkin et al., 2008). While nearly all single amplified genomes (SAGs) have been produced using the multiple displacement amplification (MDA) chemistry, a variety of different approaches to isolating cells for SC-WGS have been implemented with success (Blainey, 2013).

Molecular techniques have assumed an increasingly important role in microbial ecology in recent decades, beginning with the application of gene

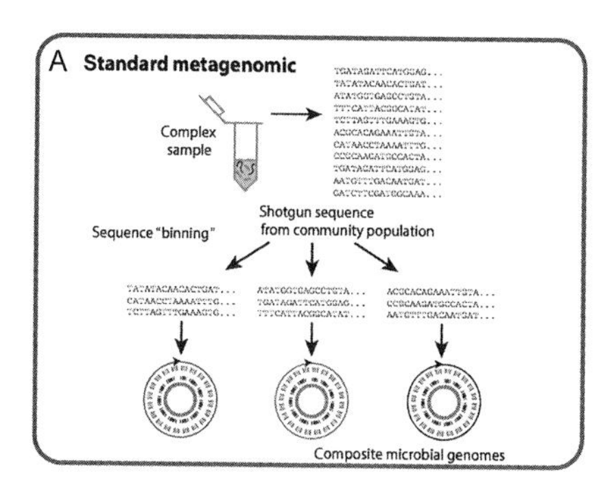

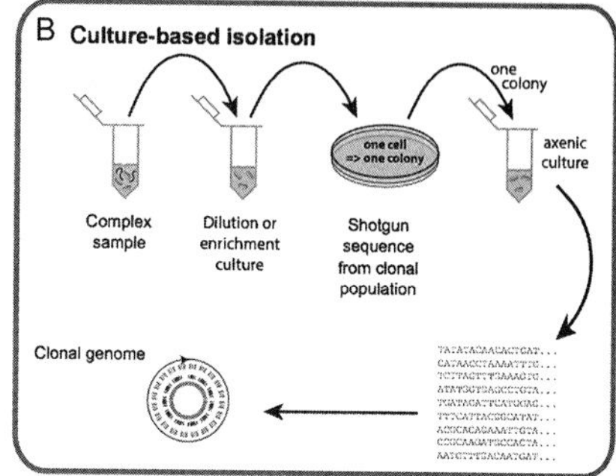

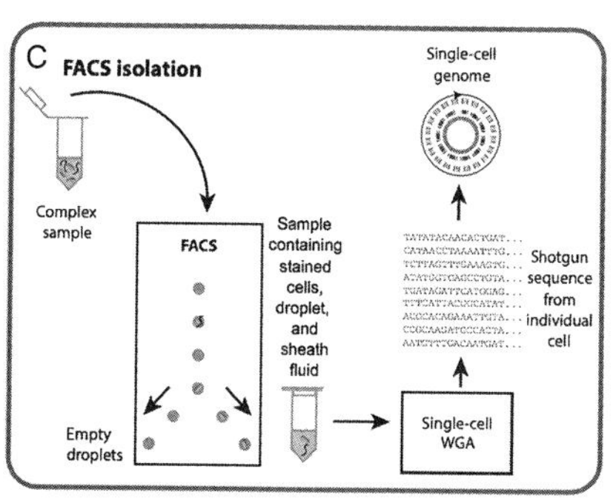

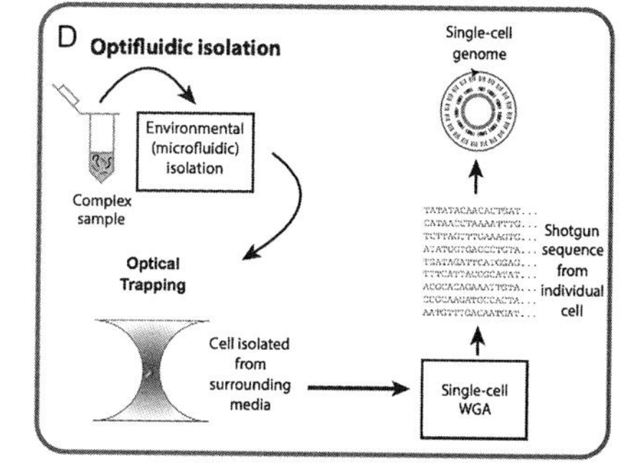

E

	Optical tweezing	Metagenomics	Fluorescence-activated cell sorting	Culturing
One benchtop apparatus	Yes	Yes	No	No
Low WGA reagent consumption	Yes (nanoliter)	Dependent on desired sequence depth	No (microliter)	N/A
Sealed device prevents extrinsic contamination	Yes	No	No	No
Amenable to automation	Yes	Yes	Yes	No
Complete genomic sequence	Fragemented genomes common	Possible in lower-complexity systems	Fragemented genomes common	Yes
Compatibility with microscopy	Yes	No	No	Yes
Compatibility with large within-sample variance of cell or aggregate size and morphology	Medium	High	Low	High
Compatability with other microfluidic devices and applications	High	Low	Medium	High
Amenable to benchtop scale *in vitro* organismal experimentation	No	No	No	Yes
Amenable to single-cell *in vitro* organismal experimentation	Yes	No	Somewhat	Yes
Level of throughput	Medium	High	High	Low

Figure 4.1 Schematic diagram contrasting of methods to obtain genome sequences from yet-uncultured microbial species. Established techniques such as metagenomics (A) or culturing (B), and the two newer methods of single-cell whole-genome amplification by fluorescence-activated cell sorting (C), or the optofluidic approach discussed in this chapter (D). Advantages and features of each methodology are summarized in the table (E). *Parts A–D were adapted with permission from Blainey (2013),* FEMS Microbiology Reviews *2013.* (For color version of this figure, the reader is referred to the online version of this chapter.)

sequencing to environmental DNA (Olsen, Lane, Giovannoni, Pace, & Stahl, 1986). Subsequently, it became clear that many species prevalent in the environment have no known relatives in culture (Rappé & Giovannoni, 2003). This presented the grand challenge of identifying the role of these unknown taxa in the environment, a task that remains unfinished today. With the rise of metagenomics, the extent of gene diversity became known for many environments (Venter et al., 2004). However, evolution acts on assemblages of genes, and the reconstruction of genomes, and thus the metabolism of cells, was hampered by the difficulty of assembling genomes from any but lower-complexity microbial systems (Tyson et al., 2004; Wrighton et al., 2012).

In recent years, SAGs have provided valuable genomic information for dominant uncultured microorganisms (Dupont et al., 2011; Pamp et al., 2012; Swan et al., 2011). In one example, SC-WGS revealed that the abundant SAR324 group of marine deltaproteobacteria may couple carbon fixation to sulfur oxidation, a prediction that was consistent with subsequent labeling of cells with radioisotope tracers (Swan et al., 2011). This information has important implications for the global ocean sulfur and carbon cycles. It is expected that similar advances in many underexplored environments will result from wider application of SC-WGS. Genome sequences have also paved the way for culturing previously uncultured strains in the laboratory by providing insight into unanticipated metabolic requirements (Renesto et al., 2003).

Fluorescence-activated cell sorting (FACS) is a popular method for producing SAGs today; however, optofluidic approaches, the topic of this chapter, offer advantages for many applications. Microscopic sorting makes it possible for the operator to sort cells visually and offers advantages in situations where (1) cells or assemblages of cells have distinctive morphology, (2) fluorescent signatures are too weak or too complex for resolution by FACS, (3) only very small volumes of sample are available, or (4) large quantities of contaminating DNA are present, such as in the isolation of intracellular parasites from cell lysates. Other examples of problems that meet these criteria are small aggregates of cells such as "marine snow" that may be too fragile to study with FACS, epiphytic interactions, and syntrophic assemblages, such as anaerobic methane-oxidizing archaeal/bacterial consortia (Boetius et al., 2000).

2. OPTICAL TWEEZING FOR CELL ISOLATION

The production of SAGs requires two fundamental steps. First, the isolation of individual cells from a bulk sample, and second, the amplification of the genomic contents of the isolated cell by WGA. Various approaches to each of these steps and the impact of different technologies on contamination, the major challenge in single-cell genomics, have been reviewed previously (Blainey, 2013). We have implemented a particular cell isolation approach, optical tweezing (Ashkin, 1992, 1997; Ashkin & Dziedzic, 1989; Ashkin, Dziedzic, & Yamane, 1987; Ashkin & Gordon, 1983), to select individual microbial cells (Ashkin et al., 1987; Huber et al., 1995) and separate them for individual analysis inside custom microfluidic devices, where WGA can be carried out in nanoliter volumes while excluding extrinsic contaminants (Blainey et al., 2011). Figure 4.2 provides a granular breakdown of the component steps in the generation and sequencing of the genome of single cells using the microfluidic/optical tweezing approach.

Optical tweezing presents two unique advantages that distinguish it from other methods of cell isolation that derive from a unique feature of the tweezing approach. First, movement of the cell is uncoupled from movement of the cell's immediate surrounding environment. This is to say that the cell *itself* rather than a voxel (volumetric pixel) from the sample is selected. This specificity for objects within the sample can be understood based on the mechanism by which laser tweezers operate.

2.1. Applying forces to objects in solution with light

Optical tweezing (the terms optical/laser and tweezing/trapping are used interchangeably here) relies on a difference in refractive index between the cell and the surrounding solution. As such, cells (or other structures with a boundary defined by a refractive index gradient), but not the solution itself, can be trapped using this method. High-performance microscope objectives can focus light very tightly to create an optical trap that has a capture radius comparable to the wavelength of light used to create the trap.

These properties allow an optical trap to be used with surgical precision to move one cell when many others are nearby, while minimizing the possibility of carrying forward an extraneous cell or foreign DNA. Contamination by environmental DNA or double sorting is a significant practical concern in single-cell genomics, particularly in *de novo* applications, where reference sequence to identify contaminating reads is not available. By contrast, other popular methods such as fluorescence-activated flow cytometry and micromanipulation work by subdividing a liquid sample, and require a high dilution of cells in clean buffer to reduce the chance of contamination.

While idealized representations often depict spherical objects being trapped, in practice, cells of essentially any morphology can be moved using

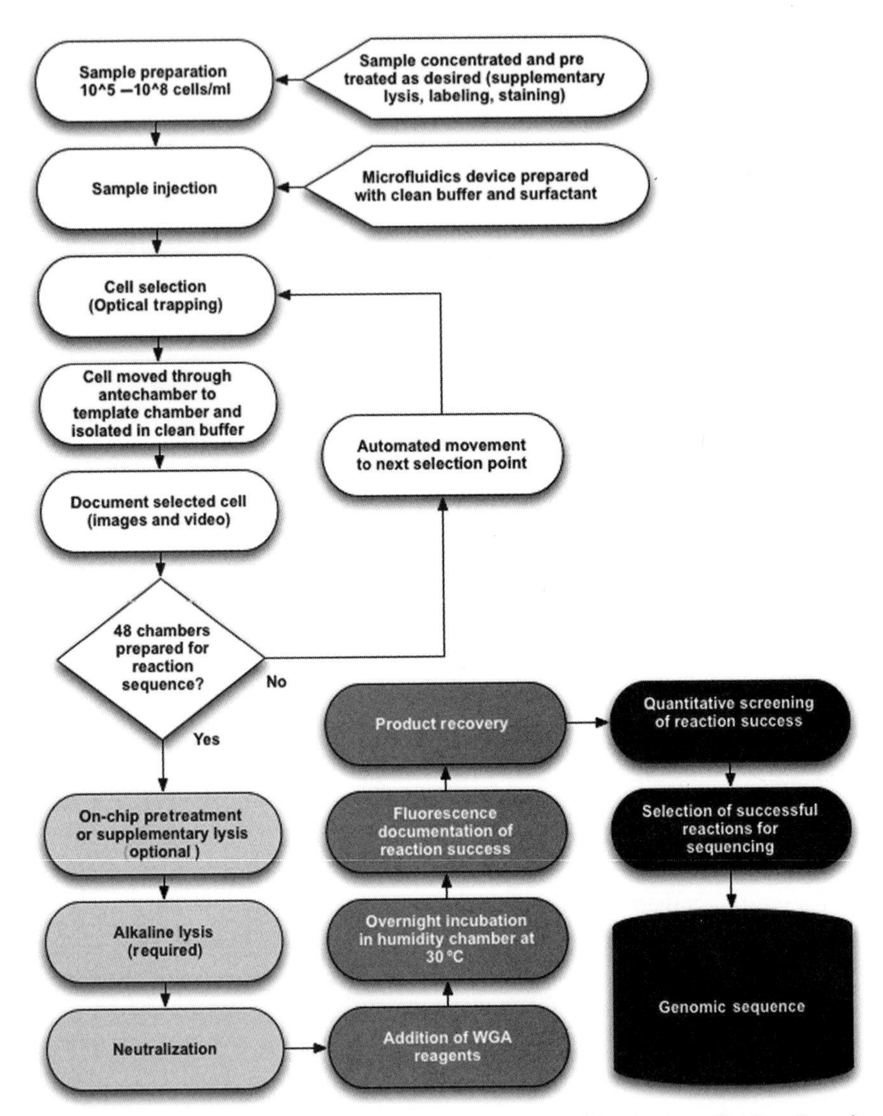

Figure 4.2 Flowchart for single-cell whole-genome amplification optofluidics. Sample preparation and cell selection steps are shown in white; lysis-related steps are shown in light gray; amplification and product recovery are shown in dark gray; postprocessing steps are shown in black. (For color version of this figure, the reader is referred to the online version of this chapter.)

an optical trap. Tiny coccoid cells with a maximum dimension of about 0.5 μm can be trapped (Blainey et al., 2011), as can high aspect ratio rods (Dodsworth et al., 2013), or even huge filaments over 100 μm in length (Pamp et al., 2012). Cells much larger than the dimensions of the optical trap are effectively moved by tugging at the cell boundary.

2.2. Action at a distance

The second unique property of the optical trap is its ability to apply forces to cells at a distance, without bringing instrumentation for manipulation into contact with the cells or the sample solution. This property enables manipulation of cells inside a closed (transparent) container. The notion of carrying out single-cell genomics inside a sealed microfluidic device is extremely appealing for the purpose of eliminating contamination from extrinsic sources and automating the subsequent lysis and whole-genome amplification of the sorted cells in tiny nanoliter volumes. The unique features of the trap allow the order of operations to be reversed compared with other common single-cell workflows. In flow cytometry and micromanipulation, the sample is first subdivided into isolate cells and then sealed in separate vessels where lysis and WGA are carried out. The optical tweezer enables an inversion of steps: the sample can first be introduced to a sealed vessel prior to the isolation of individual cells.

2.3. Practical approach to SC-WGS with lower resources

One of the largest hurdles to overcome when attempting single-cell genomics is the issue of reagent cost. Kits marked for WGA, including MDA, currently cost on the order of $1 per reaction microliter. The cost of reagents can be prohibitive for even moderate-scale studies when conventional reaction volumes of 10–50 μl are used. Although ultra-clean WGA reagents can be produced in-house at lower cost (Blainey et al., 2011), reducing reaction volumes to the nanoliter scale on a microfluidic device (1 nl = 0.001 μl) is an attractive alternative to sourcing low-cost reagents. For example, the series of devices developed at Stanford carry out WGA in 60 nl volumes, using about $0.06 worth of retail reagents per reaction (Marcy, Ishoey, et al., 2007; Marcy, Ouverney, et al., 2007). If 40 conventional-scale (20 μl) reactions normally requiring $800 worth of WGA reagents can be reduced to the nanoliter scale, the reagent cost becomes insignificant and the incremental cost for single-cell genomics becomes dominated by the cost of the microfluidic device. Two-layer custom polydimethylsiloxane (PDMS) devices of the type presented in this review cost $50–100 each in low-volume manufacturing (e.g., from the Stanford Microfluidics Foundry), yielding a net cost savings for moderate scale single-cell genomics projects.

When capabilities for sorting and reaction setup are integrated in a single benchtop microdevice, the specialized infrastructure for single-cell genomics sample prep collapses to a single apparatus that sits on a bench in a standard lab environment. An advanced laser-trapping microscope can be built from about $100,000 in parts (far less if a suitable microscope is already available, which is the case in many labs) and turn-key laser-tweezing microscopes are priced around $200,000. These costs compare favorably with those for flow cytometry equipment and cleanroom space. Paired with a modern benchtop sequencing instrument, the optofluidic approach to single-cell genomics is well suited to individual research labs, teaching labs, and portable operation. Figure 4.2 gives an overview of the single-cell genomics process using the optical trapping/microfluidic amplification approach from sample to data.

2.4. Avoiding contamination

Reducing reaction volume pays an additional dividend in the reduction of reagent-borne contamination (Blainey et al., 2011). In the example moving from 20 μl to 60 nl reactions, the amount of contamination from reagents is expected to drop more than 300-fold.

Overall, optical tweezing provides a means to isolate cells that intrinsically selects against carry-over contamination and can be used to pluck cells from concentrated cell suspensions while leaving behind sample-borne contaminants. If paired with a capability for lysis and WGA in small volumes, a great savings in reagent cost and reduced reagent contamination can also be achieved.

2.5. Integrating microscopy and cell sorting

Optical trapping approaches to single-cell genomics retain all the characteristics of classical microscopy: the ability to identify and characterize cells or cell aggregates according to morphology as well as providing the option of integrating a number of staining or labeling techniques. At present, microfluidic approaches have lower throughput rates than single-cell genomics by flow cytometry and might not be suitable for broadly assessing biodiversity. However, the unique attributes of microfluidics open the door for the application of novel screens for cell selection. When implemented in a semimanual approach where a human user selects cells under high-resolution imaging (typically 100× phase contrast microscopy), the selection of cells allows for direct interaction of the user with the sample in a notional "microbial safari" where the user observes a variety of fauna, occasionally "bagging" a trophy. Cells can straightforwardly be selected on the basis of any observable characteristic such as size, morphology, intracellular structure, or even faint fluorescence from native chromophores or reagents such as fluorescent *in situ* hybridization probes (Yilmaz, Haroon, Rabkin, Tyson, & Hugenholtz, 2010). Screens based on complex morphological characteristics such as the selection of certain microaggregates are also feasible and are likely to blur the distinction between single-cell genomics and metagenomics in ways that reveal new information about communities of interest.

2.6. Low required sample volume

Finally, microfluidic approaches are extremely parsimonious with the sample. In principle, a sample containing a single cell can be introduced to a device and intercepted with the laser trap for sorting. In practice, a 5-μl sample containing a few thousand cells in total is quite usable, indicating the suitability of this approach to single-cell genomics for especially rare or precious samples, where the volume required for other types of analyses may present a significant commitment or risk.

2.7. Technical considerations for sorting using an optical trap

For objects larger than the wavelength of impinging light (often the case when trapping cells with light, but not a necessary condition), a ray optics analysis of the optical trap is valid (Rocha, 2009). An object captured in a laser trap can be considered as a small lens that refracts (bends) and scatters the light from the optical trap. When light moves through an interface from a medium with one refractive index to a different medium with a different refractive index, the direction of the light is changed and the cell experiences a small reactive force in the opposite direction of this change. This happens because photons passing through the object or scattering off of it can exchange momentum with the trapped object, and in this way, apply force to it. When the object is centered in the trap, the optical forces on the object are balanced, but when the object is displaced from the center of the trap, the applied forces are unbalanced, and due to the optical geometry of the trap, the object experiences a force pushing it toward the center of the trap (in three dimensions). The amount of momentum carried by each photon is small, so a large number of photons must interact with the cell for any appreciable force to be applied. This necessitates the use of milliwatts of optical power for many applications, which corresponds to a very high power density when focused to a diffraction-limited spot to form a laser trap. The power density at the focus of an optical trap commonly exceeds the optical power density at the surface of the sun by several orders of magnitude. Surprisingly, this is not necessarily harmful to cells, which can easily survive continuous trapping when near-infrared light is used (Ericsson, Hanstorp, Hagberg, Enger, & Nystrom, 2000; Neuman, Chadd, Liou, Bergman, & Block, 1999).

3. OPTICAL HARDWARE SETUP

3.1. Specifying the laser and microscope

A complete list of parts for the optical setup and microfluidic controller is available in Supplement 2 (http://dx.doi.org/10.1016/B978-0-12-407863-5.00004-6). Selection of a light source for minimally perturbative manipulation of biological materials is limited to the near-infrared wavelengths where water and the stuff of life absorb only weakly. The choice of wavelength is especially important as even a small amount of absorption can be devastating to a cell exposed to the high optical power in an optical trap. Although, in principle, any collimated light source of sufficient power can be used for trapping, high-powered lasers are used almost exclusively. Recommended wavelengths are 980 nm, which has been shown experimentally to be one of the least biologically active wavelengths for use (Liang et al., 1996; Neuman & Block, 2004; Neuman et al., 1999; Zakharov & Thanh, 2008), and 1064 nm, a wavelength used in most commercially available trapping systems (also accepted to be compatible with cells). The 1064 nm lasers are readily available in high wattages with excellent beam quality from $5000 to $15,000 in the 0.5–3W range often used for optical trapping.

Considerations other than power and wavelength are relevant to trapping. Solid-state, continuous-wave lasers (as opposed to pulsed lasers) are recommended for trapping to avoid high peak powers and unwanted multiphoton photophysical phenomena. To form a single, tight focus (helpful for small cells), the beam must have a Gaussian transverse mode (single round spot brightest in the center with Gaussian intensity distribution across the beam when projected, also known as "TEM-00"). Lastly, power is a consideration. Although not the only factor in determining trapping force, a higher power light source will help increase trapping capabilities. A minimum of 100 mW reaching the sample is necessary for practical trapping, with higher powers allowing the application of more force to cells (e.g., faster movement of cells). The optics used to condition the beam for trapping attenuate the beam by absorption and back reflection. The laser setup described in this chapter outputs a 1-W, 976-nm beam from the head of the laser module, with about 250 mW reaching the sample (the largest single loss in the optical train occurs at the microscope objective). An additional safety consideration when purchasing a laser source is the laser power supply. A variable power supply is desirable, as it allows alignment of the laser source at lower (less dangerous) power levels, while prolonging the life of the laser.

When taking an automated or semiautomated microscopy approach to microfluidic single-cell genomics, microscope and objective selection are paramount. By far, the most convenient implementation entails an inverted "fluorescence" microscope with an automated (motorized) stage. Unwanted beam attenuation can be mitigated by careful selection of optical components. Although many groups build simple microscopes for optical trapping from individual components, a commercial inverted microscope is desirable when one desires to use many different imaging configurations (e.g., different magnifications and different techniques like phase, fluorescence, etc.) or would like to automate configuration changes. Typically, the laser beam will be introduced to the microscope via the "fluorescence excitation" path. Importantly, some microscopes have internal optics that may need to be accounted for and possibly modified. Specifically, any diffusing optics will need to be removed (preferably by the manufacturer) to achieve a tight focus, as was necessary in the Leica model DMI6000B automated scopes used for our applications. Preferably, the microscope presents a port with a clear path (no optics) to the dichroic mirror that reflects the fluorescence excitation up to enter the microscope objective. Figure 4.3 shows an idealized (simplest) optical diagram (part A) and two as-configured layouts specific to the Leica DMI6000B microscope (parts B and C).

We specified the Leica DMI6000B stand due to its full automation and compatibility with particular Leica objectives known at the time to be available with high numerical aperture (NA), phase contrast, and acceptable infrared (IR) transmittance (Neuman & Block, 2004). However, all major microscope manufacturers today (Leica, Nikon, Olympus, Zeiss) are optimizing objectives for improved IR transmittance (*transmittance* should not be confused with *correction* for IR chromatic aberration), and useful features

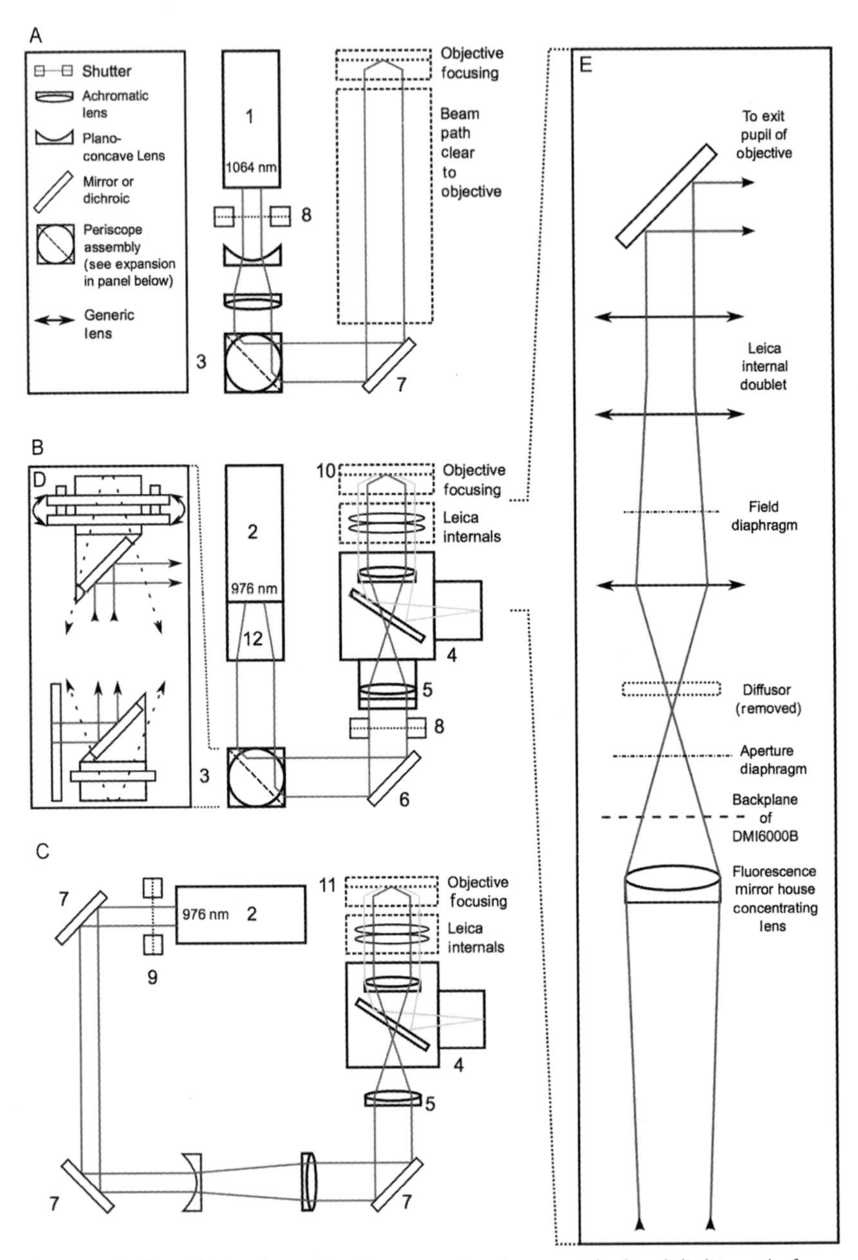

Figure 4.3 Simplified schematic diagrams showing an idealized light path for an inverted microscope with a minimum of internal components (schematic A), and two functional designs based around the Leica DMI6000B microscope stand (schematics B and C). The suggested design in schematic A includes a 1064-nm, 3-W laser source (1) chosen for its high power and ready availability, as well as its wavelength, which is suitable for biological materials. The idealized design contains a beam expander

consisting of a set of matched lenses preceded by a shutter (8; a Thorlabs part no. SH05 with a SC10 shutter controller was used in the configuration shown in B) placed at the head of the laser to enable control the beam without cycling the laser. Beam positioning and alignment is accomplished with a periscope consisting of two rotating mirrors mounted at 90 °C angles. An expanded view of the periscope used in the second configuration is presented in inset D (3; Thorlabs part no. RS99, equipped with dielectric broadband mirrors, Thorlabs part no. BB1-E03). By placing the periscope mirrors at right angles to each other, rotation of the mirrors can be used to coarsely position the beam horizontally and vertically on the third mirror, eliminating the need for three independent kinematic mounts. The periscope design used includes kinematic controls on the top mirror mount, allowing for the fine pitch-and-yaw adjustments needed for positioning at the second mirror. A third kinematic mount (6; Thorlabs part no. KM100 kinematic mount with BB1-E03 mirror) is used to adjust beam angle. The two designs utilizing the Leica DMI6000B microscope stand are somewhat more complex, largely due to the number of internal optics in the scope stand. Both of these setups utilized a 1.0–1.1-W, 976-nm laser (Crystalaser model DL-980-1W-OX/DL980-1W; 2), the wavelength chosen because of biological compatibility. The laser in the schematic B was modified at the factory with a spatial filter (12) to improve beam shape and power distribution. This spatial filter had the added effect of expanding the beam to ~8 mm, eliminating the need for additional beam expander optics. This design uses a periscope for two of the mirrors as described above. A third kinematic mount provides correction for beam angle (6). In this design, the shutter (8) is mounted outside of a customized mirror house (4) that includes a dichroic mirror to allow for simultaneous trapping and fluorescence. Due to the mirror house optics designed to bring the fluorescence excitation source into wide illumination field at the focal plane, a corrective achromatic lens (5) was installed in the mirror house at the factory to correct expansion of the trapping beam so that the beam would remain collimated upon entering the exit pupil of the trapping objective. This design utilized a 100× magnification, 1.4 numerical aperture (NA) apochromatic oil immersion objective as the trapping objective (10; Leica part no. 11504107). An alternative design shown in schematic C was also used. This design uses an independent corrective lens in the free space laser setup to offset the fluorescence optics and also includes a free space Galilean beam expander made from two matching lenses located between the second and third kinematic mounts. In this setup, the laser was mounted at a height and three individual generic mirrors and kinematic mounts (7) were used to align the laser. As in the suggested setup, a shutter (9; Uniblitz) was placed at the head of the laser. The trapping objective used here was 100× magnification, 1.3 NA apochromatic oil objective (Leica part no. 11506197; 11). A simplified diagram of the Leica DMI6000B internal components (inset E) is shown in the expansion on the right of the figure. Because the optical design inside the microscope is Leica-proprietary, the exact specifications are not indicated here. The important feature to note is that expansion and focusing optics placed external to the scope stand must be positioned so as to produce a collimated beam at the end of the light path prior to entering the exit pupil of the trapping objective. An additional, important modification is removal of the standard diffuser included in the Leica excitation light path (see inset E). The diffuser is located on a circuit board inside the scope at the position marked in inset E. Modifications to the scope internals are exacting and more accurate diagrams of the light path and scope internals can be requested from the manufacturer. (For color version of this figure, the reader is referred to the online version of this chapter.)

like external phase contrast are also now available. An additional consideration when choosing a microscope "stand" (the main body of the microscope) is competitive pricing as well as adaptability to your application and compatibility with particular software packages for automation. Prior to making a final decision on a microscope, be sure to demand IR transmission curves for the trapping objective lens and internal optics, as well as a comprehensive light path schematic indicating all internal optics the trap beam will encounter. One final consideration in choosing a microscope stand is selection of an appropriate automated stage. For our purposes, we used Leica DMI6000B stands equipped with either the standard Marhauser-Wetzlar stage available from Leica or the Applied Scientific Instrumentation MS-2000 (Eugene, OR). The stage must be able to be controlled both manually (e.g., via a joystick) and programmatically through a software interface (often drivers for RS-232, Labview or a .dll file is provided). Ensure the stage has a high level of movement precision and can be fluidly translated at speeds in the range 1–1000 μm/s. A linear encoder is also useful, as it improves the accuracy of repeated positioning.

3.2. Choosing an objective lens

By far, the single most important factors to consider when designing any optical trapping platform are IR transmission and NA of the objective lens that will be used for trapping. Higher NA provides oblique rays on which trapping in the focal dimension (z-coordinate) depends and dramatically improves overall trapping performance. A minimum of 1.0 NA is required for practical trapping, and NA of 1.3–1.5 is recommended. The objective also needs to be able to transmit IR wavelengths efficiently; however, priority should be given to maximizing the NA as low transmission can be mitigated by increased laser input power. If the trapping performance is too weak, the cell will easily escape the trap along the direction of the beam propagation. In a microfluidic channel, the upper channel surface may restrain the cell, allowing trapping of the cell in two dimensions against the upper surface of the microchannel. Even so, a better-performing trap able to hold the cell in three dimensions is strongly preferred because the cell can be moved more quickly when suspended in the center of the microchannel cross-section, and the possibility that the cell sticks to the channel surface is eliminated. Our applications utilized a 100 × magnification 1.3 or 1.4 NA apochromatic oil immersion objective (Leica part nos. 506211 and 11506197).

3.3. Alignment procedure and safety

IR light at the powers used for optical trapping is capable of quickly and permanently damaging the retina in certain exposure scenarios, and the upmost care needs to be taken in terms of safety. Alignment of laser optics should only be attempted with appropriate safety precautions. A laser safety class is recommended and is required at many research institutions. IR laser beams, invisible to the naked eye, merit extra caution because emissions are not visually apparent, and the IR light does not trigger a blink reflex. When working with lasers, appropriate shielding should be in place to ensure that laser light will not impinge on untrained or unaware individuals. Laser goggles blocking the appropriate wavelengths should be worn at all times when there is a possibility of laser emission. In general, the highly divergent laser light emitted from the microscope objective during trapping falls below a permissible exposure level after a short distance. A laser safety class should train users to calculate this distance for their particular configuration. Even so, it is good practice to place shielding above the stage to block user exposure to this divergent emission. All work should be done standing with the plane of the laser lines below the user's eye level since stray beams are most likely to occur within this plane. Care should be taken to never look directly at the laser source or at reflective surfaces (e.g., mirrors) in the path of the laser. Reflective surfaces in the work area and on the users' body (rings, watches, etc.) should be kept to a minimum. Importantly, the eyepieces on a microscope used for optical trapping with IR light must be removed or permanently disabled to eliminate the possibility that a user could directly view the laser beam in the case of user error or microscope malfunction.

The end goal of alignment is to position an expanded and collimated laser beam on the optical axis of the objective lens. To take advantage of the full NA of the objective lens, the beam should fully fill the back aperture of the objective, and thus prior to entry into the scope should be expanded to slightly overfill the objective's back aperture. A convenient way to build a simple Galilean beam expander is from two lenses, one plano-concave and one convex (a Keplerian beam expander constructed from two positive lenses will also work). The two lenses should be selected to achieve the desired degree of beam expansion according to the equation:

$$2(r)(f2)/|f1| = E$$

where r is the radius of initial beam, $f1$ is the focal length of plano-concave lens, $f2$ is the focal length of convex lens, E is the expanded beam diameter, and the space between the two lenses is the sum of the two focal lengths.

When aligning the laser setup, at least three mirrors and adjustable kinematic mounts (Thorlabs part no. KM100 kinematic mount with BB1-E03 mirror were used in our application) should be used (alignment can be achieved with two kinematic mounts, but is easier with three). Dielectric broadband mirrors (Thorlabs part no. BB1-E03) are recommended due to their very high IR reflectance (hence low attenuation). The simplest setup positions the laser aperture at the same height above the table as the entry of the beam into the microscope. Alternatively, two of the mirrors may be configured as a periscope (we used Thorlabs part no. RS99 for this purpose) to raise the path of the beam (Fig. 4.3B). If the laser lacks a modulation/shuttering capability, a shutter is best placed between the laser source and the first mirror so that laser light can be excluded from the work area without turning the laser on and off. For our purposes, we mounted the Leica DMI6000B microscope stand in an amenable fixed position on a 1 m × 1 m × 50 cm optics table prior to placement of mirror optics. Some microscopes come with accessories to directly mount the stand to a standard optical table (drilled and tapped 1/4–20 on 1 in. centers or M6 on 25 mm centers). The laser was mounted in a fixed position appropriate for our setup. Initially, all mirrors should be positioned based on measurements such that the path of the laser will follow right angles as closely as possible at each mirror. It is important to use the reflective face of the mirrors as your reference for where the beam will be redirected, as the face of the mirror may project proud of the mount. When initially placing the mirror mounts and mount stands, it is good practice to physically verify their location and angles, and not simply "eyeball" their initial position (there will be plenty to adjust even with the most exacting initial placement of optics). Good practice demands that all optics on the table are fixed in position (directly bolted or clamped with "dogs") when the laser is active to avoid unexpected redirection of the beam if an optic shifts or topples.

After the initial placement of the optics, alignment can begin. After ensuring all shielding is in place and everyone present is alert, the laser can be powered on. To align the laser, a clear way to visualize the end point of the beam and assess alignment needs to be in place. A useful tool consists of a short piece of optical tube (threaded aluminum tubing) 1–3 in. long with an adapter that allows mounting on the objective turret. The tool should have an iris next to the thread adapter, the length of the tube, then a second iris, and a cap consisting of a phosphorescent plate (allowing visualization of the invisible beam) with the center marked. For our purposes, we used Thorlabs part no. SM1A11 (adapter), SM1D12C (iris), SM1L30C (3 in. slotted lens tube), SM1D12C (iris), SM1L05 (0.5 in. lens tube), with VRC4D1 (phosphorescent disc) mounted using two retaining rings (SM1RR). The configuration of our alignment tool is detailed in supplementary figure 1 (http://dx.doi.org/10.1016/B978-0-12-407863-5.00004-6). Phosphorescent cards can be inserted into the laser path at multiple points in the beam to track the position and angle of the laser.

Initially, rotating the positions of the mirror mounts should allow for coarse positioning of the beam. When the laser is roughly aligned with the optomechanical hardware in fixed positions, adjustment using the finely threaded thumbscrews on the kinematic mounts can begin. The first two mirrors should be used to adjust the position of the beam on the third mirror. Adjustment of the third mirror should be reserved to correct the angle of the beam. Between proper positioning of the optomechanical hardware, coarse adjustment of mount position and fine alignment using the pitch-and-yaw screws of the kinematic mounts, some portion of the beam can be made to appear on the phosphorescent plate of the objective tool when it is positioned in place of the trapping objective. By continuing to use the first two mirrors for positioning, the beam can be centered. By adjusting the pitch-and-yaw of the third mirror, the angle of the beam can be corrected.

By placing an iris in between the final mirror and the entry point of the scope (centered on the optical axis of the microscope), the beam can be straightened as much as possible by ensuring that both this iris and the lower iris on the objective alignment tool close evenly around the beam (visualizing the projected beam on the alignment tool phosphorescent plate). When this is achieved, the beam is roughly aligned. The beam alignment can be fine-tuned by bringing the trapping objective into place, mounting a microscope coverslip or coverslip-mounted microfluidic device on the stage of the microscope, and visualizing the light reflected when the trap is focused on the upper surface of the coverslip with a CCD camera. By adjusting the mirrors in the beam path, the trap can be moved to the appropriate position within the field of view and the angle of the beam fine-tuned by optimizing the symmetry and concentricity of rings in the out-of-focus image of the trap reflection. The trap can be made coincident with the image plane of the microscope by adjusting the divergence of the trapping beam that enters the microscope by adjusting the distance between the lenses making up the beam expander. A diverging beam will form a trap above the imaging plane of the microscope, while a converging beam will form a trap below the imaging plane of the microscope.

4. MICROFLUIDICS CONFIGURATION AND SETUP

4.1. Microdevice design

The most current single-genome amplification device is the 48x_v4 device illustrated in Fig. 4.4. This chip can be ordered directly from the Stanford Microfluidics Foundry, and mold sets can also be obtained for in-house

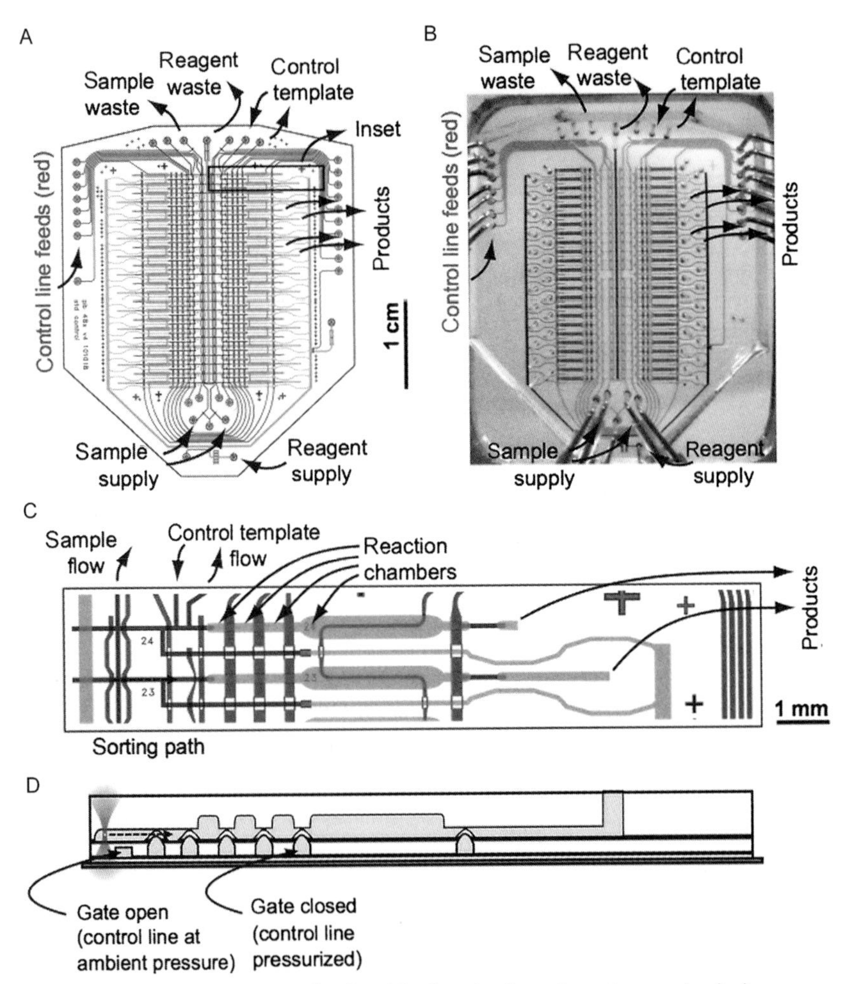

Figure 4.4 48-chamber microfluidic chip for single cell sorting and whole-genome amplification. (A) Design of 48x_v4 device with key ports marked. (B) Photograph of 48-chamber device with corresponding ports marked. (C) Inset showing detail of two reaction chambers in the 48x_v4 device. (D) Cross-sectional schematic corresponding to part (C) indicating the trapping of a cell in the (upper) flow layer and the actuation of valves by pressurization of channels in the (lower) control layer. (See color plate.)

production of microfluidic devices. Supplement S4 (http://dx.doi.org/10.1016/B978-0-12-407863-5.00004-6) contains data files with the design of the 48x_v4 microfluidic device, Supplement S5 (http://dx.doi.org/10.1016/B978-0-12-407863-5.00004-6) describes the standard operating procedure for fabricating molds for the 48x_v4 device, and Supplement S7 (http://dx.doi.org/10.1016/B978-0-12-407863-5.00004-6) is a sample order sheet for obtaining molds and/or devices from the Stanford Microfluidics Foundry.

The 48x_v4 device includes 48-arrayed independent reactor systems for cell selection, lysis, and whole-genome amplification. The device consists of two molded layers of PDMS mounted on a number one microscope coverslip. The transparent nature of the PDMS and mounting on the thin coverslip allows the application of most microscopy techniques and optical trapping. We have designed a custom mount for the 48x_v4 device that can be produced on a 3D printer and fits into the standard 110 × 160 mm stage insert opening. The holder securely grips the device, flattening any curling of the coverslip, protects the device from peeling forces, and prevents damage to the coverslip. The design files/renderings of the chip holder are provided in Supplement S3 (http://dx.doi.org/10.1016/B978-0-12-407863-5.00004-6).

The microfluidic device utilizes a "push-up" approach to the actuation of microfabricated valves. Control channels in the thin PDMS layer adjacent to the coverslip are filled with water and pressurized to actuate valves (also referred to as "gates") on the device. Where the pressurized control channels cross flow channels in the top device layer, the thin membrane separating the two stretches to pinch off the microchannel in the flow layer (Fig. 4.4D). Control lines are typically configured as dead-end channels and are filled completely with water, taking advantage of the high gas permeability of PDMS, which allows air to rapidly escape microchannels as fluid is pushed in. Under normal use, the lines are under atmospheric pressure (valves open), and 10–20 psi of air when actuated to close valves.

Each control line is coupled to an individual fluid reservoir with an overhead pressure controlled by an electronic solenoid. The microfluidic controller used to control the solenoids consists of a USB microcontroller with 24 amplified digital output bits/channels, each of which powers a single solenoid to select either atmospheric (valves on chip open) or "control" (valves on chip closed) pressures. The solenoids are mounted on three 8-port manifolds supplied with the desired "control" pressure by a gas regulator. By sending a computer signal to control the solenoid states, the valves on the chip can be opened or closed at the behest of a user or in automated routines. The Stanford Microfluidics Foundry hosts information on the capabilities, assembly, use, limitations, and alternatives to the USB chip controller, as well as in-depth workshops on the design, fabrication, and use of microfluidic devices (http://www.stanford.edu/group/foundry/).

The 48x_v4 chip is designed using two nominal control channel dimensions (both square profile produced in SU8 resist): 100 μm wide and 25 μm high, and 200 μm wide and 25 μm high. Two nominal flow channel dimensions are also specified: small channels (round profile, produced in SPR resist, valved by the narrow control channels) 100 μm wide and 7–13 μm high (taller to accommodate larger cells) and large channels (round profile, produced in AZ50 resist, valved by the wide control channels) 200 μm wide and 50 μm high.

The 48x_v4 chip has been designed for robust operation, with valves tuned (by varying control channel width from the nominal dimensions) to operate at uniform pressure, extra space between components (and component redundancy) to allow proper function in the face of fabrication defects, and a large sealing area around the device perimeter for robust lamination. We produce three 48x_v4 devices on a 10-cm wafer, which allows ample space between devices for handling in the course of manufacture.

The 48_v4 variant was in fact designed for reuse, with extra features like tuned outlet resistances that facilitate clean-out. PDMS devices are mechanically robust, with valves able to cycle many thousands of times without showing wear (Unger, 2000). The most challenging aspect of reuse is preventing the DNA products of the previous cycle from contaminating reactions in the current cycle. We validated a clean-out protocol utilizing sodium hydroxide and 0.6% hypochlorite washes, plus a UV decontamination step (PDMS is transparent in the UV). This clean-out procedure was found to be sufficient in eliminating a PCR-traceable test template from the device (data not shown). With appropriate negative controls, reuse of the 48x_v4 device is certainly a feasible option.

5. PROCEDURE FOR SORTING AND AMPLIFYING SINGLE CELLS

Each reactor on the device is intended for the processing of an individual cell, and consists of a "template chamber," two "lysis chambers," and a "reaction chamber," connected in series. Separating the template chamber from the "sample lines" is an antechamber filled with clean buffer and tapped for the addition of reagents from the "reagent line." This configuration provides a "privileged" flow path for reagents to each reactor that does not trace any channel where the bulk sample flows. A full protocol for using the 48x_v4 device is provided in Supplement S6 (http://dx.doi.org/10.1016/B978-0-12-407863-5.00004-6).

Two independent "sample lines" run the length of the device, each intersecting half of the reactors. Cells identified for selection under the microscope are sorted by moving the targeted cell from the sample line to the template chamber using the optical trap. The antechamber buffer is kept clean by slightly pressurizing the antechamber (from a 2 psi feed via the reagent line) prior to opening the gate (valve) separating it from the sample selection line, preventing cells from inadvertently crossing into the antechamber. After a cell is selected and moved into the antechamber, this gate can be closed again and the cell can be taken through the antechamber to the template chamber by transiently opening a second valve, which is closed after the cell is placed in the template chamber. Using this "air-lock" setup (the cell must pass through two valves on the chip), cells can be individually passed to the template chambers while the bulk sample solution is completely excluded from the template chambers. This works despite the design necessity that the air-lock valves operate in blocks of 12 template chambers (four blocks in all). Cells in each of the four device blocks can be treated with different lysis or reaction conditions, since the second air-lock valve controls the flow of reagents into each cell's reactor section. The ability to run different samples independently on the two halves of the chip, and to apply as many as four sets of reaction conditions, makes the device adaptable to different project configurations and allows efficient optimization of conditions for a given sample type. At the top of the device, provisions are made to inject positive control solutions (typically cells or nucleic acids) directly into reactors 24 and 48. Properly selected controls can be very useful to distinguish failures of lysis from failures of amplification.

The large surface area to volume ratios associated with microfluidic arrays present issues of "sticking" where hydrophobic or electrostatic interactions between the cells (when not held away from the walls using the optical trap) or reagent components and the walls of the device necessitate the use of surface treatments or surfactants to mediate the interaction between solution components and the walls of the device. We have found that 0.04–0.08% Pluronic F127 in the sample dilution buffer is extremely effective in preventing the sticking of most cell types to the device. Bovine serum albumin (BSA) can be supplemented if necessary, but the level of contaminating DNA should be assayed (Blainey et al., 2011), and measures taken to source a low-contamination sample or degrade contaminants (UV treatment is recommended, but can cause aggregation). Provided that amplifiable DNA can be removed or destroyed, 8 μg/ml of BSA can be used in both the sample dilution buffer as well as the sterile diluent. Many groups have developed chemical and physical coatings that prevent adsorption of cells and biomolecules to PDMS, but we find the application of dynamic coatings (soluble surfactants such as Pluronic F127 and Tween-20) convenient and effective.

Potassium hydroxide treatment (0.4 *M*, optionally with EDTA, DTT, surfactants, and elevated temperature) is the most commonly applied lysis treatment in single-cell genomics, but is not sufficient for the lysis of many environmental microorganisms. Pretreatment (of the bulk sample or individual cells on chip) with enzymes and/or detergents can be helpful in many cases, but appropriate positive controls should be used to ensure that WGA is not inhibited by the lysis reagents. As a side note, a coverslip-mounted chip was observed to survive cycling to liquid nitrogen temperatures with no ill effects, indicating the possibility of lysis by ice crystal formation within the device. Two 3.5 nl lysis chambers are provided in each reactor to allow the flexible application of sequential lysis treatments. When one reagent is changed for another, the entire reagent delivery manifold is flushed out of the central waste line to guarantee that the new reagent is applied at its full concentration.

After lysis, the third 3.5 nl chamber can be used to chemically neutralize any lysis conditions applied, for example, adding sodium acetate to neutralize an alkaline lysis step. Subsequently, the WGA reaction buffer can be prepared and applied to flush the lysate into the 50 nl reaction chamber for a total reaction volume of 60 nl. For MDA, the device is incubated at 30–31 °C on a digital hot plate with the hydration line, which runs directly under the reaction chambers, pressurized. The degree of hydration is an important parameter in optimizing reactivity. Too little hydration and the volume or the reaction will shrink as a net loss of water vapor occurs; too much, and the chamber can swell, "burping" products from the outlet and diluting the reactants. The correct amount of hydration depends on the gas permeability of the PDMS and the ambient humidity. A humidity chamber is useful and can consist of a sealed Pyrex dish that the microfluidic device is placed into with a small amount of purified water. This is placed on a hot plate with a temperature probe inserted into the chamber to regulate the supply of heat from the hot plate. (The Thermo Scientific Super-Nuova model was used for this purpose; its advantages include probe-controlled temperature as well as logging through an RS232 port.) Only the MDA chemistry has been tested in the device, although other WGA chemistries dependent on short high-temperature steps would likely work if provided increased hydration pressure.

We add an intercalating dye (0.5 × EVA Green) to the reactions and monitor product accumulation in real time by taking fluorescence images of the reaction chambers (a less informative alternative approach is endpoint fluorescence monitoring with 1 × SYBR). When complete, products are recovered in parallel by applying Tris buffer to the reagent supply and flushing the products into pipette tips inserted into the outlet of each reactor. Individual recovery of the product from each reaction eliminates the possibility of cross-contamination, while simultaneous recovery of the products in parallel (the device has been optimized for a uniform flow rate across all the chambers) enables rapid product collection in the pipette tips, with 5 μl recovered from every reaction within a few minutes. Although the products are diluted about 100-fold during this recovery step, leftover primers and considerable phi29 DNAP activity exist in the product mixture. To avoid interference with downstream assays such as PCR, it is strongly recommended that the products be heat treated (65 °C, 10 min will abolish the polymerase and exonuclease activities of phi29 DNAP) and further diluted and/or purified prior to other analyses.

6. SOFTWARE

The LabView graphical programming language was used to design an integrated software interface for the scope, stage, and microfluidic controller. Matlab, Micro-Manager, and a variety of commercial imaging programs can also be used to coordinate hardware control and image acquisition. In our implementation, the user is presented with a graphical interface (Fig. 4.5A) that shows a live view of the chip through the microscope (with fiducial marking the position of the optical trap within the field of view) and a chip schematic indicating the status of control lines. The interface provides for manual control of the on-chip valves, laser, stage, and camera, as well as automatic functions such as setting preprogrammed valve states, moving to predefined locations on the chip, time-lapse imaging across multiple fields of view, and initialization routines for calibrating the stage. Although a previous design relying on pumped flow for sorting was automated for user-independent

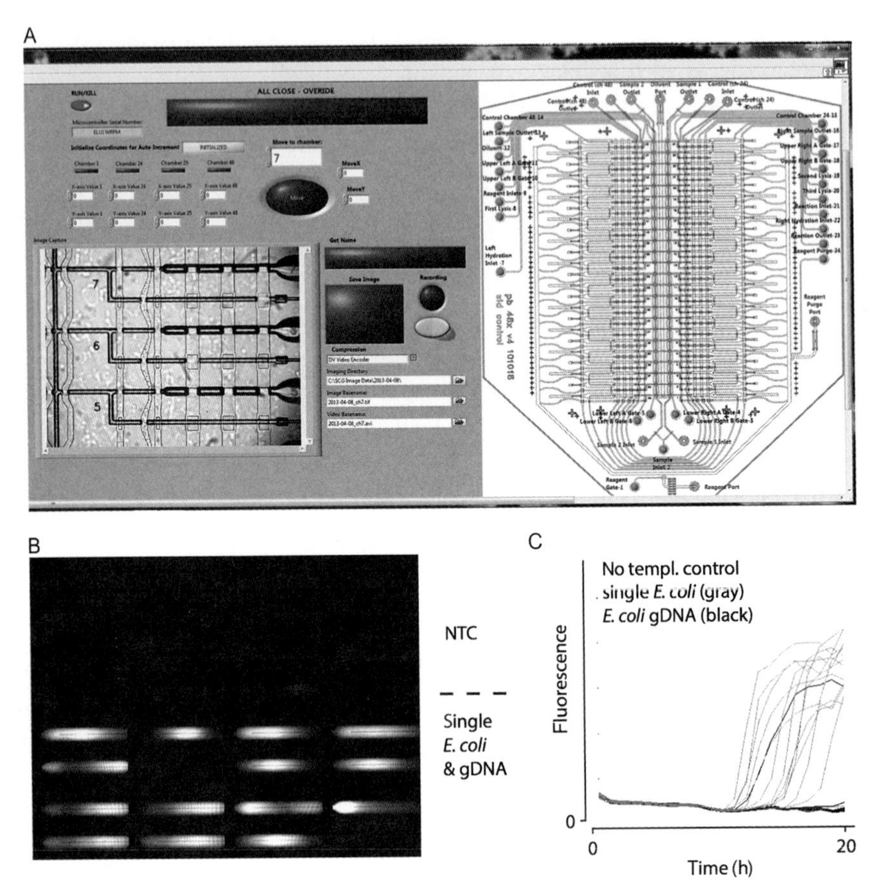

Figure 4.5 Example of dual-screen graphical user interface with integrated microfluidic controls, stage automation, and video and image recording (A). Automated fluorescence data acquisition (B) and real-time kinetics curve of double-stranded DNA formation in individual microfluidic reaction chambers (C). (See color plate.)

sorting (Marcy, Ishoey, et al., 2007), the current software runs in a semiautomated mode where the human user selects cells and initiates a programmed sorting routine. Applications where the choice of freely diffusing cells can be reduced to a machine-vision task (e.g., by a fluorescence intensity metric) may benefit from complete automation of the sorting process.

Valve control is integrated using the LabView drivers for the USB microcontroller card used in the Stanford valve manifold controller design. LabView drivers were used for control of the ASI and Marhauser-Wetzlar motorized stages. In both setups, a coordinate teaching system was developed to automatically calculate the location of each sorting intersection on the microfluidic device so as to improve the speed of cell sorting and help with navigation at higher magnifications. Our system measured focal positions as well, interpolating the z-position of the flow channel in between measurement points. For effective movement across large areas of the device, a chip-mounting system such as our 3D printed holder that flattens the coverslip and levels the device on the microscope stage is critical to prevent excessive focus run-out with high-power trapping objectives, which have a shallow depth of field. Alternatively, or in addition, an active autofocus system can be utilized to maintain the flow channels in sharp focus.

Computer-based control of the laser system is also highly recommended to improve safety and facilitate automatic control of sorting operations. The preferred solution is a laser that can be directly modulated by a digital or analog signal. Alternatively, a stand-alone electronic shutter is also suitable (e.g., Thorlabs part no. SH05 with SC10 shutter controller).

High-speed video imaging is critical to both manual and automatic manipulations of cells with the trap as well as documentation of the cells selected for analysis. A live-view frame rate of at least 25 frames per second is critical for eyestrain-free viewing over long periods of time. For optimally resolved and high-contrast fluorescence imaging of cells, the size of pixels on the detector should be between $0.5\times$ and $1\times$ the projected size of diffraction-limited objects in the field of view. For example, a $100\times$ NA 1.4 objective lens imaging green light would have a calculated resolution in the sample of $\sim 0.6 \times 525$ nm$/1.4 = 225$ nm, and the optimal pixel size would be between 225 nm$/2 \times 100 = 11$ μm and 225 nm $\times 100 = 22.5$ μm. For this system, pixels smaller than 11 μm would compromise signal-to-noise ratio for little gain in spatial resolution, while pixels larger than 22.5 μm would compromise spatial resolution. Particularly useful is the integration of application-specific imaging modalities into the GUI. These may include fluorescence or interference-based imaging of cells and automated endpoint or time-lapse fluorescence of DNA stains in each MDA reaction chamber as illustrated in Fig. 4.5B and C. Additional facilities for video and image documentation of cells with automated file naming (by day, sample designation, and chamber number) during sorting may also be built into the GUI.

7. PERFORMANCE AND LIMITATIONS

In practice, the optofluidic laser-tweezing/microfluidic WGA approach provides a number of unique advantages as a method for highly targeted single-cell genomics studies. With a well-engineered system and medium-sized cells, the speed at which cells can be manipulated can reach ~200 μm/s or more; however, as trapping forces are dependent both on the size and shape of the cell, the maximum transport speed can vary. Optimal cell sizes for fast sorting range from 1 to 10 μm in diameter, but larger cells or groups of cells can be manipulated by trapping a portion of the structure and moving at slower rates. Smaller cells of 0.5 μm can also be effectively manipulated by reducing the rate of movement; in fact, individual viruses (Ashkin et al., 1987), molecules (Katsura, Hirano, Matsuzawa, Mizuno, & Yoshikawa, 1998), and even atoms (Chu, Bjorkholm, Ashkin, & Cable, 1986) can also be confined in simple, single-beam optical traps. The limiting factor in trapping small cells is not usually the ability to achieve sufficient confinement of the cell, but rather the ability to visualize the trapped cell (highly sensitive approaches like fluorescence or scattering are typically employed for detection of small particles in optical traps). Even manual selection and sorting of cells can be performed quickly, with up to several cells per minute being placed into their respective reaction chambers. Full automation for amenable target cell types has a potential to increase this speed further, with the ultimate limit being placed by the time needed for image acquisition/processing and the maximum achievable speed of cell movement. The optical trap rejects contaminating DNA more effectively than methods based on subdivision of the input sample and does not require dilution of the sample. Laboratory experiments have shown that the trapping technique can remove cells from buffer containing 10^4 cp/μl of plasmid with no carry-over amplification of the plasmid DNA (Dodsworth et al., 2013).

The microfluidic technique produces nanogram quantities of DNA from each successful reaction (~30 ng) that are sufficient for many analyses including PCR and high-throughput sequencing (Blainey et al., 2011; Marcy, Ouverney, et al., 2007). The 48-chamber design of the microfluidic device provides ample opportunity for success. As an added benefit, the complexity of the system is not overwhelming. In the hands of an experienced user, the optical alignment is very repeatable and stable for months to years, and the relatively straightforward design can be broken down and set back up fairly easily, lending itself to travel or even field studies. As a demonstration, one of the systems used in our experiments was shipped from Oregon to Bermuda in 2012 for application as a teaching instrument in a course on microbial oceanography.

The most significant limitation of the current implementation of this optofluidic technology is throughput. The footprint of the 50 μm deep reactors and use of dedicated recovery ports limited the reactor density on the 48x_v4 device, although there are no fundamental barriers to increase the reactor density or size of the chip by reconfiguring the reactors and product recovery method to achieve batchwise improvements in throughput of several orders of magnitude. The genomic coverage achieved when sequencing single-cell WGA product is highly variable and is thought to be limited by incomplete postlysis accessibility of the genome and genome fragmentation. Coverage is limited further by amplification bias, or the uneven coverage of the genome common in single-cell WGA datasets. Single-cell WGA coverage of environmental microbial genomes has been reported at 13–41% with 160 × sequencing (Woyke et al., 2011), although much higher coverage has been observed in select cases. Since many reactions are unsuccessful or result in lower coverage than needed for desired analyses, it is advantageous to process many reactions to increase the number of high-quality reaction products obtained. Despite the current limitation of 48 reactions per chip in the optofluidic approach, applications with limited sample quantity, FACS-incompatible samples, selection criteria requiring high-resolution microscopy, or that benefit from a low-cost casual operating mode in the lab of an individual investigator may be suitable for the optical trapping approach.

Specific examples of where this approach would be particularly practical include:

- separation of known or unknown components of an enrichment culture for WGA
- separation of cells from the environment or enrichment culture for isolate culture
- selection of bacteria from tissue or intracellular parasites from a cell lysate, where contamination from host DNA would be a serious concern
- acquisition of cells with a defining or interesting morphology or visible phenotype
- manipulation of cell aggregates
- combinatorial approaches where high-quality microscopy documentation would be useful
- single-cell genomics using extremely small volumes of sample (1–20 μl).

The reagent and personnel costs for this technology are likely to be significantly lower than those associated with a facility designed for very high-throughput single-cell genomics. In this vein, the approach is suitable for laboratories where it is desirable to have a single-cell genomics capability readily available, but where they are not necessarily the central focus of the research. It is also worth noting here, that while the microfluidic device described here was designed specifically for single-genome amplification, optical trapping is a microbiological technique that is applicable to a broad variety of studies and integration with other culture-based, imaging-based microfluidic and/or molecular approaches. We eagerly anticipate an expanding role for optical trapping in a variety of novel microbiological studies in the coming years.

ACKNOWLEDGMENTS

The authors acknowledge the work of Geoffrey Schiebinger for software development and sequence data analysis, Nicholas Gobet in designing the first chip holder, and David McIntyre for assistance with optical trapping at OSU. P. C. B. was supported through the Burroughs Welcome Fund via a Career Award at the Scientific Interface. Z. C. L. and the educational laser-trapping system at OSU were supported through funding from the Gordon and Betty Moore Foundation.

REFERENCES

Ashkin, A. (1992). Forces of a single-beam gradient laser trap on a dielectric sphere in the ray optics regime. *Biophysical Journal, 61*, 569–582.

Ashkin, A. (1997). Optical trapping and manipulation of neutral particles using lasers. *Proceedings of the National Academy of Sciences of the United States of America, 94*, 4853–4860.

Ashkin, A., & Dziedzic, J. M. (1989). Internal cell manipulation using infrared laser traps. *Proceedings of the National Academy of Sciences of the United States of America, 86*, 7914–7918.

Ashkin, A., Dziedzic, J. M., & Yamane, T. (1987). Optical trapping and manipulation of single cells using infrared laser beams. *Nature, 330*, 769–771.

Ashkin, A., & Gordon, J. P. (1983). Stability of radiation-pressure particle traps: An optical Earnshaw theorem. *Optics Letters, 8*, 511–513.

Blainey, P. C. (2013). The future is now: Single-cell genomics of bacteria and archaea. *FEMS Microbiology Reviews, 37*, 407–427.

Blainey, P. C., Mosier, A. C., Potanina, A., Francis, C. A., & Quake, S. R. (2011). Genome of a low-salinity ammonia-oxidizing archaeon determined by single-cell and metagenomic analysis. *PLoS One, 6*, e16626.

Boetius, A., Ravenschlag, K., Schubert, C. J., Rickert, D., Widdel, F., Gieseke, A., et al. (2000). A marine microbial consortium apparently mediating anaerobic oxidation of methane. *Nature, 407*, 623–626.

Chu, S., Bjorkholm, J. E., Ashkin, A., & Cable, A. (1986). Experimental observation of optically trapped atoms. *Physical Review Letters, 57*, 314–317.

Dean, F. B. (2002). Comprehensive human genome amplification using multiple displacement amplification. *Proceedings of the National Academy of Sciences of the United States of America, 99*, 5261–5266.

Dean, F. B., Nelson, J. R., Giesler, T. L., & Lasken, R. S. (2001). Rapid amplification of plasmid and phage DNA using Phi29 DNA polymerase and multiply-primed rolling circle amplification. *Genome Research, 11*, 1095–1099.

Dodsworth, J. A., Blainey, P. C., Murugapiran, S. K., Swingley, W. D., Ross, C. A., Tringe, S. G., et al. (2013). Single-cell and metagenomic analyses indicate a fermentative and saccharolytic lifestyle for members of the OP9 lineage. *Nature Communications, 4*, 1854.

Dupont, C. L., Rusch, D. B., Yooseph, S., Lombardo, M.-J., Richter, R. A., Valas, R., et al. (2011). Genomic insights to SAR86, an abundant and uncultivated marine bacterial lineage. *The ISME Journal*, *6*, 1186–1199.

Ericsson, M., Hanstorp, D., Hagberg, P., Enger, J., & Nystrom, T. (2000). Sorting out bacterial viability with optical tweezers. *Journal of Bacteriology*, *182*, 5551–5555.

Fan, H. C., Wang, J., Potanina, A., & Quake, S. R. (2010). Whole-genome molecular haplotyping of single cells. *Nature Biotechnology*, *29*, 51–57.

Frumkin, D., Wasserstrom, A., Itzkovitz, S., Stern, T., Harmelin, A., Eilam, R., et al. (2008). Cell lineage analysis of a mouse tumor. *Cancer Research*, *68*, 5924–5931.

Huber, R., Burggraf, S., Mayer, T., Barns, S. M., Rossnagel, P., & Stetter, K. O. (1995). Isolation of a hyperthermophilic archaeum predicted by in situ RNA analysis. *Nature*, *376*, 57–58.

Katsura, S., Hirano, K., Matsuzawa, Y., Mizuno, A., & Yoshikawa, K. (1998). Direct laser trapping of single DNA molecules in the globular state. *Nucleic Acids Research*, *26*, 4943–4945.

Kvist, T., Ahring, B. K., Lasken, R. S., & Westermann, P. (2006). Specific single-cell isolation and genomic amplification of uncultured microorganisms. *Applied Microbiology and Biotechnology*, *74*, 926–935.

Liang, H., Vu, K. T., Krishnan, P., Trang, T. C., Shin, D., Kimel, S., et al. (1996). Wavelength dependence of cell cloning efficiency after optical trapping. *Biophysical Journal*, *70*, 1529.

Malmstrom, R. R., Rodrigue, S., Huang, K. H., Kelly, L., Kern, S. E., Thompson, A., et al. (2013). Ecology of uncultured Prochlorococcus clades revealed through single-cell genomics and biogeographic analysis. *The ISME Journal*, 7, 184–198.

Marcy, Yann, Ishoey, T., Lasken, R. S., Stockwell, T. B., Walenz, B. P., Halpern, A. L., et al. (2007). Nanoliter reactors improve multiple displacement amplification of genomes from single cells. *PLoS Genetics*, *3*, e155.

Marcy, Y., Ouverney, C., Bik, E. M., Losekann, T., Ivanova, N., Martin, H. G., et al. (2007). Inaugural article: Dissecting biological "dark matter" with single-cell genetic analysis of rare and uncultivated TM7 microbes from the human mouth. *Proceedings of the National Academy of Sciences of the United States of America*, *104*, 11889–11894.

Neuman, K. C., & Block, S. M. (2004). Optical trapping. *The Review of Scientific Instruments*, *75*, 2787.

Neuman, K. C., Chadd, E. H., Liou, G. F., Bergman, K., & Block, S. M. (1999). Characterization of photodamage to *Escherichia coli* in optical traps. *Biophysical Journal*, *77*, 2856–2863.

Olsen, G. J., Lane, D. J., Giovannoni, S. J., Pace, N. R., & Stahl, D. A. (1986). Microbial ecology and evolution: A ribosomal RNA approach. *Annual Review of Microbiology*, *40*, 337–365.

Pamp, S. J., Harrington, E. D., Quake, S. R., Relman, D. A., & Blainey, P. C. (2012). Single-cell sequencing provides clues about the host interactions of segmented filamentous bacteria (SFB). *Genome Research*, *22*, 1107–1119.

Rappé, M. S., & Giovannoni, S. J. (2003). The uncultured microbial majority. *Annual Review of Microbiology*, *57*, 369–394.

Renesto, P., Crapoulet, N., Ogata, H., La Scola, B., Vestris, G., Claverie, J.-M., et al. (2003). Genome-based design of a cell-free culture medium for Tropheryma whipplei. *The Lancet*, *362*, 447–449.

Rocha, M. S. (2009). Optical tweezers for undergraduates: Theoretical analysis and experiments. *American Journal of Physics*, *77*, 704.

Stepanauskas, R., & Sieracki, M. E. (2007). Matching phylogeny and metabolism in the uncultured marine bacteria, one cell at a time. *Proceedings of the National Academy of Sciences of the United States of America*, *104*, 9052–9057.

Swan, B. K., Martinez-Garcia, M., Preston, C. M., Sczyrba, A., Woyke, T., Lamy, D., et al. (2011). Potential for chemolithoautotrophy among ubiquitous bacteria lineages in the dark ocean. *Science, 333*, 1296–1300.

Sykes, P. J., Neoh, S. H., Brisco, M. J., Hughes, E., Condon, J., & Morley, A. A. (1992). Quantitation of targets for PCR by use of limiting dilution. *BioTechniques, 13*, 444–449.

Tyson, G. W., Chapman, J., Hugenholtz, P., Allen, E. E., Ram, R. J., Richardson, P. M., et al. (2004). Community structure and metabolism through reconstruction of microbial genomes from the environment. *Nature, 428*, 37–43.

Unger, M. A. (2000). Monolithic microfabricated valves and pumps by multilayer soft lithography. *Science, 288*, 113–116.

Venter, J. C., Remington, K., Heidelberg, J. F., Halpern, A. L., Rusch, D., Eisen, J. A., et al. (2004). Environmental genome shotgun sequencing of the Sargasso Sea. *Science, 304*, 66–74.

Vogelstein, B., & Kinzler, K. W. (1999). Digital Pcr. *Proceedings of the National Academy of Sciences of the United States of America, 96*, 9236–9241.

Wang, J., Fan, H. C., Behr, B., & Quake, S. R. (2012). Genome-wide single-cell analysis of recombination activity and de novo mutation rates in human sperm. *Cell, 150*, 402–412.

Woyke, T., Sczyrba, A., Lee, J., Rinke, C., Tighe, D., Clingenpeel, S., et al. (2011). Decontamination of MDA reagents for single cell whole genome amplification. *PLoS One, 6*, e26161.

Woyke, T., Xie, G., Copeland, A., González, J. M., Han, C., Kiss, H., et al. (2009). Assembling the marine metagenome, one cell at a time. *PLoS One, 4*, e5299.

Wrighton, K. C., Thomas, B. C., Sharon, I., Miller, C. S., Castelle, C. J., VerBerkmoes, N. C., et al. (2012). Fermentation, hydrogen, and sulfur metabolism in multiple uncultivated bacterial phyla. *Science, 337*, 1661–1665.

Yilmaz, S., Haroon, M. F., Rabkin, B. A., Tyson, G. W., & Hugenholtz, P. (2010). Fixation-free fluorescence in situ hybridization for targeted enrichment of microbial populations. *The ISME Journal, 4*, 1352–1356.

Youssef, N. H., Blainey, P. C., Quake, S. R., & Elshahed, M. S. (2011). Partial genome assembly for a candidate division OP11 single cell from an anoxic spring (zodletone spring, Oklahoma). *Applied and Environmental Microbiology, 77*, 7804–7814.

Zakharov, S. D., & Thanh, N. C. (2008). Combined absorption of light in the process of photoactivation of biosystems. *Journal of Russian Laser Research, 29*, 558–563.

Zhang, K., Martiny, A. C., Reppas, N. B., Barry, K. W., Malek, J., Chisholm, S. W., et al. (2006). Sequencing genomes from single cells by polymerase cloning. *Nature Biotechnology, 24*, 680–686.

Zong, C., Lu, S., Chapman, A. R., & Xie, X. S. (2012). Genome-wide detection of single-nucleotide and copy-number variations of a single human cell. *Science, 338*, 1622–1626.

CHAPTER FIVE

Quantifying and Identifying the Active and Damaged Subsets of Indigenous Microbial Communities

Corinne Ferrier Maurice, Peter James Turnbaugh[1]
FAS Center for Systems Biology, Harvard University, Cambridge, Massachusetts, USA
[1]Corresponding author: e-mail address: pturnbaugh@fas.harvard.edu

Contents

1. Introduction 92
2. Flow Cytometry and Fluorescent Dyes 93
   2.1 Viewing microbial communities with flow cytometry 93
   2.2 Identifying metabolically active cells 93
   2.3 Monitoring membrane integrity and polarity 95
3. Experimental Procedures 95
   3.1 Sample preparation 95
   3.2 Quantifying microbial activity and cell damage 96
   3.3 FACS-Seq 98
4. Experimental Validation 99
   4.1 Optimizing single-cell methods with isolates and fecal samples 99
   4.2 Validating the FACS-Seq protocol on the human gut microbiota 102
5. Troubleshooting 103
   5.1 Sample handling issues: Storage and oxygen exposure 103
   5.2 General troubleshooting tips 105
Acknowledgment 106
References 106

Abstract

Flow cytometry and fluorescent dyes represent valuable experimental tools for studying complex microbial communities, enabling the quantification and sorting of cells with distinct levels of activity or damage, and providing information that can be difficult to infer from metagenomic sequencing alone. Despite this potential, these single-cell methods have seldom been applied to the study of host-associated microbial communities. Here, we present our recently developed protocols utilizing four distinct fluorescent dyes that label cells based on nucleic acid content, respiratory activity, and membrane damage. These methods have been successfully applied to study the trillions of microorganisms inhabiting the human gastrointestinal tract (the gut microbiota), in addition to a collection of isolates from five common gut-associated

Methods in Enzymology, Volume 531
ISSN 0076-6879
http://dx.doi.org/10.1016/B978-0-12-407863-5.00005-8

bacterial phyla. By merging these protocols with fluorescence-activated cell sorting and downstream multiplex 16S rRNA gene sequencing, it is possible to rapidly assess the taxonomic composition of each physiological category. These methods provide an initial step toward a robust toolkit allowing a rapid, culture-independent, and comprehensive assessment of the physiology and metabolic activity of host-associated microbial communities.

1. INTRODUCTION

The human body is home to trillions of microbial cells representing thousands of species (Turnbaugh et al., 2009; Qin et al., 2010) that have a profound effect on human physiology (Dutton and Turnbaugh, 2012; Haiser and Turnbaugh, 2012). These microorganisms are widely distributed among different body habitats and densely colonize the gastrointestinal tract (referred to as the gut microbiota) (Costello et al., 2009; Maurice and Turnbaugh, 2011). Recent metagenomic surveys have extensively described the structure and dynamics of the gut microbiota and its vast array of genes, the gut microbiome, in states of health and disease (Consortium, H. M. P., 2012; Turnbaugh et al., 2009; Qin et al., 2010; Yatsunenko et al., 2012). However, these techniques provide limited information about metabolic activity, necessitating other methods for the quantification and identification of the active or damaged cells within host-associated microbial communities.

Flow cytometry (FCM) and cell sorting have been used in a variety of ecosystems to access the physiology of cells within complex microbial communities (Del Giorgio and Gasol, 2008; Shapiro, 2000). Combined with appropriate fluorescent dyes, FCM allows the quantification of cells with distinct levels of activity or damage, using the individual cell fluorescence and light scatter signals. In addition, the stained cells of interest can also be sorted and identified by sequencing the appropriate marker genes.

Here, we present an optimized protocol for the single-cell analysis of the human gut microbiota, validated with bacterial isolates and the intact fecal microbiota. First, we provide general information about the use of FCM and the fluorescent dyes available. We then detail our optimized protocols for characterizing the physiology of microbial cells from human fecal samples, as well as methods for the sorting and downstream 16S rRNA gene sequencing, referred to here as FACS-Seq. Finally, we discuss various experimental considerations and troubleshooting options.

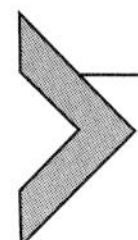

2. FLOW CYTOMETRY AND FLUORESCENT DYES

2.1. Viewing microbial communities with flow cytometry

The main benefits of FCM include speed, hundreds of thousands of events processed allowing for robust statistical analyses, information about general cellular features, and physiological information (for more information, see the excellent review by Shapiro, 1995). Briefly, microbial cells in suspension are exposed to a laser and the resulting light scatter and fluorescence emission signals for each cell are acquired (Shapiro, 1995). Thousands of events per second are recorded and characteristics such as cell abundance, size, shape, intracellular content, and granularity can be rapidly determined. Additional physiological information including nucleic acid content, membrane integrity, respiratory activity, and intracellular enzymatic activity can also be obtained by staining cells with fluorescent dyes prior to FCM acquisition (Del Giorgio and Gasol, 2008; Maurice and Turnbaugh, 2013; Shapiro, 1995; Fig. 5.1A). By using a fluorescence-activated cell sorter (FACS), each cell population can be sorted for downstream analyses (Fig. 5.1B).

2.2. Identifying metabolically active cells

Multiple dyes indicative of microbial physiology are available, but we have chosen to focus here on four dyes (two indicative of metabolic activity and two of cell damage; Fig. 5.1A) that have been extensively used in bacterial cultures, aquatic ecosystems, biofilms, and even the human gut microbiota (Del Giorgio and Gasol, 2008; Maurice, Haiser, and Turnbaugh, 2013; Novo, Perlmutter, Hunt, and Shapiro, 2000). Most of the other available dyes remain to be validated before their application to the human microbiota (Maurice and Turnbaugh, 2013).

Numerous studies indicate that cell growth and transcriptional activity can be determined using nucleic acid-binding fluorescent dyes, such as SybrGreen I. These dyes enter all microbial cells irrespective of the membrane status, with strong affinity to double- and single-stranded nucleic acids (Del Giorgio and Gasol, 2008; Martens-Habbena and Sass, 2006; Maurice and Turnbaugh, 2013).

Extensive data from aquatic microbial communities show that microbial cells tend to cluster into two groups, according to their relative green fluorescence: cells with low (LNA) and high nucleic acid (HNA) content (Bouvier, del Giorgio, and Gasol, 2007; Del Giorgio and Gasol, 2008).

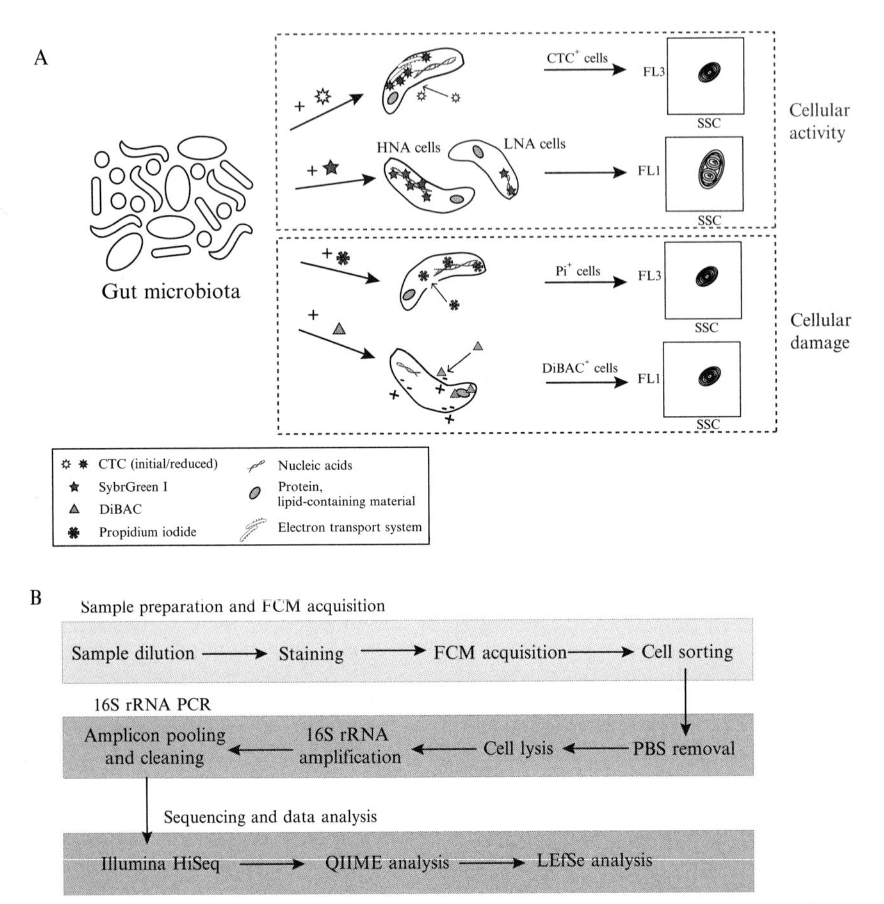

Figure 5.1 Identifying the active and damaged subsets of the human gut microbiota. (A) Fluorescent dyes, intracellular targets, and resulting FCM signals. The dyes used to determine cellular activity are highlighted in red, those for cellular damage in blue. (B) Overall workflow for FACS-Seq. (For interpretation of the references to color in this figure legend, the reader is referred to the online version of this chapter.)

Relative nucleic acid content can serve as a proxy of cellular activity, with several studies indicating that HNA cells exhibit higher rates of cell division and metabolic activity than LNA cells (Bouvier et al., 2007; Wang, Hammes, Boon, Chami, and Egli, 2009; Zubkov, Fuchs, Burkill, and Amann, 2001). Both populations have also been observed in the human gut microbiota, where the fluorescence differences could not be explained by increased cell or genome size, and most probably reflected differences in metabolic activity (Maurice et al., 2013).

Metabolically active microbial cells can also be quantified with the tetrazolium salt 5-cyano-2,3-ditolyltetrazolium chloride (CTC) (Rodriguez, Phipps, Ishiguro, and Ridgway, 1992; Sherr, del Giorgio, and Sherr, 1999; Sieracki, Cucci, and Nicinski, 1999). This salt is reduced to an insoluble red compound, formazan, during both aerobic and anaerobic respiration (Bhupathiraju, Hernandez, Landfear, and Alvarez-Cohen, 1999; Smith and McFeters, 1997; Walsh, Lappin-Scott, Stockdale, and Herbert, 1995). Preliminary data suggest that CTC can be applied to the human gut microbiota (Maurice and Turnbaugh, 2013), but more extensive validation is necessary prior to a wide application of this method.

2.3. Monitoring membrane integrity and polarity

Membrane integrity is a crucial indicator of cellular damage, and several fluorescent vital dyes will not enter microbial cells with intact membranes. For example, propidium iodide (Pi), commonly used to assess bacterial viability, is a red nucleic acid dye excluded from cells with intact membranes due to its size, molecular structure, and insolubility in the hydrophobic membrane phase (Haugland, 2005; Strauber and Muller, 2010). Pi^+ cells correspond to severely damaged cells (Fig. 5.1A).

Loss of membrane polarity is another physiological cue indicative of cell damage and can be assessed with oxonol dyes such as bis-(1,3-dibutylbarbituric acid)trimethine oxonol ($DiBAC_4$). $DiBAC_4$ is a negatively charged molecule that binds to lipid-containing material, generally excluded from cells because of their negative membrane gradient (Strauber and Muller, 2010; Suller and Lloyd, 1999). The dye is able to enter cells with a loss of membrane polarity, detectable by their green fluorescence (Fig. 5.1A).

These physiological dyes target distinct cellular processes. Thus, proxies of bacterial damage such as Pi^+ and $DiBAC^+$ cells might not always correlate (Novo et al., 2000). No single dye will provide all the physiological information about a microbial cell, and we therefore strongly recommend using several complementary dyes.

3. EXPERIMENTAL PROCEDURES

3.1. Sample preparation

Reagents and equipment

- Reduced 1× PBS (rPBS: 80 g l^{-1} NaCl, 2 g l^{-1} KCl, 14.4 g l^{-1} Na_2HPO_4, 2.4 g l^{-1} KH_2PO_4) containing L-cysteine (final concentration

1 mg ml^{-1}) and the oxygen indicator resazurin (final concentration 1 μg ml^{-1}). Filter (0.2 μm) and store anaerobically. Once the rPBS is clear again, it is oxygen free and ready to use.

- A monitored anaerobic chamber (e.g., Coy Laboratory).
- A tabletop centrifuge, with a swing-bucket or fixed-angle rotor adapted to volumes $\geq$10 ml.
- Fresh human fecal samples with limited oxygen exposure.

  Note: Additional validation may be necessary for samples from other body locations.

Experimental protocol

1. Dilute a portion of a fresh human fecal sample 1:10 g ml^{-1} in 1 × rPBS.
2. Vortex thoroughly.
3. Centrifuge 1 min at 700 × *g* to pellet organic matter and larger particles.
4. Transfer to anaerobic chamber. Take 1 ml of bacterial supernatant and transfer to a new Eppendorf tube.

   Note: This volume can be adjusted according to your experimental needs.
5. Spin 3 min at 6000 × *g* to pellet bacteria.
6. Wash bacterial pellet with 1 × rPBS and resuspend in 1 ml (or adjusted volume from step 4).
7. Spin 2 min at 6000 × *g*.
8. Repeat steps 6 and 7.
9. Repeat step 6. The bacterial suspension can now be further diluted and stained for FCM analysis.

3.2. Quantifying microbial activity and cell damage

Reagents and equipment

- SybrGreen I (Invitrogen). Prepare a stock solution at 1000 × following the provider's instructions and keep at − 20 °C.
- CTC (Invitrogen). Prepare a stock solution at 50 m*M* following the provider's instructions. This stock solution should be prepared fresh for each experiment.
- Propidium iodide (Sigma-Aldrich). Prepare a stock solution at 10 mg ml^{-1} following the provider's instructions and keep at −20 °C.
- $DiBAC_4$ (Invitrogen). Prepare a stock solution at 0.5 mg ml^{-1} following the provider's instructions and keep at −20 °C.
- Reference fluorescent beads as internal standards. This setup was originally tested with 5.8 μm fluorescent beads (Spherotech Inc.), but 3 μm Sphero Rainbow beads (BD Biosciences) can also be used.

– A flow cytometer equipped with at least one solid-state 488 nm laser, standard filter setup, and analysis software such as FlowJo (TreeStar Inc.) or CellQuest Pro (BD Biosiences).

Experimental protocol

1. Dilute your anaerobic bacterial cell suspension 120-fold in 1× rPBS.

   Note: This dilution factor can be adjusted according to your samples, but resulted in appropriate cell concentration for the samples we have analyzed to date.
2. Prepare and label FCM tubes in triplicate (technical staining replicates) for each dye.
3. Proceed with an additional 2.5-fold dilution of your samples in rPBS for bacterial physiology acquisition.
4. Pipette 500 μl of your diluted sample in triplicate FCM-labeled tubes, and add the different dyes from your prepared stocks according to the volumes and concentrations specified in Table 5.1. You should have three staining replicates/dye used. Incubate in the dark under anaerobic conditions.
5. Add 10 μl of green fluorescent beads. They should be prepared fresh, but can be kept at 4 °C for up to 2 weeks.

   Note: The bead concentration ($\sim 10^4$ beads ml^{-1}) should be verified *every time* you do a FCM experiment, either with Countbright beads (Invitrogen) or by epifluorescence microscopy.
6. FCM acquisition: Make sure the number of events per second is ≤1000 (can vary depending on the instrument). Acquisition plots should be log scaled.
7. FCM analysis: Gate the cells of interest and the reference beads using the appropriate fluorescence and light scatter channels. Do not include bead doublets or triplets in your gate. Record the number of events, the fluorescence, and light scatter values to control for any variation in the FCM signals.

Table 5.1 Staining procedure

Dye	Volume of dye (μl) in 500 μl total	Initial dye concentration	Final dye concentration	Staining time (min)
SybrGreen I	*0.5*	*1000×*	*1×*	*15*
CTC	*50*	*50 mM*	*5 mM*	*180*
DiBAC	***1***	***0.5 mg ml^{-1}***	***1 μg ml^{-1}***	***10***
Pi	***2***	***10 mg ml^{-1}***	***40 μg ml^{-1}***	***10***

The dyes in italic indicate activity, whereas those in bold italic assess cell damage.

Note: As DiBAC is sensitive to cell size, start by gating the microbial cells using a forward and side scatter plot (FSC vs. SSC) to decrease the number of false events and proceed with your analysis on this subset.

8. Convert FCM events to abundances, using the following equation:

$$(V_b \times C_b \times E_c)/(E_b \times V_s)$$

where V_b = bead volume; C_b = bead concentration; E_c = cell events; E_b = bead events; and V_s = total volume of sample. Proceed with determining the relative proportions of cells in each population relative to the total SybrGreen I positive cells.

3.3. FACS-Seq

Reagents and equipment

- All the reagents listed for Section 3.2.
- 96-well plates with strip caps (VWR).
- 5 *M* Betaine (Sigma-Aldrich).
- 10× PCR buffer (100 m*M* Tris (pH 8.3), 500 m*M* KCl, 20 m*M* $MgSO_4$).
- 10 n*M* dNTP, diluted down to 2 n*M* (New England Biolabs).
- TAQ polymerase (5 units μl^{-1}; Invitrogen).
- Barcoded primers targeting an appropriate marker gene (e.g., variable region 4 of the 16S rRNA gene; Caporaso et al., 2011, 2012). Dilute to 10 μ*M*.
- A FACS equipped with at least one solid-state 488 nm laser, standard filter setup, and a 70 μm nozzle.
- A thermocycler (e.g., MJ Research).
- A centrifuge with plate holders (e.g., Sorvall Legend RT).

Experimental protocol

1. Dilute your anaerobic bacterial cell suspension 120-fold in 1× rPBS.
   Note: This is the same step 1 as for the bacterial physiology experimental setup.
2. Proceed with an additional 1.7-fold dilution of your samples in rPBS for bacterial sorting.
3. Add the different dyes and incubate in the dark under anaerobic conditions, according to Table 5.1. Add 10 μl of green fluorescent beads.
4. FCM acquisition: Determine a narrow gate around the cells of interest, and sort at least 50,000 events into triplicate adjacent wells of a 96-well plate.

5. Centrifuge the plate at 4400 rpm for 15 min at 4 °C to pellet the bacterial cells.
6. Remove the supernatant, add 19.8 μl of PCR buffer (2 μl 5 *M* Betaine in 1 × PCR buffer) and incubate the cell pellets for 30 min at 4 °C.
7. Lyse cells: Place plate in liquid nitrogen before placing in thermocycler at 100 °C. Repeat procedure twice.
8. Add 2 μl 2 m*M* dNTP, 1 unit TAQ polymerase, and 1 μl each of the 10 μ*M* forward and reverse primers.
9. PCR amplification: 94 °C for 3 min, 35 cycles of 94 °C for 45 s, 50 °C for 30 s, and 72 °C for 90 s, with final extension at 72 °C for 10 min.
   Note: These settings should be adjusted according to your specific primer set.
10. Pool triplicate reactions and confirm amplification on 1.5% agarose gel.
11. Clean amplicons with AMPure XP beads (Agencourt).
12. Quantify each sample with the Quant-iT Picogreen dsDNA Assay kit (Invitrogen).
13. Pool equimolar amounts of each sample and dilute the final pool to 10 n*M* (for sequencing on the Illumina HiSeq; otherwise, adjust final concentration as needed).
14. Proceed with sequencing protocol.
    Note: Loading samples at 4 p*M* on the Illumina HiSeq can reduce errors due to the low complexity of amplicon samples. Alternatively, samples can be pooled with a high-complexity shotgun library.
15. Proceed with data analysis using QIIME (quantitative insights into microbial ecology; Caporaso et al., 2010), LEfSe (Segata et al., 2011), and/or additional bioinformatics tools.

4. EXPERIMENTAL VALIDATION

4.1. Optimizing single-cell methods with isolates and fecal samples

For each dye, we used unstained, stained, heat, and/or ethanol-treated samples to determine the optimal dye concentration. Cells were discriminated from noise or unstained cells using 2-parameter scatter plots of light scatter signals (SSC) and the appropriate emission filter channel (FL1 or FL3) (Fig. 5.2A).

We validated our experimental approach with representative strains from the five major phyla found in the human gut: *Eggerthella lenta* (Actinobacteria), *Bacteroides fragilis* (Bacteroidetes), *Lactobacillus casei* (Firmicutes), *Escherichia coli* (Proteobacteria), and *Akkermansia muciniphila* (Verrumicrobia), and fecal

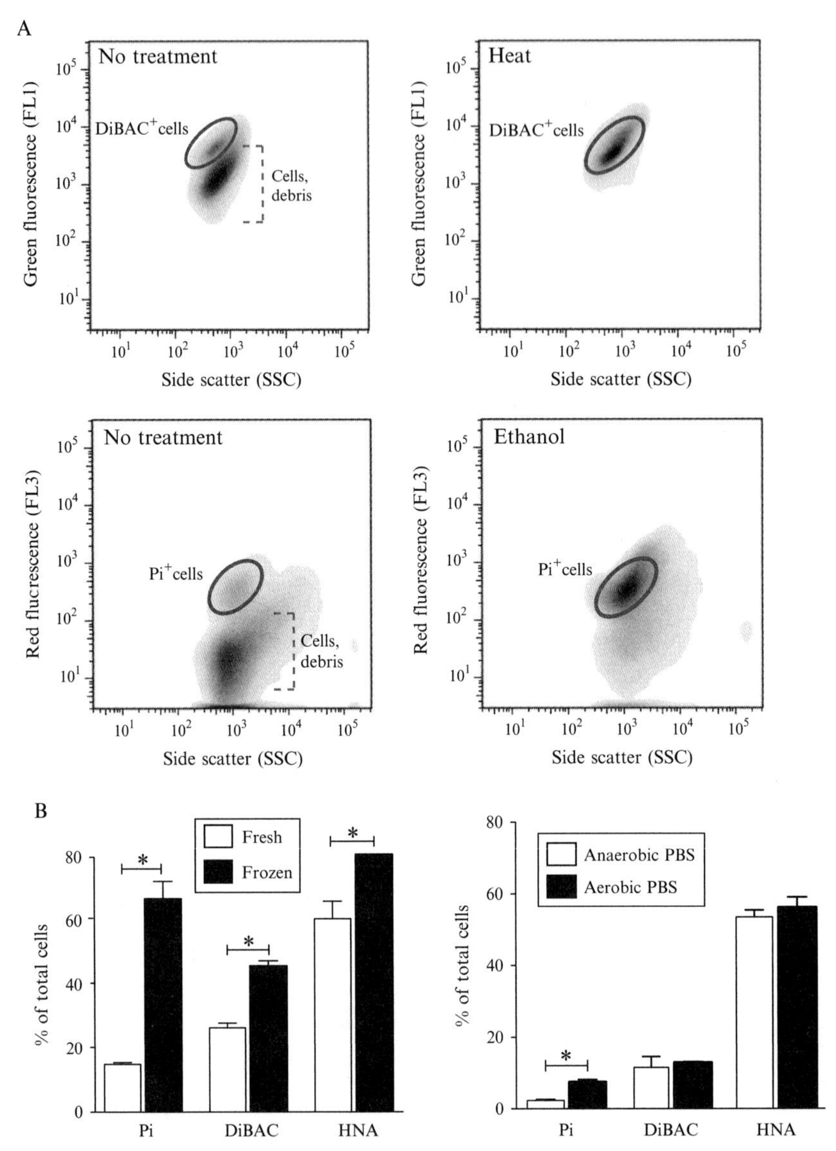

Figure 5.2 Microbial physiology in human fecal samples. (A) Heat and ethanol treatments induce cell damage. Density plots of a human fecal sample stained with DiBAC and Pi (left panel, no treatment), and the same sample heated (60 °C, 10 min, top right panel) and incubated with ethanol (70%, 10 min, bottom right panel). (B) Sample storage and handling induce cell damage. Proportions of Pi^+, $DiBAC^+$, and HNA cells from fresh versus frozen fecal samples (left panel) and fresh fecal samples immediately diluted in anaerobic versus aerobic PBS (right panel). Values represent mean ± sem. * indicates $p<0.05$, Student t-test. (For color version of this figure, the reader is referred to the online version of this chapter.)

communities from three unrelated individuals. The optimal SybrGreen I concentration of 1× was based on previously reported values and was assessed mainly with the gut isolates. Total cell abundance was obtained using the FSC and SSC signals, and the SybrGreen I-stained cells. Considering all isolates, both methods for determining cell abundance were equivalent (Fig. 5.3A), confirming that SybrGreen I can stain all bacterial cells, irrespective of their phylum. The proportion of HNA cells during exponential growth averaged 92.9±1.6% of the total cells, with the exception of *A. muciniphila* (6.8±0.7% of the total cells). Although HNA and LNA cells

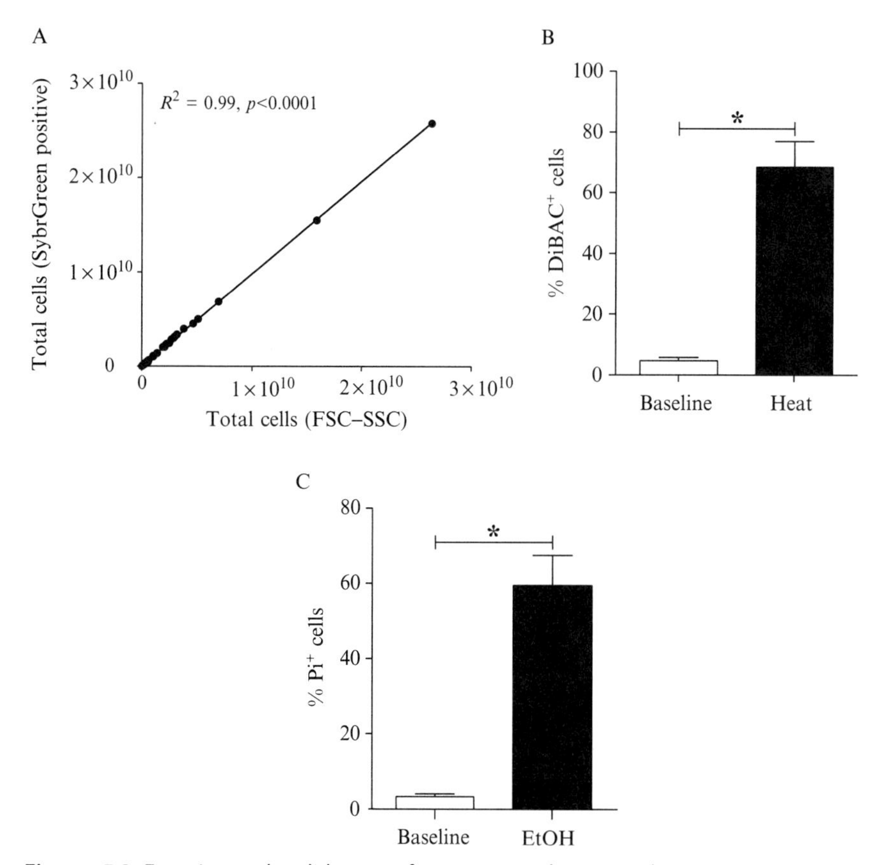

Figure 5.3 Experimental validation of staining with gut isolates. (A) Correlation between total cell counts obtained using the forward and side scatter values (FSC, SSC) and the total cell counts determined after SybrGreen I staining ($N = 5$ bacterial isolates with 2–3 replicates each). (B) Percentage of cells with altered membrane polarity ($DiBAC^{+}$) with or without heat treatment. (C) Percentage of cells with damaged membrane integrity (Pi^{+}) with or without ethanol treatment. Values in panels (B) and (C) represent mean±sem from the five isolates. * indicates $p < 0.05$, Student t-test.

have been associated with distinct levels of activity in microbial communities, more extensive analyses of representative cultured isolates and mock communities are needed.

The optimal concentrations for DiBAC and Pi were determined based on a dose–response curve of the dyes for untreated samples (gut isolates and fecal communities), heat-treated samples (60 °C, 10 min), and ethanol-treated samples (70% final concentration, 10 min) (Fig. 5.2A). This allowed us to confirm dye uptake when cells are exposed to stressful conditions. DiBAC concentrations higher than 1 μg ml^{-1} overestimated the damaged cells, both in the normal and heat-treated samples, with more damaged cells than the total cells. On average, the proportions of damaged $DiBAC^{+}$ cells remained low during exponential phase of the gut isolates ($4.5 \pm 1.2\%$ $DiBAC^{+}$) and were higher in untreated fecal communities ($27.4 \pm 2.3\%$ across three individuals, $N=21$). These values significantly increased in the heat treatments (Fig. 5.3B), reaching $68.5 \pm 8.5\%$ and $92.7 \pm 13.5\%$ of total cells for the isolates and the fecal communities, respectively (Maurice et al., 2013). We tested Pi concentrations ranging from 0.02 to 0.1 mg ml^{-1} with normal and ethanol-treated samples. We determined an optimal Pi concentration of 0.04 mg ml^{-1}; Pi^{+} cells reached nearly 100% of total cells in the ethanol-treated samples and higher concentrations did not significantly modify the proportions of Pi^{+} cells in the untreated samples. Damaged Pi^{+} cells also remained low during the exponential growth of the gut isolates ($3.4 \pm 0.7\%$ Pi^{+} cells) and averaged $17.2 \pm 2.1\%$ in human fecal samples ($N=21$). These proportions increased with the ethanol treatments to $59.5 \pm 8.0\%$ for the gut isolates and $97.7 \pm 6.8\%$ for the fecal microbiota (Fig. 5.3C).

The experimental procedure for CTC staining was only optimized with fecal samples, using previously published concentrations for anaerobic bacteria (Bhupathiraju et al., 1999). However, we did test different incubation times, using once again untreated and stained samples, untreated and unstained samples, formaldehyde-killed and stained samples, as well as rPBS incubated with CTC (no cells). Incubation times longer than 3 h increased the proportions of CTC^{+} cells in the killed samples threefold, although the values remained below 1%. On average, $12.0 \pm 2.2\%$ of the fecal microbiota was CTC positive.

4.2. Validating the FACS-Seq protocol on the human gut microbiota

We tested three crucial aspects of the FACS-Seq protocol: (i) the accurate sorting of the cell population of interest, (ii) the minimum number of sorted cells necessary for reliable PCR amplification, and (iii) microbial community structure prior to and after cell sorting.

First, sorted cells were reanalyzed by FCM, and were highly enriched within the initial sorting gate (>72% of cells, range: 67–88%), without any detectable change in fluorescence and SSC values (Fig. 5.4A). Events from each subset were also visualized by epifluorescence microscopy (data not shown), confirming the presence of the sorted cells. Next, we sorted incrementing numbers of cells, from 30,000 to 75,000 cells, that were used for PCR amplification of the 16S rRNA gene. Successful PCR amplification was obtained with $\geq$50,000 sorted cells. We also determined that the centrifugation step was necessary, as residual PBS from FACS inhibited PCR amplification. Centrifugation did not lead to significant cell loss (<5%), as confirmed by FCM.

Next, we compared the overall structure of the gut microbiota (by gating the entire SybrGreen I population) to the HNA and LNA subsets (Fig. 5.4B). As expected, the entire gut microbiota exhibited an intermediate community structure to the sorted LNA and HNA populations from the same individual. However, we did detect a clear difference between unsorted samples and the sorted gut microbiota (Maurice et al., 2013), suggesting that there may be technical biases introduced during FACS-Seq and/or that fecal samples contain substantial levels of free nucleic acids from lysed cells. Additional follow-up experiments are necessary to tease apart these, and potentially other, confounding factors.

5. TROUBLESHOOTING

5.1. Sample handling issues: Storage and oxygen exposure

In order to ensure representative measurements of bacterial physiology, it is essential to minimize sample handling prior to analysis. We tested the effect of sample storage by comparing fresh samples to those maintained at −80 °C for <3 months, from the same three unrelated individuals (Fig. 5.2B, left panel). Freezing significantly increased the proportions of all physiological categories (ANOVA, $p < 0.0001$), and we therefore recommend using fresh fecal samples.

We also tested the effects of both short- and long-term oxygen exposures. Fresh fecal samples were collected and either transferred immediately to our anaerobic chamber and processed under anaerobic conditions with rPBS, or processed under aerobic conditions. We compared

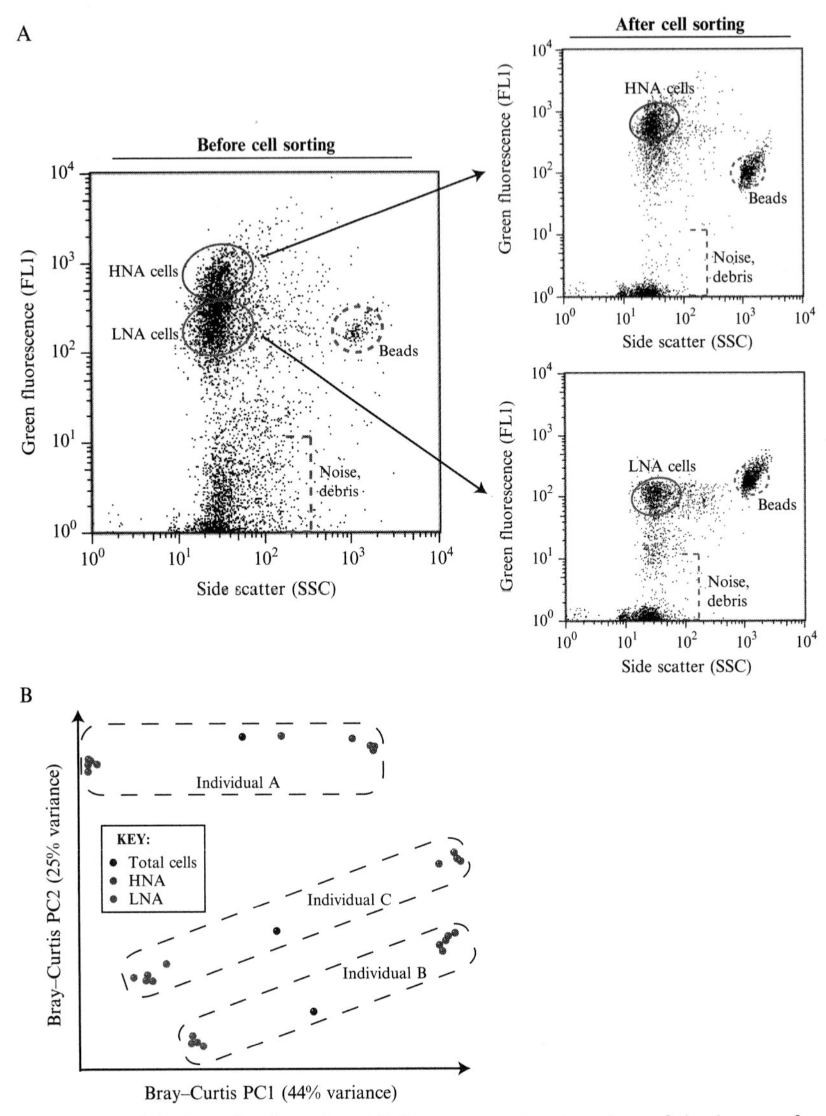

Figure 5.4 Validation of cell sorting. (A) Representative dot plots of the human fecal microbiota stained with SybrGreen I. The HNA and LNA populations are highlighted, as well as the reference beads. After sorting, beads were added and the microbial cells were restained and reanalyzed by FCM. (B) Principal coordinates analysis of microbial community structure in all SybrGreen I positive, HNA, and LNA cells from three unrelated individuals. The Bray Curtis distance metric was used to compare 16S rRNA gene sequencing profiles generated from each sorted cell population. (For color version of this figure, the reader is referred to the online version of this chapter.)

bacterial physiology immediately, 2, 4, and 24 h after sample collection. Samples processed under aerobic conditions consistently exhibited higher proportions of $DiBAC^+$, Pi^+, and HNA cells, relative to the samples kept under anaerobic conditions. This was significant even for samples processed immediately, that is, within 15 min after collection ($DiBAC^+$ aerobic $= 18.4 \pm 0.6\%$ vs. $DiBAC^+$ anaerobic $= 10.5 \pm 0.6$; Pi^+ aerobic $= 9.8 \pm 0.2\%$ vs. Pi^+ anaerobic $= 5.5 \pm 0.3$; HNA aerobic $= 85.8 \pm 3.2\%$ vs. HNA anaerobic $= 49.3 \pm 4.0$; Student's *t*-tests, all $p < 0.001$), but this increase was most obvious after 4 h of oxygen exposure. This suggested that oxygen exposure is critical for bacterial physiology. We further explored this issue by comparing the physiology of cells only processed with rPBS versus cells initially diluted and centrifuged in normal PBS before being transferred to the anaerobic chamber and further processed with rPBS (Fig. 5.2B, right panel). This very short oxygen exposure did not significantly impact the proportions of $DiBAC^+$ cells or of HNA cells, but there was a significant increase in the proportions of Pi^+ cells (Student's *t*-test, $p < 0.0001$). Samples should therefore be diluted as quickly as possible into rPBS before further downstream experimentation.

5.2. General troubleshooting tips

- *How to identify the cells of interest:* Compare your stained samples to unstained samples and stained killed samples (e.g., heat, ethanol, formaldehyde, or UV treated; Fig. 5.2A).
- *The cell counts are not consistent:* Check your reference bead concentration. Acquire enough events for your reference beads to ensure correct calculations ($\geq 1\%$ of total events).
- *Cell proportions are $>100\%$:* Ensure that you have acquired enough reference beads events (see previous note). For complex communities, make an initial FSC versus SSC gate to select only microbial cells (based on size and cellular content), before proceeding with your analysis. Cells with autofluorescence and/or chloroplasts can also be excluded from the analysis by using red versus green fluorescence plots.
- *Avoid contamination:* For the FACS-Seq protocol, make sure your 96-well plate is DNA and RNA free. Unused wells should always be capped. Ensure that the PBS used for FCM is as sterile as possible.
- *No PCR amplification:* Make sure that there is no PBS remaining in the wells with sorted cells. Modify the number of sorted events you need for correct amplification based on the specific population being sorted.

ACKNOWLEDGMENT

This work was supported by the National Institutes of Health (P50 GM068763).

REFERENCES

Bhupathiraju, V. K., Hernandez, M., Landfear, D., & Alvarez-Cohen, L. (1999). Application of a tetrazolium dye as an indicator of viability in anaerobic bacteria. *Journal of Microbiological Methods, 37*, 231–243.

Bouvier, T., del Giorgio, P. A., & Gasol, J. M. (2007). A comparative study of the cytometric characteristics of high and low nucleic-acid bacterioplankton cells from different aquatic ecosystems. *Environmental Microbiology, 9*, 2050–2066.

Caporaso, J. G., Kuczynski, J., Stombaugh, J., Bittinger, K., Bushman, F. D., Costello, E. K., et al. (2010). QIIME allows analysis of high-throughput community sequencing data. *Nature Methods, 7*, 335–336.

Caporaso, J. G., Lauber, C. L., Walters, W. A., Berg-Lyons, D., Huntley, J., Fierer, N., et al. (2012). Ultra-high-throughput microbial community analysis on the Illumina HiSeq and MiSeq platforms. *The ISME Journal, 6*, 1621–1624.

Caporaso, J. G., Lauber, C. L., Walters, W. A., Berg-Lyons, D., Lozupone, C. A., Turnbaugh, P. J., et al. (2011). Global patterns of 16S rRNA diversity at a depth of millions of sequences per sample. *Proceedings of the National Academy of Sciences of the United States of America, 108*(Suppl. 1), 4516–4522.

Costello, E. K., Lauber, C. L., Hamady, M., Fierer, N., Gordon, J. I., & Knight, R. (2009). Bacterial community variation in human body habitats across space and time. *Science, 326*, 1694–1697.

Del Giorgio, P. A., & Gasol, J. M. (2008). Physiological structure and single-cell activity in marine bacterioplankton. In D. L. Kirchman (Ed.), *Microbial ecology of the oceans* (2nd ed., pp. 243–285). Hoboken, New Jersey: Wiley-Blackwell.

Dutton, R. J., & Turnbaugh, P. J. (2012). Taking a metagenomic view of human nutrition. *Current Opinion in Clinical Nutrition and Metabolic Care, 15*, 448–454.

Haiser, H. J., & Turnbaugh, P. J. (2012). Is it time for a metagenomic basis of therapeutics? *Science, 336*, 1253–1255.

Haugland, R. P. (2005). *The handbook: A guide to fluorescent probes and labeling technologies* (10th ed.). Carlsbad, California: Invitrogen Corp.

Human Microbiome Project Consortium (2012). Structure, function and diversity of the healthy human microbiome. *Nature, 486*, 207–214.

Martens-Habbena, W., & Sass, H. (2006). Sensitive determination of microbial growth by nucleic acid staining in aqueous suspension. *Applied and Environmental Microbiology, 72*, 87–95.

Maurice, C. F., Haiser, H. J., & Turnbaugh, P. J. (2013). Xenobiotics shape the physiology and gene expression of the active human gut microbiome. *Cell, 152*, 39–50.

Maurice, C. F., & Turnbaugh, P. J. (2011). The human microbiome: Exploring and manipulating our microbial selves. In D. Marco (Ed.), *Metagenomics: Current innovations and future trends*. Norfolk, UK: Caister Academic Press.

Maurice, C. F., & Turnbaugh, P. J. (2013). Quantifying the metabolic activities of human-associated microbial communities across multiple ecological scales. *FEMS Microbiology Reviews*, , *37*(5), 830–848. http://dx.doi.org/10.1111/1574-6976.12022.

Novo, D. J., Perlmutter, N. G., Hunt, R. H., & Shapiro, H. M. (2000). Multiparameter flow cytometric analysis of antibiotic effects on membrane potential, membrane permeability, and bacterial counts of Staphylococcus aureus and Micrococcus luteus. *Antimicrobial Agents and Chemotherapy, 44*, 827–834.

Qin, J., Li, R., Raes, J., Arumugam, M., Burgdorf, K. S., Manichanh, C., et al. (2010). A human gut microbial gene catalogue established by metagenomic sequencing. *Nature, 464*, 59–65.

Rodriguez, G. G., Phipps, D., Ishiguro, K., & Ridgway, H. F. (1992). Use of a fluorescent redox probe for direct visualization of actively respiring bacteria. *Applied and Environmental Microbiology, 58*, 1801–1808.

Segata, N., Izard, J., Waldron, L., Gevers, D., Miropolsky, L., Garrett, W. S., et al. (2011). Metagenomic biomarker discovery and explanation. *Genome Biology, 12*, R60.

Shapiro, H. M. (1995). *Practical flow cytometry* (3rd ed.). New York: John Wiley and Sons.

Shapiro, H. H. (2000). Microbial analysis at the single-cell level: Tasks and techniques. *Journal of Microbiological Methods, 42*, 3–16.

Sherr, B. F., del Giorgio, P., & Sherr, E. B. (1999). Estimating abundance and single-cell characteristics of respiring bacteria via the redox dye CTC. *Aquatic Microbial Ecology, 18*, 117–131.

Sieracki, M. E., Cucci, T. L., & Nicinski, J. (1999). Flow cytometric analysis of 5-cyano-2,3-ditolyl tetrazolium chloride activity of marine bacterioplankton in dilution cultures. *Applied and Environmental Microbiology, 65*, 2409–2417.

Smith, J. J., & McFeters, G. A. (1997). Mechanisms of INT (2-(4-iodophenyl)-3-(4-nitrophenyl)-5-phenyl tetrazolium chloride) and CTC (5-cyano-2,3-ditolyl tetrazolium chloride) reduction in Escherichia coli K-12. *Journal of Microbiological Methods, 29*, 161–175.

Strauber, H., & Muller, S. (2010). Viability states of bacteria—Specific mechanisms of selected probes. *Cytometry. Part A, 77A*, 623–634.

Suller, M. T. E., & Lloyd, D. (1999). Fluorescence monitoring of antibiotic-induced bacterial damage using flow cytometry. *Cytometry, 35*, 235–241.

Turnbaugh, P. J., Hamady, M., Yatsunenko, T., Cantarel, B. L., Duncan, A., Ley, R. E., et al. (2009). A core gut microbiome in obese and lean twins. *Nature, 457*, 480–484.

Walsh, S., Lappin-Scott, H. M., Stockdale, H., & Herbert, B. N. (1995). An assessment of the metabolic activity of starved and vegetative bacteria using two redox dyes. *Journal of Microbiological Methods, 24*, 1–9.

Wang, Y., Hammes, F., Boon, N., Chami, M., & Egli, T. (2009). Isolation and characterization of low nucleic acid (LNA)-content bacteria. *The ISME Journal, 3*(8), 889–902.

Yatsunenko, T., Rey, F. E., Manary, M. J., Trehan, I., Dominguez-Bello, M. G., Contreras, M., et al. (2012). Human gut microbiome viewed across age and geography. *Nature, 486*, 222–227.

Zubkov, M. V., Fuchs, B. M., Burkill, P. H., & Amann, R. (2001). Comparison of cellular and biomass specific activities of dominant bacterioplankton groups in stratified waters of the Celtic Sea. *Applied and Environmental Microbiology, 67*, 5210–5218.

SECTION II

Microbial Community Genomics—Sampling, Metagenomic Library Preparation, and Analysis

CHAPTER SIX

Preparation of BAC Libraries from Marine Microbial Populations

Gazalah Sabehi, Oded Béjà[1]
Faculty of Biology, Technion—Israel Institute of Technology, Haifa, Israel
[1]Corresponding author: e-mail address: beja@tx.technion.ac.il

Contents

1. Introduction 111
2. BAC Library Construction Procedures 112
   2.1 Competent cell preparation for BAC transformation via electroporation 112
   2.2 BAC vector preparation for cloning 113
   2.3 Seawater sample collection and purification of HMW DNA 115
   2.4 Partial digestion of the HMW DNA for cloning 116
   2.5 HMW DNA extraction from gel 117
   2.6 HMW DNA ligation into the BAC vector 119
   2.7 Transformation of the ligation products into *E. coli* 119
   2.8 Determination of the average size of the clones 120
   2.9 Library construction and storage 121
References 121

Abstract

A protocol is presented here for the construction of BAC (bacterial artificial chromosome) libraries from planktonic microbial communities collected in marine environments. The protocol describes the collection and preparation of the planktonic microbial cells, high molecular weight DNA purification from those cells, the preparation of the BAC vector, and the special ligation and electrotransformation procedures required for successful library preparation. With small modifications, this protocol can be applied to microbes collected from other environments.

1. INTRODUCTION

Construction of bacterial artificial chromosome (BAC) libraries form marine samples has proved to be a successful tool to study microbial communities via understanding their phylogenetic affiliation, physiological and metabolic properties, and ecological function (Béjà, Aravind, et al., 2000;

Methods in Enzymology, Volume 531
ISSN 0076-6879
http://dx.doi.org/10.1016/B978-0-12-407863-5.00006-X

Béjà et al., 2002; Béjà, Suzuki, et al., 2000; de la Torre et al., 2003; Feingersch & Béjà, 2009; Feingersch et al., 2010; Loy et al., 2009; Sabehi, Béjà, Suzuki, Preston, & DeLong, 2004; Sabehi et al., 2005). This tool allows cloning of high molecular weight (HMW) genomic DNA fragments that are recovered from microbial cell populations into BAC cloning vector. BAC vector contains an origin of replication derived from the *Escherichia coli* F plasmid and allows the cloning and stable maintenance of HMW DNA fragments of up to 300 kb (Shizuya et al., 1992). Since bacteria have small genomes, only a few BAC clones are required for full coverage of a bacterium genome. Thus, this approach has advantages in the genomic analysis of microbial populations.

Construction of large insert-containing BAC libraries from marine samples involves the following steps: first, collecting a large water sample and filtration or concentration through a desired size filter mesh to concentrate the cell population; second, purification of HMW DNA from the cells, followed by subsequent enzymatic digestion and size selection for HMW DNA fragments; third, ligation of the purified size-selected fragments into a specially prepared BAC vector, and finally, transformation of the ligated recombinant BACs into *E. coli* and subsequent selection and archiving of positive recombinant BAC containing colonies for further analysis.

2. BAC LIBRARY CONSTRUCTION PROCEDURES

2.1. Competent cell preparation for BAC transformation via electroporation

The transformation of BAC clones into *E. coli* employs electroporation. For this purpose, electrocompetent cells are required. The *E. coli* strain DH10B [F^-, *mcr*A, Δ(*mrr-hsd*RMS-*mcr*BC) φ80dLacZΔM15, Δ*lac*X74, *deo*R, *rec*A1, *end*A1, *ara*D139, Δ(*ara, leu*)7697, *gal*U, *gal*K, λ^-, *rps*L, *nup*G] is recommended for BAC cloning since this strain is efficiently electroporated and transformed with large DNA clones (Sheng, Mancino, & Birren, 1995). *E. coli* DH10B competent cells are prepared as described by Sheng et al. (1995) as follows. As this strain is resistant to the antibiotic streptomycin, cultures in the following steps are grown with streptomycin sulfate salt (Sigma) with a final concentration of 10 μg/ml. *E. coli* DH10B single colony from a freshly streaked agar plate (1.5% Bacto agar) is picked and inoculated into a starter culture of 2 ml LB medium (1% Bacto tryptone, 0.5% Bacto yeast extract, and 1% NaCl) in a tube with a volume of at least five times the volume of the culture. The culture is incubated for ~12 h at 37 °C with vigorous shaking (~200 rpm). The fresh saturated starter culture is then diluted 1:1000 into 1 l of SOB medium (2% tryptone, 0.5% Bacto yeast extract, 0.05% NaCl and 2.5 m*M* KCl, pH 7, without $MgCl_2$), in Erlenmeyer with a volume of at least five times the volume of the culture and grown at 37 °C with vigorous shaking (~200 rpm) to an OD_{550} of 0.7 (no higher than 0.8). The culture is divided into 250 ml centrifuge tube. The cells are then harvested by centrifugation at 4000 × *g* for 10 min at 4 °C. The pellet is resuspended in a volume equal to the original culture volume of prechilled to 4 °C 10% sterile glycerol and centrifuged again and this step is repeated. Then, the pellet is resuspended in ~30 ml of 10% glycerol and centrifuged in one centrifuge tube at 6000 × *g* for 10 min at 4 °C. The pellet is resuspended in 2 ml of 10% glycerol per liter of initial culture volume, and aliquots of 30 μl are immediately snap-frozen in liquid nitrogen and stored at −80 °C.

2.2. BAC vector preparation for cloning

The preparation of the BAC vector for cloning includes the following steps: purification of a BAC vector from *E. coli* host culture, digestion with restriction enzyme and dephosphorylation of the BAC vector ends with alkaline phosphatase to prevent self-ligation in the ligation step, and finally, testing the efficiency of the digested and dephosphorylated BAC vector. The BAC vector pIndigo536, a derivative of pBeloBAC11 (Kim et al., 1996) is recommended for the construction of BAC libraries. pIndigo536 BAC vector encodes for chloramphenicolacetyl transferase and provides resistance to chloramphenicol. *E. coli* DH10B cells carrying the pIndigo536 vector are grown as shown in the following steps, with a final chloramphenicol concentration of 12.5 μg/ml. A single colony from a freshly streaked agar plate is picked and inoculated into a starter culture of 2 ml LB medium. The culture is incubated for ~12 h at 37 °C with vigorous shaking (~200 rpm). The fresh saturated starter culture is then diluted 1:1000 into 1 l of LB medium and grown for ~12 h at 37 °C with vigorous shaking (~200 rpm). The BAC vector is extracted from the bacterial culture using a Qiagen large-construct kit (Qiagen) according to manufacturer's protocol. This kit is suitable for the purification of a single-copy vector and allows the purification of a high quality vector DNA free of the host genomic DNA. The pIndigo536 BAC vector restriction enzyme digestion can be performed using either *Hin*dIII or *Bam*HI restriction enzymes since their sites are located within the reporter gene lacZ, which allows identification of recombinants by insertional inactivation and blue/white selection. Digestion of the BAC vector is done using *Hin*dIII restriction enzyme digestion as follows: Mix 10 μg BAC vector, 0.1 mg/ml BSA, 7.5 μl of 10× NEBuffer 2, 10 units *Hin*dIII restriction enzyme (New England BioLabs), and double-distilled (dd) H_2O in a final volume of 75 μl. The reaction mixture is incubated at 37 °C for at least 5 h. The longer reaction incubations are necessary to ensure full digestion of the vector. Following digestion, the BAC vector treated with alkaline phosphatase to dephosphorylate the vector ends. The reaction is performed as follows: First, prepare 25 μl of alkaline phosphatase reaction cocktail containing 10 units of CIP alkaline phosphatase (New England BioLabs), 2.5 μl of 10× NEBuffer 2, 0.1 mg/ml BSA, and ddH_2O to a final volume of 25 μl. Next, add 25 μl of alkaline phosphatase reaction cocktail to the 75 μl restriction enzyme BAC digest (above), and incubate the reaction for 1 h at 37 °C. The BAC vector should not be overexposed to the alkaline phosphatase because it can damage the sticky ends produced by the restriction enzyme. The *Hin*dIII digested and dephosphorylated vector (7.4 kb) is next separated in 1% agarose gel by gel electrophoresis and extracted using the QIAEX II gel extraction kit (Qiagen). The estimation of the vector concentration can be performed by spectrophotometry using a NanoDrop or similar device. The purified vector is stored at −20 °C in aliquots until use. It is recommended to prepare the digested and dephosphorylated BAC vector immediately before construction of the BAC library, since the cloning efficiency decreases when the treated vector is stored for long periods. It is recommended to test the efficiency of the digested and dephosphorylated BAC vector before starting the construction of the BAC library by performing the following reactions and quality control checks: Test Reaction 1, self-ligation of the *Hin*dIII digested BAC vector to check the efficiency of the restriction reaction. Test Reaction 2, self-ligation of the *Hin*dIII digested and dephosphorylated BAC vector to check the efficiency of the dephosphorylation reaction. Test Reaction 3, ligation of the *Hin*dIII digested and dephosphorylated BAC vector with *Hin*dIII cut phage Lambda DNA as follows: 50 ng *Hin*dIII digested and dephosphorylated BAC vector, 15 ng of Lambda DNA cut with *Hin*dIII, 10 μl of 10× T4 DNA ligase buffer, 5 units T4 DNA ligase (New England BioLabs), and ddH_2O to a final volume of 100 μl. Ligation and transformation conditions should be performed as described below. Most of the colonies from the Test Reaction 1 should be blue because of self-ligation of the vector. In Test Reaction 2, colony yield should be low. If there is a high percentage of blue colonies in Test Reaction 2, then the vector may not be fully digested or not fully dephosphorylated, and if there is a high percentage of false-positive (white) colonies, then the vector has been damaged during the dephosphorylation treatment. In those cases, the vector needs to be prepared again. In Test Reaction 3, most of the colonies should be white; if so, then the treated vector is ready for use.

Nowadays, different commercial ready-to-use BAC vectors are available from several manufacturers as well.

2.3. Seawater sample collection and purification of HMW DNA

There are on average about 5×10^5 prokaryotic cells per milliliter in the continental shelf and the upper 200 m of the open ocean water (Whitman, Coleman, & Wiebe, 1998). Because BAC library construction demands a large amount of starting material ($\sim 10^{10}$–10^{11} cells), a large volume (about 1000 l) of seawater is necessary to concentrate the required number of cells. Since the interest here is to construct a BAC library from the prokaryotic population, size filtration should be performed to remove the large eukaryotic cells, and collect the prokaryotic population. The seawater sample can be prefiltered through a GF/A glassfiber filter (Whatman) and concentrated by tangential flow filtration using the Pellicon systems (Millipore), to about 0.5–1.0 l final volume. Further concentration is performed by centrifugation at $30{,}000 \times g$ at 4 °C for 0.5 h. The cell pellet is resuspended in 0.5 ml of 0.2 μm filtered seawater. The quality of the isolated DNA is important for the construction of the BAC library, as large DNA fragments are required. HMW DNA purification from the cells is performed following the protocol described by Béjà, Suzuki et al. (2000) and Stein, Marsh, Wu, Shizuya, and DeLong (1996) in which the concentrated cells are embedded in agarose plugs, which provide physical support and prevent DNA shearing during the DNA purification: A 1% SeaPlaque GTG agarose (Lonza) is prepared in 0.2 μm filtered seawater and kept at ~42 °C. The concentrated cells are then mixed with an equal volume of 1% SeaPlaque GTG agarose drawn into 1 cc syringes and chilled on ice for 10 min to solidify the agarose. It is critical to perform this step quickly since the agarose-cell suspension may solidify before being pulled into the syringe. The cell-containing agarose plugs are then placed in 20 ml STE buffer (1 *M* NaCl, 0.1 *M* EDTA, pH 8.0 and 10 m*M* Tris–HCl, pH 8.0) and incubated in a rotating incubator for 1 h at room temperature. To lyse the cells, the agarose plug is transferred to low salt concentration buffer and treated with lysozyme. The STE buffer is replaced with 15 ml lysis buffer (0.1 *M* EDTA, pH 8.0, 10 m*M* Tris–HCl, pH 8.0, 50 m*M* NaCl, 0.2% sodium deoxycholate (deoxycholic acid) (Sigma), 1% sarkosyl (*N*-lauroylsarcosine sodium salt) (Sigma) and 1 mg/ml lysozyme (Sigma)), and incubated in a rotating incubator for 1 h at 37 °C. For protein digestion, the plug is then treated with proteinase-K. The lysis buffer is replaced with 40 ml ESP buffer (1% sarkosyl and 1 mg/ml proteinase-K (Sigma), in 0.5 *M* EDTA, pH 8.0), and incubated in a rotating incubator for 16 h at 50 °C and this step is repeated. The plug is then moved into 50 ml of 50 m*M* EDTA, pH 8.0 and stored at 4 °C until use.

Before proceeding with the plug, it is important to inactivate and to wash the proteinase-K traces from the plug since it can affect the enzymatic reactions in the following steps. The plug is washed with a solution containing 10 m*M* Tris–HCl, pH 7.5, 0.1 m*M* EDTA, pH 8.0, and 1 m*M* phenylmethylsulfonyl fluoride (PMSF) (Sigma) three times for 2 h each with slow rotation at room temperature. This is followed by three washes with the same solution without PMSF for 2 h each with slow rotation at room temperature. The plug can then be stored at 4 °C until use.

2.4. Partial digestion of the HMW DNA for cloning

HMW DNA fragments with a size range between 150 and 350 kb are prepared for cloning by enzymatic digestion. In order to determine the optimal enzymatic reaction conditions that yield fragments at this size range, reactions with different enzyme concentrations are set. The partial digestion of the HMW DNA in plug is performed following the protocol described by Béjà, Suzuki, et al. (2000) and Stein et al. (1996). A series of restriction reactions are set with different restriction enzyme concentrations. Gel slices with a thickness of ~1 mm from the plug are cut using a microscope cover glass, added to an Eppendorf tube containing 200 μl mixtures of 1× NEBuffer 2, 0.1 mg/ml BSA, 4 m*M* spermidine (Sigma), and ddH_2O to a final volume of 200 μl. The tubes are then incubated for 30 min on ice for buffer exchange. To treat the DNA in the gel slices with restriction enzyme, the solution is then replaced with a mixture plus *Hin*dIII enzyme with different concentrations (e.g., 0, 0.25, 0.5, 0.75, and 1 unit; optimal enzyme concentrations need to be determined empirically) and incubated for 1 h on ice to allow diffusion of the enzyme into the gel slice. To achieve partial restriction enzyme digestion, the DNA is exposed to the enzyme for a short incubation by incubating at 37 °C for 20 min and the reaction is immediately stopped by adding 1/10 volume of 0.5 *M* EDTA, pH 8.0 and placed on ice.

The *Hin*dIII digested DNA in the gel slices is run in a gel and separated by pulse field gel electrophoresis (PFGE), which allows separation of large fragments of DNA. The gel slices are loaded on 1% SeaPlaque GTG agarose in 1 × TAE buffer (40 m*M* Tris–acetate and 1 m*M* EDTA). The loading of the gel slices into the gel wells can be done by using a microscope cover glass. PFGE lambda Ladder DNA (New England BioLabs) is loaded in wells flanking the sample wells. The PFGE is performed on a clamp homogeneous electrical field CHEF-DR II PFGE machine (Bio-Rad). The electrophoresis conditions are: 6 V/cm, 20–40 s pulses, 18 h, and 12 °C.

It is important to avoid staining of the gel containing the HMW DNA used for the cloning by ethidium bromide or other DNA staining solutions and exposing to UV, since these damage DNA. After PFGE, the flanking gel portions containing the marker are cut, stained with a highly sensitive stain, GelStar nucleic acid gel stain (1X) (Lonza) and then examined under a UV light. The region between 150 and 350 kb is marked by making incisions on the gel. Then, the two gel pieces are aligned with the unstained gel portion that contains the digested HMW DNA. The gel portion containing the DNA fragments between 150 and 350 kb from the unstained part of the gel are cut by following the incisions on the gel pieces containing the marker. Then, the two gel pieces containing the digested DNA flanking to the gel portion containing the 150–350 kb are stained with GelStar nucleic acid gel stain (1X) and examined by UV. The optimal enzymatic reaction conditions ensure that most of the fragments at the size range of 150–350 kb are used for a large-scale digestion. A gel after PFGE of HMW DNA digested with different concentrations of *Hin*dIII is shown in Fig. 6.1. The region between 150 and 350 kb is excised and used as DNA inserts for BAC library construction. The flanking regions of the gel are stained. When the enzyme concentration increases, the HMW weight of the DNA (marked by white arrows) decreases. The optimal enzyme concentration for a large-scale digestion is 1 unit of *Hin*dIII, since using this concentration renders most of the DNA fragments between 150 and 350 kb, and most of the HMW DNA is digested.

2.5. HMW DNA extraction from gel

HMW DNA extraction from gel is done by digestion of the gel by GELase. The gel pieces containing the HMW digested DNA are washed three times

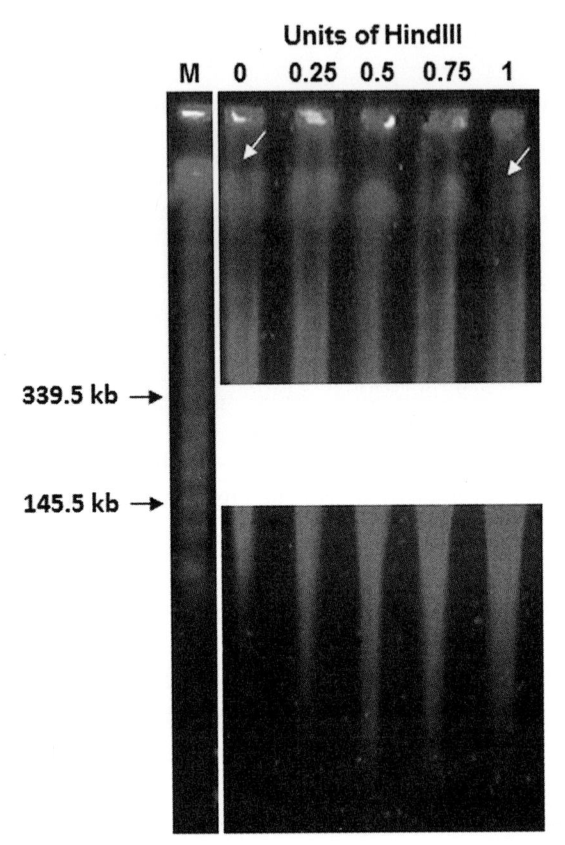

Figure 6.1 PFGE of partially digested DNA from microbial population. PFGE of microbial genomic DNA digested by different *Hin*dIII concentrations (0, 0.25, 0.5, 0.75, and 1 unit). When the enzyme concentration increases, the HMW DNA (marked by white arrows) decreases. HMW DNA restriction fragments between 150 and 350 kb are cut from the gel and used as DNA inserts for BAC library construction. Markers are shown on the left. M, PFGE lambda Ladder DNA (New England BioLabs).

for 20 min each with 1 × TE (10 m*M* Tris–HCl, pH 8.0, and 1 m*M* EDTA, pH 8.0) supplemented with 50 m*M* NaCl to exchange the 1 × TAE running buffer. The treatment with GELase is done according to the manufacturer's instructions, except omitting the GELase buffer since the enzyme is active in 1 × TE supplied with 50 m*M* NaCl. The Agarose pieces are melted at 65 °C and transferred to 45 °C. One unit of GELase (Epicentre) per 100 mg of gel is added and incubated at 45 °C for 1 h for gel digestion. A rough estimation for the concentration of the HMW DNA can be performed by running aliquots of the DNA in a gel together with different concentrations of uncut

Lambda DNA (Promega) (e.g., 5, 10, 25, 50, 75, and 100 ng). HMW digested DNA with a concentration of >5 ng/μl can be used for the BAC library construction. The HMW digestion DNA can be stored at 4 °C.

2.6. HMW DNA ligation into the BAC vector

Since the HMW DNA is digested with one restriction enzyme, the chimera between two unrelated DNA fragments can occur and can be inserted into the same clone. To avoid this problem, the ligation between the vector and the inserts is done with high vector to inset ratio. HMW DNA pipetting in the following steps has to be done by using broad-end tips (regular tips could be used after cutting their ends), to prevent DNA shearing. The ligation reaction is done by the combination of about 200 ng of digested and size-selected HMW DNA with digested and dephosphorylated pIndigo536 BAC vector in a molar ratio of 1:10, insert to vector. The mixture is then incubated at 55 °C for 10 min to denatured hybridized sticky ends and cooled on ice. Five units of T4 DNA ligase, 10 μl of 10 × T4 DNA ligase buffer and ddH_2O to a final volume of 100 μl are added. The ligation reaction conditions are as follows: incubation at 16 °C for 16 h followed by incubation at 65 °C for 20 min to heat inactivation of the T4 ligase, this step increases the transformation efficiency (Zhu & Dean, 1999).

2.7. Transformation of the ligation products into *E. coli*

The transformation of the ligation products into *E. coli* is done by electroporation. The ligation buffer has to be exchanged with 0.5 × TE. This is done by drop dialysis of the ligation reaction on nitrocellulose filters, 0.025 μm pore size (Millipore) against 0.5 × TE for 1 h. Following dialysis, 25 μl of the ligation products are combined with 30 μl DH10B electrocompetent cells on ice then transferred into prechilled ice in 1 mm cuvette. The electroporation is performed using a Gene Pulser II electroporation system (Bio-Rad) using the following parameters: voltage set to 13 kV/cm, resistance to 100 Ω, and capacitance to 25 μF (Sheng et al., 1995). The cells are immediately suspended in 0.5 ml SOC (2% Bacto tryptone, 0.5% Bacto yeast extract, 0.05% NaCl, and 20 m*M* glucose), transferred to a culture tube and incubated for 1 h at 37 °C to recover. Following incubation, the cells are plated on LB agar plates containing 12.5 μg/ml chloramphenicol, 50 μg/ml X-gal, and 25 μg/ml IPTG and incubated at 37 °C for 16 h. The efficiency of the cloning and the transformation should be determined. If the percentage of the white colonies is high (more than

60%), and the clones have large inserts (as described below) then one can proceed to large-scale cloning and transformation for construction of a large BAC library.

2.8. Determination of the average size of the clones

To determine the average size of the cloned DNA in the library, BAC clones (about 50 white colonies picked randomly) are purified from white colonies recovered from the BAC transformation. The colonies are inoculated in 2 ml LB medium with 12.5 μg/ml chloramphenicol and grown for ~16 h at 37 °C with vigorous shaking (~200 rpm). BAC clone purification is performed using a standard alkaline lysis procedure (below). Cultures are transferred to a microcentrifuge tube and centrifuged at 18,000 × *g* for 1 min in a table-top microcentrifuge. The pelleted bacterial cells are resuspended in 0.3 ml P1 solution (50 m*M* Tris–HCl, pH 8, 10 m*M* EDTA, pH 8, and 100 μg/ml RNase A (Sigma)). Then 0.3 ml of P2 (0.2 N NaOH and 1% SDS) is added and the tubes are gently inverted six times to mix. Then, 0.3 ml of P3 (3 *M* KOAc, pH 4.8–5.5) is added and tubes are inverted six times and incubated in ice for 5 min. The tubes are then centrifuged at 18,000 × *g* for 10 min. The supernatant (~750 μl) is transferred into a new tube containing equal volume of cold isopropanol. The tubes are mixed by six inversions and incubated at −20 °C for 1 h. To pellet the DNA, the tubes are centrifuged for 15 min at maximal speed (>18,000 × *g*) at 4 °C. The DNA pellet is washed with 0.5 ml cold 70% ethanol and centrifuged again at maximal speed at 4 °C. The DNA pellet is airdried to remove ethanol traces for ~10 min and resuspended in 40 μl of 10 m*M* Tris–HCl, pH 8.5.

To determine the size of the inserts in the isolated clones, the clones are digested with *Not*I, which its sites located on both sides of the *Hin*dIII cloning sitting in the BAC vector. Digestion of the clones is done as follows: 200 ng BAC clone, 2 μl of 10 × NEBuffer 3, 0.1 mg/ml BSA, 1 unit of *Not*I restriction enzyme (New England BioLabs), and ddH_2O to a final volume of 20 μl. The reaction is incubated for 1 h at 37 °C. The digested clones are then run and separated in 1% Seakem Gold agarose gel (Lonza) in 0.5 × TBE (45 m*M* Tris–borate and 1 mM EDTA, pH 8) by PFGE under the following conditions: 6 V/cm, 5–15 s pulse, 13 h, and 14 °C. The gel is stained with GelStar nucleic acid gel stain (1X). Example of *Not*I digested BAC clones with different size range inserts is shown in Fig. 6.2. By evaluating the insert sizes of the random clones examined, one can estimate the average insert size in the BAC library.

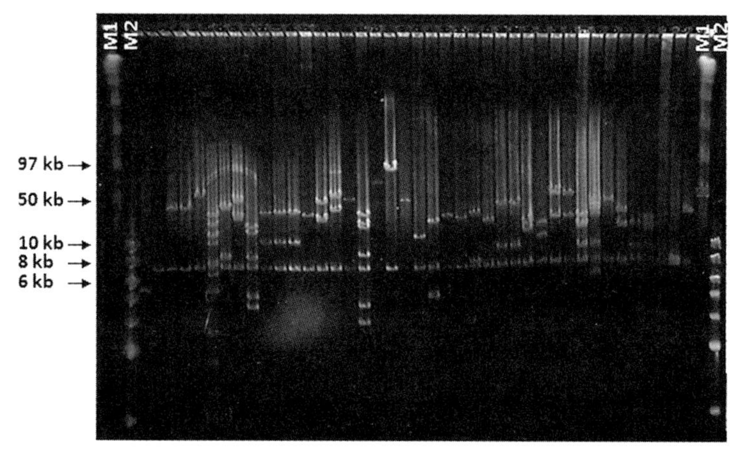

Figure 6.2 PFGE of randomly selected BAC clones form a BAC library digested with *Not*I. The BAC vector is 7.5 kb and the inserts appear in a variety of different sizes. Markers are shown in the left and the right of the gel. M1, PFGE lambda Ladder DNA; M2, 1 kb DNA ladder (New England BioLabs).

2.9. Library construction and storage

The final step is to perform a large-scale cloning and transformation to construct a large library with all the available DNA. The transformed cells should be plated in 145 mm agar plates. The white colonies are picked to 96-well or 384-well plates, filled with 180 or 60 μl, respectively, of LB with 7% glycerol (v/v) and 12.5 μg/ml chloramphenicol. The plates are incubated at 37 °C for ~16 h and stored at −80 °C.

REFERENCES

Béjà, O., Aravind, L., Koonin, E. V., Suzuki, M. T., Hadd, A., Nguyen, L. P., et al. (2000). Bacterial rhodopsin: Evidence for a new type of phototrophy in the sea. *Science*, *289*, 1902–1906.

Béjà, O., Suzuki, M. T., Heidelberg, J. F., Nelson, W. C., Preston, C. M., Hamada, T., et al. (2002). Unsuspected diversity among marine aerobic anoxygenic phototrophs. *Nature*, *415*, 630–633.

Béjà, O., Suzuki, M. T., Koonin, E. V., Aravind, L., Hadd, A., Nguyen, L. P., et al. (2000). Construction and analysis of bacterial artificial chromosome libraries from a marine microbial assemblage. *Environmental Microbiology*, *2*, 516–529.

de la Torre, J. R., Christianson, L., Béjà, O., Suzuki, M. T., Karl, D., Heidelberg, J. F., et al. (2003). Proteorhodopsin genes are widely distributed among divergent bacterial taxa. *Proceedings of the National Academy of Sciences of the United States of America*, *100*, 12830–12835.

Feingersch, R., & Béjà, O. (2009). Bias in assessments of marine SAR11 biodiversity in environmental fosmid and BAC libraries? *The ISME Journal*, *3*, 1117–1119.

Feingersch, R., Suzuki, M. T., Shmoish, M., Sharon, I., Sabehi, G., Partensky, F., et al. (2010). Microbial community genomics in eastern Mediterranean Sea surface waters. *The ISME Journal, 4*, 78–87.

Kim, U. J., Birren, B. W., Slepak, T., Mancino, V., Boysen, C., Kang, H. L., et al. (1996). Construction and characterization of a human bacterial artificial chromosome library. *Genomics, 34*, 213–218.

Loy, A., Duller, S., Baranyi, C., Mussmann, M., Ott, J., Sharon, I., et al. (2009). Reverse dissimilatory sulfite reductase as phylogenetic marker for a subgroup of sulfur-oxidizing prokaryotes. *Environmental Microbiology, 11*, 289–299.

Sabehi, G., Béjà, O., Suzuki, M. T., Preston, C. M., & DeLong, E. F. (2004). Different SAR86 subgroups harbour divergent proteorhodopsins. *Environmental Microbiology, 6*, 903–910.

Sabehi, G., Loy, A., Jung, K. H., Partha, R., Spudich, J. L., Isaacson, T., et al. (2005). New insights into metabolic properties of marine bacteria encoding proteorhodopsins. *PLoS Biology, 3*, e173.

Sheng, Y., Mancino, V., & Birren, B. (1995). Transformation of *Escherichia coli* with large DNA molecules by electroporation. *Nucleic Acids Research, 23*, 1990–1996.

Shizuya, H., Birren, B., Kim, U. J., Mancino, V., Slepak, T., Tachiiri, Y., et al. (1992). Cloning and stable maintenance of 300-kilobase-pair fragments of human DNA in *Escherichia coli* using an F-factor-based vector. *Proceedings of the National Academy of Sciences of the United States of America, 89*, 8794–8797.

Stein, J. L., Marsh, T. L., Wu, K. Y., Shizuya, H., & DeLong, E. F. (1996). Characterization of uncultivated prokaryotes: Isolation and analysis of a 40-kilobase-pair genome fragment from a planktonic marine archaeon. *Journal of Bacteriology, 178*, 591–599.

Whitman, W. B., Coleman, D. C., & Wiebe, W. J. (1998). Prokaryotes: The unseen majority. *Proceedings of the National Academy of Sciences of the United States of America, 95*, 6578–6583.

Zhu, H., & Dean, R. A. (1999). A novel method for increasing the transformation efficiency of Escherichia coli-application forbacterial artificial chromosome library construction. *Nucleic Acids Research, 27*, 910–911.

CHAPTER SEVEN

Preparation of Fosmid Libraries and Functional Metagenomic Analysis of Microbial Community DNA

Asunción Martínez[*,1], **Marcia S. Osburne**[†]
[*]Department of Civil and Environmental Engineering, Massachusetts Institute of Technology, Cambridge, Massachusetts, USA
[†]Department of Molecular Biology and Microbiology, Tufts University School of Medicine, Boston, Massachusetts, USA
[1]Corresponding author: e-mail address: chon@mit.edu

Contents

1. Introduction 124
2. General Considerations: Genomic Library Vector and Host 125
   2.1 Choice of genomic library vector 125
   2.2 Choice of heterologous screening host 126
3. Construction of Fosmid Environmental Libraries 128
   3.1 Isolation of environmental DNA from marine samples 128
   3.2 Library construction using vector pCC1FOS™ 129
   3.3 Storing fosmid libraries 131
4. General Considerations for Working with Fosmids 131
   4.1 Preparation of fosmid DNA 131
   4.2 Retransforming fosmids 132
   4.3 Sequencing fosmids 132
5. Functional Screens 133
   5.1 Use of heterologous hosts with chromosomal mutations 133
   5.2 Expression enhancement 134
   5.3 Format 134
   5.4 Examples 135
   5.5 Increasing the odds 137
   5.6 Characterization of functional screen hits 138
6. Conclusions 139
References 139

Methods in Enzymology, Volume 531
ISSN 0076-6879
http://dx.doi.org/10.1016/B978-0-12-407863-5.00007-1

Abstract

One of the most important challenges in contemporary microbial ecology is to assign a functional role to the large number of novel genes discovered through large-scale sequencing of natural microbial communities that lack similarity to genes of known function. Functional screening of metagenomic libraries, that is, screening environmental DNA clones for the ability to confer an activity of interest to a heterologous bacterial host, is a promising approach for bridging the gap between metagenomic DNA sequencing and functional characterization. Here, we describe methods for isolating environmental DNA and constructing metagenomic fosmid libraries, as well as methods for designing and implementing successful functional screens of such libraries.

1. INTRODUCTION

Community DNA and RNA sequencing have become key tools in microbial ecology. Recent technological advances, including next-generation sequencing, combined with new sampling methods and single-cell genomics, have made feasible sampling and analysis of natural microbial communities at a scale unthinkable just a few years ago (Ottesen et al., 2011; Urich et al., 2008; Zhang et al., 2006). Yet determining the functional roles of newly discovered sequences remains challenging, as a large fraction shows no significant similarity to previously annotated sequences. For example, it is not uncommon to find that 40–60% of nonribosomal RNA gene reads in marine depth profiles have no significant match in the non-redundant NCBI database, and these numbers are even higher when using functional databases like KEGG or COG (Shi, Tyson, Eppley, & DeLong, 2011; Stewart, Ulloa, & Delong, 2011). Similar numbers of unannotated sequences are found in other environments (Urich et al., 2008). Therefore, developing approaches to assign biochemical and ecological functions to novel sequences is critical.

Transcriptomic approaches have been used to associate gene expression to natural or experimentally induced environmental conditions (McCarren et al., 2010; Shi et al., 2011; Stewart et al., 2011). These studies identify genes that may be important in specific environments, but their individual biochemical functions are often elusive. An alternative approach, nucleotide sequence-based screening (Courtois et al., 2003; Okuta, Ohnishi, & Harayama, 1998) requires *a priori* knowledge of sequences of interest. In contrast, functional screens, in which environmental library DNA clones are screened for their ability to confer a function of interest to a heterologous host, require no sequence knowledge and thus can provide direct evidence of function for previously unknown genes. Functional screens are particularly powerful when combined with *in vitro* biochemical analysis of purified proteins (McSorley et al., 2012), and with traditional metagenomic and metatranscriptomic analyses that provide useful information about the occurrence and ecological distribution of newly identified genes in other microbes (Martinez, Tyson, & Delong, 2010). In this chapter, we describe methods for constructing and screening environmental DNA libraries in fosmid vectors (Fig. 7.1). While an exhaustive review of the relevant literature is beyond the aim of this chapter, we discuss general considerations, illustrate possible strategies, and provide selected examples, with the aim of empowering the reader to design successful functional screens.

2. GENERAL CONSIDERATIONS: GENOMIC LIBRARY VECTOR AND HOST

2.1. Choice of genomic library vector

One of the first considerations when designing a functional screen is the desired insert size for the genomic library. Many plasmid vectors can maintain DNA inserts of 4–10 kb, are relatively easy to use, and offer high copy number and inducible promoters to control expression levels. Libraries based on such vectors have been used successfully in functional screens when only one or few genes are required to obtain the desired phenotype (Maresca, Braff, & DeLong, 2009; Owen, Robins, Parachin, & Ackerley, 2012). However, these vectors are inadequate for cloning the larger gene clusters required to express more complex metabolic pathways.

Construction of large-insert libraries requires either multicopy cosmid vectors or, for added stability, single-copy fosmid or bacterial artificial chromosome (BAC) vectors, the latter of which are derived from the *E. coli* F-factor replicon (Kim, Shizuya, de Jong, Birren, & Simon, 1992; Shizuya et al., 1992). Cosmid and fosmid libraries, which are introduced into *E. coli* by phage infection, are packaged using bacteriophage λ components and thus their insert size is restricted to 35–40 kb. BACs on the other hand, normally introduced into *E. coli* by transformation, can accommodate inserts up to 300 kb significantly increasing the odds of containing complete gene clusters and reducing the number of clones required to obtain the desired coverage. In our view, however, the use of fosmids strikes a good compromise between the requirement for inserts large enough to include complete gene clusters and the relative ease of fosmid environmental library construction.

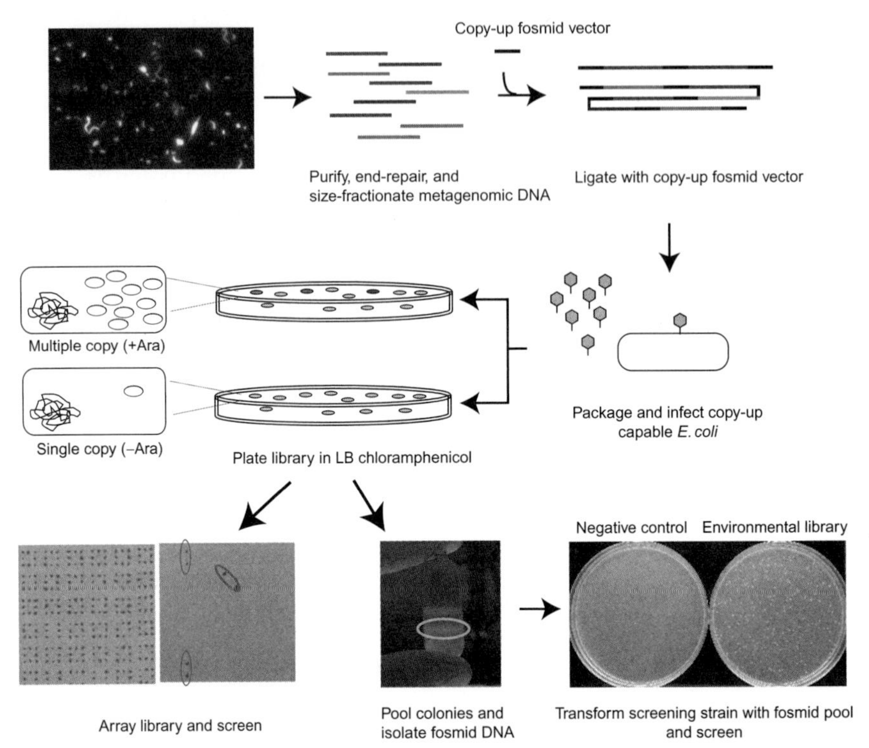

Figure 7.1 Process for extraction and cloning of environmental DNA into fosmid vector and screening of metagenomic libraries. Large genomic fragments obtained from uncultured microbial microorganisms are end-repaired, ligated with fosmid vector, and packaged *in vitro*. After infection of a copy-up *E. coli* strain, the metagenomic library can be screened directly, either in the absence of arabinose (single copy) or in the presence of arabinose which increases the copy number of the fosmid and may result in increased expression (represented here as colored colonies). Alternatively, the clones can be arrayed and frozen for screening (bottom left), or pooled and used to prepare a fosmid DNA pool to transform additional hosts (bottom right). (See color plate.)

2.2. Choice of heterologous screening host

E. coli is the most common heterologous host used in functional screens because of the vast resources available in terms of vectors, ease of manipulation, genetic tools, and biochemical knowledge. However, the success of a functional screen depends on the ability of the host to express environmental DNA and to provide precursors and modifying enzymes. Thus, the use of an alternative heterologous host may sometimes be advantageous, as when screening for natural products synthesized by polyketide and nonribosomal peptide synthases which are not easily expressed in *E. coli* (Wenzel & Muller, 2005).

Functional screening of metagenomic libraries in hosts other than *E. coli* requires specialized shuttle vectors that can be maintained in the second host either by autonomous replication or by integration into the chromosome. Autonomous replication in the alternative host can be accomplished using vectors with broad host-range origins of replication. For example, the shuttle BAC vector pGNS-BAC uses the RK2 minireplicon for high-copy replication in diverse Gram negative species (Kakirde et al., 2011). In another example, Johnston and colleagues have used a wide host-range cosmid to construct environmental gene libraries for screening in *E. coli*, *Pseudomonas aeruginosa*, and the alpha proteobacterium *Rhizobium leguminosarum* (Li, Wexler, Richardson, Bond, & Johnston, 2005; Wexler, Bond, Richardson, & Johnston, 2005). Cosmid libraries are transferred among these different species by conjugation of individual clones or *en masse* using pooled libraries. Importantly, the authors showed that the genes uncovered using functional screens for novel alcohol dehydrogenase and tryptophan biosynthetic genes were different in the different hosts, highlighting the advantage of using multiple hosts in functional screens (Li et al., 2005; Wexler et al., 2005).

Some shuttle vectors have the ability to integrate into the chromosome of the second host using engineered phage integrase-mediated site-specific DNA recombination. For example, Courtois and colleagues (Courtois et al., 2003) screened by PCR a soil library constructed in an *E. coli–Streptomyces* shuttle cosmid. Positive cosmids were transferred into *Streptomyces lividans* by spheroplast transformation and then integrated into the *S. lividans* chromosome and tested for expression. Improved shuttle BAC vectors that can be easily transferred by conjugation and have chromosome integration features are also available (Martinez et al., 2004; Miao et al., 2005).

Screening environmental libraries in multiple bacterial hosts is greatly facilitated by more versatile shuttle BAC vectors and high-throughput conjugation, described by Martinez et al. (2004). Using this system, a single BAC library can be screened in parallel in *E. coli*, *Streptomyces*, and an engineered *Pseudomonas putida* strain, allowing researchers to take advantage of the known variation in expression patterns of some small-molecule gene clusters.

The construction of BAC libraries is discussed by Beja et al. in Chapter 6. The remainder of this chapter focuses on the construction and functional screening of fosmid environmental libraries in *E. coli*. The functional screening strategies described here can be applied to all other types of metagenomic libraries and to other heterologous hosts.

3. Construction of Fosmid Environmental Libraries

3.1. Isolation of environmental DNA from marine samples

Because construction of fosmid libraries requires DNA inserts in the 35–45 kb size range, the embedding of cells in agar plugs for DNA isolation that is required for BAC library construction is unnecessary (Chapter 6). However, to avoid excessive shearing, all DNA extraction steps should be performed with gentle mixing and vortexing should be avoided. Using the method described below, we routinely obtain marine metagenomic DNA samples of the appropriate size range necessary for library construction (DeLong et al., 2006). Specialized methods to remove humic acids that interfere with library construction from soil samples, or that make use of additional cell wall hydrolases to improve DNA isolation from certain microbes, have been described elsewhere (Courtois et al., 2001; Liles et al., 2008; Rondon et al., 2000).

1. Harvest microbial biomass by filtration into 0.2-μm Sterivex filters (Millipore). Required sample volumes vary depending on the environment. While 20 L is usually sufficient for marine surface water samples, larger volumes are recommended for deep samples because of their lower microbial biomass (DeLong et al., 2006). For marine planktonic samples, prefilter samples through a 1.6-μm GFA glass filter (Whatman) to remove particles. After filtration, add 1.8 ml of DNA lysis buffer (0.75 *M* sucrose, 40 m*M* EDTA, 50 m*M* Tris–HCl, pH 8.0) to the Sterivex filter and freeze at −80 °C for later use. The sucrose is used to keep samples isosmotic with seawater, so its concentration may need to be changed depending on the salinity of the habitat sampled.
2. Thaw Sterivex filter on ice. Add 50 μl of fresh lysozyme solution (25 mg lysozyme in 200 μl of water). Incubate for 1 h at 37 °C with intermittent mixing.
3. Add 100 μl of fresh proteinase K solution (20 mg proteinase K in 200 μl of water) and 100 μl of 20% SDS. Incubate at 55 °C for 1 h with intermittent mixing.
4. Remove lysate from Sterivex filter using a 3-cc syringe and place gently in a 15-ml centrifuge tube. Rinse the filter with 1 ml lysis buffer and pool with initial lysate.
5. Add an equal volume of phenol:chloroform:isoamylalcohol (25:24:1), pH 8. Mix gently by inversion and centrifuge at 3500 × *g* for 10 min. Remove aqueous phase to new tube and repeat extraction.

6. Extract with an equal volume of chloroform:isoamylalcohol (24:1) as above.
7. Concentrate DNA by spin dialysis using an Amicon Ultra-4 filter 30k cutoff (Millipore). Wash three times with 2 ml of TE buffer (10 m*M* Tris–HCl, pH 8.0, 1 m*M* EDTA) and concentrate down to approximately 200 μl. Remove liquid with a micropipettor, using care not to damage the membrane. Rinse membrane with 30 μl of TE and pool.
8. Purify high-molecular-weight (HMW) DNA by CsCl density gradient centrifugation (this step, although not required, results in more efficient cloning). Mix 180 μl (3–5 μg) of DNA with 160 mg of CsCl and 10 μl of ethidium bromide (5 mg/ml). Spin at 100,000 rpm for 20 h in a TLA100 fixed angle rotor (Optima TL Ultracentrifuge, Beckman Coulter). Visualize HMW DNA band using a long-wave UV lamp and carefully remove band with a 1-ml syringe and a 20-gauge needle. Extract three times with equal volumes of water-saturated butanol. Wash DNA with TE and concentrate down to 52 μl using an Amicon Ultra spin filter as above.

3.2. Library construction using vector pCC1FOS™

The CopyControl™ Fosmid Library Production Kit (Epicentre, Madison, WI) is convenient for preparing environmental DNA fosmid libraries, as this system allows vector copy number to be adjusted. pCC1FOS™ contains both the single-copy F-plasmid origin of replication, and the *oriV* origin, which can be induced to allow high-copy plasmid replication (Wild, Hradecna, & Szybalski, 2002). In this system, the *trfA* gene, required for *oriV* function, is located in the *E. coli* chromosome under control of the *araC*–P_{BAD} regulatory region. Thus, clones can be maintained in single copy for stability in the absence of inducer, or induced to high copy by adding L-arabinose to the growth medium to induce *trfA* expression, which then activates *oriV*, resulting in up to 100 plasmid copies per cell. High-copy conditions can result in increased gene expression, sometimes allowing identification of library hits not detectable under single-copy conditions (Martinez, Bradley, Waldbauer, Summons, & DeLong, 2007; Martinez et al., 2010). Similar conditionally amplifiable fosmid systems are also available from Lucigen (Lucigen Corporation), but in this case vectors are designed to eliminate read-through of insert genes from vector promoters. This could have negative consequences for functional screening, as native environmental promoters may not function efficiently.

Environmental DNA is cloned using the CopyControl™ Fosmid Library Production Kit (Epicentre) following the manufacturer's protocol, with some modifications as follows:

1. End-repair 52 μl HMW DNA with 8 μl end-repair buffer, 8 μl 2.5 m*M* dNTP mix, 8 μl 10 m*M* ATP, and 4 μl end-repair enzyme mix at room temperature for 1 h. Heat inactivate at 70 °C for 10 min.
2. Size-fractionate HMW end-repaired DNA by pulsed-field gel electrophoresis on a CHEF-DR-II system (Bio-Rad, Hercules, CA) using 1% SeaPlaque low-melt agarose (Cambrex, Baltimore, MD) under the following conditions: 10 °C, 6 V/cm for 15 h and 5–15 s pulse time in 1 × TAE (40 m*M* Tris-acetate, 1 m*M* EDTA, pH 8.0) buffer.
3. Stain gel with SYBR gold (Molecular Probes, Eugene, OR) and visualize on a Dark Reader transilluminator (Clare Chemical Research, Dolores, CO). Excise gel regions containing genomic DNA in 40–50 kb region. This staining and non-UV illumination approach avoids UV damage of the DNA. If using a UV transilluminator, stain and visualize only the size markers to avoid UV damage to genomic DNA.
4. Melt gel slice at 70 °C for 10 min and digest with 5 μl of gelase (1 unit/μl, Epicentre) for 2 h at 45 °C. Inactivate gelase at 70 °C for 10 min, place tube on ice for 10 min, and spin at 13,000 rpm for 10 min to remove remaining oligosaccharides.
5. Wash three times with 2 ml TE buffer on an Amicon Ultra-4 100k (Millipore) and concentrate size-fractionated DNA to about 30 μl.
6. Combine 0.5 μg of size-fractionated DNA with 1 μl of 10 × Fast-Link ligation buffer, 1 μl 10 m*M* ATP, 1 μl of the CopyControl™ pCC1FOS™ vector (0.5 μg/μl), and 1 μl Fast-Link ligase in 10 μl final volume (for a 10:1 insert:vector molar ratio) and ligate at room temperature for at least 4 h. We routinely set up four ligations, thus using the majority of the size-fractionated DNA.
7. Package fosmid clones *in vitro* using MaxPlax™ Lambda Packaging Extracts according to the manufacturer's instructions.
8. Grow *E. coli* EPI300™ in LB containing 10 m*M* $MgSO_4$ and 0.2% maltose to an OD_{600} of 0.8. Mix 250 μl of undiluted phage mix with 10 ml of *E. coli* EPI300™, incubate 1 h at 37 °C, and plate 600 μl per Q-tray (Genetix, Christchurch, MA) containing LB agar with 12 μg/ml chloramphenicol, using sterile glass beads to spread the cells. Phage stock dilution and plating densities may need adjustment for different size agar plates or other variables.

9. Be sure that all fluid is absorbed before placing the plates at 37 °C; if necessary, first dry inoculated plates in a sterile laminar flow hood. Incubate plates at 37 °C for up to 24 h. Estimate the total number of fosmid clones.

3.3. Storing fosmid libraries

Libraries can be stored in different ways depending upon their intended use(s).

Arrays. Although arraying a library is labor intensive, arrays provide a permanent record and a reliable source of fosmid DNA for subsequent DNA sequencing, and also allows for multiple successive screens. If a functional screen is to be performed in *E. coli* EPI300™, the library can be arrayed manually or robotically (e.g., using a Genetix QPix (Genetics)) in 96- or 364-well plates with the appropriate volume of LB containing 12 μg/ml chloramphenicol and 10% glycerol. Plates are incubated at 37 °C for up to 24 h and then stored frozen at −80 °C. For functional screens, a sterile replicator can be used to inoculate screening plates from partially thawed array plates.

Colony pools. Alternatively, infected cells can be stored in pools by harvesting colonies from selective plates using a sterile spreader and resuspending them in sterile LB containing 12 μg/ml chloramphenicol and 10% glycerol. The suspension is mixed and stored in aliquots at −80 °C. Aliquots can be thawed and screened at a later time. However, because of the resulting clone amplification many more clones need to be analyzed to obtain the desired coverage.

DNA pools. If the screen is to be performed in a different *E. coli* strain, fosmid DNA can be extracted from the pooled colonies and stored in aliquots for future transformation.

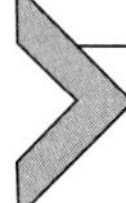

4. GENERAL CONSIDERATIONS FOR WORKING WITH FOSMIDS

4.1. Preparation of fosmid DNA

Because high-copy number occasionally leads to the loss of certain clones or accumulation of insert deletions, we routinely maintain fosmid clones in single copy by growing them in the absence of L-arabinose. For DNA isolation, fosmid clones are gown in 2 ml of LB with 12 μg/ml chloramphenicol at 37 °C for 24 h. 0.2% glucose may be added to prevent leaky expression of *trfA*, thus ensuring single copy (Wild et al., 2002), although this is not usually required. For high-copy induction, 0.02% L-arabinose is used as recommended (Wild et al., 2002). A commercial autoinduction solution is also available from Epicentre.

Fosmid DNA is best isolated using standard alkaline lysis methods (Sambrook, Fritsch, & Maniatis, 1989), with the added precaution of mixing by inversion rather than vortexing in order to minimize shearing. Commercial products such as Quiagen plasmid miniprep spin columns should be avoided, as plasmid yields are poor for plasmids larger than 10 kb. For high-throughput fosmid isolation, we use the AutoGenprep 960 (AutoGen Inc., Holliston, MA) which can extract fosmid DNA from up to four 96-well plates at a time.

While standard alkaline lysis methods yield fosmid DNA in sufficient amount and quantity for individual fosmid transformations, when preparing fosmid DNA pools for electroporation, we have found it is best to further purify fosmid DNA in a cesium chloride density gradient (Sambrook et al., 1989).

4.2. Retransforming fosmids

Because of their large size, transformation of fosmids by electroporation is less efficient than electroporation of smaller plasmids. We recommend using 1.2 kV/cm, 200 Ω, and 25 μF, as described by Sheng, Mancino, and Birren (1995), and high-efficiency electrocompetent cells. For retransforming fosmid pools, we highly recommend pilot experiments with pooled DNA to determine optimal conditions for obtaining desired coverage and plating density.

4.3. Sequencing fosmids

When using functional screens, large-insert clones of interest are best sequenced by taking advantage of *in vitro* transposition systems that generate a collection of clones with single random transposon insertion (e.g., EZ-Tn5™ KAN-2 Insertion kit (Epicentre), the GPS-1 Genome Priming System (NEB)). Fosmid inserts can be sequenced bidirectionally outward from the transposon insertion site using the primers provided in the kit. The advantage of this method is that individual transposon insertion mutants can be screened directly in the functional assay to determine the role of the disrupted gene in the phenotype of interest. For example, we have used this method to analyze the functions in retinal biosynthesis and light-driven ATP production of genes on a marine fosmid insert (Martinez et al., 2007).

For fosmid DNA sequencing, we typically pick individual transposon insertion clones into four 96-well plates and sequence two of the plates with forward and reverse primers using fluorescence-based Sanger sequencing. While this is typically sufficient for sequencing the entire insert, additional sequencing might be required to find a transposon insertion in every predicted gene of interest.

5. FUNCTIONAL SCREENS

The range of functional screens that can be performed with an environmental library is immense. As long as there is an assay for the function of interest and a bacterial heterologous host lacking that function, a functional screen is possible. For the purpose of this chapter, we will focus on general considerations and provide some examples that will illustrate various strategies that can be used when designing a functional screen. Because of space limitations, the list of cited screens is undoubtedly incomplete. We apologize for any omissions and refer the reader to the literature for more specific information.

5.1. Use of heterologous hosts with chromosomal mutations

Functional screens most often comprise screening for a gain-of-function conferred by a cloned environmental DNA fragment. If the heterologous host is already proficient in the function of interest, a mutant defective in this function is used for the gain-of-function screen. Several examples of such functional screens of environmental libraries have been published, including the identification of *trp* genes using a tryptophan auxotroph (Li et al., 2005) and the isolation of phosphonate biosynthetic clusters in an *E. coli* strain containing a deletion of the endogenous *phn* cluster (Martinez et al., 2010). For such screens, pooled metagenomic libraries are transformed into electrocompetent cells of the desired mutant host and transformants are selected in the appropriate medium.

Recently, the Keio collection of *E. coli* K-12 strains containing in-frame, single-gene knockout mutants in almost every nonessential gene has been constructed (Baba et al., 2006) and made available to the scientific community via GenoBase (http://ecoli.aist-nara.ac.jp). In this collection, transposons replace the gene of interest, but they can be easily excised (Datsenko & Wanner, 2000), leaving a clean unmarked, in-frame deletion of the desired gene. This collection is clearly an invaluable resource for obtaining functional screening hosts.

5.2. Expression enhancement

The conditions needed for optimal expression of environmental genes in heterologous hosts are context-specific and not readily generalizable. However, we and others have found that slow-growth conditions and/or lower temperatures often result in increased expression. For example, incubating colonies for 3 days at 30 °C (Rondon et al., 2000) or 1 day at 37 °C followed by 6 days at room temperature (MacNeil et al., 2001) has been successful in a variety of soil library functional screens. We have also observed that in the case of clones selected for their ability to use phosphonate, an alternative P source, real hits appeared even after incubation of screening plates for 20 days at 30 °C (Martinez et al., 2010). For extended incubations, care should be taken to incubate plates either in a humidified incubator or sealed in plastic sleeves to prevent desiccation of the agar. In addition, a negative control plate with cells transformed with empty vector should always be prepared and monitored closely.

Another strategy to increase gene expression is to increase clone copy number using a selectively amplifiable vector such as pCC1FOS™, as described above (Wild et al., 2002). For example, we have found that adding 0.02% L-arabinose to LB plates resulted in a greater than 10-fold increase in the number of colored clones in marine fosmid libraries constructed in pCC1FOS™ (unpublished observations). Further, these conditions were required for the detection of marine clones expressing a complete proteorhodopsin-based photosystem in *E. coli* (Martinez et al., 2007). On the other hand, some clones were sufficiently expressed in single copy but became toxic to the host under copy-up conditions (Martinez et al., 2010). Thus, it is worth considering performing the functional screen under both single-copy and copy-up conditions to increase the probability of success.

Since copy-up requires expression of a chromosomal *araC*–P_{BAD}-regulated *trfA* gene, we have used P1*vir* transduction to transfer this gene to an alternative screening strain to confer copy-up capability (Martinez et al., 2010). This method should be applicable to any $RecA^+$ strain, including those in the *E. coli* Keio collection (Baba et al., 2006).

5.3. Format

Functional screens can be performed in a variety of formats. Screening colonies on agar plates may be the least labor intensive, as it does not require high-throughput liquid handling and can be used to screen very large numbers of clones using relatively few plates, especially when the screen involves selecting for growth resulting from gain-of-function (Martinez et al., 2010).

Environmental libraries can also be screened using liquid cultures in microtiter plates. Although more labor intensive than colony screening, this approach is used when it is necessary to prepare high-throughput crude protein extracts for screening in enzymatic assays (Suenaga, Ohnuki, & Miyazaki, 2007) or to prepare crude small-molecule extracts for antibacterial screening (Martinez, Kolvek, Hopke, Yip, & Osburne, 2005) or other high-throughput assays.

Finally, functional screens can be performed using fluorescence-activated cell sorting (FACS) when the desired phenotype results in fluorescence. This occurs, for example, when coupling regulatory regions to *gfp* expression (Martinez, Hopke, MacNeil, & Osburne, 2005; Uchiyama, Abe, Ikemura, & Watanabe, 2005). This powerful technique allows high-complexity libraries to be grown as a single liquid culture prior to sorting and to then screen extremely high numbers of clones with ease.

5.4. Examples

Examples of functional screens are listed below.

5.4.1 Enzyme-activity-based screens

One of the pioneering studies using functional screens to identify novel functions in large-insert environmental libraries focused on identifying antibacterial, lipase, amylase, nuclease, and hemolytic activities (Rondon et al., 2000). The soil BAC library was plated on a variety of diagnostic agar plates, and positive clones were identified by characteristic phenotypes. For example, clones encoding hemolytic activity were identified by the appearance of a clear halo around the colony on blood agar plates. Some additional examples of enzyme-based screens are included below.

5.4.2 Pigmentation screens

Although not a specific target, screening libraries for pigmented colonies has led to the identification of several interesting activities in soil libraries, including the biosynthetic gene cluster encoding the antibiotic violacein (blue pigment) (Brady, Chao, Handelsman, & Clardy, 2001) and novel antibiotics turbomycin A and B (brown pigment) (Gillespie et al., 2002). Further, orange clones in a marine fosmid library contained a complete and functional proteorhodopsin-based photosystem that allowed *E. coli* to make ATP from light (Martinez et al., 2007).

5.4.3 Antibacterial and antifungal screens

Antibacterial activity can be identified by a zone of inhibition surrounding an environmental library colony overlaid with a lawn of sensitive bacteria on an agar plate (MacNeil et al., 2001; Rondon et al., 2000). Similar assays for antifungal activity were used to identify a gene cluster encoding the biosynthesis of the antifungal compound tambjamine in the marine bacterium *Pseudoalteromonas tunicata* (Burke, Thomas, Egan, & Kjelleberg, 2007).

5.4.4 Antibiotic resistance screens

Riesenfeld, Goodman, and Handelsman (2004) screened a soil metagenomic BAC library for fragments that conferred resistance to aminoglycoside antibiotics and found the soil environment to be a reservoir of novel antibiotic resistance genes.

5.4.5 Chitinase screens

Chitinases from uncultured marine microorganisms have been identified by their ability to cleave the fluorogenic chitin analogue 4-methylumbelliferyl beta-D-*N*,*N*′-diacetylchitobioside (Cottrell, Moore, & Kirchman, 1999).

5.4.6 Screens for genes that allow utilization of a specific substrate

These screens are based on the ability of a library clone expressed in a heterologous host to confer the capability to use a substrate the host cannot normally use. Examples include genes for utilization of 4-hydroxybutyrate as sole C source (Henne, Daniel, Schmitz, & Gottschalk, 1999), utilization of phosphonates as the sole P source (Martinez et al., 2010), and identification of different classes of *trp* genes (Li et al., 2005).

5.4.7 High-throughput screens of library extracts to detect enzyme activity

A particularly interesting example from the point of view of methodology was the identification of novel extradiol dioxygenases by screening crude extracts of high-copy fosmid clones for the ability to degrade catechol (Suenaga et al., 2007). In this case, HTP extracts were prepared using a bacterial lysis product (BugBuster plus Benzonase, Novagen) in a simple two-step procedure.

5.4.8 Screens for modulators of eukaryotic cell growth

Gloux and colleagues screened a gut microbiome library for clones that could modulate eukaryotic cell growth *in vitro* (Gloux et al., 2007). They used a two-step microplate screening method in which clone libraries were first grown in microtiter plates and then lysed using glass beads. Bacterial lysates were then added to fibroblast cultures, and their growth was assessed by means of crystal violet staining.

5.4.9 Intracellular biosensor screens

Intracellular biosensors are reporter systems in which the gene encoding an easily detectable protein such as GFP is placed under control of a promoter that is activated in response to an activity of interest. Positive (GFP-expressing) clones are then isolated by FACS. For example, a reporter system consisting of *luxR*–P_{luxI}–*gfp* was used to identify clones that inhibit or induce quorum sensing in *E. coli* (Williamson et al., 2005).

This approach can also be used to screen for catabolic genes by substrate-induced gene expression (Uchiyama et al., 2005) exploiting the fact that bacterial catabolic genes, which have linked regulatory regions, are often regulated by substrate concentration. Using this method, a metagenomic library was constructed in an operon-trap *gfp*-expression vector and screened by FACS for catabolic genes, based on substrate-induced fluorescence of their regulatory regions.

5.5. Increasing the odds

The chance of success of a functional screen depends in part on the representation of the desired genes in the metagenomic library. Given the complexity of natural environmental samples, identifying functions present in only a small minority of the population can be daunting. Simply increasing the number of clones screened can be one means of improving the odds, and methods that incorporate FACS, as described for biosensor screens, greatly amplify screening capacity.

Methods that increase the representation of the desired organisms in the metagenomic library can greatly increase the odds of success. One approach is to construct libraries from enrichment cultures that favor the growth of microbes containing the targeted functions. For example, in order to identify genes encoding alcohol oxidoreductase activity, environmental samples were first enriched for microorganisms able to grow in the presence of glycerol, followed by library construction and screening in *E. coli* for the production of carbonyls from polyols on indicator agar (Knietsch, Waschkowitz, Bowien, Henne, & Daniel, 2003).

Another interesting approach is to combine stable isotope probing with metagenomic analysis. Dumont, Radajewski, Miguez, McDonald, and Murrell (2006) added $^{13}CH_4$ to an environmental sample to preferentially label the DNA of methylotrophs. ^{13}C-labeled gDNA was then separated from bulk unlabeled community DNA in a density gradient and used to construct a metagenomic library enriched for DNA from organisms that had obtained the majority of their C from the labeled substrate. This approach enriches the library for organisms involved in a particular metabolic process, thus increasing the chances of identifying genes of interest.

Finally, a whole community library can be prescreened to increase the odds of finding a desired activity. This is particularly useful when the final screen has limited throughput. For example, Courtois and colleagues (Courtois et al., 2003) screened a pooled library by PCR to identify the small number of clones encoding polyketide synthase genes in a large soil library. Those clones were then transferred into *Streptomyces* to analyze small-molecule production by HPLC. Methods for prescreening libraries by assessing the activation of a variety of stress-response promoters fused to *gfp* have also been published, although their usefulness remains to be tested (Martinez, Hopke, et al., 2005).

5.6. Characterization of functional screen hits

A hit in a functional screen must first be confirmed by retransforming the plasmid, fosmid, or BAC into a fresh host. Once the phenotype is confirmed, we recommend preparing a random transposon insertion library using the EZ-Tn5™ KAN-2 Insertion kit (Epicentre) or equivalent to sequence the insert as described above. After protein-encoding genes have been predicted, selected transposon insertion mutants may be used to deduce the role of the inactivated genes. This is particularly valuable when the predicted genes lack significant similarity to proteins known to be involved in the activity of interest (Martinez et al., 2010). In this case, transposon insertions in every predicted gene were individually tested for utilization of 2-aminoethylphosphonate; two of the 25 predicted genes encoded a novel phosphonate degradation pathway. The novel enzymes were then characterized biochemically *in vitro* (McSorley et al., 2012).

6. CONCLUSIONS

One of the major challenges of contemporary microbial ecology is to interpret the vast amount of genomic and metagenomic information that is accumulating as a result of large sequencing projects (DeLong, 2009). In the marine ecosystem, for example, more than 6 million proteins have been putatively identified, many of which lack functional annotation (Yooseph et al., 2007). Because few environmentally relevant microorganisms can currently be grown in the laboratory, let alone manipulated genetically, traditional approaches for linking genes to functions are of limited use. In this context, functional screening of metagenomic libraries in heterologous hosts offers an alternative approach for assigning functions to hypothetical genes identified in metagenomic surveys (Martinez et al., 2010). A major limitation of this approach is the requirement for expression and correct protein assembly and localization in the heterologous host. Increasing the repertoire of shuttle vectors, taxonomically diverse heterologous hosts, and high-throughput library transfer methods will undoubtedly facilitate large-scale functional screening and improve the rate of functional assignment to hypothetical genes. Despite the limitations, functional screens offer a powerful means to identify genes not previously associated with a particular function of interest. When subsequently combined with traditional biochemical characterization methodologies and with analyses of relevant genomic, metagenomic, and metatranscriptomic data, this approach can yield invaluable information regarding the function, distribution, and expression of the ecologically relevant novel activities in the environment.

REFERENCES

Baba, T., Ara, T., Hasegawa, M., Takai, Y., Okumura, Y., Baba, M., et al. (2006). Construction of Escherichia coli K-12 in-frame, single-gene knockout mutants: The Keio collection. *Molecular Systems Biology*, *2*, 2006.0008.

Brady, S. F., Chao, C. J., Handelsman, J., & Clardy, J. (2001). Cloning and heterologous expression of a natural product biosynthetic gene cluster from eDNA. *Organic Letters*, *3*(13), 1981–1984.

Burke, C., Thomas, T., Egan, S., & Kjelleberg, S. (2007). The use of functional genomics for the identification of a gene cluster encoding for the biosynthesis of an antifungal tambjamine in the marine bacterium Pseudoalteromonas tunicata. *Environmental Microbiology*, *9*(3), 814–818.

Cottrell, M. T., Moore, J. A., & Kirchman, D. L. (1999). Chitinases from uncultured marine microorganisms. *Applied and Environmental Microbiology*, *65*(6), 2553–2557.

Courtois, S., Cappellano, C. M., Ball, M., Francou, F. X., Normand, P., Helynck, G., et al. (2003). Recombinant environmental libraries provide access to microbial diversity for drug discovery from natural products. *Applied and Environmental Microbiology, 69*(1), 49–55.

Courtois, S., Frostegard, A., Goransson, P., Depret, G., Jeannin, P., & Simonet, P. (2001). Quantification of bacterial subgroups in soil: Comparison of DNA extracted directly from soil or from cells previously released by density gradient centrifugation. *Environmental Microbiology, 3*(7), 431–439.

Datsenko, K. A., & Wanner, B. L. (2000). One-step inactivation of chromosomal genes in Escherichia coli K-12 using PCR products. *Proceedings of the National Academy of Sciences of the United States of America, 97*(12), 6640–6645.

DeLong, E. F. (2009). The microbial ocean from genomes to biomes. *Nature, 459*, 200–206.

DeLong, E. F., Preston, C. M., Mincer, T., Rich, V., Hallam, S. J., Frigaard, N.-U., et al. (2006). Community genomics among stratified microbial assemblages in the ocean's interior. *Science, 311*(5760), 496–503.

Dumont, M. G., Radajewski, S. M., Miguez, C. B., McDonald, I. R., & Murrell, J. C. (2006). Identification of a complete methane monooxygenase operon from soil by combining stable isotope probing and metagenomic analysis. *Environmental Microbiology, 8*(7), 1240–1250.

Gillespie, D. E., Brady, S. F., Bettermann, A. D., Cianciotto, N. P., Liles, M. R., Rondon, M. R., et al. (2002). Isolation of antibiotics turbomycin a and B from a metagenomic library of soil microbial DNA. *Applied and Environmental Microbiology, 68*(9), 4301–4306.

Gloux, K., Leclerc, M., Iliozer, H., L'Haridon, R., Manichanh, C., Corthier, G., et al. (2007). Development of high-throughput phenotyping of metagenomic clones from the human gut microbiome for modulation of eukaryotic cell growth. *Applied and Environmental Microbiology, 73*(11), 3734–3737.

Henne, A., Daniel, R., Schmitz, R. A., & Gottschalk, G. (1999). Construction of environmental DNA libraries in Escherichia coli and screening for the presence of genes conferring utilization of 4-hydroxybutyrate. *Applied and Environmental Microbiology, 65*(9), 3901–3907.

Kakirde, K. S., Wild, J., Godiska, R., Mead, D. A., Wiggins, A. G., Goodman, R. M., et al. (2011). Gram negative shuttle BAC vector for heterologous expression of metagenomic libraries. *Gene, 475*(2), 57–62.

Kim, U. J., Shizuya, H., de Jong, P. J., Birren, B., & Simon, M. I. (1992). Stable propagation of cosmid sized human DNA inserts in an F factor based vector. *Nucleic Acids Research, 20*(5), 1083–1085.

Knietsch, A., Waschkowitz, T., Bowien, S., Henne, A., & Daniel, R. (2003). Construction and screening of metagenomic libraries derived from enrichment cultures: Generation of a gene bank for genes conferring alcohol oxidoreductase activity on Escherichia coli. *Applied and Environmental Microbiology, 69*(3), 1408–1416.

Li, Y., Wexler, M., Richardson, D. J., Bond, P. L., & Johnston, A. W. (2005). Screening a wide host-range, waste-water metagenomic library in tryptophan auxotrophs of Rhizobium leguminosarum and of Escherichia coli reveals different classes of cloned trp genes. *Environmental Microbiology*, 7(12), 1927–1936.

Liles, M. R., Williamson, L. L., Rodbumrer, J., Torsvik, V., Goodman, R. M., & Handelsman, J. (2008). Recovery, purification, and cloning of high-molecular-weight DNA from soil microorganisms. *Applied and Environmental Microbiology, 74*(10), 3302–3305.

MacNeil, I. A., Tiong, C. L., Minor, C., August, P. R., Grossman, T. H., Loiacono, K. A., et al. (2001). Expression and isolation of antimicrobial small molecules from soil DNA libraries. *Journal of Molecular Microbiology and Biotechnology, 3*(2), 301–308.

Maresca, J. A., Braff, J. C., & DeLong, E. F. (2009). Characterization of canthaxanthin biosynthesis genes from an uncultured marine bacterium. *Environmental Microbiology Reports, 1*(6), 524–534.

Martinez, A., Bradley, A. S., Waldbauer, J. R., Summons, R. E., & DeLong, E. F. (2007). Proteorhodopsin photosystem gene expression enables photophosphorylation in a heterologous host. *Proceedings of the National Academy of Sciences of the United States of America, 104*(13), 5590–5595.

Martinez, A., Hopke, J., MacNeil, I. A., & Osburne, M. S. (2005). Accessing the genomes of uncultivated microbes for novel natural products. In L. Zhang & A. L. Demian (Eds.), *Natural products: Drug discovery and therapeutic medicine* (pp. 295–312). Totowa, NJ: Humana Press Inc.

Martinez, A., Kolvek, S. J., Hopke, J., Yip, C. L., & Osburne, M. S. (2005). Environmental DNA fragment conferring early and increased sporulation and antibiotic production in Streptomyces species. *Applied and Environmental Microbiology, 71*(3), 1638–1641.

Martinez, A., Kolvek, S. J., Yip, C. L., Hopke, J., Brown, K. A., MacNeil, I. A., et al. (2004). Genetically modified bacterial strains and novel bacterial artificial chromosome shuttle vectors for constructing environmental libraries and detecting heterologous natural products in multiple expression hosts. *Applied and Environmental Microbiology, 70*(4), 2452–2463.

Martinez, A., Tyson, G. W., & Delong, E. F. (2010). Widespread known and novel phosphonate utilization pathways in marine bacteria revealed by functional screening and metagenomic analyses. *Environmental Microbiology, 12*(1), 222–238.

McCarren, J., Becker, J. W., Repeta, D. J., Shi, Y., Young, C. R., Malmstrom, R. R., et al. (2010). Microbial community transcriptomes reveal microbes and metabolic pathways associated with dissolved organic matter turnover in the sea. *Proceedings of the National Academy of Sciences of the United States of America, 107*(38), 16420–16427.

McSorley, F. R., Wyatt, P. B., Martinez, A., Delong, E. F., Hove-Jensen, B., & Zechel, D. L. (2012). PhnY and PhnZ comprise a new oxidative pathway for enzymatic cleavage of a carbon-phosphorus bond. *Journal of the American Chemical Society, 134*, 8364–8367.

Miao, V., Coeffet-Legal, M. F., Brian, P., Brost, R., Penn, J., Whiting, A., et al. (2005). Daptomycin biosynthesis in Streptomyces roseosporus: Cloning and analysis of the gene cluster and revision of peptide stereochemistry. *Microbiology, 151*(Pt. 5), 1507–1523.

Okuta, A., Ohnishi, K., & Harayama, S. (1998). PCR isolation of catechol 2,3-dioxygenase gene fragments from environmental samples and their assembly into functional genes. *Gene, 212*(2), 221–228.

Ottesen, E. A., Marin, R. I., Preston, C. M., Young, C. R., Ryan, J. P., Scholin, C. A., et al. (2011). Metatranscriptomic analysis of autonomously collected and preserved marine bacterioplankton. *The ISME Journal, 5*(12), 1881–1895.

Owen, J. G., Robins, K. J., Parachin, N. S., & Ackerley, D. F. (2012). A functional screen for recovery of 4'-phosphopantetheinyl transferase and associated natural product biosynthesis genes from metagenome libraries. *Environmental Microbiology, 14*(5), 1198–1209.

Riesenfeld, C. S., Goodman, R. M., & Handelsman, J. (2004). Uncultured soil bacteria are a reservoir of new antibiotic resistance genes. *Environmental Microbiology, 6*(9), 981–989.

Rondon, M. R., August, P. R., Bettermann, A. D., Brady, S. F., Grossman, T. H., Liles, M. R., et al. (2000). Cloning the soil metagenome: A strategy for accessing the genetic and functional diversity of uncultured microorganisms. *Applied and Environmental Microbiology, 66*(6), 2541–2547.

Sambrook, J., Fritsch, E. F., & Maniatis, T. (1989). *Molecular cloning: A laboratory manual* (2nd ed.). Cold Spring Harbor, NY: Cold Spring Harbor Laboratory Press.

Sheng, Y., Mancino, V., & Birren, B. (1995). Transformation of Escherichia coli with large DNA molecules by electroporation. *Nucleic Acids Research, 23*(11), 1990–1996.

Shi, Y., Tyson, G. W., Eppley, J. M., & DeLong, E. F. (2011). Integrated metatranscriptomic and metagenomic analyses of stratified microbial assemblages in the open ocean. *The ISME Journal, 5*(6), 999–1013.

Shizuya, H., Birren, B., Kim, U. J., Mancino, V., Slepak, T., Tachiiri, Y., et al. (1992). Cloning and stable maintenance of 300-kilobase-pair fragments of human DNA in Escherichia coli using an F-factor-based vector. *Proceedings of the National Academy of Sciences of the United States of America*, *89*(18), 8794–8797.

Stewart, F. J., Ulloa, O., & Delong, E. F. (2011). Microbial metatranscriptomics in a permanent marine oxygen minimum zone. *Environmental Microbiology*, *14*(1), 23–40.

Suenaga, H., Ohnuki, T., & Miyazaki, K. (2007). Functional screening of a metagenomic library for genes involved in microbial degradation of aromatic compounds. *Environmental Microbiology*, *9*(9), 2289–2297.

Uchiyama, T., Abe, T., Ikemura, T., & Watanabe, K. (2005). Substrate-induced gene-expression screening of environmental metagenome libraries for isolation of catabolic genes. *Nature Biotechnology*, *23*(1), 88–93.

Urich, T., Lanzen, A., Qi, J., Huson, D. H., Schleper, C., & Schuster, S. C. (2008). Simultaneous assessment of soil microbial community structure and function through analysis of the meta-transcriptome. *PLoS One*, *3*(6), e2527.

Wenzel, S. C., & Muller, R. (2005). Recent developments towards the heterologous expression of complex bacterial natural product biosynthetic pathways. *Current Opinion in Biotechnology*, *16*(6), 594–606.

Wexler, M., Bond, P. L., Richardson, D. J., & Johnston, A. W. (2005). A wide host-range metagenomic library from a waste water treatment plant yields a novel alcohol/aldehyde dehydrogenase. *Environmental Microbiology*, 7(12), 1917–1926.

Wild, J., Hradecna, Z., & Szybalski, W. (2002). Conditionally amplifiable BACs: Switching from single-copy to high-copy vectors and genomic clones. *Genome Research*, *12*(9), 1434–1444.

Williamson, L. L., Borlee, B. R., Schloss, P. D., Guan, C., Allen, H. K., & Handelsman, J. (2005). Intracellular screen to identify metagenomic clones that induce or inhibit a quorum-sensing biosensor. *Applied and Environmental Microbiology*, *71*(10), 6335–6344.

Yooseph, S., Sutton, G., Rusch, D. B., Halpern, A. L., Williamson, S. J., Remington, K., et al. (2007). The Sorcerer II Global Ocean Sampling expedition: Expanding the universe of protein families. *PLoS Biology*, *5*(3), e16.

Zhang, K., Martiny, A. C., Reppas, N. B., Barry, K. W., Malek, J., Chisholm, S. W., et al. (2006). Sequencing genomes from single cells by polymerase cloning. *Nature Biotechnology*, *24*(6), 680–686.

CHAPTER EIGHT

Preparation of Metagenomic Libraries from Naturally Occurring Marine Viruses

Sergei A. Solonenko*, **Matthew B. Sullivan***,†,1

*Department of Ecology and Evolutionary Biology, University of Arizona, Tucson, Arizona, USA
†Department of Molecular and Cellular Biology, University of Arizona, Tucson, Arizona, USA
[1]Corresponding author: e-mail address: mbsulli@email.arizona.edu

Contents

1. On the Importance of Environmental Viruses and Viral Metagenomics 144
2. The DNA Viral Metagenomic Sample-to-Sequence Pipeline 147
3. The Library Preparation Process 149
   3.1 Fragmentation 149
   3.2 Insert size choices 154
   3.3 End repair and adaptor ligation: A key step in low-input DNA library construction 155
   3.4 Sizing and other options 157
   3.5 Amplification protocols for enrichment, quantity, and signal detection 157
   3.6 Library quantification 158
   3.7 Sequencing reaction and technologies 159
4. Conclusions 159
Acknowledgments 160
References 160

Abstract

Microbes are now well recognized as major drivers of the biogeochemical cycling that fuels the Earth, and their viruses (phages) are known to be abundant and important in microbial mortality, horizontal gene transfer, and modulating microbial metabolic output. Investigation of environmental phages has been frustrated by an inability to culture the vast majority of naturally occurring diversity coupled with the lack of robust, quantitative, culture-independent methods for studying this uncultured majority. However, for double-stranded DNA phages, a quantitative viral metagenomic sample-to-sequence workflow now exists. Here, we review these advances with special emphasis on the technical details of preparing DNA sequencing libraries for metagenomic sequencing from environmentally relevant low-input DNA samples. Library preparation steps broadly involve manipulating the sample DNA by fragmentation, end repair and adaptor ligation, size fractionation, and amplification. One critical area of future research and development is parallel advances for alternate nucleic acid types such

Methods in Enzymology, Volume 531
ISSN 0076-6879
http://dx.doi.org/10.1016/B978-0-12-407863-5.00008-3

as single-stranded DNA and RNA viruses that are also abundant in nature. Combinations of recent advances in fragmentation (e.g., acoustic shearing and tagmentation), ligation reactions (adaptor-to-template ratio reference table availability), size fractionation (non-gel-sizing), and amplification (linear amplification for deep sequencing and linker amplification protocols) enhance our ability to generate quantitatively representative metagenomic datasets from low-input DNA samples. Such datasets are already providing new insights into the role of viruses in marine systems and will continue to do so as new environments are explored and synergies and paradigms emerge from large-scale comparative analyses.

1. ON THE IMPORTANCE OF ENVIRONMENTAL VIRUSES AND VIRAL METAGENOMICS

Viruses infect all forms of life from the smallest microbes to the largest plants and animals. The outcomes of these infections can range from no discernible impact (some chronic or lysogenic infections) to death (lytic infections), but together viruses likely have profound impacts across all ecosystems on Earth as they number over $\sim 10^{31}$ planet-wide—approximately 10 times more viruses than prokaryotes (Wommack & Colwell, 2000). Particularly, well studied are marine bacterial viruses (phages) (Suttle, 2007), which kill ~20–40% of bacteria per day (Suttle, 2005; Weinbauer, 2004), move 10^{29} genes per day (Paul, 1999), and exist as a prophages within the genomes of about half the microbes at any given time (Paul, 2008). This implicates marine viruses in altering global biogeochemical cycling (the "viral shunt" keeps substrates from higher trophic levels, Fuhrman, 1999; Wilhelm & Suttle, 1999), structuring microbial communities (with most theory focused on "kill the winner," Thingstad, 2000; Weinbauer & Rassoulzadegan, 2004), and moving genes from one host to another, possibly driving microbial niche differentiation (e.g., Sullivan et al., 2006).

One phage–host system—cyanobacterial viruses (cyanophages) that infect abundant, marine *Prochlorococcus* and *Synechococcus* (Sullivan, Waterbury, & Chisholm, 2003)—has been relatively well studied due to its ecological importance and amenability to culturing. In fact, cyanophages harbor core "host" photosynthesis genes that are expressed during infection (Clokie, Shan, Bailey, Jia, & Krisch, 2006; Dammeyer, Bagby, Sullivan, Chisholm, & Frankenberg-Dinkel, 2008; Lindell, Jaffe, Johnson, Church, & Chisholm, 2005; Thompson et al., 2011), can recombine with host copies to alter the evolutionary trajectory of their host's photosystems (Ignacio-Espinoza & Sullivan, 2012; Lindell et al., 2004; Sullivan et al., 2006), and are modeled to improve phage fitness by boosting photosynthesis during infection (Bragg & Chisholm, 2008; Hellweger, 2009). This "photosynthetic phage" paradigm demonstrates that an infected cell is intimately controlled by its viral predator and calls for deeper investigation to document other coevolutionary paradigms in representative model systems from the diversity of viruses and hosts in nature.

Problematically, however, the bulk of microbial hosts and their viruses have not yet been cultivated. In fact, 85% of 1100 genome-sequenced phages derive from only 3 of the 45 known bacterial phyla (Holmfeldt et al., 2013), and these statistics are worse for archaeal and eukaryotic hosts. This is changing as new marine phage–host systems emerge (Holmfeldt et al., 2013; Zhao et al., 2013). However, the disparity between known potential hosts and those in culture led environmental virologists to culture-independent methods (e.g., metagenomics) to survey natural viral communities. Environmental viral metagenomes preceded those of their microbial hosts by 2 years with the development of the linker-amplified shotgun library method (Breitbart et al., 2002; Schoenfeld et al., 2008; Tyson et al., 2004; Venter et al., 2004) and even inspired Norman Anderson (Viral Defense Foundation) and N. Leigh Anderson (Plasma Proteome Institute) to propose sequencing, cataloging, and tracking viruses in human blood to treat human disease (Anderson, Gerin, & Anderson, 2003). Such efforts have not yet been realized, but in the environmental sciences, application of viral metagenomics has indeed led to a number of important discoveries (Breitbart, 2012).

Environmental viral metagenomic studies over the past decade have revealed how little we know—the bulk of viral metagenomes are (Cesar Ignacio-Espinoza, Solonenko & Sullivan, 2013) or completely new to science (reviewed in Hurwitz & Sullivan, 2013)—but new biology has emerged including evidence for recombination between ssDNA and ssRNA viruses (Rosario, Duffy, & Breitbart, 2012), delineation of compositional differences between freshwater and marine viral communities (Roux, Krupovic, Poulet, Debroas, & Enault, 2012), and the discovery of novel and diverse auxiliary metabolic genes found in viral metagenomes (Sharon et al., 2011). More recent work expands the above "photosynthetic virus" paradigm from photosynthesis genes in cyanophages to diverse host metabolic genes in a majority of phages (Hurwitz, Hallam, & Sullivan, in review-a; Hurwitz & Sullivan, in review-b). This, in combination with decades-old coliphage studies, suggests that the metabolic output of an uninfected cell drastically differs from that of a metabolically reprogrammed virus-infected cell. While few quantitative data are available, ocean virus–microbe interactions clearly impact the global carbon cycle, often dictating whether carbon in any individual microbial cell is sequestered to the deep ocean or released to the atmosphere through respiration of viral lysates (Fuhrman, 1999).

The challenge to developing a quantitative understanding of viral roles in ecosystems has been the lack of optimized tools to study viruses in a quantitative manner. For viral community sequence space, however, there is now an optimized, quantitative ocean viral metagenomic sample-to-sequence workflow (Fig. 8.1) that has been thoroughly evaluated using replicated metagenomic analyses to understand impacts of choices made in viral particle concentration and purification, nucleic acid amplification, and sequencing library preparation and platform choice (Duhaime, Deng, Poulos, & Sullivan, 2012; Duhaime & Sullivan, 2012; Hurwitz, Deng, Poulos, & Sullivan, 2013; John et al., 2011; Solonenko et al., 2013). This new quantitative data type has facilitated exciting discoveries, including uncovering the most abundant viruses in the oceans (Zhao et al., 2013) and advancing informatic solutions to organize unknown viral sequence space (*sensu* Yooseph 2007 protein clusters) (Hurwitz et al., 2013). This organization is tremendously powerful for viromic studies, as it helped reveal that the core Pacific Ocean virome (POV) is made of only 180 proteins, its pan-genome is relatively well sampled (~422k proteins), and the bulk of these proteins—even those core to all samples—are functionally

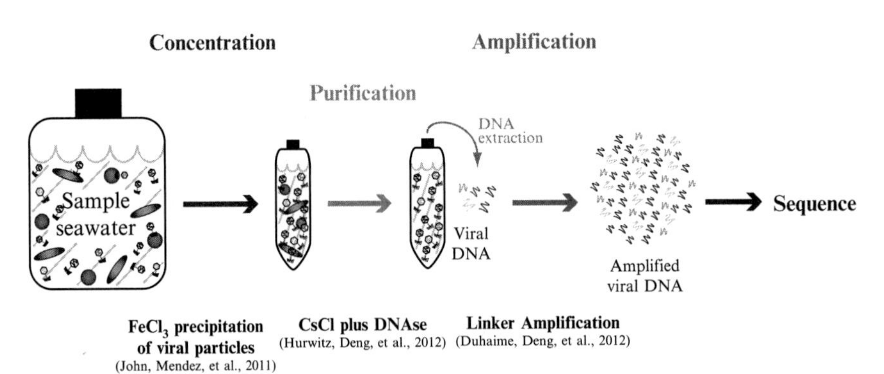

Figure 8.1 Overview of the environmental viral metagenomic sample-to-sequence workflow. The four basic steps in the creation of viral metagenomic data are illustrated, including references for suggested protocols for sequencing dsDNA viruses from marine samples. *Reprinted with permission from Duhaime and Sullivan (2012).* (For color version of this figure, the reader is referred to the online version of this chapter.)

unknown, but abundant, and presumably driving viral effects on ecosystem function. Further, the POV dataset has revealed that viral metabolic reprogramming extends far beyond cyanophage manipulation of photosynthesis, as it appears that Pacific Ocean viruses manipulate all of central microbial metabolism during infection, which profoundly alters our perception of viral roles in global carbon cycling. Specifically, Pacific Ocean virus gene content suggests that viral communities manipulate all starvation-related central metabolic pathways during infection in ways that could define viral niche space across hosts and the water column. Finally, protein clusters are powerful ecological inference tools. Specifically, they can (i) serve as a universal metric for comparing community viral diversity—something currently problematic due to reliance upon quantification derived from assembly output not yet tuned for metagenomic datasets—and (ii) offer a basis on which one can apply OTU-based ecological theory, independent of known function, using new and expanding community tools (e.g., QIIME).

Clearly, viral metagenomics will lead to myriad discoveries and, with careful optimization of the sample-to-sequence workflows, to help develop a more comprehensive understanding of the roles viruses play in the function of Earth's ecosystems.

2. THE DNA VIRAL METAGENOMIC SAMPLE-TO-SEQUENCE PIPELINE

Prior to constructing sequencing libraries, one needs to obtain a viral community concentrate and nucleic acids. This sample-to-sequence workflow (Fig. 8.1) is relatively well established now for double-stranded DNA (dsDNA) viruses and involves prefiltration to remove cellular material, concentration and purification of viral particles, and DNA extraction. While choice of prefilter is dependent upon environmental microbial concentrations and types, as well as the research questions being investigated, the remaining steps are now relatively well constrained (exceptions in the following paragraph) as follows. Viral particles are concentrated by $FeCl_3$ precipitation (John et al., 2011), with choice of purification (DNAse alone, DNAse + cesium chloride density gradient ultracentrifugation, or DNAse + sucrose density gradient ultracentrifugation) (Hurwitz et al., 2013), and the resulting limiting DNA (usually less than a few tens of nanograms) available for linker amplification techniques yielding metagenomes that are ±1.5-fold biased by %G+C content (e.g., Duhaime et al., 2012), which sharply contrasts up to ±10,000-fold biases of phi29-based whole-genome amplification methods (Yilmaz, Allgaier, & Hugenholtz, 2010; Zhang et al., 2006), although this value for phi29-based amplification may be an overestimate, since the measurements were done under the challenging conditions of single-cell amplification.

Based upon SYBR Gold particle counts, the current sample-to-sequence workflow captures the vast majority of detectable viral particles. However, there remain issues and opportunities for research and development, particularly for studies needing to document less common phage types. These include the following: (i) very large viruses are problematic because the prefilters are either too small (0.2 μm) or else coselect many microbes (e.g., 0.8 or 0.45 μm), (ii) lipid-containing viruses may require tweaks to concentration and purification protocols, (iii) the current methods are optimized for dsDNA viruses. On this latter point, it is possible that RNA viruses are missed because RNA is not commonly extracted from viral concentrates, and ssDNA viruses are missed because we cannot detect these well by staining (Holmfeldt, Odic, Sullivan, Middelboe, & Riemann, 2012) and density gradients often select against them (Thurber, Haynes, Breitbart, Wegley, & Rohwer, 2009). Notably, however, some studies have enriched for ssDNA viruses using one of the inherent systematic biases of the phi29 whole-genome amplification enzyme (Kim & Bae, 2011; Kim et al., 2008).

The nucleic acid extraction step is particularly challenging for microbial samples and thought to be one of the largest sources of bias in microbial metagenomes (Morgan, Darling, & Eisen, 2010). However, this step is unlikely to be problematic for environmental viruses because microbes have incredible diversity in cell membranes resulting in highly variable accessibility of their DNA. In contrast, viruses use a relatively simple method for protecting their DNA—protein capsids—which lends itself to nearly universally effective DNA extraction protocols. Protocols to date have largely focused on extracting DNA from viral concentrates, but there are also methods available for studying RNA and ssDNA metagenomes (Culley, Lang, & Suttle, 2006; Filiatrault et al., 2010; Roux et al., 2012). In fact, recent work suggests that RNA viruses may represent half of the viruses in the oceans (Steward et al., 2013), and methods exist to simultaneously separate ssDNA, dsDNA, and RNA from the same viral sample (Andrews-Pfannkoch, Fadrosh, Thorpe, & Williamson, 2010). Clearly, viruses with other nucleic acid types are promising targets for exploration in the environment. However, we focus here on DNA viruses since the sample-to-sequence pipeline is now well understood. Specifically, this chapter focuses on DNA library construction from natural viruses for metagenomic sequencing, including optimizations necessary for obtaining high-quality data from limiting DNA input amounts that are common to such samples.

3. THE LIBRARY PREPARATION PROCESS

Over the last decade, many variations in library preparation have emerged. However, the overall process is relatively constrained to manipulating the sample DNA by fragmentation, end repair and adaptor ligation, size fractionation, and amplification (Fig. 8.2).

3.1. Fragmentation

Obtaining the desired size of genomic DNA for sequencing library preparation requires fragmenting the DNA using a variety of options (summarized in Table 8.1 and detailed below). The overall goals of these methods are identical—to create fragments of the desired size while minimizing loss through efficient DNA recovery and narrowing the resulting fragment length distribution—but each method has strengths and weaknesses.

Traditional DNA fragmentation for genome sequencing projects was done using hydrodynamic shearing, nebulization, or enzymatic digestion, but these approaches have significant limitations for application to metagenomics. Nebulization mechanically breaks long DNA strands by forcing a nucleic acid solution through a narrow opening with varied air pressure. The advantages of nebulization are (i) random breakage with a relatively small fragment size range and (ii) no need for expensive equipment beyond pressurized air, while the disadvantages are (i) low throughput as only one sample can be fragmented per nebulizer and (ii) loss of up to 50% of total DNA which necessitates several micrograms of input DNA as starting material (Quail, 2010; Quail et al., 2008). Another mechanical shearing method, hydrodynamic shearing, uses the shear forces generated when repeatedly streaming a DNA sample through a narrow opening to generate large (>2 kb) and relatively tightly sized fragments, a great advantage for mate-pair protocols. As in nebulization, some material is lost, and a high sample minimum of several micrograms of DNA is required (see HydroShear Technical Brochure, 2009). Alternative to mechanical shearing, traditional protocols have used enzymatic digestion by endonucleases either with specific and known cleavage sites

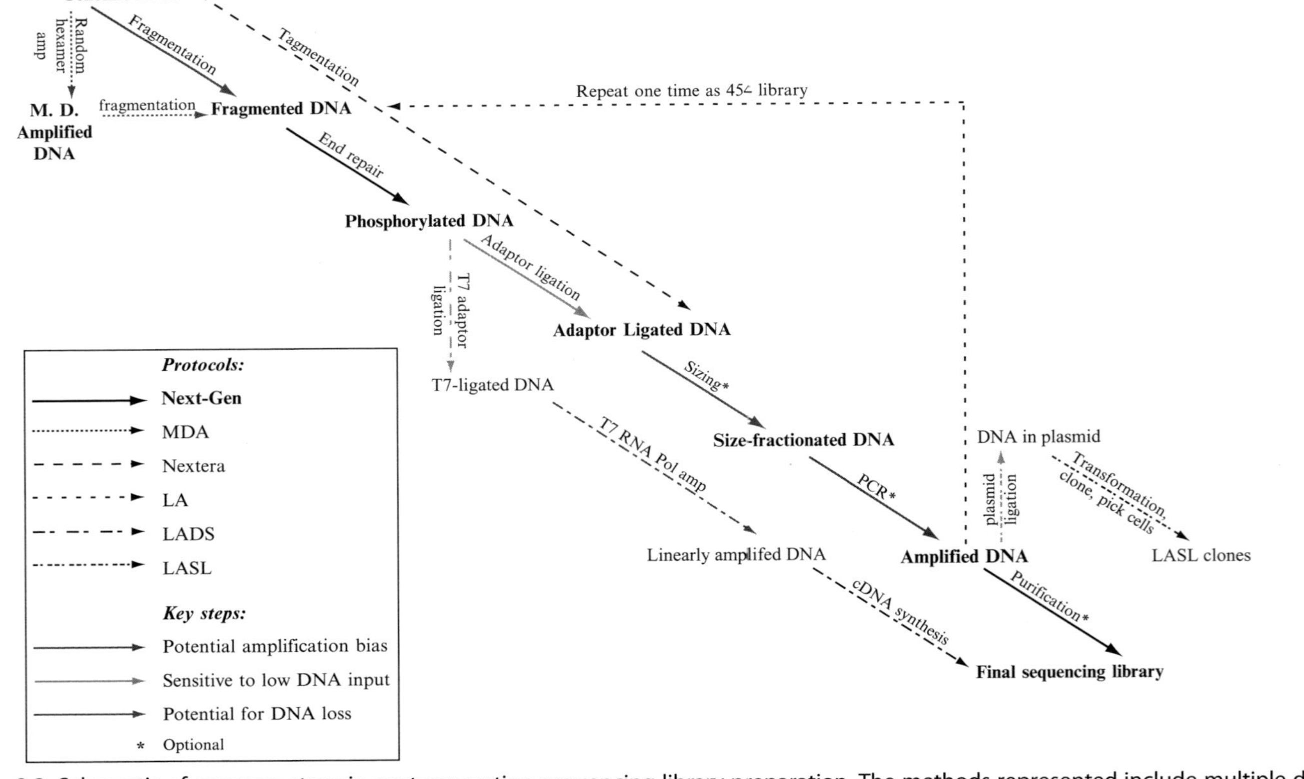

Figure 8.2 Schematic of common steps in next-generation sequencing library preparation. The methods represented include multiple displacement amplification (MDA, a phi29 whole-genome amplification method), an amplification of raw genomic DNA; linear amplification for deep sequencing (LADS), an alternative to PCR amplification for amplifying library DNA; linker-amplified shotgun library (LASL), a clone library protocol which shares many steps with sequencing library preparation; as well as several next-generation library construction methods, some of which may not require sizing, PCR, or purification (see Table 8.1). This figure highlights several steps in the procedure that are associated with issues that may impact the success or quality of the constructed library, in particular, amplification bias, ligation conditions, and choice of fragmentation method. (For color version of this figure, the reader is referred to the online version of this chapter.)

Table 8.1 A summary of several common library prep protocols available for Illumina, 454, and Ion Torrent sequencing systems

	Input DNA	Fragmentation method	DNA ends treatment	Ligation method	Adaptor type	Sizing method	Amplification	Sequencing	References
Illumina TruSeq	1 μg	Acoustic shear	End Repair & A-Tailing	T/A overhang	Y-adaptors	Gel extraction	Adaptor specific	Illumina	TruSeq Sample Prep Guide
454 GS FLX+	1 μg	Nebulization	End Repair	Blunt ended	Dual dsDNA or Y-adaptors	Bead	None	454	GS FLX+ Library Prep Manual
Ion Torrent	100 ng or 1 μg	Acoustic or enzymatic shear	End Repair	Blunt ended	Dual dsDNA	Gel extraction	Adaptor specific	Ion PGM	Ion Torrent Library Prep Manual
Multiple displacement amplification	1–100 ng	Endonuclease	LC Dependent	LC dependent	LC dependent	LC dependent	Random Hexamer	LC dependent	Yilmaz et al. (2010)
Linker-amplified library construction	>10 pg	Acoustic shear	End Repair	Blunt ended	Dual dsDNA	Bead or gel extraction	Adaptor specific	454	Duhaime et al. (2012)
Linker-amplified for deep sequencing	3–40 ng	Nebulization	End Repair & A-Tailing	T/A overhang	Identical dsDNA	Gel extraction	Transcription	Illumina	Hoeijmakers et al. (2011)

Continued

Table 8.1 A summary of several common library prep protocols available for Illumina, 454, and Ion Torrent sequencing systems—cont'd

	Input DNA	Fragmentation method	DNA ends treatment	Ligation method	Adaptor type	Sizing method	Amplification	Sequencing	References
Linker-amplified shotgun library	1 μg	HydroShear	End Repair	Blunt ended	Identical dsDNA	None	Adaptor specific	Sanger	Breitbart et al. (2002)
Nextera XT	1 ng	Simultaneous fragmentation and tagging			Dual dsDNA	Bead	Adaptor specific (limited cycle)	Illumina	Nextera XT Sample Prep Guide

DNA amounts refer to the recommended starting DNA necessary for the protocol (unsheared viral dsDNA). Four fragmentation options are represented across these protocols, but most are intercompatible except for the transposase, where fragmentation and adaptor attachment happen in one step. Adaptor types are Y-adaptor, which includes two separate adaptors that share a region of homology and form a Y structure during ligation, dual adaptors, two different adaptors ligated on either end of a genomic DNA fragment, and identical adaptors, where the same adaptor is ligated on both ends of the genomic DNA fragment. Some methods of attaching dual adaptors generate many adaptor combinations, requiring a purification/enrichment step to obtain properly ligated library fragments (ones with different adaptors on each end). MDA is done before fragmentation and is thus compatible with many different types of downstream sequencing preparation, with the affected steps marked as library construction (LC) dependent.

for controlled genomic DNA fragmentation or with more permissive cleavage sites for nonspecific shearing of DNA. Advantages of enzymatic digestion include (i) no need for equipment investment, (ii) random digestion (for nonspecific enzymes), (iii) marginally tunable sizing by adjusting the restriction reaction conditions, while the disadvantages are (i) nonrandom fragmentation (for specific cut-site restriction endonucleases), (ii) poor control for generating large fragments (e.g., NEB Fragmentase kit), and (iii) lower reproducibility (Adey et al., 2010; Linnarsson, 2010).

In contrast, newer library preparation protocols fragment DNA using acoustic shearing or tagmentation (Nextera kit, Illumina TruSeq kit, Duhaime et al., 2012). To generate fragmented DNA, acoustic shearing simply uses cavitation to randomly break up the DNA (Quail, 2010), while tagmentation combines fragmentation with adaptor attachment in one transposition reaction (Adey et al., 2010). These two methods pervade modern library protocols due to several desirable features. First, both can produce fragments with narrow size distributions that are optimal for short-read sequencing (e.g., 150–300 bp, Henn et al., 2010), which is not efficiently done with nebulization or enzymatic digestion (Quail et al., 2008). Notably, downstream sizing may not be needed for acoustic shearing but is required for tagmentation to remove small fragments where size distributions extend as low as 40 bp (Nextera XT manual; Adey et al., 2010). Second, acoustic shearing and tagmentation are high-efficiency methods: acoustic shearing because it incurs virtually no sample loss because it is performed in closed tubes, and tagmentation because it reduces sample manipulation. Third, acoustic shearing, in particular, has reduced chance of contamination because the entire process is done in a closed tube. Finally, both methods can be scaled for high-throughput work. For example, acoustic shearing can already be done in 96-well plate format and has recently been utilized in microfluidic applications (Tseng, Lomonosov, Furlong, & Merten, 2012), with development heading toward automated microfluidic μl-scale sequencing library preparation (Vyawahare, Griffiths, & Merten, 2010). The disadvantages of these methods are that acoustic shearing requires expensive equipment or fee-for-service access, while tagmentation leads to slight %G+C biases in genomes (Marine et al., 2011) and metagenomes (Solonenko et al., 2013), presumably due to insertion biases inherent to the transposase (Adey et al., 2010).

3.2. Insert size choices

Many library preparation options should be tuned to accommodate the type of sequence data best suited to the research question being addressed. For example, metagenomic sequencing has predominantly relied on data derived from a single sequencing read per DNA fragment. However, two sequencing reads per DNA fragment (paired-end sequencing) can be obtained by attachment of different sequencing adaptors to DNA fragment ends to allow directional sequencing off each end. This strategy can be used to provide longer "reads" for small-insert libraries where the two sequencing reads overlap each other. For large-insert libraries, such paired-end data can drastically increase metagenomic assembly contig sizes (e.g., Rodrigue et al., 2010). Several assembly algorithms use paired-end information for genome scaffolding, with Allpaths-lg (Gnerre et al., 2011), the most popular, and options in Velvet (Zerbino, McEwen, Margulies, & Birney, 2009), Abyss (Simpson et al., 2009), and SOAP-denovo (Luo et al., 2012) also available. Notably, these algorithms were designed for single genome assembly and have problems handling large differences in coverage (>100) present in metagenomic data, in which high coverage contigs may be mistaken for repeat regions or lead to misassembly due to heterogeneity, while low-coverage contigs may become overly fragmented due to low read overlap (Peng, Leung, Yiu, & Chin, 2012). Two recently published methods, IDBA-UD (Peng et al., 2012) and MetaVelvet (Namiki, Hachiya, Tanaka, & Sakakibara, 2012), address the above issues and are capable of analyzing metagenomic paired-end data, but either method has yet to be used on viral metagenomic data.

Currently, paired-end sequencing libraries are limited to small (<800 bp) insert sizes due to limitations in bridge amplification clustering (Illumina Paired End Sample Prep Guide, Rev. E., February 2011) and emPCR (GS FLX Ti General Preparation Method Manual, April 2009). One way to overcome this hard limit is by mate pairing (similar to long-range paired-end or paired-end tag libraries), whereby longer DNA fragments are circularized by ligating the two ends together and then fragmented down to <800 bp size compatible with paired-end library construction and sequencing. Current mate-pair library size limits are 40 kbp, with mate-pair creation efficiency decreasing and size distribution variation increasing as insert sizes increase (Asan et al., 2012). As well, a major reason why mate-pair libraries are not standard in environmental metagenomic surveys is the requirement for prohibitively large amounts of starting DNA, for example, 50 μg for a 35-kb mate-pair library (Asan et al., 2012). Successful use of mate-pair data yields a new level of organization to metagenomic data (Iverson et al., 2012). While such quantities are currently impossible for viral metagenomic studies, there is potential for creative amplification-based solutions which could augment environmental DNA to the point where environmental virologists may also benefit from mate-pair data.

3.3. End repair and adaptor ligation: A key step in low-input DNA library construction

Fragmentation commonly results in ssDNA ends which require repair to prepare for dsDNA adaptor ligation (Table 8.1). In fact, end repair is part of every protocol except tagmentation, where the transposition reaction leaves no damage to DNA ends and includes addition of adaptors. Some protocols, such as Illumina and LADS, utilize A-tailing to create an overhang to which T-tailed adapter sequences are ligated so as to leverage improved efficiency over blunt-end ligation and prevent concatenation of template DNA (Bratbak, Wilson, & Heldal, 1996). However, because A-tailing adds another step to the procedure in which DNA may be lost (i.e., DNA binding to tubes, Ellison, English, Burns, & Keer, 2006), many protocols utilize blunt-end ligation for adding adaptor sequence to the fragments (Table 8.1).

The indispensable step in library preparation is the addition of adaptors to the genomic DNA fragments, which eventually act as a primer site during the sequencing reaction. Most protocols achieve this using ligation, with the exception of tagmentation, where the transposition reaction attaches the adaptors (Table 8.1). Adaptor sequences vary by sequencing technology and application (overview in Fig. 8.3). Adaptors can contain just the sequencing primer site, commonly also with a barcode incorporated to identify pooled libraries sequenced together on one run. Custom barcodes are easy to develop for the 454 and Ion Torrent systems (examples available at http://www.eebweb.arizona.edu/faculty/mbsulli/protocols/TMPL_LAs.pdf), but more complicated for Illumina sequencing where barcoding of the first several sequenced bases disrupts the identification of clustered reads on the sequencing plate (Rohland & Reich, 2012). Particular library methods can have variations in the attached sequences, including the T7 promoter for transcription in the LADS protocol, the mosaic end sequence that is necessary for transposition in the Nextera tagmentation protocol, and sequences specific for amplifying library fragments (e.g., P5 and P7 sequences in the TruSeq Illumina protocol and LADS, or the A-linker in Linker Amplification). A Y-adaptor instead of dual dsDNA adaptors has also

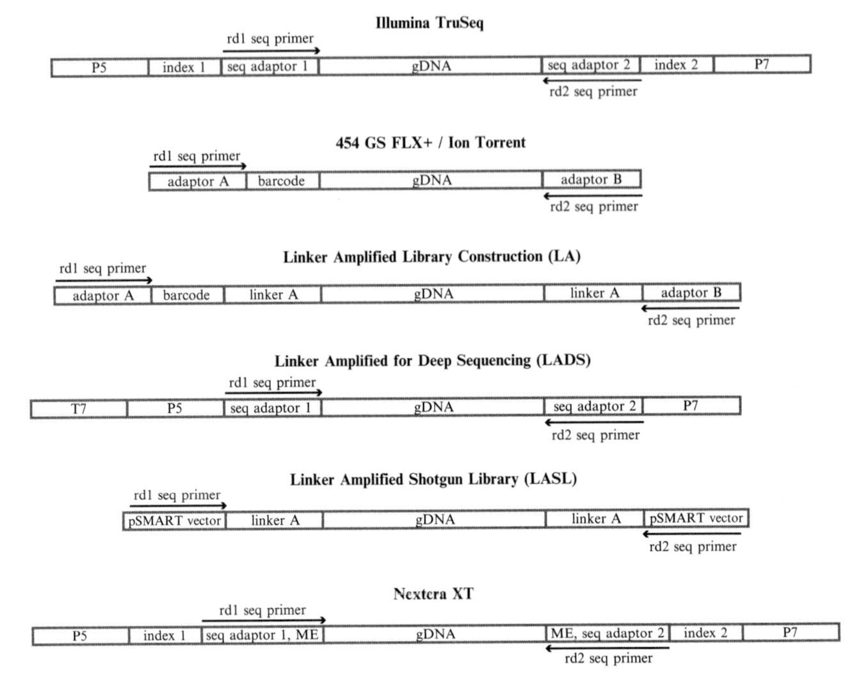

Figure 8.3 Overview schematic of adaptor sequences involved in commonly used library preparation technologies. This figure represents an overview of finished library fragments generated using each library preparation method discussed in this review. Particular focus is placed on (1) the presence of index or barcode regions that allow a library to be pooled with other libraries for efficient sequencing, (2) the location of sequencing primers illustrated as arrows here, indicating the parts of each fragment that will appear in the final sequencing output (and potentially require trimming), and (3) auxiliary sequences such as T7, P5, and P7 that are important for library amplification. ME stands for mosaic end, a subsection of the Nextera sequencing adaptor that allows transposition to occur.

been used to prevent the loss of library DNA due to the attachment of incorrect combinations of adaptors (see 454 General vs. 454 Rapid prep kits, also Zheng et al., 2010).

The success of adaptor ligation is critical to the generation of a robust sequencing library, particularly for low-input DNA samples where optimizing the adaptor-to-template (calculated as free DNA ends) ratio is critical (Solonenko et al., 2013). For particularly low-input DNA libraries, adaptors can also be used to amplify DNA prior to sequencing preparations as shown for LADS (Hoeijmakers, Bartfai, Francoijs, & Stunnenberg, 2011) and LA (Duhaime et al., 2012) in Fig. 8.3. Notably, LADS and LA amplifications are much preferable to amplifications using random hexamer primers and phi29, which result in nonquantitative and nonreproducible library composition (Yilmaz et al., 2010) where quantities can vary as much as 10,000-fold from starting concentrations (Zhang et al., 2006).

3.4. Sizing and other options

From gel sizing to beads and chip-based systems, there are many options available for controlling the size of library fragments (summarized in Table 8.1). Gel sizing has traditionally been used for DNA sizing, but it is problematic for low-input DNA samples as there may be too little DNA to visualize it on a gel and the protocols suffer from inefficient yields (~50%) and intersample contamination (Duhaime et al., 2012). Sizing is, however, critical for targeting a small range of fragment sizes so as to improve final library quality (Linnarsson, 2010; Quail et al., 2008). An overabundance of small DNA fragments may alter the stoichiometry of adaptor ligation reactions or overpopulate the library during PCR amplification steps. Tightly sized input DNA is also particularly valuable for downstream analyses (e.g., scaffolding for genome assembly) that rely upon paired-end or mate-pair information (Simpson et al., 2009). Acoustic shearing can even produce a fragment distribution that is narrow enough that sizing can be skipped (Solonenko et al., 2013). The Pippin Prep is a more accurate method of gel sizing, and while it requires more investment in equipment, this method is recommended for low-DNA viral metagenomic protocols (Duhaime et al., 2012). The LabChip XT system is another automated sizing method with greater accuracy compared to gel sizing, but currently this has a higher price point. By far, the most cost-effective and high-throughput sizing method uses carboxylic acid coated beads (SPRI, Ampure XP, or My One) to capture different sizes of DNA (Borgstrom, Lundin, & Lundeberg, 2011; Rohland & Reich, 2012). Lastly, columns commonly used to remove extra nucleotides, primers, or adaptors and adaptor dimers may also function as a sizing step, as small DNA fragments are removed (>100 bp for QiaQuick PCR Cleanup Kit).

3.5. Amplification protocols for enrichment, quantity, and signal detection

Once DNA has been processed as above, there remains only the need to amplify the resultant DNA molecules before sequencing. Amplification serves several purposes in metagenomic sequencing library protocols. First, limited amplification cycles (10 or fewer for Illumina TruSeq prep) enrich the DNA pool for molecules containing correctly ligated adaptors. Second, for low-input DNA samples, amplification can be used to augment sample DNA so as to have enough material to survive library preparation loss steps. Amplification is also used to improve signal detection when a pool of synchronized sequenced reads is required (e.g., 454, Illumina, Ion Torrent). Commonly, this is a separate, final step in library preparation before sequencing—an amplification to create the ~1000 copies that are read by the sequencer. Notably, these PCRs are done with each template isolated in some manner: 454 and Ion Torrent utilize emPCR on a primer-covered bead, while Illumina uses bridge amplification to generate localized "clusters" on a primer-covered sequencing plate (Metzker, 2010). Third, the amplification step is of critical importance and associated choices should not be made lightly. This is because whole-genome amplification methods lead to nonquantitative metagenomes (Yilmaz et al., 2010), while PCR-based amplification is prone to several biases including stochasticity of amplification, heteroduplex formation, chimeric amplicons, and %G+C bias due to the polymerase, high-temperature amplification conditions, and differential priming (reviewed in Duhaime & Sullivan, 2012). However, for PCR-based amplification methods, conditions can be optimized to yield less biased products (Adey et al., 2010), including adjustment of cycling conditions and addition of stabilizing compounds (Schwientek, Szczepanowski, Ruckert, Stoye, & Puhler, 2011), linear amplification (LADS, Hoeijmakers et al., 2011) to lower cross-amplicon competition for primers (Shaw, 2002), and leaving out the amplification step entirely when DNA amounts are not limiting (>1 μg (Kozarewa et al., 2009) for Illumina, standard 454 protocol). Because emPCR and bridge amplification physically isolate amplicons from each other, the signal amplification reactions are a minimal source of bias, with artificial duplicates being the largest issue and observed only for emPCR-based technologies (454 and Ion Torrent; Gomez-Alvarez, Teal, & Schmidt, 2009). Notably, single-molecule sequencing developments may improve these technologies further (Wanunu, 2012).

3.6. Library quantification

The final step of any library preparation procedure is quantification of the library before loading the library for sequencing by emPCR for 454 or Ion Torrent and bridge amplification for Illumina. Correct quantification prevents the library DNA from being overloaded, which can lead to mixed signals, or underloaded, which underutilizes sequencing capacity. Library concentration information also gives the user the opportunity to strategically pool several libraries when sequencing depth requires less than one run or lane. Several methods are available for this procedure including qPCR, and titration-free qPCR (Zheng et al., 2010), but typically this step is done by the sequencing center and is not a choice for the user to make.

3.7. Sequencing reaction and technologies

Ultimately, each sequencing technology differs not only in preparation (reviewed here) but also in type of sequencing data generated (reviewed in Glenn, 2011; Kircher & Kelso, 2010; Metzker, 2010). Briefly, two important features are the cost efficiency of sequencing data and the read length. Illumina sequencing is the current leader in cost with tens of millions of reads per run, with high potential to overwhelm downstream bioinformatic processing pipelines (Chiang, Clapham, Qi, Sale, & Coates, 2011). 454 GS FLX produces the longest reads available in a next-generation system, an important characteristic for assembly, as well as routine metagenomic analysis (Wommack, Bhavsar, & Ravel, 2008). Beyond these predominantly genome-centered reviews, our own previous work used replicated metagenomics to evaluate the impact of sequencing platforms on the resulting viral metagenomes and showed that the choice of sequencing technology may be less of an influence on the content of metagenomic data than choices made during library preparation (Solonenko et al., 2013).

4. CONCLUSIONS

As new library preparation methods are developed, viral metagenomics continues to become less expensive and more reproducible, as well as more accessible to an expanding diversity of viral types. While the viral metagenomic sample-to-sequence workflow is relatively well established now for dsDNA viruses, there is a need for parallel research and development toward quantitative metagenomic processing steps for accessing ssDNA and RNA viruses in the environment. Mindful of this, it is clear that modern sequencing capacity now empowers metagenomics to adopt experimental designs involving technical replicates (Knight et al., 2012) and that such designs have proven critical for understanding impacts of library preparation methods and sequencing platforms on the resulting viral metagenomes (Solonenko et al., 2013). Implied in these goals is the use of efficient, replicable methods for generating viral metagenomes, an important part of metagenomic experimental design. Making informed choices at key steps in metagenomics library preparation, such as fragmentation, ligation, and amplification, may reduce the chances of unexpected failure of library preparation or bias in metagenomic sequencing data. As such, refined metagenomic datasets coupled with myriad emerging viral ecology tools that allow access to single viral genomes, link wild viruses to their hosts, and evaluate viral community morphology (Allen et al., 2011, 2013; Brum, Schenck, & Sullivan, 2013; Deng et al., 2013; Tadmor, Ottesen, Leadbetter, & Phillips, 2011) are transforming the landscape of questions that researchers can ask. Together, these advances beckon a new era for the field where we can finally develop a mechanistic understanding of the principles governing variations in natural virus and microbial communities, one virus and one host at a time.

ACKNOWLEDGMENTS

We thank Christine Schirmer for assistance with figures and tables and technical discussions as well as Jennifer Brum and Natalie Solonenko for comments on the chapter. Funding was provided by the Gordon and Betty Moore Foundation to M. B. S. and an NSF IGERT Comparative Genomics Training Grant to S. A. S.

REFERENCES

Adey, A., Morrison, H. G., Asan, Xun, X., Kitzman, J. O., & Turner, E. H. (2010). Rapid, low-input, low-bias construction of shotgun fragment libraries by high-density in vitro transposition. *Genome Biology*, *11*(12), R119.

Allen, L. Z., Ishoey, T., Novotny, M. A., McLean, J. S., Lasken, R. S., & Williamson, S. J. (2011). Single virus genomics: A new tool for virus discovery. *PLoS One*, *6*(3), e17722.

Allers, E., Moraru, C., Duhaime, M., Beneze, E., Solonenko, N., Barerro-Canosa, J., et al. (2013). Single-cell and population level viral infection dynamics revealed by phageFISH, a method to visualize intracellular and free viruses. *Environmental Microbiology*, *15*(8), 2306–2318.

Anderson, N. G., Gerin, J. L., & Anderson, N. L. (2003). Global screening for human viral pathogens. *Emerging Infectious Diseases*, *9*(7), 768–774.

Andrews-Pfannkoch, C., Fadrosh, D. W., Thorpe, J., & Williamson, S. J. (2010). Hydroxyapatite-mediated separation of double-stranded DNA, single-stranded DNA, and RNA genomes from natural viral assemblages. *Applied and Environmental Microbiology*, *76*(15), 5039–5045.

Asan, Geng, C., Chen, Y., Wu, K., Cai, Q., Wang, Y., et al. (2012). Paired-end sequencing of long-range DNA fragments for de novo assembly of large, complex mammalian genomes by direct intra-molecule ligation. *PLoS One*, 7(9), e46211.

Borgstrom, E., Lundin, S., & Lundeberg, J. (2011). Large scale library generation for high throughput sequencing. *PLoS One*, *6*(4), e19119.

Bragg, J. G., & Chisholm, S. W. (2008). Modelling the fitness consequences of a cyanophage-encoded photosynthesis gene. *PLoS One*, *3*, e3550.

Bratbak, G., Wilson, W., & Heldal, M. (1996). Viral control of *Emiliania huxleyi* blooms? *Journal of Marine Systems*, *9*(1), 75–81.

Breitbart, M. (2012). Marine viruses: Truth or dare. *Annual Review of Marine Science*, *4*, 425–448.

Breitbart, M., Salamon, P., Andresen, B., Mahaffy, J. M., Segall, A. M., Mead, D., et al. (2002). Genomic analysis of uncultured marine viral communities. *Proceedings of the National Academy of Sciences of the United States of America, 99*(22), 14250–14255.

Brum, J. R., Schenck, R. O., & Sullivan, M. B. (2013). Global morphological analysis of marine viruses shows minimal regional variation and dominance of non-tailed viruses. *The ISME Journal* advance online publication, 2 May 2013. http://dx.doi.org/10.1038/ismej.2013.67.

Cesar Ignacio-Espinoza, J., Solonenko, S. A., & Sullivan, M. B. (2013). The global virome: Not as big as we thought? *Current Opinion in Virology*.

Chiang, G. T., Clapham, P., Qi, G., Sale, K., & Coates, G. (2011). Implementing a genomic data management system using iRODS in the Wellcome Trust Sanger Institute. *BMC Bioinformatics, 12*, 361.

Clokie, M. R. J., Shan, J., Bailey, S., Jia, Y., & Krisch, H. M. (2006). Transcription of a 'photosynthetic' T4-type phage during infection of a marine cyanobacterium. *Environmental Microbiology, 8*, 827–835.

Culley, A. I., Lang, A. S., & Suttle, C. A. (2006). Metagenomic analysis of coastal RNA virus communities. *Science, 312*(5781), 1795–1798.

Dammeyer, T., Bagby, S. C., Sullivan, M. B., Chisholm, S. W., & Frankenberg-Dinkel, N. (2008). Efficient phage-mediated pigment biosynthesis in oceanic cyanobacteria. *Current Biology, 18*(6), 442–448.

Deng, L., Gregory, A., Yilmaz, S., Poulos, B. T., Hugenholtz, P., & Sullivan, M. B. (2012). Contrasting life strategies of viruses that infect photo- and heterotrophic bacteria, as revealed by viral tagging. *mBio, 3*(6), e00373-12. http://dx.doi.org/10.1128/mBio.00373-12.

Duhaime, M., Deng, L., Poulos, B., & Sullivan, M. B. (2012). Towards quantitative metagenomics of wild viruses and other ultra-low concentration DNA samples: A rigorous assessment and optimization of the linker amplification method. *Environmental Microbiology, 14*, 2526–2537.

Duhaime, M., & Sullivan, M. B. (2012). Ocean viruses: Rigorously evaluating the metagenomic sample-to-sequence pipeline. *Virology, 434*, 181–186.

Ellison, S. L., English, C. A., Burns, M. J., & Keer, J. T. (2006). Routes to improving the reliability of low level DNA analysis using real-time PCR. *BMC Biotechnology, 6*, 33.

Filiatrault, M. J., Stodghill, P. V., Bronstein, P. A., Moll, S., Lindeberg, M., Grills, G., et al. (2010). Transcriptome analysis of Pseudomonas syringae identifies new genes, noncoding RNAs, and antisense activity. *Journal of Bacteriology, 192*(9), 2359–2372.

Fuhrman, J. A. (1999). Marine viruses and their biogeochemical and ecological effects. *Nature, 399*, 541–548.

Glenn, T. C. (2011). Field guide to next-generation DNA sequencers. *Molecular Ecology Resources, 11*(5), 759–769.

Gnerre, S., Maccallum, I., Przybylski, D., Ribeiro, F. J., Burton, J. N., Walker, B. J., et al. (2011). High-quality draft assemblies of mammalian genomes from massively parallel sequence data. *Proceedings of the National Academy of Sciences of the United States of America, 108*(4), 1513–1518.

Gomez-Alvarez, V., Teal, T. K., & Schmidt, T. M. (2009). Systematic artifacts in metagenomes from complex microbial communities. *The ISME Journal, 3*(11), 1314–1317.

Hellweger, F. L. (2009). Carrying photosynthesis genes increases ecological fitness of cyanophage *in silico*. *Environmental Microbiology, 11*, 1386–1394.

Henn, M. R., Sullivan, M. B., Stange-Thomann, N., Osburne, M. S., Berlin, A. M., Kelly, L., et al. (2010). Analysis of high-throughput sequencing and annotation strategies for phage genomes. *PLoS One, 5*(2), e9083.

Hoeijmakers, W. A., Bartfai, R., Francoijs, K. J., & Stunnenberg, H. G. (2011). Linear amplification for deep sequencing. *Nature Protocols*, *6*(7), 1026–1036.

Holmfeldt, K., Odic, D., Sullivan, M. B., Middelboe, M., & Riemann, L. (2012). Cultivated single-stranded DNA phages that infect marine Bacteroidetes prove difficult to detect with DNA-binding stains. *Applied and Environmental Microbiology*, *78*(3), 892–894.

Holmfeldt, Karin, et al. (2013). Twelve previously unknown phage genera are ubiquitous in global oceans. *Proceedings of the National Academy of Sciences*, *110*(31), 12798–12803.

Hurwitz, B. H., Deng, L., Poulos, B., & Sullivan, M. B. (2013). Evaluation of methods to concentrate and purify wild ocean virus communities through comparative, replicated metagenomics. *Environmental Microbiology*, *15*(5), 1428–1440. http://dx.doi.org/10.1111/j.1462-2920.2012.02836.x.

Hurwitz, B. H., & Sullivan, M. B. (2013). The Pacific Ocean Virome (POV): A marine viral metagenomic dataset and associated protein clusters for quantitative viral ecology. *PLoS One*, *8*, e57355.

Ignacio-Espinoza, J. C., & Sullivan, M. B. (2012). Phylogenomics of T4 cyanophages: Lateral gene transfer in the "core" and origins of host genes. *Environmental Microbiology*, *14*, 2113–2126.

Iverson, V., Morris, R. M., Frazar, C. D., Berthiaume, C. T., Morales, R. L., & Armbrust, E. V. (2012). Untangling genomes from metagenomes: Revealing an uncultured class of marine Euryarchaeota. *Science*, *335*(6068), 587–590.

John, S. G., Mendez, C. B., Deng, L., Poulos, B., Kauffman, A. K. M., Kern, S., et al. (2011). A simple and efficient method for concentration of ocean viruses by chemical flocculation. *Environmental Microbiology Reports*, *3*(2), 195–202.

Kim, K. H., & Bae, J. W. (2011). Amplification methods bias metagenomic libraries of uncultured single-stranded and double-stranded DNA viruses. *Applied and Environmental Microbiology*, 77(21), 7663–7668.

Kim, K. H., Chang, H. W., Nam, Y. D., Roh, S. W., Kim, M. S., Sung, Y., et al. (2008). Amplification of uncultured single-stranded DNA viruses from rice paddy soil. *Applied and Environmental Microbiology*, *74*(19), 5975–5985.

Kircher, M., & Kelso, J. (2010). High-throughput DNA sequencing—Concepts and limitations. *BioEssays: News and Reviews in Molecular, Cellular and Developmental Biology*, *32*(6), 524–536.

Knight, R., Jansson, J., Field, D., Fierer, N., Desai, N., Fuhrman, J. A., et al. (2012). Unlocking the potential of metagenomics through replicated experimental design. *Nature Biotechnology*, *30*(6), 513–520.

Kozarewa, I., Ning, Z., Quail, M. A., Sanders, M. J., Berriman, M., & Turner, D. J. (2009). Amplification-free Illumina sequencing-library preparation facilitates improved mapping and assembly of (G+C)-biased genomes. *Nature Methods*, *6*(4), 291–295.

Lindell, D., Jaffe, J. D., Johnson, Z. I., Church, G. M., & Chisholm, S. W. (2005). Photosynthesis genes in marine viruses yield proteins during host infection. *Nature*, *438*(7064), 86–89.

Lindell, D., Sullivan, M. B., Johnson, Z. I., Tolonen, A. C., Rohwer, F., & Chisholm, S. W. (2004). Transfer of photosynthesis genes to and from *Prochlorococcus* viruses. *Proceedings of the National Academy of Sciences of the United States of America*, *101*(30), 11013–11018.

Linnarsson, S. (2010). Recent advances in DNA sequencing methods—General principles of sample preparation. *Experimental Cell Research*, *316*(8), 1339–1343.

Luo, R., Liu, B., Xie, Y., Li, Z., Huang, W., Yuan, J., et al. (2012). SOAPdenovo2: An empirically improved memory-efficient short-read de novo assembler. *GigaScience*, *1*(1), 18.

Marine, R., Polson, S. W., Ravel, J., Hatfull, G., Russell, D., Sullivan, M., et al. (2011). Evaluation of a transposase protocol for rapid generation of shotgun high-throughput sequencing libraries from nanogram quantities of DNA. *Applied and Environmental Microbiology*, 77(22), 8071–8079.

Metzker, M. L. (2010). Sequencing technologies—The next generation. *Nature Reviews. Genetics*, *11*(1), 31–46.

Morgan, J. L., Darling, A. E., & Eisen, J. A. (2010). Metagenomic sequencing of an in vitro-simulated microbial community. *PLoS One*, *5*(4), e10209.

Namiki, T., Hachiya, T., Tanaka, H., & Sakakibara, Y. (2012). MetaVelvet: An extension of Velvet assembler to de novo metagenome assembly from short sequence reads. *Nucleic Acids Research*, *40*(20), e155.

Paul, J. H. (1999). Microbial gene transfer: An ecological perspective. *Journal of Molecular Microbiology and Biotechnology*, *1*, 45–50.

Paul, J. H. (2008). Prophages in marine bacteria: Dangerous molecular time bombs or the key to survival in the seas? *The ISME Journal*, *2*(6), 579–589.

Peng, Y., Leung, H. C., Yiu, S. M., & Chin, F. Y. (2012). IDBA-UD: A de novo assembler for single-cell and metagenomic sequencing data with highly uneven depth. *Bioinformatics*, *28*(11), 1420–1428.

Quail, M. A. (2010). *DNA: Mechanical breakage. Encyclopedia of Life Sciences*. Chichester: John Wiley & Sons, Ltd.

Quail, M. A., Kozarewa, I., Smith, F., Scally, A., Stephens, P. J., Durbin, R., et al. (2008). A large genome center's improvements to the Illumina sequencing system. *Nature Methods*, *5*(12), 1005–1010.

Rodrigue, S., Materna, A. C., Timberlake, S. C., Blackburn, M. C., Malmstrom, R. R., Alm, E. J., et al. (2010). Unlocking short read sequencing for metagenomics. *PLoS One*, *5*(7), e11840.

Rohland, N., & Reich, D. (2012). Cost-effective, high-throughput DNA sequencing libraries for multiplexed target capture. *Genome Research*, *22*(5), 939–946.

Rosario, K., Duffy, S., & Breitbart, M. (2012). A field guide to eukaryotic circular single-stranded DNA viruses: Insights gained from metagenomics. *Archives of Virology*, *157*(10), 1851–1871.

Roux, S., Krupovic, M., Poulet, A., Debroas, D., & Enault, F. (2012). Evolution and diversity of the *Microviridae* viral family through a collection of 81 new complete genomes assembled from virome reads. *PLoS One*, *7*, e40418.

Schoenfeld, T., Patterson, M., Richardson, P. M., Wommack, K. E., Young, M., & Mead, D. (2008). Assembly of viral metagenomes from Yellowstone hot springs. *Applied and Environmental Microbiology*, *74*(13), 4164–4174.

Schwientek, P., Szczepanowski, R., Ruckert, C., Stoye, J., & Puhler, A. (2011). Sequencing of high G+C microbial genomes using the ultrafast pyrosequencing technology. *Journal of Biotechnology*, *155*(1), 68–77.

Sharon, I., Battchikova, N., Aro, E. M., Giglione, C., Meinnel, T., Glaser, F., et al. (2011). Comparative metagenomics of microbial traits within oceanic viral communities. *The ISME Journal*, *5*(7), 1178–1190.

Shaw, C. A. (2002). Theoretical considerations of amplification strategies. *Neurochemical Research*, *27*, 1123–1131.

Simpson, J. T., Wong, K., Jackman, S. D., Schein, J. E., Jones, S. J., & Birol, I. (2009). ABySS: A parallel assembler for short read sequence data. *Genome Research*, *19*(6), 1117–1123.

Solonenko, S., Ignacio-Espinoza, J. C., Alberti, A., Cruaud, C., Hallam, S. J., Konstantinidis, K. T., et al. (2013). Sequencing platform and library preparation choices impact viral metagenomes. *BMC Genomics*, *14*(1), 320.

Steward, G. F., Culley, A. I., Mueller, J. A., Wood-Charlson, E. M., Belcaid, M., & Poisson, G. (2013). Are we missing half of the viruses in the ocean? *The ISME Journal*, 7(3), 672–679.

Sullivan, M. B., Lindell, D., Lee, J. A., Thompson, L. R., Bielawski, J. P., & Chisholm, S. W. (2006). Prevalence and evolution of core photosystem II genes in marine cyanobacterial viruses and their hosts. *PLoS Biology*, *4*, e234.

Sullivan, M. B., Waterbury, J. B., & Chisholm, S. W. (2003). Cyanophages infecting the oceanic cyanobacterium *Prochlorococcus*. *Nature*, *424*, 1047–1051.

Suttle, C. A. (2005). Viruses in the sea. *Nature*, *437*(7057), 356–361.

Suttle, C. A. (2007). Marine viruses—Major players in the global ecosystem. *Nature Reviews. Microbiology*, *5*, 801–812.

Tadmor, A. D., Ottesen, E. A., Leadbetter, J. R., & Phillips, R. (2011). Probing individual environmental bacteria for viruses by using microfluidic digital PCR. *Science*, *333*(6038), 58–62.

Thingstad, T. F. (2000). Elements of a theory for the mechanisms controlling abundance, diversity, and biogeochemical role of lytic bacterial viruses in aquatic ecosystems. *Limnology and Oceanography*, *45*(6), 1320–1328.

Thompson, L. R., Zeng, Q., Kelly, L., Huang, K. H., Singer, A. U., Stubbe, J., et al. (2011). Phage auxiliary metabolic genes and the redirection of cyanobacterial host carbon metabolism. *Proceedings of the National Academy of Sciences of the United States of America*, *108*(39), E757–E764.

Thurber, R. V., Haynes, M., Breitbart, M., Wegley, L., & Rohwer, F. (2009). Laboratory procedures to generate viral metagenomes. *Nature Protocols*, *4*(4), 470–483.

Tseng, Q., Lomonosov, A. M., Furlong, E. E., & Merten, C. A. (2012). Fragmentation of DNA in a sub-microliter microfluidic sonication device. *Lab on a Chip*, *12*(22), 4677–4682.

Tyson, G. W., Chapman, J., Hugenholtz, P., Allen, E. E., Ram, R. J., Richardson, P. M., et al. (2004). Community structure and metabolism through reconstruction of microbial genomes from the environment. *Nature*, *428*(6978), 37–43.

Venter, J. C., Remington, K., Heidelberg, J. F., Halpern, A. L., Rusch, D., Eisen, J. A., et al. (2004). Environmental genome shotgun sequencing of the Sargasso Sea. *Science*, *304*(5667), 66–74.

Vyawahare, S., Griffiths, A. D., & Merten, C. A. (2010). Miniaturization and parallelization of biological and chemical assays in microfluidic devices. *Chemistry & Biology*, *17*(10), 1052–1065.

Wanunu, M. (2012). Nanopores: A journey towards DNA sequencing. *Physics of Life Reviews*, *9*(2), 125–158.

Weinbauer, M. G. (2004). Ecology of prokaryotic viruses. *FEMS Microbiology Reviews*, *28*(2), 127–181.

Weinbauer, M. G., & Rassoulzadegan, F. (2004). Are viruses driving microbial diversification and diversity? *Environmental Microbiology*, *6*(1), 1–11.

Wilhelm, S. W., & Suttle, C. A. (1999). Viruses and nutrient cycles in the sea. *Bioscience*, *49*(10), 781–788.

Wommack, K. E., Bhavsar, J., & Ravel, J. (2008). Metagenomics: Read length matters. *Applied and Environmental Microbiology*, *74*(5), 1453–1463.

Wommack, K. E., & Colwell, R. R. (2000). Virioplankton: Viruses in aquatic ecosystems. *Microbiology and Molecular Biology Reviews*, *64*, 69–114.

Yilmaz, S., Allgaier, M., & Hugenholtz, P. (2010). Multiple displacement amplification compromises quantitative analysis of metagenomes. *Nature Methods*, 7(12), 943–944.

Zerbino, D. R., McEwen, G. K., Margulies, E. H., & Birney, E. (2009). Pebble and rock band: Heuristic resolution of repeats and scaffolding in the velvet short-read de novo assembler. *PLoS One*, *4*(12), e8407.

Zhang, K., Martiny, A. C., Reppas, N. B., Barry, K. W., Malek, J., Chisholm, S. W., et al. (2006). Sequencing genomes from single cells by polymerase cloning. *Nature Biotechnology*, *24*(6), 680–686.

Zhao, Y., Temperton, B., Thrash, J. C., Schwalbach, M. S., Vergin, K. L., Landry, Z. C., et al. (2013). Abundant SAR11 viruses in the ocean. *Nature*, *494*(7437), 357–360.

Zheng, Z., Advani, A., Melefors, O., Glavas, S., Nordstrom, H., Ye, W., et al. (2010). Titration-free massively parallel pyrosequencing using trace amounts of starting material. *Nucleic Acids Research*, *38*(13), e137.

SECTION III

Microbial Community Transcriptomics—Sample Preparation, cDNA Library Construction, and Analysis

CHAPTER NINE

Preparation and Metatranscriptomic Analyses of Host–Microbe Systems

Jarrad T. Hampton-Marcell*[,†], Stephanie M. Moormann*, Sarah M. Owens*[,‡], Jack A. Gilbert*[,†,1]

*Argonne National Laboratory, Argonne, Illinois, USA
[†]Department of Ecology and Evolution, University of Chicago, Chicago, Illinois, USA
[‡]Computation Institute, University of Chicago, Chicago, Illinois, USA
[1]Corresponding author: e-mail address: gilbertjack@anl.gov

Contents

1. Introduction 170
2. Materials 171
   2.1 Kits 171
   2.2 Reagents and equipment 172
3. Components 172
   3.1 DNA/RNA purification 172
   3.2 Quantification 173
   3.3 RNA integrity and size distribution 174
4. Experiments 175
   4.1 Total RNA extraction 175
   4.2 DNase treatment 177
   4.3 Eukaryotic mRNA depletion 177
   4.4 Ribosomal depletion 178
   4.5 cDNA synthesis 181
   4.6 Metatranscriptome generation using Index Barcodes 183
5. Sequencing and Data Analysis 184
Acknowledgments 185
References 185

Abstract

Metatranscriptomics has increased our working knowledge of the functional significance and genetic variability of microbial communities, yet there is still limited information concerning how gene expression and regulation in a microbiome influences interactions with a host organism. During a pathogenic infection, eukaryotic organisms are subject to invasion by bacteria and other agents, or these "pathogens" can switch from a commensal to pathogenic trophic relationship with the host. Understanding how these trophic relationships initiate and persist in the host requires deciphering

Methods in Enzymology, Volume 531
ISSN 0076-6879
http://dx.doi.org/10.1016/B978-0-12-407863-5.00009-5

the functional response of the host and the microbiome, so-called Dual RNA-Seq. This technique is both fast and relatively cheap compared to proteomics and metabolomics and provides information on the potential functional interactions that occur between microbes, and with the host. These metatranscriptomic analyses can also be coupled with metagenomic analyses and statistical models to provide an in-depth approach to systems biology. In this chapter, we detail a standardized method to process and analyze host-associated microbial metatranscriptomes independent of the associated host.

1. INTRODUCTION

Metagenomics has become an important tool for analyzing complex microbial community structure in free-living and host-associated environments. Metagenomic analysis allows researchers to identify the major and minor players in microbial communities and determine functional relationships with one another in complex microbial systems. Despite the power of this tool, the ability of researchers to model microbial functions is limited by current databases that describe the functional attributes of genes and the resultant proteins that can be predicted (Giannoukos et al., 2012). With the incorporation of metatranscriptomics, we can study the expressed functional capacity of communities by utilizing high-throughput sequencing of mRNA transcripts. Here, we explore techniques developed to sequence transcriptional responses from host-derived microbial communities (Westermann, Gorski, & Vogel, 2012). This is a particularly complex system in which samples can be used to generate RNA from both the host and the microbial community in question, for example, an intestinal epithelial biopsy will contain microbial communities associated with the tissue, as well as host cells, which must be appropriately separated using the techniques described here, and sequenced to enable characterization of the transcriptional response of both host and microbiome. Metagenomics coupled with metatranscriptomics is becoming more necessary as the scientific community generates more evidence that microbial community metabolism can significantly impact health, well-being, and behavior in animals, as well as, productivity and disease state in plants. A rapid way to assess this is to sequence the relative changes in the transcription of genes from the microbial communities associated with the organism of interest, and then under experimental conditions to determine whether these responses have the capability to generate metabolites that could impact the host. When combined with metaproteomics and metabolomics of the same community and environment, respectively, it is possible to use this tool to reconstruct microbial–host interactions at the metabolic level, enabling the elucidation of mechanisms underling host and microbial responses to given stimuli. As with any methodology, there are some stumbling blocks associated with performing metatranscriptomic analyses, that is, acquisition of sufficient amounts of high-quality RNA. It is important to note that RNA is extremely unstable and susceptible to degradation. It is essential that proper preparation methods, sterile equipment, and RNase removal products are used before any RNA work begins to maintain the integrity of the RNA while processing. As shown previously (Urich, Lanzen, Huson, Schleper, & Schuster, 2008), it is possible to sequence all the RNA in a sample and pull out the sequences that pertain to the mRNA that describes functional responses. This is an ideal strategy as it removes bias associated with RNA handling, preparation, and concentration of mRNA; however, it is limited to the number of samples that can be handled and the cost of generating enough coverage of the mRNA to observe significant changes in less abundant transcripts. This extremely deep coverage is necessary because on average 90–95% of total RNA in any cell is ribosomal RNA, which swamps the signals for the functionally relevant mRNA. Therefore, removal of rRNA has become commonplace (Frias-Lopez et al., 2008; Gilbert et al., 2008; Poretsky et al., 2009), so as to enrich for the functional mRNA signal. This means that less sequencing effort is required to generate a relevant (arbitrarily) coverage of the transcriptional response of the microbial community. In the eukaryotic host, this is not a significant problem as we can utilize nature's own tagging system of a poly-A tail at the end of each mRNA fragment to selectively isolate and sequence the mRNA. However, as microbial mRNA is rarely polyadenylated, we must enrich the mRNA signal in each sample, so that we can process thousands of samples in parallel even if we have to introduce the possibility of certain biases in data interpretation.

2. MATERIALS

2.1. Kits

1. MoBio PowerMicrobiome RNA Isolation Kit
2. Life Technologies Turbo DNA-free
3. Invitrogen Dynabead Oligo$(dT)_{25}$
4. Agilent RNA 6000 Nano Kit
5. Agilent RNA 600 Pico Kit

6. Epicentre Ribo-Zero Magnetic Gold Kit (Epidemiology)
7. Epicentre ScriptSeq v2 RNA-Seq Library Preparation Kit
8. Epicentre ScriptSeq Index PCR Primers Set 1

2.2. Reagents and equipment

1. Phenol:chloroform:isoamyl alcohol (25:24:1), pH 6.5–8.0
2. Qubit RNA Assay for Qubit Fluorometer
3. Epicentre FailSafe PCR Enzyme Mix
4. 100% Absolute Ethanol
5. Agencourt AMPure RNAClean XP beads
6. Agencourt AMPure XP beads
7. Magnetic Stand or Rack (1.5 ml or 96 well)
8. Invitrogen Qubit Fluorometer
9. Agilent 2100 Bioanalyzer

3. COMPONENTS

3.1. DNA/RNA purification

It is recommended to start this protocol with at least 1 μg of high-quality total RNA. Bacterial messenger RNA comprises less than 10% of the total RNA from host-associated tissue samples, with the majority of the RNA associated with the host and microbial ribosomal RNA. Considering the small amount of material and the instability of mRNA, it is necessary to use a purification kit that will recover as much mRNA as possible, while keeping the integrity of the mRNA intact during processing. We recommend performing three purifications in the process of depleting the ribosomal RNA and generation of metatranscriptomic libraries that can be scaled to expected outcomes. In this protocol, we suggest Agencourt AMPure XP and Agencourt AMPure RNAClean XP beads as they consistently display higher recovery rates than column filtration, regardless of fragment size. In addition, input material can be adjusted to produce maximum yields regardless of initial DNA or RNA material. It is recommended to run a 1.8 × reaction, that is, utilizing 180 μl of beads for a 100 μl reaction. This will capture all fragments in solution, for example, ribosomal RNA depletion and cDNA generation. After generating the cDNA library via PCR, we suggest a 0.9 × reaction with AMPure XP beads. This reaction removes all fragments under 125 bp, including primer dimers. It is necessary to remove small fragments and primer dimers prior to sequencing as they will preferentially amplify during the sequence library generation, for example, Illumina as described here, and outcompete your sequencing libraries during sequencing.

3.2. Quantification

Appropriate quantification of libraries is recommended when using next-generation sequencing technologies. Invitrogen's Qubit Fluorometer is highly recommended as it utilizes a fluorescent dye that binds to nucleic acids for quantification. We recommend avoiding the Nanodrop, as it consistently overestimates nucleic acid concentrations. Unlike the Nanodrop, the Qubit Fluorometer can discriminate between DNA and RNA when both templates are present, where trace amounts of DNA will contaminate an RNA-mediated extraction. It is recommended to quantitate RNA concentrations after each step, for example, total genomic extraction, DNase treatment, and eukaryotic mRNA depletion using the Qubit RNA Assay Kit for the Qubit Fluorometer. To do this, follow the below instructions:

1. Make a Qubit working solution by diluting the Qubit RNA reagent 1:200 in Qubit RNA buffer using a sterile plastic tube.
2. Load 190 μl of Qubit working solution into tubes labeled standard 1 and 2.
3. Add 10 μl of standard 1 solution and standard 2 solution to the appropriate tube and mix by vortexing for 2–3 s.
   - Note: These are positive and negative controls used to calibrate the instrument.
4. Load 198 μl of Qubit working solution into each individual assay tube.
5. Add 2 μl of DNA to each assay tube and mix by vortexing 2–3 s. The final volume of this solution should equal 200 μl.
   - Note: The amount of sample and working solution added to each assay tube can vary depending on concentration of the sample. The sample can vary between 1 and 20 μl and the working solution can vary between 199 and 180 μl with the final volume equaling 200 μl. It is recommended to use 2 μl of sample to produce the most accurate results.
6. Allow all the tubes to incubate at room temperature for 2 min.
7. Select RNA assay on the Qubit Fluorometer. Select run a new calibration.
8. Insert the tube containing standard 1, close lid, and press read.
9. Remove standard 1 and repeat step 8 for standard 2.

10. Insert sample, close lid, and press read.
11. Calculate concentration using dilution calculation on Qubit Fluorometer by selecting original volume of sample added to the assay tube.
12. Repeat steps 10 and 11 until all samples have been quantified.

3.3. RNA integrity and size distribution

The Agilent Bioanalyzer is a microfluidics platform used for sizing, quantification, and quality control for DNA, RNA, and proteins. This platform is preferred over the Qubit Fluorometer after depleting the sample of host and bacterial ribosomal RNA. Because we are expecting to recover only ~10% of the total RNA input as enriched mRNA, the Nano and Pico chips will likely be used having an input range of 25–500 and <25 ng/μl, respectively. Utilizing this platform, we are able to quantitate the concentration of an RNA sample and assess the quality of the sample through the RNA Integrity Number (RIN), which quantifies the fragmentation of the RNA sample (Schroeder et al., 2006). High-quality RNA will contain an RIN of at least 8, where partially fragmented RNA will contain an RIN within the range of 6–8. Any RNA sample that has a RIN below 5 should not be subjected to further fragmentation during the ScriptSeq protocol, as it will generate smaller than desired fragments.

- Note: In this protocol, we will assume an RNA concentration of 30 ng/μl, where we will highlight the RNA Nano protocol. The RNA Nano and Pico chip protocol only differ in the amount of gel–dye mix used and the concentrations of gel matrix and dye.

1. Check to make sure that the Chip Priming Station has its base adjusted to position C and the syringe clip is adjusted to the topmost position.
2. The electrodes on the head of the Bioanalyzer should be cleaned with RNaseZap followed by RNase-free water.
3. To prepare the gel for an RNA Nano chip, all reagents should be left at room temperature for at least 30 min to equilibrate.
4. 550 μl of Agilent RNA 6000 Nano gel matrix should be aliquoted into the top of a spin filter and centrifuged for 10 min at $1500 \times g \pm 20\%$. 65 μl of the filtered Agilent RNA 6000 Nano gel matrix should be aliquoted in a 0.5-ml RNase-free microcentrifuge tube and mixed with 1 μl of RNA 6000 Nano dye.
5. This solution should be vortexed well and spun for 10 min at $13{,}000 \times g$.
   - Note: The gel–dye mixture should be used within 24 h.

6. An RNA Nano chip should be placed on the priming station.
7. 9.0 μl of gel–dye mix should be pipetted into the well marked "G." The plunger should be positioned at 1 ml. The Chip Priming Station should be closed and the plunger should be depressed until the clip holds it. The plunger should remain pressurized for 30 s then released. If the plunger does not move back up to the 1 ml position after it is released, the plunger needs to be manually moved to the 1 ml position.
8. 9 μl of gel–dye mix should be pipetted into the remaining two gel–dye wells.
9. 5 μl of RNA 6000 Nano marker should be pipetted into all of the sample and ladder wells.
10. Before use, the RNA ladder should be heat denatured at 70 °C for 2 min and immediately cooled on ice to prevent secondary structure from forming. 1 μl of ladder should be added to the ladder well.
11. 1 μl of each RNA sample should be added into the 12 sample wells.
12. The chip should be vortexed with an IKA vortex mixer for 1 min at 2400 rpm.
13. The chip should be immediately loaded into the Bioanalyzer, the corresponding chip type selected in the software, and "Start run" selected.
14. To check the quality of the run, seven markers should be seen on the electropherogram of the ladder at 25, 200, 500, 1000, 2000, 4000, and 6000 nt.
    - Note: If these peaks are not seen, the run has failed and the data is not accurate.
15. The Bioanalyzer will produce an RNA concentration, rRNA ratios, and the RIN number.

4. EXPERIMENTS

4.1. Total RNA extraction

MoBio Laboratory's PowerMicrobiome RNA Extraction kit allows for fast and easy purification of total RNA from samples by removing PCR inhibitors that are seen in many host-associated samples. It is optimal to extract RNA from samples immediately after sampling as fast decay rate and an abundance of dead cells (from both human and bacterial sources) are commonly seen in these types of sample types (Stewart et al., 2013).

- Note: Trizol extractions are acceptable alternatives for samples that contain high degradation activity. It is necessary to remove all organic material and guanidine from extracted RNA in order to prevent problems with downstream applications.

1. Add 500 μl of phenol:chloroform:isoamyl alcohol (pH 6.5–8.0) to the bead tube.
   - Note: Beta-mercaptoethanol can be used in place of phenol:chloroform:isoamyl alcohol.
2. Add 0.25 g of sample into the glass bead tube.
3. Secure the glass bead tube using a MoBio Vortex Adapter pointing the tube cap toward the center of the adapter.
4. Vortex at maximum speed for 10 min.
5. Centrifuge at 13,000 × *g* for 1 min at room temperature. Transfer the upper aqueous layer to a clean 2-ml collection tube.
   - Note: There will be three aqueous layers. The top layer will contain the RNA. The middle layer will contain the phenol:chloroform and digestive enzymes. The bottom layer will contain all cellular debris. Be sure not to aspirate from the middle layer as your RNA will be degraded if this layer is transferred.
6. Add 150 μl of solution PM2 and vortex briefly to mix. Incubate at 4 °C for 5 min.
7. Centrifuge the tubes at 13,000 × *g* for 1 min.
8. Avoiding the pellet, transfer the supernatant to a clean, RNase-free 2-ml collection tube.
9. Add 650 μl of solution PM3 and 650 μl of solution PM4. Vortex briefly to mix.
10. Load 650 μl of supernatant onto a spin filter and centrifuge at 13,000 × *g* for 1 min. Discard the flow through and repeat until all the supernatant has been loaded onto the spin filter.
11. Shake to mix solution PM5. Add 650 μl of solution PM5 to the spin filter and centrifuge at 13,000 × *g* for 1 min.
12. Discard the flow through and centrifuge at 13,000 × *g* for 1 min to remove residual wash.
13. Place the spin filter basket into a clean 2-ml collection tube.
14. Add 50 μl of DNase I solution (prepared by mixing 45 μl of solution PM6 and 5 μl of DNase I stock solution) to the center of the spin filter.
15. Incubate at room temperature for 15 min.
16. Add 400 μl of solution PM7 and centrifuge at 13,000 × *g* for 1 min. Discard the flow through.
17. Add 650 μl of solution PM5 and centrifuge at 13,000 × *g* for 1 min. Discard flow through.

18. Add 650 μl of PM4 and centrifuge at 13,000 × g for 1 min. Discard flow through.
19. Centrifuge empty column at 13,000 × g for 2 min to remove residual wash solution.
20. Place the spin filter basket into a clean 2-ml collection tube.
21. Add 100 μl of solution PM8 (RNase-free water) to the center of the white filter membrane. Allow the water to sit on the membrane for at least 1 min.
22. Centrifuge at 13,000 × g for 1 m. Discard spin filter basket.
23. RNA is now ready for downstream applications. Store at −80 °C.

4.2. DNase treatment

As no RNA extraction method can completely remove DNA contamination from RNA preparations, it is recommended to perform an additional DNA removal treatment. Life Technologies' Turbo DNA-free kit digests DNA concentrations below known levels of detection through PCR, and it removes cations that degrade RNA. DNA carries microbes that are not actively functioning in the microbial community. If carried through the remainder of the process, an individual might analyze microbial species that are nonfunctional.

1. Add 0.1 volume 10 × Turbo DNase Buffer and 1 μl Turbo DNase to the RNA. Mix gently.
2. Incubate at 37 °C for 20–30 min.
3. Add 0.1 volume resuspended DNase Inactivation Reagent and mix well.
4. Incubate 5 min at room temperature, mixing occasionally.
5. Centrifuge at 10,000 × g for 1.5 min and transfer the RNA to a fresh tube.

4.3. Eukaryotic mRNA depletion

Eukaryotic mRNA represents a small percentage of total RNA extracted. Currently, there are no commercially available kits that isolate bacterial rRNA and mRNA from host rRNA and mRNA. We recommend employing Invitrogen's Dynabead Oligo(dT)$_{25}$ kit to remove the host mRNA. Utilizing the poly-A tail associated with eukaryotic mRNA, we will pair a poly-T tail to remove it from solution without disturbing the remaining microbial and eukaryotic RNA present. Instead of performing a series of washes on the magnetically bound eukaryotic mRNA, we will work with the unbound supernatant.

- Note: There are certain bacterial species that have an attenuated poly-A tail. Bacterial species that display this poly-A tail are normally marked for degradation. This kit has limited efficacy in binding these species.

Wash Dynabeads

1. Resuspend the Dynabeads in a vial by vortexing for at least 30 s.
2. Transfer the desired volume of Dynabeads to a 1.5-ml microcentrifuge tube.
3. Add the same volume of binding buffer, or at least 1 ml, and resuspend.
4. Place the tubes in a magnetic stand for 1 min and discard the supernatant.
5. Remove the tube from the magnetic stand and resuspend the washed Dynabeads in the same volume of binding buffer as the initial volume of Dynabeads.

Eukaryotic mRNA removal

1. Adjust the volume of the total RNA sample to 100 μl with nuclease-free water.
   - Note: This kit can handle up to 75 μg of total RNA.
2. Add 100 μl of binding buffer. If total RNA is more dilute than 75μg/100μl, add an equal volume of binding buffer to the beads.
3. Heat at 65 °C for 2 min to disrupt secondary structures. Immediately place on ice.
4. Add the 200 μl of total RNA to the 100 μl washed beads.
5. Mix thoroughly and allow binding by placing continuously on a mixer for 5 min at room temperature.
6. Place the tube on the magnetic stand for 1–2 min and carefully transfer all of the supernatant to a 1.5-ml microcentrifuge tube.
   - Note: The supernatant will contain the bacterial rRNA and mRNA as well as the host rRNA.

4.4. Ribosomal depletion

Removing the ribosomal RNA from the host and bacterial communities is an important aspect of creating metatranscriptomic libraries. Host- and bacterial-derived ribosomal RNA constitutes more than 90% of the total RNA extracted. While direct translation and sequencing of total RNA may reduce potential biases associated with the following steps, it significantly decreases the overall sequencing depth of metatranscriptomic coverage. Currently, we have seen only circumstantial evidence that these biases are anything but partially significant; however, as with any experiment, these caveats must be taken into consideration. Ribo-Zero Magnetic Gold Kit (Epidemiology) will effectively remove sequences found in host- and bacterial ribosomal RNA from solution with 99% efficiency working within a range of 500 ng to 5 μg of total RNA (Giannoukos et al., 2012; Lim et al., 2013):

Wash Beads

1. Pipette 225 μl of Magnetic Beads into a 1.5-ml RNase-free microcentrifuge tube. Pipette suspension slowly.
2. Place each 1.5-ml tube on the magnetic stand for at least 1 min.
3. With the 1.5-ml tube still on the stand, remove and discard the supernatant.
4. Remove the 1.5-ml tube from the magnetic stand and add 225 μl of RNase-free water to each tube. Mix well by repeated pipetting or vortexing at medium speed.
5. Repeat steps 2 and 3.
6. Remove the 1.5-ml tube from the magnetic stand. Add 65 μl of Magnetic Bead Resuspension Solution to each tube. Mix well by repeated pipetting or vortexing at medium speed.
7. Add 1 μl of Riboguard RNase Inhibitor to each tube to resuspended Magnetic Beads and mix briefly by vortexing.

Ribosomal RNA removal

Amount of input total RNA (μg)	Maximum volume of total RNA that can be added to each reaction (μl)	Volume of Ribo-Zero rRNA Removal Solution used per reaction (μl)
1–2.5	28	8
2.5–5	26	10

8. In a 0.2-ml RNase-free microcentrifuge tube, combine in given order below on ice:
   - Note: This kit is still optimal for samples with less than 1 μg of total RNA input. All subsequent reagents must be scaled down to mimic ratios of recommended total RNA input.

x μl	RNase-free water
4 μl	Ribo-Zero Reaction Buffer
1–5 μg	Total RNA sample
y μl	Ribo-Zero rRNA Removal Solution
40 μl	Total volume

9. Gently mix the reaction(s) by pipetting and incubate at 68 °C for 10 min. Store the remaining Ribo-Zero rRNA Removal Solution and Ribo-Zero Reaction Buffer at −70 to −80 °C.
10. Remove the reaction tube(s) and incubate each at room temperature for 5 min.
11. Using a pipette, add the treated RNA to the 1.5-ml tube containing the washed Magnetic Beads without changing the pipette tip. Immediately and thoroughly mix the contents of the tube pipetting up and down at least 10 times. Immediately vortex the tube again at a medium setting for 10 s and place at room temperature. Repeat this process for each sample.
12. Incubate the 1.5-ml tube at room temperature for 5 min.
13. Following incubation, mix the reactions by vortexing at medium speed for 5 s. Heat the tube at 50 °C for 5 min in an appropriate water bath or heating block. Avoid condensation.
14. After incubation, remove the tubes and immediately place them on magnetic stand for at least 1 min.
15. Carefully remove each supernatant (85–90 μl), this contains the bacterial mRNA, and transfer it to a labeled 1.5-ml RNAase-free tube.

Bacterial mRNA purification

1. Vortex the AMPure RNAClean XP beads until they are resuspended.
2. Add 160 μl of the mixed AMPure RNAClean XP Beads to each 1.5-ml microcentrifuge tube containing 85–90 μl of rRNA-depleted mRNA. Mix thoroughly by gently pipetting the entire volume 10 times.
3. Incubate the tubes at room temperature for 15 min. During incubation, prepare 80% ethanol solution.
4. Place the tubes on the magnetic stand at room temperature for at least 5 min, until the liquid appears clear.
5. Remove and discard the supernatant from each tube. Do not disturb the beads.
6. With the tubes still on the magnetic stand, add 200 μl of freshly prepared 80% ethanol to each tube, without disturbing the beads.
7. Incubate at room temperature for at least 30 s while still on the magnetic stand, then remove and discard all of the supernatant from each tube. Again, do not disturb the beads.
8. Repeat steps 6 and 7 one more time for a total of two 80% ethanol washes.
9. Allow the tubes to air-dry on the magnetic stand at room temperature for 15 min.

10. Add at least 15 μl of DNase/RNase-free water to each tube.
11. Thoroughly resuspend the beads by gently pipetting 10 times.
12. Incubate the tubes at room temperature for 2 min.
13. Place the tubes back onto the magnetic stand at room temperature for at least 5 min, until the liquid appears clear.
14. Transfer the clear supernatant from each tube to an appropriate collection tube. Leave at least 1 μl of the supernatant behind to avoid carry-over of Magnetic Beads.
15. Store on ice for immediate use or store at −70 to −80 °C for longer term storage.
16. Add 2 volumes of RNA Binding Buffer to each volume of the mRNA sample and mix well.

4.5. cDNA synthesis

Using random hexamers, Epicentre's ScriptSeq V2 RNA-Seq Library Preparation Kit uses rRNA-depleted samples to create metatranscriptomic libraries through reverse transcription. This kit takes advantage of directional, paired-end sequencing using Illumina's TruSeq Cluster Kit.

Fragmentation

1. In a 0.2-ml PCR tube, assemble the following reaction mixture on ice:
    - Note: It is recommended to use 2–5 μl of RNA sample when assembling reaction.

x μl	Nuclease-free water
y μl	rRNA-depleted or poly(A) RNA
1 μl	RNA fragmentation solution
2 μl	cDNA Synthesis Primer
12 μl	Total reaction volume

2. Fragment RNA: Incubate at 85 °C for 5 min in a thermocycler.
3. Stop the fragmentation reaction by placing the tube on ice.

Synthesize DNA

1. On ice, prepare the cDNA Synthesis Master Mix. For each reaction, combine on ice:

3 μl	cDNA Synthesis PreMix
0.5 μl	100 m*M* DTT
0.5 μl	StarScript Reverse Transcriptase
4 μl	Total volume

2. Add 4 μl of the cDNA Synthesis Master Mix to each reaction on ice and mix by pipetting.
3. Incubate at 25 °C for 5 min followed by 42 °C for 20 min.
4. Cool reaction to 37 °C and pause thermocycler.
5. Remove one reaction at a time from the thermocycler, add 1 μl of Finishing Solution, and gently mix thoroughly with a pipette. Return each reaction to the thermocycler before proceeding with the next.
6. Incubate at 37 °C for 10 min.

3′-Terminal tag

1. On ice, prepare the Terminal Tagging Master Mix. Note: It is important to place each reaction on ice.
2. Thoroughly mix the viscous Terminal Tagging Master Mix by pipetting or by flicking the tube.
3. Remove one reaction at a time from the thermocycler and add 8 μl of the Terminal Mix. Gently but thoroughly mix the reaction by pipetting. Return each reaction to the thermocycler before proceeding with the next.
4. Incubate each reaction at 25 °C for 15 min.
5. Incubate each reaction at 95 °C for 3 min. Then, cool the reaction to 4 °C on ice or in the thermocycler.

cDNA purification

1. Warm the AMPure XP beads to room temperature. While the beads warm, prepare 400 μl of fresh 80% ethanol at room temperature for each sample.
2. Using a 96-well plate format, transfer the 24 μl of each di-tagged cDNA independently into a well of the plate.
3. Add 45 μl of the beads to each well of the 96-well plate containing the di-tagged cDNA.
4. Mix thoroughly by gently pipetting the entire volume of each well 10 times.
5. Incubate the 96-well plate at room temperature for 15 min.
6. Place the 96-well plate in a magnetic stand at room temperature for at least 5 min, until the liquid appears clear.
7. Remove and discard the supernatant from each well using a pipette. Some liquid may remain in each well. Do not disturb the beads.
8. With the plate remaining on the magnetic stand, add 200 μl of 80% ethanol to each well without disturbing the beads.
9. Incubate the plate at room temperature for at least 30 s, then remove and discard all of the supernatant from each well. Do not disturb the beads.

10. Repeat steps 8 and 9 one more time for a total of two 80% ethanol washes.
11. Allow the plate to air-dry on their magnetic stand for 15 min at room temperature or until the pelleted beads are dry.
    - Note: It is important to not over dry the beads as this will result in a lower recovery rate.
12. Add 24.5 μl of nuclease-free water to each well and remove from the magnetic stand.
13. Thoroughly resuspend the beads by gently pipetting 10 times.
14. Incubate the plate at room temperature for 2 min.
15. Place the plate on the magnetic stand at room temperature for at least 5 min, until the liquid appears clear.
16. Transfer 22.5 μl of the clear supernatant, which contains the di-tagged cDNA, from each well to a new 0.2-ml PCR tube.
17. Place tubes on ice or place at −80 °C for longer term storage.

4.6. Metatranscriptome generation using Index Barcodes

1. In a 0.2-ml PCR tube containing 22.5 μl of di-tagged cDNA, add on ice:

25 μl	FailSafe PCR PreMix E
1 μl	Forward PCR Primer
1 μl	ScriptSeq Index PCR Primer
0.5 μl	FailSafe PCR Enzyme (1.25 units)
50 μl	Total volume

2. Perform PCR:

Denature the dsDNA at 95 °C for 1 min Followed by 15 cycles of: 95 °C for 30 s 55 °C for 30 s 68 °C for 3 min After cycles, incubate at 68 °C for 7 min

RNA-Seq purification

1. Warm the AMPure XP beads to room temperature. While the beads warm, prepare 400 μl of fresh 80% ethanol at room temperature for each sample.

2. Using a 96-well plate format, transfer each amplified library independently into a well of the plate.
3. Add 50 μl of the beads to each well of the 96-well plate.
4. Mix thoroughly by gently pipetting the entire volume up of each well 10 times.
5. Incubate the 96-well plate at room temperature for 15 min.
6. Place the 96-well plate in a magnetic stand at room temperature for at least 5 min, until the liquid appears clear.
7. Remove and discard the supernatant from each well using a pipette. Some liquid may remain in each well. Do not disturb the beads.
8. With the plate on the magnetic stand, add 200 μl of 80% ethanol to each well without disturbing the beads.
9. Incubate the plate at room temperature for at least 30 s, then remove and discard all of the supernatant. Do not disturb the beads.
10. Repeat steps 8 and 9 one more time for a total of two 80% ethanol washes.
11. Allow the plate to air-dry on their magnetic stand for 15 min at room temperature.
12. Add 20 μl of nuclease-free water to each well and remove the plate from their magnetic stand.
13. Thoroughly resuspend the beads by gently pipetting 10 times.
14. Incubate the plate at room temperature for 2 min.
15. Place the plate on the magnetic stand at room temperature for at least 5 min, until the liquid appears clear.
16. Transfer the clear supernatant, which contains the RNA-Seq library, from each well to an appropriate collection tube.

5. SEQUENCING AND DATA ANALYSIS

The output from an Illumina next-generation sequencing run contains quality files, a binary file, and a fastq file. Sequences with low-quality scores are filtered out before analysis. Metatranscriptomes will be analyzed using the available online resources (e.g., IMG/M, MGRAST, CAMERA, EBIs Metagenomics portal, etc.) providing annotation by comparing transcripts to different functional gene databases (e.g., using BLAST to assign functions against M5NR, SFams, and SEED) (Markowitz et al., 2012; Meyer et al. 2008; Sun et al., 2010). For more detailed descriptions of potential functional pipelines and analyses of these data (see Thomas, Gilbert, & Meyer 2012).

ACKNOWLEDGMENTS

We would like to acknowledge Areej A. Amaar and the Next Generation Sequencing Facility of the Institute for Genomics and Systems Biology at Argonne National Laboratory (IGSB-NGS). Their insight regarding sequencing platforms, constructs, and strategies was tremendously helpful in developing our metatranscriptomic pipeline from host-associated tissues.

REFERENCES

Frias-Lopez, J., Shi, Y., Tyson, G., Coleman, M., Schuster, S., Chisholm, S., et al. (2008). Microbial community gene expression in ocean surface waters. *Proceedings of the National Academy of Sciences of the United States of America*, *105*, 3805–3810.

Giannoukos, G., Ciulla, D., Huang, K., Haas, B., Izard, J., Levin, J., et al. (2012). Efficient and robust RNA-seq process for cultured bacteria and complex community transcriptomes. *Genome Biology*, *13*(r23), 1–13. http://dx.doi.org/10.1186/gb-2012-13-3-r23.

Gilbert, J., Field, D., Huang, Y., Edwards, R., Li, W., Gilna, P., et al. (2008). Detection of large numbers of novel sequences in the metatranscriptomes of complex marine microbial communities. *PLoS One*, *3*, e3042. http://dx.doi.org/10.1371/journal.pone.0003042.

Lim, Y., Schmieder, R., Haynes, M., Willner, D., Furlan, M., Youle, M., et al. (2013). Metagenomics and metatranscriptomics: Windows on CF-associated viral and microbial communities. *Journal of Cystic Fibrosis*, *12*, 154–164.

Markowitz, V., Chen, I., Chu, K., Szeto, E., Palaniappan, K., Yuri, G., et al. (2012). IMG/M: The integrated metagenome data management and comparative analysis system. *Nucleic Acids Research*, *40*, D123–D129.

Meyer, F., Paarmann, D., Souza, M., Olson, R., Glass, E., Kubal, M., et al. (2008). The metagenomics RAST server—A public resource for the automatic phylogenetic and functional analysis of metagenomes. *BMC Bioinformatics*, *9*(386), 1–8. http://dx.doi.org/10.1186/1471-2105-9-386.

Poretsky, R., Hewson, I., Sun, S., Allen, A., Zehr, J., & Moran, M. (2009). Comparative day/night metatranscriptomic of microbial communities in the North Pacific subtropical gyre. *Environmental Microbiology*, *11*, 1358–1375.

Schroeder, A., Mueller, O., Stocker, S., Salowsky, R., Leiber, M., Gassmann, M., et al. (2006). The RIN: An RNA integrity number for assigning integrity values to RNA measurements. *BMC Molecular Biology*, 7, 3.

Stewart, C., Nelson, A., Scribbins, D., Marrs, E., Lanyon, C., Perry, J., et al. (2013). Bacterial and fungal viability in the preterm gut: NEC and sepsis. *Archives of Disease in Childhood Fetal and Neonatal Edition*, *98*(4), F298–F303. http://dx.doi.org/10.1136/archdischild-2012-302119.

Sun, S., Chen, J., Li, W., Altinatas, I., Lin, A., Peltier, S., et al. (2010). Community cyberinfrastructure for advanced microbial ecology research and analysis: The CAMERA resource. *Nucleic Acids Research*, *39*(suppl 1), D546–D551. http://dx.doi.org/10.1093/nar/gkq1102.

Thomas, T., Gilbert, J., & Meyer, F. (2012). Metagenomics—A guide from sampling to data analysis. *Microbial Informatics and Experimentation*, *2*(3), 1–12. http://dx.doi.org/10.1186/2042-5783-2-3.

Urich, T., Lanzen, A., Huson, D., Schleper, C., & Schuster, S. (2008). Simultaneous assessment of soil microrbial community structure and function through analysis of the meta-transcriptome. *PLoS One*, *3*(6), e2527. http://dx.doi.org/10.1371/journal.pone.0002527.

Westermann, A., Gorski, S., & Vogel, J. (2012). Dual RNA-Seq of pathogen and host. *Nature Reviews Microbiology*, *10*, 618–630.

CHAPTER TEN

Preparation of Microbial Community cDNA for Metatranscriptomic Analysis in Marine Plankton

Frank J. Stewart[1]
School of Biology, Georgia Institute of Technology, Atlanta, Georgia, USA
[1]Corresponding author: e-mail address: frank.stewart@biology.gatech.edu

Contents

1. Introduction 188
2. Sample Collection and RNA Preservation 192
   2.1 Shipboard filtration and preservation 192
   2.2 *In situ* preservation 196
   2.3 Pump-cast collection 197
3. RNA Extraction and Standardization 198
   3.1 Extraction and DNA removal 198
   3.2 Standardization with exogenous RNA transcripts 201
4. rRNA Depletion 201
   4.1 Summary of rRNA depletion methods 201
   4.2 Commercial rRNA depletion methods 202
   4.3 Subtractive hybridization with sample-specific probes 205
   4.4 rRNA removal efficiency, alternative methods, and perspectives 207
5. RNA Amplification and cDNA Synthesis 209
   5.1 Amplification by *in vitro* transcription 209
   5.2 cDNA synthesis 210
   5.3 Preserving strand orientation 211
6. Summary 212
Acknowledgments 213
References 213

Abstract

High-throughput sequencing and analysis of microbial community cDNA (metatranscriptomics) are providing valuable insight into *in situ* microbial activity and metabolism in the oceans. A critical first step in metatranscriptomic studies is the preparation of high-quality cDNA. At the minimum, preparing cDNA for sequencing involves steps of biomass collection, RNA preservation, total RNA extraction, and cDNA synthesis.

Methods in Enzymology, Volume 531
ISSN 0076-6879
http://dx.doi.org/10.1016/B978-0-12-407863-5.00010-1

Each of these steps may present unique challenges for marine microbial samples, particularly for deep-sea samples whose transcriptional profiles may change between water collection and RNA preservation. Because bacterioplankton community RNA yields may be relatively low (<500 ng), it is often necessary to amplify total RNA to obtain sufficient cDNA for downstream sequencing. Additionally, depending on the nature of the samples, budgetary considerations, and the choice of sequencing technology, steps may be required to deplete the amount of ribosomal RNA (rRNA) transcripts in a sample in order to maximize mRNA recovery. cDNA preparation may also involve the addition of internal RNA standards to biomass samples, thereby allowing for absolute quantification of transcript abundance following sequencing. This chapter describes a general protocol for cDNA preparation from planktonic microbial communities, from RNA preservation to final cDNA synthesis, with specific emphasis placed on topics of sampling bias and rRNA depletion. Consideration of these topics is critical for helping standardize metatranscriptomics methods as they become widespread in marine microbiology research.

1. INTRODUCTION

Metatranscriptomic analysis has emerged as a powerful and increasingly common tool for exploring microbial community gene expression in natural environments (DeLong, 2009; Moran et al., 2013; Poretsky, Hewson, et al., 2009). In these analyses, complementary DNA (cDNA) is synthesized by reverse transcription from RNA extracted from an environmental sample and used as template for massively parallel shotgun sequencing. The first generation of marine metatranscriptome studies relied on data generated primarily by pyrosequencing on the Roche 454 platform (Table 10.1). However, Illumina sequencing using chain-termination technology may soon replace pyrosequencing as the method of choice for low-cost, high-quality sequence generation (Gifford et al., 2013). A typical metatranscriptome dataset produced using 454 or Illumina contains hundreds of thousands to millions of sequences. These sequences encompass diverse RNA pools, including structural (ribosomal) RNA (rRNA), messenger RNA (mRNA), and numerous regulatory small RNAs (Shi, Tyson, & DeLong, 2009). Collectively, these data yield valuable information on the metabolic activity and taxonomic diversity of the community, potentially revealing species-specific responses to environmental change as well as covariation in metabolism between distinct community members. Obtaining information-rich metatranscriptome datasets is of course dependent on the preparation of high-quality cDNA, which must be optimized to

Table 10.1 Percentage of rRNA in marine microbial metatranscriptomes following rRNA depletion

References	Samples	Depth (m)	Sequencing	Mb[a]	rRNA dep[b]	% rRNA[c]
Poretsky et al. (2005)	Tidal marsh; soda lake	0–23	Sanger	0.15	MEx	29
Frias-Lopez et al. (2008)	Subtropical oligotrophic ocean	75	454 GS 20	15	MAII	53
Gilbert et al. (2008)	Temperate fjord	Surface	454 GS FLX	29	MEx, MDA	<1
Poretsky, Hewson, et al. (2009)	Subtropical oligotrophic ocean	25	454 GS 20	12	Exo, MEx, MAII	37
Hewson, Poretsky, Beinart, et al. (2009)	Tropical ocean, phytoplankton bloom	37	454 GS FLX	10	Exo, MEx, MAII	76
Hewson, Poretsky, Dyhrman, et al. (2009)	Tropical ocean, phytoplankton bloom	Surface	454 GS FLX	NR	Exo, MEx, MAII	95[d]
Gilbert et al. (2010)	Temperate, coastal ocean	Surface	454 GS FLX Titanium	22	MEx, MDA	45
Hewson, Poretsky, Tripp, Montoya, and Zehr (2010)	Tropical, subtropical ocean	0–37	454 GS 20, 454 GS FLX	5	Exo, MEx, MAII	66
Poretsky, Sun, Mou, and Moran (2010)	Tidal marsh; surface	Surface	454 GS FLX	73	Exo, MEx, MAII	53
Vila-Costa et al. (2010)	Subtropical oligotrophic ocean	10	454 GS FLX	63	Exo, MEx, MEn, MAII	50
Stewart, Ottesen, and DeLong (2010)	Subtropical oligotrophic ocean; oxygen minimum zone	0–100	454 GS FLX	44	CHyb	43

Continued

Table 10.1 Percentage of rRNA in marine microbial metatranscriptomes following rRNA depletion—cont'd

References	Samples	Depth (m)	Sequencing	Mb[a]	rRNA dep[b]	% rRNA[c]
Mou et al. (2011)	Tidal marsh	Surface	454 GS FLX	15	MEx	67
Gifford, Sharma, Rinta-Kanto, and Moran (2011)	Tidal marsh	Surface	454 GS FLX	229	Exo, MEx, MEn, MAII	50
Hollibaugh, Gifford, Sharma, Bano, and Moran (2011)	Tidal marsh	Surface	454 GS FLX	257	Exo, MEx, MEn, MAII	50
Ottesen et al. (2011)	Temperate, coastal ocean	Surface	454 GS FLX	40	CHyb	38
Stewart, Sharma, Bryant, Eppley, and DeLong (2011)	Subtropical oligotrophic ocean	110, 500	454 GS FLX	109	CHyb	74[e]
Feike et al. (2012)	Oxygen minimum zone	118	454 GS FLX Titanium	12	Exo, MEx, MAII	64
Lesniewski, Jain, Anantharaman, Schloss, & Dick (2012)	Hydrothermal vent plume	1600–2000	454 GS FLX Titanium	140	MAII	86
Rinta-Kanto, Sun, Sharma, Kiene, and Moran (2012)	Gulf of Mexico, near-coastal	Surface	454 GS FLX	15	Exo, MEx, MEn, MAII	62
Shi, McCarren, and DeLong (2012)	Subtropical oligotrophic ocean	75	454 GS FLX	51	CHyb	71
Stewart, Dalsgaard, et al. (2012)	Oxygen minimum zone	80	454 GS FLX Titanium	367	CHyb	45
Stewart, Ulloa, et al. (2012)	Oxygen minimum zone	50–200	454 GS FLX	67	CHyb	51

Table 10.1 Percentage of rRNA in marine microbial metatranscriptomes following rRNA depletion—cont'd

References	Samples	Depth (m)	Sequencing	Mb[a]	rRNA dep[b]	% rRNA[c]
Burow et al. (2013)	Temperate estuary (microbial mat)	Surface	454 GS FLX Titanium	212	MEx, MEn, MAII	63
Gifford, Sharma, Booth, and Moran (2013)	Tidal marsh	Surface	Illumina GAIIx	780	Exo, MEx, MEn, MAII	63
Sanders, Beinart, Stewart, DeLong, and Girguis (2013)	Deep-sea vent symbionts	>2000	454 GS FLX Titanium	61	CHyb	54
Ottesen et al. (2013)	Temperate, coastal ocean	Surface	454 GS FLX Titanium	341	CHyb	54

[a]Megabases of sequence generated per sample (on average per study).
[b]rRNA depletion method: MAII, selective cDNA synthesis using the MessageAmp™II-Bacteria kit (Ambion); MEx, subtractive hybridization using the MICROBExpress™ kit; MEn, subtractive hybridization using the MICROBEnrich™ kit; Exo, mRNA-Only™ kit (exonuclease digestion); CHyb, subtractive hybridization with sample-specific probes (Stewart et al., 2010); MDA, selective cDNA amplification of mRNA using multiple displacement amplification (GenomiPHI™ V2 kit).
[c]Percentage of transcripts identified as rRNA by the authors following rRNA depletion (samples averaged per study).
[d]% rRNA not reported; estimated at 95% (Hewson, I., personal communication).
[e]Does not include archived datasets not directly sequenced in this study, or nonmarine datasets.

ensure that maximum information is recovered without bias from the RNA pool of interest, typically mRNA. This chapter reviews recent advancements in metatranscriptomic sample preparation, with the goal of identifying standardized methods and key considerations for obtaining cDNA from planktonic marine microbial communities.

Marine microbial metatranscriptomics is an evolving field, made possible only recently by advances in bioinformatics and sequencing in massively parallel configurations. The first metatranscriptomic study of planktonic microorganisms using high-throughput sequencing (Roche GS 20 instrument) was published only 6 years ago, describing a global analysis of community transcription in marine bacterioplankton from the oligotrophic Pacific Ocean (Frias-Lopez et al., 2008). The method represented a significant advance over array-based transcriptomics methods by eliminating the need for *a priori* knowledge of genomic diversity in the sample and thereby allowing for detection of transcripts from previously uncharacterized community members and genes. Subsequent to this study, metatranscriptomics has been applied to a diverse set of marine habitats, including phytoplankton blooms, anoxic oxygen minimum zones, deep-sea symbiotic organisms, and hydrothermal vent plumes (Anantharaman, Breier, Sheik, & Dick, 2013; Rinta-Kanto et al., 2012; Sanders et al., 2013; Stewart, Ulloa, & DeLong, 2012). These studies have identified novel regulatory processes as well as niche- and strain-specific variation in microbial metabolism. However, sampling from such habitats can itself cause changes in community transcription that could bias inferences of *in situ* metabolic activity (Feike et al., 2012), thereby imposing a special need for the development of *in situ* RNA preservation methods or rapid sample recovery. The following sections discuss key considerations for minimizing collection bias, maximizing isolation of target RNA (mRNA), and obtaining high yields of intact cDNA in preparation for sequencing. A generalized work plan highlighting key steps and recommendations for cDNA preparation is presented in Fig. 10.1. Downstream techniques relating to sequencing technology and bioinformatic analyses of gene expression are not discussed in this chapter. These important topics are covered in depth elsewhere in this volume (see Chapter 14 and Section V). Metatranscriptomics techniques will continue to evolve as they are applied to diverse marine habitats and experimental questions, and as the field as a whole converges on standard practices to facilitate comparisons across datasets.

2. SAMPLE COLLECTION AND RNA PRESERVATION

2.1. Shipboard filtration and preservation

Methods for sampling planktonic microorganisms should be optimized both to prevent transcriptional shifts in response to collection and to ensure total RNA stabilization in an appropriate preservative without the potential for downstream degradation. Transcriptional response times and RNA degradation rates are poorly characterized for most microbial communities. Comparisons involving (primarily) cultured nonmarine bacteria indicate that mRNA half-lives are generally much shorter than those for rRNA or tRNA and are typically on the order of minutes (e.g., median half-life of 2.4 min for *Prochlorococcus*; Deutscher, 2006; Steglich et al., 2010; Selinger, Saxena, Cheung, Church, & Rosenow, 2003; Taniguchi et al., 2010). However,

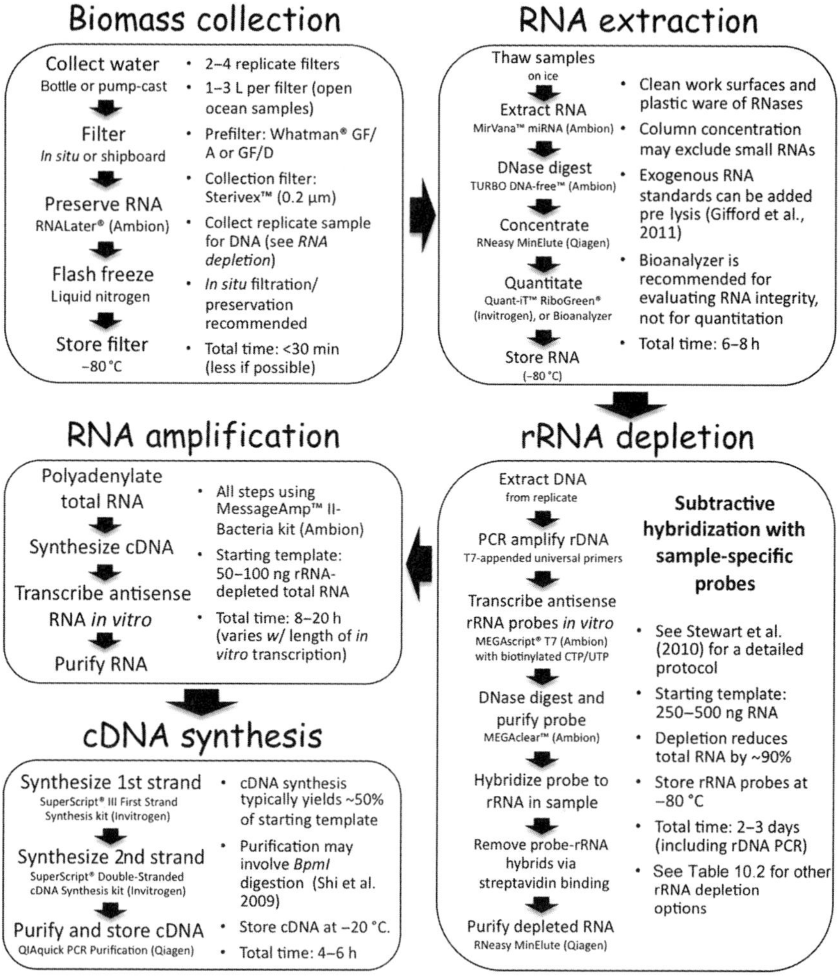

Biomass collection

Collect water
Bottle or pump-cast
Filter
In situ or shipboard
Preserve RNA
RNALater® (Ambion)
Flash freeze
Liquid nitrogen
Store filter
−80 °C

- 2–4 replicate filters
- 1–3 L per filter (open ocean samples)
- Prefilter: Whatman® GF/A or GF/D
- Collection filter: Sterivex™ (0.2 μm)
- Collect replicate sample for DNA (see *RNA depletion*)
- *In situ* filtration/preservation recommended
- Total time: <30 min (less if possible)

RNA extraction

Thaw samples
on ice
Extract RNA
MirVana™ miRNA (Ambion)
DNase digest
TURBO DNA-free™ (Ambion)
Concentrate
RNeasy MinElute (Qiagen)
Quantitate
Quant-iT™ RiboGreen® (Invitrogen), or Bioanalyzer
Store RNA
(−80 °C)

- Clean work surfaces and plastic ware of RNases
- Column concentration may exclude small RNAs
- Exogenous RNA standards can be added pre lysis (Gifford et al., 2011)
- Bioanalyzer is recommended for evaluating RNA integrity, not for quantitation
- Total time: 6–8 h

rRNA depletion

Extract DNA
from replicate
PCR amplify rDNA
T7-appended universal primers
Transcribe antisense rRNA probes *in vitro*
MEGAscript® T7 (Ambion) with biotinylated CTP/UTP
DNase digest and purify probe
MEGAclear™ (Ambion)
Hybridize probe to rRNA in sample
Remove probe-rRNA hybrids via streptavidin binding
Purify depleted RNA
RNeasy MinElute (Qiagen)

Subtractive hybridization with sample-specific probes

- See Stewart et al. (2010) for a detailed protocol
- Starting template: 250–500 ng RNA
- Depletion reduces total RNA by ~90%
- Store rRNA probes at −80 °C
- Total time: 2–3 days (including rDNA PCR)
- See Table 10.2 for other rRNA depletion options

RNA amplification

Polyadenylate total RNA
Synthesize cDNA
Transcribe antisense RNA *in vitro*
Purify RNA

- All steps using MessageAmp™ II-Bacteria kit (Ambion)
- Starting template: 50–100 ng rRNA-depleted total RNA
- Total time: 8–20 h (varies *w/* length of *in vitro* transcription)

cDNA synthesis

Synthesize 1st strand
SuperScript® III First Strand Synthesis kit (Invitrogen)
Synthesize 2nd strand
SuperScript® Double-Stranded cDNA Synthesis kit (Invitrogen)
Purify and store cDNA
QIAquick PCR Purification (Qiagen)

- cDNA synthesis typically yields ~50% of starting template
- Purification may involve *BpmI* digestion (Shi et al. 2009)
- Store cDNA at −20 °C.
- Total time: 4–6 h

Figure 10.1 Workflow for microbial community cDNA preparation prior to sequencing, based on Frias-Lopez et al. (2008), Shi et al. (2009), and Stewart et al. (2010). Additional details can be found in the text.

mRNA degradation in multispecies environmental samples is likely non-uniform, as decay rates are known to vary between species, in response to the growth state of the cell, and among genes in an operon (Barnett, Bugrysheva, & Scott, 2007; Kristoffersen et al., 2012; Steglich et al., 2010). Additionally, environmental stimuli, including changes in local conditions (e.g., oxygen, light, and temperature) due to sampling, can cause immediate transcriptional shifts in actively growing cells and therefore introduce bias into metatranscriptome datasets (Feike et al., 2012; Moran, 2009).

Consequently, recovering unbiased marine metatranscriptome samples requires special attention to minimize the time between water collection and RNA preservation.

The majority of marine metatranscriptome studies have followed similar collection protocols (Fig. 10.1). Seawater is retrieved from discrete depths in rosette-based Niskin bottles and brought immediately to the surface for shipboard filtration. Either a peristaltic or a vacuum pump is used to move sample water through a prefilter (often a glass fiber or polycarbonate membrane filter with a pore size of 1.6–2.7 μm (e.g., Whatman® GF/A or GF/D)) and then a primary collection filter (0.2 μm pore size). Collection filters are typically either hydrophilic disc filters (e.g., 25 mm Durapore® Membrane Filters, Millipore) or small cartridge filters (Sterivex™ Filters, Millipore). Relative to disc filters, Sterivex™ filters have the advantage of accommodating larger water volumes before clogging and of overall ease of use—they are sterile and individually packaged and can be easily mounted in line to filtration tubing via Luer fittings. To minimize transcriptional shifts during sampling, filtration should be done under conditions of temperature and light as close to *in situ* as possible, with total filtration times constrained to less than 10 min. In most studies, due to challenges of water recovery and sample handling, filtration times are often closer to ~20 min. These times can be reduced by preloading prefilters and collection filters prior to arrival of water on deck and by running multiple replicate filtration lines in parallel. A requirement for short filtration times typically limits the total water that can be filtered and, consequently, RNA yield per filter. However, replicate RNA extractions can later be pooled to obtain adequate RNA amounts. In our experience, the passage of ~1–3 L of seawater yields 200 ng to >1000 ng total RNA depending on water column biomass load. Immediately after filtration, filters are typically saturated in an RNA preservative and flash-frozen in liquid nitrogen and stored at −80 °C.

There has not been a systematic evaluation of how different RNA preservation and storage conditions affect marine metatranscriptome comparisons. However, flash freezing (typically in liquid nitrogen or on dry ice) followed by storage at −80 °C has been shown to yield high-quality RNA from diverse biological samples, including marine microorganisms (Gorokhova, 2005; Mutter et al., 2004; Rissanen, Kurhela, Aho, Oittinen, & Tiirola, 2010; Simister, Schmitt, & Taylor, 2011). Freezing may not be possible in field situations, however, and accidental thawing of frozen samples may activate RNases and pose a risk to RNA integrity. It is therefore recommended that microbial biomass be saturated in a storage reagent prior to freezing. Fixatives such as acetone, phenol–chloroform–isoamyl alcohol, Carnoy fixative (30% chloroform, 10% acetic acid, in ethanol), or Trizol® (phenol plus guanidine isothiocyanate; Invitrogen) have been shown to yield high-quality microbial RNA (Feike et al., 2012; Fukatsu, 1999; Rissanen et al., 2010) but are relatively toxic. Most marine community gene expression studies have instead used commercial stabilization solutions such as RNALater® (Ambion) or RNAProtect Bacteria Reagent (Qiagen).

RNALater®, in particular, has been applied widely. This reagent is relatively nontoxic but contains high amounts of ammonium sulfate, which rapidly penetrates cells to cause precipitation of soluble proteins, presumably including RNases (Allewell & Sama, 1974; Lader, 1998). RNALater® was shown to effectively preserve RNA yield and integrity in samples of cultured bacteria (*Synechococcus*, *Pseudomonas*) stored for 3 months at 4 °C (Bachoon, Chen, & Hodson, 2001). However, in a study involving nonmicrobial samples (microcrustaceans), RNALater®-preserved samples showed degradation after 4 months of storage at 5 °C and after 1 month at room temperature (Gorokhova, 2005). Recently, total RNA from marine bacterioplankton stored in RNALater® for 4 weeks at room temperature was shown to be stable with no obvious degradation (Ottesen et al., 2011). Further, metatranscriptome profiles did not differ significantly (only 0.03% of transcripts showed differential abundance) between a control sample frozen immediately after filtration and a sample stored in RNALater® on an autonomous sampling instrument for 4 weeks (see below; Ottesen et al., 2011). These results highlight the utility of RNALater® for recovering high-quality bacterioplankton RNA during short-term field operations without access to cold storage. However, immediate flash freezing of preserved samples should be done whenever possible. Although commercial reagents such as RNALater® are relatively expensive, ammonium sulfate-based recipes for RNA preservative are available online and may provide a cost-friendly solution when large volumes of preservative are required.

It is recommended that biomass samples for DNA analysis be collected jointly with samples for metatranscriptomics (Fig. 10.1). Coupled metagenomic and metatranscriptomic datasets allow for downstream standardization of transcript abundance relative to gene or taxon abundance, for example, via RNA:DNA expression ratios (Frias-Lopez et al., 2008; Shi, Tyson, Eppley, & DeLong, 2011; Stewart et al., 2011). Furthermore, downstream protocols for rRNA depletion can utilize rRNA-targeting hybridization probes that can be made sample specific based on PCR amplifications from coupled DNA samples (Fig. 10.1 and below; Stewart et al., 2010). Coupled DNA samples can be filtered and preserved following filtration for RNA samples and should, when water volumes allow, be collected from the same sample (bottle) used for RNA collection. Commercial protocols exist for the coisolation of both DNA and RNA directly from the same biomass sample (filter) (e.g., AllPrep DNA/RNA/Protein Mini kit, Qiagen) but have not been widely used for metaomic applications and may yield insufficient amounts of total DNA if sampling times are restricted to minimize transcriptional changes in the RNA pool. A recently described protocol for analysis of biomolecules (RNA, DNA, proteins, and metabolites) coisolated from the same sample showed promising results when applied to a fecal microbiome sample (Roume et al., 2013) but has not yet been applied to environmental samples with lower biomass.

2.2. *In situ* preservation

The majority of marine metatranscriptome studies following this general sampling protocol (Fig. 10.1) have focused on shallow depths from which water can be brought to the surface relatively quickly (Table 10.1). For deep-sea environments, particularly those where *in situ* environmental conditions differ markedly from those of the overlying depths (e.g., oxygen minimum zones, hydrothermal vents), the risk of sampling-induced transcriptional changes is substantial, notably given the short turnover times of microbial RNAs. Capturing unbiased metatranscriptional signals from these challenging environments requires new methods for preserving RNA *in situ*. Recent studies have demonstrated the utility of *in situ* RNA preservation using the Environmental Sample Processor (ESP) developed by the Monterrey Bay Aquarium Research Institute. The ESP is an autonomous water collection system that allows for *in situ* filtration, preservation, and storage of community biomass samples, as well as on-instrument molecular analyses and environmental monitoring (Scholin et al., 2009). Analysis of microbial community samples filtered, preserved, and stored on the ESP over a Lagrangian deployment has revealed synchronized physiological responses among diverse microbial community members in the same water mass (Ottesen et al., 2011, 2013). A related device, the Submerged Incubation Device-*In Situ* Microbial Sampler (SID-ISMS), developed by researchers at the Woods Hole Oceanographic Institution, can collect, filter, and preserve up to 48 samples *in situ* during a wire-based deployment and is currently being tested across diverse marine habitats.

Both the ESP and SID-ISMS are invaluable resources for time-series sampling of *in situ* microbial processes. However, these instruments are relatively challenging to deploy and are not broadly available to the larger research community, suggesting a need for *in situ* sampling methods scalable to individual labs. Feike et al. (2012) recently described a modified Niskin bottle called the automatic flow injection sampler (AFIS), which mixes preservative (1/10th of sample volume) into a discrete water sample immediately following bottle closure. The preserved water is then filtered aboard ship. In a comparison involving microbial communities from the Baltic Sea suboxic zone (120 m depth), the abundances of key transcripts differed by up to 30-fold between samples preserved *in situ* using the AFIS compared to those preserved after filtration aboard ship (Feike et al., 2012), indicating a significant potential for sampling-induced changes in gene expression in certain habitats (e.g., along steep redoxclines; Feike et al., 2012). Similarly, Wurzbacher, Salka, and Grossart (2012) recently introduced the *in situ* filtration and fixation sampler (IFFS), which is equivalent to the AFIS in size, portability, and scope of use (wire-based collection of single, discrete samples) but, like the larger ESP and SID-ISMS, filters seawater *in situ* and then immediately dispenses fixative to preserve the material on the filter. Dual filtration and fixation capability reduces the volume of fixative handled by the instrument, saving on cost and reducing exposure to potentially toxic preservatives. In a test of the instrument, transcript patterns of freshwater bacteria in IFFS-collected samples differed significantly from those in samples collected by Niskin bottles and preserved 45 min after sampling (Wurzbacher et al., 2012). While the IFFS is not commercially available, detailed specifications for rebuilding the instrument have been made public by the designers (Wurzbacher et al., 2012). The development and testing of these samplers highlight a growing recognition that accurately measuring gene expression in planktonic microbial communities requires easily operated, low-cost instruments for RNA preservation *in situ*.

2.3. Pump-cast collection

High-speed pumping systems can transport water to the surface from deep in the water column in a matter of minutes, potentially helping to minimize transcriptional shifts during sampling and providing an alternative to *in situ* preservation methods. Such methods have been used for collecting water from suboxic zones while minimizing oxygen intrusion into the samples (Canfield et al., 2010; Lam et al., 2007; Strady, Pohl, Yakushev,

Krüger, & Hennings, 2008). However, these methods are costly to develop and implement and may impose additional biases in the interpretation of community RNA signals. Notably, a significant increase in the transcription of stress-response genes was observed in samples collected using a high-speed pump system compared to those collected via *in situ* fixation or using a traditional Niskin bottle followed by shipboard filtration (Feike et al., 2012). It is possible that rapid passage of water (~2.9 L/min) through the narrow (18 mm) collection tube may have caused mechanical or pressure stress to the community. Transcriptional responses to physical stress may also occur during filtration *in situ* or following traditional Niskin collection, although such effects have not been thoroughly examined.

3. RNA EXTRACTION AND STANDARDIZATION

3.1. Extraction and DNA removal

The choice of RNA extraction method depends on the intended downstream analyses and the desired size range of the recovered transcripts. Multiple extraction methods and commercial kits have proved effective for obtaining high-quality RNA from marine microorganisms. These methods typically involve a cell lysis step using mechanical disruption (bead beating, vortexing) in the presence of a denaturing lysis buffer (usually guanidine-thiocyanate) that inactivates RNases, organic extraction (e.g., with phenol:chloroform) for phase separation of nucleic acids and proteins, and then nucleic acid recovery via either alcohol (ethanol or isopropyl) precipitation or adsorption of RNA molecules onto a silica column (solid-phase extraction). Prior to initiating any extraction procedure, filters should be thawed gently on ice in the presence of an appropriate RNA stabilization solution, and care must be taken to maintain an RNase-free work environment. Gloves should be worn at all times and changed often. Plasticware, glassware, and water should be treated with an appropriate RNase decontamination procedure (diethylpyrocarbonate treatment) or purchased sterile, and bench surfaces should be cleaned with an RNase-inhibiting solution (e.g., RNaseZap®, Ambion).

It is important to note that most column purifications discriminate against small RNA molecules less than ~200 bases and therefore require special binding and wash conditions for recovery of RNA at the smallest end of the size spectrum, for example, regulatory small RNAs (sRNAs), as well as tRNA and 5S rRNA. Commercial kits optimized for small RNA recovery, such as the *mir*Vana™ miRNA Isolation Kit (Ambion), have proved effective in isolating both sRNA and longer mRNA size fractions from marine plankton samples (Shi et al., 2009). The *mir*Vana™ protocol lyses cell using a denaturing lysis solution coupled with vortexing and can be easily adapted to introduce more vigorous mechanical disruption procedures (e.g., bead beating or pestle grinding of filters). This kit can be adapted for use with samples collected on Sterivex™ filters, such that initial cell lysis is conducted directly within the filter cartridge, followed by dispensing of the lysate for downstream extraction with acid-phenol:chloroform (Stewart, Dalsgaard, et al., 2012). An alternative protocol for bacterioplankton RNA isolation involves chemical and mechanical (bead beating) lysis followed by spin-column purification using the RNeasy® Mini Kit (see Poretsky, Gifford, Rinta-Kanto, Vila-Costa, and Moran (2009) for step-by-step instructions), although this method will not retain small RNA molecules.

Total RNA should be purified and accurately quantified following extraction. Purification typically involves treatment with DNase (e.g., TURBO™ DNase (Ambion)) to remove contaminating DNA, followed by concentration and cleanup using a glass fiber-based purification method, such as the RiboPure™-Bacteria Kit (Ambion) or the RNeasy MinElute Cleanup Kit (Qiagen). A loss of ~10–20% of total RNA should be expected during column cleanup, and it is recommended that two elution steps be performed to maximize recovery for samples with potentially low yields. As discussed earlier, column purifications typically discriminate against low-molecular-weight (<200 bases) RNA molecules. In one of the only studies examining sRNA in marine metatranscriptomes, a MinElute PCR Purification kit was used to purify and concentrate total RNA following extraction with the *mir*Vana™ kit (Shi et al., 2009). The MinElute kit, developed for use with double-stranded DNA molecules ranging from 70 to 4000 bp, enabled the recovery of a wide diversity of sRNAs from ~70 to 500 bases in length. However, the potential loss of RNA fragments smaller than this size range was not assessed. Future studies of sRNA in environmental microbial communities may benefit from RNA cleanup methods designed specifically to retain RNA at the smallest end of the size range, such as the RNA Clean & Concentrator™-5 Kit (Zymo Research) or the SurePrep™ RNA Cleanup and Concentration Kit (Fisher Scientific), which claim recovery of RNA across all size ranges, including sRNA.

Following DNase treatment and cleanup, total RNA should be quantitated and assessed for quality, and then placed at −80 °C for long-term storage. The Quant-iT™ RiboGreen® RNA Assay (Invitrogen) is recommended for fluorescence-based quantitation of RNA concentrations from 1 ng/mL to 1 μg/mL. This method is highly sensitive, but requires access to a fluorescent plate reader and is relatively time consuming (~1 h). Quantitation can also be done spectrophotometrically in a matter of minutes using common benchtop instruments (e.g., NanoDrop (Thermo Scientific)), although these methods tend to yield less-accurate results, notably for samples with low RNA concentration. Alternatively, the Agilent 2100 Bioanalyzer, which provides a visualization (electropherogram) of the RNA size distribution, can be used (with the RNA 6000 Pico kit) for quantitation in the range of 0.05–5.0 ng/μL, while simultaneously allowing for assessment of RNA integrity. However, in our experience, Bioanalyzer-based quantitations are less consistent than fluorescence-based methods and may require manual integration of electropherogram peak area for optimal results. RNA integrity is determined from electrophoretic traces indicating the presence or absence of RNA degradation products and is quantified via an RNA integrity number (RIN) ranging from 1 to 10, where level 10 indicates no degradation and completely intact rRNA peaks (Fig. 10.2). Measuring RNA integrity is important for estimating the

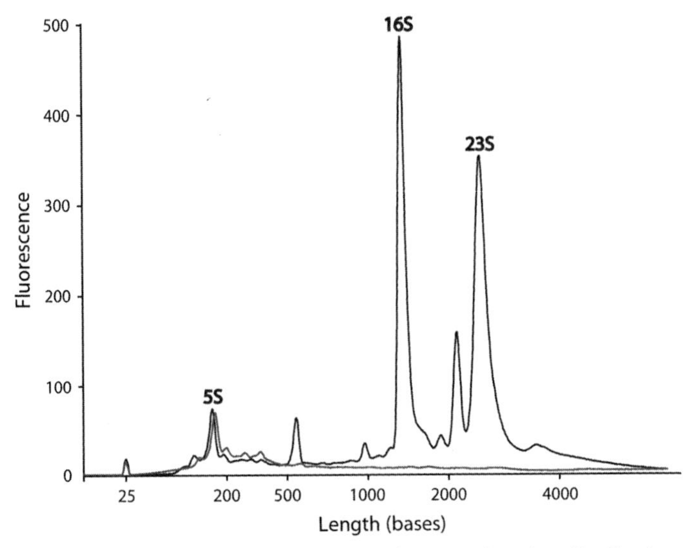

Figure 10.2 A Bioanalyzer electropherogram showing the size distribution of high-integrity total RNA (RIN ~9) from a marine bacterioplankton sample, before (blue) and after (red) rRNA depletion using sample-specific antisense probes targeting 16S and 23S rRNA transcripts. Note: despite removal of visual 16S and 23S rRNA peaks, rRNA fragments are still present in the sample. *Figure and rRNA depletion protocol are adapted from Stewart et al. (2010).* (For interpretation of the references to color in this figure legend, the reader is referred to the online version of this chapter.)

effectiveness of downstream rRNA removal methods, notably as efficient and unbiased rRNA removal may correlate with high RINs, depending on removal strategy (Giannoukos et al., 2012; see below).

3.2. Standardization with exogenous RNA transcripts

A challenge for the metatranscriptomic field is to relate measurements of relative transcript abundance inferred from sequencing to absolute transcript levels *in situ*. Recently, a standardization protocol has been developed to help bypass this limitation (Gifford et al., 2011). This technique, covered in detail in Chapter 13, involves the addition of exogenous RNA transcripts to an environmental sample, similar to approaches applied for qPCR and microarray standardization (Coyne et al., 2005; Hannah, Redestig, Leisse, & Willmitzer, 2008). RNA fragments of known length (~1 kb, approximating the average length of a bacterial protein-coding gene) are transcribed *in vitro* off of a plasmid, quantified and confirmed for the correct length, and added to samples (filtered biomass) prior to cell lysis and RNA extraction (Gifford et al., 2011). The number of standard fragments recovered by sequencing allows normalization across other genes in the RNA datasets and a direct estimate of the absolute transcript count per gene in the sample. Accurate extrapolation of standard counts to other genes assumes that both the standard RNA and the natural mRNA in the sample have uniform rates of degradation and that RNA extraction from cells is 100% efficient. Using this approach, the number of transcripts per cell has been estimated at ~200 for coastal bacterioplankton, several-fold lower than cultures of actively growing γ-proteobacteria but consistent with estimates from multiple studies based on RNA yield (Gifford et al., 2011; Lee & Kemp, 1994; Moran et al., 2013), suggesting that standardization with exogenous RNA allows a reasonable estimate of *in situ* transcript abundances. This method has thus far been applied sparingly to environmental samples, but holds considerable promise for inferring the magnitude of microbial physiological responses based on metatranscriptome data.

4. rRNA DEPLETION

4.1. Summary of rRNA depletion methods

As with any technology, it is imperative to recover the greatest amount of usable data possible. Metatranscriptomic studies are concerned primarily with transcription patterns of protein-coding genes, thereby necessitating that the ratio of mRNA to structural (ribosomal) RNA recovered per sample be maximized. This is a challenge for active microbial communities in which rRNA, including small (16S) and large (5S, 23S) subunit rRNA transcripts, may constitute greater than 90% of the total transcript pool (Ingraham, Maaløe, & Neidhardt, 1983; Neidhardt & Umbarger, 1996; Wendisch et al., 2001). As most prokaryotic mRNAs are not polyadenylated, mRNA enrichment methods developed for eukaryotes using oligo(dT)-based capture beads or selective priming (during cDNA synthesis) cannot be applied to microbial communities. However, a wide range of custom and commercially available methods have been developed to enrich for prokaryotic mRNA in total RNA samples (Tables 10.1 and 10.2). These include subtractive hybridization using custom or commercial oligonucleotide capture probes targeting rRNA (Chen & Duan, 2011; Poretsky et al., 2005; Poretsky, Hewson, et al., 2009; Stewart et al., 2010), enzymatic digestion of rRNA (Poretsky, Hewson, et al., 2009; Sharma et al., 2010), normalization methods to deplete highly abundant double-stranded molecules (primarily rRNA) in partially denatured cDNA pools (Vandernoot et al., 2012; Yi et al., 2011), selective cDNA synthesis using preferential poly(A) tailing of mRNA (Frias-Lopez et al., 2008; Wendisch et al., 2001) or not-so-random primer sets (Armour et al., 2009; Lin et al., 2010), and size separation of RNA pools by gel electrophoresis (McGrath et al., 2008). The majority of these methods have been developed and optimized for use with monocultures of bacteria. A systematic comparison of their efficiency across a range of environmental microbial communities, including those with archaeal members, has not been conducted.

4.2. Commercial rRNA depletion methods

The majority of marine metatranscriptomics studies that have included an mRNA enrichment step have used either subtractive hybridization or enzymatic digestion of rRNA, or a combination of both (Table 10.1). A recent study directly compared these methods using mixtures of RNA from cultures of nonmarine Bacteria and Archaea, and two popular commercially available kits: the mRNA-ONLY™ Prokaryotic mRNA Isolation kit (Epicentre) and the MICROBExpress™ Bacterial mRNA Enrichment kit (Ambion). The mRNA-ONLY™ method uses a $5' \rightarrow 3'$ exonuclease that preferentially degrades processed RNA molecules containing a 5′-monophosphate, the majority of which are presumably rRNA. However, secondary non-5′-monophosphate activity has been observed and may,

Table 10.2 Methods for rRNA depletion from total RNA samples

Method	Kit	Used for metatranscriptomics	Other refs; comments
Subtractive hybridization, sample specific	Custom protocol	See Table 10.1	Stewart et al. (2010); can target rRNA from all three Domains; requires probe synthesis
Subtractive hybridization, commercial	MICROBExpress™ Bacterial mRNA Enrichment kit (Ambion)	See Table 10.1	See Table 10.1 for refs; not sample specific; capture probes based on model organisms
	Ribo-Zero™ Magnetic Kit (Bacteria) (Epicentre)	Giannoukos et al. (2012)	Not sample specific; capture probes based on model organisms
	RiboMinus™ Transcriptome Isolation kit (Bacteria) (Invitrogen)	Xiong et al. (2012)	Chen and Duan (2011); not sample specific; capture probes based on model organisms
	MICROBEnrich™ kit (Ambion)	See Table 10.1	Depletion of eukaryotic rRNA and mRNA; not sample specific; capture probes based on model organisms; may remove polyadenylated prokaryotic mRNA
Size separation by gel electrophoresis	Custom protocol	McGrath et al. (2008)	mRNAs may comigrate and be removed with rRNA
Duplex removal by hydroxyapatite chromatography	Custom protocol	NA	Vandernoot et al. (2012); labor intensive

Continued

Table 10.2 Methods for rRNA depletion from total RNA samples—cont'd

Method	Kit	Used for metatranscriptomics	Other refs; comments
Exonuclease digestion of 5′-monophosphate (processed) RNA	mRNA-ONLY™ Prokaryotic mRNA Isolation kit (Epicentre)	See Table 10.1	Sharma et al. (2010); efficient to use; may degrade some mRNAs; may facilitate transcription start site detection
Duplex-specific nuclease (DSN) removal of abundant transcripts	TRIMMER-DIRECT cDNA Normalization Kit (Evrogen)	Giannoukos et al. (2012)	Yi et al. (2011); applied to cDNA (not RNA); may remove highly abundant mRNAs
Selective amplification of mRNA-derived cDNA via multiple displacement amplification	GenomiPHI™ V2 kit (GE Healthcare)	Gilbert et al. (2008)	Potential for nonuniform amplification of cDNA; not quantitative
Selective cDNA synthesis via not-so-random priming	Ovation® Prokaryotic RNA-Seq System (NuGEN), or taxon-specific primers	Giannoukos et al. (2012)	Lin, Zhang, Zhuang, Tran, and Gill (2010); requires sequence information; for example, enrichment of eukaryotic mRNA using taxon-specific leader sequences
Selective cDNA synthesis via poly(A) tailing	MessageAmp™II-Bacteria RNA Amplification Kit (Ambion)	Frias-Lopez et al. (2008)	Wendisch et al. (2001); not widely used for rRNA depletion

on occasions, lead to mRNA degradation (mRNA-ONLY™ Manual, Epicentre). The MICROBExpress™ kit removes rRNA via hybridization to 16S and 23S rRNA-targeting oligonucleotides bound to magnetic beads. A direct comparisons showed that MICROBExpress™ outperformed exonuclease digestion, reducing observed rRNA percentages from 95 to 97% of total RNA to 77–89%, compared to 86–91% rRNA following use of the Epicentre kit (He et al., 2010). This study also showed that using both kits in combination only marginally improved rRNA depletion levels. Notably, an improvement in mRNA detection sensitivity was not observed when exonuclease digestion was used upstream of hybridization, potentially because exonuclease treatment fragments the RNA pool, thereby increasing the number of rRNA fragments that lack binding sites for capture probes (He et al., 2010). Indeed, in this study, lower RINs, a measure of the level of RNA degradation in the sample, were associated with overall lower depletion efficiencies.

Several marine metatranscriptome studies have used a combination of mRNA-ONLY™ and MICROBExpress™ kits together with a third kit, the MICROBEnrich™ (Ambion) (Table 10.1). Adding the MICROBEnrich™ procedure introduces a second rRNA capture step targeting eukaryotic 18S and 28S rRNA, as well as polyadenylated mRNAs. Eukaryotic RNA removal may be especially important in microbial biomass samples that have not been prefiltered to remove larger phytoplankton or zooplankton cells. Published studies using this tripartite approach have reduced rRNA percentages in diverse marine microbial samples to an average of 55% of total RNA ($n=7$ samples (5 studies) examined here; Table 10.1). However, the capture oligonucleotides used in these commercial hybridization methods were optimized using model organisms (e.g., *Escherichia coli* RNA for MICROBExpress™; human, mouse, and rat RNA for MICROBEnrich™) and therefore may bind with lower efficiency to diverse microbial communities (prokaryotes and eukaryotes) from natural environments. Furthermore, although 3′-poly(A) tracts are typically associated primarily with eukaryotic mRNAs, mRNA polyadenylation does occur in prokaryotes (Dreyfus & Régnier, 2002), in which it has been estimated to occur on 2–60% of all molecules for a given mRNA transcript, primarily in association with RNA degradation (Sarkar, 1997). The use of poly(A)-based RNA capture via the MICROBEnrich™ kit may therefore selectively remove a fraction of prokaryotic mRNAs; this potential bias has not been explored for metatranscriptome studies.

4.3. Subtractive hybridization with sample-specific probes

The efficiency of subtractive hybridization may be improved for complex environmental samples by using customized, sample-specific rRNA probes (Stewart et al., 2010). Materials lists and a step-by-step description of a sample-specific hybridization protocol are provided as Supplementary

Material in Stewart et al. (2010). Briefly, a coupled DNA sample is used to PCR amplify rRNA genes using universal primers appended with promoter sites for a T7 RNA polymerase. The amplicon pool is then used as template for *in vitro* transcription using biotinylated nucleotides, yielding antisense rRNA probes specific to the environmental community. Biotinylated antisense probes are hybridized to complementary rRNA molecules over a step-down procedure of decreasing temperature, and probe-bound, double-stranded rRNA is then removed by binding to streptavidin-coated magnetic beads. Although marine microbial metatranscriptomes are typically dominated by bacterial RNA, this protocol contains additional broad-specificity primer sets for generating probes targeting archaeal (16S, 23S) and eukaryotic (18S, 28S) rRNA. [Primers matching the 5/5.8S rRNA are not included in this method; however, 5/5.8S transcripts are typically a minor component of the rRNA pool, based on electropherograms from environmental communities (Stewart et al., 2010). In some protocols, a low abundance of 5S rRNAs may be due to upstream purification methods, such as the RNeasy® kit (Qiagen), which discriminate against low-molecular-weight RNAs.]

An advantage of this approach is that probes are tailored specifically to the taxonomic composition of the sample and therefore have a higher likelihood of efficient binding compared to probes based on model organisms. Also, probes are synthesized across nearly the full length of the rRNA gene, helping to avoid losses in efficiency due to fragmentation of sample RNA. However, it is possible that probes matching the rRNA of rare but highly transcriptionally active members may be underrepresented in the custom probe set. To avoid this possibility, probes could instead be synthesized from amplicons generated by reverse transcription (RT)-PCR using total RNA as starting template, although RT-PCR is typically less efficient across longer sequences and could therefore reduce the level of probe coverage across target rRNA molecules. Sample-specific hybridization has been shown to have minimal effects on mRNA abundance profiles (via nonspecific hybridization) (Stewart et al., 2010) and has been used to reduce rRNA on average to 52% of total across diverse marine samples ($n=59$ samples (8 studies); Table 10.1), and to as low as 17% in a soil metatranscriptome (Stewart et al., 2011), although probes targeting all three domains were not used in all studies. Recently, a modification of the method was developed for use in analyzing insect gut metatranscriptomes and was shown to deplete rRNA to less than 11% (Kukutla, Steritz, & Xu, 2013). Modifications included two additional rounds of probe removal via biotin–streptavidin binding and the generation of probes (bacterial 16S and 23S) using multiple combinations of universal primer sets. This modified protocol has not yet been applied to marine microbial communities.

4.4. rRNA removal efficiency, alternative methods, and perspectives

A survey of published marine metatranscriptome studies suggests that the most widely used methods for rRNA depletion (primarily subtractive hybridization and exonuclease digestion, discussed earlier) reduce rRNA levels on average to only ~50% of total RNA, a seemingly low efficiency (Table 10.1). However, this statistic is misleading. Assuming that rRNA in nonmanipulated samples constitutes 90% of total RNA, a reduction to 50% of total RNA corresponds to an absolute removal of 89% of rRNA. Even if 95% of rRNA is removed, rRNA would still constitute 31% of total RNA following mRNA enrichment. These numbers assume that the sequencing preparation protocol itself does not introduce any bias for or against certain transcripts (e.g., size selection for longer transcripts dominated by rRNA). It is possible that no method will consistently yield such high efficiencies, even under the best experimental circumstances.

Other newer or less widely used rRNA depletion methods have, nonetheless, reported rRNA removal efficiencies approaching 100% for metatranscriptome samples. Gilbert et al. (2008) described a reduction of rRNA to 0.08% of total in coastal marine communities following amplification of cDNA using a multiple displacement amplification (MDA; GenomiPHI™ V2 kit, GE Healthcare), which may have discriminated against rRNA due to its secondary structure. However, MDA does not amplify uniformly across loci and its use for quantitative analysis of metagenomes and metatranscriptomes is not recommended (Kim & Bae, 2011; Yilmaz, Allgaier, & Hugenholtz, 2010). Recently, a new commercial kit for subtractive hybridization of rRNA (including 5S rRNA) using proprietary probe sets, the Ribo-Zero™ Magnetic Kit (Bacteria) (Epicentre), was shown to deplete rRNA to less than 5% of total in cultures of Gram-positive and Gram-negative bacteria and in a gut microbiome, without introducing bias in the relative abundances of mRNAs (Giannoukos et al., 2012). Ribo-Zero™ performed significantly better than other commercially available methods using subtractive hybridization (MICROBExpress™), exonuclease digestion (mRNA-ONLY™), duplex-specific nuclease removal, and enrichment of mRNAs using selective priming during cDNA synthesis (Ovation® Prokaryotic RNA-Seq System, NuGEN) (Giannoukos et al.,

2012; Peano et al., 2013). While the Ribo-Zero™ method has been shown to be effective for human microbiome samples (Lim et al., 2012), it has not yet (to the best of our knowledge) been tested on marine microbial communities. Notably, Ribo-Zero™ may cause a drop in the number of detected mRNAs when input RNA quantities are reduced below the recommended amount (1–5 μg) (Giannoukos et al., 2012), suggesting this method may require optimization for planktonic microbial samples with RNA yields in the tens to hundreds of nanograms.

Sequencing costs have dropped substantially in recent years, raising the question of whether or not it is less costly and more time efficient to skip an rRNA removal step and instead devote resources to additional sequencing. This approach has been employed in a metatranscriptome analysis of soil communities (Urich et al., 2008) and has the added benefit of providing large amounts of quantitative data on the taxonomic composition of the transcriptionally active community. The published marine metatranscriptome studies listed in Table 10.1 produced an average of ~120 megabases (Mb) of total cDNA sequence data per sample, or roughly 60 Mb of non-rRNA data, assuming 50% rRNA following depletion (Table 10.1). Many of these datasets were generated on the Roche 454 platform, which currently yields data (using the GS FLX instrument with Titanium series chemistry) at a cost of ~$10,000 per gigabase (Gb) (sequencing cost only; Liu et al., 2012). In contrast, the Illumina HiSeq 2000 instrument generates sequence at ~$40–70 per Gb (Liu et al., 2012; Quail et al., 2012), and as Illumina paired-end read lengths continue to increase, this platform will likely replace 454 as the method of choice for metatranscriptomics. Furthermore, compared to 454 pyrosequencing, new reagents and protocols developed for the Illumina platform provide added benefits such as streamlined approaches for library preparation, a reduced amount of total cDNA input required (as little as 50 ng), and information-rich strand-specific cDNA library synthesis procedures. Assuming rRNA depletion reduces rRNA to 50% of total, producing 60 Mb of non-rRNA sequence data would cost ~$8 using Illumina (at $70 per Gb) or ~$1200 using 454 (sequencing costs only). Generating a comparable amount of data from a non-rRNA-depleted sample containing 90% rRNA would cost $40 (Illumina) or $6000 (454), implying that rRNA depletion must cost less than $32 (Illumina) or $4800 (454) per sample, including materials and labor costs, to be more cost-effective than additional sequencing. Current costs of rRNA depletion are estimated at ~$15 to ~$100 per sample depending on methodology, suggesting that depletion is only narrowly (or is not) cost-effective at this depth of Illumina sequencing, but substantially more cost-effective than additional sequencing using 454 technology. However, this simplified analysis does not take into account consideration of read length differences. Mode read length for 454 ranges from 400 to 700 bp depending on sequencing kit, whereas the HiSeq 2000 generates 100 bp reads. Longer reads provide additional data to facilitate accurate gene and taxonomic predictions (Luo, Tsementzi, Kyrpides, Read, & Konstantinidis, 2012), and therefore may be a benefit when sampling environments for which few reference genomes are available. This advantage may provide incentive for choosing 454 over Illumina technology for some applications; in such analyses, rRNA depletion is recommended to offset sequencing costs.

Furthermore, for environmental metatranscriptomes representing potentially hundreds to thousands of species, current sequencing depths are almost certainly inadequate. Indeed, transcripts detected in ~115 Mb of non-rRNA cDNA data were estimated to represent only 0.00001% of the mRNA molecules in a coastal bacterioplankton sample (Gifford et al., 2011). Sequencing coverage will need to increase, potentially into the tens of Gbs per sample, in order to detect statistically significant patterns in the vast majority of transcripts, most of which are present at low percentages (<0.01%) of the total transcript pool (Haas, Chin, Nusbaum, Birren, & Livny, 2012; Stewart et al., 2010). Such coverage increases will rapidly increase the cost-effectiveness of rRNA depletion but will likely be buffered by continuing decreases in sequencing cost. Choosing an optimal cDNA preparation strategy therefore involves weighing the current cost of sequencing at desired depths and read lengths against the materials and labor costs of rRNA depletion, with the expectation that absolute RNA depletion efficiency is unlikely to be greater than 90%.

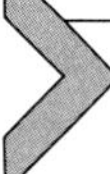

5. RNA AMPLIFICATION AND cDNA SYNTHESIS

5.1. Amplification by *in vitro* transcription

Total RNA yields are often low (<1000 ng) for marine plankton samples due to sampling and methodological constraints combined with potentially low overall community transcriptional activity. Bacterioplankton RNA yields recovered from 5 to 10 L of seawater from the oligotrophic open ocean may be less than 500 ng depending on extraction efficiency. Yields for coastal samples are greater, typically in the 1000–2000 ng range. However, efficient rRNA subtraction can reduce yields by >80%

(due both to rRNA removal and sample loss due to processing; Giannoukos et al., 2012). Consequently, to obtain sufficient quantities (typically >1000 ng) for a final cDNA synthesis step and for cDNA sequencing, it is often necessary to amplify total RNA.

RNA amplification is typically done via *in vitro* transcription following an initial round of cDNA synthesis. The MessageAmp II-Bacteria Kit (Ambion) has been used extensively for this step (Table 10.1). Briefly, this method involves polyadenylation of template RNA using *E. coli* poly(A) polymerase, followed by first-strand cDNA synthesis primed with oligo(dT) primers appended to the promoter sequence for T7 RNA polymerase. [A starting template amount of 50–100 ng RNA is recommended, although as little as 3 ng has been used successfully for amplification (Frank Stewart, personal communication).] Following second-strand synthesis, *in vitro* transcription using T7 polymerase is used to synthesize microgram quantities of antisense RNA in a linear amplification. In our experience, low yield samples may require running the transcription reaction for up to 14 h; longer times may result in minor amounts of RNA degradation. This protocol has been modified to introduce a restriction enzyme (*Bpm*I) recognition site sequence in the oligo(dT)-T7 primer, enabling digestion with *Bpm*I to remove poly(A) tails after cDNA synthesis (see below; Frias-Lopez et al., 2008). It has been suggested that the MessageAmp kit may also function to deplete rRNA due to selective polyadenylation of microbial mRNA over rRNA (Frias-Lopez et al., 2008). However, subsequent studies using this method without additional rRNA depletion steps have failed to significantly reduce rRNA concentrations (Lesniewski et al., 2012; McCarren et al., 2010), and most metatranscriptome studies use the MessageAmp technique primarily for total RNA amplification.

5.2. cDNA synthesis

The final step of cDNA preparation involves converting amplified antisense total RNA to double-stranded cDNA for sequencing. A widely used protocol (see Shi et al., 2009; Stewart et al., 2010) involves a combination of the SuperScript® III First-Strand Synthesis kit (Invitrogen) for first-strand synthesis and the SuperScript® Double-Stranded cDNA Synthesis kit (Invitrogen) for second-strand synthesis. In this protocol, first-strand synthesis via reverse transcriptase is primed with random hexamers, while second-strand synthesis by DNA polymerase is primed by RNA fragments created by RNaseH activity. Resulting double-stranded cDNA is then purified (e.g., using the QIAquick PCR purification kit (Qiagen)), digested with *Bpm*I to remove residual poly(A) tails (if using the MessageAmp protocol of Frias-Lopez et al. (2008) described earlier), and then used for sequencing. In our experience, a loss of ~50% of input antisense RNA should be expected during this cDNA step.

Although transcript abundance patterns inferred from metatranscriptome sequencing have been validated using RT-qPCR in several independent studies (Frias-Lopez et al., 2008; Hewson, Poretsky, Beinart, et al., 2009; Poretsky, Hewson, et al., 2009; Shi et al., 2009), more subtle biases potentially introduced by selective priming during cDNA synthesis are relatively unexplored. Analyses of transcriptome datasets generated on the Illumina platform have revealed nonuniformity in random hexamer priming based on nucleotide composition (Hansen, Brenner, & Dudoit, 2010). The alternative priming strategy, using oligo(dT) sequences, may also introduce artifacts due to preferential priming at internal poly(A) sites or at the 3′-end of transcripts (Brooks, Sheflin, & Spaulding, 1995; Nam et al., 2002). While attempts have been made to correct for these biases in RNA-seq experiments involving single organisms (Hansen et al., 2010; Roberts, Trapnell, Donaghey, Rinn, & Pachter, 2011), these methods are challenging for environmental metatranscriptomes with low sequence coverage per transcript and without reference genomes. Evaluating potential effects of priming strategy during cDNA synthesis should be a priority in future metatranscriptome analyses.

5.3. Preserving strand orientation

Environmental metatranscriptome studies have thus far not been concerned with preserving the strand orientation of RNA transcripts. Determining strand orientation is critical for studies of antisense RNAs, which are common in bacterial transcriptomes, although their functional significance is still not well understood (Raghavan, Sloan, & Ochman, 2012). A simple method for determining strand orientation involves incorporating dUTP (rather than dTTP) during second-strand cDNA synthesis (Parkhomchuk et al., 2009). The uridine-containing strand is then selectively degraded during the preparation of sequencing libraries by digestion with uracil-*N*-glycosylase. This approach has enabled the determination of strand-specificity using transcriptomes of cultured bacteria for which whole-genome sequences are available for transcriptome mapping (Giannoukos et al., 2012). A similar approach could presumably be integrated into the second-strand cDNA synthesis procedure described earlier, following RNA amplification. Alternatively, newer cDNA synthesis protocols developed for the Illumina platform (e.g., ScriptSeq™ v2 RNA-seq Library Preparation kit, Epicentre) that incorporate 3′ and 5′ tagging strategies of cDNAs may allow the direct identification of sense and antisense RNA sequences. Analyses of strand orientation in metatranscriptomes are likely to become more commonplace as reference genomes from environmentally representative taxa become available.

6. SUMMARY

Metatranscriptomics is now a widespread tool for exploring the physiology and ecology of microorganisms in natural environments. Application of this technology will continue to drive and be driven by advances in sequencing methods and bioinformatics. Other chapters in this volume (Section V) and several recent articles have discussed current methods and challenges relating to these important topics (Carvalhais, Dennis, Tyson, & Schenk, 2012; Filiatrault, 2011; Moran et al., 2013). First and foremost, however, robust metatranscriptome studies require unbiased sample collection and the generation of high-quality cDNA from environmental samples. For marine microbial communities, in particular, improvements in sampling for cDNA will likely involve the development and widespread application of *in situ* preservation methods and an increased understanding of the temporal response of community gene expression to both environmental fluctuations and sampling-induced biases.

As sequencing goals shift, potentially due to reductions in sequencing cost or coverage requirements for statistical inference, the general cDNA preparation workflow presented in Fig. 10.1 will likely shift as well. Notably, rRNA depletion methods may soon no longer be cost-effective for analyses targeting abundant taxa or highly expressed genes but will likely continue to be a critical need when exploring patterns in minor community members. Recent reductions in the amount of starting template required for high-throughput sequencing may eliminate the need for RNA amplification. Notably, the Illumina HiSeq and MiSeq platforms advertise starting input requirements of 50 ng, well within the range obtainable following cDNA synthesis from a typical bacterioplankton sample (however, rRNA depletion may significantly deplete total RNA, making it challenging to recover even 50 ng cDNA). Also, the potentially biasing effects of different sample storage, RNA extraction, and cDNA synthesis protocols have not been evaluated by comparative analysis of metatranscriptome profiles, suggesting that until these biases are understood, cDNA preparation protocols should be standardized to the greatest extent possible. This chapter aims to help achieve this goal, helping to create a comparative framework for exploring microbial community gene expression across diverse marine environments.

ACKNOWLEDGMENTS

The author thanks Ed DeLong for constructive input during the preparation of this chapter. This work was made possible by generous support from the Alfred P. Sloan Foundation and the National Science Foundation (Grant 1151698).

REFERENCES

Allewell, N. M., & Sama, A. (1974). The effect of ammonium sulfate on the activity of ribonuclease A. *Biochimica et Biophysica Acta*, *341*, 484–488.

Anantharaman, K., Breier, J. A., Sheik, C. S., & Dick, G. J. (2013). Evidence for hydrogen oxidation and metabolic plasticity in widespread deep-sea bacteria. *Proceedings of the National Academy of Sciences of the United States of America*, *110*, 330–335.

Armour, C. D., Castle, J. C., Chen, R., Babak, T., Loerch, P., Jackson, S., et al. (2009). Digital transcriptome profiling using selective hexamer priming for cDNA synthesis. *Nature Methods*, *6*, 647–649.

Bachoon, D. S., Chen, F., & Hodson, R. E. (2001). RNA recovery and detection of mRNA by RT-PCR from preserved prokaryotic samples. *FEMS Microbiology Letters*, *201*, 127–132.

Barnett, T. C., Bugrysheva, J. V., & Scott, J. R. (2007). Role of mRNA stability in growth phase regulation of gene expression in the group A Streptococcus. *Journal of Bacteriology*, *189*, 1866–1873.

Brooks, E. M., Sheflin, L. G., & Spaulding, S. W. (1995). Secondary structure in the 3′ UTR of EGF and the choice of reverse transcriptases affect the detection of message diversity by RT-PCR. *Biotechniques*, *19*, 806–812.

Burow, L. C., Woebken, D., Marshall, I. P., Lindquist, E. A., Bebout, B. M., Prufert-Bebout, L., et al. (2013). Anoxic carbon flux in photosynthetic microbial mats as revealed by metatranscriptomics. *The ISME Journal*, *7*, 817–829.

Canfield, D. E., Stewart, F. J., Thamdrup, B., De Brabandere, L., Dalsgaard, T., DeLong, E. F., et al. (2010). A cryptic sulfur cycle in oxygen-minimum zone waters off the Chilean Coast. *Science*, *330*, 1375–1378.

Carvalhais, L. C., Dennis, P. G., Tyson, G. W., & Schenk, P. M. (2012). Application of metatranscriptomics to soil environments. *Journal of Microbiological Methods*, *91*, 246–251.

Chen, Z., & Duan, X. (2011). Ribosomal RNA depletion for massively parallel bacterial RNA-sequencing applications. *Methods in Molecular Biology*, *733*, 93–103.

Coyne, K. J., Handy, S. M., Demir, E., Whereat, E. B., Hutchins, D. A., Portune, K. J., et al. (2005). Improved quantitative real-time PCR assays for enumeration of harmful algal species in field samples using an exogenous DNA reference standard. *Limnology and Oceanography: Methods*, *3*, 381–391.

DeLong, E. F. (2009). The microbial ocean from genomes to biomes. *Nature*, *459*, 200–206.

Deutscher, M. P. (2006). Degradation of RNA in bacteria: Comparison of mRNA and stable RNA. *Nucleic Acids Research*, *34*, 659–666.

Dreyfus, M., & Régnier, P. (2002). The poly(A) tail of mRNAs: Bodyguard in eukaryotes, scavenger in bacteria. *Cell*, *111*, 611–613.

Feike, J., Jürgens, K., Hollibaugh, J. T., Krüger, S., Jost, G., & Labrenz, M. (2012). Measuring unbiased metatranscriptomics in suboxic waters of the central Baltic Sea using a new in situ fixation system. *The ISME Journal, 6*, 461–470.
Filiatrault, M. J. (2011). Progress in prokaryotic transcriptomics. *Current Opinion in Microbiology, 14*, 579–586.
Frias-Lopez, J., Shi, Y., Tyson, G. W., Coleman, M. L., Schuster, S. C., Chisholm, S. W., et al. (2008). Microbial community gene expression in ocean surface waters. *Proceedings of the National Academy of Sciences of the United States of America, 105*, 3805–3810.
Fukatsu, T. (1999). Acetone preservation: A practical technique for molecular analysis. *Molecular Ecology, 8*, 1935–1945.
Giannoukos, G., Ciulla, D. M., Huang, K., Haas, B. J., Izard, J., Levin, J. Z., et al. (2012). Efficient and robust RNA-seq process for cultured bacteria and complex community transcriptomes. *Genome Biology, 13*, R23.
Gifford, S. M., Sharma, S., Booth, M., & Moran, M. A. (2013). Expression patterns reveal niche diversification in a marine microbial assemblage. *The ISME Journal, 7*, 281–298.
Gifford, S. M., Sharma, S., Rinta-Kanto, J. M., & Moran, M. A. (2011). Quantitative analysis of a deeply sequenced marine microbial metatranscriptome. *The ISME Journal, 5*, 461–472.
Gilbert, J. A., Field, D., Huang, Y., Edwards, R., Li, W., Gilna, P., et al. (2008). Detection of large numbers of novel sequences in the metatranscriptomes of complex marine microbial communities. *PLoS One, 3*, e3042.
Gilbert, J. A., Field, D., Swift, P., Thomas, S., Cummings, D., Temperton, B., et al. (2010). The taxonomic and functional diversity of microbes at a temperate coastal site: A 'multi omic' study of seasonal and diel temporal variation. *PLoS One, 5*, e15545.
Gorokhova, E. (2005). Effects of preservation and storage of microcrustaceans in RNAlater on RNA and DNA degradation. *Limnology and Oceanography: Methods, 3*, 143–148.
Haas, B. J., Chin, M., Nusbaum, C., Birren, B. W., & Livny, J. (2012). How deep is deep enough for RNA-Seq profiling of bacterial transcriptomes? *BMC Genomics, 13*, 734.
Hannah, M. A., Redestig, H., Leisse, A., & Willmitzer, L. (2008). Global mRNA changes in microarray experiments. *Nature Biotechnology, 26*, 741–742.
Hansen, K. D., Brenner, S. E., & Dudoit, S. (2010). Biases in Illumina transcriptome sequencing caused by random hexamer priming. *Nucleic Acids Research, 38*, e131.
He, S., Wurtzel, O., Singh, K., Froula, J. L., Yilmaz, S., Tringe, S. G., et al. (2010). Validation of two ribosomal RNA removal methods for microbial metatranscriptomics. *Nature Methods, 7*, 807–812.
Hewson, I., Poretsky, R. S., Beinart, R. A., White, A. E., Shi, T., Bench, S. R., et al. (2009). In situ transcriptomic analysis of the globally important keystone N2-fixing taxon *Crocosphaera watsonii. The ISME Journal, 3*, 618–631.
Hewson, I., Poretsky, R. S., Dyhrman, S. T., Zielinski, B., White, A. E., Tripp, H. J., et al. (2009). Microbial community gene expression within colonies of the diazotroph, *Trichodesmium*, from the Southwest Pacific Ocean. *The ISME Journal, 3*, 1286–1300.
Hewson, I., Poretsky, R. S., Tripp, H. J., Montoya, J. P., & Zehr, J. P. (2010). Spatial patterns and light-driven variation of microbial population gene expression in surface waters of the oligotrophic open ocean. *Environmental Microbiology, 12*, 1940–1956.
Hollibaugh, J. T., Gifford, S., Sharma, S., Bano, N., & Moran, M. A. (2011). Metatranscriptomic analysis of ammonia-oxidizing organisms in an estuarine bacterioplankton assemblage. *The ISME Journal, 5*, 866–878.
Ingraham, J. L., Maaløe, O., & Neidhardt, F. C. (1983). *Growth of the bacterial cell.* Sunderland, MA, USA: Sinauer Associates.
Kim, K. H., & Bae, J. W. (2011). Amplification methods bias metagenomic libraries of uncultured single-stranded and double-stranded DNA viruses. *Applied and Environmental Microbiology, 77*, 7663–7668.

Kristoffersen, S. M., Haase, C., Weil, M. R., Passalacqua, K. D., Niazi, F., Hutchison, S. K., et al. (2012). Global mRNA decay analysis at single nucleotide resolution reveals segmental and positional degradation patterns in a Gram-positive bacterium. *Genome Biology, 13*, R30.

Kukutla, P., Steritz, M., & Xu, J. (2013). Depletion of ribosomal RNA for mosquito gut metagenomic RNA-seq. *Journal of Visualized Experiments, 74*, e50093.

Lader, E. (1998). Methods and reagents for preserving RNA in cell and tissue samples. U.S. Patent 6204375, filed July 31, 1998, issued March 20, 2001.

Lam, P., Jensen, M. M., Lavik, G., McGinnis, D. F., Müller, B., Schubert, C. J., et al. (2007). Linking crenarchaeal and bacterial nitrification to anammox in the Black Sea. *Proceedings of the National Academy of Sciences of the United States of America, 104*, 7104–7109.

Lee, S., & Kemp, P. F. (1994). Single-cell RNA content of natural marine planktonic bacteria measured by hybridization with multiple 16S rRNA-targeted fluorescent probes. *Limnology and Oceanography, 39*, 869–879.

Lesniewski, R. A., Jain, S., Anantharaman, K., Schloss, P. D., & Dick, G. J. (2012). The metatranscriptome of a deep-sea hydrothermal plume is dominated by water column methanotrophs and lithotrophs. *The ISME Journal, 6*, 2257–2268.

Lim, Y. W., Schmieder, R., Haynes, M., Willner, D., Furlan, M., Youle, M., et al. (2012). Metagenomics and metatranscriptomics: Windows on CF-associated viral and microbial communities. *Journal of Cystic Fibrosis, 12*, 154–164.

Lin, S., Zhang, H., Zhuang, Y., Tran, B., & Gill, J. (2010). Spliced leader-based metatranscriptomic analyses lead to recognition of hidden genomic features in dinoflagellates. *Proceedings of the National Academy of Sciences of the United States of America, 107*, 20033–20038.

Liu, L., Li, Y., Li, S., Hu, N., He, Y., Pong, R., et al. (2012). Comparison of next-generation sequencing systems. *Journal of Biomedicine and Biotechnology, 2012*, 251364.

Luo, C., Tsementzi, D., Kyrpides, N., Read, T., & Konstantinidis, K. T. (2012). Direct comparisons of Illumina vs. Roche 454 sequencing technologies on the same microbial community DNA sample. *PLoS One, 7*, e30087.

McCarren, J., Becker, J. W., Repeta, D. J., Shi, Y., Young, C. R., Malmstrom, R. R., et al. (2010). Microbial community transcriptomes reveal microbes and metabolic pathways associated with dissolved organic matter turnover in the sea. *Proceedings of the National Academy of Sciences of the United States of America, 107*, 16420–16427.

McGrath, K. C., Thomas-Hall, S. R., Cheng, C. T., Leo, L., Alexa, A., Schmidt, S., et al. (2008). Isolation and analysis of mRNA from environmental microbial communities. *Journal of Microbiological Methods, 75*, 172–176.

Moran, M. A. (2009). Metatranscriptomics: Eavesdropping on complex microbial communities. *Microbe, 4*, 329–335.

Moran, M. A., Satinsky, B., Gifford, S. M., Luo, H., Rivers, A., Chan, L. K., et al. (2013). Sizing up metatranscriptomics. *The ISME Journal, 7*, 237–243.

Mou, X., Vila-Costa, M., Sun, S., Zhao, W., Sharma, S., & Moran, M. A. (2011). Metatranscriptomic signature of exogenous polyamine utilization by coastal bacterioplankton. *Environmental Microbiology Reports, 3*, 798–806.

Mutter, G. L., Zahrieh, D., Liu, C., Neuberg, D., Finkelstein, D., Baker, H. E., et al. (2004). Comparison of frozen and RNALater solid tissue storage methods for use in RNA expression microarrays. *BMC Genomics, 5*, 88.

Nam, D. K., Lee, S., Zhou, G., Cao, X., Wang, C., Clark, T., et al. (2002). Oligo(dT) primer generates a high frequency of truncated cDNAs through internal poly(A) priming during reverse transcription. *Proceedings of the National Academy of Sciences of the United States of America, 99*, 6152–6156.

Neidhardt, F. C., & Umbarger, H. E. (1996). Chemical composition of *Escherichia coli*. In F. C. Neidhardt (Ed.), *Escherichia coli and Salmonella: Cellular and molecular biology* (pp. 13–17). Washington, DC: ASM Press.

Ottesen, E. A., Marin, R., 3rd., Preston, C. M., Young, C. R., Ryan, J. P., Scholin, C. A., et al. (2011). Metatranscriptomic analysis of autonomously collected and preserved marine bacterioplankton. *The ISME Journal, 5*, 1881–1895.

Ottesen, E. A., Young, C. R., Eppley, J. M., Ryan, J. P., Chavez, F. P., Scholin, C. A., et al. (2013). Pattern and synchrony of gene expression among sympatric marine microbial populations. *Proceedings of the National Academy of Sciences of the United States of America, 110*, E488–E497.

Parkhomchuk, D., Borodina, T., Amstislavskiy, V., Banaru, M., Hallen, L., Krobitsch, S., et al. (2009). Transcriptome analysis by strand-specific sequencing of complementary DNA. *Nucleic Acids Research, 37*, e123.

Peano, C., Pietrelli, A., Consolandi, C., Rossi, E., Petiti, L., Tagliabue, L., et al. (2013). An efficient rRNA removal method for RNA sequencing in GC-rich bacteria. *Microbial Informatics and Experimentation, 3*, 1.

Poretsky, R. S., Bano, N., Buchan, A., LeCleir, G., Kleikemper, J., Pickering, M., et al. (2005). Analysis of microbial gene transcripts in environmental samples. *Applied and Environmental Microbiology, 71*, 4121–4126.

Poretsky, R. S., Gifford, S., Rinta-Kanto, J., Vila-Costa, M., & Moran, M. A. (2009). Analyzing gene expression from marine microbial communities using environmental transcriptomics. *Journal of Visualized Experiments, 24*, e1086.

Poretsky, R. S., Hewson, I., Sun, S., Allen, A. E., Zehr, J. P., & Moran, M. A. (2009). Comparative day/night metatranscriptomic analysis of microbial communities in the North Pacific subtropical gyre. *Environmental Microbiology, 11*, 1358–1375.

Poretsky, R. S., Sun, S., Mou, X., & Moran, M. A. (2010). Transporter genes expressed by coastal bacterioplankton in response to dissolved organic carbon. *Environmental Microbiology, 12*, 616–627.

Quail, M. A., Smith, M., Coupland, P., Otto, T. D., Harris, S. R., Connor, T. R., et al. (2012). A tale of three next generation sequencing platforms: Comparison of Ion Torrent, Pacific Biosciences and Illumina MiSeq sequencers. *BMC Genomics, 13*, 341.

Raghavan, R., Sloan, D. B., & Ochman, H. (2012). Antisense transcription is pervasive but rarely conserved in enteric bacteria. *mBio, 3*, e00156-12.

Rinta-Kanto, J. M., Sun, S., Sharma, S., Kiene, R. P., & Moran, M. A. (2012). Bacterial community transcription patterns during a marine phytoplankton bloom. *Environmental Microbiology, 14*, 228–239.

Rissanen, A. J., Kurhela, E., Aho, T., Oittinen, T., & Tiirola, M. (2010). Storage of environmental samples for guaranteeing nucleic acid yields for molecular microbiological studies. *Applied Microbiology and Biotechnology, 88*, 977–984.

Roberts, A., Trapnell, C., Donaghey, J., Rinn, J. L., & Pachter, L. (2011). Improving RNA-Seq expression estimates by correcting for fragment bias. *Genome Biology, 12*, R22.

Roume, H., Muller, E. E., Cordes, T., Renaut, J., Hiller, K., & Wilmes, P. (2013). A biomolecular isolation framework for eco-systems biology. *The ISME Journal, 7*, 110–121.

Sanders, J. G., Beinart, R. A., Stewart, F. J., DeLong, E. F., & Girguis, P. R. (2013). Metatranscriptomics reveal differences in in situ energy and nitrogen metabolism among hydrothermal vent snail symbionts. *The ISME Journal, 7*, 1556–1567. http://dx.doi.org/10.1038/ismej.2013.45.

Sarkar, N. (1997). Polyadenylation of mRNA in prokaryotes. *Annual Review of Biochemistry, 66*, 173–197.

Scholin, C., Doucette, G., Jensen, S., Roman, B., Pargett, D., Marin, R., et al. (2009). Remote detection of marine microbes, small invertebrates, harmful algae, and biotoxins using the Environmental Sample Processor (ESP). *Oceanography, 22*, 158–167.

Selinger, D. W., Saxena, R. M., Cheung, K. J., Church, G. M., & Rosenow, C. (2003). Global RNA half-life analysis in *Escherichia coli* reveals positional patterns of transcript degradation. *Genome Research*, *13*, 216–223.

Sharma, C. M., Hoffmann, S., Darfeuille, F., Reignier, J., Findeiss, S., Sittka, A., et al. (2010). The primary transcriptome of the major human pathogen *Helicobacter pylori*. *Nature*, *464*, 250–255.

Shi, Y., McCarren, J., & DeLong, E. F. (2012). Transcriptional responses of surface water marine microbial assemblages to deep-sea water amendment. *Environmental Microbiology*, *14*, 191–206.

Shi, Y., Tyson, G. W., & DeLong, E. F. (2009). Metatranscriptomics reveals unique microbial small RNAs in the ocean's water column. *Nature*, *459*, 266–269.

Shi, Y., Tyson, G. W., Eppley, J. M., & DeLong, E. F. (2011). Integrated metatranscriptomic and metagenomic analyses of stratified microbial assemblages in the open ocean. *The ISME Journal*, *5*, 999–1013.

Simister, R. L., Schmitt, S., & Taylor, M. W. (2011). Evaluating methods for the preservation and extraction of DNA and RNA for analysis of microbial communities in marine sponges. *Journal of Experimental Marine Biology and Ecology*, *397*, 38–43.

Steglich, C., Lindell, D., Futschik, M., Rector, T., Steen, R., & Chisholm, S. W. (2010). Short RNA half-lives in the slow-growing marine cyanobacterium *Prochlorococcus*. *Genome Biology*, *11*, R54.

Stewart, F. J., Dalsgaard, T., Thamdrup, B., Revsbech, N. P., Ulloa, O., Canfield, D. E., et al. (2012). Experimental incubations elicit profound changes in community transcription in OMZ bacterioplankton. *PLoS One*, *7*, e37118.

Stewart, F. J., Ottesen, E. A., & DeLong, E. F. (2010). Development and quantitative analyses of a universal rRNA-subtraction protocol for microbial metatranscriptomics. *The ISME Journal*, *4*, 896–907.

Stewart, F. J., Sharma, A. K., Bryant, J. A., Eppley, J. M., & DeLong, E. F. (2011). Community transcriptomics reveals universal patterns of protein sequence conservation in natural microbial communities. *Genome Biology*, *12*, R26.

Stewart, F. J., Ulloa, O., & DeLong, E. F. (2012). Microbial metatranscriptomics in a permanent marine oxygen minimum zone. *Environmental Microbiology*, *14*, 23–40.

Strady, E., Pohl, C., Yakushev, E. V., Krüger, S., & Hennings, U. (2008). PUMP-CTD-System for trace metal sampling with a high vertical resolution. A test in the Gotland Basin, Baltic Sea. *Chemosphere*, *70*, 1309–1319.

Taniguchi, Y., Choi, P. J., Li, G. W., Chen, H., Babu, M., Hearn, J., et al. (2010). Quantifying *E. coli* proteome and transcriptome with single-molecule sensitivity in single cells. *Science*, *329*, 533–538.

Urich, T., Lanzén, A., Qi, J., Huson, D. H., Schleper, C., & Schuster, S. C. (2008). Simultaneous assessment of soil microbial community structure and function through analysis of the meta-transcriptome. *PLoS One*, *3*, e2527.

Vandernoot, V. A., Langevin, S. A., Solberg, O. D., Lane, P. D., Curtis, D. J., Bent, Z. W., et al. (2012). cDNA normalization by hydroxyapatite chromatography to enrich transcriptome diversity in RNA-seq applications. *Biotechniques*, *53*, 373–380.

Vila-Costa, M., Rinta-Kanto, J. M., Sun, S., Sharma, S., Poretsky, R., & Moran, M. A. (2010). Transcriptomic analysis of a marine bacterial community enriched with dimethylsulfoniopropionate. *The ISME Journal*, *4*, 1410–1420.

Wendisch, V. F., Zimmer, D. P., Khodursky, A., Peter, B., Cozzarelli, N., & Kustu, S. (2001). Isolation of *Escherichia coli* mRNA and comparison of expression using mRNA and total RNA on DNA microarrays. *Analytical Biochemistry*, *290*, 205–213.

Wurzbacher, C., Salka, I., & Grossart, H. P. (2012). Environmental actinorhodopsin expression revealed by a new *in situ* filtration and fixation sampler. *Environmental Microbiology Reports*, *4*, 491–497.

Xiong, X., Frank, D. N., Robertson, C. E., Hung, S. S., Markle, J., Canty, A. J., et al. (2012). Generation and analysis of a mouse intestinal metatranscriptome through Illumina based RNA-sequencing. *PLoS One*, 7, e36009.

Yi, H., Cho, Y. J., Won, S., Lee, J. E., Jin Yu, H., Kim, S., et al. (2011). Duplex-specific nuclease efficiently removes rRNA for prokaryotic RNA seq. *Nucleic Acids Research*, *39*, e140.

Yilmaz, S., Allgaier, M., & Hugenholtz, P. (2010). Multiple displacement amplification compromises quantitative analysis of metagenomes. *Nature Methods*, 7, 943–944.

CHAPTER ELEVEN

Sequential Isolation of Metabolites, RNA, DNA, and Proteins from the Same Unique Sample

Hugo Roume, Anna Heintz-Buschart, Emilie EL Muller, Paul Wilmes[1]
Luxembourg Centre for Systems Biomedicine, University of Luxembourg, Esch-sur-Alzette, Luxembourg
[1]Corresponding author: e-mail address: paul.wilmes@uni.lu

Contents

1. Introduction 220
   1.1 From multi-omics to integrated omics: The necessity for systematic measurements 220
   1.2 Methods for the isolation of concomitant biomolecules 222
2. Sampling and Sample Preprocessing 222
   2.1 Sample fixation 222
   2.2 Preprocessing: Dealing with differing sample characteristics 223
3. Sample Processing and Small Molecule Extraction 226
   3.1 Extraction of extracellular metabolites from cell supernatant 226
   3.2 Sample homogenization by cryomilling 226
   3.3 Extraction of intracellular metabolites from cell pellet 227
4. Sequential Biomacromolecular Isolation 228
   4.1 Qiagen AllPrep DNA/RNA/Protein Mini Kit-based method 228
   4.2 Norgen Biotek All-in-One Purification Kit-based method 229
5. Quality Control 230
   5.1 Cell lysis 230
   5.2 Metabolite fractions 231
   5.3 Nucleic acids 233
   5.4 Proteins 235
6. Outlook 235
References 235

Abstract

In microbial ecology, high-resolution molecular approaches are essential for characterizing the vast organismal and functional diversity and understanding the interaction of microbial communities with biotic and abiotic environmental factors. Integrated omics, comprising genomics, transcriptomics, proteomics, and metabolomics allows

Methods in Enzymology, Volume 531
ISSN 0076-6879
http://dx.doi.org/10.1016/B978-0-12-407863-5.00011-3

conclusive links to be drawn between genetic potential and function. However, this requires truly systematic measurements. In this chapter, we first assess the levels of heterogeneity within mixed microbial communities, thereby demonstrating the need for analyzing biomolecular fractions obtained from a single and undivided sample to facilitate multi-omic analysis and meaningful data integration. Further, we describe a methodological workflow for the reproducible isolation of concomitant metabolites, RNA (optionally split into large and small RNA fractions), DNA, and proteins. Depending on the nature of the sample, the methodology comprises different (pre)processing and preservation steps. If possible, extracellular polar and nonpolar metabolites may first be extracted from cell supernatants using organic solvents. Cells are homogenized by cryo-milling before small molecules are extracted with organic solvents. After cell lysis, nucleic acids and protein fractions are sequentially isolated using chromatographic spin columns. To prove the broad applicability of the methodology, we applied it to microbial consortia of biotechnological (biological wastewater treatment biomass), environmental (freshwater planktonic communities), and biomedical (human fecal sample) research interest. The methodological framework should be applicable to other microbial communities as well as other biological samples with a minimum of tailoring and represents an important first step in standardization for the emerging field of Molecular Eco-Systems Biology.

1. INTRODUCTION

1.1. From multi-omics to integrated omics: The necessity for systematic measurements

The field of microbial ecology has advanced rapidly since the introduction of molecular biology methodologies, which have allowed the exploration of microbial diversity and function in a multitude of different environments. Over the past two decades, a wide variety of techniques have been developed to describe and characterize microbial assemblages. More recently, revolutionary improvements and advances in high-throughput "omic" technologies, ranging from metagenomics to metabolomics, are allowing resolution of community- and population-level processes from genetic potential to final phenotype. However, in order to collect such high-resolution data in a truly systematic fashion and facilitate meaningful data integration and analysis, robust sampling, sample preservation, and biomolecular isolation methodologies are essential.

In *Foundation of Systems Biology*, Kitano defined the ideal systematic measurement as the "simultaneous measurement of multiple features for a single sample" (Kitano, 2001). This consideration is particularly important in eco-systems biology as a vast microbial diversity is a hallmark of natural settings

(Caporaso et al., 2010) and even within-population genetic heterogeneity is very pronounced in natural microbial communities (Wilmes, Simmons, Denef, & Banfield, 2009). To assess the extent of heterogeneity at each biomolecular information layer, a comparative experiment was carried out on islet forming spatially resolved lipid-accumulating organisms (LAO)-enriched microbial community samples from a biological wastewater treatment plant (Roume et al., 2013). We measured the phylogenetic composition using 454 pyrosequencing of 16S rDNA amplicons (Muller et al., in preparation), the functional genetic potential using an Illumina HiSeq platform (Pinel et al., in preparation), and the community-wide metabolic profile using gas chromatography coupled to mass spectrometry (GC–MS; Roume et al., 2013). We generated data of the concomitant biomolecular fractions for four biological replicates (i.e., space resolved LAO islets) and four technical replicates of one single islet (i.e., different subsamples derived from the same islet after sample splitting), sampled at the same point in time (Fig. 11.1). Extensive sample-to-sample variation is apparent at each level of biomolecular information. Interestingly, the ability to discriminate between biological and technical replicates decreases from phylogenetic profiles (Fig. 11.1A) to metagenomes (Fig. 11.1B) and lastly metabolite profiles (Fig. 11.1C). This finding reflects

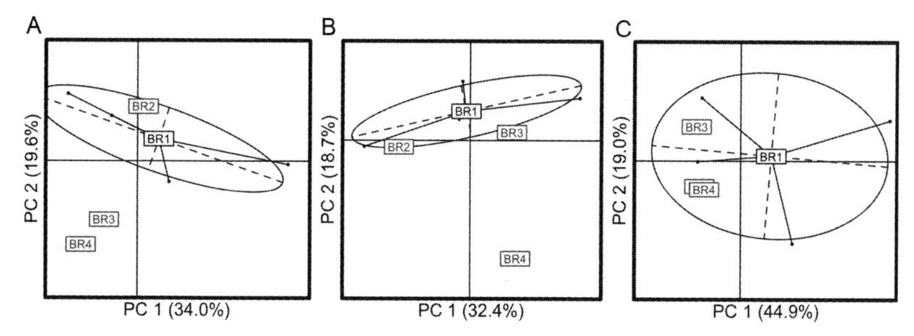

Figure 11.1 Within- and between-sample heterogeneity as determined from a biological wastewater treatment plant microbial community. Scatter plots of the results of principal coordinate analysis of Bray–Curtis dissimilarity indices of four biological replicates (BR1–BR4), from a unique sampling date, including one biological replicate (BR1) split into four technical replicates. Each biological replicate is indicated by a rectangle, and each technical replicate is represented by a dot. The ellipse covers two-thirds of the technical replicates, which are derived from the same biological replicate, with the rectangle indicating their common center of gravity. (A) Community compositions of the samples as revealed by 16S rDNA amplicon sequencing (Muller et al., in preparation), (B) corresponding functional capacities determined using KEGG orthologous group annotations of metagenomic sequence data (Pinel et al., in preparation), and (C) corresponding polar metabolomes (Roume et al., 2013). (For color version of this figure, the reader is referred to the online version of this chapter.)

increased complexity at each biomolecular information layer. The resulting fine-scale heterogeneity underlines the requirement for the isolation of concomitant biomolecules to fulfill the premise of systematic measurements and enable meaningful integration of space- and time-resolved microbial community omic datasets.

1.2. Methods for the isolation of concomitant biomolecules

Previously published methodologies for the simultaneous isolation of certain biomacromolecules, namely DNA, RNA, and proteins, from biological samples are primarily based on monophasic mixtures of water, phenol, and guanidine isothiocyanate (Chomczynski & Sacchi, 1987). The simultaneous extraction of metabolites, proteins, and RNA has previously been carried out on plant material (Weckwerth, Wenzel, & Fiehn, 2004). For safe routine laboratory use, chromatographic spin column-based methods have been developed for the combined extraction of RNA and proteins from a variety of cells and tissues (Morse, Shaw, & Larner, 2006). The additional isolation of DNA from tissues and cultured cells is the primary feature of the Qiagen AllPrep method (Tolosa, Schjenken, Civiti, Clifton, & Smith, 2007), and this has since been validated on an extensive set of medical samples (Radpour et al., 2009). Since then, several additional manufacturers, for example, GE Healthcare, Macherey-Nagel, Norgen Biotek, etc., have commercialized chromatographic spin column-based kits for the sequential isolation of DNA, RNA, and protein extraction from tissues or cultured cells.

The detailed protocols presented here provide a unique methodological framework for the reproducible isolation of high-quality genomic DNA, large and small RNA or total RNA, and proteins, as well as polar and nonpolar metabolites from single and undivided samples. This chapter focuses on biological wastewater treatment biomass, planktonic freshwater, and human fecal mixed microbial communities, which are representative of the differing sample characteristics one may encounter. However, the protocols can also be successfully applied to other samples with minor preprocessing modifications. Examples include frozen human feces, methanogenic reactor digestates, and diverse mouse tissue samples.

2. SAMPLING AND SAMPLE PREPROCESSING

2.1. Sample fixation

When collecting biological samples for systematic molecular analyses, it is crucial to immediately fix the samples in a state which best preserves the information contained within the system at the time of sampling. If care is not given to this step, the information contained within the DNA, RNA, protein, and metabolites will rapidly change due to specific and nonspecific degradation, regulation of expression, and protein modifications. Sample fixation directly after sampling is therefore a key step in the experimental workflows presented for the three different microbial communities described below:

Biomass from a biological wastewater treatment plant (LAOs)

- Collect sample in a 50 ml tube and immediately snap freeze on-site in liquid nitrogen.

Freshwater filtrate

- Collect approximately 40 l of freshwater and preserve the sample at 4 °C. Proceed immediately to the preprocessing step.

Human fecal sample

- Cover 300 mg ± 10% of fresh human fecal sample immediately with 1.5 ml of RNAlater solution (Ambion) and carry on to the preprocessing step.

2.2. Preprocessing: Dealing with differing sample characteristics

Sample preprocessing is an important step, which varies significantly depending on the overall characteristics of samples (Fig. 11.2A). If possible, this step consists of the separation of the intra- and extracellular compartments and accompanying concentration of cells.

2.2.1 Biological wastewater treatment plant biomass or similar samples

For a sample which allows recovery of the extracellular medium, the sample preprocessing begins with slowly thawing the sample on ice to allow the subsequent separation of cells from the supernatant (extracellular compartment) by centrifugation. The detailed preprocessing protocol involves the following:

1. With a sterile spatula, reduce to powder around 200 mg ±10% of frozen LAO sample. To conserve the frozen state of the powder, cool down all equipment and refreeze the material regularly by briefly dipping it into liquid nitrogen. Float a weighing boat on the surface of the liquid nitrogen and pour the powdered sludge into this.
2. Transfer the frozen powder into a precooled 2 ml sample tube and place this into a liquid nitrogen bath. Optionally, it can be stored for few days at −80 °C until further processing.

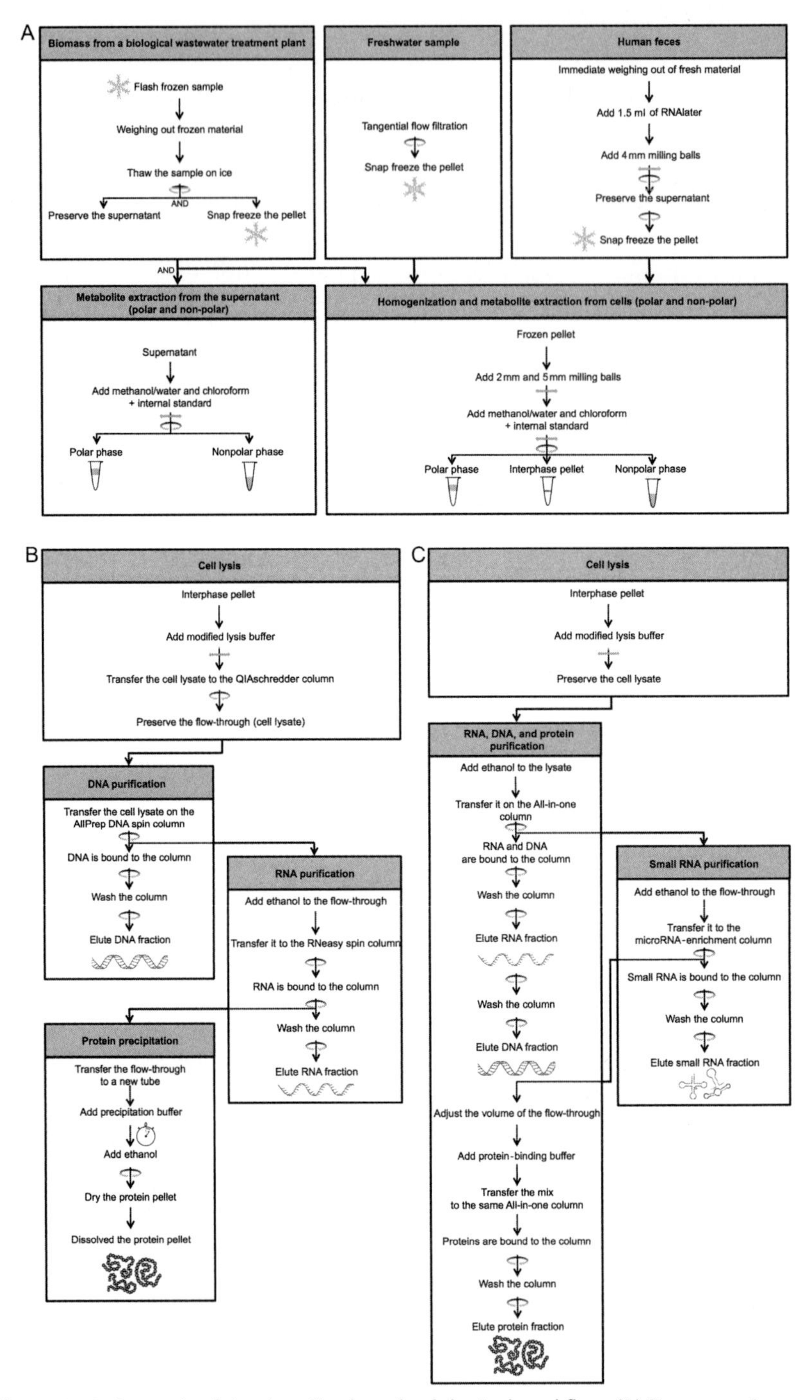

Figure 11.2 Synopsis of the described methodological workflow. (A) Preprocessing and metabolite extraction; (B) isolation of DNA, total RNA, and proteins using the Qiagen AllPrep-based procedure; and (C) isolation of DNA, small and large RNA, and protein using the Norgen Biotek All-in-One-based procedure. The circular arrows denote a centrifugation step, and the horizontal arrows symbolize a mix or bead beating step.

3. Thaw the sample on ice for approximately 10–15 min by placing the sample tube horizontally on the ice surface. Flip the tube regularly and verify the consistency of the sample every 5 min. It is highly recommended to place the thawed sample into ice immediately after the sample is completely thawed.
4. Centrifuge the thawed sample at 18,000 × *g* for 15 min at 4 °C and transfer the supernatant into a new 2 ml sample tube kept at 4 °C. Around 150 μl of supernatant for 200 mg of starting material may be expected. If visible material particles are still present in the supernatant, carry out an additional centrifugation for 5 min until a completely limpid supernatant is obtained. Proceed to Section 3.1 with the supernatant.
5. Snap freeze the cell pellet and preserve it at −80 °C until step 2 of Section 3.2.

2.2.2 Freshwater planktonic microbial communities

The preprocessing of a liquid sample with a low density of biomass consists of first concentrating the cells present within the sample before snap freezing:

1. Concentrate cells by tangential flow filtration using, for example, Sartocon silica cassettes (Satorius Stedim Biotek) with a filtration area of 0.1 m^2 and molecular weight cutoff of 10 kDa at a flow rate of ~1.5 l min^{-1}.
2. Pellet the concentrated cells by ultracentrifugation at 48,400 × *g* for 1 h at 4 °C.
3. Resuspend the resulting pellet in 1 ml of supernatant and transfer it to a 2 ml centrifugation tube.
4. Further concentrate the suspension by an additional centrifugation step at 14,000 × *g* for 5 min at 4 °C.
5. Snap freeze the resulting pellet and preserve it at −80 °C until step 2 of Section 3.2.

2.2.3 Human fecal sample

For a complex sample such as fresh feces, in which biomolecules are highly sensitive to degradation and which are rich in PCR inhibitors, the preprocessing step involves treatment of the sample to preserve the integrity of the biomolecules, especially RNA (Fitzsimons, Akkermans, de Vos, & Vaughan, 2003). The optimized workflow for fecal samples is thus as follows:

1. Add three autoclaved stainless steel milling balls with a diameter of 4 mm to the tube of fresh fecal sample overlaid with RNAlater.

2. Homogenize the mixture by cold bead beating in a Mixer Mill 400 (Retsch) for 5 min at 10 Hz. For this, the rack holding the sample tube needs to be precooled at 4 °C.
3. Centrifuge the homogenized sample at 700 × *g* for 1 min at 4 °C. This low-speed centrifugation step should result in a supernatant containing detached cells. Transfer the supernatant into a fresh 2-ml sample tube.
4. Pellet the cells contained within the supernatant at 14,000 × *g* for 5 min at 4 °C. Discard the supernatant.
5. Snap freeze the resulting pellet and preserve it at −80 °C until step 2 of Section 3.2.

3. SAMPLE PROCESSING AND SMALL MOLECULE EXTRACTION

Depending on the physical characteristics of the sample and the preprocessing, metabolite extraction can be done separately on supernatant and/or on cells (Fig. 11.2A).

3.1. Extraction of extracellular metabolites from cell supernatant

For metabolite extraction, the polar and the nonpolar solvent fractions should be present in equal volumes. The method for cold solvent extracting metabolites from the supernatant of a wastewater treatment plant biomass sample is as follows:

1. Add to one volume of the aqueous supernatant obtained in Section 2.2.1 at step 4, one volume (e.g., 150 μl) of cold methanol at −20 °C and two volumes of cold chloroform at −20 °C (e.g., 300 μl).
2. Mix the sample tube in a thermomixer (e.g., Thermomixer comfort, Eppendorf) for 30 min at 4 °C at maximum speed (e.g., 1400 rpm).
3. Centrifuge the tubes at 14,000 × *g* and 4 °C for 5 min.
4. Transfer the polar and nonpolar phases into separate 2 ml sample tubes. If required, the different phases can be preserved for a few hours at −20 °C at this step.

For a description of processing the polar and nonpolar metabolite fractions for subsequent GC–MS analyses, please see Roume et al. (2013).

3.2. Sample homogenization by cryomilling

The cryomilling (or cryogenic grinding) step involves breaking up the structure of the biological material and allows homogenization of the sample. The frozen homogenous powder obtained at the end of this step guarantees a sufficiently large surface area for the solvents to extract the small molecules. The processing, described below, is of fundamental importance for efficient metabolite extractions from cellular biomass and subsequent representative metabolic profiling thereof (Lolo et al., 2007).

1. Cool autoclaved-sterilized milling balls (5×2 mm + 2×5 mm) by dipping the tubes containing them in liquid nitrogen.
2. Add the cold milling balls to the previously obtained frozen cell pellet (Sections 2.2.1, 2.2.2, or 2.2.3 at step 5).
3. Cryomill the frozen pellet for 2 min at 25 Hz in a Mixer Mill 400 (Retsch). The adaptor rack holding the sample tube needs to be precooled in liquid nitrogen to −80 °C prior to the milling step.
4. Dip the tube immediately into liquid nitrogen to preserve the sample in a frozen state.

At the end of the cryomilling step, the sample should comprise a frozen homogenous powder. If it is not the case, repeat the cryomilling step or empirically optimize the number and size of stainless steel milling balls according to the nature of the sample.

3.3. Extraction of intracellular metabolites from cell pellet

1. Add sequentially 300 μl of cold methanol/water (1:1, v/v) and 300 μl of cold chloroform. Vortex the solution until complete dissolution of the powdered sample into the solvent mixture.
2. Mill the sample tube in a Mixer Mill 400 (Retsch) for 2 min at 20 Hz. In order to avoid solvent leakage, wrap parafilm around the rim of the cap of the sample tube before starting the milling to ensure a tight seal.
3. Centrifuge the sample tube at $14{,}000 \times g$ and 4 °C for 5 min to separate the solvent mixture into a polar (top) phase, an interphase pellet (middle), and nonpolar (lower) phase.
4. Transfer the polar and nonpolar phases into new 2 ml sample tubes. If required, the different phases can be preserved for a few hours at −20 °C at this step.
5. Preserve the interphase with the milling beads in the original centrifugation tube on ice (until proceeding with step 1 of Section 4.1 or 4.2).

For an example of the processing of polar and nonpolar metabolite fractions for subsequent GC–MS analysis, please see Roume et al. (2013).

4. SEQUENTIAL BIOMACROMOLECULAR ISOLATION

Biomacromolecules, including large, small, or total RNA, DNA, and proteins, are isolated from the interphase pellet obtained following the metabolite extraction step (Section 3.3, step 5). Two sequential biomacromolecular isolation protocols, modified from commercially available kits, are described here. The Qiagen AllPrep DNA/RNA/Protein Mini Kit-based method (Fig. 11.2B) allows for sequential extraction of total RNA, DNA, and protein. The Norgen Biotek All-in-One Purification Kit-based method (Fig. 11.2C) facilitates the sequential extraction of large RNA, small RNA, DNA, and proteins. Although these two kit-based methods for biomacromolecular purification are described here, application of alternative methods to the interphase pellet should be entirely feasible but may require some minor adjustments.

Cell disruption, comprising both sample homogenization and cell lysis, is an early and fundamental step in any biomolecular isolation methodology. Both chemical and mechanical/physical methods are available for cell disruption. However, in natural microbial communities, due to the presence of many different microorganisms with vastly differing cellular properties and due to the presence of different interfering matrixes, chemical and/or enzymatic lysis by themselves are typically ineffective at comprehensive and reproducible cell lysis. In the methodology presented here, we combine the mechanical method of cryogenic grinding with chemical lysis to result in indiscriminate cell lysis.

The efficiency of cell lysis should be assessed for each new type of sample (Section 5.1). Furthermore, following sequential extraction, an aliquot of 5 μl for small and large or total RNA and 10 μl for DNA should be stored at 4 °C for quality assessment (Sections 5.3.1 and 5.3.2). These steps are essential to guarantee the efficiency of cell lysis as well as the quality of the obtained biomolecular fractions.

4.1. Qiagen AllPrep DNA/RNA/Protein Mini Kit-based method

The interphase pellet is overlaid with lysis buffer before bead beating with the same milling balls as used previously. Importantly, a modified lysis buffer is used to prevent RNA degradation. Following addition of the modified lysis buffer, the sample is subjected to shearing of high-molecular weight cellular components in order to reduce the viscosity of the lysates, to improve the binding efficiency of nucleic acids to the chromatographic spin columns, and thereby to increase the overall yield and purity of the obtained biomolecular fractions. The methodology is summarized in Fig. 11.2B and is as follows:

1. Add 10 μl of β-mercaptoethanol per milliliter of RLT buffer (Qiagen).
2. Add 600 μl of cold (4 °C) modified RLT buffer (step 1) to the interphase pellet (step 5 of Section 3.3) and cover the rim of the closed tube cap with parafilm.
3. Resuspend the interphase in the modified lysis buffer by a quick vortexing of the sample tube.
4. Proceed to a bead beating in a Mixer Mill 400 (Retsch) for 30 s at 25 Hz. Note that the adaptor racks should be at 4 °C.
5. Transfer up to 700 μl of the lysate to a QIAshredder column (Qiagen) and centrifuge for 2 min at 12,100 × *g*. The entire lysate should pass through the QIAshredder column. If a pellet forms in the collection tube, it should be resuspended before loading the lysate onto the next column.

All further steps are described in the AllPrep DNA/RNA/Protein Mini Handbook (Qiagen, version 12/2007) in the section "Simultaneous purification of genomic DNA, total RNA, and total protein from animal and human cells" p. 22 step 4. In our own experience, the elution of DNA and total RNA in the dedicated buffers can be repeated in order to recover more nucleic acids.

Importantly, if the lysate has not passed through the AllPrep DNA spin column completely after 5 min of centrifugation at the maximum speed, transfer the liquid retained in the column to an additional AllPrep DNA column, which should then be processed along with the first tube. Both AllPrep DNA columns should be used for the subsequent DNA recovery steps.

4.2. Norgen Biotek All-in-One Purification Kit-based method

Similar to the Qiagen AllPrep-based method, the interphase pellet is resuspended in a modified lysis buffer, which prevents RNA degradation. Cell disruption is carried out in this buffer using the milling balls still present in the centrifugation tube following step 5 of Section 3.3. A few essential modifications to the manufacturer's instructions have been implemented and are detailed below. The workflow is also summarized in Fig. 11.2C.

1. Add 400 μl of cold lysis buffer supplemented with 10 μl ml^{-1} of β-mercaptoethanol and 100 μl of cold 1× Tris–EDTA buffer

(pH 7.2) to the interphase pellet. Cover the rim of the closed tube cap with parafilm.

2. Resuspend the pellet by quickly vortexing the sample tube.
3. Carry out cell lysis by bead beating the sample in the Mixer Mill 400 (Retsch) for 30 s at 25 Hz in a cold rack (4 °C).
4. Add 100 μl of pure ethanol to the lysate and mix by vortexing for 10 s.

In our own experience, increasing the proportion of ethanol in the lysate prior to loading of the column results in a higher yield of nucleic acids recovery.

Follow the procedure of the All-in-One Purification Kit protocol from (Norgen Biotek, 2008; PI24200-12) section 3: "Isolation of large RNA, Genomic DNA, microRNA, and Proteins" p. 17 3A.1.b.

Importantly, and as recommended by the manufacturer for bacterial starting material, two elutions of large RNA, DNA, and protein from the All-in-One column should be carried out in order to maximize the yield. We typically also carry out the column-based procedure for total protein purification.

5. QUALITY CONTROL

The different methods presented in this section are typically used in our laboratory to assess the efficiency of the protocol when applying it to new samples and to assess the quality of the recovered biomolecular fractions.

5.1. Cell lysis

Conservation of cell integrity until sample processing as well as representative cell lysis prior to biomolecular extraction are essential considerations for the methodology to result in reproducible and representative biomolecular fractions (Roume et al., 2013). The proportion of cells which are intact before and after cell disruption may be assessed using the Live/Dead *Bac*Light Bacterial Viability Kit (L7012, Invitrogen). Figure 11.3 shows that, in the case of LAO samples from wastewater, most cells in the snap-frozen sample still have intact cell membranes after 10 min of thawing on ice (Fig. 11.3A) and that the vast majority of the cells are lysed after cryomilling, metabolite extraction, and combined mechanical/chemical lysis in the modified Norgen Biotek lysis buffer (Fig. 11.3B). To assess the lysis efficiency:

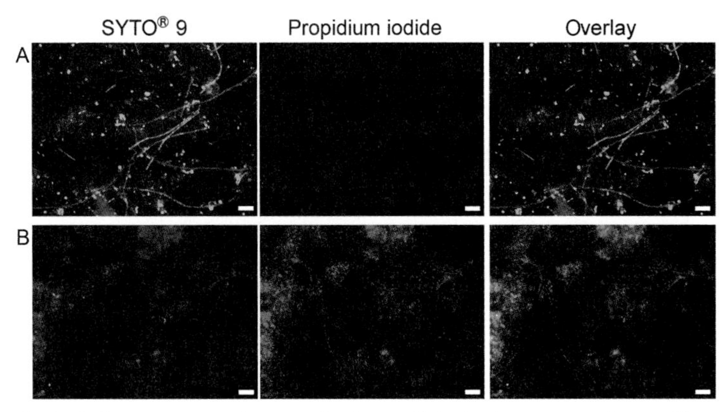

Figure 11.3 Cell integrity before and after cell lysis. Representative epifluorescent micrographs of microbial cells from a representative biological wastewater treatment plant sample stained with the Live/Dead stain. Propidium iodide specifically stains bacteria with a damaged cell membrane, whereas Syto 9 serves as the counterstain. Intact cells are highlighted in green and disrupted cells in red. Scale bar is equivalent to 10 μm. (A) Sample having undergone a single freeze–thaw cycle and (B) sample having undergone subsequent metabolite extraction followed by mechanical and chemical lysis using the Norgen Biotek All-in-One Kit's modified lysis buffer. (For interpretation of the references to color in this figure legend, the reader is referred to the online version of this chapter.)

1. Pellet cellular material from the individual protocol steps for which you want to assess cell membrane integrity by centrifugation at 14,000 × *g* and 4 °C for 5 min.
2. Wash the pellet once with phosphate buffered saline solution (1 × PBS, pH 7) and redilute into 1 ml of 1 × PBS buffer.
3. Follow the manufacturer's instructions of the Live/Dead *Bac*Light Bacterial Viability Kit (Molecular Probes, rev. July 15, 2004, p. 3).

For the determination of the red and green pixel ratio, multiple red and green fluorescence micrographs can be processed as described in Roume et al. (2013).

5.2. Metabolite fractions

Quality control in metabolomics is a challenging task. The metabolites have to be extracted without compromising either their chemical structure or their relative abundance. Some metabolites are extremely unstable and therefore prone to degradation during sample preparation. Several recommendations should be considered, from the sampling to the metabolomic analyses, to ensure the generation of reliable data.

5.2.1 Sampling and sample preprocessing

- The sample should be processed immediately or snap frozen in liquid nitrogen right away on site to stabilize the sample and stop any enzymatic activity.
- The sample should be preserved at a minimum of −80 °C, ideally −150 °C.
- The sample should be homogenized by cryomilling to facilitate efficient metabolite extraction from solid sample.

5.2.2 Validation of metabolomic analyses

- In the case of GC–MS, the analytical procedure usually needs to be optimized for each sample type with the fine tuning of several parameters, including, for example, the volume of metabolite fraction to be injected, the split or the oven program. The derivatization procedure may also need to be adapted.
- The polar and/or the nonpolar solvents used for the extraction should be supplemented with exogenous chemical compounds which should not be present endogenously within the sample and can serve as internal standards. During metabolomic analysis, the internal standard is a useful benchmark to assess metabolite recovery as well as the reproducibility of the analysis.
- When large numbers of samples are analyzed in a single analytical run, pool samples, which are mixtures of aliquots of all the samples included in the run and which, thus, reflect the entirety of metabolites detectable, should be included in the sample analysis sequence at regular intervals. For example, we typically include a pool sample as every seventh sample in a sequence, followed by a blank sample. The measurement results obtained from the pool samples can provide an indication of possible instrument drift during the run. Importantly, the pool samples, as well as the internal standards, can later also be used to normalize the data (Roume et al., 2013).

The metabolomics data obtained by GC–MS can be displayed as total ion chromatograms (TICs). We routinely assess the quality of the TICs to get an indication of the purity and representativeness of the metabolite extracts. For example, the metabolomic analysis of the three different samples, for which the extractions are described, resulted in clear TICs comprised of numerous small and well-defined peaks (Fig. 11.4A).

5.3. Nucleic acids

5.3.1 Ribonucleic acid

Because RNA is a biomacromolecule prone to degradation, the accurate assessment of its integrity is one of the most critical steps for the success of any downstream analysis (Vermeulen et al., 2011), including ribosomal RNA removal, reverse transcription, and high-throughput cDNA sequencing (RNA-Seq). A commonly accepted large and total RNA quality and integrity indicator is Agilent's RNA integrity number (RIN; Fleige & Pfaffl, 2006). A RIN $\geq$7.0 ususally satisfies accepted quality requirements for most RNA-Seq protocols (Jahn, Charkowski, & Willis, 2008) and ensures a low contamination of the small RNA fraction with degradation products. If a sample exhibits a RIN of $<$7.0, we recommend not to use all other biomolecular fractions obtained from this sample and to carry out an additional extraction.

As shown in the representative electropherograms displayed in Fig. 11.4B and C, the extraction method described here facilitates the isolation of high quality and pure RNA. The large RNA fractions of the three different sample types were analyzed using the Agilent 2100 Bioanalyzer with the RNA 6000 Nano RNA Kit and 2100 Expert software (Fig. 11.4B). In order to compare different RNA fractions, fluorescent units of the traces were normalized (Roume et al., 2013). The traces exhibit distinct peaks between 100 and 4,500 nt. The two major peaks, at around 1,500 and 2,900 nt, show distinctively the 16S and the 23S rRNA, respectively. The broad peak around 100 nt indicates the small RNA fraction.

The electropherograms of small RNA traces (Fig. 11.4C), obtained using the Agilent RNA 6000 Small RNA Kit, are dominated by peaks around 60 nt representative of the transfer RNA (tRNA). Other small RNAs are represented by multiple peaks around 120 nt, including 5S rRNA. In eukaryotic systems, the "miRNA"-like region is typically defined as the broad peak ranging from 10 to 40 nt.

5.3.2 Deoxyribonucleic acid

The size, the quality (degraded vs. intact), and semiquantitative amount of DNA extracted can be determined by agarose gel electrophoresis as highlighted in Fig. 11.4D. The methodology described herein is able to yield DNA which exhibits intense and high-molecular weight DNA bands

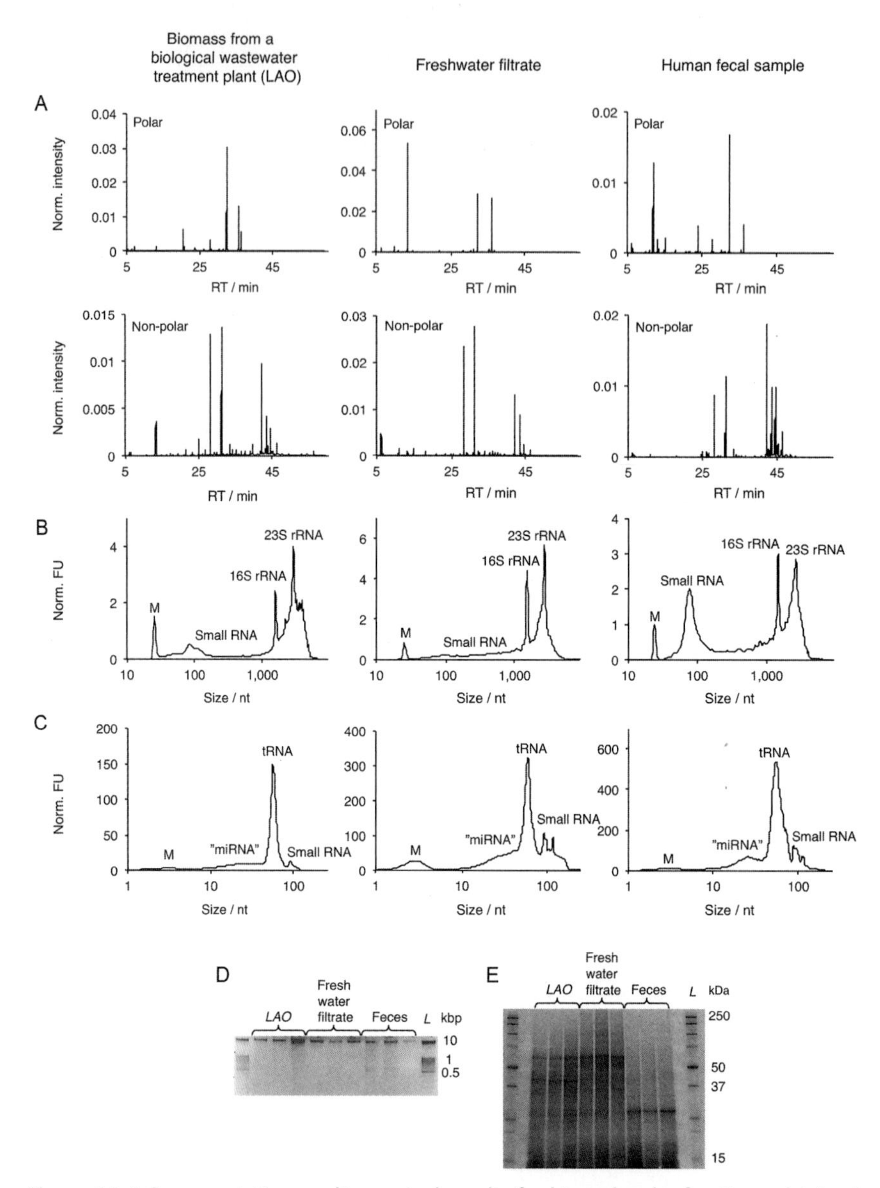

Figure 11.4 Representative quality control results for biomolecular fractions obtained using the Norgen Biotek All-in-One Purification Kit-based biomolecular isolation methodology when applied to a biological wastewater treatment sample (left hand panes), river water filtrate (middle panes), and human fecal samples (right hand panes). (A) Representative GC–MS total ion chromatograms of polar and nonpolar metabolite fractions. (B) and (C) Representative Agilent Bioanalyzer 2100 electropherograms of the obtained total RNA fractions and small RNA fractions, respectively. (D) Agarose gel electrophoresis image of genomic DNA fractions for three technical replicates for each of the samples. (E) SDS-PAGE image of the first protein elution for three technical replicates for each of the samples. Norm, normalized; FU, fluorescent unit; M, marker; L, ladder; nt, nucleotides; RT, retention time. *Figure reproduced from Roume et al. (2013) with permission from Nature Publishing Group.*

and can be directly used for library construction for high-throughput sequencing.

5.4. Proteins

The quality of obtained protein fractions is typically assessed by SDS-PAGE. The obtained gel should exhibit clear bands, as highlighted in Fig. 11.4E for the three different samples for which the biomolecular extraction protocol is described herein. The gel obtained can also be directly used for proteomics either by analysis of individually excised bands or entire lanes.

6. OUTLOOK

In microbial ecology, an increasing number of studies aim to integrate multi-omic datasets in order to obtain comprehensive high-resolution overview of microbial community structure and function. Considering recent technological improvements and the accompanying decrease in the cost of high-throughput molecular methods as well as associated progress in computational and statistical methodologies, these types of investigations will dramatically intensify in the coming years. As demonstrated in the present work, sample-to-sample variation is extensive for microbial communities at each biomolecular level. Consequently, the sequential isolation of biomolecules from an undivided sample is a clear prerequisite for meaningful multi-omic data integration and analysis (Muller, Glaab, May, Vlassis, & Wilmes, 2013). Application of the methodology presented here will allow truly systematic measurement of microbial consortia and will allow the deconvolution of microbial community-driven processes in unprecedented detail. Consequently, the described methodolgy and future iterations thereof provide the backbone of the emerging field of Eco-Systems Biology.

REFERENCES

Caporaso, J. G., Lauber, C. L., Walters, W. A., Berg-Lyons, D., Lozupone, C. A., Turnbaugh, P. J., et al. (2010). Global patterns of 16S rRNA diversity at a depth of millions of sequences per sample. *Proceedings of the National Academy of Sciences of the United States of America*, *108*, 4516–4522.

Chomczynski, P., & Sacchi, N. (1987). Single-step method of RNA isolation by acid guanidinium thiocyanate-phenol-chloroform extraction. *Analytical Biochemistry*, *162*(1), 156–159.

Fitzsimons, N. A., Akkermans, A. D. L., de Vos, W. M., & Vaughan, E. E. (2003). Bacterial gene expression detected in human faeces by reverse transcription-PCR. *Journal of Microbiological Methods*, *55*(1), 133–140.

Fleige, S., & Pfaffl, M. W. (2006). RNA integrity and the effect on the real-time qRT-PCR performance. *Molecular Aspects of Medicine*, *27*(2–3), 126–139.

Jahn, C. E., Charkowski, A. O., & Willis, D. K. (2008). Evaluation of isolation methods and RNA integrity for bacterial RNA quantitation. *Journal of Microbiological Methods, 75*(2), 318–324.

Kitano, H. (2001). Systems biology: Toward system-level understanding of biological systems. In H. Kitano (Ed.), *Foundations of systems biology* (pp. 1–36). Cambridge, MA: The MIT Press.

Lolo, M., Pedreira, S., Vázquez, B., Franco, C., Cepeda, A., & Fente, C. (2007). Cryogenic grinding pre-treatment improves extraction efficiency of fluoroquinolones for HPLC-MS/MS determination in animal tissue. *Analytical and Bioanalytical Chemistry, 387*(5), 1933–1937.

Morse, S., Shaw, G., & Larner, S. (2006). Concurrent mRNA and protein extraction from the same experimental sample using a commercially available column-based RNA preparation kit. *Biotechniques, 40*(1), 54–58.

Muller, E. E. L., Glaab, E., May, P., Vlassis, N., & Wilmes, P. (2013). Condensing the omics fog of microbial communities. *Trends in Microbiology, 21*, 325–333.

Muller, E. E. L., Pinel, N., Laczny, C., Roume, H., Lebrun, L., Hussong, R., et al. Integrated omics reveals that ecological success of a generalist is linked to fine-tuning of resource usage (in preparation).

Pinel, N., Muller, E. E. L., Narayanasamy, S., Laczny, C., Roume, H., Hussong, R., et al. Ecological dominance in the light of time-resolved, multi-omic community characterization (in preparation).

Radpour, R., Sikora, M., Grussenmeyer, T., Kohler, C., Barekati, Z., & Holzgreve, W. (2009). Simultaneous isolation of DNA, RNA, and proteins for genetic, epigenetic, transcriptomic, and proteomic analysis. *Journal of Proteome Research, 8*(11), 5264–5274.

Roume, H., Muller, E. E. L., Cordes, T., Renaut, J., Hiller, K., & Wilmes, P. (2013). A biomolecular isolation framework for eco-systems biology. *The ISME Journal, 7*(1), 110–121.

Tolosa, J. M., Schjenken, J. E., Civiti, T. D., Clifton, V. L., & Smith, R. (2007). Column-based method to simultaneously extract DNA, RNA, and proteins from the same sample. *Biotechniques, 43*(6), 799–804.

Vermeulen, J., De Preter, K., Lefever, S., Nuytens, J., De Vloed, F., Derveaux, S., et al. (2011). Measurable impact of RNA quality on gene expression results from quantitative PCR. *Nucleic Acids Research, 39*(9), e63.

Weckwerth, W., Wenzel, K., & Fiehn, O. (2004). Process for the integrated extraction, identification and quantification of metabolites, proteins and RNA to reveal their co-regulation in biochemical networks. *Proteomics, 4*(1), 78–83.

Wilmes, P., Simmons, S. L., Denef, V. J., & Banfield, J. F. (2009). The dynamic genetic repertoire of microbial communities. *FEMS Microbiology Reviews, 33*(1), 109–132.

CHAPTER TWELVE

Use of Internal Standards for Quantitative Metatranscriptome and Metagenome Analysis

Brandon M. Satinsky[*], Scott M. Gifford[†], Byron C. Crump[‡], Mary Ann Moran[§,1]

[*]Department of Microbiology, University of Georgia, Athens, Georgia, USA
[†]Department of Civil and Environmental Engineering, Massachusetts Institute of Technology, Cambridge, Massachusetts, USA
[‡]College of Earth, Ocean and Atmospheric Sciences, Oregon State University, Corvallis, Oregon, USA
[§]Department of Marine Sciences, University of Georgia, Athens, Georgia, USA
[1]Corresponding author: e-mail address: mmoran@uga.edu

Contents

1. Introduction 238
2. Method Overview 241
3. DNA Template and Vector Design for Internal RNA Standards 242
4. mRNA Standard Preparation 244
   4.1 Required materials 244
   4.2 Plasmid amplification and stock preparation 245
   4.3 Plasmid linearization and *in vitro* transcription 246
5. DNA Standard Preparation 247
   5.1 Required materials 247
   5.2 Genomic standard stock preparation 247
6. Internal Standard Addition 247
7. Internal Standard Recoveries and Quantification 248
8. Dataset Normalization Using Internal Standards 248
   8.1 Metatranscriptome normalization 248
   8.2 Metagenome normalization 249
Acknowledgments 249
References 250

Abstract

Next generation sequencing-enabled metatranscriptomic and metagenomic datasets are providing unprecedented insights into the functional diversity of microbial communities, allowing detection of the genes present in a community as well as differentiation of those being actively transcribed. An emerging challenge of meta-omics approaches is how to quantitatively compare metagenomes and metatranscriptomes collected across spatial and temporal scales, or among treatments in experimental manipulations.

Methods in Enzymology, Volume 531
ISSN 0076-6879
http://dx.doi.org/10.1016/B978-0-12-407863-5.00012-5

Here, we describe the use of internal DNA and mRNA standards in meta-omics methodologies, and highlight how data collected in an absolute framework (per L or per cell) provides increased comparative power and insight into underlying causes of differences between samples.

1. INTRODUCTION

Metagenomic and metatranscriptomic methodologies have been used with great success in generating detailed information on community-level gene abundance and transcription patterns in marine, freshwater, soil, gut, and other natural microbial systems (Damon et al., 2011; Dinsdale et al., 2008; Gifford, Sharma, Rinta-Kanto, & Moran, 2011; Maurice, Haiser, & Turnbaugh, 2013; Ottesen et al., 2013; Poretsky et al., 2005; Vila-Costa, Sharma, Moran, & Casamayor, 2013). Most studies to date have collected meta-omics data in a relative framework, in which abundance of genes or messages is calculated as percent of the sequence library (Campbell, Yu, Heidelberg, & Kirchman, 2011; Hewson et al., 2009). However, a critical limitation of relative meta-omics data from complex natural communities is that they cannot provide information on the extent or directionality of changes in any particular gene or transcript molecule in comparative analyses. For instance, an observed decrease in the percent contribution of a transcript to the community metatranscriptome may be due to a decrease in the abundance of that transcript or to an increase in the abundance of an unrelated transcript (Fig. 12.1). In the application of meta-omics technologies to ecological and biogeochemical questions in complex microbial communities, the ability to recognize which genes and transcript molecules are changing in absolute abundance is crucial information, requiring datasets that are not influenced by the myriad nontarget processes and taxa changing simultaneously in a microbial cell or ecosystem.

To circumvent the limitations of relative metagenomic and metatranscriptomic datasets, internal genomic DNA or mRNA standards can be added at the initiation of sample processing (Gifford et al., 2011; Moran et al., 2013). Because these control molecules are mixed into and processed alongside the sample-derived nucleic acids, this allows quantification of losses throughout the preparation and analysis pipeline and, based on the number of standard molecules added at the beginning of sample processing and those recovered in the sequence library, calculation of the number of molecules of each gene or transcript in the original environment

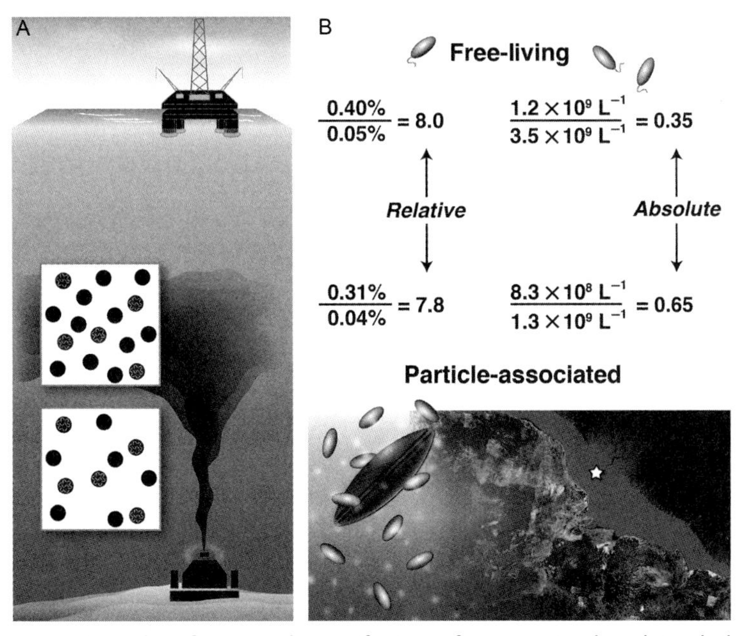

Figure 12.1 Examples of improved quantification of meta-omics data through the use of internal standards. (A) Transcripts binning to SAR11 member HTCC7211 accounted for a smaller fraction of the community metatranscriptome in the oil plume caused by the Deepwater Horizon accident compared to nonimpacted control samples below the plume, yet the absolute number of transcripts contributed by this taxon was not different. HTCC7211 is one of several bacteria taxa dominant in the prespill community that did not respond to the presence of oil, whereas other taxa greatly increased in number and activity in the hydrocarbon-impacted seawater (from Rivers et al., 2013). (B) Particle-associated bacteria in the Amazon River plume in June 2010 had twofold higher expression of proteorhodopsin genes than free-living bacteria, yet expression estimates calculated incorrectly from relative data would not have shown the differential regulation of this ecologically important gene (Satinsky B. et al., unpublished). (For color version of this figure, the reader is referred to the online version of this chapter.)

(e.g., gene copies per liter of water or average transcripts per microbial cell). Internal standards based on a known quantity of added control molecules are used routinely in quantitative PCR studies for calculating absolute gene and transcript abundance (Church, Short, Jenkins, Karl, & Zehr, 2005) and in microarray and RNA-seq analyses to normalize expression shifts in genes across different developmental stages or tissue types (Hannah, Redestig, Leisse, & Willmitzer, 2008; van de Peppel et al., 2003).

The benefits of quantitative meta-omics datasets can be illustrated by the following two examples. In the first, sequences binning to SAR11 member HTCC7211 in the bathypelagic waters of the Gulf of Mexico accounted for

0.84% of the bacterial metatranscriptome in natural seawater but only 0.07% in seawater exposed to oil and gas contamination from the Deepwater Horizon accident, indicating a 12-fold underrepresentation of HTCC7211 following the accident. Yet absolute transcript numbers for this taxon calculated based on internal standard normalization revealed that transcripts were present in equal numbers in impacted and nonimpacted seawater (2.8×10^{11} and 3.4×10^{11} transcripts/L; Fig. 12.1), and that the change in percent contribution of HTCC7211 populations was due to large increases in gammaproteobacteria groups that bloomed in response to hydrocarbon inputs (Fig. 12.1; Rivers et al., 2013). In a second example, expression ratios for proteorhodopsin genes in the near-shore Amazon River plume were nearly identical for the free-living and particle-associated bacteria when calculated on a relative basis (% of the metatranscriptome/% of the metagenome ≈ 8 for both free-living and particle-associated; Fig. 12.1). Yet on an absolute basis, the per-gene transcription level of proteorhodopsin was twofold higher for bacteria associated with particulate material compared to free-living cells in this ecosystem (Fig. 12.1). In these examples, normalization based on internal standard recovery provided insights into growth and regulation differences for bacteria in their natural environment, information that can be leveraged in comparative analyses across samples (e.g., within a time series, across a transect, or during a manipulative experiment) (Gifford et al., 2011; Moran et al., 2013; Fig. 12.1).

To generate quantitative-omics data, internal control sequences must be readily distinguished from natural microbial community sequences during bioinformatic analyses. For metatranscriptomes, artificial mRNAs produced by *in vitro* transcription from constructed DNA templates can be used as internal standards and preparation of mRNA standards with or without a poly(A) tail customizes them for bacterial/archaeal or eukaryotic studies. For metagenomes, genomic DNA obtained from a cultured microorganism not present in the studied environment can be added as an internal standard. In our marine and estuarine studies, DNA from the thermophilic bacterium *Thermus thermophilus* (ATCC) has served as the standard.

Calculations based on the internal standards assume that the natural nucleic acid (mRNA or genomic DNA) and the internal standards (artificial transcripts or exogenous genomic DNA) behave similarly throughout the sample and library preparation steps. However, the natural nucleic acid is enclosed in cell membranes at the initiation of processing while the internal standards are not, potentially resulting in underestimation of natural nucleic acid abundance due to incomplete cell lysis, or alternatively,

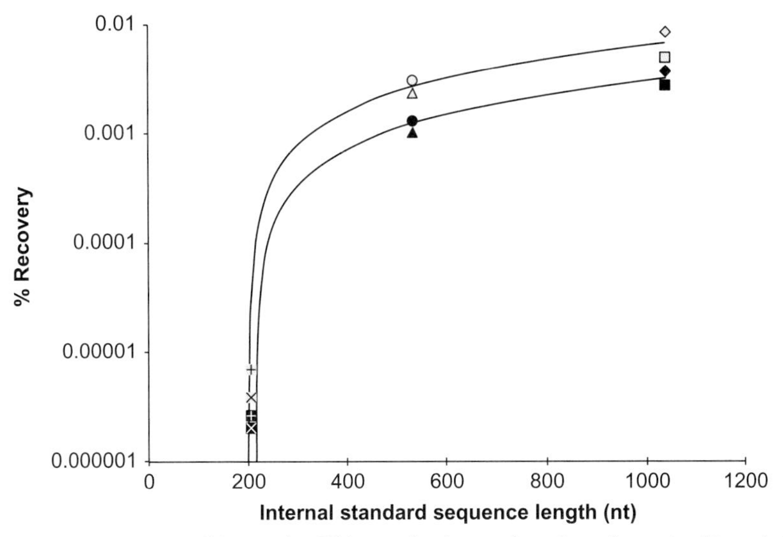

Figure 12.2 Recovery of internal mRNA standards as a function of standard length for two replicate metatranscriptome libraries (black series and gray series) from the Amazon River near Tapajos in June 2011. Two different internal standards of three different lengths (each represented as a different shaped symbol) were added to the samples at the initiation of nucleic acid extraction and percent recoveries were calculated as S_s (internal standard reads in the sequence library) $\times$ 100/S_a (internal standards added to the sample).

underestimation of standard abundance due to longer exposure time to mechanical shearing or RNAse degradation. In the case of mRNA processing, transcript length can affect recovery because of biases against small transcripts during solid-phase extraction methods and library preparation (Fig. 12.2), although this will affect both artificial and natural transcripts alike. Given an average bacterial and archaeal gene size of 924 bp (Xu et al., 2006), we use an internal standard of ~1000 nt to track recovery of mRNAs from typical prokaryotic genes. Internal standard length can be scaled down if small RNAs or short transcripts are the focus of the study; scaling up from 1000 nt does not appear to be necessary because of minimal effect on recovery for lengths >500 nt (Fig. 12.2).

2. METHOD OVERVIEW

The method given here describes the synthesis of internal mRNA standards and then the addition and quantification of mRNA and DNA internal standards for metatranscriptome and metagenome analysis.

mRNA standards are synthesized using custom templates or commercially available plasmids that are transcribed *in vitro* to RNA. A known number of standards are added to the sample of interest and metatranscriptome processing and sequencing proceeds according to the user's protocol. The number of internal standards recovered in the sequence library is quantified via BLAST homology searches. Data normalization is then based on the number of standards identified in a sequence library relative to the number of standards added. As a note of caution for working with RNA, care should be taken to avoid all contaminating nucleic acids and nucleases through the use of sterile technique and cleaning the working area with RNase*Zap*® or a similar reagent.

DNA standards can be prepared by purchasing or extracting DNA from a cultured microbe that is unrelated to microbes anticipated to be present in the system of interest and for which a complete genome sequence is available. A known number of genome copies are added to the sample, and metagenome processing and sequencing proceeds according to the user's protocol. The number of standard reads recovered in the sequence library is quantified via a two-step BLAST homology search and used for quantitative metagenomic analysis.

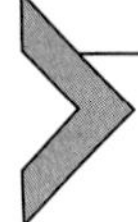

3. DNA TEMPLATE AND VECTOR DESIGN FOR INTERNAL RNA STANDARDS

Two approaches are available for obtaining the DNA template for standard synthesis. One approach involves commercially available plasmids that contain an RNA polymerase binding site. These are advantageous because of ease of use and low cost (Gifford et al., 2011; Moran et al., 2013), although the vectors make transcript length customization more difficult and they often contain regions of homology to functional proteins or to sequences deposited mistakenly into databases as functional proteins. This homology can make the subsequent identification of reads derived from standards more challenging in a high-throughput bioinformatics pipeline. A second approach involves the synthesis of custom DNA fragments that are inserted into plasmids. These fragments can easily be designed without homology to protein encoding genes, and provide optimal control of both length and composition.

For both template approaches, the final plasmid should contain the following components (in order): a T7 RNA polymerase promoter sequence, the internal standard sequence, and a restriction site targeting a unique site in the plasmid and preferably producing a blunt end (Fig. 12.3). For poly-A selective transcriptomes, a poly-A tail can be included in custom synthesized templates between the RNA polymerase promoter and the internal standard sequence. Whether using commercially available plasmids or custom synthesized internal standard templates, sequences should first be analyzed against relevant databases to identify regions of homology that could interfere with unambiguous identification of the standard in the sequence library.

Template size is also an important consideration because downstream processing steps during RNA processing and library preparation can lead to biases in the size of transcripts recovered. Based on addition of the six standards shown in Fig. 12.2 (representing two variations in base composition for each of three sizes: 200, 500, and 1000 nt), recovery efficiency in the sequence library was several orders of magnitude lower for the 200 nt mRNA standards compared to the others (Fig. 12.2). However, the duplicate standards at each size were recovered with nearly identical efficiencies, indicating that base composition is not an important factor in standard

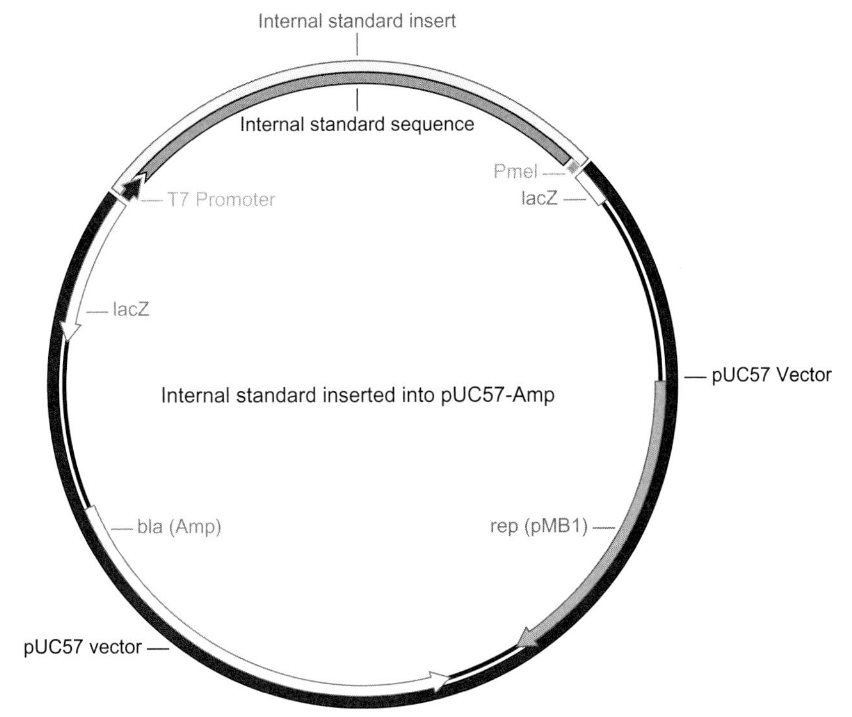

Figure 12.3 Genetic construct for *in vitro* transcription of a synthetic mRNA internal standard.

recovery. For the sequence data represented in Fig. 12.2, steps in the RNA isolation, purification, and amplification relied on solid-phase extraction, while Illumina library preparation included cDNA shearing and size selection (225 bp target size), all of which could lead to size bias for both artificial and natural mRNAs. Other extraction and library preparation methods may result in different size biases, but it is not straightforward to correct for size biases as transcript length depends on operon structure rather than individual gene length. Nonetheless, an internal standard can be selected that approximates the average size of the natural nucleic acid molecules being targeted (i.e., genomic DNA standards for metagenomes and artificial mRNAs of typical gene length for metatranscriptomes).

4. mRNA STANDARD PREPARATION

4.1. Required materials

- *Equipment*: 4 °C microcentrifuge, 10-, 20-, 200-, and 1000-μL pipettes, water bath, 37 °C shaking incubator, thermocycler, gel electrophoresis equipment and reagents, microfluidic electrophoresis instrument or fluorometry-based instrument for measuring nucleic acid concentration.
- *Media*: LB agar, LB agar + ampicillin (100 μg/mL final concentration), LB medium, LB medium + ampicillin (100 μg/mL final concentration), SOC medium (2% tryptone, 0.5% yeast extract, 10 m*M* sodium chloride, 2.5 m*M* potassium chloride, 10 m*M* magnesium chloride, 10 m*M* magnesium sulfate, 20 m*M* glucose).
- *Bacterial cell line*: One Shot® Top10 Chemically Competent *Escherichia coli* (Life Technologies, Grand Island, NY).
- *Template DNA*: Custom synthesized DNA template (T7 RNA polymerase promoter, internal standard sequence, unique restriction site) inserted into a plasmid.
- *Restriction digest and end repair*: Restriction enzyme matching unique restriction site and corresponding buffers, mung bean nuclease for end repair on digests that do not produce blunt ends.
- *Commercially available kits*: Ambion MEGAscript® T7 Kit (Life Technologies), Quant-iT™ RiboGreen® RNA Assay Kit (Life Technologies), miniPrep plasmid extraction kit.
- *Other reagents*: phenol:chloroform:isoamyl alcohol (24:24:1, pH ~7), citrate-saturated phenol:chloroform:isoamyl alcohol (24:24:1, pH 4.7),

sterile 2-propanol, ice cold 70% ethanol, nuclease-free 3 *M* sodium acetate, nuclease-free TE buffer, sterilized 100% glycerol; 1% agarose gel, nuclease-free water, RNase*Zap*® (Life Technologies).

- *Disposables*: nuclease-free 10-, 20-, 200-, and 1000-μL filter tips, nuclease-free PCR tubes, nuclease-free microcentrifuge tubes, gloves.

4.2. Plasmid amplification and stock preparation

4.2.1 Resuspension of plasmid DNA

If beginning with lyophilized plasmid DNA, spin briefly to ensure the contents are at the bottom of the tube. Resuspend the plasmid DNA in a volume of TE buffer to produce a stock concentration of 0.1 μg/μL. To prepare a working solution, add 1 μL of the stock solution to 99 μL of nuclease-free water to produce a final concentration of 1 ng/μL. The resuspended plasmid DNA can be stored at −20 °C.

4.2.2 Chemical transformation of plasmid into Top10 E. coli *cells*

Prior to beginning the transformation, ensure that all required media are prepared and sterilized. Place frozen competent cells and a prelabeled tube on ice. Prewarm a hot water bath to 42 °C. To a tube on ice, add 2 μL of (~2 ng) the plasmid working solution to 100 μL of thawed competent cells and flick the tube gently to mix. Incubate the mixture for 30 min on ice, then heat shock in the 42 °C hot water bath for 45 s. Immediately place the tube on ice for 2 min and then add 500 μL of SOC or LB liquid medium to the tube and incubate at 37 °C for 1 h with shaking (~225 rpm). During this time prewarm LB-Amp agar plates in a 37 °C incubator. From the tube, pipet and spread 10, 100, and 200 μL on three separate LB-Amp agar plates. Place the plates upside down in a 37 °C incubator for 12–24 h. Following incubation, inoculate a single, well-isolated colony from one of the plates and place into 5-mL LB-Amp media. Grow the liquid culture at 37 °C for ~8 h with vigorous shaking (~300 rpm). Remove 850 μL of the starter culture and place into a 2-mL freezer vial with 150 μL of sterilized 100% glycerol, mix thoroughly, and store at −80 °C. To work from the frozen stocks, place a loopful of stock into 10 mL of LB-Amp liquid medium and grow at 37 °C with vigorous shaking (~300 rpm) for 12–16 h. Harvest cells by centrifugation at 6000 × *g* for 15 min at 4 °C. Discard the supernatant and recover the plasmid DNA using a commercially available plasmid mini-prep kit.

4.3. Plasmid linearization and *in vitro* transcription

4.3.1 Linearization of plasmid template

Digest 2 μg of plasmid with restriction enzyme targeting the site at the end of the template sequence according to the restriction enzyme protocol. Sticky ends created by nonblunt-end cutting enzymes should be removed using mung bean nuclease. After digestion and end repair, bring the reaction to 100 μL by adding TE buffer, add 100 μL of phenol:chloroform:isoamyl alcohol (25:24:1, pH ~7), and mix by vortexing. Spin the mixture for 5 min at 12,000 × *g* in a microcentrifuge. Transfer the aqueous phase to a new tube and add 0.1 volumes (~10 μL) of 3 *M* sodium acetate and 0.7 volumes (~70 μL) of isopropanol to the tube. Mix thoroughly and incubate for 10 min at room temperature, and centrifuge for 30 min at 12,000 × *g* at 4 °C. Discard supernatant and wash pellet with 200 μL of ice cold 70% ethanol. Centrifuge for 5 min and discard the supernatant, being careful not to disturb the pellet. Air-dry the pellet to remove residual ethanol before resuspending the pellet in 5 μL of nuclease-free water. Transfer 2 μL of linearized plasmid into a new tube and add 2 μL of nuclease-free water. Use 1 μL of the diluted sample to check the concentration and analyze the remaining 3 μL on a 1% agarose gel to check for complete digestion and the presence of a single-sized product. Retain the 3 μL of undiluted DNA template for subsequent steps.

4.3.2 Synthesis and purification of mRNA internal standard

Synthesis of the internal standards from a template containing a T7 promoter is completed through the use of an *in vitro* transcription reaction using the Ambion MEGAscript® High Yield T7 Kit. In a 0.2-mL tube at room temperature, combine 2 μL of ATP solution, 2 μL of CTP solution, 2 μL of GTP solution, 2 μL of UTP solution, 2 μL of 10 × reaction buffer, 1 μg of linearized template DNA (up to 8 μl), and 2 μL of enzyme mix, and bring the total reaction volume to 20 μL with nuclease-free water. Mix thoroughly by flicking and incubate the mixture at 37 °C in a thermocycler with a heated lid for 16 h. Degrade the plasmid DNA by adding 1 μL of Turbo DNAse to the reaction tube and incubating for 15 min at 37 °C. Add 20 μL of citrate-saturated (pH 4.7) phenol:chloroform:isoamyl alcohol (25:24:1) to the tube. Vortex the mixture for 1 min and centrifuge for 2 min at 12,000 × *g* to separate the phases. Transfer the upper aqueous phase to a fresh tube and add 1 volume of chloroform:isoamyl alcohol (24:1). Vortex the mixture for 1 min and centrifuge for 2 min at 12,000 × *g*. Transfer the upper aqueous phase to a fresh tube and add 0.1 volumes of 3 *M* sodium acetate and 0.7 volumes of isopropanol. Mix by vortexing and incubate for 10 min at room temperature and then centrifuge for 30 min at 4 °C. Carefully discard the supernatant and wash the pellet with 200 μL of ice cold 70% ethanol. Centrifuge for 5 min and carefully remove the supernatant without disturbing the pellet. Air-dry the pellet until no residual ethanol remains and resuspend the dried pellet in 50 μL of nuclease-free water. Quantify the RNA fluorometrically using a Quant-iT™ RiboGreen® RNA Assay Kit and check the transcript size using a microfluidic electrophoresis instrument (e.g., Experion Automated Electrophoresis System, Agilent 2100 Bioanalyzer, or Agilent 2200 TapeStation). Store the mRNA internal standard stock at −80 °C.

5. DNA STANDARD PREPARATION

5.1. Required materials

- *Equipment*: Refrigerator, small tube rocker, 65 °C water bath or oven, fluorometry-based instrument for measuring nucleic acid concentration, 10-, 20-, 200-, and 1000-μL pipettes.
- *Materials*: Genomic DNA from a cultured, sequenced microbe unlikely to be closely related to microbes in the natural community, for example, *T. thermophilus* DSM7039 [HB27] genomic DNA (American Type Culture Collection (ATCC), Manassas, VA).
- *Commercially available kit*: Quant-iT™ PicoGreen® dsDNA Assay Kit (Life Technologies).
- *Disposables*: Sterile 10-, 20-, 200-, and 1000-μL filter tips, nuclease-free microcentrifuge tubes, gloves.

5.2. Genomic standard stock preparation

- Resuspend the genomic DNA in a volume of nuclease-free water to produce a stock concentration of 0.1 μg/μL following procedures recommended by ATCC. After rehydration incubate overnight at 4 °C while rocking, and then incubate for 1 h at 65 °C. To prepare a working solution, add 1 μL of the stock solution to 99 μL of nuclease-free water to produce a final concentration of 1 ng/μL. Check the DNA concentration of stocks fluorometrically using Quant-iT™ PicoGreen® dsDNA Assay Kit. The genomic DNA can be stored at −20 °C.

6. INTERNAL STANDARD ADDITION

Internal standards should be incorporated into the sample in a known amount just prior to RNA/DNA extraction. Prepare a tube with the desired lysis solution and add a known number of internal standard copies/genomes to the prepared lysis tube prior to the addition of the sample. The goal is to add an amount of internal standard sufficient for effective quantification in the sequence dataset, but not so high as to dominate the reads. This amount can be estimated from expected recovery of nucleic acids based on previous experience with the sample type. For example, if 5 μg of total RNA is expected from an extraction, the addition of 25 ng of internal standard (0.5% of the total RNA pool by weight) should be sufficient for a standard ~1000 nt in length. In our experience, a targeted ~0.5% addition has resulted in standards accounting for 0.1–5% of reads, depending on accuracy of our predicted RNA yield. When working with multiple standards, each standard should be added to the lysis tube independently in order to control for pipetting error.

7. INTERNAL STANDARD RECOVERIES AND QUANTIFICATION

Following sequencing, the number of mRNA internal standards recovered should be quantified by a BLASTn homology search for the template sequence using a bit score cutoff of 50, equivalent to an average percent identity of 98% in our analyses. The number of genomic internal standards should be quantified by first using a BLASTn homology search against the reference genome sequence to identify all potential standard reads, and subsequently taking any hits from the initial BLASTn homology search and performing a BLASTx search against the RefSeq Protein database to identify all protein encoding reads derived from the reference genome with a bit score cutoff of 40. The second annotation step against the RefSeq Protein database is necessary for identification of the standard reads due to a high number of false positives recruited by the BLASTn homology search. Following quantification, the internal standards should be removed from the dataset before further processing.

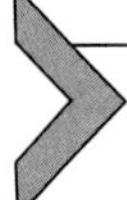

8. DATASET NORMALIZATION USING INTERNAL STANDARDS

8.1. Metatranscriptome normalization

Following identification of internal transcript standards, total transcript pool size and individual transcript abundances can be calculated as follows:

$$P_\mathrm{a} = \frac{P_\mathrm{s} \times S_\mathrm{a}}{S_\mathrm{s}}, \quad T_\mathrm{a} = \frac{T_\mathrm{s} \times P_\mathrm{a}}{P_\mathrm{s}}$$

P_a = total transcripts in the sample
P_s = protein encoding reads in the transcriptome library
S_a = molecules of internal standard added to the sample
S_s = internal standard reads in the sequence library
T_a = total molecules of any particular transcript type in the sample. This value can be divided by the mass or volume of sample collected to calculate the transcript abundance per volume or weight
T_s = number of transcripts of interest in the sequence library

8.2. Metagenome normalization

Following identification of internal genome standards, community gene pool size and individual gene abundances can be calculated as follows:

$$S_r = \frac{S_S}{S_P}$$

$$P_g = \frac{P_s \times S_a}{S_r}, \quad G_a = \frac{G_s \times P_g}{P_s}$$

S_r = no. of molecules of internal standard genome recovered from sequencing
S_S = no. of protein encoding internal standard reads in the sequence library
S_P = no. of protein encoding genes in the internal standard reference genome
P_g = total no. of protein encoding genes in the sample
P_s = no. of protein encoding sequences in the metagenome library
S_a = no. of molecules of internal standard genome added to the sample
G_a = no. of molecules of any particular gene category in the sample. This can then be divided by the mass or volume of sample collected to calculate the transcript abundance per volume or weight

ACKNOWLEDGMENTS

This research was funded by grants from the Gordon and Betty Moore Foundation (Marine Microbiology Investigator and River-Ocean Continuum of the Amazon) and the National Science Foundation (MCB-0702125).

REFERENCES

Campbell, B. J., Yu, L., Heidelberg, J. F., & Kirchman, D. L. (2011). Activity of abundant and rare bacteria in a coastal ocean. *Proceedings of the National Academy of Sciences of the United States of America, 108*(31), 12776–12781.

Church, M. J., Short, C. M., Jenkins, B. D., Karl, D. M., & Zehr, J. P. (2005). Temporal patterns of nitrogenase gene (nifH) expression in the oligotrophic North Pacific Ocean. *Applied and Environmental Microbiology, 71*(9), 5362–5370.

Damon, C., Vallon, L., Zimmermann, S., Haider, M. Z., Galeote, V., Dequin, S., et al. (2011). A novel fungal family of oligopeptide transporters identified by functional metatranscriptomics of soil eukaryotes. *ISME Journal, 5*(12), 1871–1880.

Dinsdale, E. A., Edwards, R. A., Hall, D., Angly, F., Breitbart, M., Brulc, J. M., et al. (2008). Functional metagenomic profiling of nine biomes. *Nature, 452*(7187), 629–632.

Gifford, S. M., Sharma, S., Rinta-Kanto, J. M., & Moran, M. A. (2011). Quantitative analysis of a deeply sequenced marine microbial metatranscriptome. *ISME Journal, 5*(3), 461–472.

Hannah, M. A., Redestig, H., Leisse, A., & Willmitzer, L. (2008). Global mRNA changes in microarray experiments. *Nature Biotechnology, 26*(7), 741–742.

Hewson, I., Poretsky, R. S., Beinart, R. A., White, A. E., Shi, T., Bench, S. R., et al. (2009). In situ transcriptomic analysis of the globally important keystone N2-fixing taxon Crocosphaera watsonii. *ISME Journal, 3*(5), 618–631.

Maurice, C. F., Haiser, H. J., & Turnbaugh, P. J. (2013). Xenobiotics shape the physiology and gene expression of the active human gut microbiome. *Cell, 152*(1–2), 39–50.

Moran, M. A., Satinsky, B., Gifford, S. M., Luo, H., Rivers, A., Chan, L. K., et al. (2013). Sizing up metatranscriptomics. *ISME Journal,* 7(2), 237–243.

Ottesen, E. A., Young, C. R., Eppley, J. M., Ryan, J. P., Chavez, F. P., Scholin, C. A., et al. (2013). Pattern and synchrony of gene expression among sympatric marine microbial populations. *Proceedings of the National Academy of Sciences of the United States of America, 110*(6), E488–E497.

Poretsky, R. S., Bano, N., Buchan, A., LeCleir, G., Kleikemper, J., Pickering, M., et al. (2005). Analysis of microbial gene transcripts in environmental samples. *Applied and Environmental Microbiology, 71*(7), 4121–4126.

Rivers, A., Sharma, S., Tringe, S. G., Martin, J., Joye, S. B., & Moran, M. A. (2013). Transcriptional response of bathypelagic marine bacterioplankton to the deepwater horizon oil spill. *ISME Journal,* http://dx.doi.org/10.1038/ismej.2013.129.

van de Peppel, J., Kemmeren, P., van Bakel, H., Radonjic, M., van Leenen, D., & Holstege, F. C. (2003). Monitoring global messenger RNA changes in externally controlled microarray experiments. *EMBO Reports, 4*(4), 387–393.

Vila-Costa, M., Sharma, S., Moran, M. A., & Casamayor, E. O. (2013). Diel gene expression profiles of a phosphorus limited mountain lake using metatranscriptomics. *Environmental Microbiology, 15*(4), 1190–1203.

Xu, L., Chen, H., Hu, X., Zhang, R., Zhang, Z., & Luo, Z. W. (2006). Average gene length is highly conserved in prokaryotes and eukaryotes and diverges only between the two kingdoms. *Molecular Biology and Evolution, 23*, 1107–1108.

CHAPTER THIRTEEN

Sample Processing and cDNA Preparation for Microbial Metatranscriptomics in Complex Soil Communities

Lilia C. Carvalhais, Peer M. Schenk[1]

School of Agriculture and Food Sciences, The University of Queensland, Brisbane, Queensland, Australia
[1]Corresponding author: e-mail address: p.schenk@uq.edu.au

Contents

1. Introduction 252
2. Working with RNA 253
   2.1 RNase and alien DNA-free working environment 253
3. Sampling 254
4. RNA Extraction of Complex Soil Communities 255
5. Assessing RNA Quality and Quantity 256
6. Soil Samples with Low Microbial Biomass 258
7. Removing Inhibitors 258
8. rRNA Subtraction/mRNA Enrichment 259
   8.1 Subtractive hybridization 259
   8.2 Exonuclease treatment 260
   8.3 Size separation by gel electrophoresis 261
9. mRNA Amplification 261
10. cDNA Synthesis 262
11. Current State and Future Prospects 263

References 263

Abstract

Soil presents one of the most complex environments for microbial communities as it provides many microhabitats that allow coexistence of thousands of species with important ecosystem functions. These include biomass and nutrient cycling, mineralization, and detoxification. Culture-independent DNA-based methods, such as metagenomics, have revealed operational taxonomic units that suggest a high diversity of microbial species and associated functions in soil. An emerging but technically challenging area to profile the functions of microorganisms and their activities is

Methods in Enzymology, Volume 531
ISSN 0076-6879
http://dx.doi.org/10.1016/B978-0-12-407863-5.00013-7

mRNA-based metatranscriptomics. Here, we describe issues and important considerations of soil sample processing and cDNA preparation for metatranscriptomics from bacteria and archaea and provide a set of methods that can be used in the required experimental steps.

1. INTRODUCTION

Soils are considered to be among the most complex and dynamic ecosystems on earth due to their compound structure, temporal variability, and spatial heterogeneity (White, 2006). Soil physical, chemical, and biological components enable the provision of various ecosystem services by fulfilling essential roles, including filter and reservoir, structure, fertility, climate regulation, resource and biodiversity conservation (Dominati, Patterson, & Mackay, 2010; Pare & Bedard-Haughn, 2013; Pointing & Belnap, 2012; Sheffield, Wood, & Roderick, 2012; Wohl et al., 2012). Soils are composed of fluid (water and air) and solid (particles) phases which harbor an enormous variety of macrofauna and microbes (Parker, 2010; Pulleman et al., 2012). Soil microorganisms from a multitude of taxonomic groups are densely populating soil environments and account for the majority of the diversity found in terrestrial ecosystems (Fierer et al., 2007). These microbes, together with the soil-inhabiting macrofauna, play a crucial role in delivering some of the essential ecosystem functions such as nutrient cycling, detoxification, and recycling of wastes, as well as decomposition (Barrios, 2007; Bradford & Newington, 2002; Dominati et al., 2010; Lavelle et al., 2006).

The size and diversity of soil microbial communities have hampered soil microbial ecology studies in comparison with communities from other environments. One gram of soil is estimated to contain a density of 10^8 to 10^9 bacterial cells and between few hundreds to 100,000 microbial species (Berendsen, Pieterse, & Bakker, 2012; Curtis, Sloan, & Scannell, 2002; Gans, Wolinsky, & Dunbar, 2005; Hughes, Hellmann, Ricketts, & Bohannan, 2002; Roesch et al., 2007). This extraordinarily high diversity is a consequence of the vast number of microhabitats generated within soil environments, which allow species coexistence and population persistence through resource partitioning (Ettema & Wardle, 2002; Giller, 1996; Kassen, Llewellyn, & Rainey, 2004).

Functional redundancy has been increasingly reported to be a common feature in microbial communities (Ferrer et al., 2013; Hesselsoe et al., 2009; Rineau & Courty, 2011; Schimel & Schaeffer, 2012). Therefore, development of strategies for conservation or use of soil ecosystem services is dependent on the understanding of the ongoing microbial functions. The largest portion of soil microbial biomass is comprised of bacteria and archaea (Hassink, Bouwman, Zwart, & Brussaard, 1993). However, only up to 1% of these soil microorganisms can be cultured in standard laboratory growth media (Amann, Ludwig, & Schleifer, 1995; McCaig, Grayston, Prosser, & Glover, 2001; Pham & Kim, 2012). Culture-independent methods that allow functional studies of soil communities have revolutionized our ability to investigate microbe-mediated soil ecosystem processes. Technical innovation, increasing speed, and lowering costs of high-throughput parallel sequencing platforms have made possible to assess microbial genomes much more deeply and comprehensively than ever before (Metzker, 2010; Shokralla, Spall, Gibson, & Hajibabaei, 2012). The functional investigation of microbial genomes in environmental DNA (also known as metagenomics) has been providing valuable insights into the potential functions and processes in a given environment (Delmont et al., 2012; Fierer et al., 2012; Kakirde, Parsley, & Liles, 2010; Liang et al., 2011; Uroz et al., 2013). However, extracellular DNA is not as accessible to nucleases since it can be adsorbed on soil particles and colloids, which maintain its integrity for long periods (Pietramellara et al., 2009). For this reason, DNA-based approaches most probably do not provide an accurate snapshot of the microbial activities which are taking place at the moment of the sampling. To obtain such information, metatranscriptomic approaches, which compare RNA transcript levels between samples, have been developed. However, due to the challenges imposed by the complex nature of soil samples, there has been a slow progress in comprehensive investigations of soil environments through metatranscriptomic approaches (Carvalhais, Dennis, Tyson, & Schenk, 2012; Damon et al., 2011, 2012; de Menezes, Clipson, & Doyle, 2012). Here, we describe issues and important considerations associated to soil sample processing and cDNA preparation for metatranscriptomic approaches and provide a compilation of methods that can be used in the associated steps.

2. WORKING WITH RNA

2.1. RNase and alien DNA-free working environment

RNA molecules are considerably more prone to degradation than DNA molecules, both chemically and enzymatically, and therefore have to be handled with care. The reasons for that are (1) the higher susceptibility of RNA's 2′-hydroxyl group to base-catalyzed hydrolysis in the ribose unit and (2) RNases are ubiquitous and very stable. Moreover, in bacterial

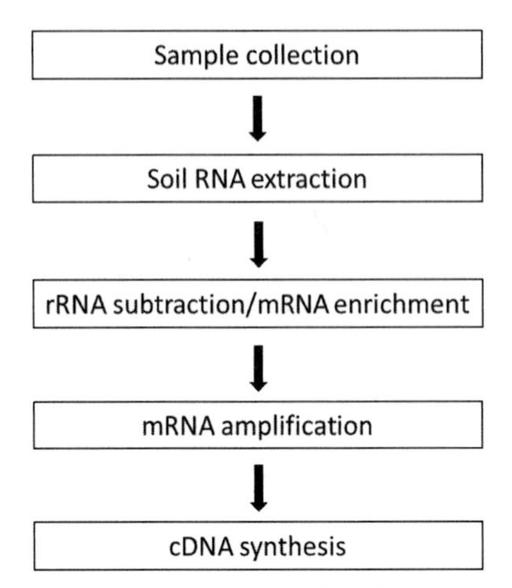

Figure 13.1 Overview of the main steps involved in the preparation of cDNA from soil.

and archaeal cells, simultaneous transcription and translation generate fragmented mRNA (Proshkin, Rahmouni, Mironov, & Nudler, 2010), and there is no addition of a poly(A) tail to the transcript for stabilization (Dreyfus & Regnier, 2002). In metatranscriptomic approaches, RNA degradation and contamination with alien DNA during sampling and RNA extraction should be avoided as some of the key steps are greatly affected when RNA is not intact and pure. All materials, plasticware, and glassware should then be decontaminated for RNases and alien DNA. Bench and table surfaces, as well as centrifuge rotors, spatulas, and pipette barrels should be wiped with a DNA and RNase surface decontaminant solution (e.g., RNase AWAY™ Reagent and RNaseZAP®, Ambion, Austin, TX, USA; RNase AWAY®, St. Louis, MO, USA; RNase AWAY Spray Bottle, Molecular BioProducts, San Diego, CA, USA). Certified RNase/DNA-free disposable plastic tubes, filter tips, and ultra-pure water can be purchased from a range of vendors, and glassware should be treated and solutions prepared with diethyl pyrocarbonate-treated water.

An overview of the main steps involved in the preparation of cDNA from soil is depicted in Fig. 13.1, each of which is described in detail below.

3. SAMPLING

Sampling is the first critical step especially for metatranscriptomic approaches. Due to the short half-life of mRNAs, which typically range from seconds to minutes (Deutscher, 2006), transcriptional profiles can be significantly altered by handling during sampling. Preferably, the collection of each sample should not take longer than a few minutes and should be made consistent between samples as well as between experiments to be compared. As soon as soil samples are collected, it is then essential to either store them in an RNA preservation solution (e.g. LifeGuard™ Soil Preservation Solution, MO BIO Laboratories Inc., Carlsbad, CA) or use liquid nitrogen to snap-freeze samples. In case further processing does not occur immediately after collection, soil samples should be stored at −70 °C or lower to prevent RNA degradation. Sieving of soil samples collected from natural environments is often used to remove large organic debris, rocks, and fine roots (mesh size 2 mm) (Bailly et al., 2007). However, ideally it should be done immediately before weighing the soil for total RNA extraction, therefore after the soil has been stored in a soil RNA preservation solution (and the sieve should also be wiped with a DNA and RNase surface decontaminant solution). If samples were stored at −70 °C without a soil preservation solution, sieving is not recommended after thawing as RNA degradation is likely to occur at room temperature. To avoid the same problem when weighing the soil immediately before RNA isolation, samples can be weighed while still frozen. A spatula can be used to scrape the surface of the frozen soil to obtain the desired amount. However, the soil should have been mixed thoroughly prior to being frozen to homogenize the sample. When the soil is mixed to a lysis solution, the RNA is relatively stable, but the lysis buffer should come into immediate contact with all parts of the soil. Alternatively, if the experiment will be conducted under controlled conditions, such as using pot experiments in growth chambers/greenhouses, sieving should be done before starting the experiment, unless results can be affected by it depending on the purpose of the experiment. Another way to extract RNA from soil is to first isolate water-suspendable microbes from soil at the site of collection by (1) adding phosphate-buffered saline (PBS) to suspend microbes, (2) letting soil debris settle, and (3) centrifuging the remaining liquid to recover a microbial pellet that can then be frozen and stored at −70 °C (McGrath et al., 2008). This method also allows concentration of soil microbes (see below).

4. RNA EXTRACTION OF COMPLEX SOIL COMMUNITIES

The main problems encountered during RNA isolation procedures are RNA adsorption to soil particles, presence of RNases, incomplete cell lysis, and coextraction of complex organic molecules that inhibit amplification steps by limiting template availability (Arbeli & Fuentes, 2007; Opel, Chung, & McCord, 2010). A number of different methods have been used for RNA extraction, including enzymatic lysis (Moran, Torsvik, Torsvik, & Hodson, 1993; Zhou, Bruns, & Tiedje, 1996), microwave-based rupture (Orsini & Romano-Spica, 2001), and liquid nitrogen grinding (Hurt et al., 2001; Volossiouk, Robb, & Nazar, 1995). However, bead-beating appears to be the most efficient to circumvent the issue of RNA adsorption to soil particles and cell lysis (Lakay, Botha, & Prior, 2007), and therefore it is the most commonly used method nowadays. Several RNA extraction kits are commercially available, including the RNA PowerSoil® Total RNA Isolation Kit (currently the most commonly used; MoBio Laboratories Inc., Carlsbad, CA, USA), the Soil Total RNA Purification kit (Thorold, Canada), ZR Soil/Fecal RNA MicroPrep™ (Irvine, CA, USA), E.Z.N.A.® Soil RNA kit (Omega Bio-tek, Norcross, GA, USA), all based on bead-beating. A widely used method was suggested by Griffiths, Whiteley, O'Donnell, and Bailey (2000), which is based on bead-beating with an extraction buffer containing hexadecyltrimethylammonium bromide (CTAB) and phenol–chlorophorm–isoamyl alcohol (25:24:1), followed by repeated centrifugation steps and precipitation of the aqueous supernatant with 1.6 *M* NaCl mixed with 2 volumes of 30% polyethylene glycol 6000 (Griffiths et al., 2000). Another bead-beating-based method includes 1% SDS, 2 *M* guanidine isothiocyanate, and β-mercaptoethanol in the extraction buffer and the precipitation solution contains Na-acetate and ethanol for total nucleic acid precipitation and isopropanol for RNA precipitation after DNase I treatment (Bailly et al., 2007).

5. ASSESSING RNA QUALITY AND QUANTITY

High RNA quality typically denotes features such as high purity and integrity. Ultraviolet–visible spectrophotometry has been traditionally used to assess quality as well as quantity by the measurement of optical density (OD) mainly at two wave lengths: 260 and 280 nm, which is specific for nucleic acids and proteins, respectively. Ideally, OD 260/280 ratios of an RNA extract should reach 1.8 or greater (Manchester, 1996). This ratio will be lowered in the presence of absorbing contaminants at 280 nm such as proteins (Gallagher, 2011). However, this method is not accurate as overestimations often occur due to the presence of contaminating molecules that absorb radiation at similar wavelengths. Another spectrometric instrument that requires very low sample volumes and no cuvettes is the NanoDrop (ND-3300, NanoDrop Technologies, USA). A combination of spectrometric approach which allows determining purity and a method that includes a RNA-specific fluorescent dye (e.g., RNA RiboGreen dye, Molecular Probes, Invitrogen, or Qubit® 2.0 Fluorometer, Life Technologies Corp., Carlsbad, CA, USA) is recommended for most downstream analysis as it provides a more sensitive quantification especially for RNA samples with low concentration (Fleige & Pfaffl, 2006).

Ethidium bromide (EtBr)-stained denaturing agarose gel electrophoresis is the most commonly used method to assess RNA integrity (Fleige & Pfaffl, 2006). Usually two bands corresponding to the largest RNA subunits are visible and should be clear and distinct as well as no smear should be detected along the lane. For RNA with high integrity, the band brightness ratio between the two largest RNA bands should be 2 or higher (23S/16S and 28S/18S for bacteria/archaea and eukaryotes, respectively). However, this approach is observer subjective and high RNA quantities ($\sim$200 ng) are needed for visualization using EtBr, which depending on the RNA source can be difficult to obtain. More sensitive stains are available, including SYBR® Green II and SYBR® Gold (Molecular Probes, Eugene, OR, EUA); both each require just 2 and 1 ng of RNA on the gel, respectively.

Two very sensitive microfluidics-based technologies can be used to assess RNA quality and quantity, the Experion (Bio-Rad Laboratories Inc., Hercules, CA, USA) and Agilent 2100 Bioanalyzer (Agilent Technologies, Palo Alto, CA, USA). Both on-chip capillary electrophoresis methods require a very small amount of RNA (at picogram levels for both Agilent RNA 6000 Pico chip and Experion RNA HighSens used in the Agilent 2100 Bioanalyzer and Experion, respectively).

RNA sizes and amounts are estimated in comparison to an RNA ladder (Fleige & Pfaffl, 2006). As mentioned above, ratios close to 2 (23S/16S for bacteria and archaea and 28S/18S for eukaryotes) of the obtained electropherogram usually depict the highest RNA integrities. In addition, a model has been developed by Agilent Technologies (Palo Alto, CA, USA) for classifying RNA samples according to its integrity (Schroeder et al., 2006). This model provides a RNA integrity number (RIN) that is calculated using features of various regions of the obtained electropherogram and ranges from 1 to 10. RIN values close to 10 signify high RNA integrity and 1 low. For samples to be used in metatranscriptomic approaches, it is typically recommended an RIN of 8 or higher.

6. SOIL SAMPLES WITH LOW MICROBIAL BIOMASS

Soils with low microbial biomass typically provide low RNA yields and may require that microbial cells are enriched in the soil sample before starting the RNA extraction. Especially when soil quantities are not limiting, starting with bigger amounts from what is usually required (e.g., 20 g instead of 2 g) captures a more representative sample of the heterogeneity that is typical of soil. An example of such a procedure has been reported previously (McGrath et al., 2008). In brief, an amount of 20 g of soil and 20 mL of PBS (137 m*M* NaCl, 10 m*M* phosphate, 2.7 m*M* KCl, pH 7.4) or saline (0.85% NaCl) is added to a 50-mL tube, and after shaking it vigorously for approximately 20 s to homogenize the sample, bigger particles are allowed to decant for 10 s and the supernatant is subsequently transferred into microcentrifuge tubes which are then centrifuged for 2 min ($14{,}000 \times g$) to recover the microbe-enriched pellet. After discarding the supernatant, this pellet should be snap frozen and stored at −70 °C or less. To avoid RNA degradation or changes in RNA profiles during transport from soil collection sites, these steps can be done on site by running the microcentrifuge from a car battery and maintaining the tubes in a 1:1 mixture of equal ratio of ice and sodium chloride (if well crushed and mixed, the temperature can go down to −10 °C, which is more suitable than room temperature). Dry ice would be obviously the best way of transporting the samples; however, it may not be easily available for purchase in remote experimental sites.

7. REMOVING INHIBITORS

Specific methods have been developed and were shown to overcome the issue of RNA coprecipitation with humic and fulvic acids, which inhibit several enzymatically catalyzed processes used in molecular biology. Such methods include addition of chemicals that adsorb polyvinylpolypyrolidone (Rajendhran & Gunasekaran, 2008), adsorption with powered activated charcoal (Desai & Madamwar, 2007), isolation of extracted nucleic acids by $CaCl_2$ (Sagova-Mareckova et al., 2008), precipitation with aluminum sulfate prior to cell lysis (Persoh, Theuerl, Buscot, & Rambold, 2008), extraction of RNA at pH 5.0 followed by purification using Q-Sepharose columns supplemented with CTAB and vitamins (Mettel, Kim, Shrestha, & Liesack, 2010), and pretreatment of soils with $CaCO_3$

(Sagova-Mareckova et al., 2008). The PowerMicrobiome™ RNA Isolation kit (MoBio Laboratories Inc., Carlsbad, CA, USA) has been especially designed for samples with high content of PCR inhibitors. If presence of inhibitors is still detected after RNA isolation, commercial kits are available for cleaning environmental RNA samples, such as RNeasy MinElute Cleanup kit (Qiagen, Valencia, CA, USA), NucleoSpin RNA Clean-up XS (Macherey-Nagel, Dueren, Germany), RNA Clean-Up and Concentration kit (Norgen Biotek Corp., Thorold, ON, Canada), and OneStep™ PCR Inhibitor Removal Kit (Zymo Research Corporation, Irvine, CA, USA).

8. rRNA SUBTRACTION/mRNA ENRICHMENT

Only a small fraction of the total RNA (up to 5%) is composed of messenger RNA (mRNA), which is the RNA type that most metatranscriptomic approaches focus on. To maximize the mRNA sequencing coverage, it is then essential to perform an mRNA enrichment step. For this purpose, a number of methods have been developed. A summary of rRNA subtraction methods are presented in Table 13.1 and discussed in more detail in the sections below.

8.1. Subtractive hybridization

This technique involves the subtraction of a specific set of nucleic acids (in this case rRNA) from a reference sample (total RNA) using probes with

Table 13.1 Currently available rRNA subtraction methods from samples containing total RNA

rRNA subtraction methods	Source references/kits
Subtractive hybridization	Stewart, Ottesen, and DeLong (2010); MICROBExpress Bacterial mRNA Enrichment (Ambion); Ribo-Zero Magnetic kit (Epicentre)
Exonuclease treatment	mRNA-only™ Prokaryotic mRNA Isolation kit (Epicentre)
Combination of subtractive hybridization and exonuclease treatment	Mettel et al. (2010), He et al. (2010)
Size separation by gel electrophoresis	McGrath et al. (2008)

sequences that are complementary to the set to be subtracted (rRNA) (Pang et al., 2004; Su & Sordillo, 1998; Winstanley, 1998). rRNA can be partially subtracted by the commercial kits MICROBExpress Bacterial mRNA Enrichment (Ambion, Austin, TX, USA) and Ribo-Zero Magnetic kit (Bacteria) (Epicentre, Madison, WI, USA) through capture 16S and 23S binding oligonucleotides immobilized on magnetic microbeads. However, insufficient complementarity of capture probes in the kit may affect the efficiency of rRNA removal. An rRNA subtraction hybridization protocol includes sample-specific biotinylated rRNA capture probes (Stewart et al., 2010). Briefly, microbial community DNA is isolated simultaneously with RNA and is used as PCR template with universal primers to amplify genes encoding the large ribosomal subunits of bacteria, archaea, and eukaryotes. The T7 RNA polymerase promoter sequence is added to the reverse primer. *In vitro* transcription is then used to produce biotinylated antisense rRNA probes from the ribosomal amplicons. After the complementary rRNA molecules present in the total RNA hybridize to the biotinylated probes, streptavidin-coated magnetic beads are then used to remove biotinylated double-stranded rRNA. In addition to being sample specific and therefore more likely to target a broader spectrum of rRNA molecules, another advantage of this method is that, unlike most available commercial kits, archaeal and eukaryotic DNA can also be subtracted (also see Chapter 10).

8.2. Exonuclease treatment

This method is based on a 5′-phosphate-dependent exonuclease that degrades only RNA molecules that possess a 5′-monophosphate, such as most bacterial rRNAs. The commercial mRNA-only™ Prokaryotic mRNA Isolation kit (Epicentre, Madison, WI, USA) uses this principle to provide mRNA enriched samples. Yet, the subtraction hybridization method has been shown to be more efficient in enriching for mRNA from soil samples when compared with exonuclease treatment (Mettel et al., 2010). This is believed to occur because soil-derived total RNA is likely to possess residual humic acids, which are known to inhibit a number of enzymatic activities (Allison, 2006). Nevertheless, for approaches that target unprocessed mRNA, it has been suggested that exonuclease treatment may be used after subtraction hybridization (Mettel et al., 2010).

Another study systematically compared the two rRNA subtraction methods described above to validate their fidelity and effectiveness using high-throughput metatranscriptome sequencing. Relative transcript abundances contained the least bias when subtractive hybridization only was used, followed by exonuclease only, and lastly by combinations of these treatments (He et al., 2010).

8.3. Size separation by gel electrophoresis

This method is among the ones that involve the least sample processing after RNA isolation (e.g., reverse transcription and *in vitro* transcription followed by cleaning up) but requires fairly large amounts of total RNA (McGrath et al., 2008). Total RNA is applied on a 1.5% TAE (40 m*M* Tris-base; 20 m*M* acetic acid; 1 m*M* ethylene diamine tetraacetic acid, pH 8.0) agarose gel, which is then run for 45 min at 100 V. Regions containing mRNA along lanes between the 23S, 16S, and 5S bands are identified using a UV transilluminator, and subsequently excised and purified using the Wizard SV Gel and PCR Clean-Up System (Promega, Madison, WI, USA). Although, mRNA that has the same size as the 23S, 16S, and 5S rRNA bands are possibly missed out, it is likely to be present in various sizes in the excised regions. This is because fragmented RNA is inevitably generated by simultaneous transcription and translation in archaeal and bacterial cells (Proshkin et al., 2010).

9. mRNA AMPLIFICATION

To obtain sufficient amounts of mRNA from soils for downstream applications on metatranscriptomic approaches can be challenging, especially for rhizosphere samples. A way to circumvent this problem is to amplify the obtained mRNA. Because bacterial mRNA is not naturally polyadenylated, the direct use of oligo(dT) primers after cDNA synthesis is not an option. Several commercial kits are available for mRNA amplification, including the MessageAmp II-Bacteria RNA Amplification (Ambion, Austin, Texas), ExpressArt Bacterial mRNA amplification Nano kit (Amsbio, Abingdon, UK), and illustra GenomiPhi HY DNA Amplification (GE Healthcare, Milwaukee, WI) kits. The most commonly used kit for mRNA amplification in environmental metatranscriptomic studies is the MessageAmp II-Bacteria RNA Amplification kit (Frias-Lopez et al., 2008; Gosalbes et al., 2011; Poretsky et al., 2009; Stewart, Ulloa, & DeLong, 2012). The methods involved usually start with the polyadenylation of the 3′-end of the total RNA with the *E. coli* Poly(A)

Polymerase, which is subsequently reverse transcribed into a first-strand cDNA with an oligo(dT) primer. A T7 promoter sequence and a restriction enzyme (*Bpm*I) recognition site can be included in the primer sequence for later removal of the tails introduced by the oligo(dT) primer (T7-*Bpm*I-$(dT)_{16}$VN oligo) (Frias-Lopez et al., 2008). The second strand of the cDNA is subsequently synthesized with a DNA polymerase and RNase H, which degrade the remaining RNA. After a purification step to remove RNA, primers, and enzymes, double-stranded cDNA is then used as templates for *in vitro* transcription by the T7 RNA polymerase. This enzyme generates high quantities of antisense RNA, which should then be cleaned up before cDNA synthesis. Initial RNA amounts for *in vitro* transcription can include a minimum of 10 and 100 ng for mRNA or total RNA, respectively. However, a 14 h incubation time is strongly recommended during *in vitro* transcription when minimum initial amounts are used. A range of commercial kits based on silica-membrane technology are available for RNA cleanup, including the RNA Clean & Concentrator (Zymo Research, Irvine, CA, USA), GeneJET RNA Cleanup and Concentration Micro Kit (Thermo Scientific, Rockford, IL, USA), NucleoSpin RNA Clean-up (Macherey-Nagel, Dueren, Germany), and RNeasy MinElute Cleanup kit (Qiagen, Valencia, CA, USA). The last one seems to be the most commonly used.

10. cDNA SYNTHESIS

The enzymatic synthesis of cDNA from mRNA requires, among other reagents, two key components: reverse transcriptase and oligonucleotides (hexamers and oligo-dT). Although enzymes and primers can be bought separately, a number of commercial kits containing all necessary reagents can be purchased, such as SuperScript III Double-Stranded cDNA Synthesis Kit (Life Technologies, Carlsbad, CA, USA), Maxima H Minus Double-Stranded cDNA Synthesis Kit (Thermo Scientific, Hudson, NH, USA), or cDNA Synthesis System (Roche Applied Science, Mannheim, Germany). A final digestion with *Bpm*I should then be performed to remove the introduced poly(A) tails, and after purification, the sample is ready to be sent for high-throughput sequencing. Although some metatranscriptomic approaches using soil samples used the traditional cDNA plasmid libraries as templates for sequencing (Bailly et al., 2007; Damon et al., 2012; Takasaki et al., 2013), this method has been rapidly superseded by high-throughput parallel sequencing, as it delivers a much greater output of reads.

11. CURRENT STATE AND FUTURE PROSPECTS

The access to metatranscriptomic approaches is still limited to laboratories that can afford the high costs of various commercial kits which are needed for most of the steps involved, as well as high-throughput sequencing necessary to compare transcript levels between samples with reasonable depth. Although sequencing technologies have become cheaper and been increasingly improved to deliver higher outputs, they are still costly. Another major constraint is data analysis, which requires high computational power, as well as skills in bioinformatics, as most softwares are still being developed or often are not user-friendly at the current stage.

Undoubtedly, soil metatranscriptomic analyses have great potential to deepen our understanding of microbial functions in the soil and help developing strategies to tackle problems associated with soil fertility and pollution, crop yield, global warming, biodiversity, ecosystem function as well as plant diseases. A few studies using soil metatranscriptomics reported interesting findings in these matters. For example, a combined metagenomic and metatranscriptomic approach assessed the potential and actual microbial functions involved in soil organic carbon degradation in Antarctic permafrost peatlands (Tveit, Schwacke, Svenning, & Urich, 2013). These environments can contribute to greater emissions of methane and carbon dioxide, and therefore to global warming. Another recent study profiled metabolic microbial processes involved in responses of soil microbial communities to the priority pollutant phenanthrene (de Menezes et al., 2012). Other approaches identified the predominant enzymes expressed in forest soils by eukaryotes as well as a novel fungal family of oligopeptide transporters (Damon et al., 2011, 2012). In addition, evolutionary patterns were revealed by microbial transcript levels in different environments including forest soils. Microbial gene expression levels were inversely related to evolutionary rates in natural environments (Stewart, Sharma, Bryant, Eppley, & DeLong, 2011). In conclusion, it is clear that the application of metatranscriptomic approaches in soil environments is still in its infancy and a lot more studies are certainly on the way as technical and analytical challenges are being overcome.

REFERENCES

Allison, S. D. (2006). Soil minerals and humic acids alter enzyme stability: Implications for ecosystem processes. *Biogeochemistry*, *81*, 361–373.

Amann, R. I., Ludwig, W., & Schleifer, K. H. (1995). Phylogenetic identification and in-situ detection of individual microbial cells without cultivation. *Microbiological Reviews*, *59*, 143–169.

Arbeli, Z., & Fuentes, C. L. (2007). Improved purification and PCR amplification of DNA from environmental samples. *FEMS Microbiology Letters*, *272*, 269–275.

Bailly, J., Fraissinet-Tachet, L., Verner, M. C., Debaud, J. C., Lemaire, M., Wesolowski-Louvel, M., et al. (2007). Soil eukaryotic functional diversity, a metatranscriptomic approach. *The ISME Journal*, *1*, 632–642.

Barrios, E. (2007). Soil biota, ecosystem services and land productivity. *Ecological Economics*, *64*, 269–285.

Berendsen, R. L., Pieterse, C. M. J., & Bakker, P. A. H. M. (2012). The rhizosphere microbiome and plant health. *Trends in Plant Science*, *17*, 478–486.

Bradford, M. A., & Newington, J. E. (2002). With the worms: Soil biodiversity and ecosystem functioning. *Biologist (London, England)*, *49*, 127–130.

Carvalhais, L. C., Dennis, P. G., Tyson, G. W., & Schenk, P. M. (2012). Application of metatranscriptomics to soil environments. *Journal of Microbiological Methods*, *91*, 246–251.

Curtis, T. P., Sloan, W. T., & Scannell, J. W. (2002). Estimating prokaryotic diversity and its limits. *Proceedings of the National Academy of Sciences of the United States of America*, *99*, 10494–10499.

Damon, C., Lehembre, F., Oger-Desfeux, C., Luis, P., Ranger, J., Fraissinet-Tachet, L., et al. (2012). Metatranscriptomics reveals the diversity of genes expressed by eukaryotes in forest soils. *PLoS One*, 7(1), e28967.

Damon, C., Vallon, L. S., Zimmermann, L., Haider, M. Z., Galeote, V., Dequin, S., et al. (2011). A novel fungal family of oligopeptide transporters identified by functional metatranscriptomics of soil eukaryotes. *The ISME Journal*, *5*, 1871–1880.

de Menezes, A., Clipson, N., & Doyle, E. (2012). Comparative metatranscriptomics reveals widespread community responses during phenanthrene degradation in soil. *Environmental Microbiology*, *14*, 2577–2588.

Delmont, T. O., Prestat, E., Keegan, K. P., Faubladier, M., Robe, P., Clark, I. M., et al. (2012). Structure, fluctuation and magnitude of a natural grassland soil metagenome. *The ISME Journal*, *6*, 1677–1687.

Desai, C., & Madamwar, D. (2007). Extraction of inhibitor-free metagenomic DNA from polluted sediments, compatible with molecular diversity analysis using adsorption and ion-exchange treatments. *Bioresource Technology*, *98*, 761–768.

Deutscher, M. P. (2006). Degradation of RNA in bacteria: Comparison of mRNA and stable RNA. *Nucleic Acids Research*, *34*, 659–666.

Dominati, E., Patterson, M., & Mackay, A. (2010). A framework for classifying and quantifying the natural capital and ecosystem services of soils. *Ecological Economics*, *69*, 1858–1868.

Dreyfus, M., & Regnier, P. (2002). The poly(A) tail of mRNAs: Bodyguard in eukaryotes, scavenger in bacteria. *Cell*, *111*, 611–613.

Ettema, C. H., & Wardle, D. A. (2002). Spatial soil ecology. *Trends in Ecology and Evolution*, *17*, 177–183.

Ferrer, M., Ruiz, A., Lanza, F., Haange, S. B., Oberbach, A., Till, H., et al. (2013). Microbiota from the distal guts of lean and obese adolescents exhibit partial functional redundancy besides clear differences in community structure. *Environmental Microbiology*, *15*, 211–226.

Fierer, N., Breitbart, M., Nulton, J., Salamon, P., Lozupone, C., Jones, R., et al. (2007). Metagenomic and small-subunit rRNA analyses reveal the genetic diversity of bacteria, archaea, fungi, and viruses in soil. *Applied and Environmental Microbiology*, *73*, 7059–7066.

Fierer, N., Lauber, C. L., Ramirez, K. S., Zaneveld, J., Bradford, M. A., & Knight, R. (2012). Comparative metagenomic, phylogenetic and physiological analyses of soil microbial communities across nitrogen gradients. *The ISME Journal*, *6*, 1007–1017.

Fleige, S., & Pfaffl, M. W. (2006). RNA integrity and the effect on the real-time qRT-PCR performance. *Molecular Aspects of Medicine*, *27*, 126–139.

Frias-Lopez, J., Shi, Y., Tyson, G. W., Coleman, M. L., Schuster, S. C., Chisholm, S. W., et al. (2008). Microbial community gene expression in ocean surface waters. *Proceedings of the National Academy of Sciences of the United States of America*, *105*, 3805–3810.

Gallagher, S. R. (2011). Quantitation of DNA and RNA with absorption and fluorescence spectroscopy. *Current Protocols in Molecular Biology*, *3*, 3D.

Gans, J., Wolinsky, M., & Dunbar, J. (2005). Computational improvements reveal great bacterial diversity and high metal toxicity in soil. *Science*, *309*, 1387–1390.

Giller, P. S. (1996). The diversity of soil communities, the 'poor man's tropical rainforest'. *Biodiversity and Conservation*, *5*, 135–168.

Gosalbes, M. J., Durban, A., Pignatelli, M., Abellan, J. J., Jimenez-Hernandez, N., Perez-Cobas, A. E., et al. (2011). Metatranscriptomic approach to analyze the functional human gut microbiota. *PLoS One*, *6*, e17447.

Griffiths, R. I., Whiteley, A. S., O'Donnell, A. G., & Bailey, M. J. (2000). Rapid method for coextraction of DNA and RNA from natural environments for analysis of ribosomal DNA- and rRNA-based microbial community composition. *Applied and Environmental Microbiology*, *66*, 5488–5491.

Hassink, J., Bouwman, L. A., Zwart, K. B., & Brussaard, L. (1993). Relationships between habitable pore-space, soil biota and mineralization rates in grassland soils. *Soil Biology and Biochemistry*, *25*, 47–55.

He, S. M., Wurtzel, O., Singh, K., Froula, J. L., Yilmaz, S., Tringe, S. G., et al. (2010). Validation of two ribosomal RNA removal methods for microbial metatranscriptomics. *Nature Methods*, *7*, 807–858.

Hesselsoe, M., Fureder, S., Schloter, M., Bodrossy, L., Iversen, N., Roslev, P., et al. (2009). Isotope array analysis of Rhodocyclales uncovers functional redundancy and versatility in an activated sludge. *The ISME Journal*, *3*, 1349–1364.

Hughes, J. B., Hellmann, J. J., Ricketts, T. H., & Bohannan, B. J. M. (2002). Counting the uncountable: Statistical approaches to estimating microbial diversity. *Applied and Environmental Microbiology*, *68*, 448.

Hurt, R. A., Qiu, X., Wu, L., Roh, Y., Palumbo, A. V., Tiedje, J. M., et al. (2001). Simultaneous recovery of RNA and DNA from soils and sediments. *Applied and Environmental Microbiology*, *67*, 4495–4503.

Kakirde, K. S., Parsley, L. C., & Liles, M. R. (2010). Size does matter: Application-driven approaches for soil metagenomics. *Soil Biology and Biochemistry*, *42*, 1911–1923.

Kassen, R., Llewellyn, M., & Rainey, P. B. (2004). Ecological constraints on diversification in a model adaptive radiation. *Nature*, *431*, 984–988.

Lakay, F. M., Botha, A., & Prior, B. A. (2007). Comparative analysis of environmental DNA extraction and purification methods from different humic acid-rich soils. *Journal of Applied Microbiology*, *102*, 265–273.

Lavelle, P., Decaens, T., Aubert, M., Barot, S., Blouin, M., Bureau, F., et al. (2006). Soil invertebrates and ecosystem services. *European Journal of Soil Biology*, *42*, S3–S15.

Liang, Y., van Nostrand, J. D., Deng, Y., He, Z., Wu, L., Zhang, X., et al. (2011). Functional gene diversity of soil microbial communities from five oil-contaminated fields in China. *The ISME Journal*, *5*, 403–413.

Manchester, K. L. (1996). Use of UV methods for measurement of protein and nucleic acid concentrations. *Biotechniques*, *20*, 968–970.

McCaig, A. E., Grayston, S. J., Prosser, J. I., & Glover, L. A. (2001). Impact of cultivation on characterisation of species composition of soil bacterial communities. *FEMS Microbiology Ecology*, *35*, 37–48.

McGrath, K. C., Thomas-Hall, S. R., Cheng, C. T., Leo, L., Alexa, A., Schmidt, S., et al. (2008). Isolation and analysis of mRNA from environmental microbial communities. *Journal of Microbiological Methods*, *75*, 172–176.

Mettel, C., Kim, Y., Shrestha, P. M., & Liesack, W. (2010). Extraction of mRNA from Soil. *Applied and Environmental Microbiology*, *76*, 5995–6000.

Metzker, M. L. (2010). Applications of next generation sequencing technologies—The next generation. *Nature Reviews Genetics*, *11*, 31–46.

Moran, M. A., Torsvik, V. L., Torsvik, T., & Hodson, R. E. (1993). Direct extraction and purification of rRNA for ecological studies. *Applied and Environmental Microbiology*, *59*, 915–918.

Opel, K. L., Chung, D., & McCord, B. R. (2010). A study of PCR inhibition mechanisms using real time PCR. *Journal of Forensic Sciences*, *55*, 25–33.

Orsini, M., & Romano-Spica, V. (2001). A microwave-based method for nucleic acid isolation from environmental samples. *Letters in Applied Microbiology*, *33*, 17–20.

Pang, X., Zhou, D. S., Song, Y. J., Pei, D. C., Wang, J., Guo, Z. B., et al. (2004). Bacterial mRNA purification by magnetic capture-hybridization method. *Microbiology and Immunology*, *48*, 91–96.

Pare, M. C., & Bedard-Haughn, A. (2013). Soil organic matter quality influences mineralization and GHG emissions in cryosols: A field-based study of sub- to high Arctic. *Global Change Biology*, *19*, 1126–1140.

Parker, S. S. (2010). Buried treasure: Soil biodiversity and conservation. *Biodiversity and Conservation*, *19*, 3743–3756.

Persoh, D., Theuerl, S., Buscot, F., & Rambold, G. (2008). Towards a universally adaptable method for quantitative extraction of high-purity nucleic acids from soil. *Journal of Microbiological Methods*, *75*, 19–24.

Pham, V. H. T., & Kim, J. (2012). Cultivation of unculturable soil bacteria. *Trends in Biotechnology*, *30*, 475–484.

Pietramellara, G., Ascher, J., Borgogni, F., Ceccherini, M. T., Guerri, G., & Nannipieri, P. (2009). Extracellular DNA in soil and sediment: Fate and ecological relevance. *Biology and Fertility of Soils*, *45*, 219–235.

Pointing, S. B., & Belnap, J. (2012). Microbial colonization and controls in dryland systems. *Nature Reviews Microbiology*, *10*, 654.

Poretsky, R. S., Hewson, I., Sun, S. L., Allen, A. E., Zehr, J. P., & Moran, M. A. (2009). Comparative day/night metatranscriptomic analysis of microbial communities in the North Pacific subtropical gyre. *Environmental Microbiology*, *11*, 1358–1375.

Proshkin, S., Rahmouni, A. R., Mironov, A., & Nudler, E. (2010). Cooperation between translating ribosomes and RNA polymerase in transcription elongation. *Science*, *328*, 504–508.

Pulleman, M., Creamer, R., Hamer, U., Helder, J., Pelosi, C., Peres, G., et al. (2012). Soil biodiversity, biological indicators and soil ecosystem services-an overview of European approaches. *Current Opinion in Environmental Sustainability*, *4*, 529–538.

Rajendhran, J., & Gunasekaran, P. (2008). Strategies for accessing soil metagenome for desired applications. *Biotechnology Advances*, *26*, 576–590.

Rineau, F., & Courty, P. E. (2011). Secreted enzymatic activities of ectomycorrhizal fungi as a case study of functional diversity and functional redundancy. *Annals of Forest Science*, *68*, 69–80.

Roesch, L. F., Fulthorpe, R. R., Riva, A., Casella, G., Hadwin, A. K. M., Kent, A. D., et al. (2007). Pyrosequencing enumerates and contrasts soil microbial diversity. *The ISME Journal*, *1*, 283–290.

Sagova-Mareckova, M., Cermak, L., Novotna, J., Plhackova, K., Forstova, J., & Kopecky, J. (2008). Innovative methods for soil DNA purification tested in soils with widely differing characteristics. *Applied and Environmental Microbiology*, *74*, 2902–2907.

Schimel, J. P., & Schaeffer, S. M. (2012). Microbial control over carbon cycling in soil. *Frontiers in Microbiology*, *3*, 348.

Schroeder, A., Mueller, O., Stocker, S., Salowsky, R., Leiber, M., Gassmann, M., et al. (2006). The RIN: An RNA integrity number for assigning integrity values to RNA measurements. *BMC Molecular Biology*, *7*, 3.
Sheffield, J., Wood, E. F., & Roderick, M. L. (2012). Little change in global drought over the past 60 years. *Nature*, *491*, 435–438.
Shokralla, S., Spall, J. L., Gibson, J. F., & Hajibabaei, M. (2012). Next-generation sequencing technologies for environmental DNA research. *Molecular Ecology*, *21*, 1794–1805.
Stewart, F. J., Ottesen, E. A., & DeLong, E. F. (2010). Development and quantitative analyses of a universal rRNA-subtraction protocol for microbial metatranscriptomics. *The ISME Journal*, *4*, 896–907.
Stewart, F. J., Sharma, A. K., Bryant, J. A., Eppley, J. M., & DeLong, E. F. (2011). Community transcriptomics reveals universal patterns of protein sequence conservation in natural microbial communities. *Genome Biology*, *12*, R26.
Stewart, F. J., Ulloa, O., & DeLong, E. F. (2012). Microbial metatranscriptomics in a permanent marine oxygen minimum zone. *Environmental Microbiology*, *14*, 23–40.
Su, C. L., & Sordillo, L. M. (1998). A simple method to enrich mRNA from total prokaryotic RNA. *Molecular Biotechnology*, *10*, 83–85.
Takasaki, K., Miura, T., Kanno, M., Tamaki, H., Hanada, S., Kamagata, Y., et al. (2013). Discovery of glycoside hydrolase enzymes in an avicel-adapted forest soil fungal community by a metatranscriptomic approach. *PLoS One*, *8*, e55485.
Tveit, A., Schwacke, R., Svenning, M. M., & Urich, T. (2013). Organic carbon transformations in high-Arctic peat soils: Key functions and microorganisms. *The ISME Journal*, *7*, 299–311.
Uroz, S., Ioannidis, P., Lengelle, J., Cebron, A., Morin, E., Buee, M., et al. (2013). Functional assays and metagenomic analyses reveals differences between the microbial communities inhabiting the soil horizons of a Norway spruce plantation. *PLoS One*, *8*, e55929.
Volossiouk, T., Robb, E. J., & Nazar, R. N. (1995). Direct DNA extraction for PCR-mediated assays of soil organisms. *Applied and Environmental Microbiology*, *61*, 3972–3976.
White, R. E. (2006). *Principles and practice of soil science: The soil as a natural resource* (4th ed.). Oxford, UK: Blackwell Publishing.
Winstanley, C. (1998). Subtractive hybridization. In J. M. Walker & R. Rapley (Eds.), *Molecular biomethods handbook* (pp. 227–238). Totowa, NJ, USA: Humana Press.
Wohl, E., Barros, A., Brunsell, N., Chappell, N. A., Coe, M., Giambelluca, T., et al. (2012). The hydrology of the humid tropics. *Nature Climate Change*, *2*, 655–662.
Zhou, J. Z., Bruns, M. A., & Tiedje, J. M. (1996). DNA recovery from soils of diverse composition. *Applied and Environmental Microbiology*, *62*, 316–322.

SECTION IV

Microbial Community Proteomics—Sampling, Sample Preparation, Spectral Analysis, and Interpretation

CHAPTER FOURTEEN

Sample Preparation and Processing for Planktonic Microbial Community Proteomics

Robert M. Morris[*,1], **Brook L. Nunn**[†,‡]
[*]School of Oceanography, University of Washington, Seattle, Washington, USA
[†]Department of Genome Sciences, University of Washington, Seattle, Washington, USA
[‡]Department of Medicinal Chemistry, University of Washington, Seattle, Washington, USA
[1]Corresponding author: e-mail address: morrisrm@uw.edu

Contents

1. Introduction 272
2. Sample Collection 273
   2.1 Sampling for planktonic community proteomics 274
3. Sample Preparation 278
   3.1 Separating soluble and membrane cell fractions 279
4. Extracting and Digesting Proteins 281
   4.1 Reduction and alkylation 282
   4.2 Trypsin digestion 282
   4.3 Desalting 282
5. Mass Spectrometry 283
References 286

Abstract

Advances in tandem mass spectrometry (tandem MS) and sequencing have enabled the field of community proteomics, which seeks to identify expressed proteins, their sequence variability, and the physiological responses of organisms to variable environmental conditions. Bottom-up tandem MS-based community proteomic approaches generate fragmentation spectra from peptides. Fragmentation spectra are then searched against genomic or metagenomic databases to deduce the amino acid sequences of peptides, providing positive identifications for proteins. Marine community proteomic studies have verified the importance of nutrient transport, energy generation, and carbon fixation functions in bacteria and archaea and revealed spatial and temporal shifts in the expressed functions of communities. Here, we discuss sample collection, preparation, and processing methods for planktonic tandem MS-based community proteomics.

Methods in Enzymology, Volume 531
ISSN 0076-6879
http://dx.doi.org/10.1016/B978-0-12-407863-5.00014-9

1. INTRODUCTION

Proteomics can verify genomic potential, confirm gene regulation, inform sequence variability, and provide information about the specific activities of proteins that have been posttranslationally modified. Here, we provide information for microbial ecologists interested in planktonic sample collection and processing for tandem MS-based community proteomics. Sections in this chapter discuss key concepts to consider when conducting a planktonic community proteomics study and provide protocols that we have used to identify the expressed functions of marine bacteria and archaea.

Tandem mass spectrometry (tandem MS)-based community proteomic studies have revealed the expressed functions of microorganisms from diverse environments. They have been applied to the microbial inhabitants of low diversity environments including acid-mine drainage, wastewater sludge, sponge symbiont, and anaerobic reductive dechlorinating microbial communities (Liu, Fan, Zhong, Kjelleberg, & Thomas, 2012; Lo et al., 2007; Morris et al., 2007; Mueller et al., 2011; Ram et al., 2005; Wilmes et al., 2008), as well as more complex microbial communities in oceans, lakes, and the human microbiome (Chen, Ryu, Gharib, Goodlett, & Schnapp, 2008; Giovannoni et al., 2005; Lauro et al., 2011; Morris et al., 2010; Sowell et al., 2011, 2009; Teeling et al., 2012; Williams et al., 2012).

Tandem MS-based community proteomic studies targeting marine bacteria and archaea have identified key functions expressed by communities in the North Atlantic, South Atlantic, North Pacific, North Sea, and Southern Ocean. Sowell and colleagues verified the importance of phosphate transporters in the Sargasso Sea, and transporters for amino acids, taurine, and polyamines in coastal upwelling waters off the Oregon Coast (Sowell et al., 2009, 2011). Morris et al. identified functional shifts along basin-scale gradients in nutrients, chlorophyll, and dissolved organic matter in the South Atlantic gyre and Benguela upwelling system. In addition, proteomics revealed the ocean-wide importance of TonB-dependent transporters in bacteria and archaea (Morris et al., 2010). Williams et al. reported the importance of chemolithoautotrophic carbon fixation in winter surface waters from the Antarctic (Williams et al., 2012) and Teeling et al. reported the importance of substrate-controlled succession in bacteria and archaea communities from the North Sea (Teeling et al., 2012). Although the potential for some of these processes had been recognized in metagenome sequence data, a functional approach like proteomics was needed to move from prediction to *in situ* verification. This is critical in ecology because not all genes are transcribed and not all transcripts are translated under the same conditions.

2. SAMPLE COLLECTION

Because marine bacteria and archaea are present in seawater at relatively low concentrations, it is important that cells are collected and stored as efficiently as possible. We recommend isolating microbial samples for tandem MS-based proteomic analyses within 1–3 h of collection (filtration start to finish). There are several key decisions to make prior to sample collection (Fig. 14.1). Start by identifying target organisms and target proteins. Consider the following: (1) location and size fraction of target organisms, (2) abundance of target organisms, and (3) availability of genomic sequence data from target organisms. Careful consideration of each of these points will significantly increase the number of proteins identified, while reducing the number of tandem MS runs needed to identify target proteins, thereby reducing cost. Water collection and filtration methods are therefore critical because they ultimately determine whether or not there is enough biomass to proceed. Both conventional and tangential flow filtration (TFF) methods have been used successfully to collect bacteria and archaea for proteomic analyses (Morris et al., 2010; Teeling et al., 2012).

With conventional filtration, seawater is passed through a membrane and particles are trapped in, or on, the filter. The main advantage of conventional filtrations is that *in situ* pumps can be used to collect samples with little sample handling in the laboratory. This can be done using the Large Volume Water Transfer System made by McLANE research laboratories, inc. (East Falmouth, MA). The disadvantage of a conventional filtration method is that biofouling occurs as a filter cake forms, blocking the flow of sample. It is also important to consider the location of target proteins because cells are often embedded in filters, and it can be difficult to extract membrane proteins. Larger volume samples can be processed by reducing pump flow rates (<1 L/min) and by carefully selecting prefilters to remove unwanted organisms that will contaminate the system and dilute the target proteins (Fig. 14.1).

With TFF the majority of seawater travels tangentially across the surface of the filter rather than into the filter. The advantage of this method is that cells are washed away from the filter surface, increasing the length of time that a filter can operate. This allows filters with much smaller pore sizes

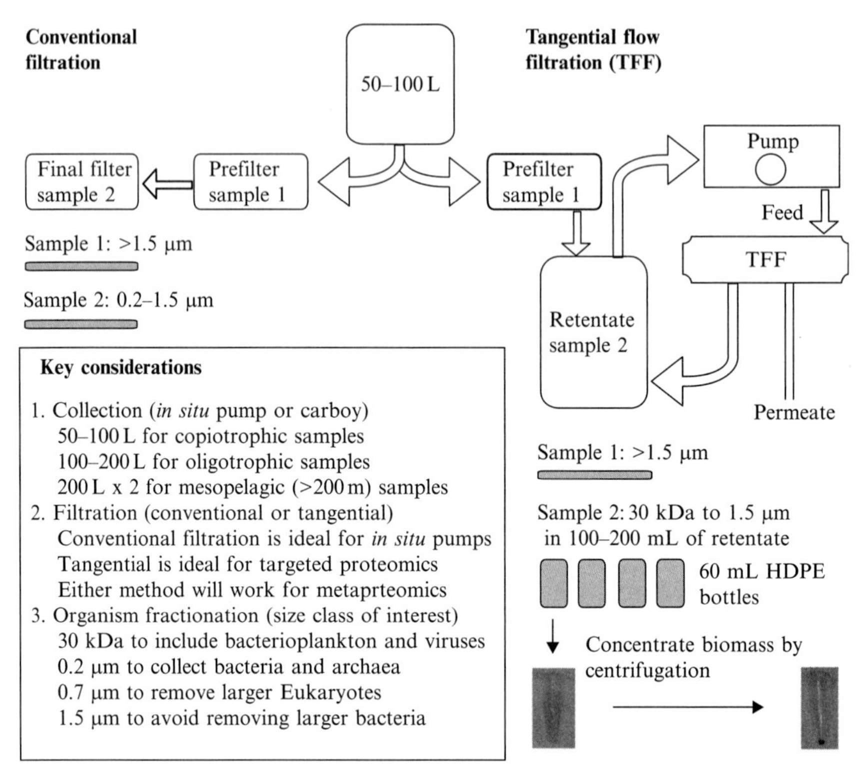

Figure 14.1 Sample collection flowchart for tandem MS-based community proteomics. Both conventional and tangential flow filtration methods have been used to collect planktonic samples for community proteomics. With conventional filtration, cells are collected *in situ*, and cells are stored directly on filters. Using TFF, cells are processed on ship or shore, concentrated, and stored in solution. Method selection depends on the organisms and proteins of interest, sample limitations, and field collection logistics.

to be used. Cells, viruses, and proteins from larger volumes of seawater can be concentrated at the same time. Concentrated samples are stored in solution and later pelleted by centrifugation. The main disadvantage of TFF is that relatively large volumes of water (50–400 L) have to be collected and processed in the laboratory. Here, we outline a detailed protocol for community proteomics of planktonic bacteria and archaea using a TFF system (Fig. 14.1).

2.1. Sampling for planktonic community proteomics

This protocol is for concentrating bacteria and archaea for tandem MS-based proteomic analyses using TFF. This system has been used successfully to collect planktonic marine samples for community proteomic analyses from 200 L of water, yielding a total of ~35 μg protein. Cells from open ocean, coastal, hydrothermal vent, and estuary waters have been previously collected and concentrated as described below for community proteomic profiling.

2.1.1 Assembling the TFF system

Assemble a Pellicon 2 TFF holder with one Biomax-30 Pellicon 2 Ultrafiltration Cassette installed (30 kDa). Refer to the holder and cassette manuals for more detailed instructions as the system can be modified depending on the application. Prepare approximately 10 L of ultrapure water, 10 L 0.1 N NaOH, and 10 L 0.1 N H_3PO_4. Begin by flushing the system with 10 L of ultrapure water. While the system is flushing, adjust the retentate flow rate to between 0.5 and 1.0 L/min and the permeate flow rate to between 0.7 and 1.4 L/min. To clean the system after it is setup and the flow rates are adjusted, between samples, and before disassembling the system and storing the filter (1) flush with 10 L of ultra pure water, then recirculate for 15–30 min; (2) flush with 1 L of 0.1 N NaOH, then recirculate for 15–30 min; (3) flush with 1 L of 0.1 N H_3PO_4, then recirculate for 15–30 min. Flush the system with 10 L of sample before each cell concentration. After all samples have been processed, disassemble the unit and store the Biomax-30 Pellicon 2 Ultrafiltration Cassette at 4 °C in 0.1 N H_3PO_4.

2.1.2 Collecting cells

Fill the desired number of 50 L carboys with seawater. The initial sample volume depends on the concentration of bacteria and archaea. Because biomass is usually a limiting factor for community proteomics, more is almost always better, although longer filtration times can reduce yield and alter community protein pools. We recommend 50–100 L for coastal surface samples, 100–200 L for open ocean surface samples, and two separate 200 L concentrations for samples below the photic zone (>200 m). Prefiltration is required for high particulate samples typical of coastal systems. A Biomax-30 Pellicon 2 Ultrafiltration Cassette (Millipore, Billerica, MA) has a 30 kDa cutoff and requires the removal of particles greater than 100 μm. Use a 0.8–1.5 μm prefilter to remove larger eukaryotic microorganisms and to enrich for bacteria and archaea (Fig. 14.1).

2.1.3 Concentrating cells

Connect the sample reservoir to the inlet of the Pellicon 2 TFF system (Fig. 14.1). Flush 10 L of sample through the system, discarding seawater passing through Silicone permeate and retentate tubing. Check the retentate clamp so that the flow rate is between 0.5 and 1.0 L/min and the permeate flow rate is between 0.7 and 1.4 L/min. Place the retentate tubing into the retentate reservoir to start concentrating sample after the system has been flushed. Retain 1 L of permeate for rinsing and cell count dilutions, the remainder of permeate can be discarded. Maintain differential flow rates throughout the concentration procedure. Do not exceed an inlet pressure of 25 psi. Record inlet and retentate pressures (psi), retentate and permeate flow rates (L/min), and initial sample volume (L). Take samples for initial cell counts 4,6-diamidino-2-phenylindole dihydrochloride (DAPI) and for DNA extractions to evaluate concentration efficiencies and community structure, respectively. For longer concentrations, closely monitor tubing for damage and adjust or replace tubing as needed.

Transfer the retentate to a smaller reservoir after it is reduced to <1 L. At this point small volumes of sample are highly enriched in cells, so take precautions not to lose any concentrated sample. Reduce tubing lengths to the minimum length needed. Continue to concentrate cells until the retentate is contained in the tubing and the Pellicon 2 TFF unit. Remove the tubing from the pump, while ensuring that everything drains into the retentate reservoir. Hold feed and retentate tubing in retentate reservoir and lift the Pellicon 2 TFF holder to drain concentrated sample from tubing and channels in the holder. Work the unit back and forth and squeeze the tubing to ensure that the entire sample is recovered. Look into the TFF unit to make sure that the sample channels are fully drained. Small volumes of sample contain large numbers of cells, so sample collection is critical at this stage. Record the final sample volume, which should be 100–200 mL (Table 14.1). Take retentate subsamples for final cell count (DAPI) and DNA for community analyses, respectively. For the DAPI count, dilute the sample 1/1000 in permeate. For the DNA sample, reserve enough retentate to allow DNA extraction from at total of 10^9–10^{10} cells or 1–10 mL. Transfer the sample to 60 mL HDPE bottles, flash freeze in liquid nitrogen, and store at -80 °C until further processing.

2.1.4 Quantifying concentration efficiency

Use total cell counts obtained by DAPI to evaluate concentration efficiency and 16S rRNA gene analyses to check for potential shifts in community structure. Cell counts in previous studies have ranged from 4.3×10^9 total cells in sample 3 from the South Atlantic to 7.9×10^{10} total cells in sample 7 from the South Atlantic (Table 14.1). Only 1/2 of the cells were used for proteomic analyses, which suggests that 2–40 billion cells are sufficient to construct community membrane metaproteomes. However, sample 3 yielded only 39 unique

Table 14.1 Planktonic marine samples concentrated for tandem MS-based proteomics

Location	Unique protein identifications	Prefilter (μm)	Initial volume (L)	Final volume (mL)	Cells recovered (counts/%)
South Atlantic gyre-1	132	0.8	200	120	3.0×10^{10}/21.4
South Atlantic gyre-2	196	0.8	200	120	1.8×10^{10}/11.4
South Atlantic gyre-3	39	0.8	200	70	4.7×10^{9}/2.3
South Atlantic coastal-4	103	0.8	100	200	6.6×10^{10}/27.4
South Atlantic coastal-5	275	0.8	100	120	6.3×10^{10}/26.6
South Atlantic coastal-6	289	0.8	100	100	3.0×10^{10}/14.4
South Atlantic coastal-7	269	0.8	100	210	7.9×10^{10}/32.7
South Atlantic coastal-8	171	0.8	100	220	1.6×10^{10}/24.5
South Atlantic gyre-9	164	0.8	100	165	5.2×10^{10}/30.9
South Atlantic gyre-10	157	0.8	100	210	5.0×10^{10}/46.9
North Pacific gyre-1	610[a]	1.5	100	260	3.2×10^{10}/35.1
North Pacific coastal-2	845[a]	1.5	100	160	8.9×10^{10}/55.2
North Pacific coastal-3	1005[a]	1.5	100	200	4.7×10^{10}/33.4
North Pacific coastal-4	670[a]	1.5	100	200	5.5×10^{10}/39.4
Hydrothermal plume-a/b	226/333	0.8	200	230	2.9×10^{10}/41.5
Puget Sound surface-1	–[a]	0.8	100	160	2.1×10^{10}/24.1

Continued

Table 14.1 Planktonic marine samples concentrated for tandem MS-based proteomics—cont'd

Location	Unique protein identifications	Prefilter (μm)	Initial volume (L)	Final volume (mL)	Cells recovered (counts/%)
Puget Sound surface-2	–[a]	0.8	100	180	6.2×10^{10}/23.4
Average		N/A	124	172	4.2×10^{10}/28.9

[a]Unpublished proteomic results.
Samples were collected from the South Atlantic in 2007 (Morris et al., 2010), Puget Sound in 2008, a hydrothermal vent plume at Axial Seamount (1450 m) in 2011 (Mattes et al., 2013), and from the North Pacific in 2012. In all cases, half of the biomass was used for tandem MS-based proteomics analyses. Additional biomass was archived.

protein identifications, while sample 7 yielded 269 unique protein identifications. This suggests that the lower limit for planktonic community proteomics using this approach is approximately 1.0×10^9 cells.

Although differences in community composition, targeted proteins, sample processing steps, and tandem MS-based analyses methods can all influence protein detection limits, we recommend collecting cells from 50 to 200 L of seawater for the best results. For example, concentrating cells from 50 L at an average concentration of 1.0×10^6 cells/mL in surface seawater will yield 1.5×10^{10} cells, assuming an average concentration efficiency of 29% (Table 14.1).

3. SAMPLE PREPARATION

Minimal manipulations, including test tube transfers and fractionations, are desired during sample preparation of low-biomass samples. There are several key decisions to make regarding sample processing (Fig. 14.2). Start by identifying target proteins. Consider the following: (1) cellular location and size of target proteins, (2) absolute and relative abundance of target proteins, and (3) availability of sequence data for target proteins. There are several approaches that can be used to enrich for target proteins, such as centrifugation, precipitation, and/or fractionation. However, biomass is lost at each stage of enrichment, so extra care should be taken to ensure that all biomass is recovered at each step. For example, we identified 226 and 333 proteins in replicate proteomes from a hydrothermal vent plume (Mattes et al., 2013; Table 14.1). 94% of the proteins identified in the first replicate were identified in the second replicate. Differences protein identification were most likely due to differences in the amount of biomass recovered after the first centrifugation step (Section 3.1.1, below).

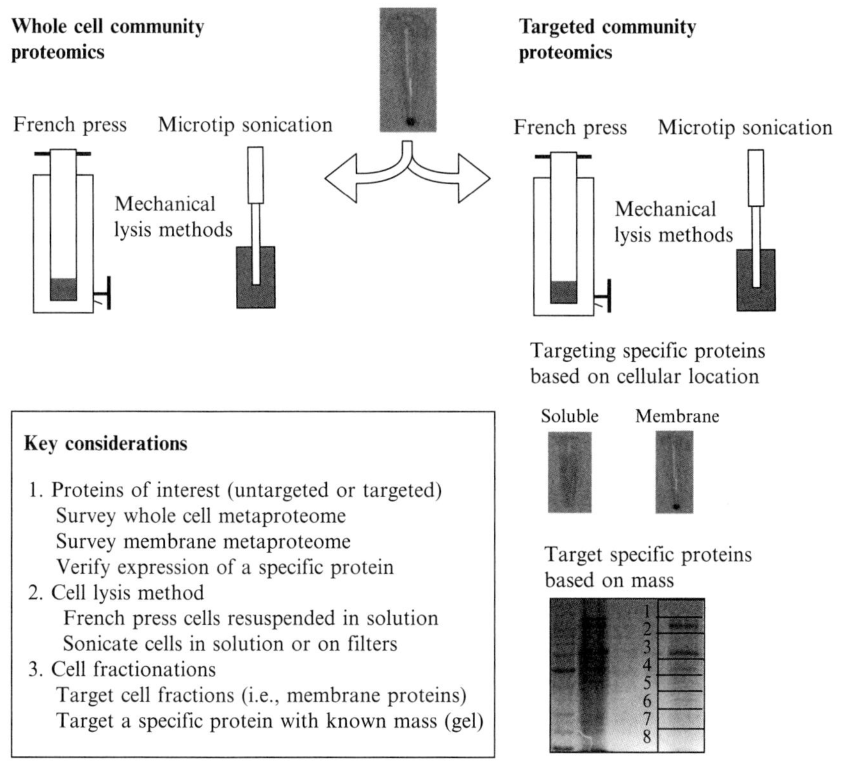

Figure 14.2 Sample processing flowchart for tandem MS-based community proteomics. Whole cell community proteomics and targeted community proteomics use the same methods to collect and lyse cells. Different fractionation methods, however, can be used to target proteins in different cell fractions (i.e., soluble versus membrane), to target specific proteins based on isoelectric point and/or mass (2D or 1D gel), or to increase the number of proteins identified by analyzing fractions separately (Morris et al., 2007, 2010; Morris, Sowell, Barofsky, Zinder, & Richardson, 2006).

3.1. Separating soluble and membrane cell fractions

This protocol is for fractionating soluble proteins from the membrane-bound proteins in bacteria and archaea samples for tandem MS-based community proteomics. It has been used successfully to produce crude membrane, enriched membrane, and soluble cell fractions for separate tandem MS analyses.

3.1.1 Pooling cells from concentrated samples

Thaw and combine a concentrated sample obtained by TFF. Distribute the concentrated sample evenly into multiple 30 mL centrifuge tubes and spin at 18,000 rpm (39,000 × *g*) for 60 min at 4 °C. Mark the orientation of each centrifuge tube so that the location of the small pellet is known. Discard the supernatant and wash the pellet from one centrifuge tube using 1 mL 4 °C 20 m*M* tris(hydroxymethyl)aminomethane (Tris) pH 7.4. The cell pellet will likely not be visible to the naked eye, so carefully wash the side of the first centrifuge tube that contains a pellet several times using a low-flow rate 1 mL pipette. Start combining the pellets by transferring the first pellet resuspended in 1 mL Tris to the second centrifuge tube that contains a pellet. Resuspend the second pellet by washing the side of the second centrifuge tube using the same 1 mL of Tris (now Tris/pellet) from the first tube. Repeat sequentially until all resuspended pellets are combined into the last centrifuge tube (using the same 1 mL of Tris). Repeat this procedure two more times with an additional 1 mL of fresh 4 °C 20 m*M* Tris–HCl pH 7.4 to rinse the sample tubes and ensure that all cells have been recovered. The final volume should be ~3 mL. This process yields a whole cell concentrate.

3.1.2 Lysing cells by French press

Transfer the whole cell concentrate (procedure above) to a thoroughly cleaned French press mini-cell that has been stored at 4 °C. Lyse the cells using a French press at 20,000 lbs/in.^{2}. Slowly release the sample (in drops, not in a stream) into an acid-washed collection bottle. French press again at 20,000 lbs/in^2 to ensure that all cells are lysed. This process yields a lysed cell extract.

3.1.3 Lysing cells by sonication

Transfer the lysed cell extract (from procedure above) into two 1.5 mL microcentrifuge tubes. Centrifuge at 14,000 rpm (18,000 × *g*) for 60 min at 4 °C to remove the Tris–HCl. Discard the supernatants. Rinse each pellet in 200 μL 4 °C 20 m*M* Tris–HCl pH 7.4. Discard the supernatant wash. To each cell pellet add 300 μL 6 *M* urea in 50 m*M* $(NH_4)HCO_3$ and sonicate cells using a microtip sonicating probe (100 W MSE 20 kHz; 20 × 10 s, 4 °C), not a sonicating bath. Ensure that the sample is completely chilled to 4 °C prior to repeating sonication by either placing in an ice bath or flash freezing with liquid nitrogen or dry ice in ethanol. The sonication probe should be submerged in the solution and not touching the sides of the centrifuge tube. This process yields a lysed cell extract.

3.1.4 Generating a membrane-enriched cell fraction

Distribute 1.5 μL of the lysed cell extract into two 1.5 mL microcentrifuge tubes. Centrifuge at 14,000 rpm (18,000 × *g*) for 60 min at 4 °C. This first centrifugation step will generate a soluble fraction supernatant and a crude membrane pellet. Remove the soluble fraction supernatants and place them into two ultrahigh-speed centrifuge tubes. Gently rinse each pellet in 200 μL 4 °C 20 m*M* Tris–HCl pH 7.4. Discard the wash, flash freeze the pellets and store at −80 °C. These are the crude membrane cell fractions. Do not add protease inhibitors to any of these samples because they will adversely influence the completeness of enzymatic digestions required for tandem MS-based proteomic analyses.

The next centrifugation step will generate a membrane-enriched cell pellet from the soluble fraction supernatants obtained above. Centrifuge the soluble fraction supernatants at 40,000 rpm (129,000 × *g*) for 60 min at 4 °C. Transfer the supernatants from this high-speed centrifugation to two new 1.5 mL microcentrifuge tubes, flash freeze, and store at −80 °C. These are the soluble cell fractions. They contain soluble proteins (and DNA sheered to ~500 bp). Wash the pellets obtained by high-speed centrifugation with 200 μL 4 °C 20 m*M* Tris pH 7.4. Discard the wash, flash freeze, and store the pellets at −80 °C. These are the membrane-enriched cell fractions. Again, do not add protease inhibitors.

4. EXTRACTING AND DIGESTING PROTEINS

For the digestion procedure presented here, we recommend starting with 30–300 μg of protein. Protein concentrations can be estimated after the cells are lysed using the Bradford assay (Bradford, 1976) or more accurately estimated from amino acid concentrations (Moore, Nunn, Goodlett, & Harvey, 2012). With each chemical assay, the list of interfering solutes needs to be considered as urea and chlorophyll commonly interfere with the wavelength used to estimate protein concentrations. If the sample size is limited, peptide concentrations can also be quantified after digestions are completed and desalted. Peptide concentrations are measured on samples using the Thermo Scientific NanoDrop 2000/2000c Spectrophotometer. The peptide bond absorbance is monitored at 205 nm UV wavelength (Grimsley & Pace, 2003; Scopes, 1974).

To increase the number of peptides generated during digestions, protein disulphide bonds are reduced and then "capped" using an alkylating reagent to ensure that the disulphide bonds do not reform. Reductions are carried out in a reducing environment to minimize the reformation of disulfide bonds and alkylating agents are added to block the sulfur groups.

4.1. Reduction and alkylation

Lyse cells accordingly to Section 3.1.2 or 3.1.3. If the sample size is limited, the following protocol can be equally divided by half, a third, or a quarter. Add 20 μL of 1.5 *M* Tris–HCl pH 8.8 to the lysed cells in 300 μL 6 *M* freshly made urea with 50 m*M* (NH_4) HCO_3. Add 7.5 μL of 200 m*M* Tris (2-carboxyethyl) phosphine, vortex and let it sit for 1 h at 37 °C in order to reduce the disulphide bonds. In the dark, add 60 μL of 200 m*M* iodoacetimide (IAA), vortex and let it sit for 1 h at 20 °C in the dark. IAA is an alkylating agent and should be made the day that it is used. Add 60 μL of 200 m*M* dithiothreitol in the dark in order to quench excess IAA, then vortex and let it sit for 1 h at 20 °C. Divide the sample into three microcentrifuge tubes (~150 μL in each tube) and add 800 μL 25 m*M* (NH_4) HCO_3 to each of the three tubes and vortex to dilute the urea. Add 200 μL methanol (HPLC grade) to each tube to assist in protein denaturation, or unfolding, allowing the digestion enzyme Trypsin to have more access to cleavage sites (Li, Schmerberg, & Ji, 2009).

4.2. Trypsin digestion

To the reduced and alkylated proteins add Trypsin at a 50:1 protein to enzyme ratio and incubate for 12 h at 37 °C (first Trypsin digestion). We recommend Promega sequencing grade modified trypsin because it is lyophilized in 20 μg aliquots that are modified by reductive methylation, reducing the rate of autolytic digestion.

Centrifuge the tubes in order to pool the condensate and add another aliquot of Trypsin at a 25:1 protein to enzyme ratio for 2 h at 37 °C (second Trypsin digestion). Speedvac samples at 4 °C, pooling samples that were split into three tubes as the volume decreases. When samples approach 100 μL, add 50 μL 1% TFA to lower the pH prior to desalting. Vortex the sample and continue to speedvac the sample to near dryness. This step should be repeated until a pH of ≤ 2 is achieved.

4.3. Desalting

We recommend desalting the peptides to remove salts and urea from the digestion buffers prior to analyzing samples on a mass spectrometer. Although desalting can be completed inline on some HPLC trapping columns, salts can clog nanoflow chromatography capillaries and when introduced into the mass spectrometer can decrease the signal to noise ratio. Using a C4, C8, C12, or C18 resin will retain nonpolar solutes, such as peptides (e.g., Nest Group, www.nestgrp.com; Southborough, MA). Dried down peptide digests are resuspended in a 5% acetonitrile (ACN) solution with 0.1% trifluoroacetic acid (TFA), making sure the final pH is <2. The size of the desalting column should be selected based on the amount of protein in the sample. The basic protocol includes conditioning the column in 80% ACN + 0.1% TFA (centrifuge 3 min, 110 × *g*; repeat twice), followed by equilibrating the column with 5% ACN, 0.1% TFA (centrifuge 3 min, 110 × *g*; repeat twice). The samples in 5% ACN, 0.1% TFA are then added to the column (centrifuge 3 min, 110 × *g*). Collect flow through and put through the column one more time. Wash the salt fraction from the column with 5% ACN, 0.1% TFA (centrifuge 3 min, 110 × *g*; repeat twice). Change the collection tube and collect flow through after adding 50–80% ACN, 0.1% TFA. Dry in a Speedvac to ~10 μL. Avoid taking the sample to complete dryness. Resuspend the peptide digest in 5% ACN, 0.1% formic acid to yield a final concentration of 1 μg/μL and store at −80 °C.

5. MASS SPECTROMETRY

For tandem MS-based community proteomics, it is important to take into consideration the type of mass spectrometers used for the analyses. Biomass is typically limited, due to the collection restraints described above, and samples are complex and comprised of many different species. The dynamic range of the instrument is therefore critical and can greatly influence the number of proteins identified. There are a wide variety of MS platforms. It is critical to consider the sensitivity of the MS and the HPLC systems, the mass accuracy of the precursor ions (MS1) and fragment ions (MS2), and the duty cycle of the instrument (how quickly the mass spectrometer selects the parent ions for fragmentation and collects fragment ion spectra). For exploratory proteomic profiling, the mass spectrometer is typically operated using a data-dependent acquisition (DDA) strategy, where the ions targeted for fragmentation are chosen from the precursor (MS1) survey scan (Fig. 14.3A). The user can design an MS experiment to fragment the top 5–20 most intense ions observed in the MS1 scan. This provides an unbiased method for collecting spectra on the most abundant ions observed in the precursor ion scan from a single analysis with a mass to charge (m/z) range of 300–2000. There are a number of other parameters that can be modified in the method file that will drive the HPLC and the mass spectrometer, allowing the experienced user multiple ways to design a relatively simple DDA experiment. For example, two-dimensional long chromatographic separations (>20 h) on very complex, protein rich samples (estimated ~25 mg)

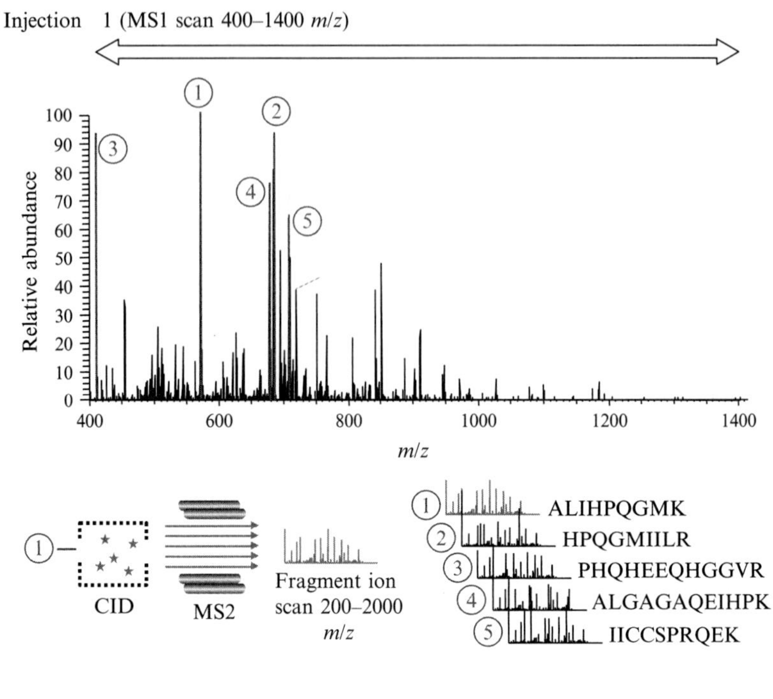

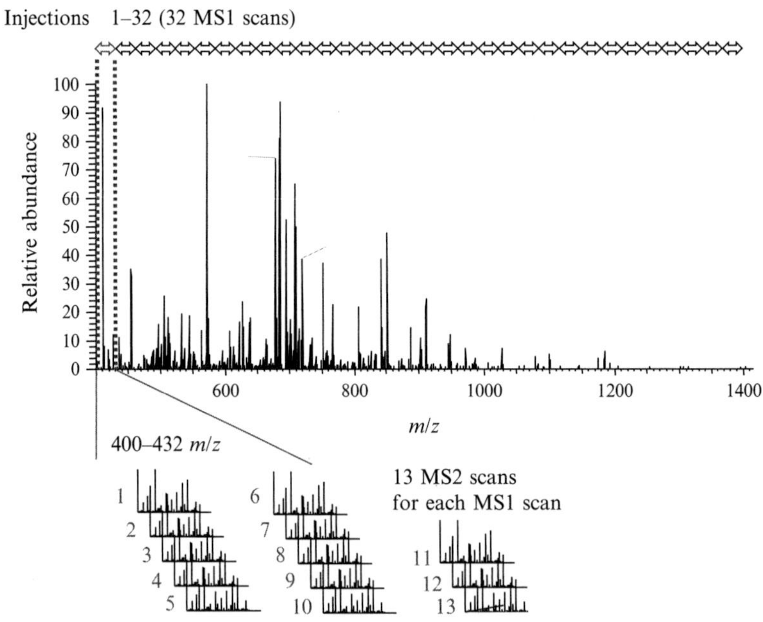

Figure 14.3 Data-dependent and data-independent tandem MS-based proteomic approaches. (A) During data-dependent analyses, the MS is programmed to collect MS2 data on the five most intense peaks observed in the MS1 precursor scan (1–5 in grey). Those ions are then isolated in the ion trap or quadrapole and undergo collision-induced dissociation (CID), creating peptide fragment ions. The resulting

using DDA can achieve a high number of protein identifications (2214 proteins; Verberkmoes et al., 2009). Alternatively, a data-independent acquisition strategy, such as the Precursor Acquisition Independent From Ion Count (PAcIFIC), has a much wider dynamic range requires less sample can yield similar numbers of protein identifications compared to a 20-h two-dimensional chromatography experiment (Fig. 14.3B). The PAcIFIC method requires that sample injections of ~ 1 µg are repeated up to 32 times in order to cover the full 300–2000 *m/z* scan range (a 48-h analysis cycle). This method allows the collection of MS2 data without the observation of an MS1 precursor prior to fragmentation, resulting in fragmentation data for very low-abundance peptides (Morris et al., 2010; Panchaud et al., 2011, 2009).

A typical DDA (top three analysis) analysis of a whole cell lysate of the diatom *Thalassiosira pseudonana* on an LTQ-DECA (scan rate: 1 Hz) using a nanoflow HPLC will yield 200–300 protein identifications with two or more peptides (20 cm analytical column; 1 injection, 90 min gradient; FDR (false discovery rate) $< 1\%$) with low mass accuracy on the precursor and fragment ions. The number of MS2 scans collected is a reflection of the duty cycle or scan rate. The LTQ-DECA is limited to collecting three fragment ion spectra across a 30 s chromatographic peak, thus the top three analyses. Analysis of the same culture on an LTQ-Orbitrap (maximum scan rate: 5 Hz) using a nanoflow HPLC system (20 cm analytical column; 1 injection, 90 min gradient; FDR $< 1\%$) and DDA (top eight analysis) will yield 400–500 protein identifications with two or more peptides (FDR $< 1\%$) and high mass accuracy collected only on MS1 precursor ions. In unknown community proteomics, high mass accuracy, such as is found on the LTQ-Orbitrap, is needed in order to increase the confidence in peptide assignments and protein identifications. The Q-Exactive is renowned for its ability to collect high mass accuracy data both in the MS1 and MS2 scans while

fragment ions are measured in the second mass spectrometer and MS2 fragmentation spectra are collected. MS2 spectra collected represent putative peptides that are identified by searching a peptide database. (B) During data-independent acquisition MS2 data are systematically collected across the 400–1400 *m/z* scan range for the purpose of identifying low-abundance peptides in samples with variable or wide protein concentration ranges (Panchaud, Jung, Shaffer, Aitchison, & Goodlett, 2011). This method, termed Precursor acquisition independent of ion count (PAcIFIC), requires that each sample be injected into the mass spectrometer a total of 32 times to cover the full *m/z* range. Each injection experiences a 60–90 min HPLC gradient and 13 MS2 scans are performed per duty cycle, each collecting fragmentation spectra in a 2.5 *m/z* channel. (See color plate.)

maintaining a superior duty cycle (maximum scan rate of 12 Hz). The same analysis of the diatom *T. pseudonana* using a Q-Exactive with DDA (top 20 analysis) yields 1100–1200 protein identifications with two or more peptides (20 cm analytical column; 1 injection, 90 min gradient; FDR < 1%).

Thus far, the Q-Exactive provides the greatest number of identifications with high mass accuracy on precursor and fragment ions in the shortest period of time. Sometimes the cost or the availability of an instrument, such as the Q-Exactive, might not allow for an experiment to be analyzed on this instrument, therefore knowledge of fractionation methods or well-designed experiments on a different instrument can yield similar results at a fraction of the cost. For more information about tandem MS-based analyses, please refer to the additional proteomics chapters in this volume.

REFERENCES

Bradford, M. M. (1976). A rapid and sensitive method for the quantitation of microgram quantities of protein utilizing the principle of protein-dye binding. *Analytical Biochemistry, 72*, 248–254.

Chen, J., Ryu, S., Gharib, S. A., Goodlett, D. R., & Schnapp, L. M. (2008). Exploration of the normal human bronchoalveolar lavage fluid proteome. *Proteomics. Clinical Applications, 2*, 585–595. http://dx.doi.org/10.1002/prca.200780006.

Giovannoni, S. J., Bibbs, L., Cho, J. C., Stapels, M. D., Desiderio, R., Vergin, K. L., et al. (2005). Proteorhodopsin in the ubiquitous marine bacterium SAR11. *Nature, 438*, 82–85. http://dx.doi.org/10.1038/nature04032.

Grimsley, G. R., & Pace, C. N. (2003). *Spectroscopic Determination of protein concentration*. In: Current Protocols in Protein Science. John Wiley & Sons, Inc. p 3.1.1–3.1.9

Lauro, F. M., DeMaere, M. Z., Yau, S., Brown, M. V., Ng, C., Wilkins, D., et al. (2011). An integrative study of a meromictic lake ecosystem in Antarctica. *ISME Journal, 5*, 879–895. http://dx.doi.org/10.1038/ismej.2010.185.

Li, F., Schmerberg, C. M., & Ji, Q. C. (2009). Accelerated tryptic digestion of proteins in plasma for absolute quantitation using a protein internal standard by liquid chromatography/tandem mass spectrometry. *Rapid Communications in Mass Spectrometry, 23*, 729–732. http://dx.doi.org/10.1002/rcm.3926.

Liu, M., Fan, L., Zhong, L., Kjelleberg, S., & Thomas, T. (2012). Metaproteogenomic analysis of a community of sponge symbionts. *ISME Journal, 6*, 1515–1525. http://dx.doi.org/10.1038/ismej.2012.1.

Lo, I., Denef, V. J., Verberkmoes, N. C., Shah, M. B., Goltsman, D., DiBartolo, G., et al. (2007). Strain-resolved community proteomics reveals recombining genomes of acidophilic bacteria. *Nature, 446*, 537–541. http://dx.doi.org/10.1038/nature05624.

Mattes, T. E., Nunn, B. L., Marshall, K. T., Proskurowski, G., Kelley, D. S., Kawka, O. E., et al. (2013). Sulfur oxidizers dominate carbon fixation at a biogeochemical hot spot in the dark ocean. *ISME Journal*, http://dx.doi.org/10.1038/ismej.2013.113.

Moore, E. K., Nunn, B. L., Goodlett, D. R., & Harvey, H. R. (2012). Identifying and tracking proteins through the marine water column: Insights into the inputs and preservation mechanisms of protein in sediments. *Geochimica et Cosmochimica Acta, 83*, 324–359. http://dx.doi.org/10.1016/j.gca.2012.01.002.

Morris, R. M., Fung, J. M., Rahm, B. G., Zhang, S., Freedman, D. L., Zinder, S. H., et al. (2007). Comparative proteomics of Dehalococcoides spp. reveals strain-specific peptides associated with activity. *Applied and Environmental Microbiology*, *73*, 320–326. http://dx.doi.org/10.1128/AEM.02129-06.

Morris, R. M., Nunn, B. L., Frazar, C., Goodlett, D. R., Ting, Y. S., & Rocap, G. (2010). Comparative metaproteomics reveals ocean-scale shifts in microbial nutrient utilization and energy transduction. *ISME Journal*, *4*, 673–685.

Morris, R. M., Sowell, S., Barofsky, D., Zinder, S., & Richardson, R. (2006). Transcription and mass-spectroscopic proteomic studies of electron transport oxidoreductases in Dehalococcoides ethenogenes. *Environmental Microbiology*, *8*, 1499–1509. http://dx.doi.org/10.1111/j.1462-2920.2006.01090.x.

Mueller, R. S., Dill, B. D., Pan, C., Belnap, C. P., Thomas, B. C., VerBerkmoes, N. C., et al. (2011). Proteome changes in the initial bacterial colonist during ecological succession in an acid mine drainage biofilm community. *Environmental Microbiology*, *13*, 2279–2292. http://dx.doi.org/10.1111/j.1462-2920.2011.02486.x.

Panchaud, A., Jung, S., Shaffer, S. A., Aitchison, J. D., & Goodlett, D. R. (2011). Faster, quantitative, and accurate precursor acquisition independent from ion count. *Analytical Chemistry*, *83*, 2250–2257. http://dx.doi.org/10.1021/ac103079q.

Panchaud, A., Scherl, A., Shaffer, S. A., von Haller, P. D., Kulasekara, H. D., Miller, S. I., et al. (2009). Precursor acquisition independent from ion count: How to dive deeper into the proteomics ocean. *Analytical Chemistry*, *81*, 6481–6488. http://dx.doi.org/10.1021/ac900888s.

Ram, R. J., Verberkmoes, N. C., Thelen, M. P., Tyson, G. W., Baker, B. J., Blake, R. C., 2nd., et al. (2005). Community proteomics of a natural microbial biofilm. *Science*, *308*, 1915–1920. http://dx.doi.org/10.1126/science.1109070.

Scopes, R. K. (1974). Measurement of protein by spectrophotometry at 205 nm. *Analytical Biochemistry*, *59*, 277–282.

Sowell, S. M., Abraham, P. E., Shah, M., Verberkmoes, N. C., Smith, D. P., Barofsky, D. F., et al. (2011). Environmental proteomics of microbial plankton in a highly productive coastal upwelling system. *ISME Journal*, *5*, 856–865. http://dx.doi.org/10.1038/ismej.2010.168.

Sowell, S. M., Wilhelm, L. J., Norbeck, A. D., Lipton, M. S., Nicora, C. D., Barofsky, D. F., et al. (2009). Transport functions dominate the SAR11 metaproteome at low-nutrient extremes in the Sargasso Sea. *ISME Journal*, *3*, 93–105. http://dx.doi.org/10.1038/ismej.2008.83.

Teeling, H., Fuchs, B. M., Becher, D., Klockow, C., Gardebrecht, A., Bennke, C. M., et al. (2012). Substrate-controlled succession of marine bacteria and archaea populations induced by a phytoplankton bloom. *Science*, *336*, 608–611. http://dx.doi.org/10.1126/science.1218344.

Verberkmoes, N. C., Russell, A. L., Shah, M., Godzik, A., Rosenquist, M., Halfvarson, J., et al. (2009). Shotgun metaproteomics of the human distal gut microbiota. *ISME Journal*, *3*, 179–189. http://dx.doi.org/10.1038/ismej.2008.108.

Williams, T. J., Long, E., Evans, F., Demaere, M. Z., Lauro, F. M., Raftery, M. J., et al. (2012). A metaproteomic assessment of winter and summer bacteria and archaea from Antarctic Peninsula coastal surface waters. *ISME Journal*, *6*, 1883–1900. http://dx.doi.org/10.1038/ismej.2012.28.

Wilmes, P., Andersson, A. F., Lefsrud, M. G., Wexler, M., Shah, M., Zhang, B., et al. (2008). Community proteogenomics highlights microbial strain-variant protein expression within activated sludge performing enhanced biological phosphorus removal. *ISME Journal*, *2*, 853–864. http://dx.doi.org/10.1038/ismej.2008.38.

Sample Handling and Mass Spectrometry for Microbial Metaproteomic Analyses

Ryan S. Mueller[*,1], **Chongle Pan**[†]
[*]Department of Microbiology, Oregon State University, Corvallis, Oregon, USA
[†]Oak Ridge National Laboratory, Oak Ridge, Tennessee, USA
[1]Corresponding author: e-mail address: ryan.mueller@oregonstate.edu

Contents

1. Introduction 290
2. Experimental Considerations and Options 291
3. Generalized Procedure 295
   3.1 Equipment and reagents for LC-MS/MS 295
   3.2 Protein extraction considerations 296
   3.3 Proteolysis and cleanup 296
   3.4 Two-Dimensional liquid chromatography separation 297
   3.5 Tandem mass spectrometry analysis 298
Acknowledgment 299
References 300

Abstract

Metaproteomic studies of whole microbial communities from environmental samples (e.g., soil, sediments, freshwater, seawater, etc.) have rapidly increased in recent years due to many technological advances in mass spectrometry (MS). A single 24-h liquid chromatograph–tandem mass spectrometry (LC–MS/MS) measurement can potentially detect and quantify thousands of proteins from many dominant and subdominant naturally occurring microbial populations. Importantly, amino acid sequences and relative abundance information for detected peptides are determined, which allows for the characterization of expressed protein functions within communities and specific matches to be made to microbial lineages, with potential subspecies resolution. Continued optimization of protein extraction and fractionation protocols, development of quantification methods, and advances in mass spectrometry instrumentation are enabling more accurate and comprehensive peptide detection within samples, leading to wider research applicability, greater ease of use, and overall accessibility. This chapter provides a brief overview of metaproteomics experimental options, including a general protocol for sample handling and LC–MS/MS measurement.

Methods in Enzymology, Volume 531
ISSN 0076-6879
http://dx.doi.org/10.1016/B978-0-12-407863-5.00015-0

1. INTRODUCTION

For decades, scientists have measured the protein complement of a biological sample with the intent of defining physiological changes and responses of living cells across various experimental treatments. Classical proteomics approaches use gel-based protein separation and detection methods (O'Farrell, 1975) to experimentally characterize proteome changes. These experiments are less than ideal for extensive proteome characterization of complex biological samples however, since they are limited in analysis throughput and proteome coverage. The advent of electrospray ionization (ESI) allowed on-line coupling of liquid chromatography separation with mass spectrometry analysis. This represented an important advance in proteomics research by enabling direct analysis of protein and peptide samples by liquid chromatograph-tandem mass spectrometry (LC–MS/MS) (Fenn, Mann, Meng, Wong, & Whitehouse, 1989; Karas & Hillenkamp, 1988). The analytical platform of LC–MS/MS has been continuously improved with the development of multidimensional nanoscale LC and high-performance hybrid MS instrumentation. Shotgun proteomics was invented to analyze total peptide digests of proteome samples with LC–MS/MS. Peptides and proteins are identified by searching MS/MS data against a protein sequence database. Current proteomics technologies provide (1) deep proteome coverage for identification and quantification of thousands of proteins from a single proteome sample or a multiplexed set of samples, (2) confident peptide identification using highly accurate mass measurements, (3) detection across large dynamic range of peptide abundances, and (4) resolution of similar peptides from complex metaproteome samples.

Based on the shotgun proteomics technology, metaproteomics was recently developed to interrogate the *in situ* physiology of microbial communities. Metaproteomics is enabled by the advent of metagenomics, which provides high-quality protein sequences for database searching and peptide/protein identification. The expected amino acid sequences from uncultured organisms in a given sample allow accurate peptide-spectrum matching of the collected MS/MS data. Significant advances in DNA sequencing technologies and their application in the field of metagenomics have greatly expanded the field of metagenomics and led to many coupled metagenomics–metaproteomics studies in environmental microbiology.

The primary goals of many metaproteomic studies are to link the expressed functions of a community to its genetic potential as defined by metagenomics, thereby characterizing physiological responses or functional changes of microbes growing in natural environments. Insights have been gained into the *in situ* physiological demands of microorganisms during bioremediation and nutrient cycling processes (Abram et al., 2011; Justice et al., 2012; Lacerda, Choe, & Reardon, 2007; Wilmes, Wexler, & Bond, 2008), and the identity of active populations within these communities and their resource preferences (Pan et al., 2011; Taubert et al., 2012). Additional studies have characterized functions of microbial communities across seasons and natural gradients within aquatic systems (Morris et al., 2010; Mueller et al., 2011; Williams et al., 2012), and in symbiotic associations with various Eukaryotic hosts (Burnhum et al., 2011; Delmotte et al., 2009; Haange et al., 2012; Klaassens, de Vos, & Vaughan, 2007; Liu, Fan, Zhong, Kjelleberg, & Thomas, 2012; Verberkmoes et al., 2009). Beyond these primary goals, metaproteomic approaches are emerging as important and versatile tools to address many questions related to microbial ecology and community assembly processes (Mueller et al., 2010; Schneider et al., 2012), natural population evolution (Denef & Banfield, 2012; Denef et al., 2010), and to aid in microbial genome annotation validation (Jaffe, Berg, & Church, 2003; Savidor et al., 2006) and characterization of protein posttranslational modifications (Erickson et al., 2010; Gupta et al., 2007; Singer et al., 2010).

Shotgun proteomics acquires a large collection of MS/MS spectra resulting from individual peptides digested from proteins (Link et al., 1999; Wolters, Washburn, & Yates, 2001). A wide variety of mass spectrometers can be used for these experiments (see Yates, Ruse, & Nakorchevsky, 2009). During each experimental run, the instrument will collect hundreds of thousands of MS/MS spectra by switching between the collection of MS1 spectra of full-length intact peptide ions and the collection of MS2 spectra of fragment ions resulting from the collision-induced dissociation (CID) of selected precursor peptide ions. Peptides are identified by searching their MS/MS spectra against a database of proteins expected to occur in a given sample. Because peptides not represented in the protein database cannot be identified, it is crucial to carefully design the database of expected proteins using existing or concurrently generated DNA sequence data in a metaproteomics experiment.

2. EXPERIMENTAL CONSIDERATIONS AND OPTIONS

Many reviews provide excellent overviews of technological principles and specific experimental protocols regarding mass spectrometry and shotgun proteomics in general (e.g., Yates et al., 2009). However, there are important factors to consider when designing and running metaproteomics experiments for environmental microbiology studies. The goal of this chapter will be to highlight significant developments and experimental options in metaproteomics and provide a general methodology detailing the steps to perform a metaproteomics measurement (Wolters et al., 2001; Fig. 15.1). This specific type of experiment analyzes complex mixtures of tryptic peptides derived from whole community protein extracts and involves the coupling of nanoscale 2-dimentional liquid chromatography (strong cation exchange (SCX) LC followed by reverse-phase (RP) LC) on-line with tandem mass spectrometry analysis through nanospray ionization. Techniques using alternative separation technologies (e.g., strong anion-exchange (Wiśniewski, Nagaraj, Zougman, Gnad, & Mann, 2010), hydrophilic interaction (Di Palma, Boersema, Heck, & Mohammed, 2011), and affinity chromatography (Ficarro et al., 2002)), or alternative ionization techniques (e.g., MALDI,

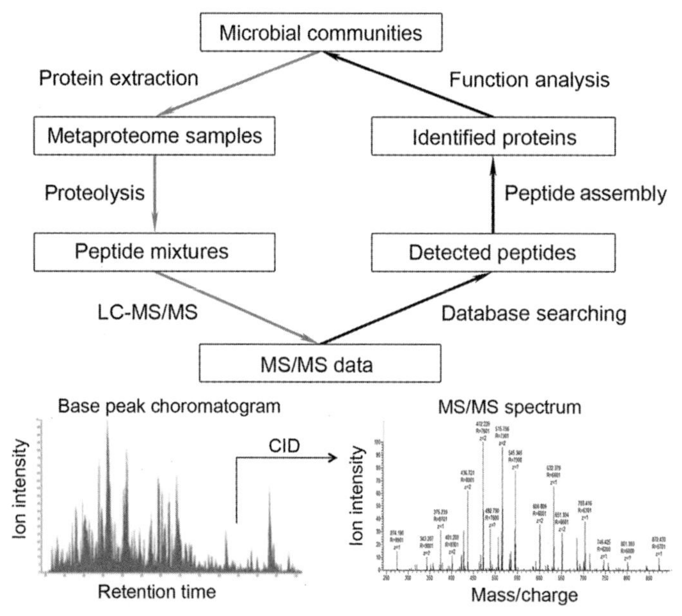

Figure 15.1 Overview of a metaproteomics workflow. The experimental procedure shown in gray includes protein extraction, proteolysis, and LC–MS/MS analysis. The obtained MS/MS data are a large collection of MS/MS spectra acquired during LC separation illustrated as a base peak chromatogram on the left and a MS/MS spectrum on the right. The MS/MS data are processed in a computational procedure shown in black, which includes database searching, peptide assembly, and functional analysis. The proteomic results provide insights into the metabolic activities of measured microbial communities.

SALDI (Chen, Shiea, & Sunner, 1998), DIOS (Thomas, Shen, Crowell, Finn, & Siuzdak, 2001), AP-MALDI (Laiko, Baldwin, & Burlingame, 2000), and SELDI (Merchant & Weinberger, 2000)) will not be described here, but should be considered depending on a given experiment's needs.

Perhaps the most important consideration for metaproteomic experiments is the biological diversity of the sample being analyzed (i.e., how many species or populations are present and their relative abundances), which will directly affect the coverage of each population's proteome obtained by each experiment (i.e., number of detected proteins of the total predicted proteome). A typical 24-h LC–MS/MS experiment acquires hundreds of thousands of MS/MS scans on the chromatographically separated peptides. These spectra will generally identify tens of thousands of peptides comprising up to a few thousand proteins. However, since most natural microbial communities will be composed of hundreds to thousands of populations each expressing a protein complement of thousands of proteins, the proteome coverage of the community is often biased toward highly abundant proteins from dominant populations. In contrast, a typical proteome run for a single sequenced isolate can identify approximately 50% of the predicted proteome being expressed if culture conditions favor high protein production and protein extraction efficiency is high.

The most common methods to expand proteome coverage rely on sample fractionation to reduce the sample complexity prior to LC–MS/MS analysis. One approach is to fractionate at the initial step of protein extraction from microbial cells utilizing differential centrifugation to separate proteins based on their general localization properties (e.g., cytosolic, membrane, extracellular, or periplasmic proteins; Ram et al., 2005). Another set of techniques fractionates extracted proteins using isoelectric focusing and gel electrophoresis (Hörth, Miller, Preckel, & Wenz, 2006; Jafari et al., 2012). All of these methods can be used alone or in combination to provide greater proteome coverage. However, it should be noted that any fractionation inevitably leads to an increased number of samples for LC–MS/MS analysis, which reduces the overall experimental throughput. Off-line fractionation can also lead to significant sample loss. Thus, one has to balance the benefits of fractionation (chiefly, deeper metaproteome coverage) with its increased requirements for instrument time and sample amount.

The biomass contained within a given environmental sample is a second important consideration for metaproteomic experiments. Typically several hundred micrograms of protein should be extracted for adequate MS analysis, allowing for repeat runs for technical replication and potential failures. If we assume that microbial cells contain 30 fg of protein per cell (based on aquatic bacterioplankton estimations; Fukuda, Ogawa, Nagata, & Koike, 1998; Zubkov, Fuchs, Eilers, Burkill, & Amann, 1999) approximately 1×10^{10} cells will be needed for a single extraction, which can create limitations for proper experimental design.

Another critical point that needs to be accounted for prior to experimentation is the design of the protein sequence database for peptide identification. Shotgun proteomic experiments are reliant on a comprehensive list of protein sequences predicted from the metagenome of the community being analyzed (see Fig. 15.1). Ideally, these databases will be comprised of high-quality assembled and annotated sequences from DNA collected from the same environmental samples. As creating specific metagenome databases for all proteome samples is not a trivial task, alternative database designs have been employed with success. These include using metagenomes derived from similar environments or sampled at different times than the proteome samples to be analyzed, or using annotated draft or closed genomes of isolates expected to populate a given proteome sample (e.g., Verberkmoes et al., 2009; Wilkins et al., 2009). These approaches each have inherent limitations, with most problems arising from a given database not capturing the proper level of protein sequence diversity, thereby resulting in poor overall numbers of peptide matches and protein identifications.

Lastly, peptide quantification is an important and active area of research within the field of proteomics. The most straightforward methods are designated "label-free" approaches, which correlate collected MS data (i.e., peptide ion peak intensity or normalized counts of spectra from a matched protein) with protein abundances that are compared relatively across sample sets (Florens et al., 2006; Griffin et al., 2010). Alternative approaches use isotopic labeling of the proteomes to detect quantitative differences between peptides containing different isotopic signatures. For example, in metabolic labeling experiments, one experimental treatment will be grown using stable isotope labeled media where ^{15}N or ^{13}C is incorporated into newly synthesized proteins, and the proteome from this treatment will be compared to a proteome collected from cells grown in label-free media (e.g., Belnap et al., 2009; Ong et al., 2002; Pan et al., 2006). Another method involves chemical labeling of peptides postextraction with isobaric tags (e.g., iTRAQ (Ross et al., 2004), TMT (Thompson et al., 2003)). Different tags can be applied to separate experimental treatments and samples can be combined and analyzed together allowing for relative quantitation to be made between peptides. If absolute quantitation is desired, the most widely used approach involves adding known quantities of labeled peptides to a sample as a standard reference (e.g., QconCAT (Benyon, Doherty, Pratt, & Gaskell, 2005), AQUA (Gerber, Rush, Stemman, Kirschner, & Gygi, 2003)). Each of these methods has inherent advantages and disadvantages (Bantscheff, Lemeer, Savitski, & Kuster, 2012; Bantscheff, Schirle, Sweetman, Rick, & Kuster, 2007). Generally, label-free approaches are the least accurate, but the easiest to perform. Labeling experiments, however, require more processing steps and additional reagents, but provide high accuracy, a wide dynamic range, and good reproducibility (see a comparative study by Li et al., 2011).

3. GENERALIZED PROCEDURE

3.1. Equipment and reagents for LC-MS/MS

- Quaternary high-performance liquid chromatograph (HPLC) system. Solvent A: 95% H_2O, 4.9% acetonitrile (ACN), and 0.1% formic acid (FA). Solvent B: 30% H_2O, 69.9% ACN, and 0.1% FA. Solvent C: 5% H_2O, 94.9% ACN, and 0.1% FA. Solvent D: 500 mM ammonium acetate in Solvent A.
- RP column: Cut a 100-μm-I.D. self-pack PicoFrit column (New Objective, Woburn, MA) to 25 cm long. Add a small amount of RP C_{18} resin (Phenomenex Inc.) to 1 ml methanol in a 1.5-ml tube and vortex to suspend resin particles. Place the RP slurry tube in a pressure cell and connect the column to the pressure cell. Pressurize the pressure cell using 800 psi helium gas and pack 15 cm RP resin into the column. Equilibrate the column with Solvent A for 20 min. Alternatively, prepacked RP C_{18} PicoFrit columns can be purchased from New Objective (Woburn, MA). Prepacked columns offer more reproducible LC performance at a higher cost than self-packed columns.
- SCX column: cut a piece of 15-cm-long 250-μm-I.D. fused silica capillary. Connect to a filter union (M-520, Upchurch Scientific). Add a small amount of SCX resin (Phenomenex Inc.) to 1 ml methanol in a 1.5-ml tube and vortex to suspend resin particles. Place the SCX slurry tube in a pressure cell and connect the column to the pressure cell. Pressurize the pressure cell using 800 psi helium gas and pack 5 cm SCX resin into the column. Equilibrate the column with Solvent A for 10 min. A new SCX column should be used for every LC–MS/MS run to minimize sample carry-over between runs.

- LTQ-Orbitrap mass spectrometer (Thermo Scientific, San Jose, CA). Hybrid MS instruments, such as LTQ-Orbitrap, allows acquiring high-resolution full scans for accurate quantification. A less expensive ion trap instrument, LTQ, can also be used with the compromise of less accurate mass measurement and abundance quantification. Time-of-flight instruments and QExactive offer good mass spectrometry performance at an intermediate instrument cost.

3.2. Protein extraction considerations

A wide range of protocols for efficient extraction of proteins from microbial cells is available, allowing for optimal extraction from a variety of samples. The appropriate extraction method will depend on a variety of intrinsic factors related to each collected sample. Once a sufficient amount of cellular material has been collected, a generalized extraction protocol will include steps for cell lysis, solubilization and chemical reduction of the proteome, separation of the proteome from residual sample components (i.e., lysed cellular debris, residual sample matrix, interfering chemical substances), and precipitation of the proteome. Different cell lysis techniques can be used depending on the cell membrane and wall structures of the members of the community within a sample (e.g., physical lysis via heat, pressure, or sonication; or chemical lysis generally using detergents and stabilizing agents). Another sample dependent consideration is removal of possible coextracted contaminants. Soil samples have proven to be particularly difficult to extract proteins from given a protein's tendency to adhere or adsorb to mineral surfaces or associate with humic colloids. Multiple modified extraction protocols are available to address these issues and should be consulted as needed (see, e.g., Chourey et al., 2010; Keiblinger et al., 2012). The protocol below will detail the steps for protein handling and MS analysis postextraction (i.e., after a proteins have been precipitated). It should be noted that approximate protein concentrations will be needed and can be performed on the sample prior to precipitation using a standard assay kit (e.g., bicinchoninic acid (BCA) protein assay kit).

3.3. Proteolysis and cleanup

1. Once a sufficient amount of proteins have been extracted, precipitated, and concentrated via centrifugation, the resulting protein pellet should be resuspended, reduced, and denatured in 1 ml of protein denaturing buffer. Iodoacetamide can be used to alkylate free cysteine residues.

Adjust the volume of added buffer depending on the size of the pellet. Incubation of the suspension for 1 h at 60 °C with vortexing every 20 min will aid in resuspension.

2. Add trypsin at a 1:100 ratio to the overall proteins (by weight). Incubate overnight at 37 °C with gentle shaking. To obtain more extensive digestion, a second aliquot of trypsin can be added at a 1:100 ratio, and the sample incubated for an additional 5 h at 37 °C with gentle shaking.
3. Clean up the digest sample with a RP solid-phase extraction (SPE) cartridge. Alternatively, sample digestion and cleanup can be performed using a filter-aided sample preparation protocol (Manza, Stamer, Ham, Codreanu, & Liebler, 2005; Wiśnewski, Zougman, Nagaraj, & Mann, 2009). Final peptide concentration should be determined using a general assay kit (i.e., BCA protein assay kit).

3.4. Two-Dimensional liquid chromatography separation

1. Add ~100 μg of the peptide sample into a 200-μl tube, place the tube in an LC pressure injection cell, and connect the SCX column to the cell. Pressurize the cell using 800 psi of helium to load the sample onto the column.
2. The two-dimensional LC system should be assembled as shown in Fig. 15.2. This LC setup shares basic configurations as the MudPIT method (McDonald, Ohi, Miyamoto, Mitchison, & Yates, 2002). The SCX column is connected directly to the RP column to achieve on-line two-dimensional separation. The RP column is integrated with the electrospray tip to eliminate postcolumn dead volume. The high voltage for ESI is applied to the waste flow to remove electrochemistry reaction products from the LC system. Compared to split-phase MudPIT, this setup eliminates the RP trap column by desalting the sample offline using SPE. The SCX column is decoupled from the RP column using a filter union such that the RP column can be reused for multiple runs and the SCX column can be replaced for every run. Adjust the length of the waste line to achieve a flow rate of 300 nl/min at the tip of the RP column at 100% Solvent A.
3. Elution of the peptides off the column and injection into the mass spectrometer is performed using HPLC. The generalized gradient profile discussed here allows elution of peptides over a 24-h time frame, allowing for continuous daily runs and minimal instrument downtime. Eleven LC elution gradient profiles are used within one run. The first 10 cycles involve the following components: (1) column equilibration

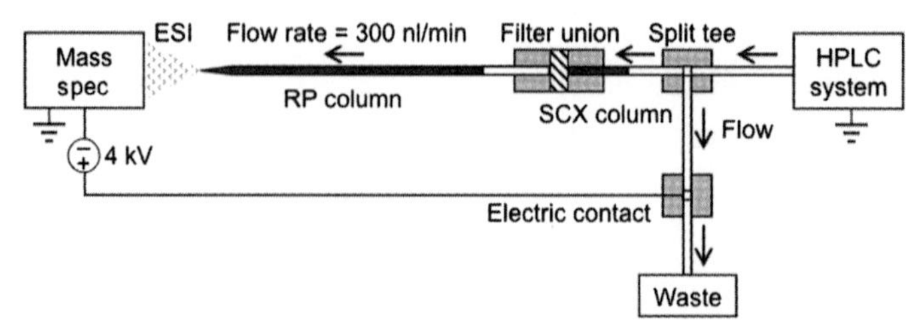

Figure 15.2 Schematic diagram of the LC-MS/MS System. The HPLC system pumps LC solvent at 100 μl/min. The split tee divides the flow, delivering a 300-nl/min flow to the LC columns and diverting the rest to the waste. The arrows indicate flow directions. Peptides are preloaded on the SCX column. During a salt pulse, an aliquot of the peptides are eluted off the SCX column and loaded onto the RP column. The RP column then resolves peptides using an RP LC gradient. The filter union functions as the frit for the SCX column. Eluent from the RP column is electrosprayed into mass spectrometer. Ionized peptides are analyzed by tandem mass spectrometry. The high voltage for electrospray ionization (ESI) is applied to the LC flow via an electric contact on the waste line.

with 100% Solvent A for 5 min, (2) pulsed addition of Solvent D at a defined percentage for each cycle for 2 min (The percentages of Solvent D used in the salt pulses of the 10 cycles are 10%, 15%, 20%, 25%, 30%, 35%, 40%, 45%, 50%, and 60% in this order.), (3) reequilibration of the column with 100% Solvent A for 3 min, and (4) a 110-min continuous gradient from 100% Solvent A to a 50:50 mixture of Solvent A and Solvent B. The final LC elution profile includes: (1) initial equilibration for 5 min with Solvent A, (2) a pulse with 100% Solvent D for 5 min, (3) reequilibration with 100% Solvent A for 5 min, and (4) and a final continuous resolution gradient of 165 min from 100% Solvent A to 100% Solvent B. The first 10 LC cycles elutes the majority of peptides using up to 60% Solvent D for the SCX step gradient and ramping up to 50% Solvent B for the RP continuous gradient. The last LC cycle elutes the remaining peptides by using 100% Solvent D for the SCX salt pulse and ramping up to 100% Solvent B for the RP gradient. SCX salt pulse steps in the 11 LC cycles are optimized to elute off approximately the same amount of peptides for each RP separation.

3.5. Tandem mass spectrometry analysis

1. Check the tune and calibration of the mass spectrometer and adjust if needed. The mass calibration of an Orbitrap instrument needs to be reset weekly to achieve consistent high mass accuracy. The ion optics of mass spectrometers should be calibrated only when necessary, and the ion

source stack should be cleaned regularly for optimum detection sensitivity.

2. Set up MS/MS methods in the MS instrument control software. Acquire MS/MS data continuously for the entire duration of the LC separation. Configure 6–12 data-dependent MS/MS scans for each full scan, targeting the precursor ions of top intensities in the full scan. Enable dynamic exclusion for the MS/MS scans. Acquire full scans using Orbitrap at resolution 30,000. Acquire MS/MS scans using ion trap or Orbitrap with CID at 35% normalized collision energy and a 3-Da isolation window. Initiate the LC separation sequence by HPLC and start concurrent data acquisition by mass spectrometer.
3. As peptides elute off the LC column, they are ionized by nanospray ionization and analyzed in real time by tandem mass spectrometry. The MS1 scans detect the *m/z* and intensity of all incoming peptide ions at regular time intervals. Every data-dependent MS2 scan following a MS1 scan automatically targets an abundant ion that has not been analyzed by a prior MS2 scan. The targeted peptide ion is isolated by its *m/z* and fragmented by CID. The *m/z* and intensity of the resultant fragment ions are recorded in the MS2 scan. Mass spectrometers can only acquire a limited number of MS2 scans per minute, whereas the LC may elute off many more peptides per minute. As a result, many peptides, particularly low-abundance ones, will not have the chance to be selected for MS2 analysis during their chromatographic peak.
4. After data acquisition is completed, it is important to manually check the quality of the LC–MS/MS data by inspecting total ion chromatograms for overall sample complexity and total signal intensity. Check base peak chromatograms for LC separation resolution. ESI is suppressed during salt pulses and should recover after salt pulses in less than 10 min. Check the quality of tandem mass spectra. Anomalies can be caused by sample problems or malfunctions of the LC–MS/MS system. The performance of the LC–MS/MS system can be verified by running standard proteome samples.
5. Copy acquired raw MS/MS data files from the MS instrument control computer to a data analysis computer. The details of data processing will be covered in the next chapter.

ACKNOWLEDGMENT

This work was supported by the Gordon and Betty Moore foundation (Grant #3302). Oak Ridge National Laboratory is managed by UT-Battelle, LLC, for the U.S. Department of Energy under contract DE-AC05-00OR22725.

REFERENCES

Abram, F., Enright, A.-M., O'Reilly, J., Botting, C. H., Collins, G., & O'Flaherty, V. (2011). A metaproteomic approach gives functional insights into anaerobic digestion. *Journal of Applied Microbiology, 110*, 1550–1560.

Bantscheff, M., Lemeer, S., Savitski, M. M., & Kuster, B. (2012). Quantitative mass spectrometry in proteomics: Critical review update from 2007 to the present. *Analytical and Bioanalytical Chemistry, 404*, 939–965.

Bantscheff, M., Schirle, M., Sweetman, G., Rick, J., & Kuster, B. (2007). Quantitative mass spectrometry in proteomics: A critical review. *Analytical and Bioanalytical Chemistry, 389*, 1017–1031.

Belnap, C. P., Pan, C., VerBerkmoes, N. C., Power, M. E., Samatova, N. F., Carver, R., et al. (2009). Cultivation and quantitative proteomic analyses of acidophilic microbial communities. *The ISME Journal, 4*, 520–530.

Benyon, R. J., Doherty, M. K., Pratt, J. M., & Gaskell, S. J. (2005). Multiplexed absolute quantification in proteomics using artificial QCAT proteins of concatenated signature peptides. *Nature Methods, 2*, 587–589.

Burnhum, K. E., Callister, S. J., Nicora, C. D., Purvine, S. O., Hugenholz, P., Warnecke, F., et al. (2011). Proteome insights into the symbiotic relationship between a captive colony of *Nasutitermes corniger* and its hindgut microbiome. *The ISME Journal, 5*, 161–164.

Chen, Y. C., Shiea, J., & Sunner, J. (1998). Thin-layer chromatography-mass spectrometry using activated carbon, surface-assisted laser desorption/ionization. *Journal of Chromatography. A, 826*, 77–86.

Chourey, K., Jansson, J., VerBerkmoes, N., Shah, M., Chavarria, K. L., Tom, L. M., et al. (2010). Direct cellular lysis/protein extraction protocol for soil metaproteomics. *Journal of Proteome Research, 9*, 6615–6622.

Delmotte, N., Knief, C., Chaffron, S., Innerebner, G., Roschitzki, B., Schlapbach, R., et al. (2009). Community proteogenomics reveals insights into the physiology of phyllosphere bacteria. *Proceedings of the National Academy of Sciences of the United States of America, 106*, 16428–16433.

Denef, V. J., & Banfield, J. F. (2012). In situ evolutionary rate measurements show ecological success of recently emerged bacterial hybrids. *Science, 336*, 462–466.

Denef, V. J., Kalnejais, L. H., Mueller, R. S., Wilmes, P., Baker, B. J., Thomas, B. C., et al. (2010). Proteogenomic basis for ecological divergence of closely related bacteria in natural acidophilic microbial communities. *Proceedings of the National Academy of Sciences of the United States of America, 107*, 2383–2390.

Di Palma, S., Boersema, P. J., Heck, A. J. R., & Mohammed, S. (2011). Zwitterionic hydrophilic interaction liquid chromatography (ZIC-HILIC and ZIC-cHILIC) provide high resolution separation and increase sensitivity in proteome analysis. *Analytical Chemistry, 83*, 3440–3447.

Erickson, B. K., Mueller, R. S., VerBerkmoes, N. C., Shah, M., Singer, S. W., Thelen, M., et al. (2010). Computational prediction and experimental validation of signal peptide cleavages in the extracellular proteome of a natural microbial community. *Journal of Proteome Research, 9*, 2148–2159.

Fenn, J. B., Mann, M., Meng, C. K., Wong, S. F., & Whitehouse, C. M. (1989). Electrospray ionization for mass spectrometry of large biomolecules. *Science, 246*, 64–71.

Ficarro, S. B., McCleland, M. L., Stukenber, P. T., Burke, D. J., Ross, M. M., Shabanowitz, J., et al. (2002). Phosphoproteome analysis by mass spectrometry and its application to *Saccharomyces cerevisiae*. *Nature Biotechnology, 20*, 301–305.

Florens, L., Carozza, M. J., Swanson, S. K., Fournier, M., Coleman, M. K., Workman, J. L., et al. (2006). Analyzing chromatin remodeling complexes using shotgun proteomics and normalized spectral abundance factors. *Methods, 40*, 303–311.

Fukuda, R., Ogawa, H., Nagata, T., & Koike, I. (1998). Direct determination of carbon and nitrogen contents of natural bacterial assemblages in marine environments. *Applied and Environmental Microbiology*, *64*, 3352–3358.

Gerber, S. A., Rush, J., Stemman, O., Kirschner, M. W., & Gygi, S. P. (2003). Absolute quantification of proteins and phosphoproteins from cell lysates by tandem MS. *Proceedings of the National Academy of Sciences of the United States of America*, *100*, 6940–6945.

Griffin, N. M., Yu, J., Long, F., Oh, P., Shore, S., Li, Y., et al. (2010). Label-free, normalized quantification of complex mass spectrometry data for proteomic analysis. *Nature Biotechnology*, *28*, 83–89.

Gupta, N., Tanner, S., Jaitly, N., Adkins, J. N., Lipton, M., Edwards, R., et al. (2007). Whole proteome analysis of post-translational modifications: Applications of mass-spectrometry for proteogenomic annotation. *Genome Research*, *17*, 1362–1377.

Haange, S.-B., Oberbach, A., Schlichting, N., Hugenholz, F., Smidt, H., von Bergen, M., et al. (2012). Metaproteome analysis and molecular genetics of rat intestinal microbiota reveals section and localization resolved species distribution and enzymatic functionalities. *Journal of Proteome Research*, *11*, 5406–5417.

Hörth, P., Miller, C. A., Preckel, T., & Wenz, C. (2006). Efficient fractionation and improved protein identification by peptide OFFGEL electrophoresis. *Molecular & Cellular Proteomics*, *5*, 1968–1974.

Jafari, M., Primo, V., Smejkal, G. B., Moskovets, E. V., Kuo, W. P., & Ivanov, A. R. (2012). Comparison of in-gel protein separation techniques commonly used for fractionation in mass spectrometry-based proteomic profiling. *Electrophoresis*, *33*, 2516–2526.

Jaffe, J. D., Berg, H. C., & Church, G. M. (2003). Proteogenomic mapping as a complementary method to perform genome annotation. *Proteomics*, *4*, 59–77.

Justice, N. B., Pan, C., Mueller, R. S., Spaulding, S. E., Shah, V., Sun, C. L., et al. (2012). Heterotropic archaea contribute to carbon cycling in low-pH, suboxic biofilm communities. *Applied and Environmental Microbiology*, *78*, 8321–8330.

Karas, M., & Hillenkamp, F. (1988). Laser desorption ionization of proteins with molecular masses exceeding 10,000 daltons. *Analytical Chemistry*, *60*, 2299–2301.

Keiblinger, K. M., Wilhartitz, I. C., Schneider, T., Roschitzki, B., Schmid, E., Eberl, L., et al. (2012). Soil metaproteomics—Comparative evaluation of protein extraction protocols. *Soil Biology & Biogeochemistry*, *54*, 14–24.

Klaassens, E. S., de Vos, W. M., & Vaughan, E. E. (2007). Metaproteomics approach to study the functionality of the microbiota in the human infant gastrointestinal tract. *Applied and Environmental Microbiology*, *73*, 1388–1392.

Lacerda, C. M. R., Choe, L. H., & Reardon, K. F. (2007). Metaproteomic analysis of a bacterial community response to cadmium exposure. *Journal of Proteome Research*, *6*, 1145–1152.

Laiko, V. V., Baldwin, M. A., & Burlingame, A. L. (2000). Atmospheric pressure matrix-assisted laser desorption/ionization mass spectrometry. *Analytical Chemistry*, *72*, 652–657.

Li, Z., Adams, R. M., Chourey, K., Hurst, G. B., Hettich, R. L., & Pan, C. (2011). Systematic comparison of label-free, metabolic labeling, and isobaric labeling for quantitative proteomics on LTQ orbitrap velos. *Journal of Proteome Research*, *11*, 1582–1590.

Link, A. J., Eng, J., Schieltz, D. M., Carmack, E., Mize, G. J., Morris, D. R., et al. (1999). Direct analysis of protein complexes using mass spectrometry. *Nature Biotechnology*, *17*, 676–682.

Liu, M., Fan, L., Zhong, L., Kjelleberg, S., & Thomas, T. (2012). Metaproteomic analysis of a community of sponge symbionts. *The ISME Journal*, *6*, 1515–1525.

Manza, L. L., Stamer, S. L., Ham, A. J., Codreanu, S. G., & Liebler, D. C. (2005). Sample preparation and digestion for proteomic analyses using spin filters. *Proteomics*, *5*, 1742–1745.

McDonald, W. H., Ohi, R., Miyamoto, D. T., Mitchison, T. J., & Yates, J. R., III. (2002). Comparison of three directly coupled HPLC MS/MS strategies for identification of

proteins from complex mixtures: Single-dimension LC-MS/MS, 2-phase MudPIT, and 3-phase MudPIT. *International Journal of Mass Spectrometry*, *219*, 245–251.
Merchant, M., & Weinberger, S. (2000). Recent advancements in surface-enhanced laser desorption/ionization-time of flight-mass spectrometry. *Electrophoresis*, *21*, 1164–1167.
Morris, R. M., Nunn, B. L., Frazar, C., Goodlett, D. R., Ting, Y. S., & Rocap, G. (2010). Comparative metaproteomics reveals ocean-scale shifts in microbial nutrient utilization and energy transduction. *The ISME Journal*, *4*, 673–685.
Mueller, R. S., Denef, V. J., Kainejais, L. H., Suttle, B., Thomas, B. C., Wilmes, P., et al. (2010). Ecological distribution and population physiology defined by proteomics in a natural microbial community. *Molecular Systems Biology*, *6*, 1–12.
Mueller, R. S., Dill, B. D., Pan, C., Belnap, C., Thomas, B. C., VerBerkmoes, N. C., et al. (2011). Proteome changes in the initial bacterial colonist during ecological succession in an acid mine drainage biofilm community. *Environmental Microbiology*, *13*, 2279–2292.
O'Farrell, P. H. (1975). High resolution two-dimensional electrophoresis of proteins. *The Journal of Biological Chemistry*, *250*, 4007–4021.
Ong, S.-E., Blagoev, B., Kratchmarova, I., Kristensen, D. B., Steen, H., Pandey, A., et al. (2002). Stable isotope labeling by amino acids in cell culture, SILAC, as a simple and accurate approach to expression proteomics. *Molecular & Cellular Proteomics*, *1*, 376–386.
Pan, C., Fischer, C., Hyatt, D., Bowen, B., Hettich, R. L., & Banfield, J. F. (2011). Quantitative tracking of isotope flows in proteomes of microbial communities. *Molecular & Cellular Proteomics*, *10*, 1–11.
Pan, C., Kora, G., McDonald, W. H., Tabb, D. L., VerBerkmoes, N. C., Hurst, G. B., et al. (2006). ProRata: A quantitative proteomics program for accurate protein abundance ratio estimation with confidence interval evaluation. *Analytical Chemistry*, *78*, 7121–7131.
Ram, R. J., VerBerkmoes, N. C., Thelen, M. P., Tyson, G. W., Baker, B. J., Blake, R. C., II., et al. (2005). Community proteomics of a natural microbial biofilm. *Science*, *308*, 1915–1920.
Ross, P. L., Huang, Y. N., Marchese, J. N., Williamson, B., Parker, K., & Hattan, S. (2004). Multiplexed protein quantitation in *Saccharomyces cerevisiae* using amine-reactive isobaric tagging reagents. *Molecular & Cellular Proteomics*, *3*, 1154–1169.
Savidor, A., Donahoo, R. S., Hurtado-Gonzales, O., VerBerkmoes, N. C., Shah, M. B., Lamour, K. H., et al. (2006). Expressed peptide tags: An additional layer of data for genome annotation. *Journal of Proteome Research*, *5*, 3048–3058.
Schneider, T., Kieblinger, K. M., Schmid, E., Sterflinger-Gleixner, K., Ellersdorfer, G., Roschitzki, B., et al. (2012). Who is who in litter decomposition? Metaproteomics reveals major microbial players and their biogeochemical functions. *The ISME Journal*, *6*, 1749–1762.
Singer, S. W., Erickson, B. E., VerBerkmoes, N. C., Hwang, M., Shah, M. B., Hettich, R. L., et al. (2010). Posttranslational modification and sequence variation of redox-active proteins correlate with biofilm life cycle in natural microbial communities. *The ISME Journal*, *4*, 1398–1409.
Taubert, M., Vogt, C., Wubet, T., Kleinsteuber, S., Tarkka, M. T., Harms, H., et al. (2012). Protein-SIP enables time-resolved analysis of the carbon flux in a sulfate-reducing, benzene-degrading microbial consortium. *The ISME Journal*, *6*, 2291–2301.
Thomas, J. J., Shen, Z., Crowell, J. E., Finn, M. G., & Siuzdak, G. (2001). Desorption/ionization on silicon (DIOS): A diverse mass spectrometry platform for protein characterization. *Proceedings of the National Academy of Sciences of the United States of America*, *98*, 4932–4937.
Thompson, A., Schäfer, J., Kuhn, K., Kienle, S., Schwarz, J., Schmidt, G., et al. (2003). Tandem mass tags: A novel quantification strategy for comparative analysis of complex protein mixtures by MS/MS. *Analytical Chemistry*, *75*, 1895–1904.

Verberkmoes, N. C., Russell, A. L., Shah, M., Godzik, A., Rosenquist, M., Halfvarson, J., et al. (2009). Shotgun metaproteomics of the human distal gut microbiota. *The ISME Journal*, *3*, 179–189.

Wilkins, M. J., VerBerkmoes, N. C., William, K. H., Callister, S. J., Mouser, P. J., Elifantz, H., et al. (2009). Proteogenomic monitoring of *Geobacter* physiology during stimulated uranium bioremediation. *Applied and Environmental Microbiology*, *75*, 6591–6599.

Williams, T. J., Long, E., Evans, F., DeMaere, M. Z., Lauro, F. M., Raferty, M. J., et al. (2012). A metaproteomic assessment of winter and summer bacterioplankton from Antarctic Peninsula coastal surface waters. *The ISME Journal*, *6*, 1883–1900.

Wilmes, P., Wexler, M., & Bond, P. L. (2008). Metaproteomics provides functional insight into activated sludge wastewater treatment. *PloS One*, *3*, e1778.

Wiśnewski, J. R., Nagaraj, N., Zougman, A., Gnad, F., & Mann, M. (2010). Brain phosphoproteome obtained by FASP-based method reveals plasma membrane protein topology. *Journal of Proteome Reasearch*, *9*, 3280–3289.

Wiśnewski, J. R., Zougman, A., Nagaraj, N., & Mann, M. (2009). Universal sample preparation method for proteome analysis. *Nature Methods*, *6*, 359–363.

Wolters, D. A., Washburn, M. P., & Yates, J. R., III. (2001). An automated multidimensional protein identification technology for shotgun proteomics. *Analytical Chemistry*, *73*, 5683–5690.

Yates, J. R., III., Ruse, C. I., & Nakorchevsky, A. (2009). Proteomics by mass spectrometry: Approaches, advances, and applications. *Annual Review of Biomedical Engineering*, *11*, 49–79.

Zubkov, M. V., Fuchs, B. M., Eilers, H., Burkill, P. H., & Amann, R. (1999). Determination of total protein content of bacterial cells by SYPRO staining and flow cytometry. *Applied and Environmental Microbiology*, *65*, 3251–3257.

CHAPTER SIXTEEN

Molecular Tools for Investigating Microbial Community Structure and Function in Oxygen-Deficient Marine Waters

Alyse K. Hawley[*], Sam Kheirandish[*], Andreas Mueller[*], Hilary T.C. Leung[*], Angela D. Norbeck[†], Heather M. Brewer[‡], Ljiljana Pasa-Tolic[‡], Steven J. Hallam[*,§,1]

[*]Department of Microbiology & Immunology, University of British Columbia, Vancouver, British Columbia, Canada
[†]Pacific Northwest National Laboratory, Richland, Washington, USA
[‡]Environmental and Molecular Sciences Laboratory at PNNL, Richland, Washington, USA
[§]Graduate Program in Bioinformatics, University of British Columbia, Vancouver, British Columbia, Canada
[1]Corresponding author: e-mail address: shallam@mail.ubc.ca

Contents

1. Introduction 306
2. Exploring Oxygen Minimum Zone Community Structure 307
   2.1 Sample processing and environmental DNA extraction 308
   2.2 Small-subunit ribosomal RNA gene primer design 310
   2.3 Quantifying small-subunit ribosomal RNA gene copy number 312
3. Detecting Oxygen Minimum Zone Proteins 314
   3.1 Sample processing and protein extraction 316
   3.2 Protein digestion and sample cleanup 317
   3.3 Tandem mass spectrometry and peptide identification 318
   3.4 Taxonomic binning and visualization of expressed proteins 320
Acknowledgments 325
References 325

Abstract

Water column oxygen (O_2)-deficiency shapes food-web structure by progressively directing nutrients and energy away from higher trophic levels into microbial community metabolism resulting in fixed nitrogen loss and greenhouse gas production. Although respiratory O_2 consumption during organic matter degradation is a natural outcome of a productive surface ocean, global-warming-induced stratification intensifies this process leading to oxygen minimum zone (OMZ) expansion. Here, we describe useful tools for detection and quantification of potential key microbial players and processes in OMZ community metabolism including quantitative polymerase chain

Methods in Enzymology, Volume 531
ISSN 0076-6879
http://dx.doi.org/10.1016/B978-0-12-407863-5.00016-2

reaction primers targeting Marine Group I Thaumarchaeota, SUP05, Arctic96BD-19, and SAR324 small-subunit ribosomal RNA genes and protein extraction methods from OMZ waters compatible with high-resolution mass spectrometry for profiling microbial community structure and functional dynamics.

1. INTRODUCTION

Oxygen minimum zones (OMZs) are widespread and expanding oceanographic features arising from microbial respiratory demand during organic matter degradation in stratified waters (Diaz & Rosenberg, 2008; Paulmier & Ruiz-Pino, 2009; Whitney, Freeland, & Robert, 2007). Operationally defined by oxygen (O_2) concentrations less than 20 μM, OMZs support thriving microbial communities that directly influence ocean nutrient and energy cycling (Ulloa, Canfield, Delong, Letelier, & Stewart, 2012; Wright, Konwar, & Hallam, 2012). In the absence of O_2, alternative respiratory substrates are utilized in microbial energy metabolism in a defined order based on free energy yield. These include metals such as iron and manganese, nitrate (NO_3^-), nitrite (NO_2^-), sulfate (SO_4^{2-}), and carbon dioxide (CO_2). Within OMZs, the use of NO_3^- and NO_2^- as terminal electron acceptors culminates in greenhouse gas production and fixed nitrogen loss in the forms of nitrous oxide (N_2O) and dinitrogen gas (N_2), respectively. Both dissimilatory nitrate reduction (denitrification) and anaerobic ammonium oxidation (anammox) processes are known to mediate these transformations within OMZs (Bruckner et al., 2013; Jayakumar, O'Mullan, Naqvi, & Ward, 2009; Lam et al., 2009). Recent studies also posit an essential role for sulfur cycling in OMZs, coupling the production and consumption of reduced sulfur compounds to dissimilatory nitrate reduction and dark CO_2 fixation (Canfield et al., 2010; Lavik et al., 2009; Stewart, Ulloa, & DeLong, 2011; Walsh et al., 2009). The integration of C, N, and S cycles represents a recurring theme within the O_2-deficient water column, where electron donors and acceptors are actively recycled between lower and higher oxidation states by metabolically coupled microorganisms (Ulloa et al., 2012).

Over the past four years, taxonomic and plurality sequencing studies have begun to describe microbial community structure and function of O_2-deficient waters in enclosed and semi-enclosed basins and coastal and open-ocean OMZs (Ulloa, Wright, Belmar, & Hallam, 2013; Wright et al., 2012). Several microbial groups have emerged as potential key players in mediating C, N, and S cycles within these ecosystems. Here, we describe known and novel polymerase chain reaction (PCR) primers targeting the small-subunit ribosomal RNA (SSU rRNA) genes from several of these groups including Marine Group I Thaumarchaeota, SUP05, Arctic96BD-19, and SAR324 SSU rRNA (or 16S) genes for use in quantitative PCR (qPCR) assays and protein extraction methods from O_2-deficient waters compatible with high-resolution mass spectrometry for profiling microbial community structure and functional dynamics.

2. EXPLORING OXYGEN MINIMUM ZONE COMMUNITY STRUCTURE

Numerous methods for investigating microbial community structure have been used with varying degrees of taxonomic resolution and quantitative power in O_2-deficient waters. These include amplicon-based methods such as terminal restriction length polymorphism (T-RFLP) or denaturing gradient gel electrophoresis (DGGE) of SSU rRNA genes (Lin, Scranton, Chistoserdov, Varela, & Taylor, 2008; Lin, Scranton, Varela, Chistoserdov, & Taylor, 2007; Rodriguez-Mora, Scranton, Taylor, & Chistoserdov, 2013; Vetriani, Tran, & Kerkhof, 2003; Zaikova et al., 2010), SSU rRNA gene clone library sequencing (Edgcomb et al., 2009; Fuchs, Woebken, Zubkov, Burkill, & Amann, 2005; Orsi, Song, Hallam, & Edgcomb, 2012; Stevens & Ulloa, 2008; Stoeck, Hayward, Taylor, Varela, & Epstein, 2006; Walsh & Hallam, 2011; Zaikova et al., 2010), and massively parallel tag sequencing (Allers et al., 2013; Herlemann et al., 2011; Stoeck et al., 2009, 2010). While T-RFLP and DGGE are inexpensive and amenable to automation, peak resolution is limited and taxonomic identification requires secondary purification and sequencing steps. Clone libraries can provide significantly more taxonomic information per read. However, quantitative power is limited by the cost of paired-end Sanger sequencing. Conversely, tag sequencing currently provides less taxonomic information per read than clone libraries but significantly more quantitative power. Catalyzed reporter deposition (CARD) and fluorescent *in situ* hybridization (FISH) have also been used in community composition (Fuchs et al., 2005; Lin et al., 2006, 2008; Wakeham et al., 2007) and group-specific studies targeting SUP05 (Glaubitz, Kiesslich, Meeske, Labrenz, & Jurgens, 2013; Lavik et al., 2009), Marine Group A (Allers et al., 2013), and Planctomycetes (Woebken, Fuchs, Kuypers, & Amann, 2007; Woebken et al., 2008). While CARD–FISH provides effective group-specific quantitation, probe development and optimization can be cost prohibitive and technically demanding when profiling large numbers of samples. Finally, metagenomic and metatranscriptomic sequencing approaches have been used to assess microbial community structure using phylogenetic anchors such as the SSU rRNA gene and conserved clusters of orthologous groups in open-ocean OMZs (Bryant, Stewart, Eppley, & DeLong, 2012, Canfield et al., 2010, Stewart et al., 2011, Ulloa et al., 2012) and semi-enclosed basins experiencing seasonal O_2-deficeincy (Walsh et al., 2009). While plurality sequencing approaches provide direct insight into microbial community structure and function, taxonomic resolution is dependent on database matching to defined reference genomes.

qPCR using dye assay chemistry such as SYBR® Green or EvaGreen can be a specific alternative or adjunct to sequencing and CARD–FISH methods. For example, qPCR using 5′-endonuclease probe-based chemistry (Taqman) or dye assay chemistry (SYBR® Green) has been successfully adapted for rapid and high-throughput quantification of microbial populations in seawater (Mincer et al., 2007; Suzuki, Taylor, & DeLong, 2000; Takai & Horikoshi, 2000). In O_2-deficient waters, qPCR has been used successfully to quantify functional gene expression of nitrogen cycling genes (Lam et al., 2009; Ward et al., 2009). Moreover, domain-specific primers targeting total Bacteria and Archaea, and group-specific primers targeting SUP05 and Arctic96BD-19 SSU rRNA gene copy number have been used in SYBR® Green-based qPCR assays to monitor secular changes in microbial community structure (Walsh & Hallam, 2011; Zaikova et al., 2010). The use of group-specific primers provides quantitative assessments of taxon abundance needed to accurately describe and monitor population dynamics in response to changing levels of water column O_2-deficiency.

2.1. Sample processing and environmental DNA extraction

Multiple methods for environmental DNA (eDNA) extraction from seawater exist and no single method will unfailingly provide ultraclean nucleic acids in sufficient quantity and quality to support qPCR applications without prior optimization. Here, we describe a method using a peristaltic pump to concentrate biomass from seawater onto a 0.2-μm Sterivex filter that has been used successfully to recover eDNA from O_2-deficient waters in the northeast subarctic Pacific Ocean (NESAP). Sample collection and filtration protocols can be viewed as visualized experiments at http://www.jove.com/video/1159/ (Zaikova et al., 2010) and http://www.jove.com/video/1161/ (Walsh et al., 2009), respectively.

1. Using a peristaltic pump (Cole-Parmer), seawater is filtered through a 2.7-μm GF/D prefilter to reduce particle and eukaryotic cell loading. Flow through biomass is concentrated in-line onto a 0.2-μm Sterivex filter (Millipore). Filter volumes will vary with cell densities ranging between 1 l in surface ocean waters and up to 200 l in dark ocean waters. Following biomass concentration, a syringe is used to purge remaining seawater from the filter cartridge prior to lysis buffer addition.
2. 1.8 ml of lysis buffer (0.75 *M* sucrose, 40 m*M* EDTA, 50 m*M* Tris, pH 8.3) is added to the Sterivex filter, sealed at both ends with parafilm, frozen on dry ice, and stored at −80 °C until extraction.
3. Prior to eDNA extraction, the Sterivex filter is thawed on ice followed by the addition of 100 μl lysozyme (125 mg in 1000 μl TE) and 20 μl RNase A (10 μg/ml) and resealed. The Sterivex filter is then incubated at 37 °C for 1 h in a rotating incubator or hybridization oven.
4. After 1 h, 100 μl Proteinase K and 100 μl of 20% SDS is added to the Sterivex filter and incubated at 55 °C for 1–2 h in a rotating incubator or hybridization oven.
5. The resulting lysate is extruded from the Sterivex filter into a 15-ml tube using a 10-cc syringe. Rinse the filter with 1 ml of lysis buffer, extrude, and combine with lysate.
6. Add an equal volume (~3 ml) of phenol:chloroform:IAA (25:24:1), pH 8.0, to lysate, invert to mix, and centrifuge at 2500 × *g* at room temperature for 5 min or until aqueous layer is clear. Transfer aqueous layer to a new 15-ml tube.
7. Add equal volume of chloroform:IAA (24:1) to tube with aqueous layer, invert to mix, centrifuge at 2500 × *g* at room temperature for 5 min until aqueous layer is clear, transfer aqueous layer to a new 15-ml tube, and add 1–3 ml TE buffer (pH 8.0).
8. For buffer exchange, transfer aqueous layer to Amicon® Ultra filter with 10K nominal molecular weight limit cutoff (Millipore) and centrifuge at 3500 × *g* for 10 min at 4 °C or until there is less the 1 ml remaining in the Amicon filter. Keep samples on ice during buffer exchange steps.
9. Buffer exchange two more times with 1–3 ml TE. In the final spin, bring volume down to 200–500 μl. Record the final extraction volume and transfer to 1.5-ml tube.
10. Quantify eDNA using the Quant-iT™ PicoGreen© dsDNA reagent (Invitrogen) and standards in 96- or 384-well microtiter plates on a

Varioskan Flash spectral scanning plate reader (Thermo Scientific) according to manufacturer's instructions. To determine eDNA quality, mix 2–5 μl of eDNA with loading dye and run on a 0.8% agarose gel with ethidium bromide (or other nucleic acid stain). Include additional wells with 10 μl of 1 kb + ladder and three lanes of 20 ng/μl λ *Hind*III ladder (2, 5, and 10 μl) as size standards. Run the gel at 15 V for 12–16 h and visualize using UV gel dock.

2.2. Small-subunit ribosomal RNA gene primer design

Quantitative PCR primers for environmental microorganisms need to be designed with both sequence specificity and target group diversity in mind. Here, we describe a simple approach to qPCR primer design extensible to both taxonomic and functional gene markers using the SAR324 SSU rRNA gene as an example. Additional qPCR primers targeting total Bacteria and Archaea, Marine Group I Thaumarchaeota, SUP05, and ArcticBD96-19 are represented in Table 16.1 with original citations. SSU rRNA gene sequences for SAR324 primer design were recovered from clone libraries sourced from NESAP waters and clustered at 97% identity with reference sequences available in the SILVA 102 full release database (Pruesse et al., 2007) using MOTHUR (Schloss et al., 2009). From a total of 1089 sequences, 179 operational taxonomic units (OTUs) were resolved. Representative sequences from each OTU were aligned using the SILVA aligner (http://www.arb-silva.de/aligner/) and subsequently imported into the ARB reference tree to verify taxonomic placement based on secondary structure alignment. Primer specificity and coverage were determined using BLAST (Altschul, Gish, Miller, Myers, & Lipman, 1990) searches against 1231 SAR234 sequences in the SILVA 111 full release database and Test-Prime (http://www.arb-silva.de/search/testprime/) (Klindworth et al., 2013). From a coverage perspective, both primers hit 93% of identified OMZ sequences and 53% of all SAR324 sequences in the database permitting a single mismatch. From a specificity perspective, both primers hit 24% of sequences sourced from oxygenated surface, mesopelagic and bathypelagic waters and less than 1% of sequences sourced from sediments, soils, or host-associated environments. Results using TestPrime were similar to BLAST with 41% of SAR324 sequences matching the primer pair with zero mismatches, 51% with one mismatch, and 52% with two mismatches. The majority of missed sequences originated in oxygenated surface, mesopelagic and bathypelagic waters, sediments, and soils.

Table 16.1 Summary table indicating qPCR primer sequences, reaction concentrations, annealing temperatures, expected amplicon lengths in base pairs, standards used, primer efficiencies, and detection limits

Target group	Primer name	Primer sequence (5′ to 3′)[a]	Final primer concentration (n*M*)	Annealing (°C)	Amplicon size (bp)	Primer efficiency	Detection limit (copies/ml)	Reference
Total Bacteria	27F	AGAGTTTGATCCTGGCTCAG	300	55	492	100.5 ± 0.78%	7.77×10^2	Zaikova et al. (2010)
	DW519R	GNTTTACCGCGGCKGCTG	300					
SAR324	SAR324_625F	CCGGAGGGTCTTTCGAAACT	300	65	140	98.14 ± 15.3%	1.14×10^2	Primer pair described in this study
	SAR324_764R	CGCCTCAGCGTCAGAGTG	300					
Arctic96BD-19	Ba519F	CAGCMGCCGCGGTAANWC	300	59	529	85.3 ± 7.72%	1.29×10^2	Walsh and Hallam (2011)
	Arcticl048R	CTATTTCTAGAAAGTTCGCAGG	300					
SUP05	Ba519F	CAGCMGCCGCGGTAANWC	300	63	529	100.2 ± 0.26%	2.67×10^2	Zaikova et al. (2010)
	1048R	CCATCTCTGGAAAGTTCCGTCT	300					
Total Archaea	20F	TTCCGGTTGATCCYGCCRG	300	65	499	95.5 ± 4.65%	1.95×10^2	Zaikova et al. (2010)
	DW519R	GNTTTACCGCGGCKGCTG	300					
Crenarchaeota	GI_751F	GTCTACCAGAACAYGTTC	500	58	205	93.2 ± 8.63%	2.30×10^2	Mincer et al. (2007)
	GI-956R	HGGCGTTGACTCCAATTG	500					

[a]International Union of Pure and Applied Chemistry (IUPAC) degenerate base symbols; N = A, G, C, or T; K = G or T; M = A or C; W = A or T; R = A or G; Y = C or T; H = A, C, or T.

2.3. Quantifying small-subunit ribosomal RNA gene copy number

Quality controls were performed against plasmid standards recovered from NESAP waters to empirically determine optimal annealing temperatures, efficiencies, and detection limits for all primer pairs (Table 16.1 and Fig. 16.1).

1. Target group SSU rRNA gene copy number is determined by qPCR. Each 20 μl reaction contains 10 μl Ssofast EvaGreen Supermix (BioRad), 2 μl of template DNA, specified primer concentration (Table 16.1), and remaining volume of sterile, nuclease-free water. Reactions are carried out in low tube strips or 96-well white qPCR plates and run on a CFX96 Real-Time PCR detection system (BioRad) under the following PCR conditions: initial denaturation at 98 °C for 2 min, followed by 40 cycles of primer annealing at specified temperature (Table 16.1) for 30 s and a plate read. Real-time data is analyzed with the CFX Manager Analysis Software (BioRad) (Fig. 16.1A). (Note: Ssofast EvaGreen does not require extension time at 72 °C. Optimal primer concentrations and annealing temperatures should be tested using primer dilution series across a range of annealing temperatures and template concentrations.)
2. Plasmid standards are prepared using the Plasmid Midi kit (Qiagen) followed by plasmid-safe DNase treatment (Epicentre) according to manufacturer's instructions. After enzyme treatment and heat inactivation, DNA is purified with phenol:chloroform:IAA (25:24:1) followed by buffer exchange (3 ×) with Tris–EDTA (pH 8.0) using an Amicon Ultra 1 10K filter (Millipore).
3. Standard DNA is quantified using Quant-iT PicoGreen dsDNA kit (Invitrogen) and copies of standard per μl is determined by Eq. (16.1). A 10-fold dilution series is prepared ranging from approximately 10 to 10^7 copies/μl. The Log gene copy number in reaction versus quantification cycle (C_q) threshold is graphed to produce a standard curve (Fig. 16.1B).

$$\text{Standard concentration}\,(\text{ng DNA}/\mu\text{l}) = \frac{\text{DNA concentration}\,(\text{ng}/\mu\text{l})}{\text{Fragment size}\,(\text{bp}) \times (1.096 \times 10^{-21})(\text{g}/\text{bp}) \times 10^{9}(\text{ng}/\text{g})} \qquad (16.1)$$

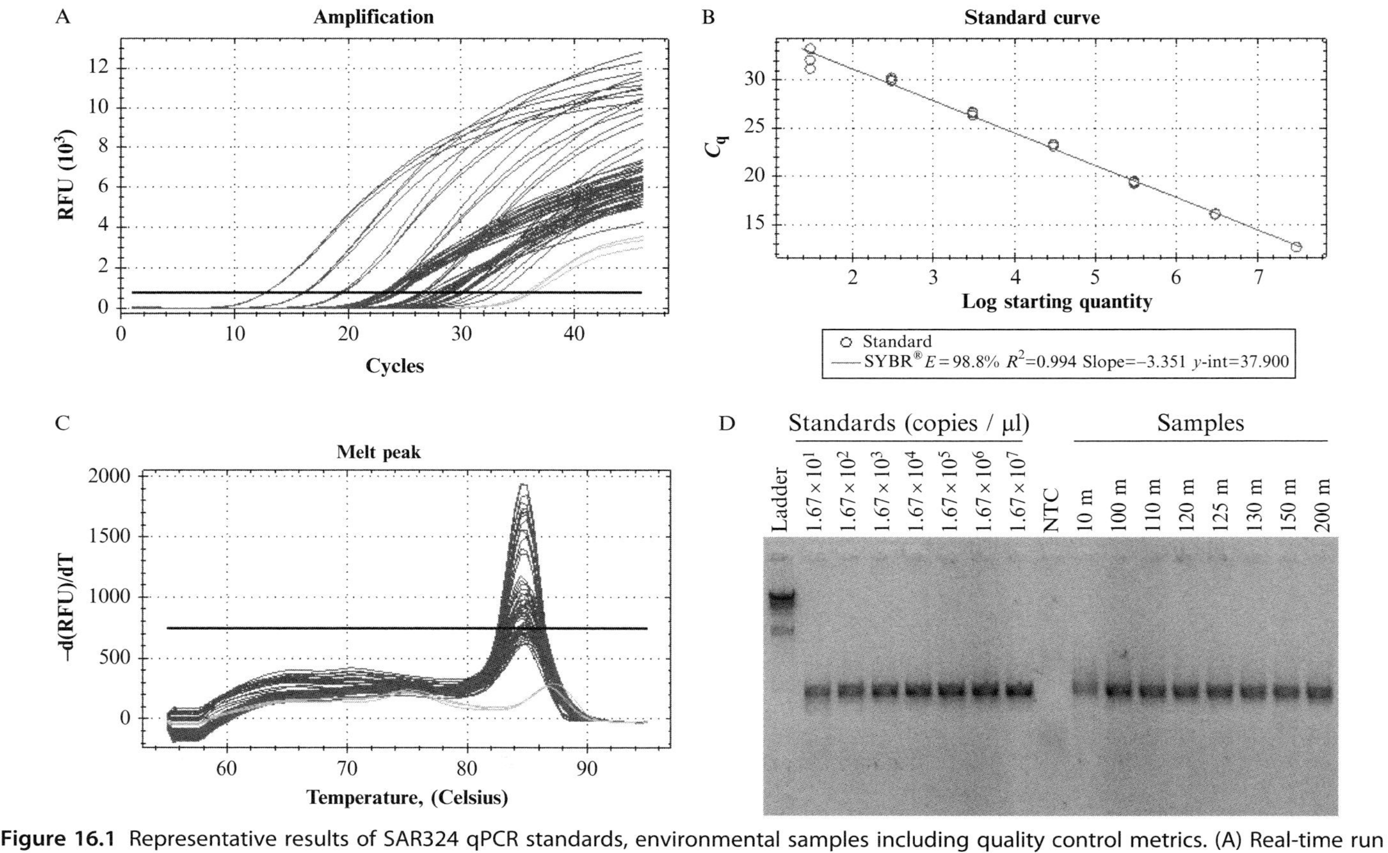

Figure 16.1 Representative results of SAR324 qPCR standards, environmental samples including quality control metrics. (A) Real-time run curves for standards (green), samples (red), and no template control (gray). (B) Standard curve of quantification cycle (C_q) versus Log of standard concentration. (C) Melting curve for standards (green), samples (red), and no template control (gray), showing single melting temperature for each sample and standard, indicating a single amplification product. (D) Representative gel image for qPCR product showing single band to confirm single amplification product observed in melting curve profiles. (See color plate.)

4. The number of gene copies in each qPCR reaction is determined from the standard curve. The number of gene copies in a ml of seawater is determined by Eq. (16.2).

$$\text{Copies/ml seawater} = \text{Copies/}\mu\text{l in reaction} \times \text{DNA extraction volume } (\mu\text{l/ml seawater filtered}) \quad (16.2)$$

5. The limit of detection is established based on a comparison of dissociation curves and C_q threshold for all sample replicates (Table 16.1). Each assay produces logarithmic amplification curves with 85–100% amplification efficiency over the standard range.
6. Following reaction cycles, a melting curve from 55 to 95 °C, held at 0.5 °C increments for 2 s is performed to check for reaction specificity (Fig. 16.1C). Standard and sample amplicons are visualized as described in DNA extraction section running the gel at 80 V for 90 min to verify expected product length and the extent of primer dimer formation (Fig. 16.1D).
7. Processed qPCR results are represented graphically in relation to environmental parameter data (Fig. 16.2). (Note: Although developed for quantifying SSU rRNA gene copy number, the primers should be extensible to reverse transcription qPCR assays to quantify SSU rRNA gene expression.)

3. DETECTING OXYGEN MINIMUM ZONE PROTEINS

Environmental proteomics also known as metaproteomics was first used to describe microbial community gene expression in an acid mine drainage ecosystem (Ram et al., 2005). Reduced community complexity in the acid mine milieu enabled the identification of key metabolic activities and metabolic partitioning between community members. Since that time, metaproteomic approaches have been successfully applied to a wide range of natural and human-engineered ecosystems including soils (Keiblinger et al., 2012; Schneider et al., 2012), leaf surfaces (Delmotte et al., 2009), human guts (Verberkmoes et al., 2009), napthalene-degrading enrichment cultures (Guazzaroni et al., 2013), and wastewater treatment plants (Kuhn et al., 2011; Wilmes, Wexler, & Bond, 2008). Although no metaproteomes for

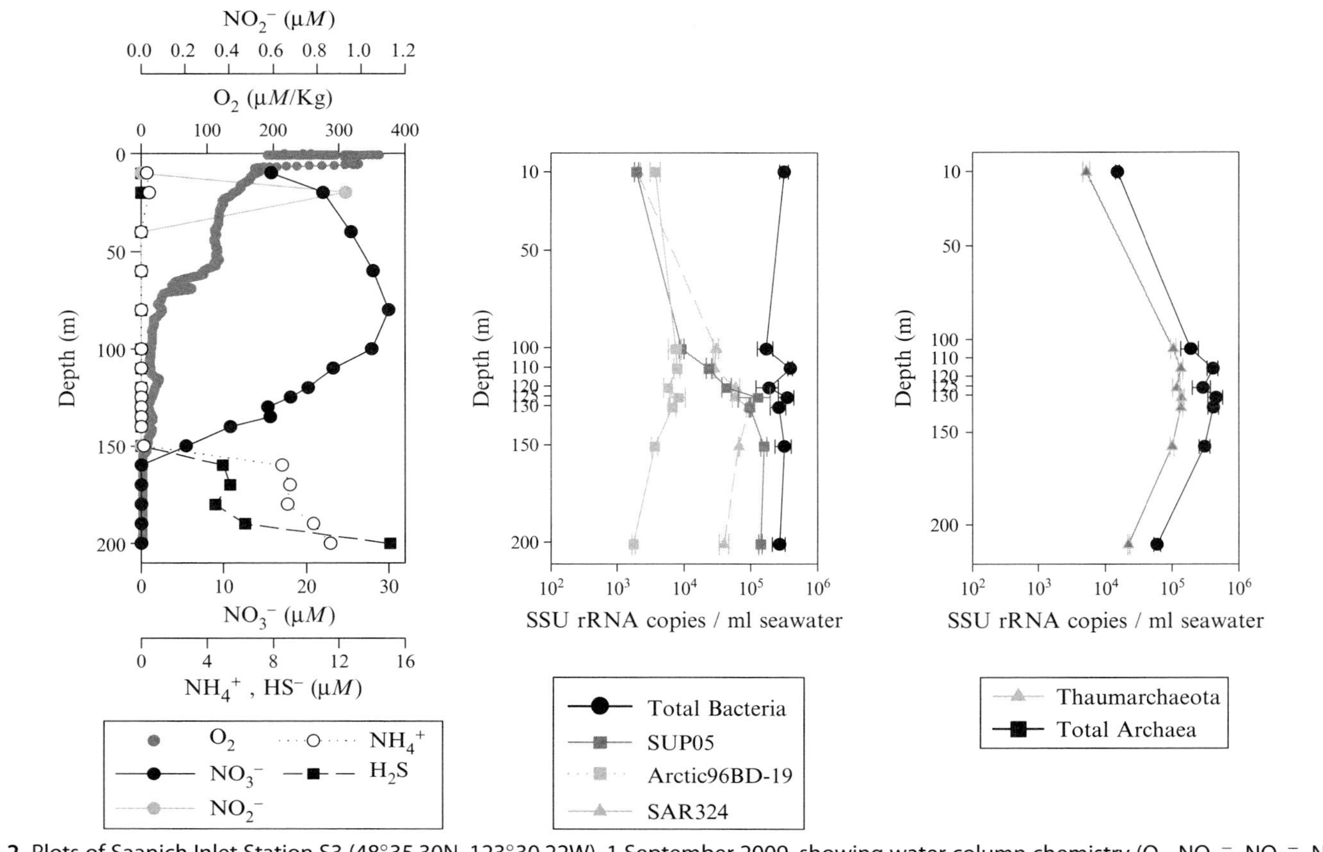

Figure 16.2 Plots of Saanich Inlet Station S3 (48°35.30N, 123°30.22W), 1 September 2009, showing water column chemistry (O_2, NO_2^-, NO_3^-, NH_4^-, and HS^-), and qPCR results of SSU rRNA copies/ml of sea water for total Bacteria, SUP05, Arctic96BD-19, SAR324, and total Archaea and Thaumarchaeota.

OMZ microbiota have been reported, surface ocean surveys have provided insight into microbial community responses to nutrient conditions along a coastal to open-ocean transect in the South Atlantic (Morris et al., 2010), coastal northeast Pacific Ocean upwelling (Sowell et al., 2011), and winter to summer transitions off the Antarctic Peninsula (Williams et al., 2012). Indeed, metaproteomics opens a functional window into microbial community metabolism and coupled biogeochemical cycles needed to more accurately monitor microbial community responses to changing levels of water column O_2-deficiency.

3.1. Sample processing and protein extraction

Protein extraction yields per unit volume of seawater need to be considered prior to large-scale sample collection to ensure sufficient biomass is filtered for downstream processing and detection steps. Here, we provide a protein extraction and peptide detection protocol optimized for community gene expression profiling in O_2-deficient marine waters in the NESAP. Empirical observations suggest that a minimum of 10^8 cells is needed to reliably detect abundant proteins under these water column conditions using nano-high-performance liquid chromatography coupled to a Thermo Electron LTQ-Orbitrap mass spectrometer with electrospray ionization.

1. Using a peristaltic pump (Cole-Parmer), seawater is filtered through a 2.7-μm GF/D prefilter to reduce particle and eukaryotic cell loading. Flow through biomass is concentrated in-line onto a 0.2-μm Sterivex filter (Millipore). Filter volumes required will vary with corresponding cell densities and typically range between 1 l in surface ocean waters and up to 200 l in dark ocean waters. Following biomass concentration, a syringe is used to purge remaining seawater from the filter cartridge prior to lysis buffer addition.
2. 1.8 ml of lysis buffer (0.75 *M* sucrose, 40 m*M* EDTA, 50 m*M* Tris, pH 8.3) is added to the Sterivex filter, sealed at both ends with parafilm, frozen on dry ice, and stored at −80 °C until extraction.
3. Prior to protein extraction, the Sterivex filter is thawed on ice followed by the addition of 200 μl of 10 × Bugbuster (Novagen). The Sterivex filter is then incubated at room temperature with rocking or rolling for 20–30 min to lyse cells.
4. The lysate is extruded from the filter into a 15-ml tube using a 10-cc syringe and put on ice prior to centrifugation at 3500 × *g* for 10 min at

4 °C to pellet cellular debris. Rinse the filter with 1 ml of lysis buffer, extrude, and combine with lysate.

5. For buffer exchange, transfer aqueous layer to Amicon® Ultra filter with 10K nominal molecular weight limit cutoff (Millipore), increase volume to 4 ml with 100 m*M* NH_4HCO_3, and centrifuge at 3500 × *g* for 10 min at 4 °C or until there is less the 1 ml remaining in the Amicon filter. Keep samples on ice during buffer exchange steps.
6. Buffer exchange two more times with 1–3 ml of 100 m*M* NH_4HCO_3. In the final spin, bring volume down to 200–500 μl. Record the final extraction volume and transfer to 1.5-ml tube.
7. Protein concentration is determined with 2-(4-carboxyquinolin-2-yl) quinoline-4-carboxylic acid (Bicinchoninic acid or BCA) assay.
8. Add powdered urea to a final concentration of 8 M (780 mg/ml). NOTE: each mg of Urea added will add 0.8 ml of volume.
9. A 50 m*M* working stock of the reducing agent Dithiothreitol (DTT) is added to a final concentration of 5 m*M* and the sample is incubated at 60 °C for 30 min.
10. Following DTT incubation, the sample is diluted 10-fold with 100 m*M* NH_4HCO_3 and 1 *M* $CaCl_2$ is added to a final concentration of 1 m*M*.
11. The sample can now be flash-frozen in liquid nitrogen and stored at −80 °C until trypsin digestion.

3.2. Protein digestion and sample cleanup

To remove residual salts from seawater samples as well as detergents used in protein extraction both a C18 column and strong cation exchange (SCX) column are used following trypsin digest.

1. Trypsin digest is carried out using 1 unit of mass spectrometry grade trypsin to 50 units protein at 37 °C for 6 h.
2. A 1 ml/50 mg bed volume C18 Solid Phase Extraction (SPE) (Sigma-Aldrich, Supelco Supelclean) column is conditioned with 3 ml of methanol and rinsed with 2 ml of 0.1% Trifluoroacetic acid (TFA) using a vacuum manifold.
3. After conditioning, the sample is added to the column and washed with 4 ml of 95:5 0.1%TFA:Acetonitrile (ACN) and allowed to dry. Peptides are eluted with 1 ml of 80:20 0.1%TFA:ACN using vacuum and concentrated to 50–100 μl in a speed-vac.

4. A 1 ml/50 mg bed volume SCX SPE column is used to clean remaining detergents from the sample. Condition the column by following steps 5–10 on a vacuum manifold. (Note: A SCX SPE 1 ml/ 50 mg tube is sufficient for up to 400 μg of protein, use a 1 ml/ 100 mg tube for larger protein amounts.)
5. Condition column with 2 ml of methanol.
6. Rinse column with 2 ml 10 m*M* ammonium formate (NH_4HCO_2), 25% ACN, pH 3.0.
7. Rinse column with 2 ml of 500 m*M* NH_4HCO_2, 25% ACN, pH 6.8.
8. Rinse column with 2 ml of 10 m*M* NH_4HCO_2, 25% ACN, pH 3.0.
9. Rinse column with 2 ml of Nanopure water.
10. Rinse column with 4 ml of 10 m*M* NH_4HCO_2, 25% ACN, pH 3.0.
11. Acidify sample by adding 10% TFA in Nanopure water to a final sample concentration of 1% and centrifuge for 5 min at 15,000 × *g* at room temperature to pellet any precipitates. Slowly pass the supernatant through column.
12. Wash the column with 4 ml of 10 m*M* NH_4HCO_2, 25% ACN, pH 3.0, and elute to dryness. Blot ends of manifold tubing below columns dry.
13. Place fresh 2.0-ml microcentrifuge tubes below columns, and with the vacuum turned off, add 1.0 ml MeOH:H_2O:NH_4OH (80:15:5) to each column.
14. Turn on vacuum, slowly elute sample from columns, and when columns are dry add an additional 500 μl of MeOH:H_2O:NH4OH (80:15:5) for a total elution volume of 1.5 ml. Concentrate the sample in a speed-vac to a final volume of 50–100 μl, adding small volumes of H_2O to dissolve particulate matter on the side of the tube (if needed). Perform BCA protein assay.
15. The sample can now be flash-frozen in liquid nitrogen and stored at −80 °C until needed for MS analysis.

3.3. Tandem mass spectrometry and peptide identification

While the detection and quantification of potential key microbial players in O_2-defcient waters provides insight into community structure and dynamics, additional methods for profiling environmental gene expression are needed for gene model and pathway validation. The application of high-pressure liquid chromatography (HPLC) coupled tandem mass spectrometry to identify expressed protein sequences from O_2-defcient waters offers a rapid and high-throughput profiling solution. The most effective peptide matching relies on the availability of environmental sequence information derived from the ecosystem under study although a standard reference database compiled from cultured isolates and publically available marine metagenomes can also be utilized. Here, we use a database of conceptually translated protein sequences from paired-end fosmid and whole-genome shotgun sequences from the Saanich Inlet water column to profile community gene expression across gradients of O_2 and hydrogen sulfide.

1. Aliquots containing $\sim$5 μg of protein are analyzed by online capillary liquid-chromatography–tandem mass spectrometry (Thermo, LTQ ion trap mass spectrometer or Thermo LTQ-Orbitrap mass spectrometer) using data-dependent fragmentation on the top 10 ions per duty cycle and a 100-min LC gradient from 0.1% formic acid in water to 0.1% formic acid acetonitrile. (Note: Reverse-phase capillary HPLC column used was made in-house at Environmental Molecular Sciences Laboratory at Pacific Northwest National Laboratories by slurry packing 3 μm Jupiter C_{18} stationary phase into a 60 cm length of 360 μm o.d. × 75 mm i.d. fused silica capillary tubing using a 1 cm sol-gel frit for retention of the packing material.)
2. Peptides are identified from MS/MS spectra using SEQUEST™ allowing for a potential oxidation of the methionine residues. Each search is performed using an environmental database of predicted protein sequences generated from the source location. For ion trap data, a mass error window of 3 m/z units is used for the precursor mass. A mass error window of 1 m/z unit is used for Orbitrap data, given the higher resolution of the instrument. In both cases, a 0 m/z tolerance is used for the fragmentation mass. Peptides identifications are permitted if they have a mass spectra generating functions value of less than 10^{-11}, which corresponds to a false discovery rate below 2% (Kim, Gupta, & Pevzner, 2008). Identifications are allowed for all possible peptide termini, that is, not limited by tryptic-only termini.
3. The number of peptide observations (scans matching to a peptide) from each protein is used as a rough measure of relative abundance and multiple charge states of a single peptide are considered as individual observations, as are the same peptides detected in different mass spectral analyses. However, it is important to note that while abundant proteins tend to produce more spectra, not all peptides ionize equally well. The most accurate quantitation requires some form of metabolic, isotopic, or isobaric tagging, or in the case of targeted proteomics, selected reaction monitoring or multiple reaction monitoring using stable isotope-labeled synthetic peptides (Walther & Mann, 2010).

3.4. Taxonomic binning and visualization of expressed proteins

There are many ways to visualize taxonomic distributions in environmental sequence data including heat maps, histograms, bubble plots, and trees. Here, we describe a composite visualization method using BLAST and the least common ancestor (LCA) algorithm implemented in MEGAN (Huson, Auch, Qi, & Schuster, 2007) superimposed on the Interactive Tree of Life (iTOL) (http://itol.embl.de/) (Letunic & Bork, 2007, 2011) to visualize the taxonomic distribution of expressed proteins from the Saanich Inlet water column (Fig. 16.3). Protein abundance information is mapped onto these tree structures using the normalized spectral abundance factor (NSAF) (Zybailov et al., 2006). The NSAF values for a given protein can be then compared for a more accurate representation of gene expression between environmental samples.

1. Amino acid sequences of detected proteins are compared to known protein sequences in genomic databases such as NCBI RefSeq via BLAST. A bit score ratio of 0.4 is used as a cutoff for confidence and the top hit is assigned to that protein sequence.
2. To increase taxonomic resolution and include genomes or metagenomes that are not yet in public database (such as RefSeq), the target database can be amended with the user-defined sequence information. For example, protein sequences for SUP05 uncultured bacterium and *Candidatus* Kuenenia stuttgartiensis were included in the BLAST against the NCBI RefSeq database by amending the RefSeq database with the additional genomic sequence information from desired organisms.
3. Peptide scan counts are summed for each protein (with $PPP > 0.95$). For peptides mapping to more than one protein, scan counts are divided between the total number of identified proteins.
4. The spectral abundance factor (SAF) (Eq. 16.3) is calculated using the sum of all scan counts for a given protein divided by the number of amino acids making up the protein sequence. The NSAF (Eq. 16.4) is the SAF for a given protein divided by the sum of all SAFs for a given sample.

$$\mathrm{SAF} = \frac{\text{Sum of scan counts for a given protein}}{\text{Length of given protein}} \tag{16.3}$$

$$\mathrm{NSAF} = \frac{\mathrm{SAF}}{\text{Sum of all SAF in a given sample}} \tag{16.4}$$

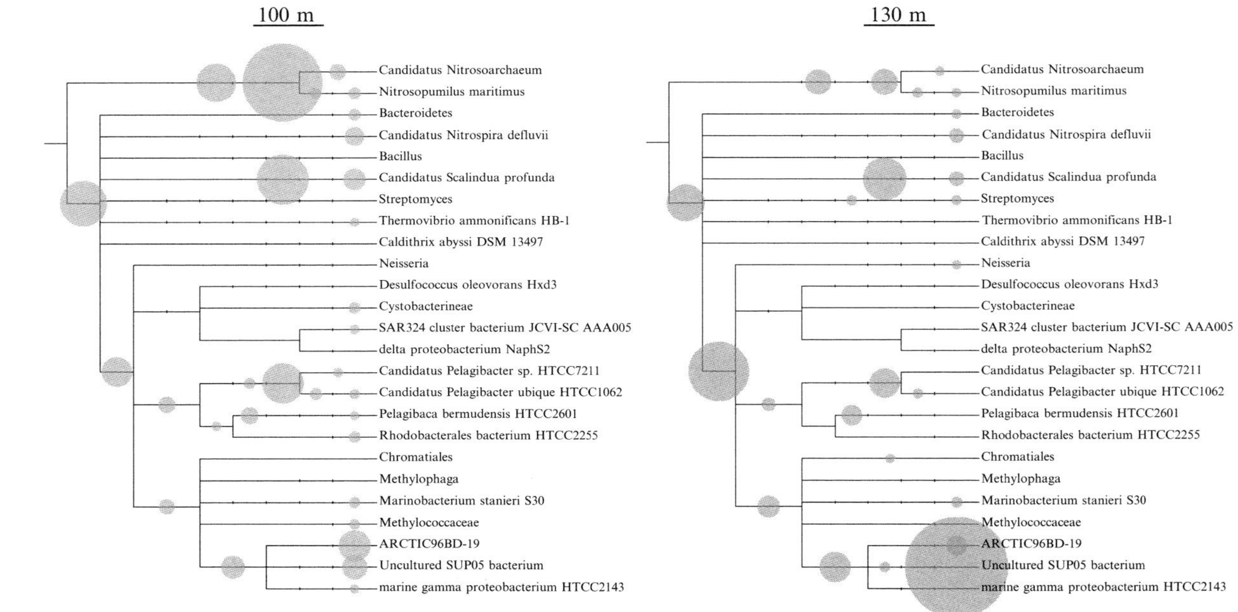
100 m
Candidatus Nitrosoarchaeum
Nitrosopumilus maritimus
Bacteroidetes
Candidatus Nitrospira defluvii
Bacillus
Candidatus Scalindua profunda
Streptomyces
Thermovibrio ammonificans HB-1
Caldithrix abyssi DSM 13497
Neisseria
Desulfococcus oleovorans Hxd3
Cystobacterineae
SAR324 cluster bacterium JCVI-SC AAA005
delta proteobacterium NaphS2
Candidatus Pelagibacter sp. HTCC7211
Candidatus Pelagibacter ubique HTCC1062
Pelagibaca bermudensis HTCC2601
Rhodobacterales bacterium HTCC2255
Chromatiales
Methylophaga
Marinobacterium stanieri S30
Methylococcaceae
ARCTIC96BD-19
Uncultured SUP05 bacterium
marine gamma proteobacterium HTCC2143
No hits/not assigned
130 m
Candidatus Nitrosoarchaeum
Nitrosopumilus maritimus
Bacteroidetes
Candidatus Nitrospira defluvii
Bacillus
Candidatus Scalindua profunda
Streptomyces
Thermovibrio ammonificans HB-1
Caldithrix abyssi DSM 13497
Neisseria
Desulfococcus oleovorans Hxd3
Cystobacterineae
SAR324 cluster bacterium JCVI-SC AAA005
delta proteobacterium NaphS2
Candidatus Pelagibacter sp. HTCC7211
Candidatus Pelagibacter ubique HTCC1062
Pelagibaca bermudensis HTCC2601
Rhodobacterales bacterium HTCC2255
Chromatiales
Methylophaga
Marinobacterium stanieri S30
Methylococcaceae
ARCTIC96BD-19
Uncultured SUP05 bacterium
marine gamma proteobacterium HTCC2143
No hits/not assigned

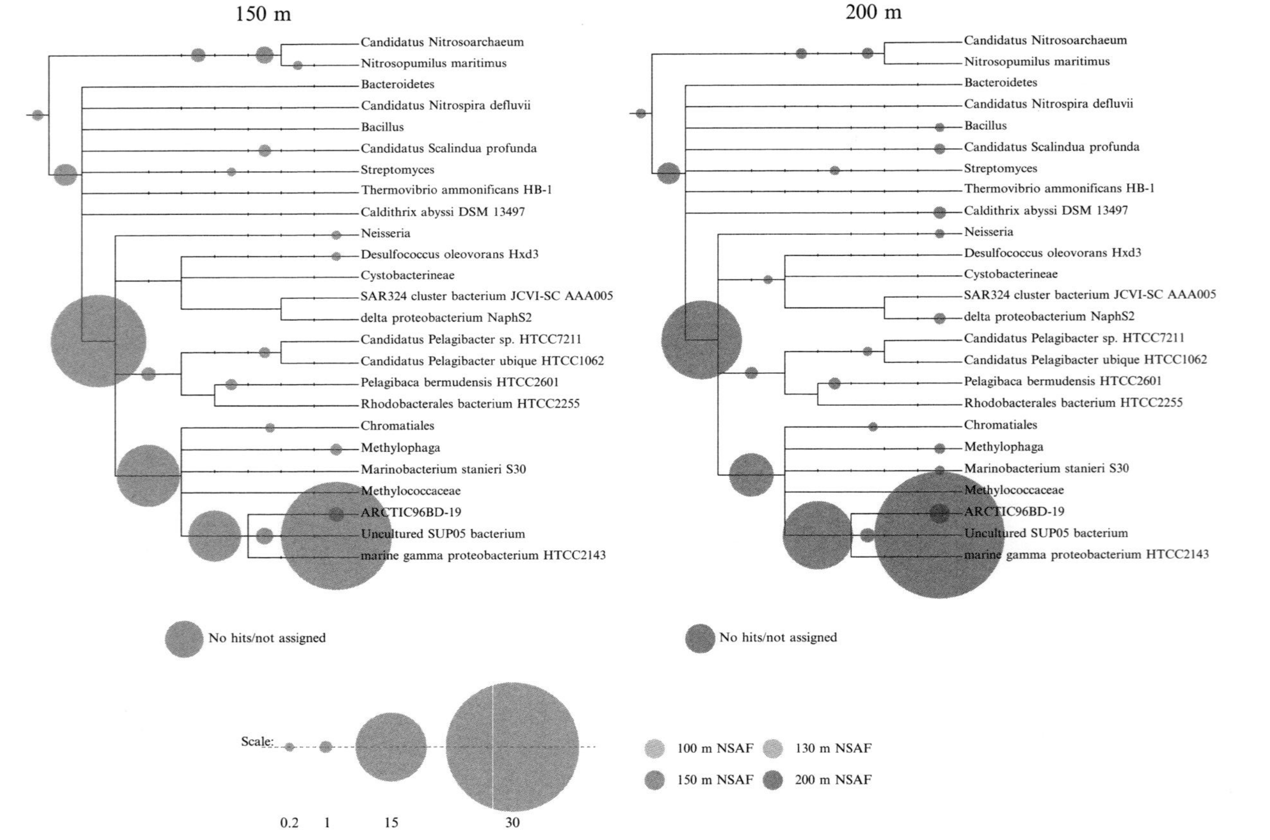

Figure 16.3 The taxonomic distribution of detected protein sequences in Saanich Inlet Station S3 (48°35.30N, 123°30.22W) determined by MEGAN least common ancestor algorithm at 100 (green), 130 (teal), 150 (blue), and 200 (purple) m sampling intervals. Taxon abundance in the metaproteome is shown using NSAF values. The NSAF value for detected proteins not assigned to the NCBI hierarchy or with no BLAST hit are shown at the base of the tree for each depth interval. (For interpretation of the references to color in this figure legend, the reader is referred to the online version of this chapter.)

5. MEGAN is run using the BLAST output for all identified protein sequences in a given sample by using the "Import from BLAST" option in the "File" menu. In the "Import" tab of the import dialogue box, select the BLAST output file, in the "'Content" tab, deselect SEED and KEGG options, in the "LCA Params" tab, change "Min Support" to 1 and deselect "Use Min-Complexity Filter." (Note: These parameters are user defined and can be altered based on user preferences or specific data requirements.)
6. Users can include taxa missing from the NCBI taxonomy or alter the structure of the NCBI taxonomy, for example, SUP05 uncultured bacterium by downloading the NCBI taxonomy structure from http://ab.inf.uni-tuebingen.de/data/software/megan4/download/welcome.html under "Updates of NCBI taxonomy." Unzip the file in the MEGAN/class/resources/files directory. Open the "names.dmp" file in a text editor, append an unused taxon ID number (left most field) for the new "species" and enter "scientific name" in the far right field maintaining the syntax present in the rest of the file. Repeat this process at the genus or family level (parent nodes) as needed. For example, in the case of SUP05 uncultured bacterium, the "names.dmp" file was amended with the following lines:

```
805819 |SUP05 cluster| |scientific name|
805820 |uncultured SUP05 cluster bacterium| |scientific name|
```

To place SUP05 in context with existing parent nodes for higher order taxonomic structure, determine the taxon ID number present in the "names.dmp" file, for example, the Gammaproteobacteria taxon ID number is 1236. Open the "nodes.dmp" file, locate the position of your new taxon ID number in the far left field, and enter the new taxon ID number in the far left field and the taxon ID number for the parent node in the second field position, for the remainder of the fields copy from an existing line (the NCBI download site explains all these fields in taxdump_readme.txt at ftp://ftp.ncbi.nlm.nih.gov/pub/taxonomy/). For example, in the case of SUP05 uncultured bacterium, the "nodes.dmp" file was amended with the following lines:

```
805819 | 135619 | order | | 1 | 1 | 1 | 1 | 5 | 1 | 0 | 0 | |
805820 | 805819 | species | | 1 | 1 | 1 | 1 | 5 | 1 | 0 | 0 | |
```

Save both of these files and relaunch MEGAN to use the newly assigned taxonomy and import your BLAST output file as described in step 1.

7. In the MEGAN tree view, open all the desired nodes by selecting a leaf on the far right, and selecting "Uncollapse Subtree" from the "Tree" menu to open the internal nodes for that leaf, repeat for all leaves. From the "Select" menu, select "All Nodes" and then from the "Export" menu select "Reads" in the "File" menu to export a list of sequences belonging to each taxon level.
8. On your desktop, open a terminal window. Using the cd command, enter the directory containing the MEGAN output files. Combine these files into a single file with sequence name and taxon ID using the following gawk scripts:

```
user$ gawk '{f=FILENAME".new";sub(">","",$0);print $0 "\t"
FILENAME> f}' *.fasta user$
cat *.new >combine_taxanames.txt user$
gawk '! a[$0]++' combine_taxanames.txt >
combine_taxonnames_clean.txt
```

9. To sum the NSAF values for all proteins assigned to a given taxon node, copy the "combine_taxonnames_clean.txt" file into Excel and sum the NSAF values for all sequences with a unique taxon ID using the SUMIF function.
10. Tree generation using iTOL requires the NCBI taxon ID assigned by MEGAN. A user account in iTOL is required to enter and store created trees. Enter a list of all taxon IDs (obtained from the "names.dmp" file for identified taxa) into iTOL at "other trees" (no number below 1 is permitted) to generate a Newick formatted output tree. Copy the tree and upload it to a new project in iTOL using "advanced options" with internal node IDs selected. View tree in iTOL to evaluate information content and visual balance. If there are more nodes than are possible to view in a single figure, consider adjusting the MEGAN parameters, or use a cutoff of minimum NSAF value, excluding all underrepresented nodes, to reduce to total number of nodes and repeat the import process. Once satisfied with your tree, export in Newick format including internal nodes.
11. Compare the taxon names in the exported tree to the taxon names of the nodes from MEGAN. It is important that names match exactly. Change spaces and "|" into "_."

12. For each sample, make a .csv file in the format “taxa” “value” “value” “value.” Use the same value in the first two “value” columns with an “R” in front of the first value and a zero for the third value (e.g., Gammaproteobacteria, R28, 28, 0). See iTOL “Uploading and working with your own trees” for more information on file headers. To provide a scale for your tree, add values to nodes which are “0” and manipulate the graphics file later to move scale bubbles into tree positions in the figure.
13. The .csv file is uploaded with a new tree in iTOL using the “Multivalue bar or pie chart” data type. Additional settings such as the minimum and maximum radii are user defined. Occasionally, the tree or the data may not be displayed in the iTOL user interface but can still be exported. Use the export function to generate an .svg file.
14. The output of multiple .svg files can be composited together using graphic design software (e.g., Adobe Illustrator) to visualize NSAF values, sequence read count, and taxonomic structure in a unified perspective (Fig. 16.3).

ACKNOWLEDGMENTS

We would like to thank David Walsh, Elena Zaikova, and Kendra Maas for help in primer design and testing. This work was performed under the auspices of the U.S. Department of Energy Joint Genome Institute under Contract No. DE-AC02-05CH11231, the Natural Sciences and Engineering Research Council (NSERC) of Canada, Canada Foundation for Innovation (CFI), and the Canadian Institute for Advanced Research (CIFAR) through grants awarded to S. J. H. Proteomic methods were performed using EMSL, a national scientific user facility sponsored by the Department of Energy's Office of Biological and Environmental Research and located at Pacific Northwest National Laboratory.

REFERENCES

Allers, E., Wright, J. J., Konwar, K. M., Howes, C. G., Beneze, E., Hallam, S. J., et al. (2013). Diversity and population structure of Marine Group A bacteria in the Northeast subarctic Pacific Ocean. *The ISME Journal*, 7(2), 256–268.

Altschul, S. F., Gish, W., Miller, W., Myers, E. W., & Lipman, D. J. (1990). Basic local alignment search tool. *Journal of Molecular Biology*, *215*(3), 403–410.

Bruckner, C. G., Mammitzsch, K., Jost, G., Wendt, J., Labrenz, M., & Jurgens, K. (2013). Chemolithoautotrophic denitrification of epsilonproteobacteria in marine pelagic redox gradients. *Environmental Microbiology*, *15*(5), 1505–1513.

Bryant, J. A., Stewart, F. J., Eppley, J. M., & DeLong, E. F. (2012). Microbial community phylogenetic and trait diversity declines with depth in a marine oxygen minimum zone. *Ecology*, *93*(7), 1659–1673.

Canfield, D. E., Stewart, F. J., Thamdrup, B., De Brabandere, L., Dalsgaard, T., Delong, E. F., et al. (2010). A cryptic sulfur cycle in oxygen-minimum-zone waters off the Chilean coast. *Science*, *330*(6009), 1375–1378.

Delmotte, N., Knief, C., Chaffron, S., Innerebner, G., Roschitzki, B., Schlapbach, R., et al. (2009). Community proteogenomics reveals insights into the physiology of phyllosphere bacteria. *Proceedings of the National Academy of Sciences of the United States of America*, *106*(38), 16428–16433.

Diaz, R. J., & Rosenberg, R. (2008). Spreading dead zones and consequences for marine ecosystems. *Science*, *321*(5891), 926–929.

Edgcomb, V., Orsi, W., Leslin, C., Epstein, S. S., Bunge, J., Jeon, S., et al. (2009). Protistan community patterns within the brine and halocline of deep hypersaline anoxic basins in the eastern Mediterranean Sea. *Extremophiles*, *13*(1), 151–167.

Fuchs, B. M., Woebken, D., Zubkov, M. V., Burkill, P., & Amann, R. (2005). Molecular identification of picoplankton populations in contrasting waters of the Arabian Sea. *Aquatic Microbial Ecology*, *39*(2), 145–157.

Glaubitz, S., Kiesslich, K., Meeske, C., Labrenz, M., & Jurgens, K. (2013). SUP05 dominates the Gammaproteobacterial sulfur oxidizer assemblages in pelagic Redoxclines of the central baltic and black seas. *Applied and Environmental Microbiology*, *79*(8), 2767–2776.

Guazzaroni, M. E., Herbst, F. A., Lores, I., Tamames, J., Pelaez, A. I., Lopez-Cortes, N., et al. (2013). Metaproteogenomic insights beyond bacterial response to naphthalene exposure and bio-stimulation. *The ISME Journal*, 7(1), 122–136.

Herlemann, D. P., Labrenz, M., Jurgens, K., Bertilsson, S., Waniek, J. J., & Andersson, A. F. (2011). Transitions in bacterial communities along the 2000 km salinity gradient of the Baltic Sea. *The ISME Journal*, *5*(10), 1571–1579.

Huson, D. H., Auch, A. F., Qi, J., & Schuster, S. C. (2007). MEGAN analysis of metagenomic data. *Genome Research*, *17*(3), 377–386.

Jayakumar, A., O'Mullan, G. D., Naqvi, S. W. A., & Ward, B. B. (2009). Denitrifying bacterial community composition changes associated with stages of denitrification in oxygen minimum zones. *Microbial Ecology*, *58*(2), 350–362.

Keiblinger, K. M., Wilhartitz, I. C., Schneider, T., Roschitzki, B., Schmid, E., Eberl, L., et al. (2012). Soil metaproteomics—Comparative evaluation of protein extraction protocols. *Soil Biology and Biochemistry*, *54*(15–10), 14–24.

Kim, S., Gupta, N., & Pevzner, P. A. (2008). Spectral probabilities and generating functions of tandem mass spectra: A strike against decoy databases. *Journal of Proteome Research*, 7, 3354–3363.

Klindworth, A., Pruesse, E., Schweer, T., Peplies, J., Quast, C., Horn, M., et al. (2013). Evaluation of general 16S ribosomal RNA gene PCR primers for classical and next-generation sequencing-based diversity studies. *Nucleic Acids Research*, *41*(1), e1.

Kuhn, R., Benndorf, D., Rapp, E., Reichl, U., Palese, L. L., & Pollice, A. (2011). Metaproteome analysis of sewage sludge from membrane bioreactors. *Proteomics*, *11*(13), 2738–2744.

Lam, P., Lavik, G., Jensen, M. M., van de Vossenberg, J., Schmid, M., Woebken, D., et al. (2009). Revising the nitrogen cycle in the Peruvian oxygen minimum zone. *Proceedings of the National Academy of Sciences of the United States of America*, *106*(12), 4752–4757.

Lavik, G., Stuhrmann, T., Bruchert, V., Van der Plas, A., Mohrholz, V., Lam, P., et al. (2009). Detoxification of sulphidic African shelf waters by blooming chemolithotrophs. *Nature*, *457*(7229), 581–584.

Letunic, I., & Bork, P. (2007). Interactive Tree Of Life (iTOL): An online tool for phylogenetic tree display and annotation. *Bioinformatics*, *23*(1), 127–128.

Letunic, I., & Bork, P. (2011). Interactive Tree Of Life v2: Online annotation and display of phylogenetic trees made easy. *Nucleic Acids Research*, *39*, W475–W478, Web Server issue.

Lin, X. J., Scranton, M. I., Chistoserdov, A. Y., Varela, R., & Taylor, G. T. (2008). Spatiotemporal dynamics of bacterial populations in the anoxic Cariaco Basin. *Limnology and Oceanography, 53*(1), 37–51.

Lin, X. J., Scranton, M. I., Varela, R., Chistoserdov, A., & Taylor, G. T. (2007). Compositional responses of bacterial communities to redox gradients and grazing in the anoxic Cariaco Basin. *Aquatic Microbial Ecology, 47*(1), 57–72.

Lin, X., Wakeham, S. G., Putnam, I. F., Astor, Y. M., Scranton, M. I., Chistoserdov, A. Y., et al. (2006). Comparison of vertical distributions of prokaryotic assemblages in the anoxic Cariaco Basin and Black Sea by use of fluorescence in situ hybridization. *Applied and Environmental Microbiology, 72*(4), 2679–2690.

Mincer, T. J., Church, M. J., Taylor, L. T., Preston, C., Karl, D. M., & DeLong, E. F. (2007). Quantitative distribution of presumptive archaeal and bacterial nitrifiers in Monterey Bay and the North Pacific Subtropical Gyre. *Environmental Microbiology, 9*(5), 1162–1175.

Morris, R. M., Nunn, B. L., Frazar, C., Goodlett, D. R., Ting, Y. S., & Rocap, G. (2010). Comparative metaproteomics reveals ocean-scale shifts in microbial nutrient utilization and energy transduction. *The ISME Journal, 4*(5), 673–685.

Orsi, W., Song, Y. C., Hallam, S., & Edgcomb, V. (2012). Effect of oxygen minimum zone formation on communities of marine protists. *The ISME Journal, 6*(8), 1586–1601.

Paulmier, A., & Ruiz-Pino, D. (2009). Oxygen minimum zones (OMZs) in the modern ocean. *Progress in Oceanography, 80*(3–4), 113–128.

Pruesse, E., Quast, C., Knittel, K., Fuchs, B. M., Ludwig, W. G., Peplies, J., et al. (2007). SILVA: A comprehensive online resource for quality checked and aligned ribosomal RNA sequence data compatible with ARB. *Nucleic Acids Research, 35*(21), 7188–7196.

Ram, R. J., VerBerkmoes, N. C., Thelen, M. P., Tyson, G. W., Baker, B. J., Blake, R. C. I., et al. (2005). Community proteomics of a natural microbial biofilm. *Science, 208*, 1915–1920.

Rodriguez-Mora, M. J., Scranton, M. I., Taylor, G. T., & Chistoserdov, A. Y. (2013). Bacterial community composition in a large marine anoxic basin: A Cariaco Basin time-series survey. *FEMS Microbiology Ecology, 84*(3), 625–639.

Schloss, P. D., Westcott, S. L., Ryabin, T., Hall, J. R., Hartmann, M., Hollister, E. B., et al. (2009). Introducing mothur: Open-source, platform-independent, community-supported software for describing and comparing microbial communities. *Applied and Environmental Microbiology, 75*(23), 7537–7541.

Schneider, T., Keiblinger, K. M., Schmid, E., Sterflinger-Gleixner, K., Ellersdorfer, G., Roschitzki, B., et al. (2012). Who is who in litter decomposition? Metaproteomics reveals major microbial players and their biogeochemical functions. *The ISME Journal, 6*(9), 1749–1762.

Sowell, S. M., Abraham, P. E., Shah, M., Verberkmoes, N. C., Smith, D. P., Barofsky, D. F., et al. (2011). Environmental proteomics of microbial plankton in a highly productive coastal upwelling system. *The ISME Journal, 5*(5), 856–865.

Stevens, H., & Ulloa, O. (2008). Bacterial diversity in the oxygen minimum zone of the eastern tropical South Pacific. *Environmental Microbiology, 10*(5), 1244–1259.

Stewart, F. J., Ulloa, O., & DeLong, E. F. (2011). Microbial metatranscriptomics in a permanent marine oxygen minimum zone. *Environmental Microbiology, 14*(1), 23–40.

Stoeck, T., Bass, D., Nebel, M., Christen, R., Jones, M. D., Breiner, H. W., et al. (2010). Multiple marker parallel tag environmental DNA sequencing reveals a highly complex eukaryotic community in marine anoxic water. *Molecular Ecology, 19*(Suppl. 1), 21–31.

Stoeck, T., Behnke, A., Christen, R., Amaral-Zettler, L., Rodriguez-Mora, M. J., Chistoserdov, A., et al. (2009). Massively parallel tag sequencing reveals the complexity of anaerobic marine protistan communities. *BMC Biology, 7*, 72.

Stoeck, T., Hayward, B., Taylor, G. T., Varela, R., & Epstein, S. S. (2006). A multiple PCR-primer approach to access the microeukaryotic diversity in environmental samples. *Protist*, *157*(1), 31–43.

Suzuki, M. T., Taylor, L. T., & DeLong, E. F. (2000). Quantitative analysis of small-subunit rRNA genes in mixed microbial populations via 5'-nuclease assays. *Applied and Environmental Microbiology*, *66*(11), 4605–4614.

Takai, K., & Horikoshi, K. (2000). Rapid detection and quantification of members of the archaeal community by quantitative PCR using fluorogenic probes. *Applied and Environmental Microbiology*, *66*(11), 5066–5072.

Ulloa, O., Canfield, D. E., Delong, E. F., Letelier, R. M., & Stewart, F. J. (2012). Microbial oceanography of anoxic oxygen minimum zones. *Proceedings of the National Academy of Sciences of the United States of America*, *109*(40), 15996–16003.

Ulloa, O., Wright, J. J., Belmar, L., & Hallam, S. J. (2013). Pelagic oxygen minimum zone microbial communities. In E. Rosenberg (Ed.), *The prokaryotes: Prokaryotic communities and ecophysiology*. Berlin Heidelberg: Springer-Verlag.

Verberkmoes, N. C., Russell, A. L., Shah, M., Godzik, A., Rosenquist, M., Halfvarson, J., et al. (2009). Shotgun metaproteomics of the human distal gut microbiota. *The ISME Journal*, *3*(2), 179–189.

Vetriani, C., Tran, H. V., & Kerkhof, L. J. (2003). Fingerprinting microbial assemblages from the oxic/anoxic chemocline of the Black Sea. *Applied and Environmental Microbiology*, *69*(11), 6481–6488.

Wakeham, S. G., Amann, R., Freeman, K. H., Hopmans, E. C., Jorgensen, B. B., Putnam, I. F., et al. (2007). Microbial ecology of the stratified water column of the Black Sea as revealed by a comprehensive biomarker study. *Organic Geochemistry*, *38*(12), 2070–2097.

Walsh, D. A., & Hallam, S. J. (2011). Bacterial community structure and dynamics in a seasonally anoxic fjord: Saanich Inlet, British Columbia. In F. J. de Burijn (Ed.), *Handbook of molecular microbial ecology II: Metagenomics in different habitats*. Hoboken, New Jersey: Wiley-Blackwell.

Walsh, D. A., Zaikova, E., Howes, C. G., Song, Y. C., Wright, J. J., Tringe, S. G., et al. (2009). Metagenome of a versatile chemolithoautotroph from expanding oceanic dead zones. *Science*, *326*(5952), 578–582.

Walther, T. C., & Mann, M. (2010). Mass spectrometry-based proteomics in cell biology. *The Journal of Cell Biology*, *190*(4), 491–500.

Ward, B. B., Devol, A. H., Rich, J. J., Chang, B. X., Bulow, S. E., Naik, H., et al. (2009). Denitrification as the dominant nitrogen loss process in the Arabian Sea. *Nature*, *461*(7260), 78–81.

Whitney, F., Freeland, H., & Robert, M. (2007). Persistently declining oxygen levels in the interior waters of the eastern subarctic Pacific. *Progress in Oceanography*, *75*(2), 179–199.

Williams, T. J., Long, E., Evans, F., DeMaere, M. Z., Lauro, F. M., Raftery, M. J., et al. (2012). A metaproteomic assessment of winter and summer bacterioplankton from Antarctic Peninsula coastal surface waters. *The ISME Journal*, *6*(10), 1883–1900.

Wilmes, P., Wexler, M., & Bond, P. L. (2008). Metaproteomics provides functional insight into activated sludge wastewater treatment. *PLoS One*, *3*(3), e1778.

Woebken, D., Fuchs, B. M., Kuypers, M. M., & Amann, R. (2007). Potential interactions of particle-associated anammox bacteria with bacterial and archaeal partners in the Namibian upwelling system. *Applied and Environmental Microbiology*, *73*(14), 4648–4657.

Woebken, D., Lam, P., Kuypers, M. M., Naqvi, S. W., Kartal, B., Strous, M., et al. (2008). A microdiversity study of anammox bacteria reveals a novel Candidatus Scalindua phylotype in marine oxygen minimum zones. *Environmental Microbiology*, *10*(11), 3106–3119.

Wright, J. J., Konwar, K. M., & Hallam, S. J. (2012). Microbial ecology of expanding oxygen minimum zones. *Nature Reviews Microbiology*, *10*(6), 381–394.
Zaikova, E., Walsh, D. A., Stilwell, C. P., Mohn, W. W., Tortell, P. D., & Hallam, S. J. (2010). Microbial community dynamics in a seasonally anoxic fjord: Saanich Inlet, British Columbia. *Environmental Microbiology*, *12*(1), 172–191.
Zybailov, B., Mosley, A. L., Sardiu, M. E., Coleman, M. K., Florens, L., & Washburn, M. P. (2006). Statistical analysis of membrane proteome expression changes in *Saccharomyces cerevisiae*. *Journal of Proteome Research*, *5*, 2339–2347.

SECTION V

Microbial Community Informatics: Computational Platforms, Comparative Analyses, and Statistics

CHAPTER SEVENTEEN

Assembling Full-Length rRNA Genes from Short-Read Metagenomic Sequence Datasets Using EMIRGE

Christopher S. Miller[*,1]

[*]Department of Integrative Biology, University of Colorado Denver, Campus Box 171, P.O. Box 173364, Denver, Colorado, USA

[1]Corresponding author: e-mail address: christopher.s.miller@ucdenver.edu

Contents

1. Introduction: The Utilitarian Small Subunit rRNA Gene in Microbial Ecology 334
2. Overview of the EMIRGE algorithm 337
   2.1 Sources of uncertainty in the rRNA assembly problem 338
   2.2 The expectation–maximization algorithm in EMIRGE 338
   2.3 Iterations, algorithm convergence, and stabilization of reconstructed sequences 340
   2.4 Splitting and merging candidate SSU rRNA sequences: algorithmic warts 342
3. EMIRGE Outputs: Reconstructed Sequences with Estimated Abundances 342
4. Practical Considerations When Running EMIRGE 343
   4.1 Strategies for quality control and filtering of input reads 343
   4.2 Running EMIRGE on metagenome shotgun sequencing data 344
   4.3 Running EMIRGE on SSU rRNA gene amplicons or transcriptome data 345
   4.4 Strategies for quality control and filtering of reconstructed sequences 346
5. Choosing a Candidate Database of rRNA Genes 347
6. Conclusions and Future Outlook 348
References 348

Abstract

Microbial ecologists have reaped enormous benefit from advances in high-throughput DNA sequencing. However, the short read lengths of currently dominant technologies have made a seemingly simple question about shotgun metagenomic experiments difficult to answer: what small subunit ribosomal RNA (SSU rRNA) genes are present in a sequenced biological sample? Without these gene sequences, it is difficult to interpret a sample within the rich context of ribosomal rRNA databases accumulated over decades. This chapter presents specialized software, EMIRGE, for the assembly of SSU rRNA genes. EMIRGE is optimized to deal with strain similarity and the fluctuating levels

Methods in Enzymology, Volume 531
ISSN 0076-6879
http://dx.doi.org/10.1016/B978-0-12-407863-5.00017-4

of conservation within the SSU rRNA gene that make assembly difficult. It has been used to successfully assemble genes from shotgun metagenomes, long PCR amplicons, and total-RNA transcriptomes. A detailed discussion of how EMIRGE works and how it deals with the uncertainty inherent in the assembly problem is presented. Practical suggestions are given for understanding and optimizing parameter choice, data preprocessing and postprocessing, and creation of a candidate SSU rRNA gene database. When high-throughput sequencing data are available, EMIRGE can serve as a valuable tool for interpreting microbial community structure.

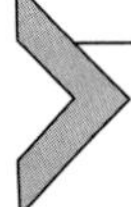

1. INTRODUCTION: THE UTILITARIAN SMALL SUBUNIT rRNA GENE IN MICROBIAL ECOLOGY

For over 35 years, microbial ecologists have used the nucleic acid sequences of ribosomal RNA genes or transcripts to document and explore the seemingly boundless diversity of the planet. Although early studies sometimes used 5S rRNA (Stahl, Lane, Olsen, & Pace, 1984), the small subunit (SSU; 16S) rRNA and its associated gene quickly became the preferred phylogenetic marker owing to several key qualities: (i) SSU genes are universally distributed; (ii) SSU genes are highly conserved, allowing for meaningful alignments and broad PCR priming sites; and (iii) SSU genes contain enough variable nucleotides to be information rich (Tringe & Hugenholtz, 2008). Woese and colleagues used these properties for building quantitative molecular phylogenies of life, shepherding taxonomists through the paradigm shift to a three-domain tree that included archaea (Fox et al., 1980). These early studies laid the foundation for reference databases of ribosomal sequences within which newly sequenced isolates could be interpreted. It was subsequently realized that molecular sequencing of ribosomal genes, eventually with "universal" primers (Lane et al., 1985) targeting conserved SSU regions, allowed for profiling the taxonomic makeup of entire environments, without the prerequisite of isolation in culture (Olsen, Lane, Giovannoni, Pace, & Stahl, 1986; Pace, Stahl, Lane, & Olsen, 1986; Stahl et al., 1984). The usefulness of this approach is self-evident: contemporary databases that catalog partial and full SSU gene sequences today contain well over three million entries (Cole et al., 2007; DeSantis et al., 2006; Pruesse et al., 2007).

A second "revolution" in the use of amplified SSU gene fragments for microbial ecology arrived with the release of massively parallel sequencing technologies. The cloning bottleneck that limited most Sanger-sequencing-based approaches to hundreds or, at most, thousands of clones has given way to inexpensive experiments that sample tens of thousands

(Sogin et al., 2006) to millions (Caporaso et al., 2011) of amplification products per biological sample. This has primarily been achieved by building adapter sequences required for sequencing via 454/Roche (Margulies et al., 2005) or Illumina (Bentley et al., 2008) technology directly into SSU-targeting "universal" primers (Caporaso et al., 2012; Sogin et al., 2006). Library preparation of amplified SSU fragments for these sequencing technologies becomes as simple as setting up a polymerase chain reaction. The ease and cost-effectiveness of this strategy have led to several ambitious projects to catalog microbial diversity, including from the human body (Huse, Ye, Zhou, & Fodor, 2012) and from hundreds of thousands of samples from across the planet (www.earthmicrobiome.org; Gilbert et al., 2010).

Despite its popularity and longevity, the technique of sequencing PCR-amplified SSU fragments has technical complications that have stubbornly persisted through the decades. Quantification of relative taxon abundances (cell numbers) is often inferred from the number of sequencing reads attributed to that taxon. However, variable genome copy number of the ribosomal RNA operon (from 1to 15 copies, but unknown for the vast majority of genomes) (Lee, Bussema, & Schmidt, 2009) can confound this inference (Farrelly, Rainey, & Stackebrandt, 1995). Attempts have been made to correct for this bias by inferring copy number from known genomes and a phylogeny (Kembel, Wu, Eisen, & Green, 2012), but even closely related species can have different rRNA operon copy number. Partly for this reason, arguments have been made for using alternate phylogenetic marker genes (Roux, Enault, Bronner, & Debroas, 2011; Vos, Quince, Pijl, De Hollander, & Kowalchuk, 2012), but these alternatives do not benefit from the vast historical accumulation of SSU reference sequences and associated metadata, nor do they provide the taxonomic breadth of PCR access that is afforded by the highly conserved priming regions in rRNA genes.

The choice of PCR primers to use in small subunit ribosomal RNA (SSU rRNA) studies has effects on the taxonomic specificity and sensitivity of the resulting amplicon. Limits in sequencing technology led to shorter sequenced fragments in early studies (Lane et al., 1985), but increasing Sanger-sequencing read lengths, and the development of PCR approaches, soon allowed for widely adopted primers that amplified the nearly full-length SSU gene (Lane, 1991). For all technologies, unfortunately, even "universal" degenerate primers were shown to have amplification biases that limit the utility of PCR-based methods for relative quantification of some taxa, especially if many cycles of PCR are performed (Fan, McElroy, & Thomas, 2012; Polz & Cavanaugh, 1998). With the advent of new sequencing technology, we have returned to short (though abundant) reads, and several recent studies have reexamined the theoretical taxonomic coverage of various primer pairs, with sometimes conflicting results about the optimal set of primers and read lengths to minimize bias (Hao & Chen, 2012; Liu, DeSantis, Andersen, & Knight, 2008; Ong et al., 2013; Soergel, Dey, Knight, & Brenner, 2012; Walters et al., 2011; Youssef et al., 2009). Chimeric amplification products, representing artificial sequences not present in the sample, can also confuse the interpretation of both richness and relative abundance calculations (Haas et al., 2011; Hugenholtz, 2003). Even with the shorter amplification products favored by high-throughput sequencing studies, a sizable fraction of reads—especially from low-abundance taxa—are from such chimeric artifacts, and chimeras can form reproducibly from reaction to reaction (Haas et al., 2011; Schloss, Gevers, & Westcott, 2011).

Recovering full or partial SSU rRNA genes directly from metagenomic reads or assembled contigs should avoid some of the biases inherent in PCR-based approaches to microbial community phylogenetic characterization. Though not without technical limitations (Morgan, Darling, & Eisen, 2010), sequencing and assembling partial or full genomes via DNA extracted directly from the environment avoids PCR primer bias, amplification bias, copy-number errors, and chimeric artifacts. Ideally, complete assembled genomes could be used to place communities in a taxonomic context. However, with notable exceptions, assembly of near-complete (Denef & Banfield, 2012; Hess et al., 2011; Tyson et al., 2004) or complete (Sharon et al., 2012) microbial genomes from community sequencing is difficult, and the catalog of complete bacterial reference genomes is in relative terms small and biased (Wu et al., 2009).

Many strategies have been proposed to infer community phylogenetic structure from metagenomic reads or contigs, only some of which exclusively rely on the SSU rRNA. When using entire read sets or assemblies, most programs assign taxonomy via sequence homology to known genomes and differ in the way they assess uncertainty. One common approach is to assign the taxonomy of the BLAST-based least common ancestor of all sequences homologous to a metagenomic fragment of interest (Huson, Mitra, Ruscheweyh, Weber, & Schuster, 2011). Another approach uses more sophisticated and sensitive hidden Markov models for homology-based assignment (Brady & Salzberg, 2009). Using just a subset of "clade-specific" genes mined from existing genomes for homology-based comparison has been reported to improve accuracy and computational speed (Segata et al., 2012).

Alternatively, nucleotide composition can be used, either alone or in combination with homology approaches (Liu et al., 2013; MacDonald, Parks, & Beiko, 2012; Parks, MacDonald, & Beiko, 2011).

Because of the breadth and depth of SSU rRNA databases, many strategies have been proposed to mine metagenomic data sets for SSU fragments, and use those fragments for taxonomic characterization. SSU fragments, rather than full-length assembled genes, are often used because modern k-mer-based assembly programs for high-throughput sequencing reads fail to properly assemble the unique alternately conserved and divergent domain structure of the SSU gene (Miller, Baker, Thomas, Singer, & Banfield, 2011). Multiple strategies have been proposed to identify metagenomic reads originating from SSU genes and use these to infer a community's taxonomic profile, via BLAST or placement in a reference tree (Bengtsson et al., 2012; Hao & Chen, 2012; Sharpton et al., 2011). For 454 data, assembled fragments, even if imperfect and/or shorter than full length, have been shown to outperform unassembled reads in taxonomic classification (Fan et al., 2012; Radax et al., 2012).

2. OVERVIEW OF THE EMIRGE ALGORITHM

Our approach to SSU rRNA-based characterization of microbial communities is to assemble near-full-length SSU genes from high-throughput Illumina sequencing data with specialized, open-source software (https://github.com/csmiller/EMIRGE). The input data can be generated via shotgun metagenomics (Gladden et al., 2011; Handley et al., 2013; Miller et al., 2011; Wrighton et al., 2012), targeted sequencing of sheared, near-full-length SSU amplicons (Miller et al., 2013; Ong et al., 2013), or total-RNA metatranscriptomic data. The assembly algorithm, expectation–maximization reconstruction of genes from the environment (EMIRGE), leverages the large number of previously reported SSU rRNA sequences as candidate sequences in a form of template-guided assembly. Over several iterations, EMIRGE maps reads to these candidate sequences, corrects them to reflect the true gene sequences sampled by the reads, and produces accurate estimates of assembled SSU rRNA gene abundances based on a probabilistic accounting of read alignment depth. The remainder of this chapter further describes the underlying algorithm behind EMIRGE, argues for the advantages of maintaining and using the uncertainty inherent in high-throughput sequencing, and discusses several practical issues that arise when using EMIRGE that point out both the strengths and limitations of the approach.

2.1. Sources of uncertainty in the rRNA assembly problem

The fundamental assumption made by EMIRGE is that one candidate reference SSU rRNA gene generated each observed read (here and below, the term "read" is used interchangeably with "paired read," as EMIRGE treats paired reads essentially as single concatenated reads). Put another way, each read is a sampled fragment from exactly one candidate SSU rRNA gene present in the biological sample. If it were certain that all SSU rRNA gene sequences in a biological sample were present in the candidate database, and if each SSU rRNA read could be unambiguously mapped to exactly one candidate, calculating read-depth abundances and correcting any errors in the candidate SSU rRNA sequence would be trivial. However, because in a natural community reads often are sampled from closely related species, and because shorter reads inherently have less information, there is usually uncertainty about exactly which reference sequence generated each read (Fig. 17.1). In addition, reads are not error free. Base-call errors are accompanied by a quality score representing the probability of an erroneous call, but a high error probability only implies uncertainty about the called base, and not certainty about any of the three alternative bases. Thus, base-call errors in reads can also confuse the issue of which reference sequence generated a given read. Of course, in a real sample, the generating SSU rRNA gene may not be present in the candidate database at all, and so there is also uncertainty about the base identity at each position in a candidate sequence. When combined, these sources of uncertainty preclude a simplistic strategy for template-guided assembly and abundance estimations.

2.2. The expectation–maximization algorithm in EMIRGE

The expectation–maximization (EM) algorithm (Dempster, Laird, & Rubin, 1977) allows for the estimation of hidden parameters in a probabilistic model from known observations, including for models of biological phenomena such as gene expression or protein domain interactions (Do & Batzoglou, 2008). In the case of EMIRGE, the observed data at hand are the short high-throughput sequencing reads and the starting database of candidate SSU rRNA genes. Hidden to us are the true sequences and relative abundances of each candidate gene in the biological sample, as well as information about which candidate gene generated each read. If it were certain which

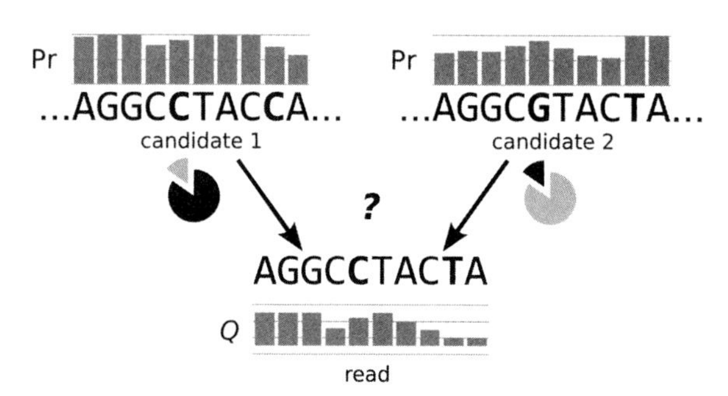

Figure 17.1 Sources of uncertainty in template-guided SSU rRNA assembly. Any given read (bottom) will have uncertain base calls at each position, as indicated by varying Sanger quality values estimated by the sequencing machine (*Q* in lower bar chart). This quality value represents only the probability that the called base is correct and indicates nothing about the individual probabilities of other possible bases being correct. A read will have been generated by exactly one of many possible candidate SSU rRNA genes (top), each of which will have different and unknown relative abundances in the biological sample (pie charts). Reads may not perfectly align with candidate SSU rRNA genes (bold bases). The sequence of each actual SSU rRNA gene in a biological sample is also uncertain and can only be described probabilistically, with the most probable base at each position being chosen for the consensus sequences (Pr in top bar charts). EMIRGE also keeps track of the probability of each of the three other possible bases at each position in a candidate SSU rRNA gene, based on information from aligned reads (not shown). All of these factors make unambiguous mapping and attribution of reads to candidate SSU rRNA genes complex. (For color version of this figure, the reader is referred to the online version of this chapter.)

candidate sequence generated which read, then calculating relative read-depth abundances would be trivial, and we could easily compute a corrected consensus for each candidate gene with mapped reads. If, on the other hand, the correct sequences for candidate genes were known alongside their relative abundances, then we could make an educated decision as to which candidate sequence generated each read.

The EM algorithm proceeds exactly by alternating between these two assumptions. First, a probabilistic guess is calculated describing which candidate SSU rRNA sequences may have generated each SSU rRNA read, under the assumption that something is known about the identity and relative abundance of the candidate sequences. Then this probabilistic description of which sequence generated each read is used to refine what is known about the consensus sequences and relative abundances of the candidate SSU rRNA genes. With multiple flip-flops between these two steps, the probabilities reflecting each of the estimated parameters begin to accurately reflect the most likely explanation for the observed set of all sequencing reads (Fig. 17.2).

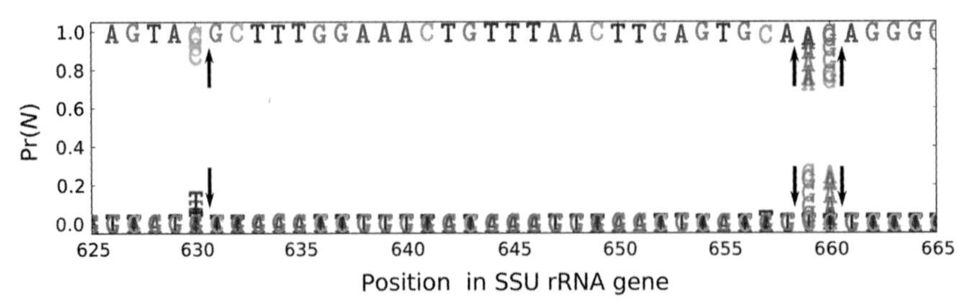

Figure 17.2 Iterative adjustment of individual base probabilities in the consensus sequences of candidate SSU rRNA genes. Shown are the probabilities (Pr(*N*); *y*-axis) for each of four possible bases at each position (*x*-axis) for selected nucleotides of a representative candidate gene. Probabilities for iterations 1–8 are overlaid with increasing opacity. Most positions in the gene are unambiguous: a single base has a probability near one, while the remaining three bases have probabilities near zero, and this does not change from iteration to iteration. For other positions (highlighted with arrows indicating increasing iteration number), EMIRGE gradually adjusts originally less certain bases to reflect increased evidence from confidently mapped reads in later iterations, ultimately revealing a highly probable consensus across all positions. (See color plate.)

2.3. Iterations, algorithm convergence, and stabilization of reconstructed sequences

In each iteration, EMIRGE performs the following steps (Miller et al., 2011), each of which is documented and timed in the output log file:

i. Map reads against the current most probable consensus sequence for each candidate SSU rRNA.

ii. Calculate the new probability of each base at each position in the candidate sequences, based on the current probabilistic distribution describing where each read could have come from, the base calls within those reads, and base quality scores.

iii. Calculate the probability that a given candidate SSU rRNA gene generated a given read, for all possible pairs of candidate genes and reads.

iv. Calculate the estimated abundances of each of the candidate genes, given the just-calculated individual probabilities that each read generated each sequence.

v. Write out a new consensus sequence for each candidate sequence for the next iteration of read mapping, based on the most probable base at each position (Fig. 17.2).

After a number of iterations, EMIRGE's estimates of the various probabilities calculated stabilize to the point where changes become minute. This is algorithm convergence. At the moment, EMIRGE does not automatically detect this, but in practice, this usually ranges from approximately 30 iterations (for simple communities) to 80–120 iterations (for complex communities). One way to observe algorithm convergence is by plotting the number of nucleotide changes made in the consensus candidate sequences over time (Fig. 17.3). Early in the process, the total number of nucleotide changes made in all candidate sequences per iteration is large. In later iterations, if the algorithm has converged, several iterations will pass with no or very few changes being made to the candidate sequences, as each position in each candidate sequence gains a clearly most probable base supported by multiple reads (Fig. 17.2). EMIRGE stores calculations and temporary files for each iteration in a separate directory. Most users will be most interested in EMIRGE's assemblies and abundance estimates after the final iteration, but extracting information about the evolution of given candidate SSU rRNA consensus sequences through the iterations can easily be done by comparing the iter.N.cons.fasta files in each directory, where *N* is the iteration number (see Section 3).

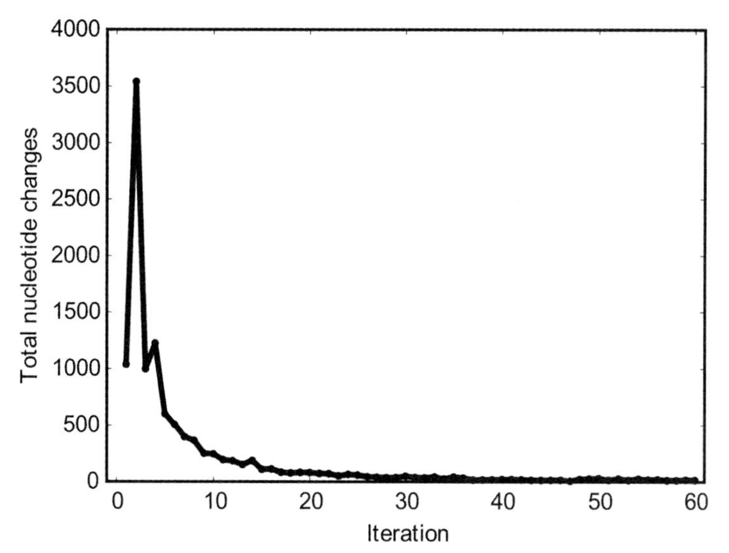

Figure 17.3 Algorithm convergence. EMIRGE was run on a supragingival plaque shotgun metagenomics sample from the Human Microbiome Project (SRS017139; Huttenhower et al., 2012). Plotted for each iteration are the total number of nucleotides in all consensus candidate SSU rRNA genes that were modified from the previous iteration. In early rounds, multiple nucleotide changes are made to consensus sequences as the candidate SSU rRNA genes are adjusted to reflect biological reality in the sample, based on information contained in the reads. In later iterations, very few or no changes are made, as EMIRGE reaches an optimal solution and assembly.

2.4. Splitting and merging candidate SSU rRNA sequences: algorithmic warts

During each iteration, EMIRGE makes two necessary but awkward rule-based decisions that diminish the elegance of the EM algorithm as it is currently implemented. If the consensus sequences of two candidate SSU rRNA sequences evolve over the course of several iterations to be highly similar, then those two sequences are *merged*. If a single sequence accumulates enough high-probability reads that indicate a minor strain with alternate bases at several positions, then it is assumed that the consensus sequence really represents two or more "strains" in the data set (a rarer minor strain is masked by a more abundant strain). That candidate sequence is then *split* into two sequences to allow differential read-mapping in further iterations. Candidate sequences which are minor strains split off in some iteration are marked with a unique identifier (_mXX) in EMIRGE-produced FASTA files. It should be noted that when to split and merge is completely user defined (via the options --snp_fraction_thresh, --variant_fraction_thresh, and --join_threshold), because it is a somewhat arbitrary choice. It may be advantageous with certain use cases to be more or less stringent when merging similar sequences. Currently, with default parameters, EMIRGE attempts to merge two candidate SSU rRNA genes when their most probable consensus sequences in any iteration share $\geq 97\%$ identity, in line with the most common definition of an operational taxonomic unit. Similarly, by default EMIRGE will split a candidate sequence into major and minor strains if more than 4% of positions in the gene show evidence of a minor strain (as secondary bases with significant probability, indicating underlying read support).

3. EMIRGE OUTPUTS: RECONSTRUCTED SEQUENCES WITH ESTIMATED ABUNDANCES

At each iteration, EMIRGE produces files which store the most probable candidate SSU rRNA consensus sequences, an estimate of their relative abundances, and the full probabilistic description of each possible base at each position in the candidate sequences. Most users are interested in the first two items (who is there and at what levels?) after the final iteration. For convenience, a script is provided with EMIRGE to produce a formatted FASTA file that contains abundance information in the header. Example usage on the 80th iteration's directory would be:

```
emirge_rename_fasta.py iter.80 > renamed.fasta
```

The resulting file, renamed.fasta, will be an estimated-abundance-sorted file of consensus sequences. The header consists of a unique internal EMIRGE sequence identifier, the accession of the original candidate sequence used to template assembly, and the estimated relative abundance of that sequence, as the Prior probability calculated by the EM algorithm (Miller et al., 2011). As implemented, the EM calculations assume that all candidate sequences are the same length, and thus any differences in the number of reads generated by each candidate is only related to relative abundance. In practice, candidate sequences can be slightly different lengths (especially, if the community contains longer eukaryotic SSU rRNA genes). Left uncorrected, longer sequences would have slightly overestimated relative abundances, simply because there are more potential bases for reads to be sampled from. The headers for each sequence thus also include a length-corrected estimate of relative abundance labeled "NormPrior." The postprocessing script will optionally remove or mask any bases left over from the original candidate gene sequence that were not supported by sufficient read depth, which can happen with very low abundance sequences with few mapping reads. These sequences with low read support, however, should generally be treated with suspicion, as they have been shown to be less reproducible in bootstrapping experiments (see below and Miller et al., 2013).

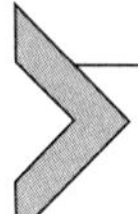

4. PRACTICAL CONSIDERATIONS WHEN RUNNING EMIRGE

EMIRGE is freely available at https://github.com/csmiller/EMIRGE and runs on any linux-based operating system. Detailed installation and usage instructions are distributed with the code. In the following sections, additional practical advice is given with considerations about how to run EMIRGE under various use cases.

4.1. Strategies for quality control and filtering of input reads

EMIRGE's underlying read mapper, bowtie (Langmead, Trapnell, Pop, & Salzberg, 2009), is parameterized to allow for multiple mismatches in alignments, mapping reads to candidate SSU rRNA sequences that may not be reflective of what actually exists in a biological sample. However, if entire sections of reads are unreliable, those reads will not be mapped. Thus, it is recommended that reads undergo some minimal quality trimming before being input into EMIRGE. In practice, it is sufficient to trim off bases from the 3′ end of reads until a base with a minimum PHRED quality of 3 is encountered (Wrighton et al., 2012). If one or both reads in a pair are shorter than some minimum threshold after trimming (e.g., 60 bp), that read pair is discarded.

Because roughly 90% of run time on metagenomic or amplicon datasets is spent in the read mapping step, which scales linearly with input data size, one strategy to reduce run time is to prefilter reads to select a high-quality representative subsample for EMIRGE. For example, by selecting the 100,000 amplicon reads with the highest average quality scores per sample for EMIRGE, Ong et al. (2013) reconstructed sequences from defined communities with high genus- and species-level precision and recall for sequences >0.1% estimated abundance. For shotgun metagenomic data of bacterial communities, less than 0.25% of all reads are typically from an SSU rRNA gene. Identifying and selecting *a priori* which reads are from the SSU rRNA gene (via an initial homology search) can vastly speed EMIRGE run times, as the 99.75% of reads not aligning to the candidate database with each iteration are not searched for mappings. However, it should be noted that EMIRGE tends to gradually recruit additional SSU rRNA reads with each iteration, presumably as candidate SSU rRNA consensus sequences are corrected to reflect the reality defined by the sample. These reads may be especially relevant for genes more distant from those in the starting candidate database. A strategy that exclusively prefilters for SSU rRNA reads with low sensitivity may exclude some of the informative data. No matter if read prefiltering is done or not, select steps in each iteration (most crucially the time-consuming read mapping step) can be carried out concurrently on multiprocessor machines, as specified with the `--processors` option.

4.2. Running EMIRGE on metagenome shotgun sequencing data

EMIRGE was originally designed to reconstruct SSU rRNA genes from shotgun metagenomic data (Gladden et al., 2011; Miller et al., 2011). The script provided with EMIRGE for this use case is emirge.py. This version of the program is optimized for the situation where the number of non-SSU rRNA reads is much larger than the number of SSU rRNA reads, and typically runs in <1 GB of RAM. Besides files containing the (paired) reads and files containing the candidate database of SSU rRNA genes, the user is also required to provide an estimate of the insert size mean and standard deviation. These values are used to constrain the mapping of paired reads, and can be approximate, or learned from an initial test mapping.

For each read, a practical decision is made to set the probability that a candidate sequence generated that read to zero if there is no alignment. This allows the data structure storing these probabilities to be a sparse matrix, rather than an expensive N read by M candidate sequence matrix (1.5×10^{11} matrix cells for a typical amplicon experiment). Similarly, EMIRGE chooses to discard candidate SSU rRNA sequences that do not have a minimum expected number of reads attributed to them. Choosing this parameter (--min_depth) represents a tradeoff between algorithm speed (consider fewer possible candidates) and potential sensitivity (keep sequences with low coverage with the hopes that more reads are recruited in later algorithm iterations). By default, sequences with $<3\times$ expected coverage depth are discarded.

4.3. Running EMIRGE on SSU rRNA gene amplicons or transcriptome data

EMIRGE has also been used to reconstruct near-full-length SSU rRNA genes from sheared long PCR amplicons. This strategy is in contrast to those that produce sequencing-ready libraries via direct incorporation of adapters into PCR primers targeting selected short hypervariable regions (e.g., Caporaso et al., 2011; Sogin et al., 2006). Two sets of broad-range SSU rRNA primers have been reported for generating amplicons for EMIRGE assembly: 27F/1492R (Miller et al., 2013) and 338F/1061R (Ong et al., 2013). With standard primers and library preparation, there is a preference for sequencing reads to initiate at amplicon ends (Miller et al., 2013), probably due to the increased availability of free ligation targets during library preparation (Harismendy & Frazer, 2009). This ligation-mediated bias can be reduced or removed by using either transposon-based library preparation (Parkinson et al., 2012) or by using standard library preparation with products amplified with amino-modified primers blocked on their 5′ end by an NH_2 group (unpublished data).

The script provided with EMIRGE for the use case of amplicon sequencing is emirge_amplicon.py. This version of the program is optimized for the situation where all reads are expected to be generated from SSU rRNA genes. The script pre-caches all reads and associated quality scores in memory to shorten computation time, for the use case of having a limited (currently, a few million) amplicon-generated reads per sample. The RAM footprint for an emirge_amplicon.py run of a few million reads is typically $\approx$5 GB. This script may also be appropriate for subsamples of reads generated from total RNA, if the expected number of SSU rRNA reads in the dataset is appropriate. The same read-filtering strategies discussed in Section 4.1 for shotgun metagenomic data also shorten execution time for transcriptome reads generated from total or rRNA-depleted RNA.

4.4. Strategies for quality control and filtering of reconstructed sequences

As discussed earlier, the outputs of EMIRGE are near-full-length reconstructed SSU rRNA gene sequences and their estimated relative abundances. As with any experimental protocol generating SSU rRNA sequences (and especially with PCR-generated data), basic quality control and filtering of the output is good practice. EMIRGE-generated chimeras can be reported in amplicon experiments if two conditions are satisfied: (i) the polymerase chain reaction generated a chimera and (ii) an appropriate chimeric candidate template SSU rRNA sequence is present in the reference database. In practice, sequences reconstructed by EMIRGE from metagenomic data appear to contain few chimeras (Gladden et al., 2011; Handley et al., 2013; Miller et al., 2011), but sequences reconstructed by EMIRGE from amplicon studies can contain as much as 40% chimeras (unpublished data). Thus, as with traditional clone libraries, filtering potential chimeric sequences with appropriate software is crucial (Haas et al., 2011; Schloss et al., 2011).

It is also necessary to remove low-abundance sequences generated by EMIRGE that are unlikely to contain sufficient actual coverage to support accurate consensus sequence reconstruction. Although EMIRGE only calculates probabilistic attributions of any read to any candidate SSU rRNA gene sequence, the expected coverage depth of any sequence can be estimated via a simple calculation (Miller et al., 2013). If N reads of length L are mapped to a SSU rRNA gene of length G with roughly uniform length coverage, average depth coverage (C) can be calculated in a standard way

$$C = \frac{L * N}{G}.$$

The only variable not readily available is the number of mapped reads, N. However, the expected number of reads for any sequence can be calculated by simply multiplying the total number of successfully mapped SSU reads by the EMIRGE estimate of abundance. As an example, for a dataset with 100,000 total mapped 100 bp reads, for a reconstructed near-full-length gene of length 1500 bp, a gene sequence with an estimated abundance of 0.1% would have an expected $0.001 \times 100{,}000 = 100$ reads sampled from it, for an estimated depth coverage of $(100 \times 100)/1500 \approx 6.7\times$. Low coverage depth can manifest itself as incomplete sampling of reads across the length of a reconstructed sequence. Bases with no mapped reads are impossible to probabilistically assign. The postprocessing script emirge_rename_fasta.py (see Section 3) has the option of filtering out final EMIRGE sequences below a certain probability/estimated abundance and of marking regions of reconstructed sequences with no read support (as *N* bases) for later removal. Both of these strategies have been employed to remove low abundance sequences, which have been shown to be less accurately reconstructed than high abundance sequences in bootstrapping experiments (Miller et al., 2013).

5. CHOOSING A CANDIDATE DATABASE OF rRNA GENES

EMIRGE has primarily been used with a filtered, clustered version of the SILVA SSU database (Pruesse et al., 2007) for candidate template sequences. SILVA includes sequences from all three domains of life, making metagenomic EMIRGE runs with this database especially useful. However, EMIRGE will run with any starting database of templates (including, in theory, non-SSU rRNA gene templates), and there may be an advantage with exclusively bacterial and archaeal communities or amplicons to using a database with more careful chimera screening, such as the greengenes core database (DeSantis et al., 2006). Preparing SILVA or any starting database for EMIRGE is straightforward. First, abnormally short (<1200 nt) or long (>2000 nt) sequences are removed to target only near-full-length sequences. Next, sequences are normalized to contain only letters within the standard set of {A,C,T,G}. This is a requirement of the read mapper. A script is distributed with EMIRGE to do this: fix_nonstandard_chars.py. Replacement bases within {A,C,T,G} are chosen at random within the set specified by the ambiguous IUPAC code; erroneous bases introduced by this procedure are quickly corrected by the sequencing reads after a handful of iterations. Candidate sequences are then clustered with USEARCH/UCLUST (Edgar, 2010) to remove highly similar SSU rRNA genes, typically at ≥97% identity. Finally, a bowtie (Langmead et al., 2009) index of the remaining SSU rRNA genes is preconstructed for read mapping with parameters optimized for speed (--offrate=1). Clustering the candidate database at other identity levels may be beneficial. If candidate sequences are *too* divergent from genes in a biological sample (e.g., with novel candidate divisions), then EMIRGE can struggle to accurately reconstruct SSU rRNA genes with sufficient taxonomic detail, although with simulated data candidate sequences <90% identical to correct sequences regularly served to template correctly assembled sequences (Miller et al., 2011). A more aggressively clustered candidate database will shorten EMIRGE's computation time.

6. CONCLUSIONS AND FUTURE OUTLOOK

As sequencing technology continues to rapidly change, a need for specialized software for specialized analysis tasks will remain. EMIRGE is one such piece of software, targeting the narrowly defined but difficult challenge of assembly of SSU rRNA genes from short, error-prone sequencing reads. Although inexpensive newer technologies with long reads and low error rates will some day obviate the need for such software, for the moment the most widely used sequencing technologies coupled to the most widely used assembly algorithms do a poor job of assembling these genes. EMIRGE will continue to be useful to reconstruct full-length SSU genes from metagenomic datasets (Gladden et al., 2011; Handley et al., 2013; Miller et al., 2011; Wrighton et al., 2012), but the most powerful use of the software may be to assemble long PCR amplicons, both of the SSU rRNA gene (Miller et al., 2013; Ong et al., 2013; Wilkins et al., 2009) as well as other targeted genes of interest. The general probabilistic model underlying EMIRGE should extend to any gene of interest with suitable primers and a sizable reference database of candidate template sequences.

REFERENCES

Bengtsson, J., Hartmann, M., Unterseher, M., Vaishampayan, P., Abarenkov, K., Durso, L., et al. (2012). Megraft: A software package to graft ribosomal small subunit (16S/18S) fragments onto full-length sequences for accurate species richness and sequencing depth analysis in pyrosequencing-length metagenomes and similar environmental datasets. *Research in Microbiology*, *163*(6–7), 407–412. http://dx.doi.org/10.1016/j.resmic.2012.07.001.

Bentley, D. R., Balasubramanian, S., Swerdlow, H. P., Smith, G. P., Milton, J., Brown, C. G., et al. (2008). Accurate whole human genome sequencing using reversible terminator chemistry. *Nature*, *456*(7218), 53–59. http://dx.doi.org/10.1038/nature07517.

Brady, A., & Salzberg, S. L. (2009). Phymm and PhymmBL: Metagenomic phylogenetic classification with interpolated Markov models. *Nature Methods*, *6*(9), 673–676. http://dx.doi.org/10.1038/nmeth.1358.

Caporaso, J. G., Lauber, C. L., Walters, W. A., Berg-Lyons, D., Huntley, J., Fierer, N., et al. (2012). Ultra-high-throughput microbial community analysis on the Illumina HiSeq and MiSeq platforms. *The ISME Journal*, *6*(8), 1621–1624. http://dx.doi.org/10.1038/ismej.2012.8.

Caporaso, J. G., Lauber, C. L., Walters, W. A., Berg-Lyons, D., Lozupone, C. A., Turnbaugh, P. J., et al. (2011). Global patterns of 16S rRNA diversity at a depth of millions of sequences per sample. *Proceedings of the National Academy of Sciences of the United States of America*, *108*(Suppl.), 4516–4522. http://dx.doi.org/10.1073/pnas.1000080107.

Cole, J. R., Chai, B., Farris, R. J., Wang, Q., Kulam-Syed-Mohideen, A. S., McGarrell, D. M., et al. (2007). The ribosomal database project (RDP-II): Introducing myRDP space and quality controlled public data. *Nucleic Acids Research*, (35 Database), D169–D172.

Dempster, A. P., Laird, N. M., & Rubin, D. B. (1977). Maximum likelihood from incomplete data via the EM algorithm. *Journal of the Royal Statistical Society. Series B (Methodological)*, *39*(1), 1–38.

Denef, V. J., & Banfield, J. F. (2012). In situ evolutionary rate measurements show ecological success of recently emerged bacterial hybrids. *Science*, *336*(6080), 462–466. http://dx.doi.org/10.1126/science.1218389.

DeSantis, T. Z., Hugenholtz, P., Larsen, N., Rojas, M., Brodie, E. L., Keller, K., et al. (2006). Greengenes, a chimera-checked 16S rRNA gene database and workbench compatible with ARB. *Applied and Environmental Microbiology*, *72*(7), 5069–5072. http://dx.doi.org/10.1128/AEM.03006-05.

Do, C. B., & Batzoglou, S. (2008). What is the expectation maximization algorithm? *Nature Biotechnology*, *26*(8), 897–899. http://dx.doi.org/10.1038/nbt1406.

Edgar, R. C. (2010). Search and clustering orders of magnitude faster than BLAST. *Bioinformatics (Oxford, England)*, *26*(19), 2460–2461. http://dx.doi.org/10.1093/bioinformatics/btq461.

Fan, L., McElroy, K., & Thomas, T. (2012). Reconstruction of ribosomal RNA genes from metagenomic data. *PLoS One*, 7(6), e39948. http://dx.doi.org/10.1371/journal.pone.0039948.

Farrelly, V., Rainey, F. A., & Stackebrandt, E. (1995). Effect of genome size and rrn gene copy number on PCR amplification of 16S rRNA genes from a mixture of bacterial species. *Applied and Environmental Microbiology*, *61*(7), 2798–2801.

Fox, G. E., Stackebrandt, E., Hespell, R. B., Gibson, J., Maniloff, J., Dyer, T. A., et al. (1980). The phylogeny of prokaryotes. *Science*, *209*(4455), 457–463.

Gilbert, J. A., Meyer, F., Antonopoulos, D., Balaji, P., Brown, C. T., Brown, C. T., et al. (2010). Meeting report: The terabase metagenomics workshop and the vision of an Earth microbiome project. *Standards in Genomic Sciences*, *3*(3), 243–248. http://dx.doi.org/10.4056/sigs.1433550.

Gladden, J. M., Allgaier, M., Miller, C. S., Hazen, T. C., VanderGheynst, J. S., Hugenholtz, P., et al. (2011). Glycoside hydrolase activities of thermophilic bacterial consortia adapted to switchgrass. *Applied and Environmental Microbiology*, 77(16), 5804–5812. http://dx.doi.org/10.1128/AEM.00032-11.

Haas, B. J., Gevers, D., Earl, A. M., Feldgarden, M., Ward, D. V., Giannoukos, G., et al. (2011). Chimeric 16S rRNA sequence formation and detection in Sanger and 454-pyrosequenced PCR amplicons. *Genome Research*, *21*(3), 494–504. http://dx.doi.org/10.1101/gr.112730.110.

Handley, K. M., Verberkmoes, N. C., Steefel, C. I., Williams, K. H., Sharon, I., Miller, C. S., et al. (2013). Biostimulation induces syntrophic interactions that impact C, S and N cycling in a sediment microbial community. *The ISME Journal*, 7, 800–816. http://dx.doi.org/10.1038/ismej.2012.148.

Hao, X., & Chen, T. (2012). OTU analysis using metagenomic shotgun sequencing data. *PLoS One*, 7(11), e49785. http://dx.doi.org/10.1371/journal.pone.0049785.

Harismendy, O., & Frazer, K. (2009). Method for improving sequence coverage uniformity of targeted genomic intervals amplified by LR-PCR using Illumina GA sequencing-by-synthesis technology. *BioTechniques*, *46*(3), 229–231. http://dx.doi.org/10.2144/000113082.

Hess, M., Sczyrba, A., Egan, R., Kim, T.-W., Chokhawala, H., Schroth, G., et al. (2011). Metagenomic discovery of biomass-degrading genes and genomes from cow rumen. *Science*, *331*(6016), 463–467. http://dx.doi.org/10.1126/science.1200387.

Hugenholtz, P. (2003). Chimeric 16S rDNA sequences of diverse origin are accumulating in the public databases. *International Journal of Systematic and Evolutionary Microbiology*, *53*(1), 289–293. http://dx.doi.org/10.1099/ijs.0.02441-0.

Huse, S. M., Ye, Y., Zhou, Y., & Fodor, A. A. (2012). A core human microbiome as viewed through 16S rRNA sequence clusters. *PLoS One*, 7(6), e34242. http://dx.doi.org/10.1371/journal.pone.0034242.

Huson, D. H., Mitra, S., Ruscheweyh, H.-J., Weber, N., & Schuster, S. C. (2011). Integrative analysis of environmental sequences using MEGAN4. *Genome Research*, *21*(9), 1552–1560. http://dx.doi.org/10.1101/gr.120618.111.

Huttenhower, C., Gevers, D., Knight, R., Abubucker, S., Badger, J. H., Chinwalla, A. T., et al. (2012). Structure, function and diversity of the healthy human microbiome. *Nature*, *486*(7402), 207–214. http://dx.doi.org/10.1038/nature11234.

Kembel, S. W., Wu, M., Eisen, J. A., & Green, J. L. (2012). Incorporating 16S gene copy number information improves estimates of microbial diversity and abundance. *PLoS Computational Biology*, *8*(10), e1002743. http://dx.doi.org/10.1371/journal.pcbi.1002743.

Lane, D. J., Pace, B., Olsen, G. J., Stahl, D. A., Sogin, M. L., & Pace, N. R. (1985). Rapid determination of 16S ribosomal RNA sequences for phylogenetic analyses. *Proceedings of the National Academy of Sciences of the United States of America*, *82*(20), 6955–6959. http://dx.doi.org/10.1073/pnas.82.20.6955.

Lane, D. J. (1991). 16S/23S rRNA sequencing. In E. Stackebrandt & M. Goodfellow (Eds.), *Nucleic Acid Techniques in Bacterial Systematics* (pp. 115–175). New York: Wiley.

Langmead, B., Trapnell, C., Pop, M., & Salzberg, S. L. (2009). Ultrafast and memory-efficient alignment of short DNA sequences to the human genome. *Genome Biology*, *10*(3), R25. http://dx.doi.org/10.1186/gb-2009-10-3-r25.

Lee, Z. M.-P., Bussema, C., & Schmidt, T. M. (2009). rrnDB: Documenting the number of rRNA and tRNA genes in bacteria and archaea. *Nucleic Acids Research*, *37*(Database issue), D489–D493. http://dx.doi.org/10.1093/nar/gkn689.

Liu, J., Wang, H., Yang, H., Zhang, Y., Wang, J., Zhao, F., & Qi, J. (2013). Composition-based classification of short metagenomic sequences elucidates the landscapes of taxonomic and functional enrichment of microorganisms. *Nucleic Acids Research*, *41*(1), e3. http://dx.doi.org/10.1093/nar/gks828.

Liu, Z., DeSantis, T. Z., Andersen, G. L., & Knight, R. (2008). Accurate taxonomy assignments from 16S rRNA sequences produced by highly parallel pyrosequencers. *Nucleic Acids Research*, *36*(18), e120. http://dx.doi.org/10.1093/nar/gkn491.

MacDonald, N. J., Parks, D. H., & Beiko, R. G. (2012). Rapid identification of high-confidence taxonomic assignments for metagenomic data. *Nucleic Acids Research*, *40*(14), e111. http://dx.doi.org/10.1093/nar/gks335.

Margulies, M., Egholm, M., Altman, W. E., Attiya, S., Bader, J. S., Bemben, L. A., & Chen, Z. (2005). Genome sequencing in microfabricated high-density picolitre reactors. *Nature*, *437*(7057), 376–380. http://dx.doi.org/10.1038/nature03959.

Miller, C. S., Baker, B. J., Thomas, B. C., Singer, S. W., & Banfield, J. F. (2011). EMIRGE: Reconstruction of full-length ribosomal genes from microbial community short read sequencing data. *Genome Biology*, *12*(5), R44. http://dx.doi.org/10.1186/gb-2011-12-5-r44.

Miller, C. S., Handley, K. M., Wrighton, K. C., Frischkorn, K. R., Thomas, B. C., & Banfield, J. F. (2013). Short-read assembly of full-length 16S amplicons reveals bacterial diversity in subsurface sediments. *PLoS One*, *8*(2), e56018. http://dx.doi.org/10.1371/journal.pone.0056018.

Morgan, J. L., Darling, A. E., & Eisen, J. A. (2010). Metagenomic sequencing of an in vitro simulated microbial community. *PLoS One*, *5*(4), e10209.

Olsen, G. J., Lane, D. J., Giovannoni, S. J., Pace, N. R., & Stahl, D. A. (1986). Microbial ecology and evolution: A ribosomal RNA approach. *Annual Review of Microbiology*, *40*, 337–365. http://dx.doi.org/10.1146/annurev.mi.40.100186.002005.

Ong, S. H., Kukkillaya, V. U., Wilm, A., Lay, C., Ho, E. X. P., Low, L., et al. (2013). Species identification and profiling of complex microbial communities using shotgun illumina sequencing of 16s rRNA amplicon sequences. *PLoS One*, *8*(4), e60811. http://dx.doi.org/10.1371/journal.pone.0060811.

Pace, N. R., Stahl, D. A., Lane, D. J., & Olsen, G. J. (1986). The analysis of natural microbial populations by ribosomal RNA sequences. *Advances in Microbial Ecology*, *9*, 1–55.

Parkinson, N. J., Maslau, S., Ferneyhough, B., Zhang, G., Gregory, L., Buck, D., et al. (2012). Preparation of high-quality next-generation sequencing libraries from picogram quantities of target DNA. *Genome Research*, *22*(1), 125–133. http://dx.doi.org/10.1101/gr.124016.111.

Parks, D. H., MacDonald, N. J., & Beiko, R. G. (2011). Classifying short genomic fragments from novel lineages using composition and homology. *BMC Bioinformatics*, *12*(1), 328. http://dx.doi.org/10.1186/1471-2105-12-328.

Polz, M. F., & Cavanaugh, C. M. (1998). Bias in template-to-product ratios in multitemplate PCR. *Applied and Environmental Microbiology*, *64*(10), 3724–3730.

Pruesse, E., Quast, C., Knittel, K., Fuchs, B. M., Ludwig, W., Peplies, J., et al. (2007). SILVA: A comprehensive online resource for quality checked and aligned ribosomal RNA sequence data compatible with ARB. *Nucleic Acids Research*, *35*(21), 7188–7196. http://dx.doi.org/10.1093/nar/gkm864.

Radax, R., Rattei, T., Lanzen, A., Bayer, C., Rapp, H. T., Urich, T., & Schleper, C. (2012). Metatranscriptomics of the marine sponge Geodia barretti: Tackling phylogeny and function of its microbial community. *Environmental Microbiology*, *14*(5), 1308–1324. http://dx.doi.org/10.1111/j.1462-2920.2012.02714.x.

Roux, S., Enault, F., Bronner, G., & Debroas, D. (2011). Comparison of 16S rRNA and protein-coding genes as molecular markers for assessing microbial diversity (Bacteria and Archaea) in ecosystems. *FEMS Microbiology Ecology*, *78*(3), 617–628. http://dx.doi.org/10.1111/j.1574-6941.2011.01190.x.

Schloss, P. D., Gevers, D., & Westcott, S. L. (2011). Reducing the effects of PCR amplification and sequencing artifacts on 16S rRNA-based studies. *PLoS One*, *6*(12), e27310. http://dx.doi.org/10.1371/journal.pone.0027310.

Segata, N., Waldron, L., Ballarini, A., Narasimhan, V., Jousson, O., & Huttenhower, C. (2012). Metagenomic microbial community profiling using unique clade-specific marker genes. *Nature Methods*, *9*(8), 811–814. http://dx.doi.org/10.1038/nmeth.2066.

Sharon, I., Morowitz, M. J., Thomas, B. C., Costello, E. K., Relman, D. A., & Banfield, J. F. (2012). Time series community genomics analysis reveals rapid shifts in bacterial species, strains, and phage during infant gut colonization. *Genome Research*, *23*(1), 111–120. http://dx.doi.org/10.1101/gr.142315.112.

Sharpton, T. J., Riesenfeld, S. J., Kembel, S. W., Ladau, J., O'Dwyer, J. P., Green, J. L., et al. (2011). PhylOTU: A high-throughput procedure quantifies microbial community diversity and resolves novel taxa from metagenomic data. *PLoS Computational Biology*, 7(1), e1001061. http://dx.doi.org/10.1371/journal.pcbi.1001061.

Soergel, D. A. W., Dey, N., Knight, R., & Brenner, S. E. (2012). Selection of primers for optimal taxonomic classification of environmental 16S rRNA gene sequences. *The ISME Journal*, *6*(7), 1440–1444. http://dx.doi.org/10.1038/ismej.2011.208.

Sogin, M. L., Morrison, H. G., Huber, J. A., Welch, D. M., Huse, S. M., Neal, P. R., et al. (2006). Microbial diversity in the deep sea and the underexplored "rare biosphere" *Proceedings of the National Academy of Sciences*, *103*(32), 12115–12120. http://dx.doi.org/10.1073/pnas.0605127103.

Stahl, D. A., Lane, D. J., Olsen, G. J., & Pace, N. R. (1984). Analysis of hydrothermal vent-associated symbionts by ribosomal RNA sequences. *Science*, *224*(4647), 409–411. http://dx.doi.org/10.1126/science.224.4647.409.

Tringe, S. G., & Hugenholtz, P. (2008). A renaissance for the pioneering 16S rRNA gene. *Current Opinion in Microbiology*, *11*(5), 442–446. http://dx.doi.org/10.1016/j.mib.2008.09.011.

Tyson, G. W., Chapman, J., Hugenholtz, P., Allen, E. E., Ram, R. J., Richardson, P. M., et al. (2004). Community structure and metabolism through reconstruction of microbial genomes from the environment. *Nature*, *428*(6978), 37–43. http://dx.doi.org/10.1038/nature02340.

Vos, M., Quince, C., Pijl, A. S., De Hollander, M., & Kowalchuk, G. A. (2012). A comparison of rpoB and 16S rRNA as markers in pyrosequencing studies of bacterial diversity. *PLoS One*, 7(2), e30600. http://dx.doi.org/10.1371/journal.pone.0030600.

Walters, W. A., Caporaso, J. G., Lauber, C. L., Berg-Lyons, D., Fierer, N., & Knight, R. (2011). PrimerProspector: De novo design and taxonomic analysis of barcoded polymerase chain reaction primers. *Bioinformatics*, *27*(8), 1159–1161. http://dx.doi.org/10.1093/bioinformatics/btr087.

Wilkins, M. J., Verberkmoes, N. C., Williams, K. H., Callister, S. J., Mouser, P. J., Elifantz, H., et al. (2009). Proteogenomic monitoring of Geobacter physiology during stimulated uranium bioremediation. *Applied and Environmental Microbiology*, *75*(20), 6591–6599. http://dx.doi.org/10.1128/AEM.01064-09.

Wrighton, K. C., Thomas, B. C., Sharon, I., Miller, C. S., Castelle, C. J., VerBerkmoes, N. C., et al. (2012). Fermentation, hydrogen, and sulfur metabolism in multiple uncultivated bacterial phyla. *Science*, *337*(6102), 1661–1665. http://dx.doi.org/10.1126/science.1224041.

Wu, D., Hugenholtz, P., Mavromatis, K., Pukall, R., Dalin, E., Ivanova, N. N., et al. (2009). A phylogeny-driven genomic encyclopaedia of Bacteria and Archaea. *Nature*, *462*(7276), 1056–1060. http://dx.doi.org/10.1038/nature08656.

Youssef, N., Sheik, C. S., Krumholz, L. R., Najar, F. Z., Roe, B. A., & Elshahed, M. S. (2009). Comparison of species richness estimates obtained using nearly complete fragments and simulated pyrosequencing-generated fragments in 16S rRNA gene-based environmental surveys. *Applied and Environmental Microbiology*, *75*(16), 5227–5236. http://dx.doi.org/10.1128/AEM.00592-09.

CHAPTER EIGHTEEN

Computational Methods for High-Throughput Comparative Analyses of Natural Microbial Communities

Sarah P. Preheim[*], Allison R. Perrotta[†], Jonathan Friedman[‡,§], Chris Smilie[§], Ilana Brito[*], Mark B. Smith[¶], Eric Alm[*,1]

[*]Biological Engineering, Massachusetts Institute of Technology, Cambridge, Massachusetts, USA
[†]Civil and Environmental Engineering, Massachusetts Institute of Technology, Cambridge, Massachusetts, USA
[‡]Physics, Massachusetts Institute of Technology, Cambridge, Massachusetts, USA
[§]Computational and Systems Biology, Massachusetts Institute of Technology, Cambridge, Massachusetts, USA
[¶]Microbiology, Massachusetts Institute of Technology, Cambridge, Massachusetts, USA
[1]Corresponding author: e-mail address: ejalm@mit.edu

Contents

1. Introduction 354
2. Sequencing Terminology 355
3. Sequence Processing 356
 3.1 Quality filtering, dereplication, and trimming of sequences 357
 3.2 Creating OTUs 360
 3.3 Artifact removal and classification of OTUs 361
4. Community Analysis 362
 4.1 Diversity estimates 363
 4.2 Community distance metrics 364
 4.3 Associations of OTUs with metadata 365
 4.4 Correlation analysis 367
5. Summary 368
References 368

Abstract

One of the most widely employed methods in metagenomics is the amplification and sequencing of the highly conserved ribosomal RNA (rRNA) genes from organisms in complex microbial communities. rRNA surveys, typically using the 16S rRNA gene for prokaryotic identification, provide information about the total diversity and taxonomic affiliation of organisms present in a sample. Greatly enhanced by high-throughput sequencing, these surveys have uncovered the remarkable diversity of uncultured organisms and revealed unappreciated ecological roles ranging from nutrient cycling

Methods in Enzymology, Volume 531
ISSN 0076-6879
http://dx.doi.org/10.1016/B978-0-12-407863-5.00018-6

to human health. This chapter outlines the best practices for comparative analyses of microbial community surveys. We explain how to transform raw data into meaningful units for further analysis and discuss how to calculate sample diversity and community distance metrics. Finally, we outline how to find associations of species with specific metadata and true correlations between species from compositional data. We focus on data generated by next-generation sequencing platforms, using the Illumina platform as a test case, because of its widespread use especially among researchers just entering the field.

1. INTRODUCTION

Prokaryotic cells make up the majority of biomass on the planet (Whitman, Coleman, & Wiebe, 1998) and influence every ecosystem from the world's oceans to the human gut. Metagenomics approaches, including amplicon-based techniques targeting the conserved small subunit ribosomal RNA genes (commonly 16S rRNA in prokaryotes), facilitate research of the structure, function, and stability of microbial communities (Amann, Ludwig, & Schleifer, 1995; Pace, 1997; Woese, Kandler, & Wheelis, 1990).

Although 16S rRNA surveys of microbial communities are widely used to characterize the composition and diversity of microorganisms present in a sample, there are many problems associated with transforming 16S rRNA sequences into proxies for species, given the ambiguity of bacterial species. Additionally, although there is no consensus for what constitutes a microbial species, the operational taxonomic unit (OTU) is a widely used construct of clustered sequence data that approximates "species" in subsequent analysis steps. OTUs are typically formed by grouping together sequences that are within a defined genetic cut-off, even though such grouping does not accurately reflect current opinion about microbial species (Gevers et al., 2005), typically overestimates sample diversity (Huse, Welch, Morrison, & Sogin, 2010), and inappropriately assumes that diverse organisms evolve at similar rates. Single base changes and chimeras can create artificial diversity during both PCR and sequencing (Qiu et al., 2001; Zhou et al., 2011). Furthermore, prior to sequencing, the preparation of samples incorporates many known biases due to differential access to microbial DNA for amplification (Forney, Zhou, & Brown, 2004). Steps should be taken to reduce these errors and biases whenever possible—from library preparation to computational processing and analysis.

Despite these limitations, 16S rRNA surveys have emerged as the most popular approach to study microbial communities, largely due to the speed, convenience, and low cost of this analysis as a result of next-generation sequencing technologies. These technologies allow researchers to compare hundreds of community profiles marked with molecular barcodes in a single sequencing run. As such, rRNA surveys have been used to complete comprehensive body site sampling in healthy adults through the Human Microbiome project (Huttenhower, Gevers, Knight, Abubucker, Badger, Chinwalla et al., 2012) and to sequence hundreds of thousands of environmental samples from across the world through the Earth Microbiome project (http://www.earthmicrobiome.org/), as well as countless other projects characterizing microbial community variation in space or time.

This chapter outlines the bioinformatics approaches used to compare microbial communities using high-throughput sequencing technologies, focusing on data generated by the Illumina sequencing platform (San Francisco, CA). We use an example of a 16S rRNA survey generated using Illumina sequencing, although many of the principles are relevant to other platforms. We focus on the following topics in the comparative analyses of microbial communities:

- Background
- Sequence processing
- Comparative analyses

2. SEQUENCING TERMINOLOGY

Identifying microorganisms present in a natural community using a sequencing-based 16S rRNA survey begins with the construction of a library. A library is a collection of DNA fragments that represents the sequence diversity in a sample. These fragments are enriched from the rest of the community genomic DNA by PCR using primers which match the microbial population or gene of interest. For Illumina sequencing, the complete molecular construct contains the amplified genomic DNA, sequences that identify the sample it originated from (i.e., index or barcode sequences) and sequences required by the platform to adhere library fragments to the solid matrix and provide a priming site for the sequencing reaction. Adding a barcode sequence to the molecular construct identifying which sample the library originated from allows for hundreds of libraries to be sequenced in the same reaction, commonly called multiplexing. Illumina offers paired-end sequencing, which provides a sequence for the forward and reverse strands of a template, enabling resequencing of very short template sequences for improved accuracy, or longer effective reads with enhanced positional information for longer template sequences.

3. SEQUENCE PROCESSING

Methodological artifacts must be minimized through quality filtering, while computational analyses require sequences to be organized into appropriate groups, and biological interpretation is facilitated by assignment of a meaningful taxonomic label to each group. Errors at any of these steps can lead to inappropriate interpretation. There are many different methods for analyzing and grouping sequence data. We will outline our approach, highlighting alternatives to commonly applied techniques when appropriate. Both the mothur (Schloss et al., 2009) and QIIME (Caporaso et al., 2010) software packages provide a comprehensive suite of tools for 16S rRNA analysis, including fastq quality filtering, OTU assignment, and classification for comparative analyses. The standard protocols associated with each package include many of the steps outlined below to process 16S rRNA sequence data.

We will use an example dataset throughout this chapter for illustration purposes. These libraries were prepared using a two-step PCR approach with primers targeting rRNA sequence positions 515 and 786 of the *Escherichia coli* genome (region V4) similar to a previously published method (Knight et al., 2011). The raw data and library construction protocol can be downloaded from the distribution-based clustering (DBC) Web site (https://github.com/spacocha/Distribution-based-clustering/blob/master/MIE_dir/). Libraries were sequenced on an Illumina MiSeq instrument (Illumina, San Diego, CA) with 2×251 paired end reads with an 8 base indexing run. Our samples were multiplexed with additional samples, including a defined mock community. 1.1 million reads were obtained from 20 freshwater samples and approximately 230,000 from the mock community.

Below, we will outline steps necessary to process the sequencing data into manageable units for further analysis, although our most recent protocol for sequence processing and OTU calling with DBC can be found at https://github.com/spacocha/Distribution-based-clustering/. The basic outline includes:

- Quality filtering, dereplication, and trimming of sequences
- Creating OTUs
- Removing artifacts and assigning informative labels to OTUs

3.1. Quality filtering, dereplication, and trimming of sequences

The overall quality of the run was visualized from the fastq file with the program FastQC (http://www.bioinformatics.babraham.ac.uk/projects/fastqc/) (Fig. 18.1). Across any sequencing platform, the average quality decreases with the length of DNA that is sequenced. One way to increase the quality of bases toward the end of a read and increase the read length is to use overlapping paired-end sequencing, commonly applied with Illumina. The observation of the same base in the reverse orientation can be used to reduce the error rate. Paired-end sequences can be joined with programs such as SHE-RA (Rodrigue et al., 2010), PandaSeq (Masella, Bartram, Truszkowski, Brown, & Neufeld, 2012), or within mothur using make.contigs. Alternatively, nonoverlapping, paired-end sequences have been concatenated and analyzed together (Werner, Zhou, Caporaso, Knight, & Angenent, 2012). Otherwise, two separate analyses can be done independently on the 5′ and 3′ reads. Although the information content is typically increased with overlapping reads, our protocol is standardized to use information from each end independently.

Raw data are first quality filtered, dereplicated, and trimmed. Improvement in diversity estimates can be gained after quality filtering (Bokulich et al., 2013) although overfiltering can lead to a loss of information. We use the mock community to determine the optimal amount of quality filtering. Our mock community is created from purified, linearized plasmids containing 16S rRNA sequences from a clone library. Thus, the concentration and sequence of each template is known, which can be used to evaluate quality-filtering performance. We use QIIME's split_libraries_fastq.py to demultiplex and quality filter the sequences, although other programs can be used to accomplish a similar filtering. [Note: we used a custom perl script (fastq2Qiime_barcode.pl) to modify the output of the Illumina data to correspond with the format required by split_libraries_fastq.py. This may not be necessary in all cases, depending on the form of the data from the sequencer and the position of the barcode sequence.]

Overfiltering can skew the abundance ratios in 16S-rRNA amplicon data and underfiltering allows many more errors into the analysis. We have identified two parameters in QIIME's split_libraries_fastq.py that increase the quality of the resulting data: (1) the maximum number of bad bases allowed before trimming (–max_bad_run_length) and (2) the threshold quality score (–last_bad_quality_char).

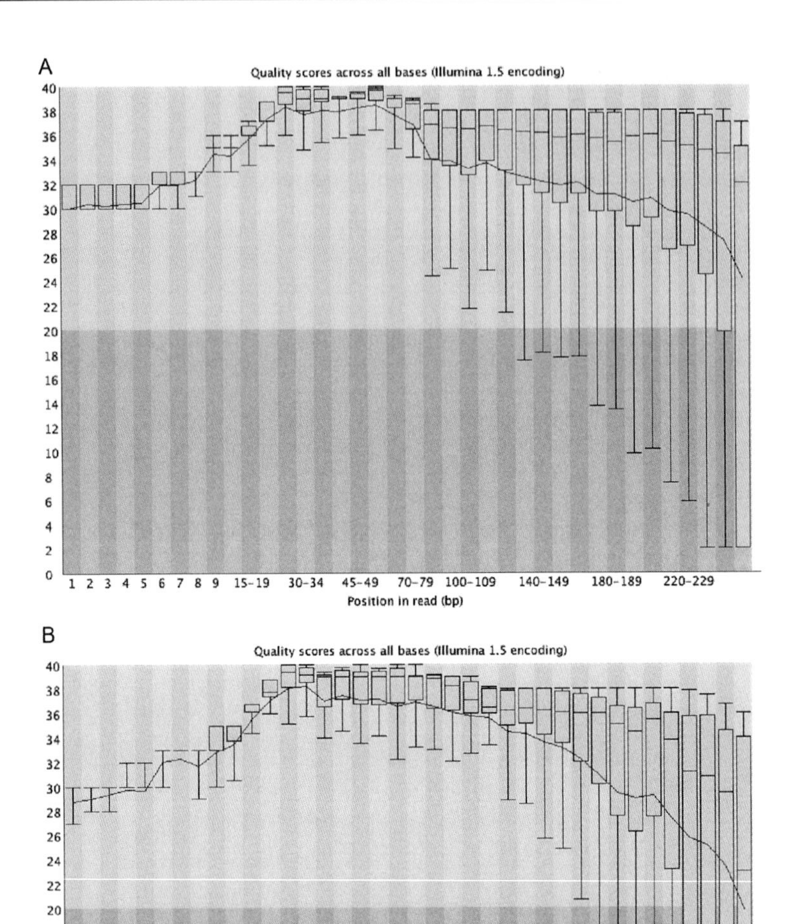

Figure 18.1 The per base sequence quality as determined by FastQC for (A) the 5′ (forward) read and (B) 3′ (reverse) end of a paired-end 250 × 250 MiSeq (Illumina) sequencing run. Raw base qualities are highest from approximately 20 (after the primer) to 70 bps or so. (For color version of this figure, the reader is referred to the online version of this chapter.)

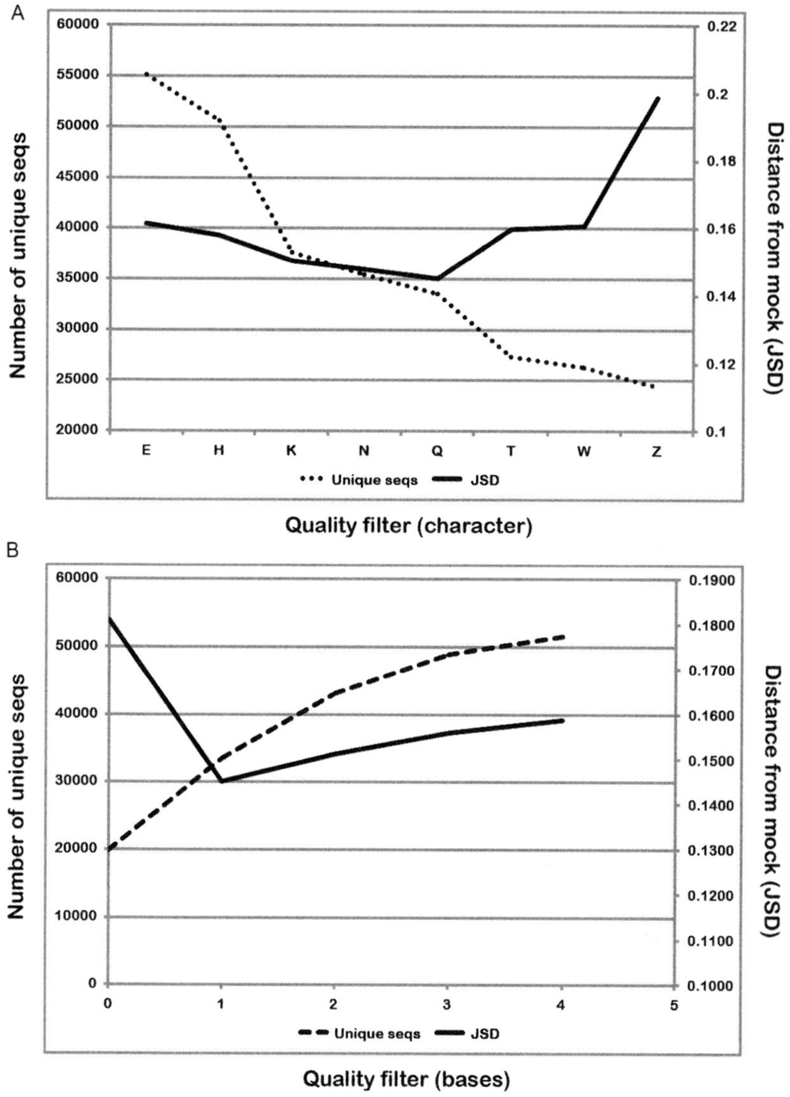

Figure 18.2 Intermediate quality filtering results in a dataset that is most similar to the input (defined, mock) community. (A) The solid line shows the distance of the resulting data from the input sequences after quality filtering using various quality thresholds (right axis). The dotted line shows the total number of unique sequences remaining after every filtering step (left axis). X-axis is the ASCII characters (Illumina's version 1.5 encoding) used in the split_library_fastq.py program of QIIME (–last_bad_char), representing different filtering stringencies ranging from E to Z, corresponding to Phred scores of 5 to 26 and probabilities of error from 0.3165 to 0.0025, respectively. (B) The solid line shows the distance of the resulting data from the input sequencing after allowing a specified number of bases to fall below the quality threshold before truncating the sequence (right axis). The dashed line shows the total number of unique sequences remaining after each step (left axis). X-axis is the number of consecutive bases with quality scores below the quality threshold that are allowed before truncating the sequence in the split_libraries_fastq.py program of QIIME (–max_bad_run_length). JSD, Jensen Shannon divergence.

We vary these two parameters and compare the resulting sequences to the known, mock community to identify the optimal filtering criteria. We found that with a quality threshold of Phred 17 (–last_bad_char Q; error rate ~0.02) and truncating the sequence when two bases fall below the quality filter (–max_bad_run_length 2) results in a mock community that has the smallest distance from our input (Fig. 18.2).

3.2. Creating OTUs

Grouping sequences into OTUs is important, because all downstream analyses are dependent on them, but there is little consensus on how this should be done. Some methods group sequences into clusters with sequences from a well-curated database as the "seed" (e.g., QIIME's "closed-reference" clustering). While this method is quick and convenient, it is not recommended because it can discard novel sequences in the dataset, even when they are abundant. This can be overcome with the "open-reference" approach (http://qiime.org/tutorials/open_reference_illumina_processing.html), where novel sequences are instead retained. Other programs cluster sequences *de novo*, forming OTUs based on their relation to other sequences in the dataset [e.g., average-linkage clustering in mothur (Schloss et al., 2009), heuristics such as USEARCH (http://www.drive5.com/usearch/) and ESPRIT (Sun et al., 2009)].

We favor an approach called DBC. DBC is an alternative method of forming OTUs *de novo* that is accurate and discriminating. This approach is different from other clustering methods because it uses the additional information contained in the distribution of sequences across libraries. Using the distribution and genetic information, DBC can reduce much of the false diversity created by sequencing error for a mock community. It can also provide the power to identify differentially distributed sequences that would otherwise be clustered together because of their sequence similarity. A manuscript describing the details of DBC will appear in the near future (Preheim, Perrotta, Martin-Platero, & Gupta, submitted for publication) and the most up-to-date, detailed protocol for running this algorithm can be found at the Web site (https://github.com/spacocha/Distribution-based-clustering).

DBC works to group genetically similar sequences that have a similar distribution across samples. The sequence by library matrix is used along with the pairwise genetic distance file to inform the clustering. The DBC algorithm can accept both aligned and unaligned phylogenetic distances, using the lowest of both to improve OTU calling. We used mothur align.seqs to create an alignment to a reference dataset (http://www.mothur.org/wiki/Silva_reference_alignment), generated from the mothur formatted Silva alignment and trimmed to the positions of our amplicon. Typically, we generate a distance matrix using FastTree—make_matix (Price, Dehal, & Arkin, 2009), although other distance programs can be used. The distribution similarity is evaluated for two sequences using the chi-squared test and the Jensen–Shannon divergence.

DBC is typically run in parallel for datasets of any significant size. This requires that the data are preclustered to a very low identity (~90% identity clusters), and each cluster is analyzed independently to form the final OTUs. Thus, the DBC algorithm evaluates whether to divide sequences found within each 90% cluster into additional OTUs based on the distribution and genetic information provided. OTUs called within the 90% clusters independently are merged into a final cluster list.

3.3. Artifact removal and classification of OTUs

Artifacts can inflate the total number of OTUs and diversity estimates. DBC can reduce the impact of sequencing errors, but non-rRNA sequences and chimeras may still be present which will inflate total diversity. We follow the recommendations on the mothur Web site (Schloss et al., 2009; http://www.mothur.org/wiki/MiSeq_SOP) and use align.seqs command with the modified reference alignment (Schloss, 2009). We discard sequences that do not start and end at the same position in the alignment using mothur: screen.seqs command with start and end options. Chimeras create false diversity, but are not easily distinguished from true rRNA sequences. They should be identified with specialized software such as UCHIME (Edgar, Haas, Clemente, Quince, & Knight, 2011) and removed from further analysis. In the example dataset, 6.2% of the resulting OTUs were chimeric, using the forward read only (35% from the overlapped reads).

Assigning a taxonomic label to each OTU created during clustering can help to interpret downstream analyses, especially when comparing against the published literature. Overlapped, paired-end sequences facilitate taxonomy assignment because their greater length increases the power of taxonomic classification. When overlapping is not possible and short reads must be used, taxonomic classification can be optimized for the specific design and read length (Mizrahi-Man, Davenport, & Gilad, 2013). The Ribosomal Database Project (RDP) naive Bayesian Classifier (Wang, Garrity, Tiedje, & Cole, 2007) is one of the most popular approaches, although others are available, such as Greengenes (DeSantis et al., 2006). The RDP classifier typically performs well on assigning short sequences at the family level (Claesson et al., 2010). However, when applying to short sequences, it is best to refine the training set that the RDP classifier uses for classification to include only the sequenced region as other regions may be characterized by quite different compositions (Werner et al., 2012). Concatenated, nonoverlapping sequences can be classified with Rtax (Soergel, Dey, Knight, & Brenner, 2012), which can be implemented through QIIME.

We typically match OTU representatives with Sanger-sequenced environmental clones from the same sample to increase the phylogenetic information associated with the most abundant organisms. This may be particularly useful in environments that are not well studied (i.e., the bottom of a lake), as opposed to more commonly analyzed samples (i.e., human microbiome). Although this type of analysis is not high-throughput, it does enhance the information gained from either the Sanger-sequenced or the Illumina-sequenced libraries alone and can be a powerful additional tool for the analysis of the most abundant sequences.

After completing these steps, the representative sequences, an OTU by library matrix and associated phylogenetic information from either the matching Sanger clones or the RDP classification can be used in downstream analysis.

4. COMMUNITY ANALYSIS

Because there are many options available to compare microbial communities, it is critical to match the tools utilized with the specific biological questions being pursued. Data generated from most high-throughput sequencing platforms (e.g., Illumina or 454) will be similar after OTU calling and classification, so many of the tools and reviews of community analyses in relation to pyrosequencing data can be applied to Illumina datasets as well. QIIME community analysis (http://qiime.org/tutorials/tutorial.html) can be used to guide the user through common types of analyses. Therefore, the goal of this section is to identify caveats of a few commonly applied analyses and provide an implementation of appropriate alternatives in areas such as:

- Diversity estimates
- Community distance metrics
- Associations of OTUs with metadata
- Correlation analysis

4.1. Diversity estimates

Estimating the total diversity (often called alpha diversity) within a community is a common first step to summarize community composition. True diversity reflects both the number of species present and the evenness of their distribution. However, because molecular surveys only capture a fraction of all species actually present in a community, all metrics reflect only an estimate of true diversity and will be sensitive to the sampling depth used. This complicates comparisons across samples with different sampling depths as more deeply sequenced samples may spuriously appear to be more diverse, independently of their actual composition. There are several metrics used to measure diversity, including the Shannon and Simpson indices (Hill, 1973), which measure both richness and evenness, and the Chao-1 estimator, which estimates species richness from the rarefaction of observed sequences (Chao, 1984; Chao, Colwell, Lin, & Gotelli, 2009). Shannon and Simpson indices are strongly preferred due to their reduced sensitivity to sampling depth (Haegeman et al., 2013).

To demonstrate the potential for artifacts of sequencing depth to confound meaningful analysis of diversity, we compared the sensitivity of richness and Shannon and Simpson diversity to sampling depth (Fig. 18.3). Using the OTU versus library matrix generated from the DBC algorithm above, we chose a library with 80,047 reads to compute the Chao-1 richness and Shannon and Simpson diversity metrics in PySurvey (sample_diversity with indices=['richness'], methods=['Chao1'] and indices=['shannon', 'simpson']). We then take random subsamples of the full dataset to simulate sequencing experiments with fewer reads, ranging from 1000 reads up to the full set 80,047 reads at 500 read intervals. We compute estimates of the three metrics in 100 iterations of random samples at each sequencing depth. While the Shannon and Simpson estimates of effective species quickly stabilize around 101 and 26, respectively, the species richness estimated by Chao-1 continues to rise with sequencing depth. Thus, while indices based on both species richness and abundance are relatively insensitive to sequencing depth, indices such as the Chao-1 estimator that measure only species richness remain highly sensitive to sequencing depth, even at high coverage. Although the Simpson index is less dependent on sequencing depth, it is also less sensitive to rare OTUs. For this reason, Shannon is the preferred metric for rRNA community analysis—achieving a balance of sensitivity to rare OTUs without undue dependency on sequencing depth.

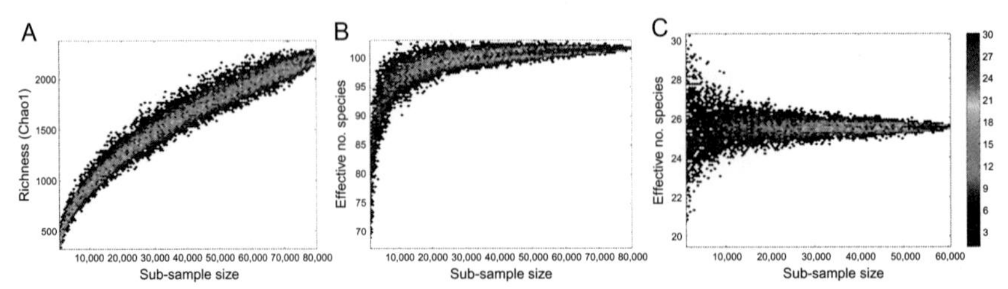

Figure 18.3 Richness is a function of sequencing depth, whereas Shannon and Simpson diversity indices stabilize with sequencing depth. (A) Chao-1 estimator (Chao1) of richness for the same library subsampled to different depths. (B) Effective number of species calculated with the Shannon diversity index (SDI) for the same library subsampled to different depths. Because SDI is computed on a log scale, we transform the SDI to the effective number of species present, which is on linear scale, by computing e^{SDI}. (C) Effective number of species calculated as the inverse of the Simpson diversity index (1/Simpsons diversity index) for the same library subsampled to different depths. (See color plate.)

4.2. Community distance metrics

Many applications call for estimating the distance between two distinct communities, a task that is impacted by sampling noise and "compositional" effects, which refer to biases that arise from the fact that only relative abundances rather than absolute values are measured (i.e., relative frequencies for all species must sum to 100%). Measures of distance between communities can be categorized based on the properties of the communities they account for: incidence (presence/absence), abundance (absolute or relative), and/or phylogenetic relatedness. Generally, incidence-based distances, where all components contribute equally, are especially challenging to estimate reliably (Chao, Chazdon, Colwell, & Shen, 2005), whereas measures which give larger weights to more abundant components are more readily estimated (Bent & Forney, 2008). This is analogous to the alpha diversity example in Fig. 18.3, where species richness is much more challenging to estimate than the Shannon entropy or the Simpson diversity.

To overcome the sampling noise-associated 16S rRNA survey data, we propose adopting the square root of the Jensen–Shannon divergence $\mathrm{JSD}^{1/2}$, defined as:

$$d_{\mathrm{JS}}^2\left(\underline{x},\underline{y}\right) = \frac{1}{2}\sum_{i=1}^{D}\left[x_i \log_2 \frac{x_i}{M_i} + y_i \log_2 \frac{y_i}{M_i}\right],$$

$$M_i = \frac{x_i + y_i}{2}$$

JSD has a clear information theoretic interpretation and is commonly used to measure differences between distributions. In this context, it gives the logarithm base 2 of the effective number of distinct communities (communities that have no species in common) and is the beta component of the Shannon entropy (Jost, 2006). The square root is of JSD is used, as it transforms JSD into a proper metric (Endres & Schindelin, 2003), enhancing interpretability. JSD accounts for the problem of compositional effects, as it is a measure of divergence between distributions, and maintains accurate and robust performance on datasets that confound Euclidean distances.

JSD was used to compare the samples from a stratified lake. Counts were normalized [fracs_f=ps.normalize(counts)] and the $JSD^{1/2}$ was used to compare communities from different extraction techniques and different depths in a stratified lake [D=ps.dist_mat (fracs_f , axis='rows', metric='JSsqrt')] and plotted with PySurvey [ps.plot_dist_heatmap (D, plot_rlabels=True, rlabel_width=0.06)]. Two major zones were identified, as the longest branches in the tree, dividing the upper (1–7 m) and lower (9–19 m) samples. Additionally, four total subzones were identified (sub1, 1–5 m; sub2, 7 m; sub3, 9–11 m; sub4, 13–19 m). Although the different extraction techniques (MoBio vs. Qiagen) tended to cluster with other samples from the same extraction technique within subzones sub1 and sub4, samples otherwise clustered by depth (Fig. 18.4).

4.3. Associations of OTUs with metadata

Multivariate statistical techniques have been developed specifically to handle high-dimensional data, including Principal Coordinates Analysis. These techniques work by first reducing the dimensionality of the data by identifying principal "axes" of differentiation among communities, which represent distinct combinations of bacterial species. These tools are particularly effective when the variable of interest (e.g., pH, disease state) is associated with major changes in community structure, but are less effective at detecting subtle variations in community structure. Furthermore, they have trouble pinpointing the specific bacterial species that drive these associations. Recently, statistical learning techniques have been employed to detect associations between bacterial species and environmental metadata. Statistical learning has many advantages over multivariate statistical approaches, including the ability to detect both major and minor variations in microbial community structure, the ability to detect nonlinear associations involving combinations of bacterial species, and the ability to pinpoint the specific bacterial species that underlie these associations.

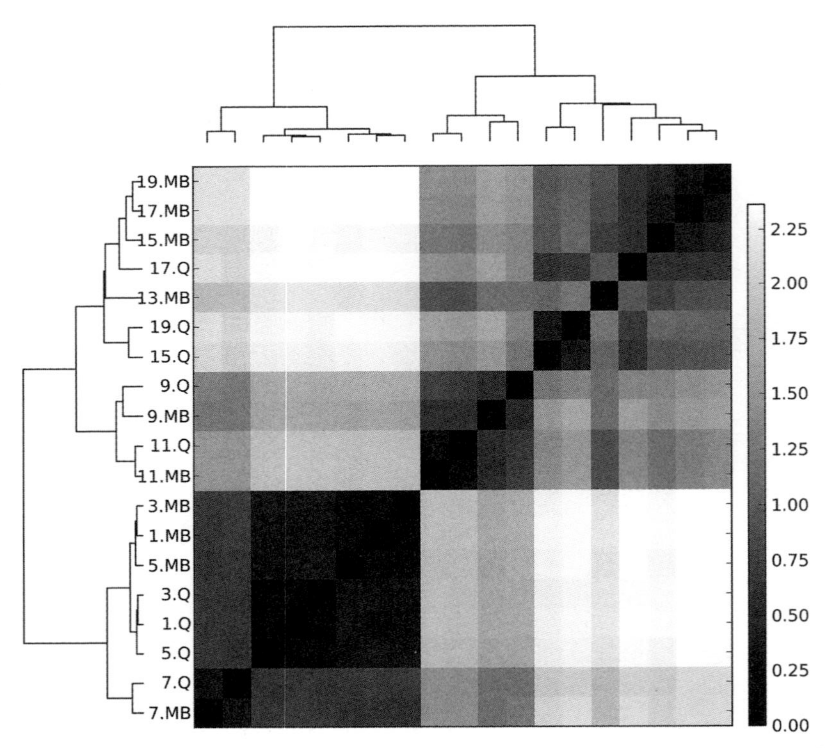

Figure 18.4 A heatmap depicting different biogeochemical zones of a stratified lake. The bacterial community for each depth was compared to all other libraries using the square root of the Jensen–Shannon divergence ($JSD^{1/2}$) and plotted as a heatmap. Bacterial communities cluster into two major zones (upper, oxic and lower, anoxic zone) as shown by the largest branch on the dendrogram (i.e., tree) and four subzones. Libraries are labeled with the depth (in meters) and the extraction technique. MB, MoBio Power Water DNA extraction kit; Q, Qiagen Blood and Tissue Kit. For example, 1.MB refers to 1 m depth in the water extracted with the MoBio kit and 19.Q refers to 19 m depth with the Qiagen kit. The dendrogram above and to the left of the heatmap are identical. (See color plate.)

One statistical learning technique that is well suited to microbial community analysis is Random Forest classification, which is implemented in the SLIME software package. For example, SLIME was recently used to discriminate patients with inflammatory bowel disease from healthy controls and to identify the specific bacterial taxa underlying this association (Papa et al., 2012). SLIME can detect associations using categorical or quantitative metadata and runs permutation tests to assess statistical significance. It is best used for relatively large datasets where the total number of samples exceeds 50. The input for this analysis is a table of OTU counts and corresponding metadata. It will proceed to test the significance of every column of metadata, reporting the *P*-value of each association, the most important bacterial taxa, and advanced metrics such as the area under the curve and other classification results.

4.4. Correlation analysis

Microorganisms interact with each other in various ways, through competition for shared resources, antagonistic mechanisms (e.g., antibiotic production), or in mutual or symbiotic associations. Correlation analysis can identify possibly interacting OTUs by identifying groups that change across a large sample in a similar way. However, correlations from low-diversity samples are often artifacts related to the compositional nature of sampling and sequencing (Friedman & Alm, 2012). For example, variation in a single, dominant OTU causes all other OTUs to have a spurious negative correlation to this dominant strain, although their absolute abundance may not have changed. As a result, standard correlation metrics (Pearson, Spearman, etc.) should not be used to compare OTUs. Instead, SparCC (https://bitbucket.org/yonatanf/sparcc) can be used to infer correlations from compositional data with high accuracy, even in low-diversity samples, and can be thought of as a replacement for Pearson correlations. Nonetheless, correlations do not imply causality and can arise when two OTUs respond to a third unmeasured environmental factor. Thus, additional experiments are needed to verify results from SparCC or other similar methods.

We used SparCC to identify correlated OTUs using the OTU by library matrix (created as stated earlier) after filtering out OTUs with fewer than 100 counts (-c 100) across at least two libraries (-s 2) using QIIME's filter_otu_table.py (removing the first line of the resulting table). We then ran SparCC on the data, using 50 bootstraps and a "one-sided" test. Interestingly, of the two pairs with correlations of 0.7, one is classified as *Saprospiraceae* (ID0000005M), which have been found attached to filamentous bacteria (Xia, Kong, Thomsen, & Nielsen, 2008). The other bacteria could not be classified (ID0000388M). This analysis provides the underlying information to identify important interactions in the environment.

5. SUMMARY

In this chapter, we provide an overview of the methods used to interpret high-throughput microbial community analysis, with a focus on data from the widely used Illumina sequencing platform. We explain how to transform Illumina sequence data into meaningful units for further analysis, including a specialized OTU calling method (DBC). We discuss why Shannon diversity is a more appropriate measure of diversity than richness and use the $JSD^{1/2}$ as a metric for distance between microbial communities. Finally, we use Random Forest classification to identify associations between specific OTUs and metadata and use SparCC to identify true correlations from proportional composition data. Armed with these tools and the increasing power of high-throughput sequencing, we expect that researchers will continue to discover exciting new connections between bacterial communities and their diverse environments.

REFERENCES

Amann, R. I., Ludwig, W., & Schleifer, K. H. (1995). Phylogenetic identification and in-situ detection of individual microbial-cells without cultivation. *Microbiological Reviews*, *59*(1), 143–169.

Bent, S. J., & Forney, L. J. (2008). The tragedy of the uncommon: Understanding limitations in the analysis of microbial diversity. *ISME Journal*, *2*(7), 689–695.

Bokulich, N. A., Subramanian, S., Faith, J. J., Gevers, D., Gordon, J. I., Knight, R., et al. (2013). Quality-filtering vastly improves diversity estimates from illumina amplicon sequencing. *Nature Methods*, *10*(1), 57–59.

Caporaso, J. G., Kuczynski, J., Stombaugh, J., Bittinger, K., Bushman, F. D., Costello, E. K., et al. (2010). QIIME allows analysis of high-throughput community sequencing data. *Nature Methods*, 7(5), 335–336.

Chao, A. (1984). Nonparametric-estimation of the number of classes in a population. *Scandinavian Journal of Statistics*, *11*(4), 265–270.

Chao, A., Chazdon, R. L., Colwell, R. K., & Shen, T. J. (2005). A new statistical approach for assessing similarity of species composition with incidence and abundance data. *Ecology Letters*, *8*(2), 148–159.

Chao, A., Colwell, R. K., Lin, C. W., & Gotelli, N. J. (2009). Sufficient sampling for asymptotic minimum species richness estimators. *Ecology*, *90*(4), 1125–1133.

Claesson, M. J., Wang, Q. O., O'Sullivan, O., Greene-Diniz, R., Cole, J. R., Ross, R. P., et al. (2010). Comparison of two next-generation sequencing technologies for resolving highly complex microbiota composition using tandem variable 16S rRNA gene regions. *Nucleic Acids Research*, *38*(22), e200.

DeSantis, T. Z., Hugenholtz, P., Larsen, N., Rojas, M., Brodie, E. L., Keller, K., et al. (2006). Greengenes, a chimera-checked 16S rRNA gene database and workbench compatible with ARB. *Applied and Environmental Microbiology*, *72*(7), 5069–5072.

Edgar, R. C., Haas, B. J., Clemente, J. C., Quince, C., & Knight, R. (2011). UCHIME improves sensitivity and speed of chimera detection. *Bioinformatics*, *27*(16), 2194–2200.

Endres, D. M., & Schindelin, J. E. (2003). A new metric for probability distributions. *IEEE Transactions on Information Theory*, *49*(7), 1858–1860.

Forney, L. J., Zhou, X., & Brown, C. J. (2004). Molecular microbial ecology: Land of the one-eyed king. [Review]. *Current Opinion in Microbiology*, 7(3), 210–220.

Friedman, J., & Alm, E. J. (2012). Inferring correlation networks from genomic survey data. *Plos Computational Biology*, *8*(9), e1002687.

Gevers, D., Cohan, F. M., Lawrence, J. G., Spratt, B. G., Coenye, T., Feil, E. J., et al. (2005). Re-evaluating prokaryotic species. *Nature Reviews Microbiology*, *3*(9), 733–739.

Haegeman, B., Hamelin, J., Moriarty, J., Neal, P., Dushoff, J., & Weitz, J. S. (2013). Robust estimation of microbial diversity in theory and in practice. *The ISME Journal*, 7, 1092–1101.

Hill, M. O. (1973). Diversity and evenness—Unifying notation and its consequences. *Ecology*, *54*(2), 427–432.

Huse, S. M., Welch, D. M., Morrison, H. G., & Sogin, M. L. (2010). Ironing out the wrinkles in the rare biosphere through improved OTU clustering. *Environmental Microbiology*, *12*(7), 1889–1898.

Huttenhower, C., Gevers, D., Knight, R., Abubucker, S., Badger, J. H., Chinwalla, A. T., et al. (2012). Structure, function and diversity of the healthy human microbiome. *Nature*, *486*(7402), 207–214.

Jost, L. (2006). Entropy and diversity. [Editorial Material]. *Oikos*, *113*(2), 363–375.

Knight, R., Caporaso, J. G., Lauber, C. L., Walters, W. A., Berg-Lyons, D., Lozupone, C. A., et al. (2011). Global patterns of 16S rRNA diversity at a depth of millions of sequences per sample. *Proceedings of the National Academy of Sciences of the United States of America*, *108*, 4516–4522.

Masella, A. P., Bartram, A. K., Truszkowski, J. M., Brown, D. G., & Neufeld, J. D. (2012). PANDAseq: PAired-eND Assembler for Illumina sequences. *BMC Bioinformatics*, *13*, 31.

Mizrahi-Man, O., Davenport, E. R., & Gilad, Y. (2013). Taxonomic classification of bacterial 16S rRNA genes using short sequencing reads: Evaluation of effective study designs. *PLoS One*, *8*(1), e53608.

Pace, N. R. (1997). A molecular view of microbial diversity and the biosphere. *Science*, *276*(5313), 734–740.

Papa, E., Docktor, M., Smillie, C., Weber, S., Preheim, S. P., Gevers, D., et al. (2012). Non-invasive mapping of the gastrointestinal microbiota identifies children with inflammatory bowel disease. *PLoS One*, 7(6), e39242.

Preheim, S. P., Perrotta, A. R., Martin-Platero, A. M., Gupta, A., & Alm, E. J. Distribution-based clustering: Using ecology to refine the operational taxonomic unit (Accepted, *Applied and Environmental Microbiology*).

Price, M. N., Dehal, P. S., & Arkin, A. P. (2009). FastTree: Computing large minimum evolution trees with profiles instead of a distance matrix. *Molecular Biology and Evolution*, *26*(7), 1641–1650.

Qiu, X. Y., Wu, L. Y., Huang, H. S., McDonel, P. E., Palumbo, A. V., Tiedje, J. M., et al. (2001). Evaluation of PCR-generated chimeras: Mutations, and heteroduplexes with 16S rRNA gene-based cloning. *Applied and Environmental Microbiology*, *67*(2), 880–887.

Rodrigue, S., Materna, A. C., Timberlake, S. C., Blackburn, M. C., Malmstrom, R. R., Alm, E. J., et al. (2010). Unlocking short read sequencing for metagenomics. *PLoS One*, *5*(7), e11840.

Schloss, P. D. (2009). A high-throughput DNA sequence aligner for microbial ecology studies. *PLoS One, 4*(12).

Schloss, P. D., Westcott, S. L., Ryabin, T., Hall, J. R., Hartmann, M., Hollister, E. B., et al. (2009). Introducing mothur: Open-source, platform-independent, community-supported software for describing and comparing microbial communities. *Applied and Environmental Microbiology, 75*(23), 7537–7541.

Soergel, D. A. W., Dey, N., Knight, R., & Brenner, S. E. (2012). Selection of primers for optimal taxonomic classification of environmental 16S rRNA gene sequences. *ISME Journal, 6*(7), 1440–1444.

Sun, Y., Cai, Y., Liu, L., Yu, F., Farrell, M. L., McKendree, W., et al. (2009). ESPRIT: Estimating species richness using large collections of 16S rRNA pyrosequences. *Nucleic Acids Research, 37*(10), e76.

Wang, Q., Garrity, G. M., Tiedje, J. M., & Cole, J. R. (2007). Naive Bayesian classifier for rapid assignment of rRNA sequences into the new bacterial taxonomy. *Applied and Environmental Microbiology, 73*(16), 5261–5267.

Werner, J. J., Koren, O., Hugenholtz, P., DeSantis, T. Z., Walters, W. A., Caporaso, J. G., et al. (2012). Impact of training sets on classification of high-throughput bacterial 16s rRNA gene surveys. *ISME Journal, 6*(1), 94–103.

Werner, J. J., Zhou, D., Caporaso, J. G., Knight, R., & Angenent, L. T. (2012). Comparison of illumina paired-end and single-direction sequencing for microbial 16S rRNA gene amplicon surveys. *ISME Journal, 6*(7), 1273–1276.

Whitman, W. B., Coleman, D. C., & Wiebe, W. J. (1998). Prokaryotes: The unseen majority. *Proceedings of the National Academy of Sciences of the United States of America, 95*(12), 6578–6583.

Woese, C. R., Kandler, O., & Wheelis, M. L. (1990). Towards a natural system of organisms—Proposal for the domains archaea, bacteria, and eucarya. *Proceedings of the National Academy of Sciences of the United States of America, 87*(12), 4576–4579.

Xia, Y., Kong, Y. H., Thomsen, T. R., & Nielsen, P. H. (2008). Identification and ecophysiological characterization of epiphytic protein-hydrolyzing Saprospiraceae ("Candidatus epiflobacter" spp.) in activated sludge. *Applied and Environmental Microbiology, 74*(7), 2229–2238.

Zhou, H. W., Li, D. F., Tam, N. F. Y., Jiang, X. T., Zhang, H., Sheng, H. F., et al. (2011). BIPES, a cost-effective high-throughput method for assessing microbial diversity. *ISME Journal, 5*(4), 741–749.

CHAPTER NINETEEN

Advancing Our Understanding of the Human Microbiome Using QIIME

José A. Navas-Molina[*], Juan M. Peralta-Sánchez[†], Antonio González[†], Paul J. McMurdie[‡], Yoshiki Vázquez-Baeza[†], Zhenjiang Xu[†], Luke K. Ursell[†], Christian Lauber[††], Hongwei Zhou[§], Se Jin Song[¶], James Huntley[†], Gail L. Ackermann[†], Donna Berg-Lyons[†], Susan Holmes[‡], J. Gregory Caporaso[||,#], Rob Knight[†,,1]**

[*]Department of Computer Science, University of Colorado, Boulder, Colorado, USA
[†]Biofrontiers Institute, University of Colorado, Boulder, Colorado, USA
[‡]Department of Statistics, Stanford University, Stanford, California, USA
[§]Department of Environmental Health, School of Public Health and Tropical Medicine, Southern Medical University, Guangzhou, China
[¶]Department of Ecology and Evolutionary Biology, University of Colorado, Boulder, Colorado, USA
[||]Department of Biological Sciences, North Arizona University, Flagstaff, Arizona, USA
[#]Institute for Genomics and Systems Biology, Argonne National Laboratory, Argonne, Illinois, USA
[**]Howard Hughes Medical Institute, University of Colorado, Boulder, Colorado, USA
[††]Cooperative Institute for Research in Environmental Sciences, Boulder, Colorado, USA
[1]Corresponding author: e-mail address: Rob.Knight@colorado.edu

Contents

1. Introduction 372
2. QIIME as Integrated Pipeline of Third-Party Tools 373
3. PCR and Sequencing on Illumina MiSeq 375
4. QIIME Workflow for Conducting Microbial Community Analysis 377
   4.1 Upstream analysis steps 379
   4.2 Downstream analysis steps 391
5. Other Features 427
   5.1 Testing linear gradients, including time series analysis 427
   5.2 Processing 454 data 429
   5.3 18S rRNA gene sequencing 430
   5.4 Shotgun metagenomics 431
   5.5 Support for QIIME in R 431
6. Recommendations 438
7. Conclusions 438
Acknowledgments 439
References 439

Methods in Enzymology, Volume 531
ISSN 0076-6879
http://dx.doi.org/10.1016/B978-0-12-407863-5.00019-8

Abstract

High-throughput DNA sequencing technologies, coupled with advanced bioinformatics tools, have enabled rapid advances in microbial ecology and our understanding of the human microbiome. QIIME (Quantitative Insights Into Microbial Ecology) is an open-source bioinformatics software package designed for microbial community analysis based on DNA sequence data, which provides a single analysis framework for analysis of raw sequence data through publication-quality statistical analyses and interactive visualizations. In this chapter, we demonstrate the use of the QIIME pipeline to analyze microbial communities obtained from several sites on the bodies of transgenic and wild-type mice, as assessed using 16S rRNA gene sequences generated on the Illumina MiSeq platform. We present our recommended pipeline for performing microbial community analysis and provide guidelines for making critical choices in the process. We present examples of some of the types of analyses that are enabled by QIIME and discuss how other tools, such as phyloseq and R, can be applied to expand upon these analyses.

1. INTRODUCTION

Advances in DNA sequencing technologies, together with the availability of culture-independent sequencing methods and software for analyzing the massive quantities of data resulting from these technologies, have vastly improved our ability to characterize microbial communities in many diverse environments. The human microbiota, the collection of microbes living in or on the human body, is of considerable interest: microbial cells outnumber human cells in our bodies by a ratio of up to 10 to 1 (Savage, 1977). These microbial communities contribute to healthy human physiology (De Filippo et al., 2010; Dethlefsen & Relman, 2011; Spencer et al., 2011) and development (Dominguez-Bello et al., 2010; Koenig et al., 2011), and dysbiosis (or imbalance in these communities) is now known to be associated with disease, including obesity (Turnbaugh et al., 2009) and Crohn's disease (Eckburg & Relman, 2007). More recently, evidence from transplants into germ-free mice suggests that some of these associations may be causal, because certain phenotypes can be transmitted by transmitting the microbiota (Carvalho et al., 2012; McLean, Bergonzelli, Collins, & Bercik, 2012; Turnbaugh et al., 2009), even including transmission of human phenotypes into mice (Diaz Heijtz et al., 2011; Koren et al., 2012; Smith et al., 2013).

Illumina's MiSeq and HiSeq DNA sequencing instruments, respectively, sequence tens of millions, or billions, of DNA fragments in a single sequencing run (Kuczynski et al., 2012). The rapidly increasing data volumes typical of recent studies drive a need for more efficient and scalable tools to study the human microbiome (Gonzalez & Knight, 2012). QIIME (Quantitative Insights Into Microbial Ecology) (Caporaso, Kuczynski, et al., 2010) is an open-source pipeline designed to provide self-contained microbial community analyses, from interacting with raw sequence data through publication-quality statistical analyses and visualizations.

QIIME integrates commonly used third-party tools and implements many diversity metrics, statistical methods, and visualization tools for analyzing microbial data. Consequently, most individual steps in the microbial community analysis can be performed in multiple ways. Here, we describe how samples are prepared for an Illumina MiSeq run, the QIIME pipeline, and our view of the current best practices for analyzing microbial communities with QIIME. Although there are other pipelines available, including mothur (Schloss et al., 2009), the RDP tools (Olsen, Larsen, & Woese, 1991; Olsen et al., 1992), ARB (Ludwig et al., 2004), VAMPS (Sogin, Welch, & Huse, 2009), and other platforms, in this review, we focus on analysis with the MiSeq platform and QIIME as this combination is increasingly popular as a method for analyzing microbial communities and a detailed comparison of other available pipelines and sequencing platforms is beyond the scope of the present work.

2. QIIME AS INTEGRATED PIPELINE OF THIRD-PARTY TOOLS

An early barrier to adoption of QIIME was that it was difficult to install, in part because of the large number of software dependencies (third-party packages that need to be installed before QIIME is operational). The large number of dependencies was, however, a deliberate choice made during QIIME development. To build a pipeline for sequence analysis that encompasses the many steps from sequence collection, curation, and statistical analysis, the user must consider many existing tools that have been developed to perform specific functions and extensively benchmarked on their ability to perform these functions, such as the uclust program for clustering sequences into Operational Taxonomic Units (OTUs) (Edgar, 2010). A pipeline thus has two options: either reimplement the algorithm or use the existing software (by creating a "wrapper" that allows its input and output to be incorporated into the pipeline). The QIIME developers choose to wrap all the algorithms rather than reimplement them. This choice preserves the integrity of the programs that make up the pipeline, as there is no doubt that the tool being used is the one designed, created, and tested by the original authors, and, in most cases, peer-reviewed by the scientific community. The reuse of existing software also allows the QIIME pipeline to include and distribute newly developed and improved algorithms more rapidly than would be possible if each algorithm had to be reimplemented and retested to check that it matched the original. Thus QIIME users can be sure that they have the most up-to-date tools for their analysis and can credit the authors of the component software packages appropriately.

One important, but sometimes poorly understood, aspect of the QIIME pipeline is that it wraps algorithms and tools produced by other researchers into a single pipeline for sequence analysis. It is therefore important to cite the individual tools that you use as well as QIIME itself. For example, an analysis using the default QIIME parameters (Caporaso, Kuczynski, et al., 2010) would use uclust (Edgar, 2010) to cluster the sequences against the GreenGenes database (DeSantis et al., 2006), assign taxonomy using the RDP classifier (Wang, Garrity, Tiedje, & Cole, 2007), and build principal coordinate analysis (PCoA) beta-diversity plots using UniFrac (Lozupone & Knight, 2005). It is important for researchers who are considering contributing to the QIIME pipeline to recognize that their contributions will be cited so that they can continue to expand upon their work. For example, the `pick_otus.py` script alone offers a choice of nine different clustering algorithms, each developed by researchers who should be acknowledged if their particular algorithm is used.

For taxonomy databases and other reference databases, including GreenGenes, it is also important to cite the release version that you are using (DeSantis et al., 2006), not least because the results will change depending on which release you used, and others may not be able to reproduce your results without this information. For GreenGenes, the default taxonomy database in QIIME, the version is named after the release date, such as the 12_10 release. The latest version of GreenGenes can always be downloaded from the qiime.org Web site. Using the same GreenGenes, reference database version is critical for comparisons of taxonomy assignments and OTUs across different studies. For this reason, all the studies in the QIIME database are always processed against the same release version of GreenGenes.

An overview of some of the key tools used by the default QIIME pipeline follows:

- uclust (Edgar, 2010). Used for OTU picking.
- usearch (Edgar, 2010). Used for OTU picking and chimera checking.

- RDP classifier (Wang et al., 2007). Used for taxonomy assignment.
- GreenGenes database (DeSantis et al., 2006). Used as a reference database for taxonomy assignment and reference-based OTU picking (see below).
- PyNAST (Caporaso, Bittinger, et al., 2010). Used for multiple sequence alignment.
- UniFrac (Lozupone & Knight, 2005). Used as a phylogenetic metric for beta-diversity analysis.

3. PCR AND SEQUENCING ON ILLUMINA MiSeq

Microbial community analysis typically begins with the extraction of DNA from primary samples (note that although most of this DNA comes from cells in the sample, some may consist of dead cells or extracellular DNA, so the representation of the active community from these sources is not perfect). Although methods for DNA extraction vary, several large initiatives such as the Earth Microbiome Project (Gilbert, Meyer, Antonopoulos, et al., 2010; Gilbert, Meyer, Jansson, et al., 2010) and the Human Microbiome Project (HMP) (Human Microbiome Project, 2012a, 2012b; Turnbaugh et al., 2007) have standardized on the MOBIO PowerSoil DNA extraction kit (www.mobio.com) to efficiently recover DNA from a wide range of sample types. After extraction, samples are PCR amplified under permissive conditions with primers containing the MiSeq sequencing adapters, a 12-nucleotide Golay barcode (first introduced in Fierer, Hamady, Lauber, & Knight, 2008) on the forward primer, followed by the bases matching the 16S rRNA gene; the reverse primer is not barcoded (Caporaso et al., 2012). The annealing temperature is set to 50 °C, which in our hands minimizes PCR artifacts (both primer dimer and background "smear") while encouraging the primers to anneal to the largest diversity of sequences possible. Similarly, we believe that including sequencing adaptors and barcodes in the PCR step has advantages over multiple enzymatic treatments of the 16S amplicon that are otherwise needed to introduce adaptors and barcodes after PCR. The first and most important consideration is the reduction of sample handling, which lowers the chance of contamination, mislabeling, and loss of small-volume samples during preparation. Combining the adapters and barcodes in the PCR step allows for exact well-to-well mapping of samples to primers, providing a standardized way to track sample-barcode combinations through library preparation, an important consideration when sequencing hundreds to thousands of samples using 96- or 384-well sample preparation formats.

Because the MiSeq can generate a large number of sequences per run, many samples can be multiplexed on each single sequencing run. The choice of barcodes thus deserves some attention. For instance, homebrew "barcodes" can be as simple as using an arbitrary sequence of known nucleotides placed at the front of the amplicon and fed into an informatics pipeline for detection. Although simple, this approach has limited ability to detect sequencing error (Caporaso et al., 2012) and increases the risk of misassignment of a sequence to the wrong sample. The use of error-correcting barcodes, such as Hamming (Hamady, Walker, Harris, Gold, & Knight, 2008) or Golay codes (Caporaso et al., 2012), allows the user to detect and correct errors in the barcode, decreasing the chances that a sequence is assigned to the wrong sample. Error-correcting barcodes also allow the user to retain more sequences because 8-nucleotide Hamming codes can detect and correct 2 and 1 bit errors, respectively (Hamady et al., 2008), and 12-nucleotide Golay codes can detect and correct 4 and 3 bit errors, respectively (Hamady & Knight, 2009). With the unique Golay codes described in Caporaso et al. (2012), up to 2167 samples could be multiplexed on a single MiSeq run at a depth of 4600 per sample, certainly sufficient to detect the effects of many biological phenomena of interest (Kuczynski, Costello, et al., 2010; Kuczynski, Liu, et al., 2010). As the QIIME default settings detect Golay barcodes, we encourage the use of these codes when possible to maximize sequence retention and assignment accuracy.

Detailed instructions for loading the MiSeq for amplicon runs with custom barcodes can be found on the Earth Microbiome Project Web site (www.earthmicrobiome.org). Briefly, pooled libraries are analyzed by Bioanalyzer (Agilent Technologies) and diluted to 2 η*M* quantitated by use of a Qubit Fluorometer (Life Technologies, high-sensitivity reagents). The phiX spike-in library (Illumina Inc.) is also diluted to 2 η*M* prior to use. Denaturation of the pooled 16S rRNA gene amplicon libraries and the phiX control is performed according to manufacturer's instructions (Illumina Inc.), giving a denatured template concentration of 20 ρ*M*. Denatured templates are further diluted to 5 ρ*M* (using Illumina HT1 buffer) and subsequently combined to give an 85% 16S rRNA gene amplicon library and 15% phiX control pool (1000 μL total volume). Improvements in the Illumina analysis software may allow reduction of this phiX spike-in, allowing more of the sequences to be used for 16S rRNA gene amplicons.

MiSeq reagent cartridges are prepared according to the manufacturer's instructions (Illumina Inc.). The sample pool (1000 μL total volume) is loaded into cartridge position 17. Custom 16S rRNA gene Read 1, Index

Read, and Read 2 sequencing primers are added directly to cartridge wells containing the standard Illumina Read 1, Index Read, and Read 2 sequencing primers (wells 12, 13, and 14, respectively, 5 μL each primer at 100 μ*M* concentration (Caporaso et al., 2012)). Primers are added to wells using a long gel loading tip and gently mixed using a plastic Pasteur pipette. Care must be taken to assure that reagents in the cartridge are localized to the bottom of the wells and that no bubbles are present.

The spike-in of PhiX, at least at low levels, is still critical for obtaining usable amplicon data because the optics requires diversity at each nucleotide position, which is not possible with absolutely conserved nucleotides within the 16S rRNA gene (or most other genes of interest). Many users have had difficulty mixing this protocol for custom amplicons with Illumina's own indexing protocol, which allows a maximum of 96 samples to be multiplexed per run at the time of writing. It is critical to use either this protocol exactly (allowing arbitrary numbers of custom barcodes) or Illumina's barcoding protocol, but not to mix and match steps and reagents.

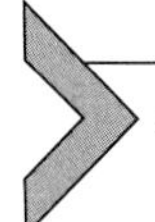

4. QIIME WORKFLOW FOR CONDUCTING MICROBIAL COMMUNITY ANALYSIS

The Illumina MiSeq technology can generate up to 10^7 sequences in a single run (Kuczynski et al., 2012). QIIME takes the instrument output and generates useful information about the community represented in each sample. At a coarse-grained level, we divide this process into "upstream" and "downstream" stages (Fig. 19.1). The upstream step includes all the processing of the raw data (sequencing output) and generating the key files (OTU table and phylogenetic tree) for microbial analysis. The downstream step uses the OTU table and phylogenetic tree generated in the upstream step to perform diversity analysis, statistics, and interactive visualizations of the data. Additionally, QIIME increasingly interfaces with other packages such as IPython and R, allowing additional analyses to be conducted.

To illustrate some of the main features of QIIME, together with some of the analyses that can be performed outside QIIME, we use an example dataset consisting of samples from different body sites of 12 mice: the oral cavity, ileum, cecum, colon, fecal pellet, skin, and whole mouse sample by homogenizing the mouse carcass. Seven mice were wild-type genotype (WT from here so on), while the five remaining mice were transgenic (TG from here so on). The samples were collected by students during the IQ-Bio course taught by Manuel Lladser and Rob Knight during Spring 2013 at

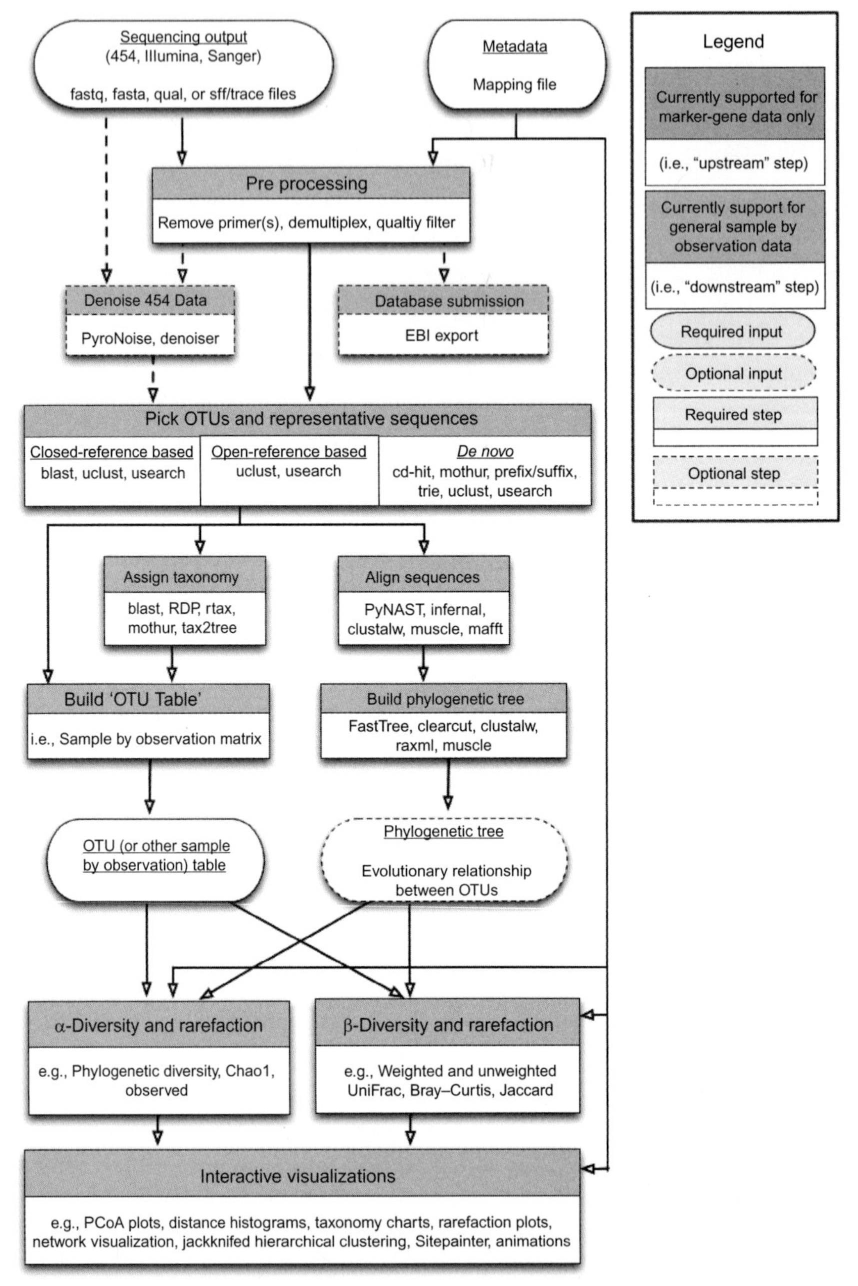

Figure 19.1 QIIME workflow overview. The upstream process (brown boxes) includes all the steps that generate the OTU table and the phylogenetic tree. This step starts by preprocessing the sequence reads and ends by building the OTU table and the phylogenetic tree. The downstream process (blue boxes) includes steps involved in analysis and interpretation of the results, starting with the OTU table and the phylogenetic tree and ending with alpha- and beta-diversity analyses, visualizations, and statistics. (See color plate.)

University of Colorado at Boulder (course identifiers: APPM5720-001-2013, CHEM4751-001-2013, CHEM5751-001-2013, CSCI4830-006-2013, CSCI7000-006-2013, MCDB6440-001-2013).

4.1. Upstream analysis steps

The QIIME analysis workflow starts with the sequencing output (fastq files) and a user-generated mapping file. The mapping file contains information for understanding what is in each sample and is therefore critical for performing the rest of the analyses; it is in tab-delimited text format. The main information in this file is a unique identifier for each sample, the barcode used for each sample, the primer sequence used, and a description for each sample, together with additional user-defined information that is necessary for understanding the results such as which species the sample was taken from, which site on the body is being studied, and clinical variables relevant to the study. The sample identifier, barcode, and primer sequence information are required for the first step of the QIIME workflow. This preprocessing step combines sample demultiplexing, primer removal, and quality-filtering. Additional information provided about the samples in the mapping file is helpful for later steps, especially for analyses that aggregate the samples by these fields (e.g., comparing lean to obese subjects). We therefore recommend including as much additional data about the samples as possible (often called "sample metadata"). This auxiliary information is also very useful for identifying contaminated samples. For example, SourceTracker (Knights, Kuczynski, Charlson, et al., 2011) is a package included in QIIME that identifies the proportion of different community sources, including contamination, in each sample based on a database of samples from known communities.

4.1.1 Demultiplexing and quality-filtering

As mentioned earlier, high-throughput sequencing allows multiple samples to be combined in a single sequencing run (Kuczynski et al., 2012). However, each sequence must then be linked back to the individual sample that it came from via a DNA barcode. The barcodes, which are short-DNA sequences unique to each sample, are incorporated into each sequence from a given sample during PCR. QIIME uses the barcodes in the mapping file to demultiplex, that is, to assign the sequences back to the samples they are derived from, using error-correcting codes where available (as noted earlier). QIIME is also able to demultiplex variable-length barcodes such as those used in the HMP, see Section 5.2.1.

During demultiplexing, QIIME removes the barcodes and primer sequences because they are not needed in later steps. Thus, the result after demultiplexing is a sequence matching the amplified 16S rRNA gene.

The third part of preprocessing is quality-filtering. Quality-filtering improves diversity estimates with Illumina sequencing substantially (Bokulich et al., 2013). Illumina instruments, like most sequencing instruments, generate a quality score for each nucleotide (Phred), related to the probability that each nucleotide was read incorrectly. QIIME uses the Phred score and user-defined parameters to remove sequence reads that do not meet the desired quality. These user-defined parameters are the percentage of consecutive high-quality base calls (p), the maximum number of consecutive low-quality base calls (r), the maximum number of ambiguous bases (typically coded as N) (n), and the minimum Phred quality score (q). For a detailed discussion of how these parameters affect diversity results, see Bokulich et al. (2013). This study recommends standard values for these parameters as $r=3$, $p=75\%$, $q=3$, and $n=0$, which are the default values in the QIIME pipeline. However, the optimal values for these parameters can vary both for individual sequencing runs and for different downstream analyses, for example, analyses such as machine-learning benefit from larger numbers of low-quality sequences, whereas accurate counts of OTUs from a mock community require fewer, higher-quality sequences. Table 19.1 contains an overview of the guidelines presented in Bokulich et al. (2013) for tuning these parameters to a given dataset.

Table 19.1 Overview of the guidelines to tune up the quality-filtering parameters

Dataset characteristics	***q***	***p***	***r***	**Results**
Majority of high-quality, full-length sequences	Increase	Increase	–	Retrieving full-length sequences with low error rates, increasing the discovery rate of rare OTUs
Short reads or reads truncated by early low-quality base calls	–	Lower	Increase	Retain lower quality but taxonomic useful reads
Maximize read count for machine-learning tools, cross-metadata OTU counts comparison, etc.	–	Lower	–	Increased sample size

Adapted from Bokulich et al. (2013).

The Illumina quality-filtering approach differs in its fundamental principles from 454 denoising (Quince et al., 2009; Reeder & Knight, 2010). 454 denoising is based on flowgram clustering (Quince et al., 2009; Quince, Lanzen, Davenport, & Turnbaugh, 2011) and is primarily targeted at reducing homopolymer runs, which are not a problem on the Illumina platform to the same extent. In contrast, the Illumina quality-filtering is based on a per-base Phred quality score and does not target indels.

The QIIME quality-filtering process works as follows. Starting at the beginning of the sequence, QIIME checks that the next *r* Phred values exceed the user-defined quality threshold *q*. If the test is positive, it continues slicing the window of *r* bases until the test fails, or the end of the sequence is reached. The sequence is then trimmed to the last base that met the quality threshold. The next test determines whether the length of the trimmed sequence exceeds *p*, expressed as the percentage of length of the raw sequence. If this check fails, the sequence is excluded. Otherwise, QIIME performs the last check on the sequence, which counts the number of ambiguous characters (*N*) in the trimmed sequence and checks that it is less than *n*. If the test fails, the sequence is rejected. QIIME combines the demultiplexing, primer removal, and quality-filtering processes in a single script, `split_libraries_fastq.py`:

```
split_libraries_fastq.py -i $PWD/IQ_Bio_16sV4_L001_sequences.fastq.
   gz -b $PWD/IQ_Bio_16sV4_L001_sequences_barcodes.fastq.gz -m $PWD/
   IQ_Bio_16sV4_L001_map.txt -o $PWD/slout --rev_comp_mapping_barcodes
```

In our example dataset, we used the `--rev_comp_mapping_barcodes` option in order to indicate that the barcodes present in the mapping file are reverse complements of Golay 12 barcodes. We used the recommended default parameters for quality-filtering on this dataset. However, to change the values for the *r*, *p*, *n*, and *q* quality-filtering parameters, we can use the -r, -p, -n, and -q options to the script. This command will write a fasta-formatted file in the *slout* folder, called *seqs.fna*, which contains the demultiplexed sequences that pass the quality filter. Each sequence contains the information about which sample it came from encoded in the name of the sequence.

Multiple lanes of Illumina fastq data can be processed together in a single call to the script, just by passing the sequence files, the barcode files, and the mapping files in the same order to the -i, -b, and -m options, respectively. For example, with two lanes, the command would look like:

```
split_libraries_fastq.py -i sequences1.fastq,sequences2.fastq
   -b sequences1_barcodes.fastq,sequences2_barcodes.fastq
   -m mapping1.txt,mapping2.txt -o slout
```

The user can check how many sequences have been demultiplexed and passed quality-filtering by using the `count_seqs.py` command. This command also shows the mean and standard deviation of the sequence length:

```
count_seqs.py -i $PWD/slout/seqs.fna
   12687021: slout/seqs.fna (Sequence lengths (mean +/- std): 150.9989
   +/- 0.1715)
   12687021: Total
```

4.1.2 OTU picking

The next step is clustering the preprocessed sequences into OTUs, which in traditional taxonomy represent groups of organisms defined by intrinsic phenotypic similarity that constitute candidate taxa (Sneath & Sokal, 1973; Sokal & Sneath, 1963). For DNA sequence data, these clusters, and hence the OTUs, are formed based on sequence identity. In other words, sequences are clustered together if they are more similar than a user-defined identity threshold, presented as a percentage (*s*). This level of threshold is traditionally set at 97% of sequence similarity, conventionally assumed to represent bacterial species (Drancourt et al., 2000); other levels approximately represent other taxa, although the fit between molecular data and traditional taxonomy varies for different taxa. QIIME supports three approaches for OTU picking (*de novo*, closed-reference, and open-reference) and multiple algorithms for each of these approaches (Table 19.2). The *de novo* approach (Fig. 19.2A) groups sequences based on sequence identity. The closed-reference approach (Fig. 19.2B) matches sequences to an existing database of reference sequences. If a sequence fails to match the database, it is discarded. The open-reference approach (Fig. 19.2C) also starts with an existing database and tries to match the sequences against them. However, if a sequence does not match the database, it is added to the database as a new reference sequence.

The OTU picking strategies shown in Fig. 19.2 are built on top of algorithms for *de novo* clustering. Of the various algorithms available, the furthest-neighbor, average-neighbor, or nearest-neighbor in mothur (Schloss & Handelsman, 2005; Schloss et al., 2009) (also named complete linkage, average linkage, and single linkage, respectively) are the most

Table 19.2 Supported OTU picking methods in QIIME with a brief description of the algorithm employed and in which OTU picking approach can be used

Method	Picking approach			Description	References
	De novo	**Closed-reference**	**Open-reference**		
cd-hit	Yes	–	–	Applies a "longest-sequence-first list removal algorithm" to cluster sequences	Li and Godzik (2006) and Li, Jaroszewski, and Godzik (2001)
Mothur	Yes	–	–	Takes an aligned set of sequences and clusters them using a nearest-neighbor, furthest-neighbor, or average-neighbor algorithm	Schloss et al. (2009)
Prefix/suffix	Yes	–	–	Clusters sequences which are identical in their first and/or last bases	QIIME team, unpublished
Trie	Yes	–	–	Clusters sequences which are identical sequences and sequences which are subsequences of other sequences	QIIME team, unpublished
blast	–	Yes	–	Compares and clusters each sequence against a reference database of sequences	Altschul et al. (1990)
uclust	Yes	Yes	Yes	Creates seed sequences which generate clusters based on percent identity	Edgar (2010)
usearch	Yes	Yes	Yes	Creates seed sequences which generate clusters based on percent identity, filters low-abundance clusters, and performs *de novo* and reference-based chimera detection	Edgar (2010)

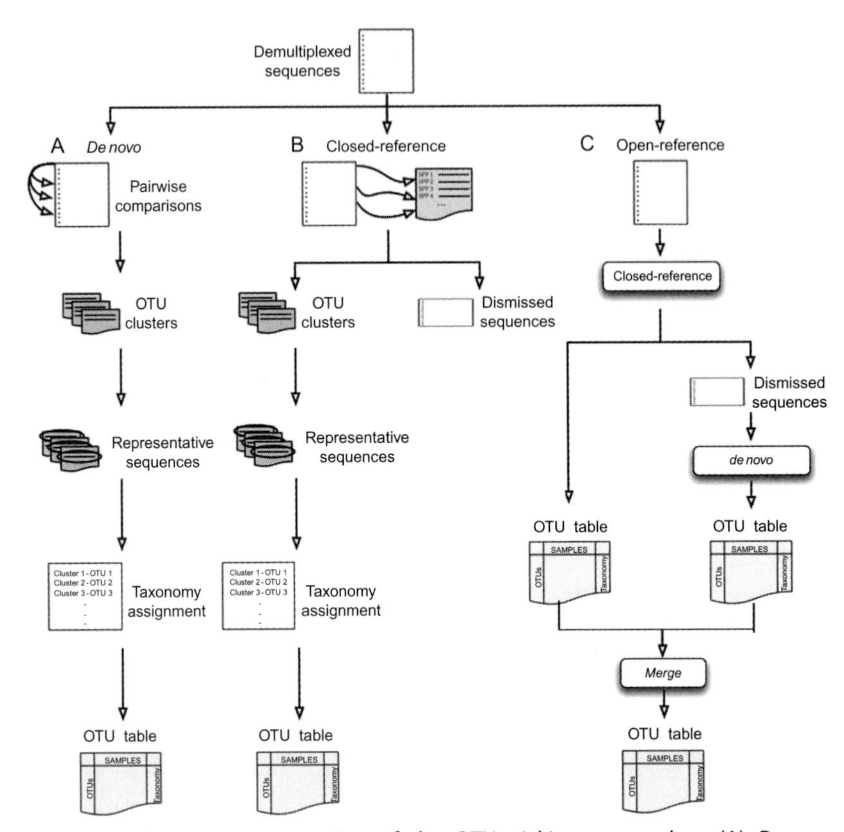

Figure 19.2 Cartoon representation of the OTU picking approaches. (A) *De novo*, (B) closed-reference, and (C) open-reference OTU picking, respectively. In the *de novo* method, sequences are compared to each other and then clusters are formed. In the closed-reference method, sequences are compared directly to a reference dataset (e.g., GreenGenes). Sequences that match a reference sequence are clustered; the remaining sequences are discarded. In both OTU picking methods, once clusters are formed, a representative sequence is selected and then taxonomy is assigned to that sequence (and applied to the rest of the sequences that make up the OTU). Open-reference combines the closed-reference and open-reference methods. The first step is identical to closed-reference, sequences discarded in the first step are clustered into OTUs by the *de novo* method, and both OTU tables are merged into a single final OTU table. *De novo* and open-reference cluster all the sequences, but closed-reference allows better comparisons between studies, especially those using different primers, because all OTUs occur in a common reference space. (For color version of this figure, the reader is referred to the online version of this chapter.)

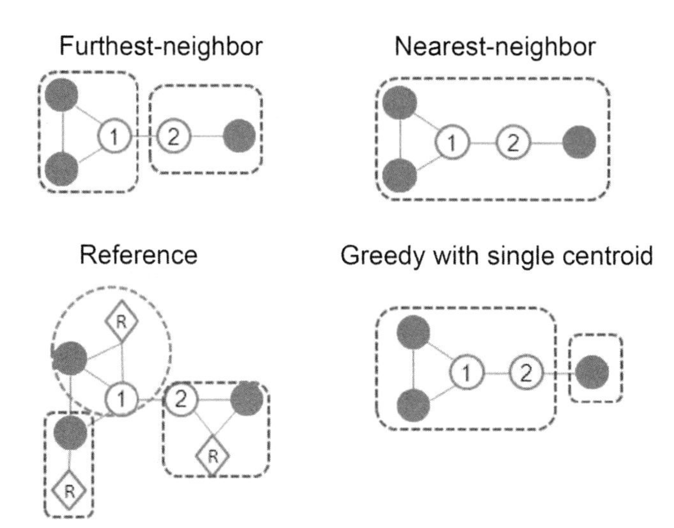

Figure 19.3 Cartoon demonstrating different clustering algorithms. Circles representing sequences linked with lines are within the distance threshold. The two numbered sequences are the first and second sequences in order in the file. The reference algorithms only consider the distance between reference (R) and sequences.

widely used. Furthest-neighbor requires that each sequence is closer than the distance threshold to every other sequence already in the OTU (Fig. 19.3). Average-neighbor requires that the average pairwise distance of all sequences in the OTU is closer than the distance threshold. Nearest-neighbor requires that each sequence is closer than the distance threshold to any sequence already in the OTU. Because these three algorithms are variants on hierarchical clustering, they require loading the distance matrix (proportional to the square of the number of dereplicated sequences) into memory and are therefore challenging to apply to large datasets (e.g., larger than 10^5 sequences). The OTUs yield by these three algorithms also change their memberships at different sequencing depths (i.e., the number of sequences chosen for clustering), which can be a problem for estimates of total OTU numbers (Roesch et al., 2007).

A solution to the distance matrix problem comes from uclust and usearch, which are greedy algorithms based on using a single centroid in each OTU (Edgar, 2010). The centroid could be either from a reference database (usearch) or identified *de novo* from the sequence dataset (both uclust and usearch) (Fig. 19.3). Sequences are serially compared to centroids in a user-defined order (usually decreasing abundance). If a sequence falls within the distance threshold of more than one centroid, the new sequence can be grouped with either the first centroid encountered or the one with the closest distance. Both uclust and usearch are much more efficient than the hierarchical methods, and they do not need to load a large distance matrix into memory (although recent versions of mothur also avoid the constraint of loading the full distance matrix). Usearch is the default *de novo* OTU picking method in QIIME. Note that it is essential to note both your OTU picking strategy, and, if *de novo* OTU picking is used, which algorithm you used to do it: it is not sufficient simply to state that you used a 97% threshold.

Because the OTU picking approach selection is a critical point in microbial community analysis, the QIIME team has produced a detailed document that describes the OTU picking protocols, their advantages, and limitations (https://github.com/qiime/qiime/blob/master/doc/tutorials/otu_picking.rst). Table 19.3 compares the different OTU picking approaches and gives guidelines for choosing an appropriate OTU picking strategy.

The recommended OTU picking approach is open-reference OTU picking, because this approach provides the best trade-off between the time taken to complete the analysis and the ability to discover novel diversity.

Once the sequences have been clustered into OTUs, a representative sequence is picked for each OTU. The entire cluster will thus be represented by a single sequence, speeding up subsequent steps (because redundant sequences need not be considered). QIIME allows the representative sequence to be selected using several techniques: choosing a sequence at random, choosing the longest sequence, and the most-abundant sequence or the first sequence. If using uclust or usearch (Edgar, 2010), the cluster seed will be used as the representative sequence. The default behavior in QIIME is to use the most abundant sequence in each OTU as the representative sequence, because these sequences are least likely to represent sequencing errors (for other applications, such as clustering with near-full-length Sanger sequences, it may be more desirable to pick the longest sequence instead). In case of closed-reference OTU picking, sequences from the reference collection should be used as the representative sequences, which is the default behavior when the closed-reference approach is selected.

4.1.3 Identify chimeric sequences

During the PCR amplification process, some of the amplified sequences can be produced from multiple parent sequences, generating sequences known as chimeras. Although these sequences are technical artifacts rather than

Table 19.3 OTU picking approaches comparison

	De novo	Closed-reference	Open-reference
Must use if	There is no reference sequence collection to cluster against (e.g., infrequently used marker gene)	Comparing nonoverlapping amplicons. The reference set of sequences must span both of the regions being sequenced	–
Cannot use if	Comparing nonoverlapping amplicons (e.g., V2 and V4 regions of 16S rRNA)	There is no reference sequence collection to cluster against (e.g., infrequently used marker gene)	Comparing nonoverlapping amplicons (e.g., V2 and V4 regions of 16S rRNA) There is no reference sequence collection to cluster against (e.g., infrequently used marker gene)
Pros	All reads are clustered	Fast, as it is fully parallelizable (useful for extremely large datasets) Better tree and taxonomy quality since the OTUs are already defined on the reference set	All reads are clustered. Fast, as is partially run on parallel
Cons	Time consuming since it runs in serial	Inability to detect novel diversity with respect to the reference set because the reads that do not hit the reference sequence collection are discarded, so the analysis focus on the "already known" diversity If the studied environment is not well characterized, a large fraction of the reads can be thrown away	There are still some steps performed in serial. If the data set contains a lot of novel diversity with respect to the reference set, this can still be slow

The table shows when each of the OTU picking approaches should be used and when they cannot be applied. It briefly describes the advantages and disadvantages of using each of the OTU picking approaches.

representing actual members of the community, chimeric sequences are important for alpha-diversity estimates (although they are less important for cross-sample comparisons, because each chimera is relatively rare and the same chimera is unlikely to be generated systematically in different samples; Ley et al., 2008). However, the same chimera can sometimes be generated in multiple PCR reactions, for example, Haas et al. (2011) reported that chimeric sequences formed from *Streptococcus* and *Staphylococcus* occurred multiple times independently, so the presence of the same sequence in multiple PCRs does not mean that it is not chimeric.

QIIME currently supports three different methods for detecting chimeras: blast fragments, a taxonomy-assignment-based approach using BLAST (Altschul, Gish, Miller, Myers, & Lipman, 1990); ChimeraSlayer (Haas et al., 2011), which uses BLAST to identify potential chimera parents; and usearch 6.1 (Edgar, 2010), which can perform *de novo* chimera detection based on abundances as well as reference-based chimera detection. The recommended method for identifying chimeric sequences is uchime (Edgar, Haas, Clemente, Quince, & Knight, 2011), which is integrated in the usearch 6.1 (Edgar, 2010) pipeline. Uchime is the fastest method for detecting chimeric sequences, and it is executed by default if the usearch method is selected for picking OTUs.

4.1.4 Taxonomy assignment

The next step in the QIIME workflow is to assign the taxonomy to each sequence of the representative set. This step connects the OTUs to named organism, which is useful for inferring likely functional roles for members of the community. When using a closed-reference approach for OTU picking, the taxonomy of the sequences can be pulled out from the reference set. In case of the open-reference and *de novo* approaches, because the clusters are not created from any reference database (as a reminder, in the open-reference approach, sequences that fail to cluster to the reference database form new clusters), the taxonomy should be assigned using a reference dataset. We recommend the GreenGenes database (DeSantis et al., 2006; McDonald, Price, et al., 2012) as the default reference data set for assigning taxonomy, although the RDP (Cole et al., 2009) and Silva (Quast et al., 2013) databases also have strengths and weaknesses relative to GreenGenes and should be considered for some analyses. Silva includes microbial eukaryotes and has invested substantial effort in cleaning up marine taxa; RDP has close links to formally recognized names in taxonomy, which can be especially useful for medical microbiology. QIIME can assign taxonomy against any of the given databases, or against a custom database, using several methods: BLAST (Altschul et al., 1990), RDP classifier (Wang et al., 2007), rtax (Soergel, Dey, Knight, & Brenner, 2012), mothur (Schloss et al., 2009), and tax2tree (McDonald, Price, et al., 2012). The QIIME team recommends the RDP classifier method (Wang et al., 2007) with a confidence value of 0.8. However, if the user has paired-end reads, the best method to use is the rtax (Soergel et al., 2012), and the user should provide the fasta files with both the first and the second reads from the paired-end sequencing. Note that the taxonomy assignment method and the reference database must both be described in order for an analysis to be reproducible, and that these methods can have a larger effect on taxonomy than the underlying biological sample, so it is important to be consistent (Liu, DeSantis, Andersen, & Knight, 2008).

4.1.5 Sequence alignment

The next step in the QIIME workflow is to align the sequences. The sequences must be aligned to infer a phylogenetic tree, which is used for diversity analyses and to understand the relationships among the sequences in the sample. Currently, QIIME supports the following methods for performing sequence alignment: PyNAST (Caporaso, Bittinger, et al., 2010), Infernal (Nawrocki, Kolbe, & Eddy, 2009), clustalw (Larkin et al., 2007), muscle (Edgar, 2004), and mafft (Katoh, Misawa, Kuma, & Miyata, 2002). The recommended (and default) method is PyNAST (Caporaso, Bittinger, et al., 2010). This method aligns the sequences against a template sequence alignment, for which we recommend the GreenGenes core set (DeSantis et al., 2006).

When sequences do not align well using PyNAST, the Infernal package (Nawrocki et al., 2009) should be used. Like PyNAST, it requires a template alignment, but unlike PyNAST, it uses stochastic context-free grammars to align incorporating secondary structure. Although this method is slow compared to other methods, it does takes advantage of RNA secondary structure (provided in a Stockholm-format file) and can be useful for aligning more variable rRNAs. For marker genes other than rRNA genes, the best strategy for building phylogenetic trees is to align the protein sequences (if available) using muscle.

4.1.6 Phylogeny construction

This step in the QIIME workflow infers a phylogenetic tree from the multiple sequence alignment generated by the previous step. The phylogenetic tree represents the relationships among sequences in terms of the amount of sequence evolution from a common ancestor. This phylogenetic tree is used in many downstream analyses, such as the UniFrac metric (Lozupone et al., 2005) for beta-diversity.

The current methods supported for inferring the phylogenetic tree in QIIME are FastTree (Price, Dehal, & Arkin, 2009), clearcut (Evans, Sheneman, & Foster, 2006), clustalw (Larkin et al., 2007), raxml (Stamatakis, Ludwig, & Meier, 2005), and muscle (Edgar, 2004). The default and recommended method in QIIME is the FastTree (Price et al., 2009) method because it shows the best trade-off between run time and reliability of the inferred tree.

4.1.7 Make OTU table

The last part of the upstream stage in QIIME is to construct the OTU table. The OTU table is a sample by observation matrix that also includes the taxonomic prediction for each OTU. For the OTU table representation, QIIME uses the Genomics Standards Consortium *candidate standard* Biological Observation Matrix (BIOM) format (McDonald, Clemente, et al., 2012). The OTU table, the mapping file, and the phylogenetic tree are the main files for performing the downstream analysis.

QIIME can perform all the steps for generating the OTU table and the phylogenetic tree from the preprocessed data in a single command. There is a separate command for each OTU picking approach. In the following commands, we assume that the GreenGenes reference files (DeSantis et al., 2006) are located in the current directory. As a remainder, our seqs.fna has 12.687.021 sequences of length 150.9989 ± 0.1715:

- For *de novo* (run time ~80 h on 1 processor (not parallelizable)):

```
pick_de_novo_otus.py -i $PWD/slout/seqs.fna -o $PWD/denovo_otus
```

- For closed-reference (run time ~2 h on 20 processors):

```
pick_closed_reference_otus.py -i $PWD/slout/seqs.fna -o $PWD/
    closed_ref_otus -r $PWD/gg_12_10_otus/rep_set/97_otus.fasta -t
    $PWD/gg_12_10_otus/taxonomy/97_otu_taxonomy.txt -a -O 20
```

- For open-reference (run time ~27 h on 20 processors):

```
pick_open_reference_otus.py -o $PWD/open_ref_otus -i $PWD/slout/
    seqs.fna -r $PWD/gg_12_10_otus/rep_set/97_otus.fasta -a -O 20
```

Because the closed-reference and open-reference OTU picking approaches can be run in parallel, we use the -a and -O 20 options in order to run them using 20 processors.

4.2. Downstream analysis steps

Once we have generated the OTU table and the phylogenetic tree, we can start the downstream analysis. At this point, we strongly recommend performing a second level of quality-filtering based on OTU abundance. The recommended procedure is to discard those OTUs with a number of sequences <0.005% of the total number of sequences (see Bokulich et al., 2013 for a detailed description of the effect of this parameter in further downstream analyses). QIIME executes the OTU abundance quality-filtering step through the script `filter_otus_from_otu_table.py`:

```
filter_otus_from_otu_table.py -i $PWD/open_ref_otus/
   otu_table_mc2_w_tax_no_pynast_failures.biom -o $PWD/
   open_ref_otus/otu_table_filtered.biom --min_count_fraction 0.00005
```

This step greatly reduces the problem of spurious OTUs, most of which are present at very low abundance.

QIIME 1.7.0 allows a first-pass view of common diversity analyses using a single command: `core_diversity_analysis.py`. One of the parameters required by this command is the sampling depth, the number of sequences that should be included in each sample for diversity analyses. This is required, because many of the commonly used diversity metrics are very sensitive to the number of sequences obtained per sample, such that samples that are similar in the number of sequences that were obtained appear similar to one another. This is bad because the number of sequences per sample is typically a methodological artifact, not reflective of biological reality. The sampling depth defines the size of the random subset of sequences that will be selected for each sample for all subsequent diversity analyses.

The optimal sampling depth is data dependent. There is no universal way of choosing a rarefaction level, although heuristics can be applied. For example, if most samples have more than 10,000 sequences and the rest range from 500 to 50 sequences per sample, it would be recommended to use 10,000 as the rarefaction level. Although many studies show marked variation in sequence depth with only a few "bad" samples, it is not always easy to choose the rarefaction level. We strongly recommend rarefying over 1000 sequences per sample for Illumina MiSeq, because samples below this level often suffer from other quality issues as well.

The information needed to choose the rarefaction level can be obtained from the script `print_biom_table_summary.py`, which shows summary information on the OTU table such as the number of sequences, the number of

OTUs, the number of samples, and the number of counts per sample, among others:

```
print_biom_table_summary.py -i $PWD/open_ref_otus/otu_table_filtered.
    biom
Num samples: 90
Num observations: 783
Total count: 10637688.0
Table density (fraction of non-zero values): 0.4289
Table md5 (unzipped): eb0f1d7fbb50bc31695dade31db1e198

Counts/sample summary:
Min: 1.0
Max: 493427.0
Median: 99111.0
Mean: 118196.533333
Std. dev.: 94277.5956531
Sample Metadata Categories: None provided
Observation Metadata Categories: taxonomy

Counts/sample detail:
BLANK4.732555: 1.0
BLANK5.732537: 1.0
Joshua.Jose.WTAbd.732533: 1.0
Nick.Krishna.TG.Fec.732513: 2.0
TH.CVA.WT.Oral.732491: 2.0
BLANK2.732552: 3.0
BLANK3.732479: 5.0
BLANK6.732470: 7.0
Elizabeth.Chris.WT.Abd.732490: 10.0
Uri.Jake.TGAbd.732468: 10.0
TH.CVA.WT.Abd.732477: 13.0
BLANK10.732524: 812.0
Elizabeth.Chris.WT.Oral.732520: 7410.0
Elizabeth.Chris.WT.Col.732481: 21746.0
Jordan.Lisette.TG.Ile.732463: 27149.0
...
TH.CVA.WT.Fec.732553: 372327.0
Wang.TG.Cec.732527: 396391.0
TH.CVA.WT.Ile.732517: 493427.0
```

In the above output, we can see the information contained in the OTU table resulting from applying the open-reference OTU picking. Some of the relevant information contained in this output is the total number of samples (90), the total number of OTUs (783), the number of reads (10,637,688), and the number of OTUs per sample. Applying the above heuristic, we could select a subsampling depth of 7410 sequences. However, because we have run three different OTU picking approaches and we want to compare them, we must search for the rarefaction level that best fits the three OTU tables. Below are the summarized information for the *de novo* OTU table and the closed-reference OTU table, respectively:

```
print_biom_table_summary.py -i $PWD/denovo_otus/otu_table_filtered.
   biom
Num samples: 93
Num observations: 600
Total count: 11122386.0
Table density (fraction of non-zero values): 0.4344
Table md5 (unzipped): b002dd85c93fd9d0571ff23b05d21dde

Counts/sample summary:
Min: 0.0
Max: 497234.0
Median: 108322.0
Mean: 119595.548387
Std. dev.: 93487.3335598
Sample Metadata Categories: None provided
Observation Metadata Categories: taxonomy

Counts/sample detail:
BLANK7.732497: 0.0
BLANK8.732522: 0.0
Jordan.Lisette.TG.Abd.732467: 0.0
BLANK4.732555: 1.0
BLANK5.732537: 1.0
Joshua.Jose.WTAbd.732533: 1.0
BLANK2.732552: 3.0
Nick.Krishna.TG.Fec.732513: 3.0
TH.CVA.WT.Oral.732491: 3.0
BLANK3.732479: 5.0
```

```
BLANK6.732470: 9.0
Elizabeth.Chris.WT.Abd.732490: 10.0
Uri.Jake.TGAbd.732468: 10.0
TH.CVA.WT.Abd.732477: 13.0
BLANK10.732524: 825.0
Elizabeth.Chris.WT.Oral.732520: 7376.0
Joey.Aaron.Kyle.WT.Abd.732541: 35655.0
...
Wang.TG.Cec.732527: 394351.0
TH.CVA.WT.Ile.732517: 497234.0
print_biom_table_summary.py -i $PWD/closed_ref_otus/
   otu_table_filtered.biom
Num samples: 90
Num observations: 673
Total count: 9434459.0
Table density (fraction of non-zero values): 0.4250
Table md5 (unzipped): 257b528478a2700c72f979ce8d9a9a1c

Counts/sample summary:
Min: 1.0
Max: 347785.0
Median: 90092.0
Mean: 104827.322222
Std. dev.: 78560.4683831
Sample Metadata Categories: None provided
Observation Metadata Categories: taxonomy

Counts/sample detail:
BLANK4.732555: 1.0
BLANK5.732537: 1.0
Joshua.Jose.WTAbd.732533: 1.0
BLANK3.732479: 2.0
Nick.Krishna.TG.Fec.732513: 2.0
TH.CVA.WT.Oral.732491: 2.0
BLANK2.732552: 3.0
Uri.Jake.TGAbd.732468: 5.0
BLANK6.732470: 7.0
Elizabeth.Chris.WT.Abd.732490: 10.0
TH.CVA.WT.Abd.732477: 12.0
BLANK10.732524: 710.0
```

```
Elizabeth.Chris.WT.Oral.732520: 7205.0
Elizabeth.Chris.WT.Col.732481: 22652.0
...
TH.CVA.WT.Fec.732553: 329988.0
TH.CVA.WT.Ile.732517: 347785.0
```

From the above output, we see that a reasonable rarefaction level for the three tables is 7205 counts per sample, derived from the closed-reference OTU picking.

Once the subsampling depth is chosen, we can execute the `core_diversity_analyses.py` command over the three OTU tables. We provide the subsampling depth via the -e parameter, the OTU table via the -i parameter, the mapping file through the -m parameter, and the metadata categories to use in categorical analyses through the -c parameter. The -o parameter is used to provide the output directory and the -a -O 64 are used to run the command in parallel using 64 processes.

```
mkdir $PWD/diversity_analysis

core_diversity_analyses.py -i $PWD/open_ref_otus/otu_table_filtered.
   biom -m $PWD/IQ_Bio_16sV4_L001_map.txt -t $PWD/open_ref_otus/
   rep_set.tre -e 7205 -c GENOTYPE,BODY_SITE -o $PWD/
   diversity_analysis/open_ref -a -O 64

core_diversity_analyses.py -i $PWD/denovo_otus/otu_table_filtered.
   biom -m $PWD/IQ_Bio_16sV4_L001_map.txt -t $PWD/denovo_otus/
   rep_set.tre -e 7205 -c GENOTYPE,BODY_SITE -o $PWD/
   diversity_analysis/denovo -a -O 64

core_diversity_analyses.py -i $PWD/closed_ref_otus/
   otu_table_filtered.biom -m $PWD/IQ_Bio_16sV4_L001_map.txt -t $PWD/
   gg_12_10_otus/trees/97_otus.tree -e 7205 -c GENOTYPE,BODY_SITE -o
   $PWD/diversity_analysis/closed_ref -a -O 64
```

The `core_diversity_analyses.py` command filters the OTU table before executing the diversity analyses. The filter removes samples from the OTU table that do not have at least the user-defined subsampling depth (7205 in our case). This filtering removes low-coverage samples from the OTU table, because they are not informative enough to be included in the study. After these samples have been filtered, the script performs the rarefaction step at the given subsampling depth.

The output of this script is an HTML file that can be opened in a Web browser (Fig. 19.4). This HTML file gives access to the results of the different diversity analysis performed (taxa summaries, α-diversity, β-diversity, and category significance) which will be explained in further sections.

For the following downstream analysis, we have used the OTU table and phylogenetic tree resulting from the open-reference OTU picking approach. In cases where we are performing comparisons between OTU picking approaches, we will specify which approaches we have used.

4.2.1 Taxa summaries

One way to visualize the OTUs in each sample is a taxa summary, which summarizes the relative abundance of the taxa present in a set of samples on multiple taxonomic levels (e.g., phylum, order, etc.) (see Fig. 19.5). This provides a quick way to identify samples that may be drastically different from others (i.e., outliers) and visually identify expected patterns and differences between and among samples. For example, this tool can be used to identify patterns such as differences in the relative abundance of Firmicutes and Bacteroidetes in the gut microbiomes of lean versus obese mice, e.g., Ley, Backhed, Turnbaugh, Lozupone, Knight, and Gordon (2005). In our example, the taxa summary shows that the fecal, colon, and cecum samples appear to be similar in composition in that their dominant phyla are present in similar relative abundances. These patterns can then be statistically tested using other methods, either within QIIME or elsewhere. QIIME contains a workflow called `summarize_taxa_through_plots.py` that generates user-specified plot types, including bar, pie, and area graphs. These graphs provide a way to visually compare the composition of each sample or of groups of samples. An OTU table with assigned taxonomies is the only required input file, and options allow the user to summarize across categories (using the metadata file), at different taxonomic levels, or only using OTUs that are present at abundances higher or lower than user-defined thresholds. The Web interface allows a scroll-over feature that identifies the taxonomy of the separate taxa. Additional output files include image files of the charts and legends, and tab-delimited files of the calculated abundances, which can then be further filtered and manipulated for downstream statistical analyses.

4.2.2 Diversity analysis

Microbial ecology studies the diversity of microorganisms by characterizing bacterial communities in different environments and determining the factors that drive diversity in these communities (Atlas & Bartha, 1998). Whittaker (1960) and Whittaker (1972) define different types of measurements of

Run summary data	
Master run log	log_20130607115410.txt
BIOM table statistics	biom_table_summary.txt
Filtered BIOM table (minimum sequence count: 7205)	table_mc7205.biom.gz
Beta diversity results (even sampling: 7205)	
Distance boxplots (weighted_unifrac)	GENOTYPE_Distances.pdf
Distance boxplots statistics (weighted_unifrac)	GENOTYPE_Stats.txt
Distance boxplots (weighted_unifrac)	BODY_SITE_Distances.pdf
Distance boxplots statistics (weighted_unifrac)	BODY_SITE_Stats.txt
3D plot (weighted_unifrac, continuous coloring)	weighted_unifrac_pc_3D_PCoA_plots.html
3D plot (weighted_unifrac, discrete coloring)	weighted_unifrac_pc_3D_PCoA_plots.html
2D plot (weighted_unifrac, continuous coloring)	weighted_unifrac_pc_2D_PCoA_plots.html
2D plot (weighted_unifrac, discrete coloring)	weighted_unifrac_pc_2D_PCoA_plots.html
Distance matrix (weighted_unifrac)	weighted_unifrac_dm.txt
Principal coordinate matrix (weighted_unifrac)	weighted_unifrac_pc.txt
Distance boxplots (unweighted_unifrac)	GENOTYPE_Distances.pdf
Distance boxplots statistics (unweighted_unifrac)	GENOTYPE_Stats.txt
Distance boxplots (unweighted_unifrac)	BODY_SITE_Distances.pdf
Distance boxplots statistics (unweighted_unifrac)	BODY_SITE_Stats.txt
3D plot (unweighted_unifrac, continuous coloring)	unweighted_unifrac_pc_3D_PCoA_plots.html
3D plot (unweighted_unifrac, discrete coloring)	unweighted_unifrac_pc_3D_PCoA_plots.html
2D plot (unweighted_unifrac, continuous coloring)	unweighted_unifrac_pc_2D_PCoA_plots.html
2D plot (unweighted_unifrac, discrete coloring)	unweighted_unifrac_pc_2D_PCoA_plots.html
Distance matrix (unweighted_unifrac)	unweighted_unifrac_dm.txt
Principal coordinate matrix (unweighted_unifrac)	unweighted_unifrac_pc.txt
Taxonomic summary results	
Taxa summary bar plots	bar_charts.html
Taxa summary area plots	area_charts.html
Taxonomic summary results (by BODY_SITE)	
Taxa summary bar plots	bar_charts.html
Taxa summary area plots	area_charts.html
Taxonomic summary results (by GENOTYPE)	
Taxa summary bar plots	bar_charts.html
Taxa summary area plots	area_charts.html
Category results	
Category significance (GENOTYPE)	category_significance_GENOTYPE.txt
Category significance (BODY_SITE)	category_significance_BODY_SITE.txt
Alpha diversity results	
Alpha rarefaction plots	rarefaction_plots.html
Alpha diversity statistics (GENOTYPE, PD_whole_tree)	GENOTYPE_PD_whole_tree.txt
Alpha diversity statistics (GENOTYPE, observed_species)	GENOTYPE_observed_species.txt
Alpha diversity statistics (GENOTYPE, chao1)	GENOTYPE_chao1.txt
Alpha diversity statistics (BODY_SITE, PD_whole_tree)	BODY_SITE_PD_whole_tree.txt
Alpha diversity statistics (BODY_SITE, observed_species)	BODY_SITE_observed_species.txt
Alpha diversity statistics (BODY_SITE, chao1)	BODY_SITE_chao1.txt

Need help?
You can get answers to your questions on the QIIME Forum.
See the QIIME tutorials for examples of additional analyses that can be run.
You can find documentation of the QIIME scripts in the QIIME script index.

Figure 19.4 HTML result from `core_diversity_analyses.py`. This HTML file summarizes and gives access to the results of the diversity analyses conducted on the given OTU table. (For color version of this figure, the reader is referred to the online version of this chapter.)

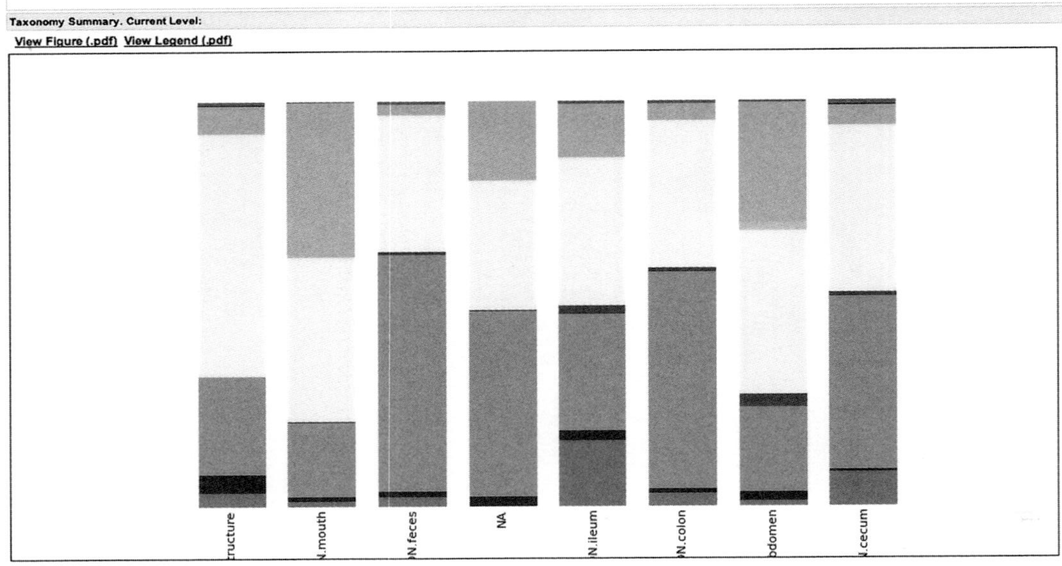

View Table (.txt)

Legend	Taxonomy	Total count	Total %	UBERON.multi.tissue.structure %	UBERON.mouth %	UBERON.feces %	NA %	UBERON.ileum %	UBERON.colon %	UBERON.skin.of.abdomen %	UBERON.cecum %
	k__Bacteria;Other	0	4.5%	3.3%	1.3%	2.5%	0.0%	16.2%	3.0%	1.3%	8.2%
	k__Bacteria;p__Actinobacteria	0	1.9%	4.5%	1.1%	1.1%	2.3%	2.3%	1.2%	2.2%	0.7%
	k__Bacteria;p__Bacteroidetes	3	36.5%	24.2%	18.2%	58.4%	45.6%	28.7%	53.4%	20.9%	42.4%
	k__Bacteria;p__Cyanobacteria	0	0.3%	0.1%	0.1%	0.0%	0.1%	0.3%	0.2%	1.3%	0.0%
	k__Bacteria;p__Deferribacteres	0	0.8%	0.0%	0.2%	0.7%	0.0%	1.7%	0.7%	1.7%	1.0%
	k__Bacteria;p__Firmicutes	3	40.2%	60.0%	40.4%	33.9%	32.4%	37.1%	36.5%	40.3%	41.4%
	k__Bacteria;p__Fusobacteria	0	0.3%	0.0%	0.0%	0.0%	0.0%	0.0%	0.0%	2.2%	0.0%
	k__Bacteria;p__Proteobacteria	1	14.9%	6.9%	38.4%	2.5%	19.5%	12.9%	4.1%	29.7%	5.1%
	k__Bacteria;p__Spirochaetes	0	0.0%	0.0%	0.0%	0.0%	0.0%	0.0%	0.0%	0.0%	0.0%
	k__Bacteria;p__TM7	0	0.1%	0.3%	0.0%	0.1%	0.0%	0.1%	0.1%	0.0%	0.1%
	k__Bacteria;p__Tenericutes	0	0.5%	0.6%	0.2%	0.8%	0.0%	0.7%	0.7%	0.4%	1.0%
	k__Bacteria;p__Verrucomicrobia	0	0.0%	0.1%	0.0%	0.0%	0.0%	0.0%	0.1%	0.0%	0.1%

Figure 19.5 A snapshot of the taxa summary of the example dataset using the web interface. Samples have been grouped and averaged by body site, and taxonomic composition is shown on the phylum level. Each column in the plot represents a body site, and each color in the column represents the percentage of the total sample contributed by each taxon group at phylum level. The taxa summaries plot help us to see which taxon groups are more prevalent in a sample. For example, the fecal samples are dominated by Bacteroidetes, while mouth and skin samples are dominated by Proteobacteria. We can also see that Fusobacteria is only present at appreciable levels in the skin samples. (See color plate.)

diversity as alpha-, beta-, and gamma-diversities. Alpha-diversity is defined as the diversity of organisms in one sample or environment. Beta-diversity is the difference in diversities across samples or environments. Finally, gamma-diversity (γ-diversity) measures the diversity at a broader scale, such as a province or region. Several different metrics of alpha- and beta-diversity are implemented in QIIME pipeline. Diversity measurements and their applications in microbial have been discussed in detail elsewhere (Jost, 2007; Kuczynski, Liu, et al., 2010; Lozupone & Knight, 2008), and here, we focus on examples of their application.

4.2.3 Alpha-diversity analysis

QIIME can generate plots showing the results of alpha-diversity, allowing the user to choose the diversity metric and different rarefaction levels (Fig. 19.6). These images are often used to estimate the true species richness of a community.

QIIME implements dozens of the most widely used alpha-diversity indices, including both phylogenetic indices (which require a phylogenetic tree) and nonphylogenetic indices. Users can obtain a list of the alpha-diversity indices implemented in QIIME by passing the parameter -s to the `alpha_diversity.py` script. Phylogenetic metrics have been especially useful in our experience because they provide additional power by accounting for the degrees of phylogenetic divergence between sequences within each sample. Thus, for alpha-diversity, we recommend phylogenetic distance (PD) (Faith, 1992) over OTU counts; however, the choice of metric will depend on the question. In particular, one might be interested in pure estimates of community richness (such as the observed number of OTUs or the Chao1 estimator of the total number that would be observed with infinite sampling), in pure estimates of evenness, or of measures that combine richness and evenness such as the Shannon entropy (if there is no hypothesis in advance about which richness measure is appropriate, remember to correct for multiple comparisons if many are applied to the same dataset). Here is an example of how to compute rarefaction curves for three different alpha-diversity metrics using a QIIME parameters file:

```
echo "alpha_diversity:metrics shannon,PD_whole_tree,observed_species"
   > alpha_params.txt
alpha_rarefaction.py -i $PWD/open_ref_otus/otu_table_filtered.biom -m
   $PWD/IQ_Bio_16sV4_L001_map.txt -o $PWD/diversity_analysis/
   alpha_rare_open_ref_uneven -a -O 64 -n 20 --min_rare_depth 1000 -e
   340000 -p $PWD/alpha_params.txt -t $PWD/open_ref_otus/rep_set.tre
```

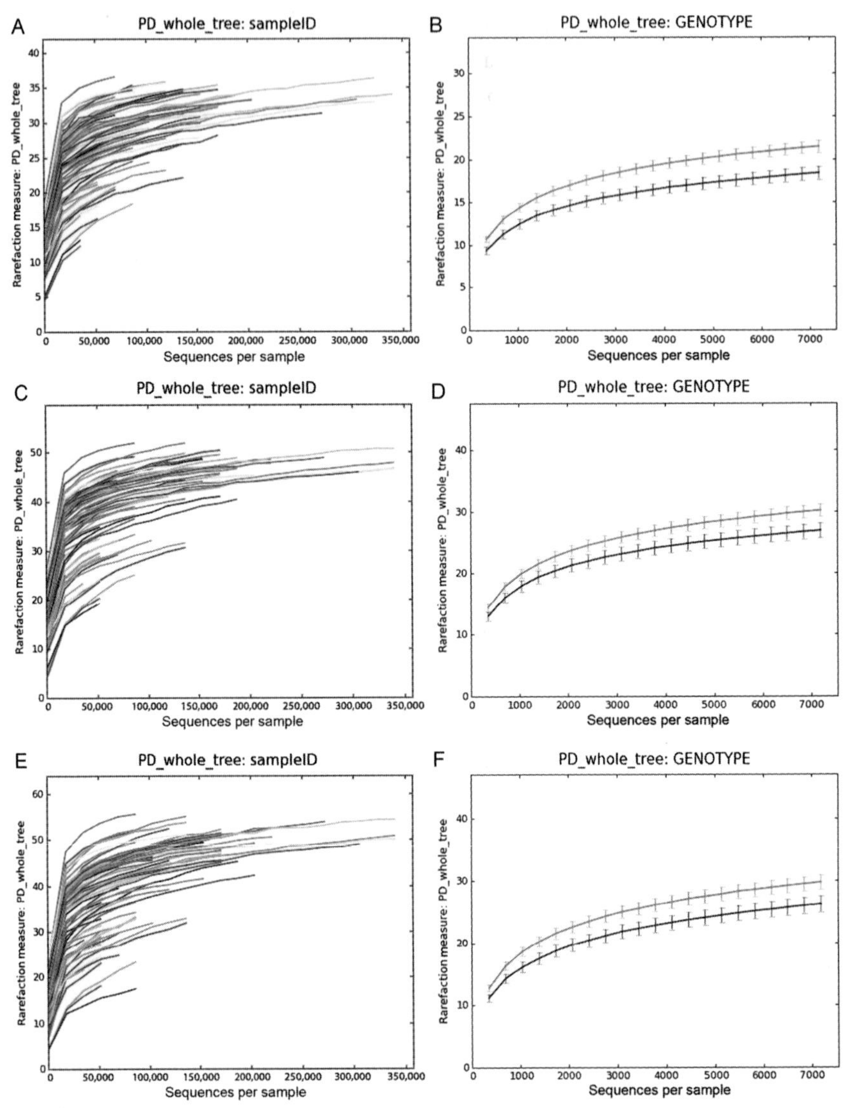

Figure 19.6 Alpha-diversity curves at different rarefaction depths using different OTU picking methods. Each line represents the results of the alpha-diversity phylogenetic diversity whole tree metric (PD whole tree in QIIME). (A), (C), and (E) represent alpha-diversity of each sample at a different sequence depth in each of the OTU picking protocols (closed-, open-reference, and *de novo*). In closed-reference, the diversity plateaus (reaches an asymptote) because only OTUs in the reference database already can be considered, greatly reducing the OTU number over what is possible if the sequences are clustered *de novo*. Comparing these curves is difficult because the sequencing depth differs among samples. (B), (D), and (F) show differences in alpha-diversity between the two mouse genotypes, wild type (WT: orange) and transgenic (TG: blue), using the

This step generates an interactive HTML document with figures showing the results for each alpha-diversity metric and for each group of samples. Curves reach asymptotes when the sequencing effort (sequencing depth) does not contribute additional OTUs. In this sense, curves would differ in their shape as a function of the selected OTU picking method.

Comparisons should be adjusted to the same depth of sequencing. Rarefaction curves can be useful for assessing the sequencing effort sufficient for representing and comparing the microbial communities (Fig. 19.6). However, although rarefaction curves were widely used during the era of Sanger sequencing, when most communities were undersampled, it is often more useful today to report the coverage and the estimated and observed numbers of OTUs at one rarefaction depth rather than to use a figure for rarefaction curves.

4.2.4 Beta-diversity analysis

Beta-diversity can also be calculated from the rarefied OTU tables, comparing the microbial communities based on their compositional structures. As with alpha-diversity, QIIME can compute many phylogenetic and nonphylogenetic beta-diversity metrics (shown by the command `beta_diversity.py -s`). Of these, we have found UniFrac to be most generally useful in revealing biologically meaningful patterns. Unifrac measures the amount of unique evolution within each community with respect to another by calculating the fraction of branch length of the phylogenetic tree that is unique to either one of a pair of communities (Lozupone et al., 2005). QIIME implements several variants of Unifrac, including weighted and unweighted Unifrac. The weighted Unifrac metric is weighted by the difference in probability mass of OTUs from each community for each branch, whereas unweighted Unifrac only considers the absence/presence of the OTUs (Lozupone, Hamady, Kelley, & Knight, 2007). Weighted Unifrac

---

different OTU picking approaches. Both curves show the same rarefaction levels, allowing easier comparisons between categories. The curves again level off, showing that the sequencing effort is sufficient to detect most of the OTUs (this saturation can be confirmed using Good's coverage, or conditional uncovered probability, or other formal coverage statistics). The error bars show the standard error of the mean diversity at each rarefaction level across the multiple iterations. (For interpretation of the references to color in this figure legend, the reader is referred to the online version of this chapter.)

is thus recommended for detecting community differences that arise from differences in relative abundance of taxa, rather than in which taxa are present. Like other metrics considering taxon abundance, weighted Unifrac is sensitive to the bias from DNA extraction efficiency, PCR amplification, etc.; this may explain why, in our hands at least, unweighted UniFrac has often provided results that correlate better with clinical or environmental variables than does weighted UniFrac. The choice of metrics is critical in beta-diversity analysis as metrics differ substantially in their ability to detect clustering or gradient patterns among microbial communities on the same dataset (Arumugam et al., 2011; Ravel et al., 2012; Schloss & Handelsman, 2006). See Kuczynski, Liu, et al. (2010) for a detailed discussion of the performance of different nonphylogenetic metrics.

QIIME calculates the beta-diversities between each pairs of input samples, forming a distance matrix. The distance matrix then can be visualized with methods such as PCoA (Mardia, Kent, & Bibby, 1979) and hierarchical clustering (Tryon, 1939), both of which have been widely used for data visualization for decades. PCoA transforms the original multidimensional matrix to a new set of orthogonal axes that explain the maximum amount of inertia in the dataset (Gower, 1966; Mardia et al., 1979) and the current implementation in QIIME scales to thousands of samples. We are currently evaluating approximate methods that will allow scaling to millions of samples (Gonzalez, Stombaugh, Lauber, Fierer, & Knight, 2012). QIIME allows the PCoA plots to be visualized interactively in three-dimensions, currently using the KiNG viewer (Chen, Davis, & Richardson, 2009). To assess the stability of the PCoA plot, jackknife resampling can be performed on the OTU table, repeating the PCoA procedure for each resampled table and plotting the aggregate results as confidence ellipsoids around the sample points (Fig. 19.7). Jackknifing is recommended because many diversity metrics, including UniFrac, are sensitive to the number of sequences per sample (Lozupone, Lladser, Knights, Stombaugh, & Knight, 2011).

Taxonomic information can be displayed on top of the PCoA using biplots (Fig. 19.8) (this analysis requires the output file from previous taxon summary step). The coordinates of a given taxon are computed as the weighted average of the coordinates of all samples, where the weights are the relative abundances of the given taxon in the set of samples. This plot is particularly suited for identifying taxa that drive the differentiation between groups of microbial communities.

Another popular method for finding relationships among samples is hierarchical clustering, which groups samples together into a tree. Although

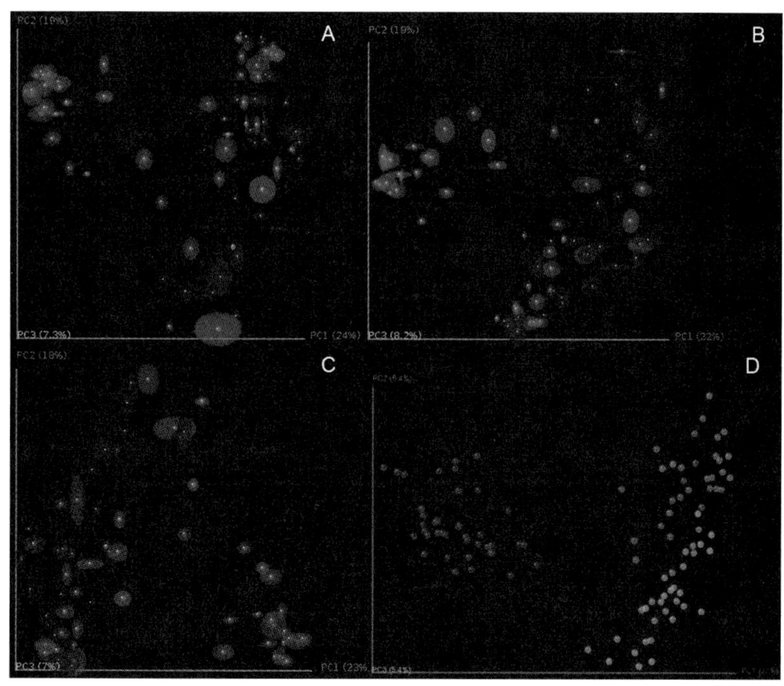

Figure 19.7 PCoA plots of unweighted Unifrac beta-diversity. Panels A–C show jackknifed replicate results for the example data set using *de novo* OTU picking, closed-reference OTU picking, and open-reference OTU picking, illustrating different results from the three OTU picking approaches (Table 19.3). Each dot represents a sample, either from a WT mouse (orange) or TG mouse (blue). The two groups are not clearly separated, probably because the data set is contaminated (recall that this is a class project and different participants varied in their dissection skills). The size of the ellipsoids shows the variation for each sample calculated from jackknife analysis. These plots are generated by the command `jackknifed_beta_diversity.py -i $PWD/denovo_otus/otu_table_filtered.biom -t $PWD/denovo_otus/rep_set.tre -m $PWD/IQ_Bio_16sV4_L001_map.txt -o $PWD/diversity_analysis/jk_denovo -e 7205 -a -O 64` (the input parameters should be adapted for using the OTU tables from different OTU picking approaches). Panel D shows the beta-diversity PCoA plot of a data set from the "keyboard" data set (Fierer et al., 2010) which links individuals to their computer keyboard through microbial community similarity. Each dot represents a microbial community sampled from either fingertips or keyboard keys from three individuals, annotated by the three colors shown in the plot. In contrast to panels A–C, panel D shows the microbial communities well separated by individual in the PCoA plot. (See color plate.)

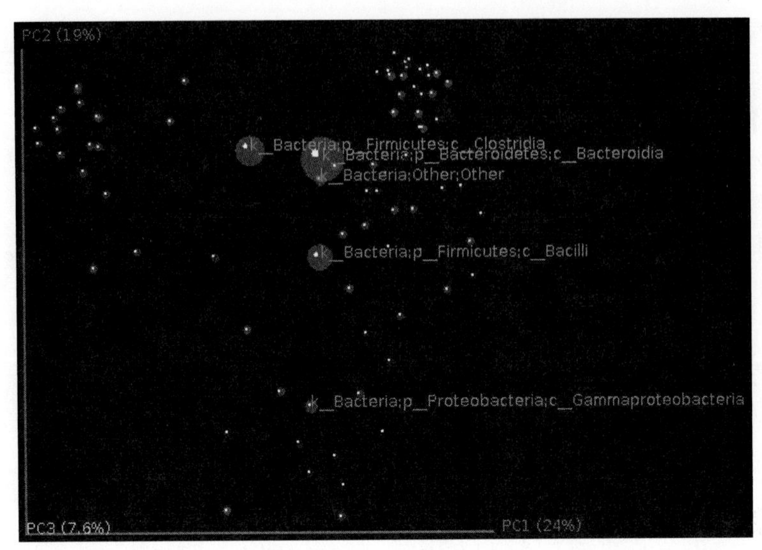

Figure 19.8 Biplot of the example data set. This is the unweighted Unifrac beta-diversity plot, similar to Fig. 19.7, with labels for the most five abundant phylum-level taxa added. The size of the sphere for each taxon is proportional to the mean relative abundance of that taxon across all samples. This plot is created by the command `make_3d_plots.py -i $PWD/diversity_analysis/open_ref/bdiv_even7205/unweighted_unifrac_pc.txt -m $PWD/IQ_Bio_16sV4_L001_map.txt -t $PWD/diversity_analysis/open_ref/taxa_plots/table_mc7205_sorted_L3.txt -n_taxa_keep 5 -o $PWD/diversity_analysis/3d_biplot`. (See color plate.)

hierarchical clustering can be effective in some cases, it should be used with caution because the eye can easily be drawn to incorrect relationships (such as samples that are adjacent in terms of the order of their labels but are topologically far apart in the tree). In general, we recommend using PCoA as a method of detecting grouping in the data but demonstrate hierarchical clustering here as an example. Here, we analyze the beta-diversity distance matrix using UPGMA, which forces the samples into an ultrametric tree (i.e., a tree in which the distance from the roots to the tips is the same for every tip) (Fig. 19.9). The resulting tree file is in Newick format and can be visualized by programs including TopiaryExplorer (Pirrung et al., 2011), the R package ape (Paradis, Claude, & Strimmer, 2004), and the package distory (Chakerian & Holmes, 2012). UPGMA can also be applied to the jackknifed subsamples to provide an estimate of the statistical confidence in the clustering, by showing the frequency of each nodes in the original full data set cluster that are supported by the jackknife replicates. We generally recommend against the use of hierarchical clustering as a method

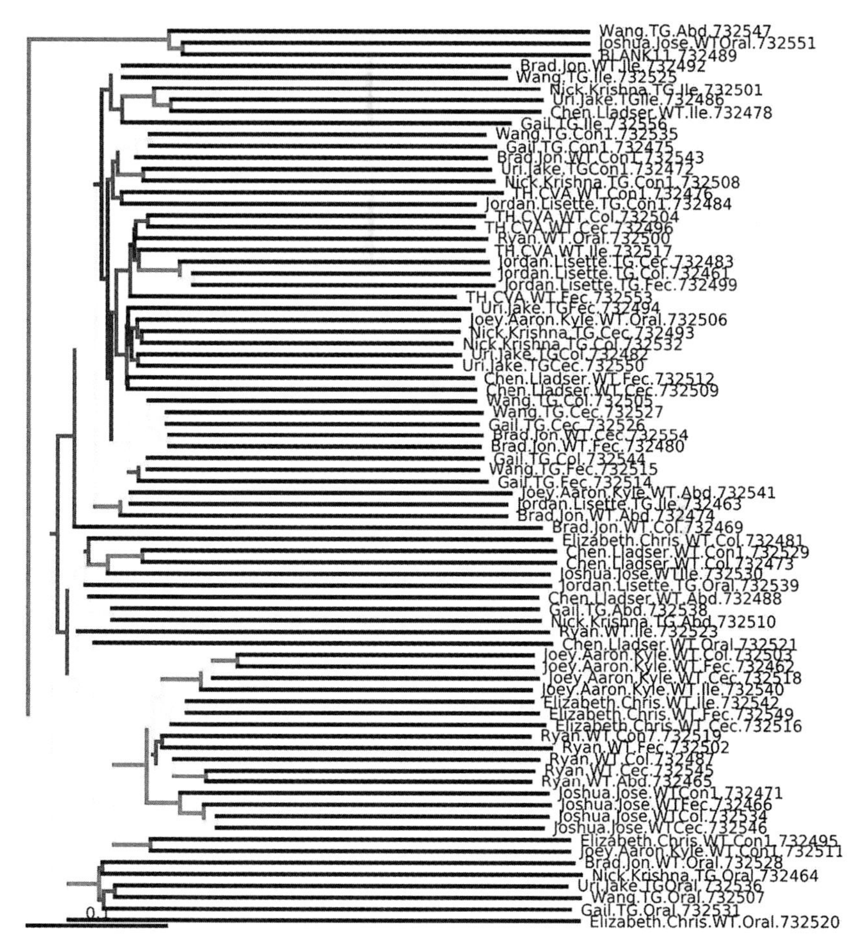

Figure 19.9 Bootstrapped UPGMA clustering on the example data set. The tree is shown with the internal nodes colored by bootstrap support (red: 75–100%, yellow: 50–75%, green: 25–50%, and blue: <25%). Although this visualization is popular in the literature, we generally recommend alternatives such as PCoA. (For interpretation of the references to color in this figure legend, the reader is referred to the online version of this chapter.)

for identifying and visualizing sample groupings, so have not invested as much effort in enabling this technique in QIIME as has been invested in other visualizations. However, if you do plan to use hierarchical clustering, it is important to be aware that substantial work has been done on more effective visualization methods, for example, in distory (Chakerian & Holmes, 2012), and performing additional analyses outside QIIME may allow improvements over the default visualizations.

4.2.5 Statistical significance of differences in alpha- and beta-diversity

Which statistical tests should be applied depends on the particular hypotheses and predictions defined *a priori* in a given research study. QIIME implements several scripts that perform a broad range of statistical tests between samples and groups of samples using both alpha- and beta-diversity measurements. For alpha-diversity, the `compare_alpha_diversity.py` script performs comparisons between groups of samples. The script uses the alpha-diversity measurements of samples standardized to a given number of sequences per sample and performs nonparametric two-sample *t*-tests (i.e., using Monte Carlo permutations to calculate the *p*-value), comparing each pair of groups of samples. Rarefaction is a critical step in these analyses, as noted earlier, because typically diversity estimates depend on the number of sequences per sample. At the maximum rarefaction depth, WT and TG mice did not show differences in alpha-diversity as measured by PD metric (WT: (mean $\pm$ s.d.) $=45.19 \pm 10.6$; TG: 40.01 ± 9.5; $t=-2.17$, $p=0.102$). We also tested for differences in alpha-diversity between body sites. We found differences between cecum and ileum (cecum (mean $\pm$ s.d.) $=51.1 \pm 3.6$; ileum: 36.72 ± 8.2; $t=5.35$, $p=0.028$), cecum and mouth (mouth: 29.54 ± 10.1; $t=6.62$, $p=0.028$), and feces and mouth (feces: 48.4 ± 4.0; $t=5.47$, $p=0.028$). None of the other pairs of comparisons between body sites showed significant differences in alpha-diversity (colon: 46.0 ± 9.2; multitissue: 46.26 ± 9.1; skin: 42.13 ± 7.4; all *p*-values >0.056).

The appropriate statistical tests of beta-diversity also depend on the research question being asked. These tests compare sets of distances between samples in the distance matrix. Careful attention must be paid to both Type I error (rejecting the null hypothesis when it is actually true) and Type II error (accepting the null hypothesis when it is actually false, i.e., lack of statistical power). Type I error is more likely when variance is unequal between groups and when many comparisons are performed on the same data (although multiple comparison corrections correct for the increased Type I error, they often raise the Type II error rate instead). As always, results should be interpreted with caution and common sense. A highly statistically significant result stemming from data with a low-correlation coefficient may indicate that a relationship has little biological meaning, and examining the scatterplot to see if the result is driven by a few outliers would be prudent. Further theoretical validation (especially of the multivariate statistical tests) is also needed, especially because the distributions underlying microbial community data have in general not yet been well characterized.

Comparisons between distance matrices are performed in QIIME using the `compare_distance_matrices.py` script. This script can perform analyses including the Mantel test, the partial Mantel test, and the Mantel Correlogram. The Mantel test is a nonparametric test that compares two distance matrices and calculates a correlation coefficient and a significant *p*-value using permutations that preserve the rows and columns. For the purpose of showing some examples (because the mouse data do not include a time series component), we will use the sequence dataset published by Caporaso, Kuczynski, et al. (2010), where the authors studied variation in the bacterial community in the human gut over time series. We will compare the Unifrac distance matrix and a distance matrix as differences in days since the treatment started. Both distance matrices showed a significant correlation (Mantel test: $p=0.035$), showing that bacterial communities were more similar as they were close in sampling. The Mantel test measures the overall correlation between distance matrices, but Mantel Correlograms measure this effect when taking into account the distances between samples marked by specific metadata variables. Essentially, the second distance matrix (in our case, days since the treatment started) is divided into classes. The classes into which the second distance matrix (days after experiment started) is determined by Sturge's rule, a method for determining the width of bars in a histogram based on the binomial formula. Then Mantel tests are run between these distance classes and the beta-diversity distance matrix. We found that none of the distance classes were significantly related to the bacterial community (Fig. 19.10: all comparisons $p>0.120$, after Bonferroni correction for multiple comparisons). The Mantel test showed us that there is an overall correlation between bacterial community and "days after the experiment started" (samples collected closer in time had more similar bacterial communities), and Mantel Correlogram showed that there is no significant correlation between the bacterial community and any of the classes into which the "days after the experiment started" matrix was divided. In other words, in this case, discretization of the data into a few timepoint classes led to an undetectable pattern; in contrast, use of the whole time series yielded an interpretable result. However, in other datasets, the reverse is often true, especially if the variation is not monotonic (e.g., in the case of seasonal variation).

The partial Mantel test is similar to the Mantel test, except that the analysis is controlled by a third variable. When we compare the beta-diversity distance matrix with days after the experiment started by controlling by sampling date, we find the same trend noted before (partial Mantel test:

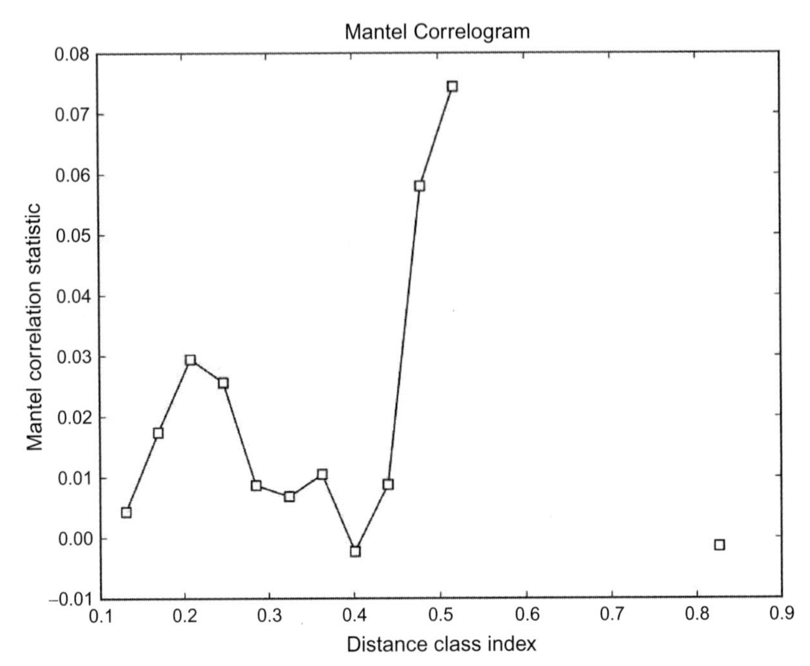

Figure 19.10 Mantel Correlogram showing the Mantel correlation statistics between unweighted Unifrac distance matrix and each class in the days after experiment started distance matrix. Classes in the second distance matrix are determined by Sturge's rule. White dots show nonsignificant relationship since black dots would show significant ones.

$p=0.010$). Samples collected close in time have similar bacterial communities and this effect is independent of the date of collection.

Several visual and statistical tests have been implemented in QIIME in order to compare between and within beta-diversity distances. Distance histograms are an easy way to compare both types of distances graphically (`make_distance_histograms.py`). The output is an html file that shows as many histograms as categories. It is very useful to compare all-within "category" against all-between "category" or the distribution of distances within each group (Fig. 19.11). Probably a more useful tool to compare these beta-diversity distances is by means of box-plots (`make_distance_boxplots.py`, Fig. 19.12). The box-plot script generates a box-plot graph and performs a *t*-test. Box-plots showed that there were no differences between the distances within mouse type and between types. However, the statistical test shows highly significant differences ($p<0.001$) when comparing within and between distances. Once again, we recommend caution and common sense when the *p*-values are interpreted. It is likely to get a significant

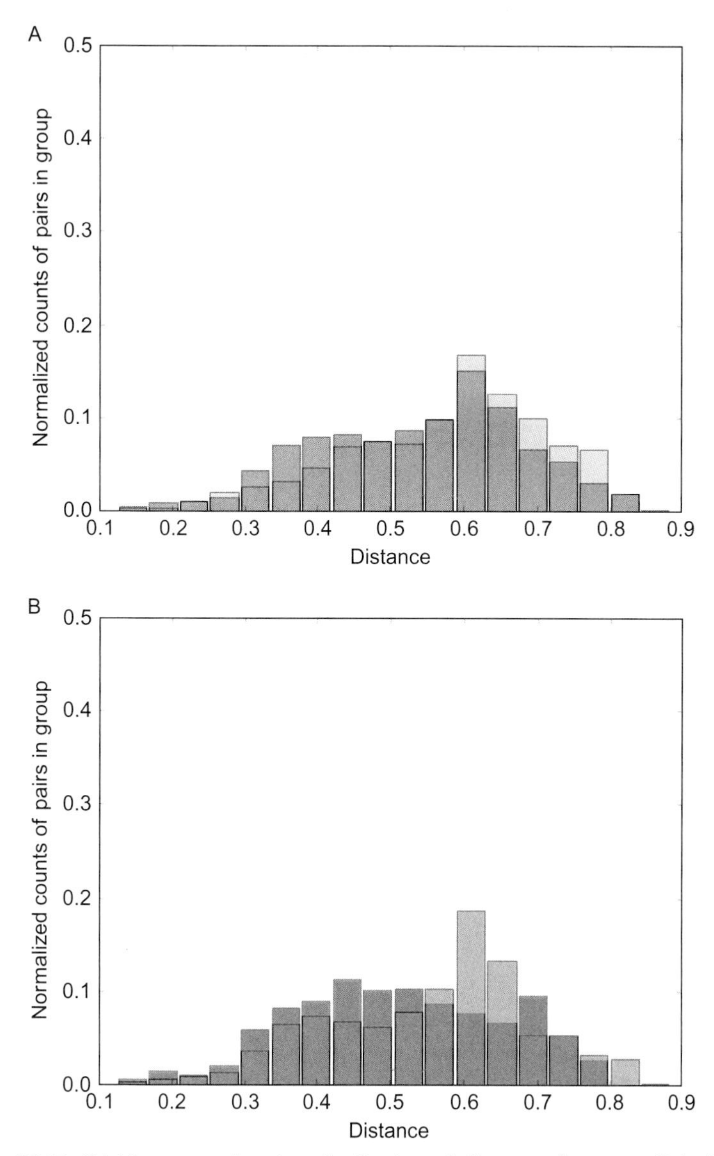

Figure 19.11 (A) Histogram showing distribution of distances between (light brown) and within (dark brown) mice gut microbiota taking into account both wild-type and transgenic mouse groups. (B) Distribution of within distances in gut bacterial community of wild-type mice (light orange) and transgenic ones (blue). (See color plate.)

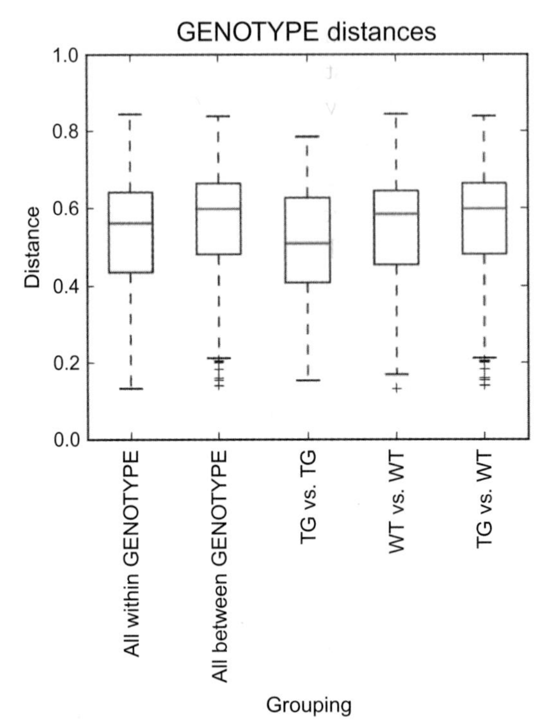

Figure 19.12 Box-plots of the unweighted UniFrac distances for bacterial gut microbiota in both mouse type (WT: wild type; TG: transgenic). "Within" distances represent distances within any of the two groups since "between" distances show distances between both groups. "TG versus TG" and "WT versus WT" represent within distances in transgenic and wild-type groups, respectively. Although averages are different, standard error overlaps in all cases. (For color version of this figure, the reader is referred to the online version of this chapter.)

p-value, although a close inspection of the box-plot reveals that standard error bars overlap. Basically, this result is due to the large number of comparisons: a small Student *t*-statistic (obtained when differences between two data sets are small) and these large degrees of freedom may be highly significant (i.e., the two data sets are very different) even with conservative multiple test corrections (as Bonferroni).

Other multivariate analyses provide additional powerful tools for exploring significant relationships between the beta-diversity distance matrix and factors or covariates. `compare_categories.py` offer different statistical tests, where ANOSIM and adonis are usually employed. ANOSIM is a nonparametric statistical test that compares ranked beta-diversity distances between different groups and calculates a *p*-value and a correlation coefficient by permutation. Adonis partitions the variance in a similar way to the analysis of variance (ANOVA) family of tests, specifically testing variation within a category is smaller or greater than variation between categories. It calculates a pseudo *F*-value, a *p*-value, and a correlation coefficient (R^2). Significant *p*-values must be interpreted together with their R^2 values to infer biological meanings from the results. It is worth mentioning here that PERMANOVA and adonis are similar statistical methods and usually provide equivalent results. However, PERMANOVA only allows categorical factors, whereas both categorical and continuous variables may be used in adonis. Both ANOSIM and adonis analyses indicate that bacterial communities in WT and TG mice significantly differ from one another (ANOSIM: $R^2=0.134$, $p<0.001$; adonis, $R^2=0.046$, $p<0.001$). However, the correlation coefficients are low, so the significant *p*-values need to be interpreted cautiously because this result may not be biologically relevant.

4.2.6 OTU networks

Network-based analysis can sometimes be very useful for displaying how OTUs are partitioned between samples, and how samples are related each other, although we have found that this analysis only works well for datasets in which the samples are not all equally connected. Networks are therefore a powerful way for visually displaying certain large and complex datasets to emphasize similarities and differences among samples. Network analyses are implemented in QIIME through the script `make_otu_network.py`. This script generates the OTU-network files to be passed into Cytoscape (Shannon et al., 2003) and statistics for those networks (specifically, a bipartite graph in which nodes represent either OTUs or samples, and edges represent a connection between an OTU and a sample; Ley et al., 2008). Cytoscape is not wrapped in the QIIME pipeline, and it is run as a separate program. The files used by Cytoscape 2.8.2 are the real edge table (real_edge_table.txt) which contains the columns "from," "to," "eweight," and "consensus_lin," among others dictated by the headers in the mapping file; and the real node file (real_node_table.txt) which contains a node for each OTU and each sample in the study. It uses the OTU file and the user metadata mapping file.

The visual output of this analysis is a clustering of samples according to their shared OTUs (i.e., samples that share more OTUs cluster closer together, as do OTUs shared by more samples): samples and OTUs are represented as dots in the space (nodes) and connected by lines (edges). The degree to which samples cluster is based on the number of OTUs shared between samples, and this is weighted according to the number of sequences within an OTU.

In the network diagram, both types of nodes, OTU nodes and sample nodes, can be easily modified using Cytoscape's graphical user interface, with symbols such as filled circles for OTUs and filled squares for samples. If an OTU is found within a sample, both nodes are connected with a line (an edge). The nodes and edges can then be colored to emphasize certain aspects of the data.

This method is not simply used for descriptive visualizations: the connections within the network can also be analyzed statistically to provide support for the clustering patterns displayed in the network. A *G*-test for independence is used to test whether sample nodes within categories (such as within a genotype, in our example mouse study) are more connected within than a group than expected by chance. Each pair of samples is classified according to whether its members shared at least one OTU, and whether they share a category. Pairs are then tested for independence in these categories (this asks whether pairs that share a category are also equally likely to share an OTU). This statistical test can also provide support for an apparent lack of clustering when it appears that a parameter is not contributing to the clustering.

In our example dataset, mouse samples show some degree of clustering in the space depending on whether the genotype is WT or TG (Fig. 19.13). These clusters in the network were significantly different (*G*-test: $p < 0.001$). Surprisingly, bacterial communities of mice did not visually cluster by body site, although the statistical test shows highly significant differences in samples from different body sites. These results must be interpreted cautiously. The degrees of freedom in the statistical test depend on the number of comparisons, so highly significant results might be obtained even when differences between clusters are slight. In other cases, these differences are obvious and easy to interpret. In the first application of this analysis in microbial ecology, the gut bacteria of a variety of mammals was surveyed and the network diagrams were colored according to the diets of the animals, which highlighted the clustering of hosts by diet category (herbivores, carnivores, omnivores). In a later meta-analysis of bacterial surveys across habitat types, the networks were colored in such a way that the phylogenetic classification of the OTUs was highlighted: this analysis revealed the dominance of shared Firmicutes in vertebrate gut samples versus a much higher diversity of phyla represented among OTUs shared among environmental samples (Ley et al., 2008).

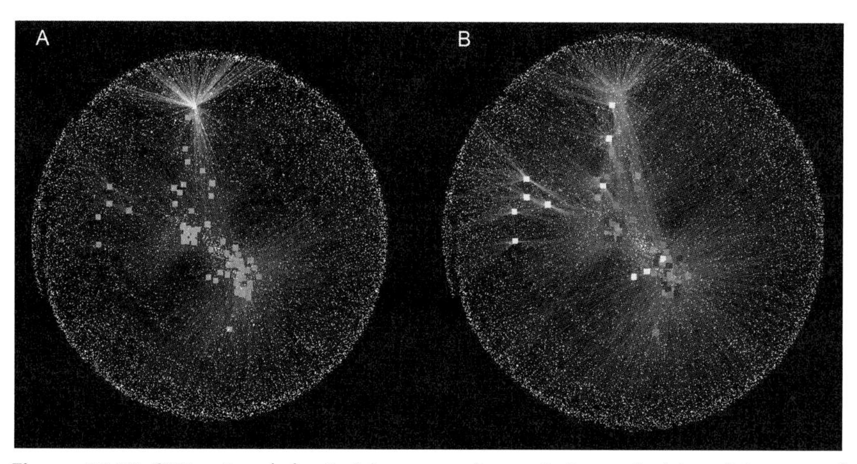

Figure 19.13 OTU-network bacterial community analysis applied in wild-type and transgenic mice. (A) Network colored by genotype (wild type: blue; transgenic: red). Control sample (yellow dot) is external in the network and several OTUs are not shared with mice. Although we can see some degree of clustering, discrimination by genotypes is difficult to assess. (B) Network colored by body site (mouth: yellow; skin: red; ileum: blue; colon: pink; cecum: orange; feces: brown; and multitissue samples: green). A control sample is colored in gray. (See color plate.)

There is no clear sample clustering by body site, suggesting that there is not a core set of OTUs that differentiates one site from another.

This OTU-based approach to comparisons between samples provides a counterpoint to the tree-based PCoA graphs derived from the UniFrac analyses. In most studies, the two approaches reveal the same patterns. They can, however, reveal different aspects of the data. The network analysis can provide taxonomic connections among samples in a visual manner, whereas PCoA–UniFrac clustering can reveal subclusters that may be obscured in the network. The principal coordinates can be pulled out individually and regressed against other metadata; the network analysis can provide a visual display of shared versus unique OTUs. Thus, together these tools can be used to draw attention to different aspects of a dataset.

4.2.7 OTU heatmaps

Another method to visualize the relationships between OTUs and samples is the heatmap, which is widely used for other applications in molecular biology (Wilkinson & Friendly, 2009). This method was initially developed by Loua (1873) to visualize population characteristics of 20 districts of Paris.

In our case, heatmaps can be used for exploratory analysis of microbiomes by mapping abundance values to a color scale in a condensed, pattern-rich format, in which each row corresponds to an OTU and each column corresponds to a sample. A good heatmap graphic can generate hypotheses about sample and/or OTU clustering in the data, which can then be followed up with additional more formal analyses. Two key structural aspects of a heatmap graphic greatly affect whether it will reveal interpretable patterns: (1) the ordering of the axes and (2) the color scaling.

QIIME can create OTU heatmaps using two different scripts: `make_otu_heatmap.py` and `make_otu_heatmap_html.py`. The first script generates a heatmap in which OTUs are represented in rows and samples in columns. OTUs and samples can be sorted and clustered by the phylogenetic tree and by the UPGMA hierarchical clustering, respectively. However, the visualizations of both trees (phylogenetic and hierarchical) in the final heatmap are not currently implemented directly in QIIME, and these hierarchical displays must be prepared using external software such as R. QIIME also supports sample clustering by a metadata category if the user provides a mapping file. The samples will be clustered within each category level using Euclidean UPGMA. The script `sort_otu_table.py` allows sorting the OTU table by a category in the mapping file, allowing defining the order of the samples in the heatmap. Figure 19.14 shows the output of `make_otu_heatmap.py`. There we can see a drawback to heatmaps: when the number of samples or OTUs included in the graphic is too high, the density of the graphic can be overwhelming. Thus, we recommend that the OTU table be filtered to a smaller number of samples (or categories) and taxa to identify the most important patterns, as we will show later in this section.

The second script (`make_otu_heatmap_html.py`) creates an interactive OTU heatmap from an OTU table (Fig. 19.15). This script parses the OTU count table and filters the table by counts per OTU (user specified). It then converts the table into a javascript array, which can be loaded into a Web browser. The OTU heatmap displays raw OTU counts per sample, where the counts are colored based on the contribution of each OTU to the total OTU count present in the sample (blue: contributes low percentage of OTUs to sample; red: contributes high percentage of OTUs). This Web application allows the user to filter the OTU table by number of counts per OTU. The user also has the ability to view the table based on taxonomy assignment. Additional features include the ability to drag rows up and down by clicking and dragging on the row headers and the ability to zoom-in on parts of the heatmap by clicking on the counts within the heatmap.

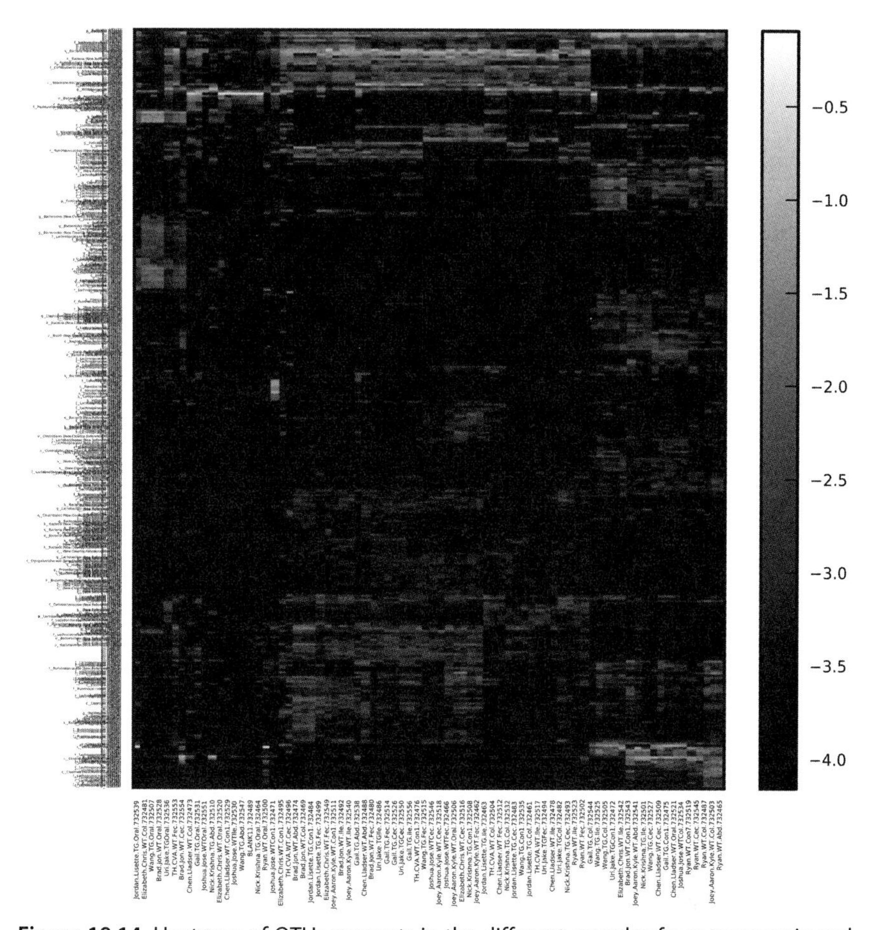

Figure 19.14 Heatmap of OTUs presents in the different samples from transgenic and wild-type mice. The intensity of black shows the abundance of certain OTU in each sample. Both samples and OTUs are sorted by UPGMA tree and the OTU phylogenetic tree, respectively.

Improved OTU heatmap visualizations can be generated using the `plot_heatmap()` command in the phyloseq package for R (McMurdie & Holmes, 2013). This package takes a similar approach to NeatMap (Rajaram & Oono, 2010), in that it uses ordination results rather than hierarchical clustering to determine the index order of each axis. For `plot_heatmap`, the default color scaling maps a particular shade of blue to a log transformation of abundance that generally works well for microbiome data, although the user can select alternative transformations.

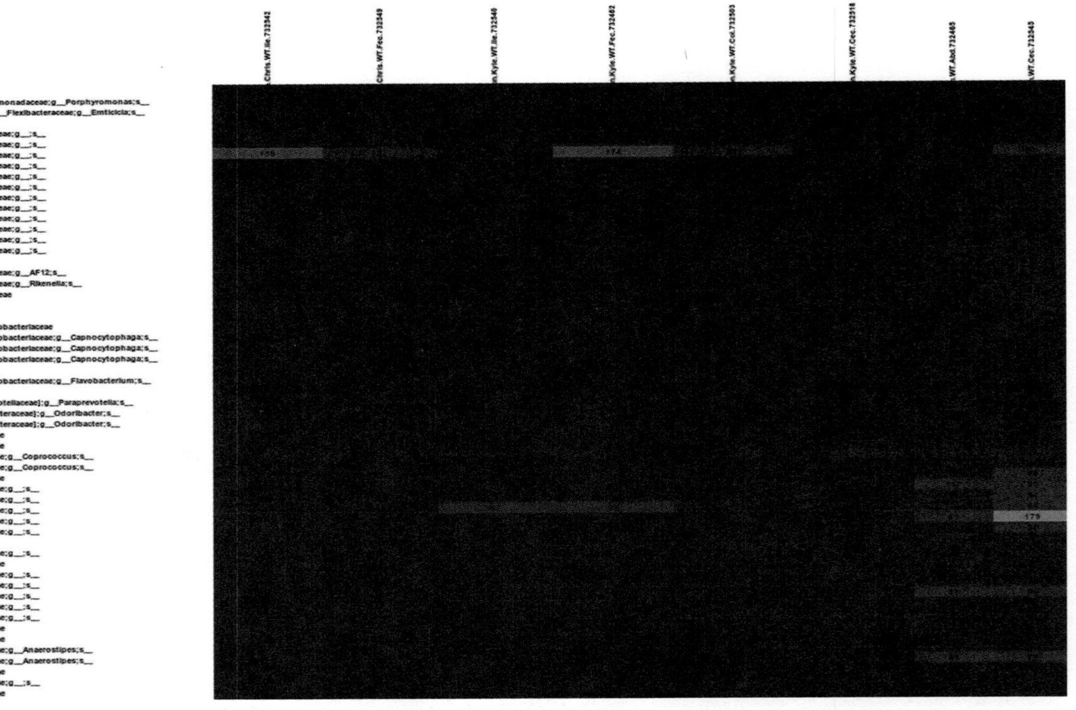

Figure 19.15 Interactive heatmap of OTUs presents in the different samples from transgenic and wild-type mice. This visualization is a result of an HTML file that can be opened in any Web browser. The advantage of this heatmap is that it is easy to manipulate the abundance level for coloring, or transpose samples and OTUs between columns and rows. (For color version of this figure, the reader is referred to the online version of this chapter.)

In this example, a key step was proper filtering of the data. We removed OTUs that appear in only a few samples. The possible contribution to the graphic of these infrequent OTUs is limited, more often contributing to "noise" that causes the heatmap to look dark, empty, and uninterpretable (see Supplemental File 1, http://dx.doi.org/10.1016/B978-0-12-407863-5.00019-8 and Fig. 19.14). We used a nonmetric multidimensional scaling (NMDS) of the Bray–Curtis distance to determine the order of the OTUs and samples. From this representation, it is possible to distinguish high-level patterns and simultaneously note the samples and OTUs involved. For instance, all but a few of the mouth samples are in a cluster towards the middle of the heatmap. One of the key features of this group is an obvious relative overabundance of three Firmicutes OTUs, which are among the most abundant in this subset of the data. Similarly, another clear pattern is a distinction between a group of WT samples from various body sites on the left of the heatmap that appear to have higher proportions of a number of different Firmicutes OTUs, as well as a few specific Bacteroidetes OTUs. This is distinct from the largest cluster of samples on the right-hand side of the heatmap, in which many of the most-abundant OTUs are a different subset of Bacteroidetes and Firmicutes OTUs. We also found it helpful to further pursue these high-level patterns by splitting the data into Firmicutes-only and Bacteroidetes-only subsets, and then plotting new heatmaps with finer-scale taxonomic labels. This required essentially the same commands and limited additional effort, well tailored for exploratory interactive analysis, much of which we have documented in Supplemental File 1 (http://dx.doi.org/10.1016/B978-0-12-407863-5.00019-8) (Fig. 19.16).

Although heatmaps have been deployed widely in molecular biology, especially in protein expression studies, some of the other displays we have discussed such as principal coordinates plots and taxonomy plots often provide more easily interpretable results. However, summarizing relations between taxa through ordination plots or network analyses have been shown to be powerful tools for highlighting similarities and differences among samples and taxa in our OTU table, and a carefully constructed heatmap (though not, in most cases, the default output) can be a useful guide to understanding and hypothesis generation.

4.2.8 OTU category significance

The experimental design of a microbial study will often involve comparing two or more groups for differences in the abundance of OTUs, for example, are there taxa that significantly differ between the control group and the

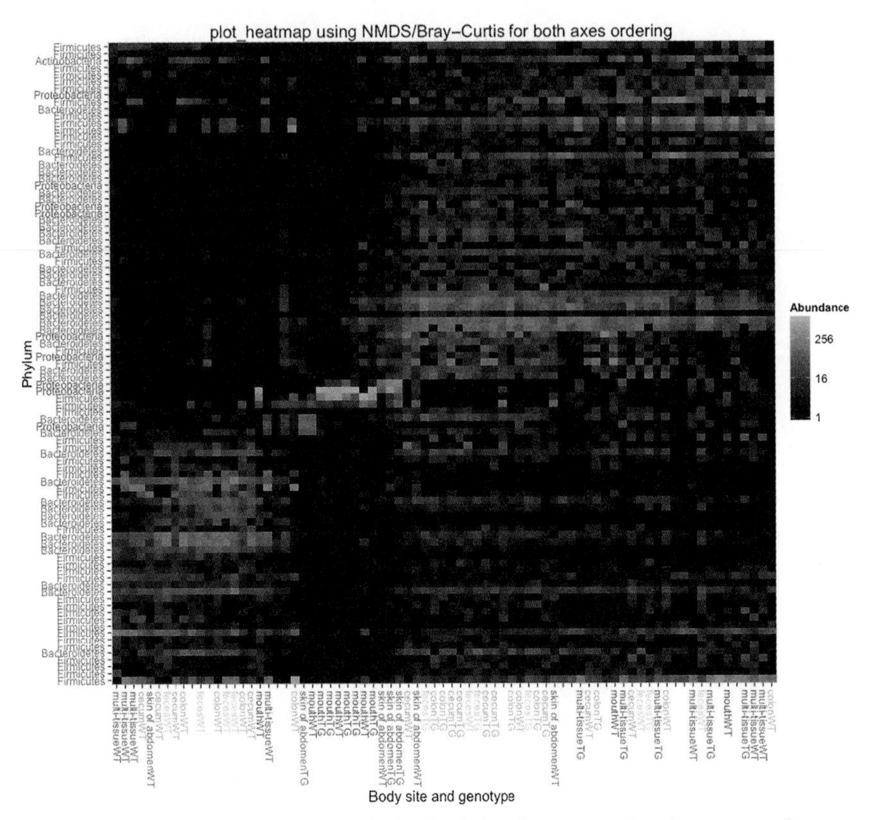

Figure 19.16 Example heatmap of the high-level patterns in the open-reference dataset. The graphic was produced by the `plot_heatmap()` function in phyloseq implemented in R after subsetting the data to the most-prevalent 100 OTUs (see Supplemental File 1, http://dx.doi.org/10.1016/B978-0-12-407863-5.00019-8). The order of sample and OTU elements was determined by the radial position of samples/OTUs in the first two aces of a nonmetric multidimensional scaling (NMDS) of the Bray–Curtis distance. Other choices for distance and ordination method can also be useful. The horizontal axis represents samples, with the genotype and body site labeled, while the vertical axis represents OTUs, labeled by phyla. Both axes are further color coded to emphasize the different categories of labels. The blue-shade color scale indicates the abundance of each OTU in each sample, from black (zero, not observed) to very light blue (highly abundant, >1000 reads). The call used to create this figure was the following, omitting some details to improve the axis labels for publication: "`plot heatmap(openfpp, "NMDS", "bray", taxa.label="Phylum", sample.label="bsgt", title="plot heatmap using NMDS/Bray-Curtis for both axes ordering")`. (For interpretation of the references to color in this figure legend, the reader is referred to the online version of this chapter.)

experimental group? One way to assess this question is to compare the relative abundances of each microbial member between the two groups. This functionality is built into a script called otu_category_significance.py. We can test if there are significant differences in OTU abundance between mouse genotypes either WT or TG. We can assess differences between these groups using the following command:

```
otu_category_significance.py -i $PWD/diversity_analysis/open_ref/
   table_mc7205.biom -m $PWD/IQ_Bio_16sV4_L001_map.txt -o $PWD/
   open_ref_otu_categ_sig_output -c GENOTYPE -s ANOVA
```

Here, we run an ANOVA to assess the relative abundance of each taxon in the OTU table between our two genotype groups. The output will be written to a user-specified file called otu_cat_sig.txt. This document will list the OTU ID, the raw *p*-value, the Bonferroni-corrected *p*-value, the false discovery rate (FDR) *p*-value, as well as the relative average abundance for each of the groups in the selected category (genotype in our case), and the OTU-taxonomy string (if provided in the initial OTU table). While many of these taxa may be significantly different between groups according to the raw *p*-value, it is extremely important that only *p*-values that have been corrected for against multiple comparisons, using either Bonferroni or FDR, be considered as significant. Many times a user's OTU table will contain hundreds or thousands of OTUs, and thus a *p*-value is likely to reach significance based solely on the large number of statistical comparisons being computed (for a probability threshold of 0.05, 1 of 20 comparisons results significant just by chance). It is often very helpful to open the .txt files produced by otu_category_significance.py in a spreadsheet so that columns can be sorted according to *p*-values.

The otu_category_significance.py script also contains several other statistics for comparing groups. The *G*-test can be used to determine if the presence or absence of a given taxa is significantly different between groups and can be specified by passing the option -s g_test in the command. The user can also run a paired *t*-test to determine whether there are taxa that significantly differ between two paired points. For example, imagine the experimental design sampled a group of mice before and after a dietary intervention. Using the paired *t*-statistic in otu_category_significance.py would then compare each mouse's after timepoint to the before timepoint, and test for differences that were consistent across mice, rather than grouping all the before and after timepoints together. For continuous variables, QIIME can calculate the Pearson correlations of OTU abundance with those variables. QIIME is also capable of longitudinal data analysis, which is suitable for the samples tracking the same subjects at multiple points in time, for example, the oral microbiota of six persons after meals in a day. Specifically, longitudinal Pearson correlation can be calculated, accounting for intra-subject correlation of measurements.

4.2.9 Machine learning

QIIME can also take advantage of several machine-learning algorithms to solve two important issues in high-throughput metagenomic studies: correction of mislabeling and quantifying sample contamination.

This mislabeling problem is an increasing issue as the number of processed and pooled sequences increases (Knights, Kuczynski, Koren, et al., 2011). This mislabeling can be addressed using supervised classifiers, a machine-learning technique that is able to fix incorrect metadata. QIIME uses the random forest (Breiman, 2001) supervised classifier implemented in R (Liaw and Wiener, 2002) to recover the mislabeled samples by training the classifier with the relative abundance taxa (Knights, Costello, & Knight, 2011). Knights, Kuczynski, Koren, et al. (2011) show that this approach can even recover up to 30–40% mislabeled samples when the biological patterns are especially clear.

This same technique can also be applied to find taxa that play a key role in differentiating groups of samples, as is done in OTU category significance. However, the difference between OTU category significance and the machine-learning technique is the type of model the construct. While the OTU category significance creates an explanatory model (i.e., it gives a model that best fits the current dataset), the machine-learning technique creates a predictive model (Knights, Costello, et al., 2011). That is, it creates a model that is able to generalize future data, minimizing the expected prediction error.

Since the supervised learning trains a classifier, it is important to provide useful predictors (OTUs in our case). Thus, it is highly recommended to filter the input OTU table to remove those OTUs that are present in few samples (e.g., <10 samples). As in previous analyses, a rarified OTU table should be used so that artificial diversity induced due to different sampling effort is removed. In our example dataset, we can use the subsampled OTU table generated for previous analyses and remove the low-abundance OTUs:

```
filter_otus_from_otu_table.py -i $PWD/diversity_analysis/open_ref/
    table_mc7205.biom -o $PWD/diversity_analysis/open_ref/
    otu_table_filtered10.biom -s 10
```

Running the following command, will run the supervised learning algorithm using the *GENOTYPE* category and 10-fold cross-validation, providing mean and standard deviation of errors:

```
supervised_learning.py -i $PWD/diversity_analysis/open_ref/
    otu_table_filtered10.biom -m $PWD/IQ_Bio_16sV4_L001_map.txt -c
    GENOTYPE -o $PWD/open_ref_supervised_learning_output -e cv10
```

This script will store several files on the output folder. The most important file is *summary.txt*:

```
cat $PWD/open_ref_supervised_learning_output/summary.txt
Model Random Forest
Error type 10-fold cross validation
Estimated error (mean +/- s.d.) 0.23373 +/- 0.15058
Baseline error (for random guessing) 0.42308
Ratio baseline error to observed error 1.81011
Number of trees 500
```

The important information in this file is the *Ratio baseline error to observed error*, which shows the ratio between the expected error of the random forest classifier and the expected error of a classifier that always guesses the most-abundant class (*Baseline error*). Our recommendation is that a ratio of at least 2 shows a good classification. In our example data set, this value is 1.81011, which is close to 2 but not enough to be considered a good classification.

The contamination quantification problem is addressed in QIIME using SourceTracker (Knights, Kuczynski, Charlson, et al., 2011). Given a list of known source environments and a sink (or set of sinks) environment(s), SourceTracker uses a Bayesian approach jointly with Gibbs sampling to predict the quantity of taxa that each source, or an unknown source, contributes to the taxa that makes up the sink environment. For a more detailed description of the algorithm, see Knights, Kuczynski, Charlson, et al. (2011).

The first step to use SourceTracker in QIIME is to modify the mapping file of our example dataset and add two columns: *SourceSink* and *Env*. The *SourceSink* column tells SourceTracker which sample is a source and which sample is a sink, while the *Env* column provides the environment. In our example, we have defined samples from mouth, ileum, cecum, colon, fecal pellet, and skin as sources and the whole mouse homogenization as a sink. In the *Env* column, we have defined the environments as the body site (mouth, ileum, cecum, colon, feces, skin, and homogenization).

As a machine-learning algorithm, SourceTracker needs useful OTUs (predictors) as inputs for training the algorithm. Here, we will use the same OTU table as used for the `supervised_learning.py` script. However, SourceTracker does not yet accept BIOM tables, so we have to transform them into to a tab-delimited OTU table (note that this table can also be opened in Excel or other popular tools):

```
convert_biom.py -i $PWD/diversity_analysis/open_ref/
   otu_table_filtered10.biom -o $PWD/diversity_analysis/open_ref/
   otu_table_filtered10.txt -b
```

Then, we can call SourceTracker using the following command (the $SOURCETRACKER_PATH variable should be defined if you have successfully install SourceTracker):

```
R --slave --vanilla --args -i $PWD/diversity_analysis/open_ref/
   otu_table_filtered10.txt -m $PWD/IQ_Bio_16sV4_L001_map_ST.txt -o
   $PWD/open_ref_sourcetracker_output < $SOURCETRACKER_PATH/
   sourcetracker_for_qiime.r
```

The output from the SourceTracker algorithm is a set of pdf files that shows the mixture of the sources that makes up the sink (see Fig. 19.17).

4.2.10 Procrustes analysis

When we want to compare samples in PCoA space that were processed in different ways, such as different ribosomal RNA subunits, primer sets, or algorithmic choices for processing, we can use procrustes analysis (Gower, 1966; Muegge et al., 2011; Vinten et al., 2011). Procrustes analysis is a statistical shape algorithm that allows us to compare different distributions by rescaling and applying a rotation matrix, that is, if the group of samples have the same shape but are in different sizes or orientation, the algorithm will resize and rotate them to make the shapes fit. As an example, we present the results of comparing the different OTU picking algorithms, see Section 4.2.2, where we can see that even as the number of OTU clusters change the distribution described is similar with a confidence of MC p-value: 0.00 and M^2: 0.097 for closed-reference versus *de novo* and MC p-value: 0.00 and M^2: 0.035 for closed-reference versus open-reference. Both cases used the first three axes (i.e., the axes displayed in

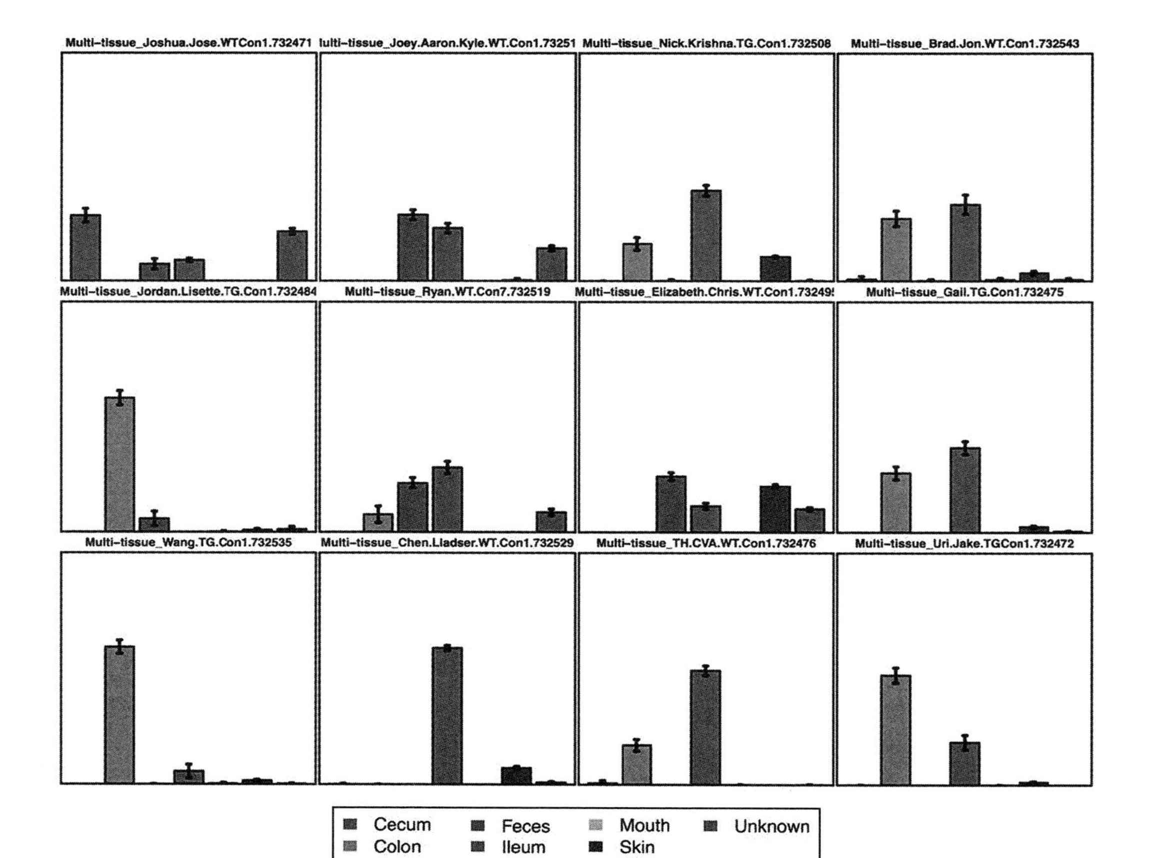

Figure 19.17 SourceTracker output showing a bar plot for each sink (mouse) present in the dataset. Each bar is a potential source (body site) and the height of each bar represents the percentage of taxa the source contributes to the taxa in the sink. The advantage of this visualization over the other two (area and pie chart) is that it shows error bars that allow to see the variance of the prediction. (For color version of this figure, the reader is referred to the online version of this chapter.)

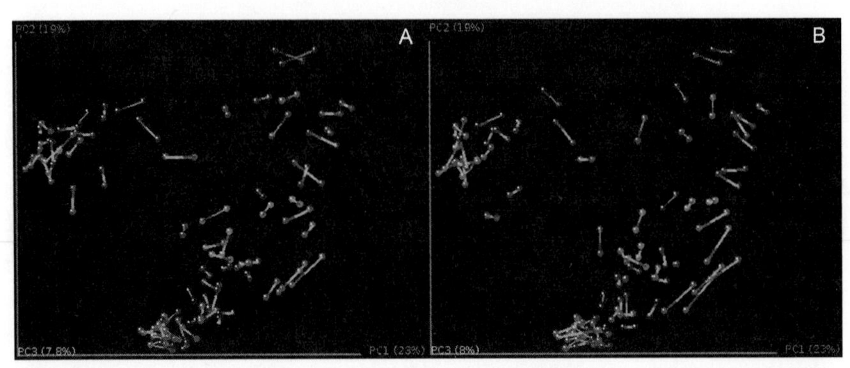

Figure 19.18 Procrustes analysis of different picking algorithms, where we can see that different OTU-clustering methods yield similar PCoA distributions. PCoA plots are colored by BODY_HABITAT. (A) Comparing samples with clusters picked using the *de novo* picking protocol against the closed-reference. (B) Comparing samples with clusters picked using the open-reference picking protocol against the closed-reference. (See color plate.)

the plot) and 100 repetitions, Fig. 19.18. To generate these plots, we ran these commands:

```
transform_coordinate_matrices.py -i $PWD/diversity_analysis/
   closed_ref/bdiv_even7205/unweighted_unifrac_pc.txt,$PWD/
   diversity_analysis/denovo/bdiv_even7205/unweighted_unifrac_pc.
   txt -r 100 -o $PWD/procrustes/closed_ref-denovo

compare_3d_plots.py -i $PWD/procrustes/closed_ref-denovo/
   pc1_transformed.txt,$PWD/procrustes/closed_ref-denovo/
   pc2_transformed.txt -o $PWD/procrustes/closed_ref-denovo/plot -m
   $PWD/IQ_Bio_16sV4_L001_map.txt

transform_coordinate_matrices.py -i $PWD/diversity_analysis/
   closed_ref/bdiv_even7205/unweighted_unifrac_pc.txt,$PWD/
   diversity_analysis/open_ref/bdiv_even7205/unweighted_unifrac_pc.
   txt -r 100 -o $PWD/procrustes/closed_ref-open_ref

compare_3d_plots.py -i $PWD/procrustes/closed_ref-open_ref/
   pc1_transformed.txt,$PWD/procrustes/closed_ref-open_ref/
   pc2_transformed.txt -o $PWD/procrustes/closed_ref-open_ref/plot -m
   $PWD/IQ_Bio_16sV4_L001_map.txt
```

4.2.11 SitePainter

Spatial data poses unique challenges, and the types of statistical analyses described earlier often obscure spatial patterns (Gevers et al., 2012; Hewitt et al., 2013). SitePainter (Gonzalez et al., 2012) is a Web-based tool that creates images representing the geographical (spatial) distribution of our samples, and then color them based on taxonomy summaries (defining which taxa occur where) and PCoA axes (defining how similar the patches are along the principal axes).

To create a new image, we suggest using Adobe Illustrator, Inkscape, or SitePainter. This list is in descending order of usability. In any of these tools, we need to create a SVG (scalable vector graphics) image that has closed paths, ellipsoids, and rectangles for any path that we want to color; and open paths, lines, or text for those that we want SitePainter to ignore. The latter are useful for static images and give a nice background for the image. Note that SVG images are text files, so they can be opened in any graphics program in the list above or in any text editor. The difference between an open and a closed path is that the element in has a letter z at the end of the definition of the lines of the path, so, for example, <path d="M 10 10 L 30 10 L 20 30 z"> is a closed path but <path d="M 10 10 L 30 10 L 20 30"> is an open one.

There are two main QIIME-generated inputs that should be loaded into SitePainter: taxa summaries and multidimensional scaling (MDS) technique results, including NMDS and PCoA. To exemplify the creation and usage of images in SitePainter, we will filter the OTU table and the beta-diversity file to only have one mouse. Filtering and summarizing the OTU table:

```
filter_samples_from_otu_table.py -i $PWD/diversity_analysis/
   open_ref/bdiv_even7205/table_mc7205_even7205.biom -m $PWD/
   IQ_Bio_16sV4_L001_map.txt -o $PWD/forSitePainter/otu_table_Gail.
   biom -s 'GROUP:Gail'

summarize_taxa.py -i $PWD/forSitePainter/otu_table_Gail.biom -o $PWD/
   forSitePainter/taxa_sum -t
```

Filtering the beta-diversity file and then recalculating PCoA is necessary every time we add or remove samples of our analyses, because PCoA results depend on the samples included in the analysis. Thus it is not sufficient to simply remove samples from PCoA results calculated on a larger set of samples:

```
filter_distance_matrix.py -i $PWD/diversity_analysis/open_ref/
   bdiv_even7205/unweighted_unifrac_dm.txt -m IQ_Bio_16sV4_L001_map.
   txt -o $PWD/forSitePainter/unweighted_unifrac_dm.txt -s 'GROUP:
   Gail'

principal_coordinates.py -i $PWD/forSitePainter/
   unweighted_unifrac_dm.txt -o $PWD/forSitePainter/
   unweighted_unifrac_pc.txt
```

Then we create an image in Adobe Illustrator that represents the mice and its gastrointestinal tract, Fig. 19.19A. Once this figure is created and saved in SVG format (this example uses version 1.1 of SVG), we open the image in any text editor and replace any letter "z" with nothing; this will destroy all the closed paths and will facilitate manipulation in SitePainter.

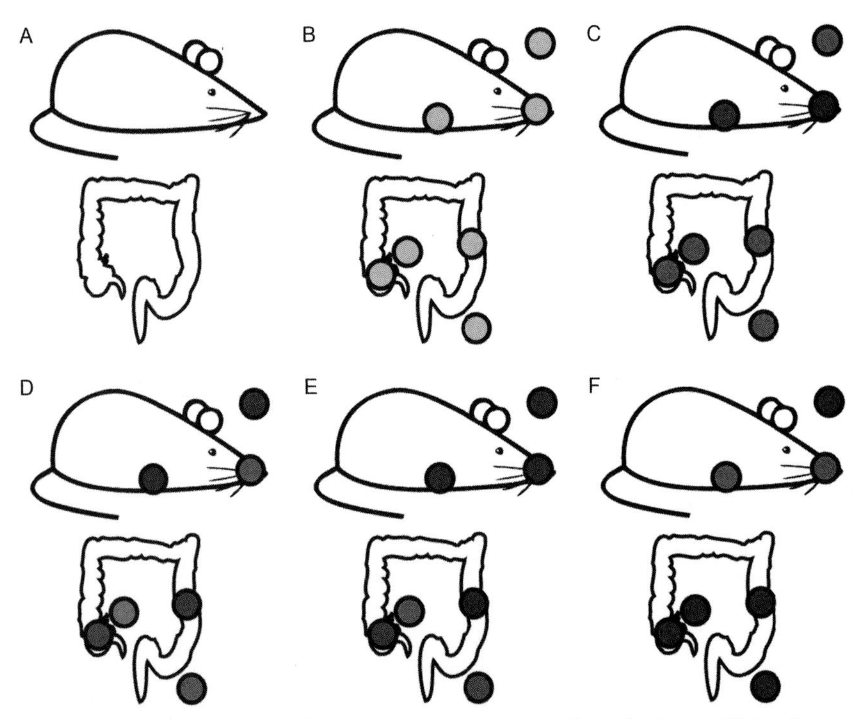

Figure 19.19 Image representing the mouse and its gastrointestinal tract. (A) Raw image without samples. (B) Image in SitePainter with samples. (C) and (D) PCoA axis 1 and 2, in red high values, in blue low values, similar colors represent similar communities. (E) and (F) Taxonomic distributions of (E) Betaproteobacteria and (F) Gammaproteobacteria, in red high abundance, in blue low abundance. (See color plate.)

Now, we can open this image in SitePainter by clicking on the pencil/flower image on the right corner, choosing "Open Image," and select our file. Then we add the places that we want to color using the rectangle or ellipsoid tool, Fig. 19.19B. Now we need to make our samples in the image match the names of the sample names from our files; for this, we need to click on "Elem. -> Click to update" on the right menu, and this will show us the current sample names in the image; then, we double-click on each one and change the name to make it match the sample name in the mapping file. Note that SitePainter does not accept sample names with dots (.), so if the sample name has this character, we need to replace it with an underscore (_). We do not need to change the QIIME files, as this will happen automatically in SitePainter. When we hover over each name, the sample will change color, facilitating the identification of the image we are selecting. If different sites have the same name, they will be colored with the same value from the QIIME output files.

The final step is to load the resulting QIIME files. To do this, we use the Metadata loader on the top left of the menu. This opens the file. We then move the right menu to the "Meta." tab. Here, we can select which column we want to use for coloring and then click "Color elements," to select more, Fig. 19.19C–F. For detailed instructions about changing colors and other details, visit http://sitepainter.sourceforge.net/tutorials/index.html.

5. OTHER FEATURES

5.1. Testing linear gradients, including time series analysis

Recent microbiome surveys have started integrating gradients (commonly over time) in their study design. We will discuss a first and general approach for those cases, using the Moving Pictures of the Human Microbiome Dataset (Caporaso et al., 2011), where two subjects were sampled daily for up to 396 days in three different body sites (sebum, saliva, and feces). Note that the mouse dataset that we use as a primary example lacks a natural temporal ordering in the study design, so we cannot use it as an example for this analysis.

PCoA plots provide a snapshot about the relative communities of many samples condensed in a single figure. However, coloring the points in PCoA space according to a color gradient can be very difficult to understand. A first approach in this case is to connect the samples belonging to the same subject/treatment subsequently sorted using the values in the gradient, that is,

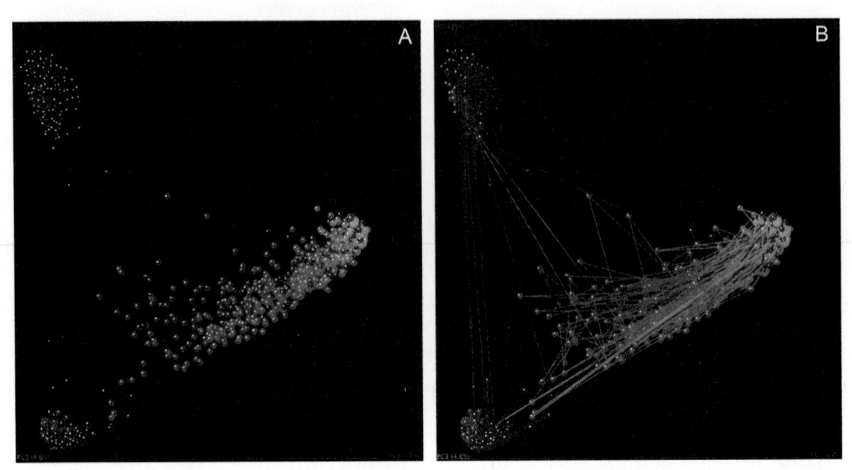

Figure 19.20 Beta-diversity plots for the moving pictures dataset using unweighted UniFrac as the dissimilarity metric (Caporaso et al., 2011). (A) PCoA plot colored by the body site and subject. (B) PCoA plot colored by the body site and subject with connecting lines between samples. Note in (B) that these lines allow us to track the individual body sites with a different approach. (See color plate.)

one timepoint after the other (see Fig. 19.20B). An interactive plot like this can be generated using the following command:

```
make_3d_plots.py -i $PWD/moving_pictures/unweighted_unifrac_pc.txt -m
   $PWD/moving_pictures/merged_columns_mapping_file.txt -o $PWD/
   moving_pictures/vectors --
   add_vectors=BODY_SITEHOST_SUBJECT_ID,DAYS_SINCE_EPOCH
```

An important thing to note here is that because we want to track each of the three body sites (SampleTypes) for the two subjects (Subject), we need a column in our mapping file that allows us to make that distinction. Hence we need to concatenate those two columns in our metadata mapping file using an external spreadsheet editor or another tool. Also note that the gradient used is a category named DAYS_SINCE_EPOCH (i.e., the number of days since January 1, 1970). The idea here is to have a common reference for the collection date of each of the samples.

Although a visualization like the one created in the previous example is often sufficient, replacing one of the axes in the PCoA plot with the data explaining the gradient provides a different insight into the analyzed data (see Fig. 19.21).

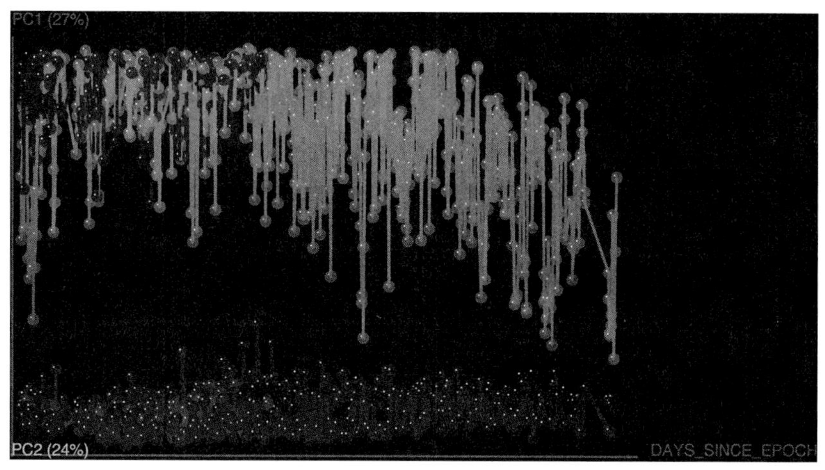

Figure 19.21 Three-dimensional plots in which two of the axes are PC1 and PC2 and the other is the day when that sample was collected in reference to the epoch time. Although this is not explicitly a beta-diversity plot, this representation allows differentiation of the individual trajectories over time. (See color plate.)

```
make_3d_plots.py -i $PWD/moving_pictures/unweighted_unifrac_pc.txt -m
    $PWD/moving_pictures/merged_columns_mapping_file.txt -o $PWD/
    moving_pictures/vectors --add_vectors=BODY_SITEHOST_SUBJECT_ID,
    DAYS_SINCE_EPOCH -a DAYS_SINCE_EPOCH
```

These visual representations can often identify meaningful patterns. To statistically support these assertions, ANOVA can be used over the values grouped by a category of interest. In a case where user wants to test for independence between the variation of one group of trajectories and another, this command could be used:

```
make_3d_plots.py -i unweighted_unifrac_pc.txt -m mapping_file.txt -o
    vectors -add_vectors=SampleTypeAndSubject,days_since_epoch -a
    days_since_epoch

--vectors_algorithm avg --vectors_path anova_stats.txt
```

5.2. Processing 454 data

We have described the recommended workflow for conducting microbial community analysis on an Illumina MiSeq dataset. However, QIIME can also perform microbial community analysis on the 454 platform. The main advantage of 454 over Illumina is that 454 generates longer sequences, which can allow a better taxonomy assignment. However, the 454 technology produces fewer reads per dollar or per sequencing run (Kuczynski et al., 2012).

The 454 processing workflow differs from the Illumina workflow in the sequence preprocessing. In this case, the output file from the sequencing facility is a fasta file containing the reads and a quality score file which contains the score for each base in each sequence included in the fasta file. In this case, the command used for the 454 preprocessing is `split_libraries.py`:

```
split_libraries.py -m Fasting_map.txt -f Fasting_Example.fna -q
   Fasting_Example.qual -o slout
```

Similar to the Illumina processing, this script also performs a quality-filtering. In this case, the quality-filtering is based on cutoffs for sequence length, end-trimming, or minimum quality score. However, to successfully remove the read artifacts, a denoising process has to be performed (Reeder & Knight, 2010) to reduce the impact of homopolymer runs (runs of the same base). The 454 denoising process is a slow, computationally intensive problem that does not scale to large datasets, as it is based on flowgram clustering (Quince et al., 2011).

5.2.1 Variable-length barcodes

Variable-length barcodes are used for two reasons: to make the number of flows (rather than the number of bases) constant (Frank, 2009) or to stagger the reads to reduce bad signal from low complexity at a given position in the set of amplicons being sequenced. This approach is not recommended today because such samples are not easily demultiplexed, and there is checksum, like Hamming or Golay, that allows error correction and improved sample assignment (Hamady et al., 2008). However, the HMP used variable-length barcodes to identify their samples within sequencing runs. Thus, QIIME allows demultiplexing such files by using the parameter -b in `split_libraries.py`, as follows:

```
split_libraries.py -m map_file_with_variable_length_barcodes.txt -f
   your_fna.fna -q your_qual.qual -o. split_library_output_ vari-
   able_length/ -b variable_length,
```

5.3. 18S rRNA gene sequencing

QIIME can also be used to perform analysis on 18S rRNA gene sequence data (in eukaryotes), as well as other markers such as ITS. The main difference between performing analyses with 18S rRNA gene data instead of 16S rRNA gene data (or ITS data) is the reference database used for OTU picking, the taxonomic assignments, and the template-based alignment building, since it must contain eukaryotic sequences.

The recommended database to use as a reference for 18S rRNA sequences is the Silva database (Quast et al., 2013). At the time of writing, the most recent QIIME-compatible Silva database is the 108 release. Since this database contains the three domains of life, it can be used as a reference for 18S rRNA data sets.

When conducting studies mixing 18S rRNA data and 16S rRNA data, you should take into account that picking OTUs against the Silva database will assign taxa to all three domains of life. In this case, it is recommended to split the OTU table by domain, generating an OTU for each domain (Archaea, Bacteria, and Eukarya). At this point, each of these tables can be used in downstream analysis in the same way as performed for 16S rRNA data.

5.4. Shotgun metagenomics

Shotgun metagenomics is also supported in QIIME, although it is still experimental and it should be used at the user's own risk. Currently, the QIIME team recommends the blat method (Kent, 2002) for searching nucleic acid sequence reads in a reference database, although usearch (Edgar, 2010) is also supported. The main reason for preferring blat against usearch is that protein reference database often requires 64-bit applications, and blat is free of charge, while the 64 bit version of usearch is not.

There are many reference databases (IMG, KEGG, M5nr, among others), and they all supported by QIIME, since the user only needs to supply a single fasta file containing the sequence records. The command that QIIME provides for mapping reads against the reference database is `map_reads_to_reference.py`, and it can be performed in parallel using the `parallel_map_reads_to_reference.py` script.

5.5. Support for QIIME in R

First published in 1996, "R" is an integrated software application and programming language designed for interactive data analysis (R Core Team). It is available for Linux, Mac OS, and Windows free of charge under an open-source license (GPL2). Since its inception, R has found a niche as a tool for interactive statistical analysis through functional programming. Primary investigation and inference are performed by writing a series of repeatable commands as "scripts" that can be recorded and published. This paradigm lends itself well to reproducible research and is enhanced substantially by R's integration with tools for literate programming such as Sweave (Leisch, 2002), knitr (Xie, 2013), and R markdown (Allaire, Horner, Marti, & Porte, 2013), as well as data graphics. There are thousands of free and open-source extensions to R (packages) available from the main R repository, CRAN, further organized by volunteer experts into 31 task "views" (which are in fact workflow inventories). Among these are dedicated package lists relevant to microbiome data, including phylogenetics, clustering, environmetrics, machine learning, multivariate, and spatial statistics, as well as a separate reviewed and curated repository dedicated to biological statistics called Bioconductor (over 600 packages).

At present, support for QIIME in R is predominantly achieved through a package called "phyloseq" (McMurdie & Holmes, 2013) dedicated to the reproducible analysis of microbiome census data in R. phyloseq defines an object-oriented data class for the consistent representation of related (heterogenous) microbiome census data that is independent of the sequencing- or OTU-clustering method (storing OTU abundance, taxonomy classification, phylogenetic relationships, representative biological sequences, and sample covariates). The package supports QIIME by including functions for importing data from biom-format files derived from more recent versions of QIIME (import_biom) as well as legacy OTU-taxonomy delimited files (import_qiime and related user accessible subfunctions). Later editions of phyloseq (>1.5.15) also include an API for importing data directly from the microbio.me/qiime data repository. In all cases, these API functions return an instance of the "phyloseq" class that contains the available heterogenous components in "native" R classes. phyloseq includes a number of tools for connecting with other microbiome analysis functions available in other R packages, as well as its own functions for flexible graphics production built using ggplot2 (Wickham, 2009), demonstrated in supplemental files (Supplemental File 1, http://dx.doi.org/10.1016/B978-0-12-407863-5.00019-8) and online tutorials. For researchers interested in developing or using methods not directly supported by phyloseq, nor its data infrastructure, the biom-format-specific core functions in phyloseq have been migrated to an official API in the biom-format project as an installable R package called "biom," now released on CRAN. This also includes some biom-format-specific functionality that is beyond the scope of phyloseq, though support for QIIME is still likely best achieved using phyloseq.

As with some of the earlier examples of QIIME commands with corresponding output and figures, in this section, we have included some key R commands potentially useful during interactive analysis in the R environment. For simplicity, show only results related to the open-reference OTU data, stored in an object in our examples named `open`, and imported into R using the phyloseq command import_biom.

```
open = import_biom("path-to-file.biom", ...)
```

Additional input data files can also be provided to import_biom or merged with open after its instantiation. For clarity, subsets and transformations of the data in `open` are stored in objects having names that begin with "`open.`" As with the remainder of the examples highlighted in this section, the complete code sufficient for reproducing all results and figures is included in the R Markdown originated document, Supplemental File 1 (http://dx.doi.org/10.1016/B978-0-12-407863-5.00019-8), which includes several additional examples not shown here and is available with supporting files on GitHub (https://github.com/joey711/navasetal).

Although not always very illuminating, a comparison of OTU richness between samples and groups of samples can easily be achieved with the `plot_richness` command. For the most precise estimates of richness for most samples, this should be performed *before* random subsampling or other transformations of the abundance data. Here, `open` contains data that has already been randomly subsampled. In Fig. 19.22, we can see that the WT samples are generally more diverse (higher richness) and somewhat more variable than the TG samples for essentially all body sites, though the differences between the two mice genotypes are small.

```
plot_richness(open, x= "BODY_SITE", color = "GENOTYPE") +
  geom_boxplot()
```

This plot command also illustrates the use of a function in ggplot2, `geom_boxplot`, that instructs the ggplot2 graphics engine to add an additional graphical element—in this case, a box-plot for each of the natural groups in the graphic. These available additional graphical instructions (called "layers" in the grammar of graphics nomenclature) are embedded with the returned plot object for subsequent rendering, inspection, or further modification, allowing for powerfully customized representations of the data.

Here is an example leveraging the abundance bar plot function from phyloseq, `plot_bar`, in order to compare the relative abundances of key phyla between the WT and TG mice across body sites. The first step was

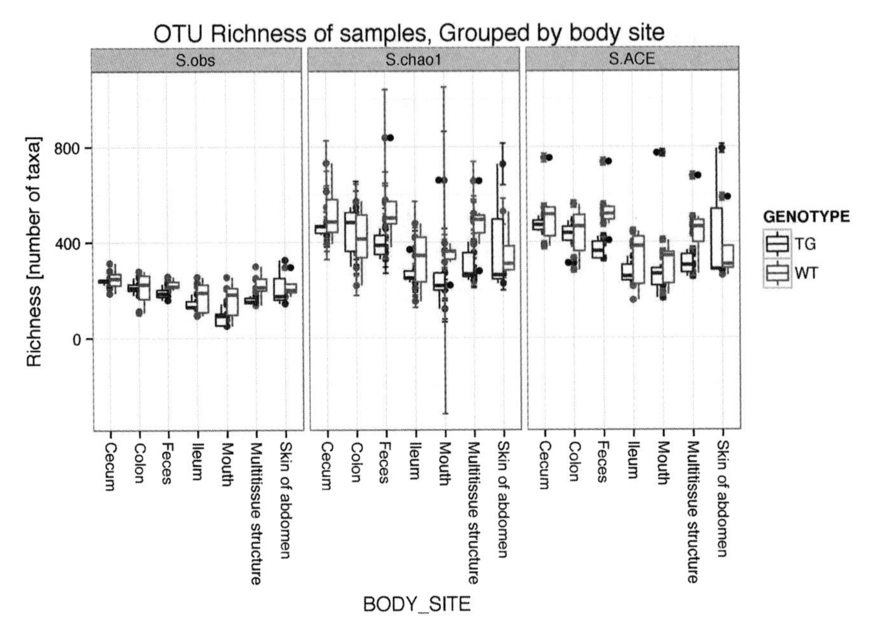

Figure 19.22 Categorically summarized OTU richness estimates using the `plot_richness` function. Samples are grouped on the horizontal axis according to body site and color shading indicates the mouse genotype. The vertical axis indicates the richness estimates in number of distinct OTUs, and a separate box-plot is overlaid on the points for each combination of genotype and body site. The "S.obs," "S.chao1," and "S.ACE" panels show the "rarefied" observed richness, Chao-1 richness, and ACE richness estimates, respectively. (For interpretation of the references to color in this figure legend, the reader is referred to the online version of this chapter.)

actually some additional data transformations (not shown, see Supplemental File 1, http://dx.doi.org/10.1016/B978-0-12-407863-5.00019-8) in order to subset the data to only major expected phyla (`subset_taxa`), merge OTUs from the same phyla as one entry (`merge_taxa`), and merge samples from the same body site and mouse genotype (`merge_samples`) (Fig. 19.23).

```
p2 = plot_bar(openphyab, "bodysite", fill = "phyla", title = title)
p2 + facet_gird(~GENOTYPE)
```

From this first bar plot, it is clear that all body sites from the average WT mouse have Firmicutes as their phylum of largest cumulative proportion, except for the "feces," where it is anyway a close call between Firmicutes and Bacteroidetes. By contrast, some of the average TG mice samples have a much higher proportion of Proteobacteria or Bacteroidetes than the corresponding WT samples. One drawback to this type of stacked bar

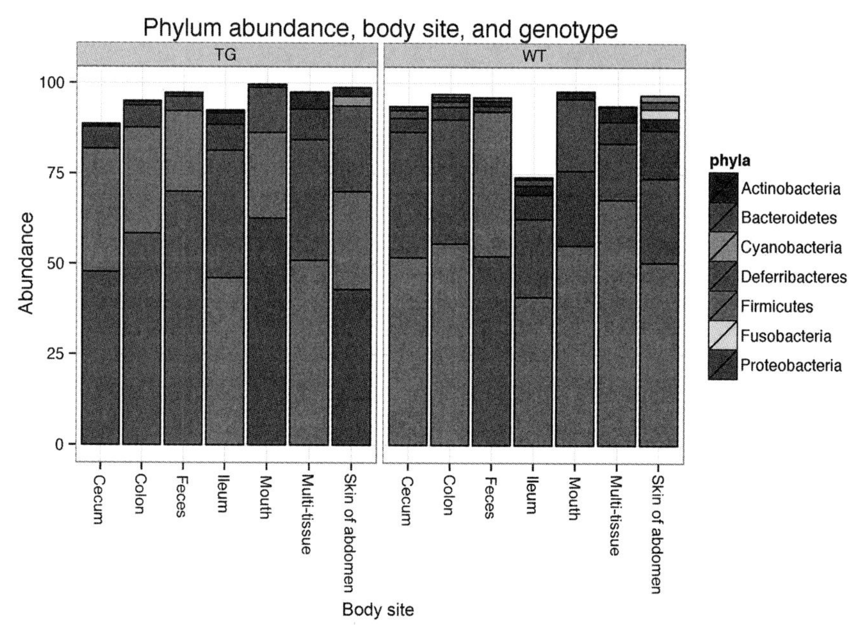

Figure 19.23 Stacked bar plot of the abundance values in the open-reference dataset. The bars are shaded according to phyla with each rectangle representing the relative abundance of a phylum in a particular sample group. The OTU rectangle in each stack is ordered according to abundance. The horizontal and vertical axes indicate the body site of the samples and the average fractional abundance of the OTU within the sample group, respectively. The separate panels "TG" and "WT" indicate the mouse genotype, achieved automatically by the `facet_grid(~GENOTYPE)` layer in the command. (See color plate.)

representation is that it is difficult to compare any of the subbars except for those at the bottom. If needed, this can be alleviated by changing the `facet_grid` call such that a separate panel is made for each phyla in the dataset, as follows (Fig. 19.24):

```
p2 + facet_grid(phyla ~ GENOTYPE) + ylim(0, 100)
```

With essentially the same effort to produce, the 14 panels of this second bar plot graphic allow an easy and quantitative comparison of the relative abundances of each phylum across body sites and genotype.

Microbiome datasets can be highly multivariate in nature, and dimensional reduction (ordination) methods can be a useful form of exploratory analysis to better understand some of the largest patterns in the data. Many ordination methods are wrapped in phyloseq by the ordinate function, and many more are offered in available R packages. Here, we show an example

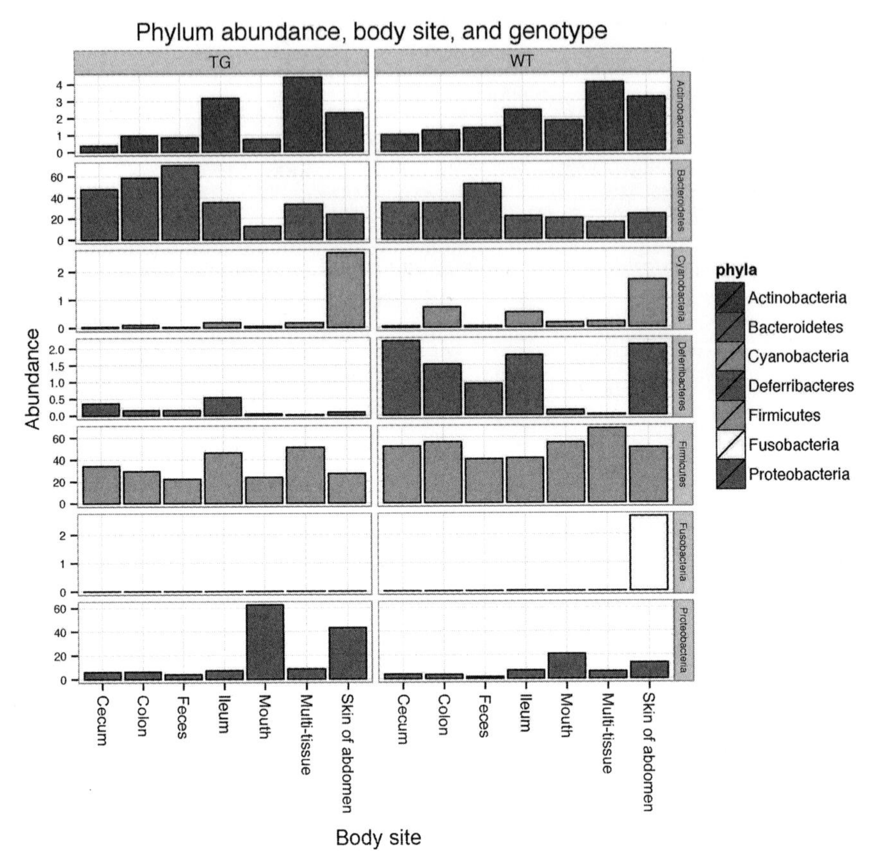

Figure 19.24 Alteration of the stacked bar plot shown in Fig. 19.23 with an additional facet dimension. In this case, an additional argument has been added to the faceting formula so that the data are separated by a row of panels for each phyla, as well as a column of panels for each mouse genotype. The color shading and other attributes generally remain the same with the average cross-category changes for each phylum more discernible. (For interpretation of the references to color in this figure legend, the reader is referred to the online version of this chapter.)

performing MDS on the precomputed unweighted UniFrac distance matrix for the open-reference dataset. The ordination result (`openUUFMDS`) is first passed to `plot_scree` in order to explore the "scree plot" representing the relative proportions of variability represented by each successive axis. Both the ordination result and the original data are then passed to `plot_ordination` with sufficient parameters to shade the sample points by genotype and create separate panels for each body site (Fig. 19.25).

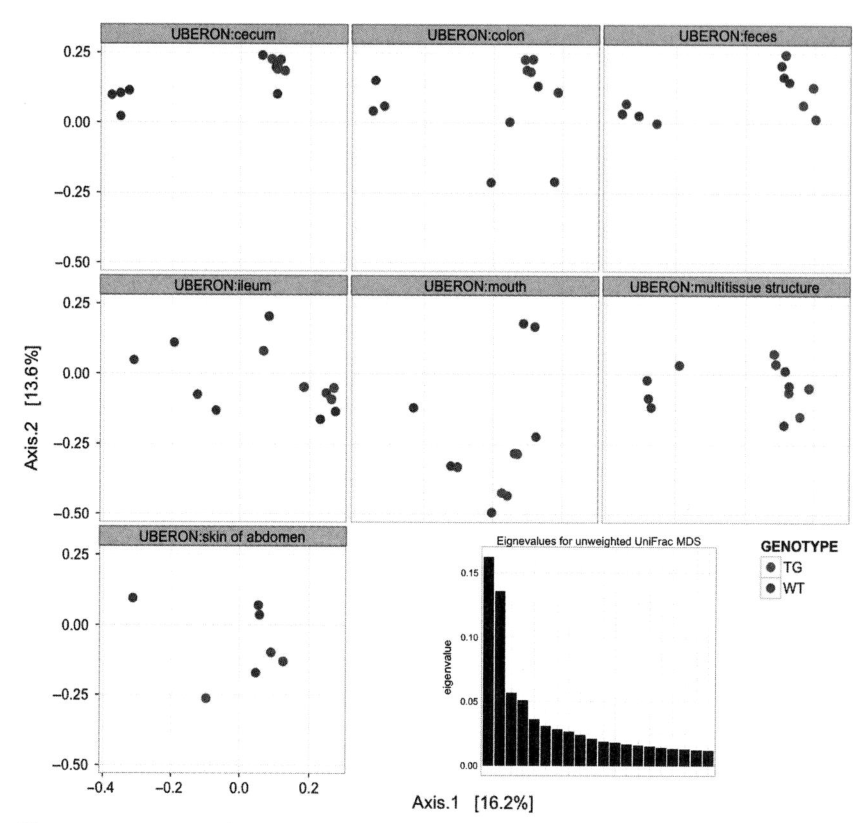

Figure 19.25 MDS ordination results on the unweighted UniFrac distances of the open-reference dataset. The samples are separated into different panels according to body site and shaded red or blue if they were from transgenic or wild-type mice, respectively. The horizontal and vertical axis of each panel represents the first and second axis of the ordination, respectively, with the relative fraction of variability indicated in brackets. (Inset) A scree plot showing the distribution of eigenvalues associated with each ordination axis. (See color plate.)

```
openUUFMDS = ordinate(open, "MDS, distance = UniFrac[["unweighted"]]
   [["open""]])
plot_scree(openUUFMDS, "Unweighted Unifrac MDS")
plot_ordination(open, openUUFMDS, color = "GENOTYPE") + geom_point(-
   size = 5) + facet_wrap(~BODY_SITE)
```

It appears that a subset of the WT samples from all but the mouth and abdomen-skin body sites cluster towards the left of the plot. This appears to be the major pattern along the axis that also comprises the greatest proportion of variability in the dataset. At this stage of analysis, it seems worthwhile to try to identify which OTU abundances are most different between these groups, and then perform some formal validation/testing of these differences.

6. RECOMMENDATIONS

Here, we highlight some of the main aspects to take into account when performing microbial community analysis:

- Use the open-reference OTU picking approach if your data allow it. It will reduce the running time and will recover all the diversity in your samples.
- Perform an OTU quality-filtering based on abundance, by removing singletons, for instance. See Bokulich et al. (2013) for further discussion on how to tune this quality-filtering and its effects on downstream analysis. Quality-filtering is critical for obtaining reasonable numbers of OTUs from a sample.
- Consider whether you need to remove specific taxa from your study, such chloroplast or host DNA sequences when analyzing microbial datasets.
- Remove samples from your study that have low coverage (i.e., low OTU counts). They are likely uninformative and usually indicate low-quality reads.
- Rarefy your OTU table in order to mitigate the differences on the sequencing effort, so the downstream diversity analyses would not be biased by the artificial diversity generated due to the difference in sequencing depth.

7. CONCLUSIONS

QIIME is a powerful tool for the analysis of bacterial community allowing researchers to recapitulate the necessary steps in the processing of sequences from the raw data to the visualizations and interpretation of the results. Two advantages make QIIME very useful: fidelity to the algorithms used and consistency in the analysis. Fidelity is obtained because QIIME wraps existing software, preserving the integrity of the original programs and algorithms designed, created, and tested by the original authors. Consistency is obtained because QIIME can be applied to sequences from different platforms, and once the upstream process is done; the analysis (downstream) process is the same independent of the sequencing platform used. These characteristics, together with the fact that QIIME is open-source software with continuous support to users via QIIME forum, have promoted the rapid increase in the QIIME user community since its publication (Caporaso, Kuczynski, et al., 2010).

Downstream and upstream processes are implemented in QIIME in a way that offers several options to perform the analyses. In this review, we discuss and demonstrate the principles for each step, what the scripts do and how to choose between options. Independent of the use of QIIME, this review also provides an overview of many of the typical steps in a microbial community analysis based on analysis of 16S rRNA sequences produced by high-throughput sequencing. Some of these tools are well developed with a long history in general ecology, whereas others are still in rapid development; we encourage microbial ecologists and bioinformaticians to work together to create, develop, and implement new strategies and tools that allow further exploration of this fascinating field.

ACKNOWLEDGMENTS

We thank William A. Walters and Jessica Metcalf for productive discussion and their useful comments about QIIME. We also acknowledge Manuel Lladser for helping collect the dataset and allowing us to use it, and the IQBio IGERT grant for funding data collection. J.A.N.M. is supported by a graduate scholarship funded jointly by the Balsells Foundation and by the University of Colorado at Boulder. S.H. is partially supported by NIH Grant R01 GM086884. This work was partially supported by the Howard Hughes Medical Institute.

REFERENCES

Allaire, J., Horner, J., Marti, V., & Porte, N. (2013). *Markdown: Markdown rendering for R*. From http://CRAN.R-project.org/package=markdown.

Altschul, S. F., Gish, W., Miller, W., Myers, E. W., & Lipman, D. J. (1990). Basic local alignment search tool. *Journal of Molecular Biology*, *215*(3), 403–410.

Arumugam, M., Raes, J., Pelletier, E., Le Paslier, D., Yamada, T., Mende, D. R., et al. (2011). Enterotypes of the human gut microbiome. *Nature*, *473*(7346), 174–180.

Atlas, R. M., & Bartha, R. (1998). *Microbial ecology: Fundamentals and applications* (4th ed.). Menlo Park, CA/Harlow: Benjamin/Cummings.

Bokulich, N. A., Subramanian, S., Faith, J. J., Gevers, D., Gordon, J. I., Knight, R., et al. (2013). Quality-filtering vastly improves diversity estimates from Illumina amplicon sequencing. *Nature Methods*, *10*(1), 57–59.

Breiman, L. (2001). Random forests. *Machine Learning*, *45*(1), 5–32.

Caporaso, J. G., Bittinger, K., Bushman, F. D., DeSantis, T. Z., Andersen, G. L., & Knight, R. (2010). PyNAST: A flexible tool for aligning sequences to a template alignment. *Bioinformatics*, *26*(2), 266–267.

Caporaso, J. G., Kuczynski, J., Stombaugh, J., Bittinger, K., Bushman, F. D., Costello, E. K., et al. (2010). QIIME allows analysis of high-throughput community sequencing data. *Nature Methods*, 7(5), 335–336.

Caporaso, J. G., Lauber, C. L., Costello, E. K., Berg-Lyons, D., Gonzalez, A., Stombaugh, J., et al. (2011). Moving pictures of the human microbiome. *Genome Biology, 12*(5), R50.
Caporaso, J. G., Lauber, C. L., Walters, W. A., Berg-Lyons, D., Huntley, J., Fierer, N., et al. (2012). Ultra-high-throughput microbial community analysis on the Illumina HiSeq and MiSeq platforms. *ISME Journal, 6*(8), 1621–1624.
Carvalho, F. A., Koren, O., Goodrich, J. K., Johansson, M. E., Nalbantoglu, I., Aitken, J. D., et al. (2012). Transient inability to manage proteobacteria promotes chronic gut inflammation in TLR5-deficient mice. *Cell Host & Microbe, 12*(2), 139–152.
Chakerian, J., & Holmes, S. (2012). Computational tools for evaluating phylogenetic and hierarchical clustering trees. *Journal of Computational and Graphical Statistics, 21*(3), 581–599.
Chen, V. B., Davis, I. W., & Richardson, D. C. (2009). KING (Kinemage, Next Generation): A versatile interactive molecular and scientific visualization program. *Protein Science, 18*(11), 2403–2409.
Cole, J. R., Wang, Q., Cardenas, E., Fish, J., Chai, B., Farris, R. J., et al. (2009). The Ribosomal Database Project: Improved alignments and new tools for rRNA analysis. *Nucleic Acids Research, 37*(Database issue), D141–D145.
De Filippo, C., Cavalieri, D., Di Paola, M., Ramazzotti, M., Poullet, J. B., Massart, S., et al. (2010). Impact of diet in shaping gut microbiota revealed by a comparative study in children from Europe and rural Africa. *Proceedings of the National Academy of Sciences of the United States of America, 107*(33), 14691–14696.
DeSantis, T. Z., Hugenholtz, P., Larsen, N., Rojas, M., Brodie, E. L., Keller, K., et al. (2006). Greengenes, a chimera-checked 16S rRNA gene database and workbench compatible with ARB. *Applied and Environmental Microbiology, 72*(7), 5069–5072.
Dethlefsen, L., & Relman, D. A. (2011). Incomplete recovery and individualized responses of the human distal gut microbiota to repeated antibiotic perturbation. *Proceedings of the National Academy of Sciences of the United States of America, 108*(Suppl. 1), 4554–4561.
Diaz Heijtz, R., Wang, S., Anuar, F., Qian, Y., Bjorkholm, B., Samuelsson, A., et al. (2011). Normal gut microbiota modulates brain development and behavior. *Proceedings of the National Academy of Sciences of the United States of America, 108*(7), 3047–3052.
Dominguez-Bello, M. G., Costello, E. K., Contreras, M., Magris, M., Hidalgo, G., Fierer, N., et al. (2010). Delivery mode shapes the acquisition and structure of the initial microbiota across multiple body habitats in newborns. *Proceedings of the National Academy of Sciences of the United States of America, 107*(26), 11971–11975.
Drancourt, M., Bollet, C., Carlioz, A., Martelin, R., Gayral, J. P., & Raoult, D. (2000). 16S ribosomal DNA sequence analysis of a large collection of environmental and clinical unidentifiable bacterial isolates. *Journal of Clinical Microbiology, 38*(10), 3623–3630.
Eckburg, P. B., & Relman, D. A. (2007). The role of microbes in Crohn's disease. *Clinical Infectious Diseases, 44*(2), 256–262.
Edgar, R. C. (2004). MUSCLE: Multiple sequence alignment with high accuracy and high throughput. *Nucleic Acids Research, 32*(5), 1792–1797.
Edgar, R. C. (2010). Search and clustering orders of magnitude faster than BLAST. *Bioinformatics, 26*(19), 2460–2461.
Edgar, R. C., Haas, B. J., Clemente, J. C., Quince, C., & Knight, R. (2011). UCHIME improves sensitivity and speed of chimera detection. *Bioinformatics, 27*(16), 2194–2200.
Evans, J., Sheneman, L., & Foster, J. (2006). Relaxed neighbor joining: A fast distance-based phylogenetic tree construction method. *Journal of Molecular Evolution, 62*(6), 785–792.
Faith, D. P. (1992). Conservation evaluation and phylogenetic diversity. *Biological Conservation, 61*(1), 1–10.
Fierer, N., Hamady, M., Lauber, C. L., & Knight, R. (2008). The influence of sex, handedness, and washing on the diversity of hand surface bacteria. *Proceedings of the National Academy of Sciences of the United States of America, 105*(46), 17994–17999.

Fierer, N., Lauber, C. L., Zhou, N., McDonald, D., Costello, E. K., & Knight, R. (2010). Forensic identification using skin bacterial communities. *Proceedings of the National Academy of Sciences of the United States of America, 107*(14), 6477–6481.

Frank, D. N. (2009). BARCRAWL and BARTAB: Software tools for the design and implementation of barcoded primers for highly multiplexed DNA sequencing. *BMC Bioinformatics, 10*, 362.

Gevers, D., Knight, R., Petrosino, J. F., Huang, K., McGuire, A. L., Birren, B. W., et al. (2012). The Human Microbiome Project: A community resource for the healthy human microbiome. *PLoS Biology, 10*(8), e1001377.

Gilbert, J. A., Meyer, F., Antonopoulos, D., Balaji, P., Brown, C. T., Brown, C. T., et al. (2010). Meeting report: The terabase metagenomics workshop and the vision of an Earth microbiome project. *Standards in Genomic Sciences, 3*(3), 243–248.

Gilbert, J. A., Meyer, F., Jansson, J., Gordon, J., Pace, N., Tiedje, J., et al. (2010). The Earth Microbiome Project: Meeting report of the "1 EMP meeting on sample selection and acquisition" at Argonne National Laboratory October 6 2010. *Standards in Genomic Sciences, 3*(3), 249–253.

Gonzalez, A., & Knight, R. (2012). Advancing analytical algorithms and pipelines for billions of microbial sequences. *Current Opinion in Biotechnology, 23*(1), 64–71.

Gonzalez, A., Stombaugh, J., Lauber, C. L., Fierer, N., & Knight, R. (2012). SitePainter: A tool for exploring biogeographical patterns. *Bioinformatics, 28*(3), 436–438.

Gower, J. C. (1966). Some distance properties of latent root and vector methods used in multivariate analysis. *Biometrika, 53*, 325–338.

Haas, B. J., Gevers, D., Earl, A. M., Feldgarden, M., Ward, D. V., Giannoukos, G., et al. (2011). Chimeric 16S rRNA sequence formation and detection in Sanger and 454-pyrosequenced PCR amplicons. *Genome Research, 21*(3), 494–504.

Hamady, M., & Knight, R. (2009). Microbial community profiling for Human Microbiome Projects: Tools, techniques, and challenges. *Genome Research, 19*(7), 1141–1152.

Hamady, M., Walker, J. J., Harris, J. K., Gold, N. J., & Knight, R. (2008). Error-correcting barcoded primers for pyrosequencing hundreds of samples in multiplex. *Nature Methods, 5*(3), 235–237.

Hewitt, K. M., Mannino, F. L., Gonzalez, A., Chase, J. H., Caporaso, J. G., Knight, R., et al. (2013). Bacterial diversity in two Neonatal Intensive Care Units (NICUs). *PLoS One, 8*(1), e54703.

Human Microbiome Project Consortium (2012a). A framework for human microbiome research. *Nature, 486*(7402), 215–221.

Human Microbiome Project Consortium (2012b). Structure, function and diversity of the healthy human microbiome. *Nature, 486*(7402), 207–214.

Jost, L. (2007). Partitioning diversity into independent alpha and beta components. *Ecology, 88*(10), 2427–2439.

Katoh, K., Misawa, K., Kuma, K., & Miyata, T. (2002). MAFFT: A novel method for rapid multiple sequence alignment based on fast Fourier transform. *Nucleic Acids Research, 30*(14), 3059–3066.

Kent, W. J. (2002). BLAT—The BLAST-like alignment tool. *Genome Research, 12*(4), 656–664.

Knights, D., Costello, E. K., & Knight, R. (2011). Supervised classification of human microbiota. *FEMS Microbiology Reviews, 35*(2), 343–359.

Knights, D., Kuczynski, J., Charlson, E. S., Zaneveld, J., Mozer, M. C., Collman, R. G., et al. (2011). Bayesian community-wide culture-independent microbial source tracking. *Nature Methods, 8*(9), 761–763.

Knights, D., Kuczynski, J., Koren, O., Ley, R. E., Field, D., Knight, R., et al. (2011). Supervised classification of microbiota mitigates mislabeling errors. *ISME Journal, 5*(4), 570–573.

Koenig, J. E., Spor, A., Scalfone, N., Fricker, A. D., Stombaugh, J., Knight, R., et al. (2011). Succession of microbial consortia in the developing infant gut microbiome. *Proceedings of the National Academy of Sciences of the United States of America, 108*(Suppl. 1), 4578–4585.

Koren, O., Goodrich, J. K., Cullender, T. C., Spor, A., Laitinen, K., Backhed, H. K., et al. (2012). Host remodeling of the gut microbiome and metabolic changes during pregnancy. *Cell, 150*(3), 470–480.

Kuczynski, J., Costello, E. K., Nemergut, D. R., Zaneveld, J., Lauber, C. L., Knights, D., et al. (2010). Direct sequencing of the human microbiome readily reveals community differences. *Genome Biology, 11*(5), 210.

Kuczynski, J., Lauber, C. L., Walters, W. A., Parfrey, L. W., Clemente, J. C., Gevers, D., et al. (2012). Experimental and analytical tools for studying the human microbiome. *Nature Reviews. Genetics, 13*(1), 47–58.

Kuczynski, J., Liu, Z., Lozupone, C., McDonald, D., Fierer, N., & Knight, R. (2010). Microbial community resemblance methods differ in their ability to detect biologically relevant patterns. *Nature Methods,* 7(10), 813–819.

Larkin, M. A., Blackshields, G., Brown, N. P., Chenna, R., McGettigan, P. A., McWilliam, H., et al. (2007). Clustal W and Clustal X version 2.0. *Bioinformatics, 23*(21), 2947–2948.

Leisch, F. (2002). Sweave: Dynamic generation of statistical reports using literate data analysis. In B. R. Wolfgang Härdle (Ed.), *Compstat: Proceedings in computational statistics* (pp. 575–580). Heidelberg, Germany: Physica-Verlag.

Ley, R. E., Backhed, F., Turnbaugh, P., Lozupone, C. A., Knight, R. D., & Gordon, J. I. (2005). Obesity alters gut microbial ecology. *Proceedings of the National Academy of Sciences of the United States of America, 102*(31), 11070–11075.

Ley, R. E., Hamady, M., Lozupone, C., Turnbaugh, P. J., Ramey, R. R., Bircher, J. S., et al. (2008). Evolution of mammals and their gut microbes. *Science, 320*(5883), 1647–1651.

Li, W., & Godzik, A. (2006). Cd-hit: A fast program for clustering and comparing large sets of protein or nucleotide sequences. *Bioinformatics, 22*(13), 1658–1659.

Li, W., Jaroszewski, L., & Godzik, A. (2001). Clustering of highly homologous sequences to reduce the size of large protein databases. *Bioinformatics, 17*(3), 282–283.

Liaw, A., & Wiener, M. (2002). Classification and Regression by randomForest. *R News: The Newsletter of the R Project, 2*(3), 4.

Liu, Z., DeSantis, T. Z., Andersen, G. L., & Knight, R. (2008). Accurate taxonomy assignments from 16S rRNA sequences produced by highly parallel pyrosequencers. *Nucleic Acids Research, 36*(18), e120.

Loua, T. (1873). *Atlas statistique de la population de Paris.* Paris: J. Dejey & Cie.

Lozupone, C. A., Hamady, M., Kelley, S. T., & Knight, R. (2007). Quantitative and qualitative beta diversity measures lead to different insights into factors that structure microbial communities. *Applied and Environmental Microbiology, 73*(5), 1576–1585.

Lozupone, C., & Knight, R. (2005). UniFrac: A new phylogenetic method for comparing microbial communities. *Applied and Environmental Microbiology, 71*(12), 8228–8235.

Lozupone, C. A., & Knight, R. (2008). Species divergence and the measurement of microbial diversity. *FEMS Microbiology Reviews, 32*(4), 557–578.

Lozupone, C., Lladser, M. E., Knights, D., Stombaugh, J., & Knight, R. (2011). UniFrac: An effective distance metric for microbial community comparison. *ISME Journal, 5*(2), 169–172.

Ludwig, W., Strunk, O., Westram, R., Richter, L., Meier, H., Yadhukumar, et al. (2004). ARB: A software environment for sequence data. *Nucleic Acids Research, 32*(4), 1363–1371.

Mardia, K. V., Kent, J. T., & Bibby, J. (1979). *Multivariate analysis.* London: Academic Press.

McDonald, D., Clemente, J. C., Kuczynski, J., Rideout, J. R., Stombaugh, J., Wendel, D., et al. (2012). The Biological Observation Matrix (BIOM) format or: How I learned to stop worrying and love the ome-ome. *Gigascience*, *1*(1), 7.

McDonald, D., Price, M. N., Goodrich, J., Nawrocki, E. P., DeSantis, T. Z., Probst, A., et al. (2012). An improved Greengenes taxonomy with explicit ranks for ecological and evolutionary analyses of bacteria and archaea. *ISME Journal*, *6*(3), 610–618.

McLean, P. G., Bergonzelli, G. E., Collins, S. M., & Bercik, P. (2012). Targeting the microbiota-gut-brain axis to modulate behavior: Which bacterial strain will translate best to humans? *Proceedings of the National Academy of Sciences of the United States of America*, *109*(4), E174, author reply E176.

McMurdie, P. J., & Holmes, S. (2013). phyloseq: An R package for reproducible interactive analysis and graphics of microbiome census data. *PLoS One*, *8*(4), e61217.

Muegge, B. D., Kuczynski, J., Knights, D., Clemente, J. C., Gonzalez, A., Fontana, L., et al. (2011). Diet drives convergence in gut microbiome functions across mammalian phylogeny and within humans. *Science*, *332*(6032), 970–974.

Nawrocki, E. P., Kolbe, D. L., & Eddy, S. R. (2009). Infernal 1.0: Inference of RNA alignments. *Bioinformatics*, *25*(10), 1335–1337.

Olsen, G. J., Larsen, N., & Woese, C. R. (1991). The ribosomal RNA database project. *Nucleic Acids Research*, *19*(Suppl.), 2017–2021.

Olsen, G. J., Overbeek, R., Larsen, N., Marsh, T. L., McCaughey, M. J., Maciukenas, M. A., et al. (1992). The Ribosomal Database Project. *Nucleic Acids Research*, *20*(Suppl.), 2199–2200.

Paradis, E., Claude, J., & Strimmer, K. (2004). APE: Analyses of phylogenetics and evolution in R language. *Bioinformatics*, *20*(2), 289–290.

Pirrung, M., Kennedy, R., Caporaso, J. G., Stombaugh, J., Wendel, D., & Knight, R. (2011). TopiaryExplorer: Visualizing large phylogenetic trees with environmental metadata. *Bioinformatics*, *27*(21), 3067–3069.

Price, M. N., Dehal, P. S., & Arkin, A. P. (2009). FastTree: Computing large minimum evolution trees with profiles instead of a distance matrix. *Molecular Biology and Evolution*, *26*(7), 1641–1650.

Quast, C., Pruesse, E., Yilmaz, P., Gerken, J., Schweer, T., Yarza, P., et al. (2013). The SILVA ribosomal RNA gene database project: Improved data processing and web-based tools. *Nucleic Acids Research*, *41*(D1), D590–D596.

Quince, C., Lanzen, A., Curtis, T. P., Davenport, R. J., Hall, N., Head, I. M., et al. (2009). Accurate determination of microbial diversity from 454 pyrosequencing data. *Nature Methods*, *6*(9), 639–641.

Quince, C., Lanzen, A., Davenport, R. J., & Turnbaugh, P. J. (2011). Removing noise from pyrosequenced amplicons. *BMC Bioinformatics*, *12*, 38.

Rajaram, S., & Oono, Y. (2010). NeatMap—non-clustering heat map alternatives in R. *BMC Bioinformatics*, *11*, 45.

Ravel, J., Gajer, P., Fu, L., Mauck, C. K., Koenig, S. S., Sakamoto, J., et al. (2012). Twice-daily application of HIV microbicides alter the vaginal microbiota. *MBio*, *3*(6), e00370-12.

Reeder, J., & Knight, R. (2010). Rapidly denoising pyrosequencing amplicon reads by exploiting rank-abundance distributions. *Nature Methods*, *7*(9), 668–669.

Roesch, L. F., Fulthorpe, R. R., Riva, A., Casella, G., Hadwin, A. K., Kent, A. D., et al. (2007). Pyrosequencing enumerates and contrasts soil microbial diversity. *ISME Journal*, *1*(4), 283–290.

Savage, D. C. (1977). Microbial ecology of the gastrointestinal tract. *Annual Review of Microbiology*, *31*, 107–133.

Schloss, P. D., & Handelsman, J. (2005). Introducing DOTUR, a computer program for defining operational taxonomic units and estimating species richness. *Applied and Environmental Microbiology*, *71*(3), 1501–1506.

Schloss, P. D., & Handelsman, J. (2006). Introducing SONS, a tool for operational taxonomic unit-based comparisons of microbial community memberships and structures. *Applied and Environmental Microbiology, 72*(10), 6773–6779.

Schloss, P. D., Westcott, S. L., Ryabin, T., Hall, J. R., Hartmann, M., Hollister, E. B., et al. (2009). Introducing mothur: Open-source, platform-independent, community-supported software for describing and comparing microbial communities. *Applied and Environmental Microbiology, 75*(23), 7537–7541.

Shannon, P., Markiel, A., Ozier, O., Baliga, N. S., Wang, J. T., Ramage, D., et al. (2003). Cytoscape: A software environment for integrated models of biomolecular interaction networks. *Genome Research, 13*(11), 2498–2504.

Smith, M. I., Yatsunenko, T., Manary, M. J., Trehan, I., Mkakosya, R., Cheng, J., et al. (2013). Gut microbiomes of Malawian twin pairs discordant for kwashiorkor. *Science, 339*(6119), 548–554.

Sneath, P. H. A., & Sokal, R. R. (1973). *Numerical taxonomy: The principles and practice of numerical classification.* San Francisco: Freeman.

Soergel, D. A., Dey, N., Knight, R., & Brenner, S. E. (2012). Selection of primers for optimal taxonomic classification of environmental 16S rRNA gene sequences. *ISME Journal, 6*(7), 1440–1444.

Sogin, M., Welch, D. M., & Huse, S. (2009). *The visualization and analysis of microbial population structures.* From http://vamps.mbl.edu.

Sokal, R. R., & Sneath, P. H. A. (1963). *Principles of numerical taxonomy.* San Francisco: W.H. Freeman.

Spencer, M. D., Hamp, T. J., Reid, R. W., Fischer, L. M., Zeisel, S. H., & Fodor, A. A. (2011). Association between composition of the human gastrointestinal microbiome and development of fatty liver with choline deficiency. *Gastroenterology, 140*(3), 976–986.

Stamatakis, A., Ludwig, T., & Meier, H. (2005). RAxML-III: A fast program for maximum likelihood-based inference of large phylogenetic trees. *Bioinformatics, 21*(4), 456–463.

Tryon, R. C. (1939). *Cluster analysis.* Ann Arbor, MI: Edwards Bros.

Turnbaugh, P. J., Hamady, M., Yatsunenko, T., Cantarel, B. L., Duncan, A., Ley, R. E., et al. (2009). A core gut microbiome in obese and lean twins. *Nature, 457*(7228), 480–484.

Turnbaugh, P. J., Ley, R. E., Hamady, M., Fraser-Liggett, C. M., Knight, R., & Gordon, J. I. (2007). The Human Microbiome Project. *Nature, 449*(7164), 804–810.

Vinten, A. J., Artz, R. R., Thomas, N., Potts, J. M., Avery, L., Langan, S. J., et al. (2011). Comparison of microbial community assays for the assessment of stream biofilm ecology. *Journal of Microbiological Methods, 85*(3), 190–198.

Wang, Q., Garrity, G. M., Tiedje, J. M., & Cole, J. R. (2007). Naive Bayesian classifier for rapid assignment of rRNA sequences into the new bacterial taxonomy. *Applied and Environmental Microbiology, 73*(16), 5261–5267.

Whittaker, R. H. (1960). Vegetation of the Siskiyou Mountains, Oregon and California. *Ecological Monographs, 30*(3), 280–338.

Whittaker, R. H. (1972). Evolution and measurement of species diversity. *Taxon, 21*(2/3), 213–251.

Wickham, H. (2009). ggplot2: Elegant graphics for data analysis. *Use R!* (Vol. 6991). New York: Springer.

Wilkinson, L., & Friendly, M. (2009). The history of the cluster heat map. *The American Statistician, 63*(2), 179–184.

Xie, Y. (2013). *Dynamic documents with R and knitr.* London, United Kingdom: Chapman and Hall/CRC.

CHAPTER TWENTY

Disentangling Associated Genomes

Daniel B. Sloan, Gordon M. Bennett, Philipp Engel, David Williams, Howard Ochman[1]
Department of Ecology and Evolutionary Biology, Yale University, New Haven, Connecticut, USA
[1]Corresponding author: e-mail address: howard.ochman@yale.edu

Contents

1. Introduction 446
2. Sequencing 446
   2.1 Obtaining genomic DNA 447
   2.2 Choosing among sequencing technologies 448
   2.3 Minimizing nucleotide composition bias in library preparation 450
   2.4 Multiplexed sequencing and the problem of library cross-contamination 450
3. Data Analysis and Genome Assembly 451
   3.1 Quality assessment of read data 452
   3.2 Presorting reads by genome 453
   3.3 Less is more: The pitfalls of too much data in a high-throughput world 454
   3.4 Assemblies as graphs: Embracing sequence diversity 454
   3.5 Methods for improving assemblies 456
4. Postassembly Analysis 457
   4.1 Methods to group assembled fragments into phylogenetic bins 457
   4.2 Binning metagenomes—simple communities 458
   4.3 Binning metagenomes—complex communities 458
   4.4 From bins to genomes 459
   4.5 Analyzing taxonomic bins or draft genomes for completeness 460
5. Summary 460
Acknowledgments 461
References 461

Abstract

The recovery and assembly of genome sequences from samples containing communities of organisms pose several challenges. Because it is rarely possible to disassociate the resident organisms prior to sequencing, a major obstacle is the assignment of sequences to a single genome that can be fully assembled. This chapter delineates many of the decisions, methodologies, and approaches that can lead to the generation of complete or nearly complete microbial genome sequences from heterogeneous samples—that is, the procedures that allow us to turn metagenomes into genomes.

Methods in Enzymology, Volume 531
ISSN 0076-6879
http://dx.doi.org/10.1016/B978-0-12-407863-5.00020-4

1. INTRODUCTION

New sequencing technologies have made it possible, at least in theory, to characterize the genomic contents of microbial communities to near completion at a fairly reasonable cost. To researchers in many areas of biology, this is a highly attractive prospect; it will finally become possible to understand the roles, functions, and interactions of all members of a community. Whereas the generation of the vast amounts of sequence necessary to investigate such matters is now within reach, methodological questions of more immediate concern have surfaced: how is it possible to determine which sequences originate from the same genome? And, how is it possible to assemble complete genomes for each of the constituent organisms or species?

Natural microbial communities span a wide range in organismal complexity and in experimental tractability. For the last decade, research in our laboratory examined the genomic characteristics of perhaps the simplest communities, the endosymbionts of insects, which often consist of a single (but sometimes multiple) bacterial species residing within an insect host. These symbionts cannot be cultivated, so they must be extracted and analyzed directly from natural sources, which complicates determination of their genome sequences. By studying progressively more diverse and complex communities, we have learned, mostly through the rote application and adjustment of available technologies and software, what assists and what hinders the genomic analysis of these communities.

In this chapter, we introduce and discuss the issues that must be considered when attempting to resolve full-genome sequences from heterogeneous samples of co-occurring organisms (Fig. 20.1). Although the goal of most projects is to recover complete genomic information for all members of the community, this is rarely achieved; moreover, there is no single protocol that guarantees the best results. Our objectives are to describe the methodological options and alternatives for the genomic characterization of microbial communities of varying degrees of complexity, and to show how experiments and analyses might be tailored to a particular system—in short, we present the information that we wish we had prior to undertaking such studies.

2. SEQUENCING

Many of the key decisions in genome analysis are made before the sequence data are even obtained. In this section, we discuss some of the most relevant considerations for generating raw sequence data.

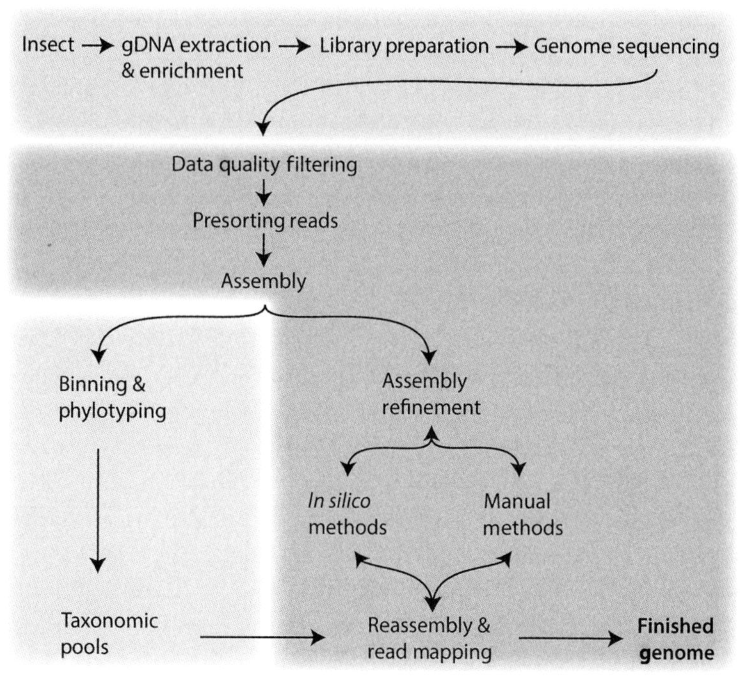

Figure 20.1 Flow diagram of experimental procedures and informatic steps used to obtain complete endosymbiont genomes from total insect DNA. Boxes are color-coordinated according to text sections: Red, Section 2; Blue, Section 3; and Green, Section 4. (For interpretation of the references to color in this figure legend, the reader is referred to the online version of this chapter.)

2.1. Obtaining genomic DNA

Obtaining sufficient quantities of DNA for genome sequencing is sometimes difficult. Ideally, DNA is available in large quantities from a single clonal population that is free of contamination from other organisms, but this is rarely the case when DNA is being prepared from environmental samples or noncultivable organisms. In the case of insect-associated endosymbionts, there are a number of techniques to increase DNA yield of symbiotic organisms by pooling multiple hosts and enriching symbiont communities through dissection, centrifugation, filtration, and/or selective amplification steps (e.g., Burke & Moran, 2011). These approaches can potentially yield microgram quantities of highly enriched microbial DNA; however, the disadvantage of pooled material is that it introduces polymorphism into the sample, complicating subsequent assembly efforts.

Attempts to minimize the extent of polymorphism within samples focus on the recovery and sequencing of DNA isolated from the bacteria of a single host or even a single bacterium (Woyke et al., 2010). But because contemporary sequencing technologies require substantially higher amounts of source DNA than can be isolated from individual cells, they often necessitate a whole genome amplification (WGA) step using ϕ29 polymerase to increase the starting material. Amplification of low-concentration DNA samples may result in the biased enrichment and low sequencing coverage of AT- and GC-rich regions. Although WGA methods are effective in increasing DNA concentration, they often result in uneven sequencing coverage due to the preferential amplification of some regions of the genome (Lasken & Egholm, 2003).

An alternative approach to WGA and sample pooling is to sequence total (i.e., host and symbiont) DNA without any enrichment or amplification step. This strategy comes with drawbacks because often the vast majority of the sequencing represents DNA from the host or other co-occurring microbial species and assembly efforts must contend with high levels of contamination from nontarget genomes. Nevertheless, the rapid decrease in sequencing costs makes this an attractive alternative in some systems, and recent efforts have been successful in assembling complete endosymbiont genomes from total insect DNA (Sloan & Moran, 2012a,b). In these cases, we have found that the assembly challenges associated with polymorphism are far more vexing than the problem of distinguishing between target and nontarget sequences. For that reason, when targeting highly reduced endosymbiont genomes, we suggest using total DNA from a single host individual rather than purified or enriched endosymbiont DNA derived from multiple unrelated individuals.

Having obtained a suitable sample of DNA, several procedures are used to enumerate its component organisms. The most common approach is to use genome/species/taxon-specific qPCR to quantitate sample constituents, which helps determine how much sequencing is required to obtain each desired genome. A small-scale test sequencing run can also be helpful in estimating the repetitiveness, nucleotide composition, size, and relative abundance of target genomes.

2.2. Choosing among sequencing technologies

Selecting the appropriate sequencing method(s) requires balancing trade-offs between cost, accuracy, and read length and understanding the biological characteristics of the target genome(s). The array of current sequencing technologies has been reviewed extensively (e.g., Niedringhaus, Milanova, Kerby,

Snyder, & Barron, 2011), so we will highlight only a few considerations that are useful when working with heterogeneous DNA samples.

Current sequencing technologies offer a range of read lengths. Methods that produce short reads (<250 base pairs (bp)) can generate large volumes of data with minimal costs. *Illumina* is currently the dominant platform and has been subject to constant development that has improved accuracy, read lengths, and total read number (Bentley et al., 2008; Minoche, Dohm, & Himmelbauer, 2011; Pandey, Nutter, & Prediger, 2008). The drawback of short-read technologies is that they are not well suited to *de novo* genome assembly of heterogeneous DNA samples, especially when target genomes contain large repeats or associated organisms share large stretches of sequence identity.

Several technologies exist that can produce longer reads. For example, *Ion Torrent* and *Roche* 454 offer read lengths of up to 400 and 1000 bp, respectively, but they are more costly per bp and are vulnerable to generating sequencing errors in single-nucleotide repeats, that is, homopolymers (Margulies et al., 2005; Merrimen, Ion Torrent R&D Team, & Rothberg, 2012; Metzker, 2005). In addition, providers such as *Pacific Biosciences* (*PacBio*) and *Oxford Nanopore Technologies* (*ONT*) have developed the so-called third-generation sequencing methods that offer substantially longer reads. To date, only *PacBio* is commercially available, boasting an average read length of ~5 kb with many reads exceeding 10 kb (English et al., 2012). The long *PacBio* reads can overcome many of the assembly problems associated with shorter read data, but they come at the cost of an error rate as high as 15% in individual reads (English et al., 2012; Koren et al., 2012). However, these errors can be corrected by either mapping short-read *Illumina* data onto long *PacBio* reads (Koren et al., 2012) or by applying a self-correction pipeline (HGAP1.4) in which *PacBio* reads are aligned to each other to reach a consensus (Chin et al., 2013).

Technological modifications for short-read methods are available to provide additional structural information for genome assembly. For example, paired-end sequencing produces short-reads at both ends of longer size-selected DNA fragments. The resulting read-pairs can be used to link contigs into scaffolds and to estimate the size of assembly gaps. With the *Illumina* platform, it is straightforward to generate paired-end libraries from linear DNA fragments of up to 500 bp or even 1 kb in length, which is preferable to single-read sequencing for *de novo* assembly. It is also possible to generate larger-insert paired-end libraries (known as mate-pair libraries) that require the circularization of fragments that are many kilobases in length. These are particularly valuable for assembling genomes with large repeats, but they are more costly and require increased quality and quantity of DNA. Methods are also in development by *Illumina* (*Moleculo Technology*) to perform highly multiplexed sequencing of separate libraries generated from individual DNA molecules. Assembly of each individual library has the potential to generate the equivalent of long single-molecule reads (10 kb or more) with very high accuracy, but this technology is not yet widely available.

2.3. Minimizing nucleotide composition bias in library preparation

Avoiding nucleotide composition bias during library construction is particularly important when working with heterogeneous DNA samples. The depth of sequencing coverage offers one way of identifying the particular genome from which each contig originates. This can only be effective, however, if sequence coverage is a reliable indication of the relative abundance of each sequence in the original DNA sample. With the exception of single-molecule sequencing technologies (e.g., *PacBio* and *ONT*), high-throughput sequencing platforms generally involve one or more PCR amplification steps. As a result, sequence coverage across a genome can vary due to local differences in nucleotide composition because regions that are exceptionally GC- or AT-rich do not amplify efficiently (Aird et al., 2011).

Modified protocols for *Illumina* library construction have been developed to ameliorate the effects of nucleotide composition bias. Elimination of the PCR amplification step used in library construction is the best way to circumvent such biases, but producing amplification-free libraries requires substantial modification of standard *Illumina* protocols and increases in the amount of input DNA (Kozarewa & Turner, 2011). It is also possible to significantly reduce nucleotide composition bias with only minor adjustments to library-construction protocols. For example, replacing the DNA polymerase provided with standard *Illumina* library-construction kits with one that is less sensitive to nucleotide composition (e.g., Accuprime Taq HiFi or KAPA Bio HiFi) can yield dramatic improvements (Aird et al., 2011; Sloan & Moran, 2012b). In addition, it is preferable to use the smallest number of PCR cycles necessary to generate sufficient library concentration. In our experience, most sequencing facilities use no more than 10 cycles, and it is possible to use as few as four with good starting material.

2.4. Multiplexed sequencing and the problem of library cross-contamination

As of this writing, a single lane on an *Illumina* HiSeq2000 produces 400 million 100-bp read-lengths, yielding a total of 40 Gb of sequence, or roughly

10,000× coverage of a typical bacterial genome. Given this enormous throughput, it is now commonplace for researchers to multiplex sequencing runs by barcoding and pooling individual samples. Barcoded libraries include a unique identifier for each sample (i.e., an index-sequence) that is incorporated into one of the adapters used in library construction. Multiplexing can greatly reduce costs, but it comes with the risk of cross-contamination between samples. Even with no physical contamination between DNA samples or barcoded adapters, *Illumina* sequencing of multiplexed libraries still results in a cross-contamination rate of ~0.3% due to the misassignment of barcodes to samples (Kircher, Sawyer, & Meyer, 2012). This occurs because the sequencing clusters, which correspond to individual library fragments, are so densely distributed within an *Illumina* lane that index-sequences can be mistakenly mapped onto sequencing reads from overlapping or mixed clusters. Note that for many applications, this rate of cross-contamination is negligible, since low-frequency contaminating sequences are easily recognized and excluded as assembly "debris." However, low rates of cross-contamination can be problematic when analyzing heterogeneous DNA samples, in which the relative abundance of different genomes can vary by orders of magnitude. In such cases, cross-contamination from a genome present at high abundance can be difficult to distinguish from legitimate sequences that are present at low abundance in the sample.

The problem of cross-contamination can be partially mitigated by clustering *Illumina* library fragments at lower densities, but this reduces the total number of sequence-reads from a run. An alternative strategy is to use a modified library-construction protocol, in which index-sequences are incorporated into both library adapters, allowing for the exclusion of any read for which either of the barcodes fails to match. This dual indexing approach has been shown to reduce multiplex cross-contamination rates by more than 100-fold (Kircher et al., 2012).

3. DATA ANALYSIS AND GENOME ASSEMBLY

The *in silico* reconstruction of complete genomes from short, overlapping sequence-reads is a complex task, made more challenging when the sample contains DNA from more than one organism. This section describes a number of important considerations for sorting sequence-reads according to their origins, assembling them into contigs, and refining the resulting assemblies and provides examples of some commonly used computational tools for these steps (Table 20.1).

Table 20.1 Examples of software for disentangling, assembling, and refining complete genomes

Task	Program	References
Quality Filtering Data	Trimmomatic	Lohse et al. (2012)
	FASTX-Toolkit	http://hannonlab.cshl.edu
	SolexaQA	Cox, Peterson, and Biggs (2010)
Genome Assembly	Velvet	Zerbino and Birney (2008)
	MIRA	Chevreux, Wetter, and Suhai (1999)
	SOAPdenovo	Luo et al. (2012)
	ALLPATHS-LG	Gnerre et al. (2011)
	Ray Meta	Boisvert, Raymond, Godzaridis, Laviolette, and Corbeil (2012)
	SPAdes	Bankevich et al. (2012)
Assembly Improvement	SEQuel	Ronen, Boucher, Chitsaz, and Pevzner (2012)
	IMAGE	Tsai, Otto, and Berriman (2010)
	PAGIT	Swain et al. (2012)
Metagenomic Binning	MEGAN	Huson, Mitra, Weber, Ruscheweyh, and Schuster (2011)
	iClaMS	Pati, Heath, Kyrpides, and Ivanova (2011)
	Phylothia	McHardy, Martin, Tsirigos, Hugenholtz, and Rigoutsos (2007)
	SPHINX	Mohammed, Ghosh, Singh, and Mande (2011)
	PhymmBL	Brady and Salzberg (2009)
	TETRA	Teeling, Waldmann, Lombardot, Bauer, and Glöckner (2004)
	S-GSOM	Chan, Hsu, Halgamuge, and Tang (2008)

3.1. Quality assessment of read data

On all sequencing platforms, the frequency of sequencing errors and systematic artifacts can vary within reads, among reads from the same run, and between sequencing runs using the same protocol. Data quality affects the performance of all downstream procedures for assigning sequences to genomes and assembling contigs. Thus, it is wise to assess the quality of sequencing reads with respect to their error rates and systematic artifacts to determine which data should be retained.

All commonly used sequencing platforms assign probabilities that each position along a read has been called correctly. These quality-scores usually decrease as a sequencing run progresses, reflecting the deterioration in sequencing accuracy toward the ends of reads. It is important to note that quality-scores do not reveal systematic artifacts and, therefore, do not describe all aspects of read quality. Whereas many assembly programs, such as *MIRA* (Chevreux et al., 1999), incorporate quality-scores to improve accuracy, programs that ignore quality-scores, such as *Velvet* (Zerbino & Birney, 2008), produce more complete assemblies if the terminal regions of reads with low-quality-scores are trimmed. Software packages for trimming sequences prior to assembly include *Trimmomatic* (Lohse et al., 2012), the *FASTX-Toolkit* (http://hannonlab.cshl.edu/fastx_toolkit), and *SolexaQA* (Cox et al., 2010). Algorithms for inferring and correcting errors in reads have been developed but are not recommended when analyzing heterogeneous mixtures of genomes as no error correction algorithms have been designed to accommodate high levels of sequence polymorphism that often occur in such samples.

3.2. Presorting reads by genome

When working with heterogeneous DNA samples, it is common to produce a single metagenome assembly and sort the resulting contigs into bins according to their inferred genome of origin. An alternative first step, however, is to sort individual reads into bins before assembly. Presorting allows for the separate treatment of reads that are derived from a single genome, thereby improving assembly accuracy. Assemblers typically assume even coverage across the target genome, and base-calling error rates falling within a specific range. Uneven coverage is often interpreted as repeated sequences rather than a true deviation from even coverage due to the presence of genomes from different organisms, resulting in assembly fragmentation. Therefore, it can be helpful to use sequence features to sort reads into bins that can then be assembled independently (Brady & Salzberg, 2009; Segata et al., 2012). In Section 4.2, we discuss the sorting of contigs assembled from mixed genome samples, and several of these approaches can also be applied to sequence-reads. Presorting of reads can avoid the construction of chimeric contigs composed of sequences from different genomes, but it is more error-prone because reads are shorter than contigs and contain less information for binning. For this reason, we typically prefer to perform phylogenetic binning at the postassembly stage rather than with individual reads.

3.3. Less is more: The pitfalls of too much data in a high-throughput world

The rapid proliferation of DNA-sequencing capacity poses an paradoxical challenge for genome assembly—too much coverage. Although obtaining more sequence data should, in principle, aid the assembly process, genome assemblers actually perform optimally at intermediate levels of coverage. At very high depths of sequencing, rare errors and polymorphisms occur frequently enough to be interpreted as alternative genome sequences, structures, or repeats that result in assembly ambiguity and fragmentation. One of the most common mistakes made with deep sequencing datasets is the failure to recognize that poor assembly quality may be caused by too much coverage rather than too little. In our experience with short-read sequencing technologies, an average read-depth in the range of 30–100 × is suitable for high-quality, gap-free coverage, while avoiding the assembly artifacts that accompany more extreme sequencing depths (Fig. 20.2).

The issue of optimal read depth is particularly relevant for heterogeneous DNA samples that contain genomes at very different relative abundances. For example, in insects, obligate and facultative bacterial endosymbionts can co-occur, with the latter often orders of magnitude less abundant than the former. As a result, assembling facultative endosymbiont genomes from total DNA requires an amount of sequence data that would produce excessive coverage for the obligate endosymbionts. In such cases, an effective strategy is to perform separate assemblies with small subsets of the sequence data to target the more abundant genomes. It is also possible to take advantage of recently developed assembly algorithms that improve handling of datasets generated from heterogeneous samples with skewed abundances (*Ray Meta*: Boisvert et al., 2012) or from single-cell amplification, which suffers from amplification biases (*SPAdes*: Bankevich et al., 2012).

3.4. Assemblies as graphs: Embracing sequence diversity

One of the main challenges associated with genome assembly from heterogeneous DNA is dealing with the sequence and structural polymorphism that exists within a sample. Closely related genomes often share enough similarity to be indistinguishable in some regions but clearly distinct in others.

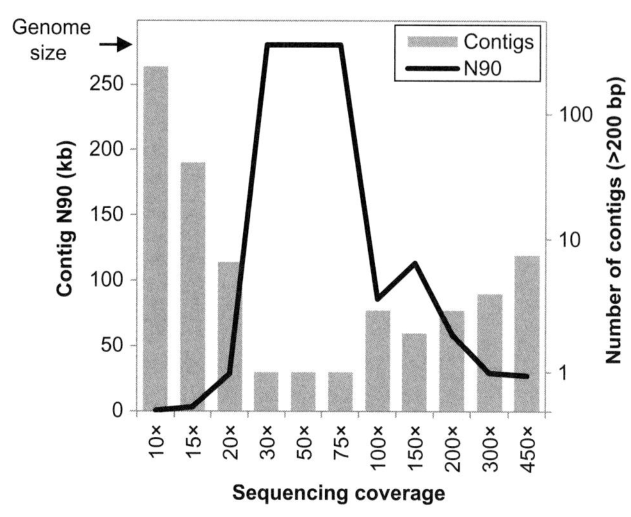

Figure 20.2 Intermediate levels of sequencing coverage are optimal for genome assembly. Plot shows statistics for assemblies at different levels of average sequencing coverage for the whitefly endosymbiont *Candidatus* Portiera aleyrodidarum TV. Assemblies of *Illumina* paired-end sequence were performed in *Velvet* v1.1.06 (Zerbino & Birney, 2008) using a *k*-mer of 39. N90 is a measure of assembly connectivity, defined as the maximum threshold length for which contigs of that length or longer constitute 90% of the total assembly length. *Data from Sloan and Moran (2013).*

This results in assemblies that may be best thought of as networks or graphs, in which identical regions are collapsed into single sequences that branch out into alternative sequences and structural connections that differ between genomes.

Because most tools used for analysis of genome assemblies only work with linear DNA sequences (i.e., contigs), the nature of metagenomes often requires that information is essentially thrown away by breaking assembly graphs into linear fragments or only reporting the most abundant sequence variant. However, a growing number of assembly tools are more actively reporting the diversity of genome sequence and structure found within samples. For example, as its name indicates, the program *ALLPATHS-LG* seeks to retain information about all the alternative sequence paths through an assembly graph (Gnerre et al., 2011). The developers of *ALLPATHS-LG* utilize modified versions of the familiar *FASTA* format (*eFASTA* and *FASTG*) for reporting sequence data (The *FASTG* Format Specification Working Group, 2012). These data formats have the advantage of providing a standardized syntax for retaining information about sequence and structural diversity within assemblies while being easily convertible to *FASTA* format for use with conventional analysis tools. At this point, there are few tools for downstream analysis that are designed to take advantage of the extra information, but the development of such tools will be particularly valuable for metagenomics and for working with heterogeneous samples.

3.5. Methods for improving assemblies

Seldom does a *de novo* assembly program produce a complete genome with a single click, although this could happen given the small size and high coding content of a bacterial genome. Assemblies are typically broken or misassembled as a result of large repetitive sequences and underrepresentation of AT- or GC-rich regions (Durham, 2003)—problems that are only compounded when working with heterogeneous DNA samples. After first-pass assembly of sequence data, it is important to evaluate the quality of the assembly for possible errors, including miscalling of single-nucleotide polymorphisms and insertions and deletions (indels), collapsing of repeated sequences, creation of structural chimeras, and even the exclusion of entire genomic regions that are underrepresented in the data. The simplest way to determine which if any of these errors exist in a genome assembly is to visualize read-coverage across the genome to identify areas of unexpectedly high or low coverage that need to be reevaluated.

Several methods are available to recognize and correct problematic regions, and to fill gaps within and between scaffolds. The most efficient are *in silico* methods that harness information from initial assemblies and read data to refine, correct, and close assemblies. Several options exist that work to identify misassemblies (e.g., Phillippy, Schatz, & Pop, 2008), correct indel and substitution errors (e.g., Ronen et al., 2012), or fill and improve gaps (e.g., Tsai et al., 2010). Some more recent methods refine whole assemblies to correct erroneous substitutions and indels, using initial assembly results to simplify the *de Bruijn* graph approach employed by many assemblers (Luo et al., 2012; Pevzner, Tang, & Waterman, 2001; Zerbino & Birney, 2008). For example, *SEQuel* uses information from already assembled contigs and the reads that map to them to locally refine assemblies by correcting indels and base pair substitutions (Ronen et al., 2012).

The *SEQuel* method, however, does not attempt to bridge broken contigs. One tool for connecting fragmented assemblies is the Iterative Mapping and Assembly for Gap Elimination (*IMAGE*) algorithm that is incorporated in the *PAGIT* assembly improvement suite (Swain et al., 2012; Tsai et al., 2010). *IMAGE* employs an iterative approach of local assemblies, taking paired-end information to build outward into the gapped regions from the flanking ends.

After these *in silico* procedures are implemented, some assembly ambiguities and contig breaks might still need to be addressed by PCR and direct sequencing to obtain a completely closed genome. In some regions, sequencing of a single PCR product will be sufficient to close a gap, but given the $<$ 1-kb read lengths of Sanger sequencing, large gaps and repetitive regions will remain. One proposed approach for resolving such regions combines long PCRs with a high-fidelity polymerase (up to 15 kb in length) with *PacBio*'s long-read sequencing approach (Zhang et al., 2012).

4. POSTASSEMBLY ANALYSIS

Although the goal is to obtain complete genomes, this is often not possible from metagenomic or heterogeneous datasets. Once an optimal assembly has been obtained, sequences can be classified according to their taxonomic identity, a process referred to as metagenomic binning (Kunin, Copeland, Lapidus, Mavromatis, & Hugenholtz, 2008). In this section, we describe some of the available methods for performing metagenomic binning and discuss the circumstances in which it is possible to move from metagenomes to genomes.

4.1. Methods to group assembled fragments into phylogenetic bins

The existing methods for placing contigs into phylogenetic bins rely on information that can be classified into three categories: homology-, composition-, and assembly-based binning (Dröge & McHardy, 2012). The principle of homology-based binning is to assign sequences based on local sequence similarity to known taxa. In contrast, composition-based approaches utilize the taxonomic signal contained in sequence features such as codon usage, GC content, or frequency patterns of short oligonucleotides (4–6-bp long). Finally, assembly-based binning classifies contigs according to similarity in read-coverage and information from read-pairs spanning different contigs. Several extensive reviews on metagenomic binning are available (Dröge & McHardy, 2012; Mande, Mohammed, & Ghosh, 2012). Below, we provide recommendations on how to apply binning strategies to metagenomic data, as selection of the proper strategy depends on the complexity of the DNA sample and the availability of reference data.

4.2. Binning metagenomes—simple communities

Metagenomic data originating from insects containing only a few bacterial endosymbionts should be relatively easy to analyze, and binning has often resulted in completely assembled bacterial genomes. The assignment of contigs to different taxonomic bins with composition-based information is straightforward because coexisting endosymbionts often have different population sizes within an insect, and their genomic base compositions generally differ from other symbionts and the host insect.

When the coexisting endosymbionts belong to different phylogenetic groups, homology-based methods can also prove successful for binning. As symbiont genomes are typically tiny (<1 Mb), they lack large repeat regions and assemble into a small number of large contigs; binning them is relatively easy and may not require any specific binning software. A simple *BLAST* analysis against the NCBI nonredundant database is often sufficient for unambiguous assignment of assembled sequences to different taxa or phylotypes. There are, however, potential pitfalls that may influence simple homology or composition-based methods: (1) the extreme base composition of certain regions may result in different oligonucleotide patterns or dips in read-coverage (Section 2.3); (2) reads originating from repeat copies tend to collapse into one contig, increasing local read-coverage; and (3) base composition and read-coverage of contigs representing extrachromosomal sequences, such as plasmids, can differ from the rest of the genome.

4.3. Binning metagenomes—complex communities

When analyzing complex microbial communities, such as those occurring within the gut of animals, taxonomic assignment of sequences will be more difficult since tens or hundreds of distinct species often coexist. This makes sequence binning solely based on read-coverage difficult because many species will likely be sequenced at similar depth. Similarly, homology- and composition-based methods also have several limitations. Sequences from closely related species may not be distinguishable, and horizontally transferred genes can have a different phylogenetic signal or base composition than the rest of the genome.

The binning method most suitable for a given metagenome depends largely on the specific characteristics of the sampled community, the type of sequencing data, and the availability of reference genomes for binning. For example, if the community consists of a limited number of species of known identity (e.g., a dozen species identified with 16S rRNA), a supervised binning approach may be the best strategy. The premise of this approach is to first assign the metagenomic sequences to predefined taxonomic bins that represent each species present in the community, and then use homology-based assignment to recruit related sequences to their respective bins. This strategy has been applied to a metagenome of the honey bee gut microbiota, which consists of eight bacterial taxa, all belonging to branches of the bacterial phylogeny for which sufficient reference data are available (Engel, Martinson, & Moran, 2012). The program, *MEGAN4*, offers a popular homology-based binning tool for this purpose (Huson et al., 2011).

When analyzing microbial communities in the absence of suitable reference genomes, supervised composition-based methods, such as *ClaMS* or *Phylopythia* (McHardy et al., 2007; Pati et al., 2011), offer more accurate results than homology-based approaches. These methods require a training dataset for each species to be binned—typically 100 kb of representative sequence for each taxon extracted from the metagenome itself. Supervised composition-based binning was used to taxonomically classify the foregut microbiota of the Tammar wallaby, which provided a draft genome of a novel bacterial species that was subsequently cultivated in the lab based on metabolic information gleaned from its genome (Pope et al., 2010, 2011).

The accuracy of binning increases with fragment length, particularly for composition-based approaches, in which fragments smaller than 1–2 kb do not provide enough information to distinguish sequences from different genomes (McHardy et al., 2007; Pati et al., 2011). Therefore, the lengths of assembled sequences will be an important criterion when deciding which binning method to use, and often the application of multiple approaches and criteria will prove best. Some software options include *SPHINX* and *PhymmBL* (Brady & Salzberg, 2009; Mohammed et al., 2011) that implement combinatorial approaches by employing homology and composition-based binning criteria. In cases where there is virtually no information *a priori* about the taxonomic composition of a metagenomic sample, unsupervised binning algorithms, such as *TETRA* (Teeling et al., 2004), *S-GSOM* (Chan et al., 2008), or *iClaMS* (Pati et al., 2011), can categorize sequences based on compositional features.

4.4. From bins to genomes

It is possible to use complex metagenomic data to assemble relatively complete or draft genomes. Several different approaches have been implemented, and the most effective uses existing databases to recruit or assign metagenomic sequences to reference genomes (Rusch, Martiny, Dupont, Halpern, & Venter, 2010; Walsh et al., 2009). For example, multiple genomes were partially assembled from a cow rumen metagenome (Hess et al., 2011), which used binning approaches (e.g., tetra-nucleotide frequency and read-coverage) and gene-linkage information to generate scaffolds. In another example, Iverson and colleagues (2012) used mate-paired short-read sequencing information to reconstruct genomes from complex metagenomic assembly graphs, which were split into linear scaffolds and grouped on the basis of tetra-nucleotide frequency and 16S rRNA sequence information. This resulted in the assembly of one complete and 13 draft genomes (Iverson et al., 2012). These examples demonstrate that the extraction of nearly complete genomes from complex metagenomes is possible; however, several important prerequisites must be met for the assembly of a given genome from a complex mixture: (1) selection of a sequencing technology that includes good linkage information for scaffolding contigs into genome map, (2) sequencing depth must be enough so that individual genotypes are sufficiently covered, (3) there should not be extensive heterogeneity in the genome population to be assembled, and (4) the binning method(s) used should provide high-resolution signatures that specifically identify the genome in question. Whether all these criteria can be met for any given target will depend on the specific environmental context and microorganisms in question.

4.5. Analyzing taxonomic bins or draft genomes for completeness

After taxonomic bins have been generated, it is necessary to analyze the extent to which they represent the complete genomic content of a given species or species group. To this end, sequences can be mapped against closely related reference genomes. A more general approach is to measure completeness based on the presence of a minimal set of core genes (Gil, Silva, Peretó, & Moya, 2004; Lerat, Daubin, & Moran, 2003). These genes ($\sim$200) are conserved in almost all bacteria, and the proportion of this gene set that is present allows for estimates of genome completeness, even if no other information is available.

5. SUMMARY

The rapid rate of technological advance in DNA sequencing has made it feasible to perform genomic analysis on complex and heterogeneous DNA samples. Successful genome sequencing projects of heterogeneous microbial communities rely on understanding the biological and genomic characteristics of the target organisms, such as their anticipated genome sizes, relative numbers, degree of polymorphism, GC contents, and sequence repeat contents. Therefore, some careful planning and prior understanding of the constituents and characteristics of a sample provides realistic perceptions about the sequencing strategy and about the amount and type of information that can be extracted from a particular sample. In our experience, the strategies and potential pitfalls that we have described here represent some of the important considerations when undertaking a project.

ACKNOWLEDGMENTS

D. B. S. was supported by a NIH Postdoctoral Fellowship (1F32GM099334). P. E. was supported by the Swiss National Science Foundation and the European Molecular Biology Organization. H. O. and D. W. acknowledge support from the Templeton Foundation (award number 23536) and from the NIH (award number GM101209).

REFERENCES

Aird, D., Ross, M. G., Chen, W. S., Danielsson, M., Fennell, T., Russ, C., et al. (2011). Analyzing and minimizing PCR amplification bias in Illumina sequencing libraries. *Genome Biology*, *12*, R18.

Bankevich, A., Nurk, S., Antipov, D., Gurevich, A. A., Dvorkin, M., Kulikov, A. S., et al. (2012). SPAdes: A new genome assembly algorithm and its applications to single-cell sequencing. *Journal of Computational Biology*, *19*, 455–477.

Bentley, D. R., Balasubramanian, S., Swerdlow, H. P., Smith, G. P., Miltion, J., Brown, C. G., et al. (2008). Accurate whole human genome sequencing using reversible terminator chemistry. *Nature*, *456*, 53–59.

Boisvert, S., Raymond, F., Godzaridis, E., Laviolette, F., & Corbeil, J. (2012). Ray meta: Scalable de novo metagenome assembly and profiling. *Genome Biology*, *13*, R122.

Brady, A., & Salzberg, S. L. (2009). Phymm and PhymmBL, metagenomic phylogenetic classification with interpolated Markov models. *Nature Methods*, *6*, 673–676.

Burke, G. R., & Moran, N. A. (2011). Massive genomic decay in *Serratia symbiotica*, a recently evolved symbiont of aphids. *Genome Biology and Evolution*, *3*, 195–208.

Chan, C.-K. K., Hsu, A. L., Halgamuge, S. K., & Tang, S.-L. (2008). Binning sequences using very sparse labels within a metagenome. *BMC Bioinformatics*, *9*, 215.

Chevreux, B., Wetter, T., & Suhai, S. (1999). Genome sequence assembly using trace signals and additional sequence information. *Computer Science and Biology: Proceedings of the German Conference on Bioinformatics (GCB)*, *99*, 45–56.

Chin, C.-S., Alexander, D. H., Marks, P., Klammer, A. A., Drake, J., Heiner, C., et al. (2013). Nonhybrid, finished microbial genome assemblies from long-read SMRT sequencing data. *Nature Methods*, *10*, 563–569.

Cox, M., Peterson, D., & Biggs, P. (2010). SolexaQA: At-a-glance quality assessment of Illumina second-generation sequencing data. *BMC Bioinformatics*, *11*, 485.

Dröge, J., & McHardy, A. C. (2012). Taxonomic binning of metagenome samples generated by next-generation sequencing technologies. *Briefings in Bioinformatics*, *13*, 646–655.

Durham, I. (2003). *Genome mapping and sequencing*. Norwich, UK: Horizon Scientific Press.

Engel, P., Martinson, V. G., & Moran, N. A. (2012). Functional diversity within the simple gut microbiota of the honey bee. *Proceedings of the National Academy of Sciences of the United States of America*, *109*, 11002–11007.

English, A. C., Richards, S. R., Han, Y., Wang, M., Vee, V., Qu, J., et al. (2012). Mind the gap: Upgrading genomes with pacific biosciences RS long-read sequencing technology. *PLoS One*, *11*, e47768.

Gil, R., Silva, F. J., Peretó, J., & Moya, J. (2004). Determination of the core of a minimal bacterial gene set. *Microbiology and Molecular Biology Reviews*, *68*, 518–537.

Gnerre, S., Maccallum, I., Przybylski, D., Ribeiro, F. J., Burton, J. N., Walker, B. J., et al. (2011). High-quality draft assemblies of mammalian genomes from massively parallel sequence data. *Proceedings of the National Academy of Sciences of the United States of America*, *108*, 1513–1518.

Hess, M., Sczyrba, A., Egan, R., Kim, T.-W., Chokhawala, H., Schroth, G., et al. (2011). Metagenomic discovery of biomass-degrading genes and genomes from cow rumen. *Science*, *331*, 463–467.

Huson, D. H., Mitra, S., Weber, N., Ruscheweyh, H., & Schuster, S. C. (2011). Integrative analysis of environmental sequences using MEGAN4. *Genome Research*, *21*, 1552–1560.

Iverson, V., Morris, R. M., Frazar, C. D., Berthiaume, C. T., Morales, R. L., & Armbrust, E. V. (2012). Untangling genomes from metagenomes: Revealing an uncultured class of marine Euryarchaeota. *Science*, *335*, 587–590.

Kircher, M., Sawyer, S., & Meyer, M. (2012). Double indexing overcomes inaccuracies in multiplex sequencing on the Illumina platform. *Nucleic Acids Research*, *40*, e3.

Koren, S., Schatz, M. C., Walenz, B. P., Martin, J., Howard, J. T., Ganapathy, G., et al. (2012). Hybrid error correction and *de novo* assembly of single-molecular sequencing reads. *Nature Biotechnology*, *30*, 693–701.

Kozarewa, I., & Turner, D. J. (2011). Amplification-free library preparation for paired-end Illumina sequencing. *Methods in Molecular Biology*, *733*, 257–266.

Kunin, V., Copeland, A., Lapidus, A., Mavromatis, K., & Hugenholtz, P. (2008). A bioinformatician's guide to metagenomics. *Microbiology and Molecular Biology Reviews*, *72*, 557–578.

Lasken, R. S., & Egholm, M. (2003). Whole genome amplification: Abundant supplies of DNA from precious samples or clinical specimens. *Trends in Biotechnology*, *21*, 531–535.

Lerat, E., Daubin, V., & Moran, N. A. (2003). From gene trees to organismal phylogeny in prokaryotes: The case of the gamma-Proteobacteria. *PLoS Biology*, *1*, E19.

Lohse, M., Bolger, A. M., Nagel, A., Fernie, A. R., Lunn, J. E., Stitt, M., et al. (2012). RobiNA: A user-friendly, integrated software solution for RNA-Seq-based transcriptomics. *Nucleic Acids Research*, *40*, W622–W627.

Luo, R., Liu, B., Xie, Y., Li, Z., Huang, W., Yuan, J., et al. (2012). SOAPdenovo2: An empirically improved memory-efficient short-read de novo assembler. *GigaScience*, *1*, 18.

Mande, S. S., Mohammed, M. H., & Ghosh, T. S. (2012). Classification of metagenomic sequences: Methods and challenges. *Briefings in Bioinformatics*, *13*, 669–681.

Margulies, M., Engholm, M., Altman, W. E., Attiya, S., Bader, J. S., Bemben, L. A., et al. (2005). Genome sequencing in microfabricated high-density picolitre reactors. *Nature*, *437*, 376–380.

McHardy, A. C., Martin, H. G., Tsirigos, A., Hugenholtz, P., & Rigoutsos, I. (2007). Accurate phylogenetic classification of variable-length DNA fragments. *Nature Methods*, *4*, 63–72.

Merrimen, B., Ion Torrent R&D Team, & Rothberg, J. M. (2012). Progress in Ion Torrent semidonductor chip based sequencing. *Electrophoresis*, *33*, 3397–3417.

Metzker, M. L. (2005). Emerging technologies in DNA sequencing. *Genome Research*, *15*, 1767–1776.

Minoche, A. E., Dohm, J. C., & Himmelbauer, H. (2011). Evaluation of genomic high-throughput sequencing data generated on Illumina HiSeq and genome analyzer system. *Genome Biology, 12*, R112.

Mohammed, M. H., Ghosh, T. S., Singh, N. K., & Mande, S. S. (2011). SPHINX–an algorithm for taxonomic binning of metagenomic sequences. *Bioinformatics, 27*, 22–30.

Niedringhaus, T. P., Milanova, D., Kerby, M. B., Snyder, M. P., & Barron, A. E. (2011). Landscape of next-generation sequencing technologies. *Analytical Chemistry, 83*, 4327–4341.

Pandey, V., Nutter, G., & Prediger, E. (2008). Applied Biosystems SOLiD system: Ligation-based sequencing. In M. Janitz (Ed.), *Next-generation genome sequencing: Towards personalized medicine* (pp. 29–41). Weihnheim: Wiley-VCH.

Pati, A., Heath, L. S., Kyrpides, N. C., & Ivanova, N. (2011). ClaMS: A classifier for metagenomic sequences. *Standards in Genomic Sciences, 5*, 248–253.

Pevzner, P. A., Tang, H., & Waterman, M. S. (2001). An Eulerian path approach to DNA fragment assembly. *Proceedings of the National Academy of Sciences of the United States of America, 98*(17), 9748–9753.

Phillippy, A., Schatz, M., & Pop, M. (2008). Genome assembly forensics: Finding the elusive mis-assembly. *Genome Biology, 9*, r55.

Pope, P. B., Denman, S. E., Jones, M. J., Tringe, S. G., Barry, K., Malfatti, S. A., et al. (2010). Adaptation to herbivory by the wallaby includes bacterial and glycoside hydrolase profiles different from other herbivores. *Proceedings of the National Academy of Sciences of the United States of America, 107*, 14793–14798.

Pope, P. B., Smith, W., Denman, S. E., Tringe, S. G., Barry, K., Hugenholtz, P., et al. (2011). Isolation of *Succinivibrionaceae* implicated in low methane emissions from Tammar wallabies. *Science, 333*, 646–648.

Ronen, R., Boucher, C., Chitsaz, H., & Pevzner, P. (2012). SEQuel: Improving the accuracy of genome assemblies. *Bioinformatics, 28*, 188–196.

Rusch, D. B., Martiny, A. C., Dupont, C. L., Halpern, A. L., & Venter, J. C. (2010). Characterization of *Prochlorococcus* clades from iron-depleted oceanic regions. *Proceedings of the National Academy of Sciences of the United States of America, 107*, 16184–16189.

Segata, N., Waldron, L., Ballarini, A., Narasimhan, V., Jousson, O., & Huttenhower, C. (2012). Metagenomic microbial community profiling using unique clade-specific marker genes. *Nature Methods, 9*, 811–814.

Sloan, D. B., & Moran, N. A. (2012a). Endosymbiotic bacteria as a source of carotenoids in whiteflies. *Biology Letters, 8*, 986–989.

Sloan, D. B., & Moran, N. A. (2012b). Genome reduction and co-evolution between the primary and secondary bacterial symbionts of psyllids. *Molecular Biology and Evolution, 29*, 3781–3792.

Sloan, D. B., & Moran, N. A. (2013). The evolution of genomic instability in the obligate endosymbionts of whiteflies. *Genome Biology and Evolution, 5*, 783–793.

Swain, M. T., Tsai, I. J., Assefa, S. A., Newbold, C., Berriman, M., & Otto, T. D. (2012). A post-assembly genome-improvement toolkit (PAGIT) to obtain annotated genomes from contigs. *Nature Protocols, 7*, 1260–1284.

Teeling, H., Waldmann, J., Lombardot, T., Bauer, M., & Glöckner, F. O. (2004). A web-service and a stand-alone program for the analysis and comparison of tetranucleotide usage patterns in DNA sequences. *BMC Bioinformatics, 5*, 163.

The FASTG Format Specification Working Group (2012). The FASTG Format Specification (v1.00): An expressive representation for genome assemblies. *Broad Institute.*

Tsai, I. J., Otto, T. D., & Berriman, M. (2010). Improving draft assemblies by iterative mapping and assembly of short reads to eliminate gaps. *Genome Biology, 11*, R41.

Walsh, D. A., Zaikova, E., Howes, C. G., Song, Y. C., Wright, J. J., Tringe, S. G., et al. (2009). Metagenome of a versatile chemolithoautotroph from expanding oceanic dead zones. *Science*, *326*, 578–582.
Woyke, T., Tighe, D., Mavromatis, K., Clum, A., Copeland, A., Schackwitz, W., et al. (2010). One bacterial cell, one complete genome. *PLoS One*, *5*, e10314.
Zerbino, D. R., & Birney, E. (2008). Velvet: Algorithms for de novo short read assembly using de Bruijn graphs. *Genome Research*, *18*, 821–829.
Zhang, X., Davenport, K. W., Gu, W., Faligault, H. E., Munk, A. C., Hazuki, M., et al. (2012). Improving genome assemblies by sequencing PCR products with PacBio. *BioTechniques*, *53*, 61–62.

CHAPTER TWENTY-ONE

Microbial Community Analysis Using MEGAN

Daniel H. Huson[1], Nico Weber

Center for Bioinformatics, University of Tübingen, Tübingen, Germany

[1]Corresponding author: e-mail address: daniel.huson@uni-tuebingen.de

Contents

Abstract

Metagenomics, the study of microbes in the environment using DNA sequencing, depends upon dedicated software tools for processing and analyzing very large sequencing datasets. One such tool is MEGAN (MEtaGenome ANalyzer), which can be used to interactively analyze and compare metagenomic and metatranscriptomic data, both taxonomically and functionally.

To perform a taxonomic analysis, the program places the reads onto the NCBI taxonomy, while functional analysis is performed by mapping reads to the SEED, COG, and KEGG classifications. Samples can be compared taxonomically and functionally, using a wide range of different charting and visualization techniques. PCoA analysis and clustering methods allow high-level comparison of large numbers of samples. Different attributes of the samples can be captured and used within analysis. The program supports various input formats for loading data and can export analysis results in different text-based and graphical formats. The program is designed to work with very large samples containing many millions of reads. It is written in Java and installers for the three major computer operating systems are available from http://www-ab.informatik.uni-tuebingen.de.

1. INTRODUCTION

Environmental metagenomics involves the study of microbial organisms in their native environment using DNA sequencing (Handelsman, Rondon, Brady, Clardy, & Goodman, 1998). Metagenomic samples contain

Methods in Enzymology, Volume 531
ISSN 0076-6879
http://dx.doi.org/10.1016/B978-0-12-407863-5.00021-6

large number of organisms, for example, 4×10^7 prokaryotic cells can be found in 1 g of forest soil (Richter & Markewitz, 1995). Research in this area has benefited from the rise of second-generation sequencing technologies and hopes to benefit further from the third generation of sequencing techniques. Sampling and sequencing different environmental niches can now be done very cheaply and efficiently.

Given a dataset of DNA sequencing reads obtained from an environmental sample, there are three initial computational challenges to address. The first task is to estimate the taxonomic content, that is the qualitative and, if possible, quantitative distribution of organisms in the sample. The second problem is to determine the functional content of the sample. The third challenge is to compare different samples of interest. In many cases, the aim is to detect changes in taxonomic and/or function composition that correlate to external parameters or properties of the samples.

To address these challenges, the first step is often to align the set of sequencing reads against a database of known reference protein sequences such as NCBI-NR or RefSeq (Wheeler et al., 2008) using a pairwise alignment tool such as BLASTX (Altschul, Gish, Miller, Myers, & Lipman, 1990), RAPSearch2 (Zhao, Tang, & Ye, 2012), or PAUDA (Huson & Xie, 2013). A read is said to *hit* a given reference sequence, if a significant alignment is found during this process. The comparison of the sequencing reads against a reference database is usually considered the computationally most expensive step of the analysis and subsequent steps are based on the obtained alignments, however, new tools such as PAUDA may change this.

As an alternative to reference-based methods, one can use alignment-free taxonomic predictors to estimate the taxonomic content or profile of a sample. Such tools often employ machine-learning techniques such as naïve Bayesian classifiers or support vector machines, based on k-mer counts (McHardy, Martin, Tsirigos, Hugenholtz, & Rigoutsos, 2007). The advantages of machine-learning techniques are their speed and their (limited) ability to classify reads even when alignments to sequences in the reference database do not exist. However, such methods are usually not able to assess functional content and they do not produce pairwise alignments, which have many uses beyond general profiling.

Unfortunately, current databases only represent a small percentage of the microbial diversity encountered on earth and are strongly biased toward model organisms. It will be some time before projects such as GEBA (Wu et al., 2009) will have a significant impact on this problem.

Given the result of the alignment step or similarity search of a set of metagenomic reads against a reference database, an analysis program such as MEGAN is then required to explore and analyze the data. MEGAN is a tool for analyzing metagenomic sequence data, allowing the user to interactively explore the taxonomic and functional content of a sample. It also supports the comparison of multiple samples on taxonomic and functional levels. The program was originally published in Huson, Auch, Qi, and Schuster (2007) and the version 4 is presented in Huson, Mitra, Weber, Ruscheweyh, and Schuster (2011). Here, we describe version 5 of the software. Written in Java, the program runs on all major operating systems. The program can be downloaded from http://www-ab.informatik.uni-tuebingen.de/software/megan.

We would like to emphasize that those technical issues such as sample-preparation protocol, DNA extraction method, and sequencing technology all have a marked effect on the resulting metagenome data. It is current practice to keep all these variables constant within a given project, especially when sample comparison is of interest.

The basic input to MEGAN is a set of sequencing reads and the result of a pairwise alignment of the reads to a database of appropriate reference sequences. MEGAN supports a number of different input formats, such as BLAST (text, tabular, and XML) (Altschul et al., 1990), SAM (Li et al., 2009), RapSearch2 (Zhao et al., 2012), RDP (Wang, Garrity, Tiedje, & Cole, 2007), NBC (Rosen, Reichenberger, & Rosenfeld, 2011), QIIME (Caporaso et al., 2010) as well as a number of different CSV (comma-separated value) formats. Processed data is stored in a so-called RMA (read-match archive) file that contains all reads and matches in a compressed and indexed format. The results of analyses produced by MEGAN can be exported in a number of CSV formats and all visualizations provided by the program can be exported in a wide range of graphics formats. The program provides search tools for locating taxa and genes of interest.

In this chapter, we discuss how to analyze and compare metagenomic samples and address some of the typical questions encountered along the way. Sampling, library preparation, sequencing, and quality control are beyond the scope of this work and are not discussed here. As sequence comparison software, we recommend use of RAPSearch2 or PAUDA.

We use a permafrost soil metagenome dataset that was published in Mackelprang et al. (2011) as a running example. This dataset was sequenced from multiple permafrost cores originating from Hess Creek, Alaska. The samples were acquired in two cores, each containing soil from two different layers representing the active and the permafrost layer. Samples were sequenced immediately after collection as well as 2 and 7 days after thawing of the cores at 5 °C. Hence, there are 12 sequenced samples in total, each associated with one of two cores, *core 1* or *core 2*, one of two layers, *active layer* or *permafrost*, and one of three time points, *frozen* (time point 0), *2 days*, or *7 days*. MEGAN files for all 12 samples can be downloaded from http://ab.inf.uni-tuebingen.de/software/megan/.

The first step in a MEGAN analysis is to parse the reads and sequence alignment (BLAST or similar) files, using the *Import from Blast* menu item. MEGAN stores the result of this step in an RMA file. The initial parsing of a dataset may take a number of hours and is often performed on a server using MEGAN in command-line mode. However, it takes MEGAN only seconds to open or reopen an RMA file.

2. TAXONOMIC ANALYSIS

One popular approach to assess the taxonomic content of a sample is to focus on specific phylogenetic markers such as 16S rRNA. This type of analysis is supported by MEGAN and one can easily import the result of an analysis of such data obtained, for example, using the RDP classifier (Wang et al., 2007) or by performing a BLASTN comparison of the reads against the Silva database (Pruesse et al., 2007). The taxonomy used by MEGAN can be modified to suit the purposes of 16S rRNA, see Lanzén et al. (2012) for details.

Metagenome sequencing proper employs environmental shotgun sequencing of reads from genomic DNA or cDNA from RNA. To perform a taxonomic analysis of a metagenomic shotgun dataset, MEGAN aims at placing each read onto a node in the NCBI taxonomy, based on an analysis of its hits against a reference database. A key idea is to use all ranks of the taxonomy so as to assign reads specific to a particular species near the leaves of the taxonomy and to map sequences that are conserved across a wider range of organisms to higher-level nodes. For example, a read that comes from a gene than is only present in *Escherichia coli* will be placed on the *E. coli* node, whereas a read that comes from a gene that is shared widely across different *Proteobacteria* will be assigned to the node labeled *Proteobacteria*.

The input to MEGAN usually consists of a file of DNA reads and a file containing all their hits to a reference database, usually in BLAST or SAM format, but other formats are also supported. In addition, at startup, MEGAN reads in the whole NCBI taxonomy, which is a rooted tree with over one million nodes. To perform a taxonomic analysis of a metagenome sample, MEGAN processes each DNA read in turn, assigning each read to the node in the NCBI taxonomy that is the *lowest common ancestor* of the set of species associated with all reference sequences that were hit by the read. This approach is known as the LCA algorithm. In essence, the LCA algorithm places reads by gene content and is thus quite conservative and unlikely to be overly effected by horizontal gene transfer.

The LCA algorithm has a number of parameters. The *minScore* parameter sets a minimum threshold for the bit score that an alignment must achieve to be considered. For reads of length 100, a value of 35 is appropriate, while for longer reads this threshold should be increased accordingly. In addition, the *maxExpected* parameter can be used to filter matches by their expected value, which is especially useful when reads have varying read lengths. The *topPercent* parameter (by default 10%) provides additional filtering of matches. Only those matches are kept whose score lies within the given percentage of the best score for the given read. The *minSupport* parameter specifies the minimum number of reads that a node in the NCBI taxonomy must attract before it is shown in the final output. The reads assigned to a node for which this requirement is not met are pushed up the taxonomy until a node with a sufficient number of assigned reads is reached.

Most sequencing protocols support sequencing of *paired reads*, that is, pairs of reads that are guaranteed to come from the same DNA insert of a specified length. MEGAN can make use of read pairing information during the LCA assignment. For a given pair of reads, the program determines the set of taxa for which both reads have significant matches and the scores of these matches are boosted by a factor of 20%.

In consequence, the LCA algorithm will base the placement of reads mainly on those taxa, for which both reads of a pair have a good match, often leading to a more specific placement.

In the analysis of 16S rRNA sequences, it is often required that a specific level of DNA identity between two 16S rRNA sequences is achieved, before they can be assigned to the same taxon at a specific taxonomic rank, for example, more than 99% of identity is required to be counted as the same species. To address this, MEGAN provides a *Percent Identity Filter* that can be used to enforce the following levels of percentage sequence identities for an assignment at a given taxonomic level, namely, Species 99%, Genus 97%, Family 95%, Order 90%, Class 85%, and Phylum 80%.

Reads that have no hits are assigned to a special node labeled *No Hits*, whereas reads that have hits, but cannot be assigned to a taxon are mapped to a special *Unassigned* node. In addition, reads consisting of repetitive sequence are assigned to a *Low Complexity* node.

The part of the NCBI taxonomy to which reads are assigned is displayed in the Taxonomy Viewer of MEGAN, and by default, each node is scaled by the square root of the number of reads associated with it (see Fig. 21.1). Nodes can also be scaled linearly, by square root, or logarithmically; the latter two options are useful when many nodes have very low counts.

The number of reads assigned to a node, as well as the sum of reads up to this node, are displayed when the node is selected or when to user places the mouse over the node. Nodes can be interactively collapsed or expanded to show more or fewer details of the classification. The user can select nodes of interest, and then can either inspect the associated reads and alignments, or export the associated reads to a file. The user can also summarize data in a number of ways using different types of charts. Additionally, the user can chart the number of reads assigned to different microbial attributes to get additional information about the selected species. Nodes can be selected in a number of different ways and many of the menu items provided by the program apply to all selected nodes. This includes various charting options such as bar, bubble, radial space-filling and word-cloud charts (see Fig. 21.2).

MEGAN allows one to compute a rarefaction curve for a given sample (see Fig. 21.3). The underlying algorithm repeatedly samples replicates of size 10%, 20%, . . . 100% from the original sample and then plots the number of leaves produced for each replicate. The steepness of the curve will give an indication how close to saturation the sequencing is. (However, because organisms can be arbitrarily rare in a sample, the rarefaction curve cannot help decide whether all organisms present have been seen.)

Depending on the sampling location, the sequencing technique used and other considerations, the researcher will have to interactively explore the data and adapt the parameters of the LCA algorithm and other options to suit their needs. Typically, after initial parsing of an input file, one will first look at a taxonomic analysis of the data. In a second step, one might then explore in the functional content of the sample.

3. FUNCTIONAL ANALYSIS

MEGAN currently supports functional analysis using three different functional classification schemes, namely, SEED, KEGG, and COG (eggNOG).

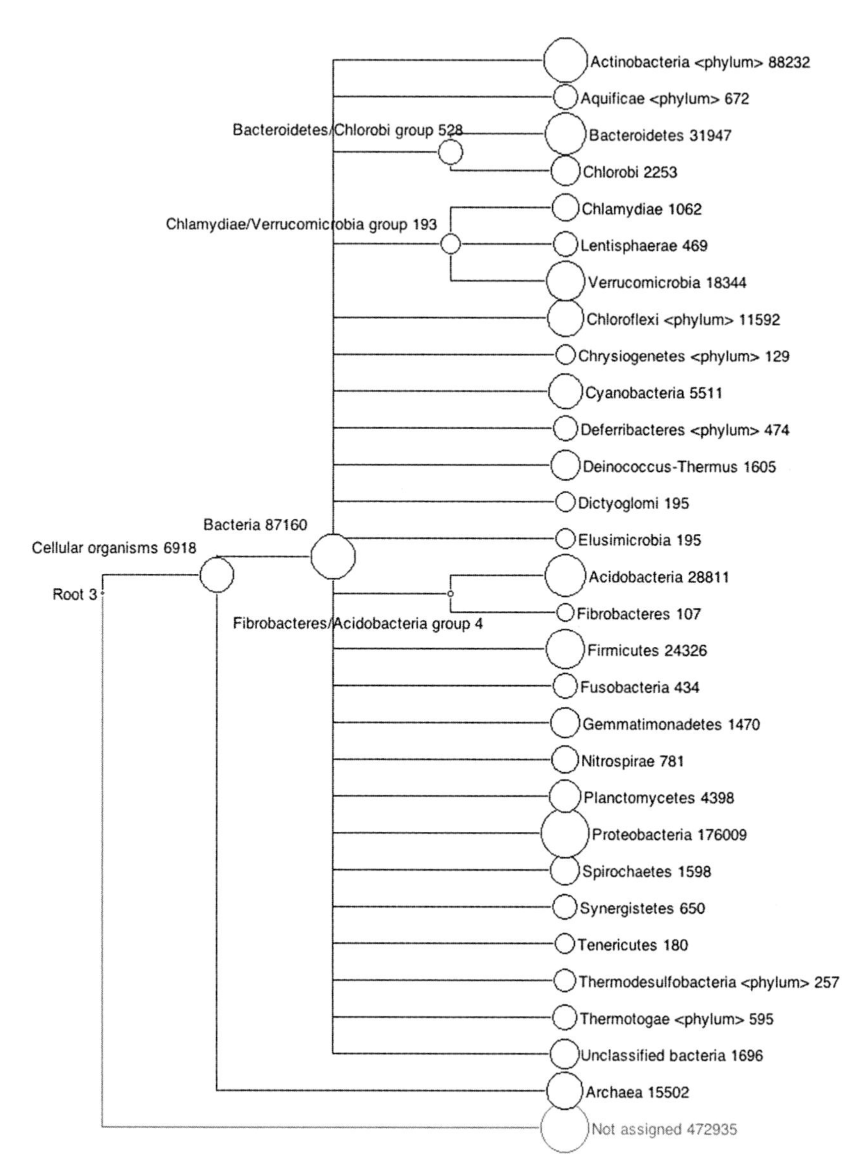

Figure 21.1 Taxonomy analysis of ≈990,000 reads originating from a permafrost sample published by Mackelprang et al. (2011). Each circle represents a taxon in the NCBI taxonomy and is scaled logarithmically to indicate how many reads have been assigned to it. In addition to the taxon name, each node is also labeled by the cumulative number of reads assigned to, or below, that node.

Figure 21.2 Word cloud of genus rank analysis. This type of representation makes it easy for the human eye to identify whether specific taxa have a significant presence in a sample. Fonts are scaled by number of reads assigned. (For color version of this figure, the reader is referred to the online version of this chapter.)

The SEED Viewer in MEGAN is based on the SEED classification (Overbeek et al., 2005) and the concept of a subsystem, which consists of a set of functional roles that implement a specific biological process or structural complex. To perform a SEED-based analysis, for each read in the input, MEGAN identifies the highest scoring hit to a reference sequence for which the corresponding functional role is known and then maps the read to that functional role.

The KEGG Viewer in MEGAN is based on the concepts of enzymes and pathways (Kanehisa & Goto, 2000). To perform a KEGG analysis, MEGAN maps reads to genes, which are then mapped to KEGG orthology groups and these appear in one or more pathways.

The COG Viewer in MEGAN is based on the Clusters of Orthologous Groups (COGs) (Tatusov, Galperin, Natale, & Koonin, 2000) and the extension of this system, called eggNOG (Powell et al., 2012). Reads are

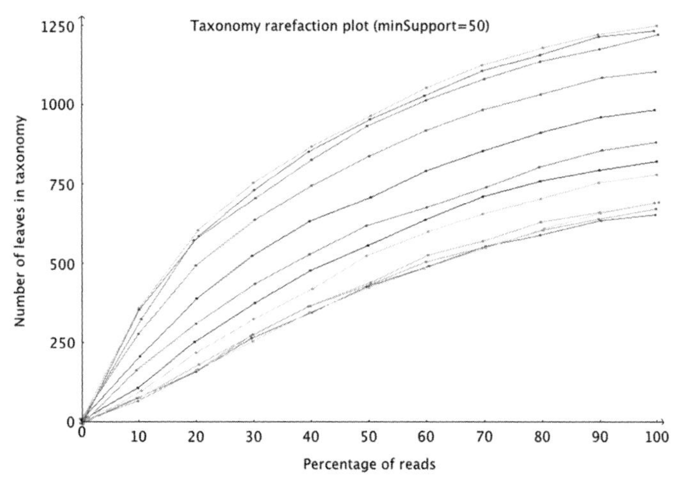

Figure 21.3 Rarefaction curve for taxonomic assignment of all 12 samples at the taxonomic rank of genus. MEGAN generates this plot by randomly and repeatedly subsampling 10%, 20%, ... 100% of the reads and counting the number of leaves in the resulting graph. Such curves are used to estimate how many additional taxa may be discovered if one increases the amount of sequencing by a certain amount. (See color plate.)

mapped to so-called COGs and NOGs, in a manner similar to the other two functional classification schemes.

MEGAN uses built-in RefSeq mappings for all three classifications. In addition, users can supply their own mapping files in a number of different formats, which is necessary when comparing reads against a reference database that does not contain RefSeq ids or using a comparison method that does not pass through all information from the reference database to its output (such as RAPSearch2). Updated SEED and COG mappings will be included on a regular basis in new releases of MEGAN. As of July 2011, the KEGG classification system is no longer freely available for academic usage. MEGAN is shipped with the last free version of KEGG.

The SEED, KEGG, and COG classifications are displayed as trees in MEGAN and the three corresponding viewers each provide the same interactive features as the Taxonomy Viewer. In addition to the tree view, a separate navigation pane allows the user to browse the structure by subsystem or groups for easy access. Moreover, the number of reads assigned to classes can be displayed as a heat map. The KEGG Viewer allows one to see how reads map to different enzymes in a given pathway, see Figs. 21.4–21.6. If only interested in specific functions, then one can select and extract all assigned reads into a new document for further analysis.

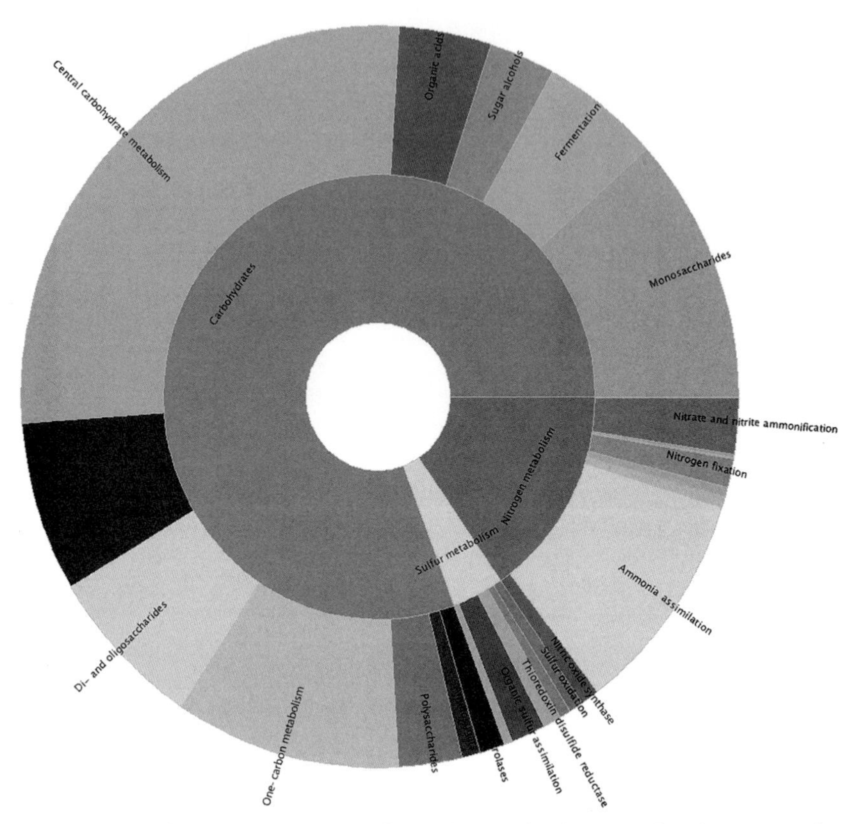

Figure 21.4 High-level SEED analysis of a single sample. The SEED classification tree has been partially selected and is displayed as radial space-filling chart. The angle covered by the region associated with a term is proportional to the number of reads assigned. The chart displays the selected nodes carbohydrates, nitrogen and sulfur metabolism. (For color version of this figure, the reader is referred to the online version of this chapter.)

4. SEQUENCE ALIGNMENT

As mentioned above, the main computational step is to determine all pairwise alignments between the set of DNA reads and all sequences in an appropriate reference database. Based on the result of this computation, it is possible to construct a reference-guided multiple sequence alignment between all reads that hit the same reference sequence. MEGAN provides access to such multiple sequence alignments in its Alignment Viewer. Once the user has selected a node in the Taxonomy, SEED, KEGG, or COG

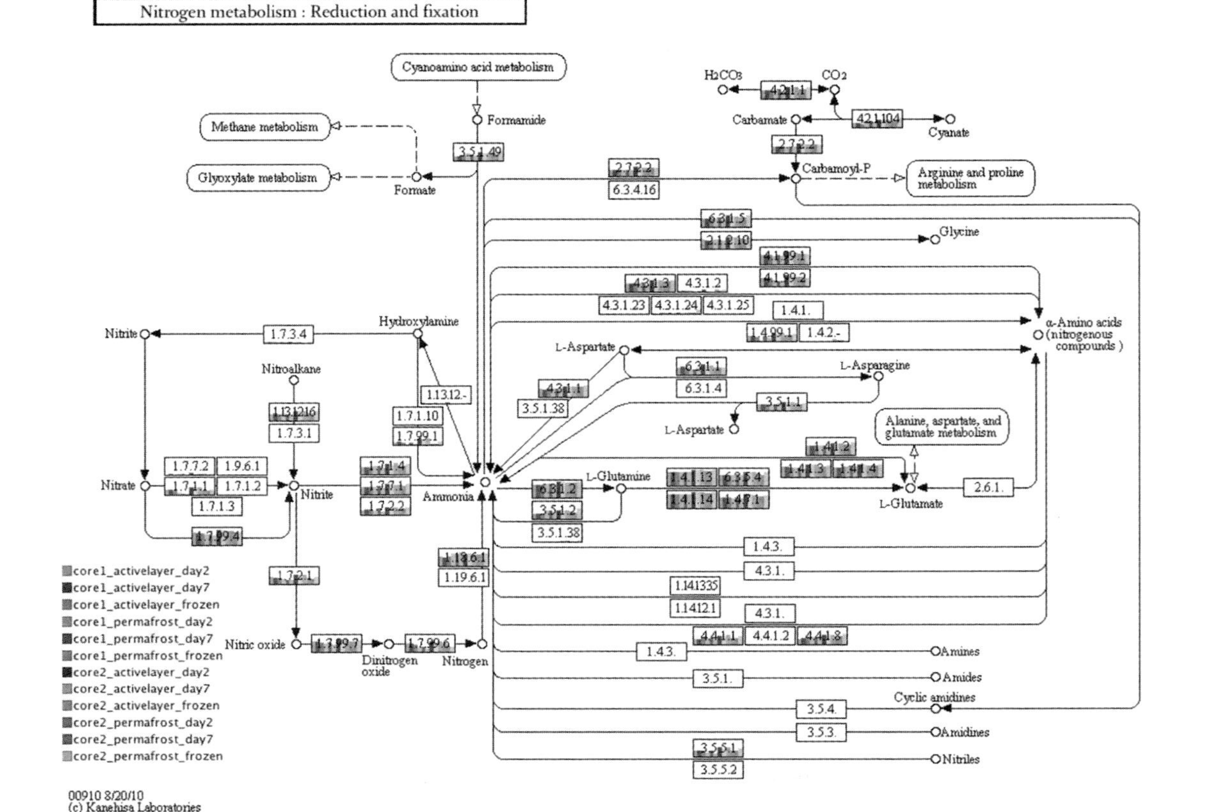

Figure 21.5 KEGG-based functional analysis of all of 12 permafrost samples (Mackelprang et al., 2011). The example shows the number of assigned reads as bar charts for each enzyme involved in the nitrogen-metabolism pathway. (For color version of this figure, the reader is referred to the online version of this chapter.)

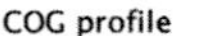

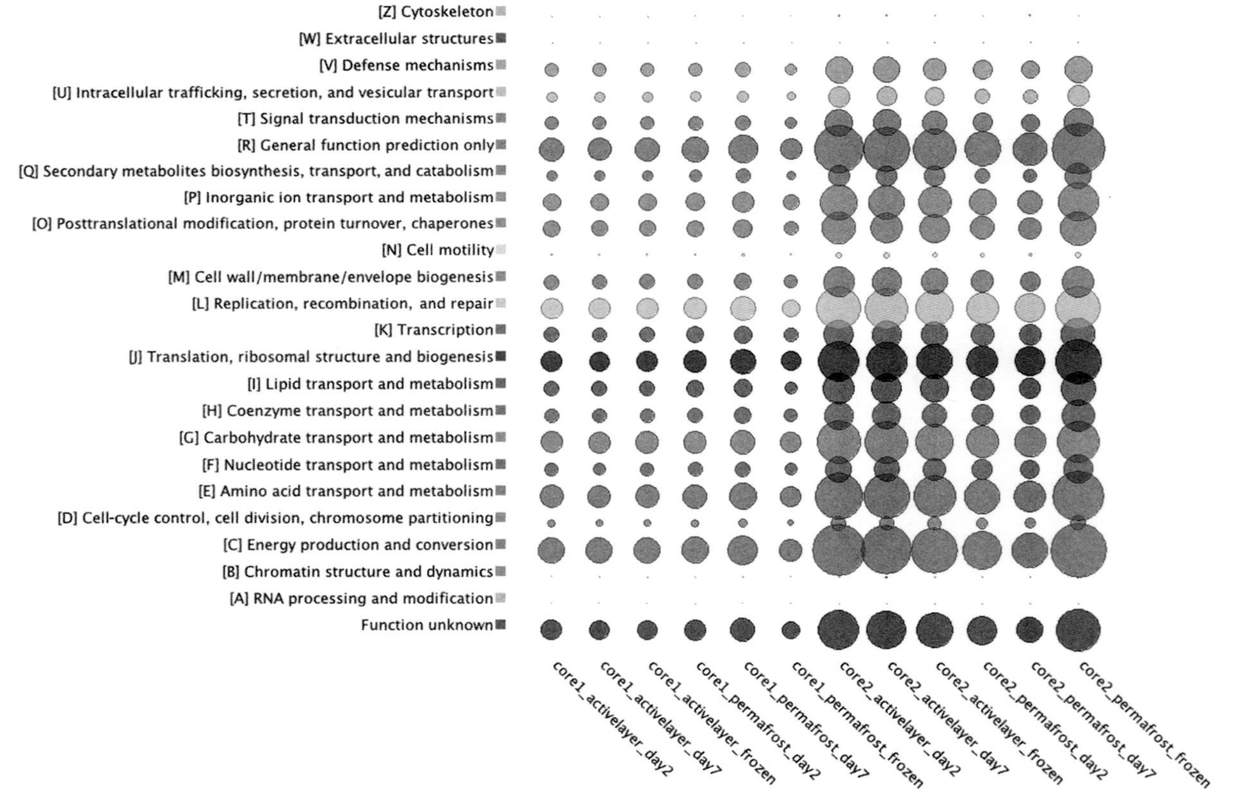

Figure 21.6 Visual output of the COG analyzer. The information is charted here as a bubble chart. Samples are displayed along the *X*-axis. The *Y*-axis represents different COG categories. Nodes represent the number of assigned reads and can be scaled linearly by square root or logarithmically. (See color plate.)

Viewer for which the Alignment Viewer is to be launched, the program collects all reference sequences that correspond to the given node and then, for each such reference sequence, the program determines all reads that align to it. The user can then select a reference sequence and the corresponding sequence alignment is subsequently displayed. The resulting alignment and/or consensus sequence can be exported to a text file.

This feature allows the user to create gene-specific multiple alignments without having to first extract reads and then align them using an external multiple alignment program. Average coverage of different genes as well as the layout of reads that match a given reference sequence may help to determine the reliability of the assignment of reads to a specific node. Functional analysis may also benefit from inspecting the alignment. The Alignment Viewer provides a function for performing a "diversity analysis" for protein-guided assemblies, which aims at estimating how many distinct genomes contribute to a specific alignment.

5. COMPARISON OF SAMPLES

Metagenome projects usually comprise multiple samples taken from different environments, experimental settings, time points, or locations. The comparison of tens or hundreds of large samples is a challenging task. Depending on the project, the goal of a comparison can vary between the detection of simple changes in taxonomic composition to complex functional shifts. To facilitate the comparison of samples, MEGAN allows the user to open multiple samples simultaneously, showing each sample in a different window. The user can then select a number of open samples to be combined into a single new comparison document.

To take different sample sizes into account, one can select absolute, relative, or subsampled reads for comparison. When comparing samples of very different sizes, the subsample mode is recommended, in which all samples are randomly and repeatedly subsampled down to the size of the smallest sample.

The resulting comparison document offers the same Taxonomy, SEED, KEGG, and COG viewers as in the above described single-sample view. In all of the viewers, the tree indicates how many reads were assigned to each node for each original input sample by drawing the node as a pie chart or bar chart (see Fig. 21.7).

All the features available for the analysis of a single sample can also be applied to the comparison of multiple samples, except for the possibility

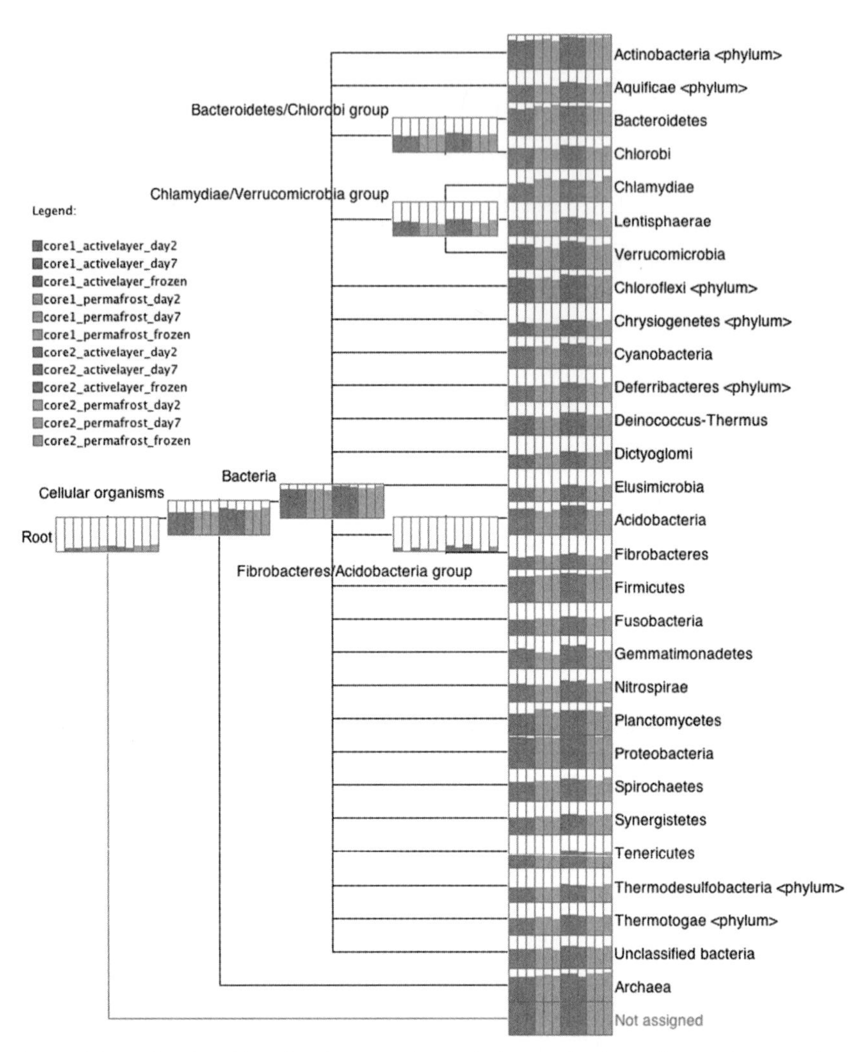

Figure 21.7 High-level (phylum) comparison of all samples from the original publication (Mackelprang et al. (2011)). Here, we have colored samples by layer (permafrost or active) that they originated from. The number of reads is represented by a bar chart next to the nodes label. (For color version of this figure, the reader is referred to the online version of this chapter.)

to save, inspect, or align reads assigned to a given node. The number of reads assigned to given nodes can be exported as a CSV file for additional analysis with external tools. Attribute charts can also be generated within the comparison view. A general shift or difference in the microbial community can be seen easily without inspecting each node. Examples range from aerobic to anaerobic oxygen requirements or critical changes in distribution of non-pathogenic to pathogenic microorganisms.

When comparing multiple samples, one often observes that the abundances for a fixed taxonomic rank vary quite strongly even between closely related samples, whereas the functional content appears to be much more conserved. Figure 21.8 illustrated this by using all 12 permafrost samples.

MEGAN's new Sample Viewer window allows one to place multiple samples into a single document and also provides methods for combining multiple samples in different ways (Fig. 21.9). In more detail, for a selected set of samples, one can compute a *core biome* containing all taxa and functions present in more than P percent of all selected samples, with $P=50$, say. The *total biome* is simply the union of taxa and functions of the selected samples. The *shared biome* is the intersection of taxa and functions present in the selected samples. The *rare biome* is the set of taxa and functions that are present in less than P percent of all selected samples.

The Sample Viewer also provides a table of attributes or *metadata* associated with the different samples. The attributes can be used to select samples by value or to color samples and also to control other features of their representation in PCoA plots and other similar comparisons.

MEGAN also supports the calculation of standard ecological indices for a comparison document (Mitra, Gilbert, Field, & Huson, 2010), such as Goodall, Chi-Square, Hellinger, Bray–Curtis, Kulczynski as well as normalized functions. Distances between samples can be calculated from taxonomic or functional data. The calculation only operates on selected nodes, so the user can easily control which parts of the sample to take into account. Resulting distances can be represented as a tree, network, or PCoA plot. Additionally, the distance matrix can be exported to a text file and then analyzed using other software packages.

To illustrate MEGAN's support of sample comparison, we imported all 12 permaforst samples into MEGAN. We used the Sample Viewer to color the samples based on their core (e.g., core 1 or core 2). We performed a PCoA analysis of the KEGG profiles of the 12 samples using MEGAN's Cluster Viewer. As shown in Fig. 21.10, samples form two major clusters that separate into four smaller ones. The two major clusters separate both layers (permafrost and active) from each other. Within those clusters, the cores are also separated. While the samples from the active layer start out quite similar and stay similar during thawing, the two permafrost samples

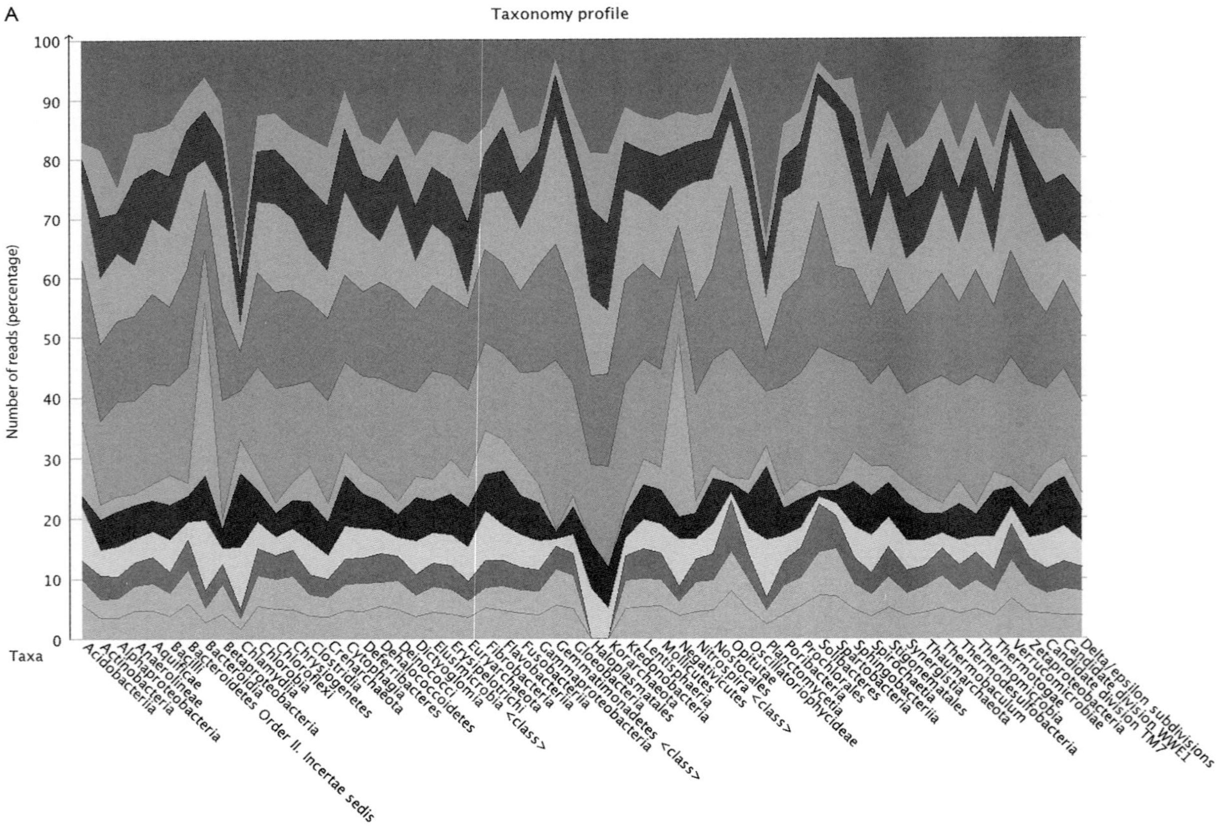

A

Taxonomy profile

Number of reads (percentage)

100
90
80
70
60
50
40
30
20
10
0

Taxa

Acidobacteria
Actinobacteria
Alphaproteobacteria
Anaerolineae
Aquificae
Bacilli
Bacteroidetes Order II. Incertae sedis
Bacteroidia
Betaproteobacteria
Chlamydiia
Chlorobia
Chloroflexi
Chrysiogenetes
Clostridia
Crenarchaeota
Cytophagia
Deferribacteres
Dehalococcoidetes
Deinococci
Dictyoglomia
Elusimicrobia <class>
Erysipelotrichi
Euryarchaeota
Fibrobacteria
Flavobacteriia
Fusobacteriia
Gammaproteobacteria
Gemmatimonadetes <class>
Gloeobacteria
Haloplasmatales
Korarchaeota
Ktedonobacteria
Lentisphaeria
Mollicutes
Negativicutes
Nitrospira <class>
Nostocales
Opitutae
Oscillatoriophycideae
Planctomycetia
Poribacteria
Prochlorales
Solibacteres
Spartobacteria
Sphingobacteriia
Spirochaetia
Stigonematales
Synergistia
Thaumarchaeota
Thermobaculum
Thermodesulfobacteria
Thermomicrobia
Thermotogae
Verrucomicrobiae
Zetaproteobacteria
Candidate division TM7
Candidate division WWE1
Delta/epsilon subdivisions

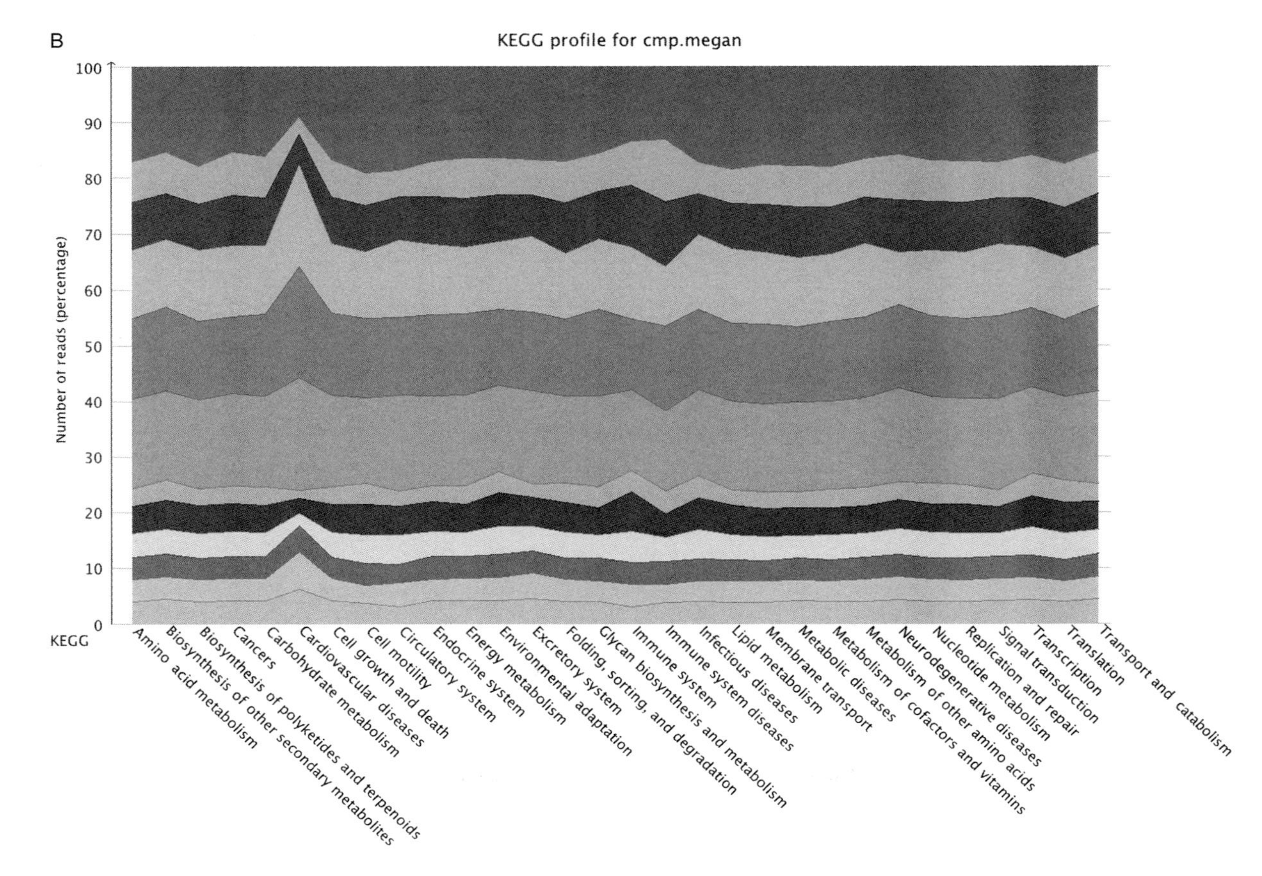

Figure 21.8 Relative abundance of the taxonomic content (A) and the KEGG functional content (B) of multiple samples. The taxonomic content is much more unstable and therefore more difficult to interpret. Function on the other hand is more conserved over multiple samples. Significant changes may therefore be easier to detect, because overall noise is less present. (See color plate.)

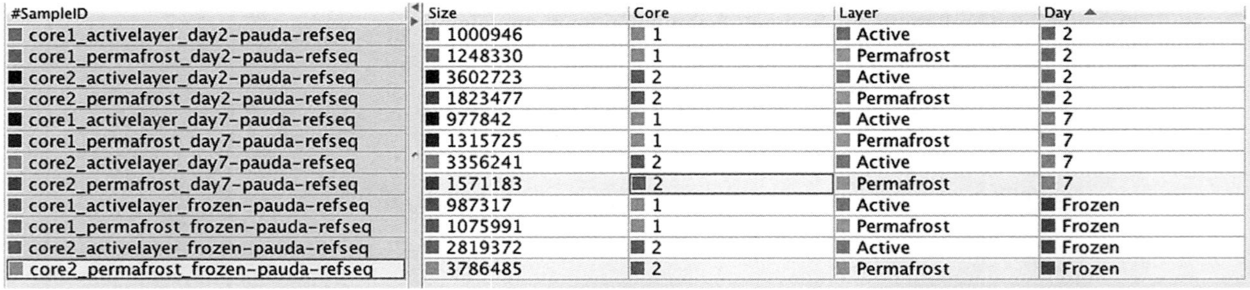

#SampleID	Size	Core	Layer	Day
core1_activelayer_day2-pauda-refseq	1000946	1	Active	2
core1_permafrost_day2-pauda-refseq	1248330	1	Permafrost	2
core2_activelayer_day2-pauda-refseq	3602723	2	Active	2
core2_permafrost_day2-pauda-refseq	1823477	2	Permafrost	2
core1_activelayer_day7-pauda-refseq	977842	1	Active	7
core1_permafrost_day7-pauda-refseq	1315725	1	Permafrost	7
core2_activelayer_day7-pauda-refseq	3356241	2	Active	7
core2_permafrost_day7-pauda-refseq	1571183	2	Permafrost	7
core1_activelayer_frozen-pauda-refseq	987317	1	Active	Frozen
core1_permafrost_frozen-pauda-refseq	1075991	1	Permafrost	Frozen
core2_activelayer_frozen-pauda-refseq	2819372	2	Active	Frozen
core2_permafrost_frozen-pauda-refseq	3786485	2	Permafrost	Frozen

Figure 21.9 The Sample Viewer shows all samples contained in a given comparison file. It can be used to select, color, or assign shapes to samples based on the values of different attributes (or *metadata*) associated with samples. (For color version of this figure, the reader is referred to the online version of this chapter.)

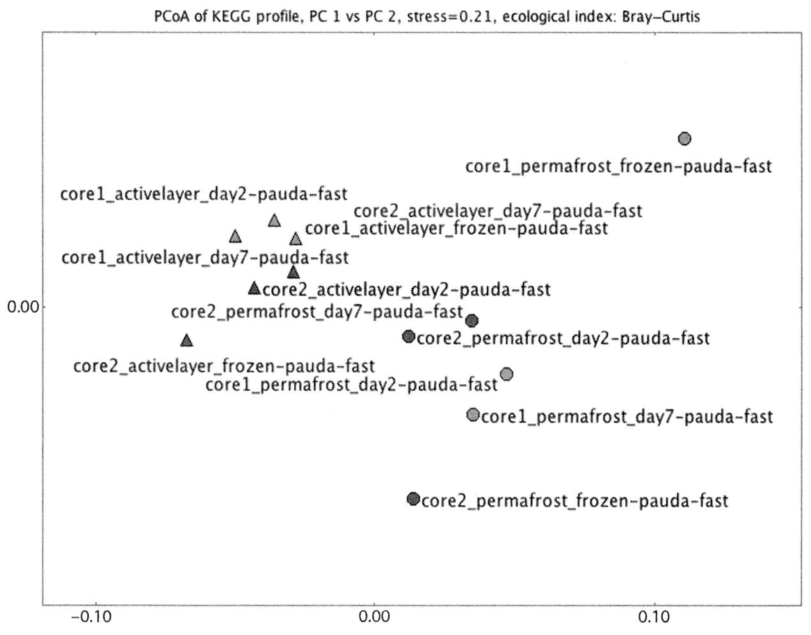

Figure 21.10 PCoA analysis of the KEGG functional content of 12 permafrost samples (Mackelprang et al., 2011). Samples from core 1 and 2 are shown in orange and blue, respectively. Samples from the active layer are shown as triangles, whereas samples from the permafrost layer are shown as circles. (For interpretation of the references to color in this figure legend, the reader is referred to the online version of this chapter.)

initially have very different functional profiles, which become much more similar during thawing.

6. CONCLUSION

MEGAN is an interactive program for analyzing the taxonomic and functional content of metagenomic (and metatranscriptomic) samples. With MEGAN, we hope to provide a versatile tool for analyzing single or groups of metagenomes on a desktop computer, aimed at the biologist in the field (or lab) rather than the trained bioinformatician, and thus we try to keep usability as simple as possible. Input is a set of DNA reads and the result of comparing the reads against a reference database.

MEGAN supports various standard file formats to make it easier to import data from different types of alignment or mapping tools. After sequence comparison, the taxonomic analysis is performed by placing

DNA reads onto nodes of the NCBI taxonomy based on the LCA approach. Offering various options, the user can tune the result to match specific needs of the analysis. Functional analysis is based on mapping reads to SEED and KEGG categories using the NCBI RefSeq identifiers.

The program also supports comparative analysis of multiple samples. An Alignment Viewer allows one to obtain multiple alignments of reads against reference sequences. Information generated by MEGAN can be exported in standard text and graphical files.

MEGAN is written in Java and runs on all major operating systems. When run in command-line mode, the program can also be integrated into larger bioinformatics analysis pipelines. As sequencing technologies continue to improve, the size of analyzed samples continues to increase. MEGAN was reportedly used to perform the taxonomic analysis of 124 human gut samples involving around 600 GB of sequence (Qin et al., 2010). This work only shows parts of MEGAN capabilities to analyze metagenomic data. The program is under active development and new features are added on a regular basis. We routinely use MEGAN to analyze large samples consisting of many million reads.

As samples continue to increase in both size and number, increasingly efficient methods for analyzing and comparing data will be needed. We have developed a new sequence alignment method called PAUDA that is a companion program to MEGAN and performs read alignment at a speed of more than 10,000 times that of BLASTX. Another advantage of PAUDA is that it can produce RMA files as output, thus eliminating the time-consuming step of initial parsing.

REFERENCES

Altschul, S. F., Gish, W., Miller, W., Myers, E. W., & Lipman, D. J. (1990). Basic local alignment search tool. *Journal of Molecular Biology, 215*, 403–410.

Caporaso, J. G., Kuczynski, J., Stombaugh, J., Bittinger, K., Bushman, F. D., Costello, E. K., et al. (2010). Qiime allows analysis of high-throughput community sequencing data. *Nature Methods*, 7(5), 335–336.

Handelsman, J., Rondon, M., Brady, S., Clardy, J., & Goodman, R. (1998). Molecular biological access to the chemistry of unknown soil microbes: A new frontier for natural products. *Chemistry and Biology, 5*, 245–249.

Huson, D. H., Auch, A. F., Qi, J., & Schuster, S. C. (2007). MEGAN analysis of metagenomic data. *Genome Research, 17*(3), 377–386.

Huson, D. H., Mitra, S., Weber, N., Ruscheweyh, H., & Schuster, S. C. (2011). Integrative analysis of environmental sequences using megan4. *Genome Research, 21*, 1552–1560.

Huson, D. H., & Xie, C. (2013). A poor man's BLASTX—High-throughput metagenomic protein database search using PAUDA. *Bioinformatics, 4*, 1236–1242.

Kanehisa, M., & Goto, S. (2000). KEGG: Kyoto encyclopedia of genes and genomes. *Nucleic Acids Research, 28*(1), 27–30.

Lanzén, A., Jørgensen, S. L., Huson, D. H., Gorfer, M., Grindhaug, S. H., et al. (2012). CREST—Classification resources for environmental sequence tags. *PLoS One*, 7(11), e49334.

Li, H., Handsaker, B., Wysoker, A., Fennell, T., Ruan, J., Homer, N., et al. (2009). The sequence alignment/map (SAM) format and SAMtool. *Bioinformatics*, *25*, 2078–2079.

Mackelprang, Rachel, Waldrop, Mark P., DeAngelis, Kristen M., David, Maude M., Chavarria, Krystle L., Blazewicz, Steven J., et al. (2011). Metagenomic analysis of a permafrost microbial community reveals a rapid response to thaw. *Nature*, *480*, 368–373.

McHardy, A. C., Martin, H. G., Tsirigos, A., Hugenholtz, P., & Rigoutsos, I. (2007). Accurate phylogenetic classification of variable-length DNA fragments. *Nature Methods*, *4*(1), 63–72.

Mitra, S., Gilbert, J., Field, D., & Huson, D. (2010). Comparison of multiple metagenomes using phylogenetic networks based on ecological indices. *The ISME Journal*, *4*, 1236–1242. http://dx.doi.org/10.1038/ismej.2010.51.

Overbeek, R., Begley, T., Butler, R. M., Choudhuri, J. V., Chuang, H.-Y., Cohoon, M., et al. (2005). The subsystems approach to genome annotation and its use in the project to annotate 1000 genomes. *Nucleic Acids Research*, *33*(17), 5691–5702.

Powell, S., Szklarczyk, D., Trachana, K., Roth, A., Kuhn, M., Muller, J., et al. (2012). eggNOG v3.0: Orthologous groups covering 1133 organisms at 41 different taxonomic ranges. *Nucleic Acids Research*, *40*, D284–D289.

Pruesse, E., Quast, C., Knittel, K., Fuchs, B., Ludwig, W., Peplies, J., et al. (2007). SILVA: A comprehensive online resource for quality checked and aligned ribosomal RNA sequence data compatible with ARB. *Nucleic Acids Research*, *35*(21), 7188–7196.

Qin, J., Li, R., Raes, J., Arumugam, M., Burgdorf, K. S., Manichanh, C., et al. (2010). A human gut microbial gene catalogue established by metagenomic sequencing. *Nature*, *464*(7285), 59–65.

Richter, D., Jr., & Markewitz, D. (1995). How deep is soil? *Bioscience*, *45*, 600–609.

Rosen, G. L., Reichenberger, E., & Rosenfeld, A. (2011). NBC: The naive Bayes classification tool webserver for taxonomic classification of metagenomic reads. *Bioinformatics*, *27*, 127–129.

Tatusov, R. L., Galperin, M. Y., Natale, D. A., & Koonin, E. V. (2000). The COG database: A tool for genome-scale analysis of protein functions and evolution. *Nucleic Acids Research*, *28*, 33–36.

Wang, Q., Garrity, G. M., Tiedje, J. M., & Cole, J. R. (2007). Naive bayesian classifier for rapid assignment of rRNA sequences into the new bacterial taxonomy. *Applied and Environmental Microbiology*, *73*(16), 5261–5267.

Wheeler, D. L., Barrett, T., Benson, D. A., Bryant, S. H., Canese, K., Chetvernin, V., et al. (2008). Database resources of the National Center for Biotechnology Information. *Nucleic Acids Research*, *36*, D13–D21.

Wu, D., Hugenholtz, P., Mavromatis, K., Pukall, R., Dalin, E., Ivanova, N. N., et al. (2009). A phylogeny-driven genomic Encyclopaedia of Bacteria and Archaea. *Nature*, *462*(7276), 1056–1060.

Zhao, Y., Tang, H., & Ye, Y. (2012). RAPSearch2: A fast and memory-efficient protein similarity search tool for next- generation sequencing data. *Bioinformatics*, *28*, 125–126.

CHAPTER TWENTY-TWO

A Metagenomics Portal for a Democratized Sequencing World

Andreas Wilke[*,†], Elizabeth M. Glass[*,†], Daniela Bartels[*,†], Jared Bischof[*,†], Daniel Braithwaite[*,†], Mark D'Souza[*,†], Wolfgang Gerlach[*,†], Travis Harrison[*,†], Kevin Keegan[*,†], Hunter Matthews[*], Renzo Kottmann[‡], Tobias Paczian[*,†], Wei Tang[*,†], William L. Trimble[*,†], Pelin Yilmaz[‡], Jared Wilkening[*,†], Narayan Desai[*,†], Folker Meyer[*,†,1]

[*]Argonne National Laboratory, Lemont, Illinois, USA
[†]University of Chicago, Chicago, Illinois, USA
[‡]Max-Planck Institute für Marine Biologie, Bremen, Germany
[1]Corresponding author: e-mail address: folker@anl.gov

Contents

1. Introduction 488
2. Pipeline and Technology Platform 489
   2.1 Details on the new MG-RAST pipeline 490
   2.2 Additional improvements 494
3. Web Interface 495
   3.1 The upload and metadata pages 496
   3.2 Metadata-enabled data discovery 496
   3.3 The overview page 496
   3.4 The analysis page 503
4. How to Drill Down Using the Workbench 510
   4.1 2.6 Downloads 513
   4.2 Viewing evidence 514
5. MG-RAST Downloads 515
6. Discussion 519
7. Future Work 521
Acknowledgments 521
References 521

Abstract

The democratized world of sequencing is leading to numerous data analysis challenges; MG-RAST addresses many of these challenges for diverse datasets, including amplicon datasets, shotgun metagenomes, and metatranscriptomes. The changes from version 2 to version 3 include the addition of a dedicated gene calling stage using FragGenescan, clustering of predicted proteins at 90% identity, and the use of BLAT for the

Methods in Enzymology, Volume 531
ISSN 0076-6879
http://dx.doi.org/10.1016/B978-0-12-407863-5.00022-8

computation of similarities. Together with changes in the underlying software infrastructure, this has enabled the dramatic scaling up of pipeline throughput while remaining on a limited hardware budget. The Web-based service allows upload, fully automated analysis, and visualization of results. As a result of the plummeting cost of sequencing and the readily available analytical power of MG-RAST, over 78,000 metagenomic datasets have been analyzed, with over 12,000 of them publicly available in MG-RAST.

1. INTRODUCTION

The growth in data enabled by next-generation sequencing platforms provides an exciting opportunity for studying microbial communities, of which ~99% of the microbes have not yet been cultured (Riesenfeld, Schloss, & Handelsman, 2004). To support user-driven analysis of metagenomic data, we have provided MG-RAST (Meyer et al., 2008). The MG-RAST portal offers automated quality control, annotation and comparative analysis services, and archives over 78,000 datasets contributed by over 10,000 researchers.

While the previous version of MG-RAST (v2) was widely used, it was limited to datasets smaller than a few 100 Mbases, and comparison of samples was limited to pairwise comparisons. In the new version, datasets of tens of gigabases can be annotated and comparison of taxa or functions that differed between samples is now limited by the available screen real estate. Figure 22.1 shows a comparison of the analytical and computational approaches used in MG-RAST v2 and v3. The major changes are the inclusion of a dedicated gene calling stage using FragGenescan (Rho, Tang, & Ye, 2010), clustering of predicted proteins at 90% identity using UCLUST (Edgar, 2010), and the use of BLAT (Kent, 2002) for the computation of similarities. Together with changes in the underlying infrastructure, this has allowed dramatic scaling of the analysis with the limited hardware available.

The new version of MG-RAST represents a rethinking of core processes and data products, as well as new user-interface metaphors and a redesigned computational infrastructure. MG-RAST supports a variety of user-driven analyses, including comparisons of many samples, previously too computationally intensive to support for an open user community.

Scaling to the new workload required changes in two areas: the underlying infrastructure needed to be rethought and the analysis pipeline needed to be adapted to address the properties of the newest sequencing technologies.

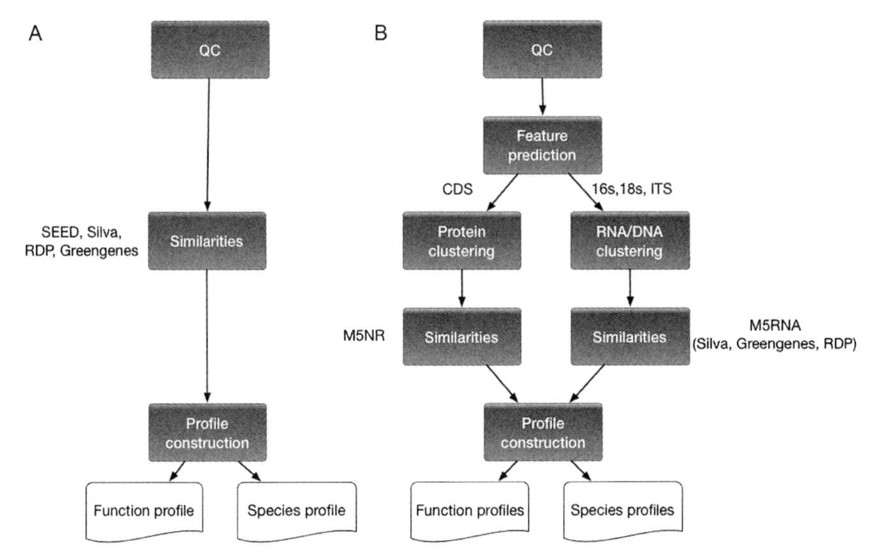

Figure 22.1 Overview of processing pipeline in (A) MG-RAST 2 and (B) MG-RAST 3. In the old pipeline, metadata was rudimentary, compute steps were performed on individual reads on a 40-node cluster that was tightly coupled to the system, and similarities were computed by BLAST to yield abundance profiles that could then be compared on a per-sample or per-pair basis. In the new pipeline, rich metadata can be uploaded, normalization and feature prediction are performed, faster methods such as BLAT are used to compute similarities, and the resulting abundance profiles are fed into downstream pipelines on the cloud to perform community and metabolic reconstruction and to allow queries according to rich sample and functional metadata. (For color version of this figure, the reader is referred to the online version of this chapter.)

2. PIPELINE AND TECHNOLOGY PLATFORM

One key aspect of scaling MG-RAST to large numbers of modern NGS datasets is the use of cloud computing which decouples MG-RAST from its previous dedicated hardware resources. Using our task server AWE (Wilke, Wilkening, Glass, Desai, & Meyer, 2011) and the SHOCK data management tool developed alongside it, we have updated our underlying computational platform using purpose-built software platform optimized for large-scale sequence analysis.

The new analytical pipeline for MG-RAST version 3 (Fig. 22.2) is encapsulated and separated from the data store, enabling far greater scalability.

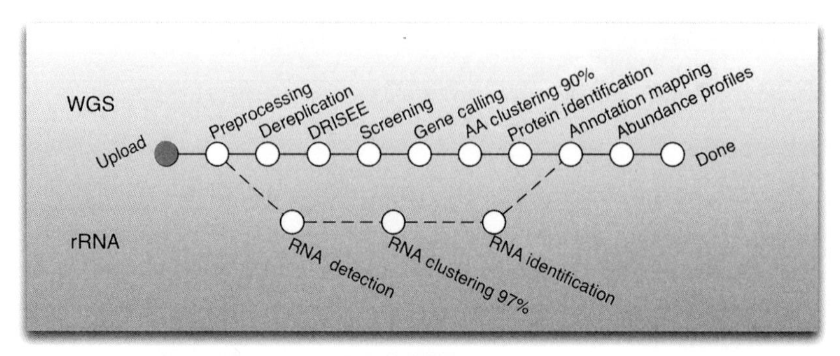

Figure 22.2 Details of the analysis pipeline for MG-RAST version 3. After initial quality control, the pipeline diverges into steps for Amplicon and shotgun reads (WGS). The duplicate read inferred sequencing error estimation (DRISEE) is based on read clusters created by dereplication step and only applicable for metagenomic shotgun sequences. (For color version of this figure, the reader is referred to the online version of this chapter.)

2.1. Details on the new MG-RAST pipeline

Several key algorithmic improvements were needed to support the flood of user-generated data.

Preprocessing: First, we replaced the read-centric approach, no longer performing a search for each independent read. The new pipeline actually consists of two independent flows. After upload, data are preprocessed using SolexaQA (Cox, Peterson, Biggs, & Solexa, 2010) to trim low-quality regions from FASTQ data. Platform-specific approaches are used for 454 data submitted in FASTA format: reads more than two standard deviations away from the mean read length are discarded as described previously (Huse, Huber, Morrison, Sogin, & Welch, 2007).

RNA detection, clustering, and identification: rRNA reads are identified using a simple rRNA detection pipeline and are searched in a separate flow in the pipeline. An initial BLAT (Kent, 2002) search against a reduced RNA database efficiently identifies RNA. The RNA-similar reads are then clustered at 97% identity, and a BLAT similarity search is performed for the longest cluster representative.

Dereplication: For the remaining data after quality filtering (which we assume to be protein coding), the processing starts with a dereplication step used to remove artificial duplicate reads (ADRs) (Gomez-Alvarez, Teal, & Schmidt, 2009). Instead of discarding the ADRs, we use these technical duplicates to estimate an error score for the entire dataset based on the variations in the sets of *nearly* identical reads stemming from artificial duplication (Keegan et al., 2012).

Quality assessment: The MG-RAST pipeline offers a variety of summaries of technical aspects of the sequence quality to enable sequence data triage. These tools include duplicate read inferred sequencing error estimation (DRISEE) for estimating sequence error, summaries of the spectra of long k-mers, and visualizations of the base caller output.

2.1.1 Duplicate read inferred sequencing error estimation

DRISEE (Keegan et al., 2012) is a method to provide a measure for sequencing error for whole genome shotgun metagenomic sequence data that are independent of sequencing technology, and accounts for many of the shortcomings of Phred. It utilizes ADRs (artifactual/artificial duplicate reads) to generate internal sequence standards from which an overall assessment of sequencing error in a sample is derived. DRISEE values are normally reported as percent error.

DRISEE values can be used to assess the overall quality of sequence samples. DRISEE data are presented on the overview page for each MG-RAST sample for which a DRISEE profile can be determined. Total DRISEE error presents the overall DRISEE-based assessment of the sample as a percent error:

$$\text{Total DRISEE Error} = \text{base\_errors}/\text{total\_bases} * 100$$

where "base_errors" refers to the sum of DRISEE detected errors and total_bases refers to the sum of all bases considered by DRISEE.

The current implementation of DRISEE is not suitable for amplicon sequencing data or other samples that may contain natural duplicated sequences (e.g., eukaryotic DNA where gene duplication and other forms of highly repetitive sequences are common) in high abundance.

2.1.2 k-mer profiles

k-mer digests are an annotation-independent method to describe sequence datasets that can support inferences about genome size and coverage. Here, the overview page presents several visualizations of the k-mer spectrum of each dataset, evaluated at $k = 15$.

Three visualizations provided of the k-mer spectrum are the k-mer spectrum, k-mer rank abundance, and ranked k-mer consumed. All three graphs represent the same spectrum, but in different ways. The k-mer spectrum plots the number of distinct k-mers against k-mer coverage. The k-mer coverage is equivalent to number of observations of each k-mer. The k-mer rank abundance plots the relationship between k-mer coverage and the k-mer rank—answering the question "what is the coverage of the *n*th most abundant k-mer." Ranked k-mer consumed plots the largest fraction of the data explained by the *n*th most abundant k-mers only.

2.1.3 Nucleotide histograms

These graphs show the fraction of base pairs of each type (A, C, G, T, or ambiguous base "N") at each position starting from the beginning of each read. Amplicon datasets (see Fig. 22.3) should show biased distributions of bases at each position, reflecting both conservation and variability in the recovered sequences.

Shotgun datasets should have roughly equal proportions of A, T, G, and C base calls, independent of position in the read as shown in Fig. 22.4.

Vertical bars at the beginning of the read indicate untrimmed (see Fig. 22.5), contiguous barcodes. Gene calling via FragGeneScan (Rho et al., 2010) and RNA similarity searches is not impacted by the presence of barcodes. However, if a significant fraction of the reads is consumed by barcodes, it reduces the biological information contained in the reads.

If a shotgun dataset has clear patterns in the data (see Fig. 22.6), this indicates likely contamination with artificial sequences. This dataset had a large fraction of adapter dimers.

Screening: The pipeline provides the option to remove reads that are near-exact matches to the genomes of a handful of model organisms, including fly, mouse, cow, and human. The screening stage uses bowtie (Langmead, Trapnell, Pop, & Salzberg, 2009) and only reads that do not match the model organisms pass into the next stage of the annotation pipeline.

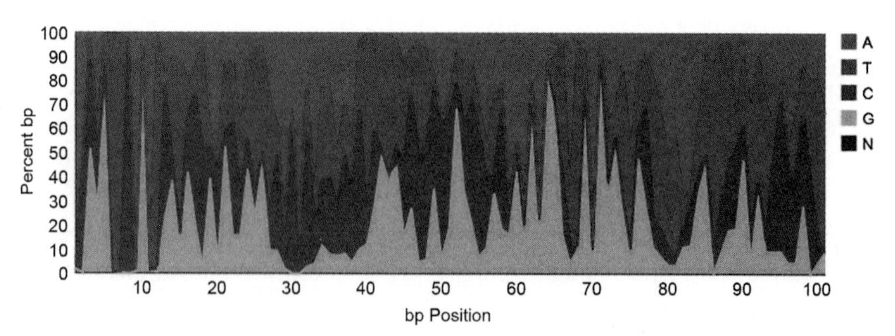

Figure 22.3 Nucleotide histogram with biased distributions. (See color plate.)

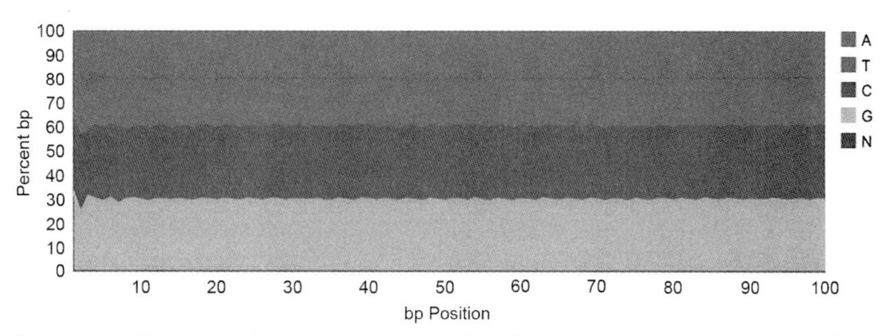

Figure 22.4 Nucleotide histogram showing ideal distributions. (For color version of this figure, the reader is referred to the online version of this chapter.)

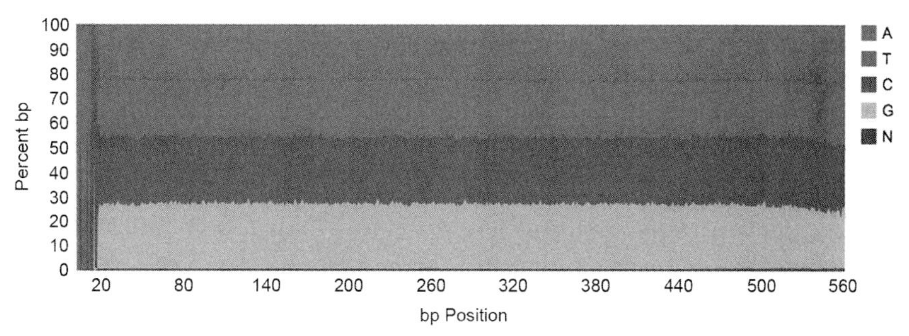

Figure 22.5 Nucleotide histogram with untrimmed barcodes. (For color version of this figure, the reader is referred to the online version of this chapter.)

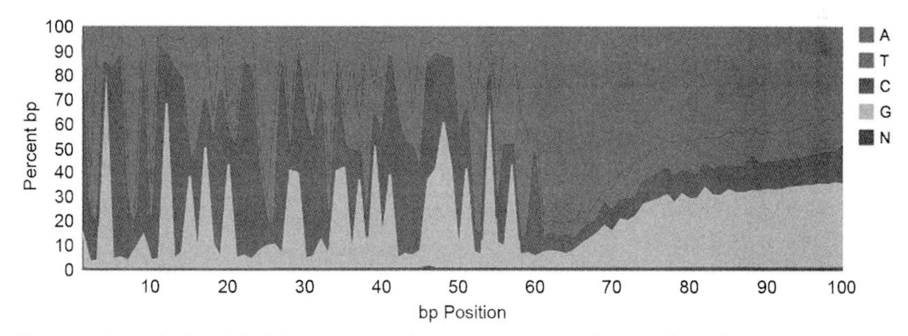

Figure 22.6 Nucleotide histogram with contamination. (See color plate.)

Gene prediction and AA clustering: While the previous version of MG-RAST used similarity-based gene predictions, this approach is significantly more expensive computationally than *de novo* gene prediction. After an in-depth investigation of tool performance (Trimble et al., 2012), we have moved to a machine learning approach: FragGeneScan (Rho et al., 2010). Using this approach, we can now predict coding regions in DNA sequences of 75 bp and longer. Our novel approach also enables the analysis of user-provided assemblies. MG-RAST builds clusters of proteins at the 90% identity level using the UCLUST (Edgar, 2010) implementation in QIIME (Caporaso et al., 2010) preserving the relative abundances. These clusters greatly reduce the computational burden of comparing all pairs of short reads, while clustering at 90% identity preserves sufficient biological signal. Once created, a representative (the longest sequence) for each cluster is subjected to similarity analysis; instead of BLAST, we use sBLAT, an implementation of the BLAT algorithm (Kent, 2002), which we parallelized using OpenMPI (Gabriel et al., 2004) for this work.

Once the similarities are computed, we present reconstructions of the species content of the sample based on the similarity results. We reconstruct the putative species composition of the sample by looking at the phylogenetic origin of the database sequences hit by the similarity searches.

Protein identification and annotation mapping: Sequence similarity searches are computed against a protein database derived from the M5NR (Wilke et al., 2011), which provides a nonredundant integration of many databases (GenBank (Benson et al., 2012), SEED (Overbeek et al., 2005), IMG (Markowitz et al., 2012), KEGG (Kanehisa, Goto, Sato, Furumichi, & Tanabe, 2012), and eggNOGs). Unlike MG-RAST 2, which relied solely on SEED, MG-RAST now supports many complementary views into the data with one similarity search, including different functional hierarchies: SEED subsystems, IMG terms, COG (Tatusov et al., 2003)/eggNOGs (Jensen et al., 2008), and ontologies such as GO (Gene Ontology Consortium, 2013). Users can easily change views without recomputation. For example, COG and KEGG views can be displayed, which both show the relative abundances of histidine biosynthesis in a dataset of four cow rumen metagenomes.

After similarity computation, a number of additional pipeline stages are executed, transforming the data into several representations that enable rapid query and or comparison.

2.2. Additional improvements

Adding the ability for users to encode rich information about each sample is another key improvement in MG-RAST 3. By using the standards developed by the Genomics Standards Consortium, we have enabled users to contribute GSC (Field et al., 2011) standard formatted metadata. Specifically, we use MIxS (minimum information about any (*x*) sequence) and MIMARKS (minimum information about a MARKer gene survey) specifications (Yilmaz et al., 2011) to store metadata and to search for related datasets in terms of geographic location, biochemical environment, or other contextual data.

This enables data discovery by end users using contextual metadata using searches like "retrieve soil samples from the continental United States." If the users have added additional metadata (domain-specific extension), additional queries are enabled, for example, "restrict the results to soils with a specific pH."

We have also enabled users to extract abundance profile data via the use of the BIOM format (McDonald et al., 2012). This enables downstream processing with BIOM compliant tools, for example, QIIME (Caporaso et al., 2010).

3. WEB INTERFACE

The MG-RAST system provides a rich Web user interface that covers all aspects of the metagenome analysis from data upload to ordination analysis. The Web interface can also be used for data discovery. Metagenomic datasets can be easily selected individually or on the basis of filters such as technology (including read length), quality, sample type, and keyword, with dynamic filtering of results based on similarity to known reference proteins or taxonomy. For example, a user might want to perform a search so as (phylum eq "actinobacteria" and function in "KEGG pathway Lysine Biosynthesis" and sample in "Ocean") to extract sets of reads matching the appropriate functions and taxa across metagenomes. The results can be displayed in familiar formats, including bar charts, trees that incorporate abundance information, heatmaps, or principal components analyses, or exported in tabular form. The raw or processed data can be recovered via download pages. Metabolic reconstructions based on mapping to KEGG pathways are also provided.

Sample selection is crucial for understanding large-scale patterns when multiple metagenomes are compared. Accordingly, MG-RAST supports MIxS and MIMARKS (Yilmaz et al., 2011) (as well as domain-specific plug-ins for specialized environments not extending the minimal GSC standards); several projects, including TerraGenome, HMP, TARA, and EMP, use these GSC standards, enabling standardized queries that integrate new samples into these massive datasets. An example query, using the metadata browser, shows how users can interrogate the existing pool of public datasets for a Biome of interest (e.g., Hot springs), and performing comparisons and a search for organisms encoding a specific gene function (e.g., Beta-lactamase or Aldo/keto reductase; see Fig. 22.7).

3.1. The upload and metadata pages

Data and metadata can be uploaded in the form of spreadsheets along with the sequence data using both the ftp and the http protocols. The Web uploader will automatically split larger files and allow parallel uploads.

MG-RAST supports datasets that are augmented with rich metadata using the standards and technology developed by the GSC.

Each user has a temporary storage location inside the MG-RAST system. This "inbox" provides temporary storage for data and metadata to be submitted to the system. Using the *inbox*, users can extract compressed files, convert a number of vendor-specific formats to MG-RAST submission compliant formats, and obtain an MD5 checksum for verifying that transmission to MG-RAST has not altered the data.

The Web uploader has been optimized for large datasets of over 100 GBp (gigabasepairs) often resulting in file sizes in excess of 150 GB.

3.2. Metadata-enabled data discovery

The metagenome browse page lists all datasets visible to the user. Datasets in MG-RAST are private by default, but the submitting user has the option to share datasets with specific users or to make datasets public. This page also provides an overview of the nonpublic datasets submitted by the user or shared with them. Figure 22.8 shows the metagenome browse table, which provides an interactive graphical means to discover data based on technical data (e.g., sequence type or dataset size) or metadata (e.g., location or biome).

3.3. The overview page

MG-RAST automatically creates an individual summary page for each dataset. This "metagenome overview page" provides a summary of the annotations for a single dataset. The page is made available by the automated pipeline once the computation is finished.

The page is intended as a single point of reference for metadata, quality, and data. It also provides an initial overview of the analysis results for individual datasets with default parameters. Further analysis is available on the analysis page.

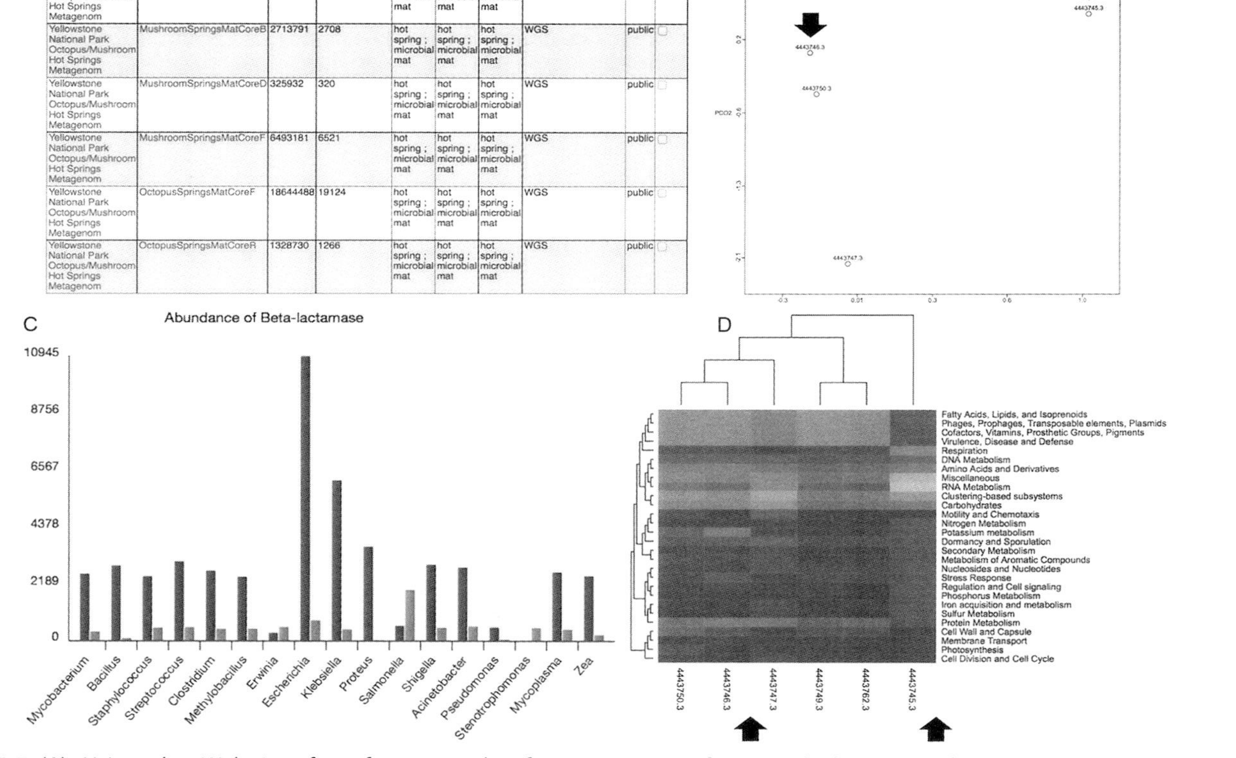

Figure 22.7 (A) Using the Web interface for a search of metagenomes for microbial mats in hotsprings (GSC-MIMS-Keywords Biome="hotspring; microbial mat"), we find six metagenomes (refs: 4443745.3, 4443746.3, 4443747.3, 4443749.3, 4443750.3, 4443762.3). (B) Initial comparison reveals some differences in protein functional class abundance (using SEED subsystems level 1). (C) From the PCoA plot using normalized counts of functional SEED subsystem-based functional annotations (level 2) and Bray-Curtis as metric, we attempt to find differences between two similar datasets (MG-RAST-IDs: 444749.3, 4443762.3). (D) Using exported tables with functional annotations and taxonomic mapping, we analyze the distribution of organisms observed to contain Beta-lactamase and plot the abundance per species for two distinct samples. (For color version of this figure, the reader is referred to the online version of this chapter.)

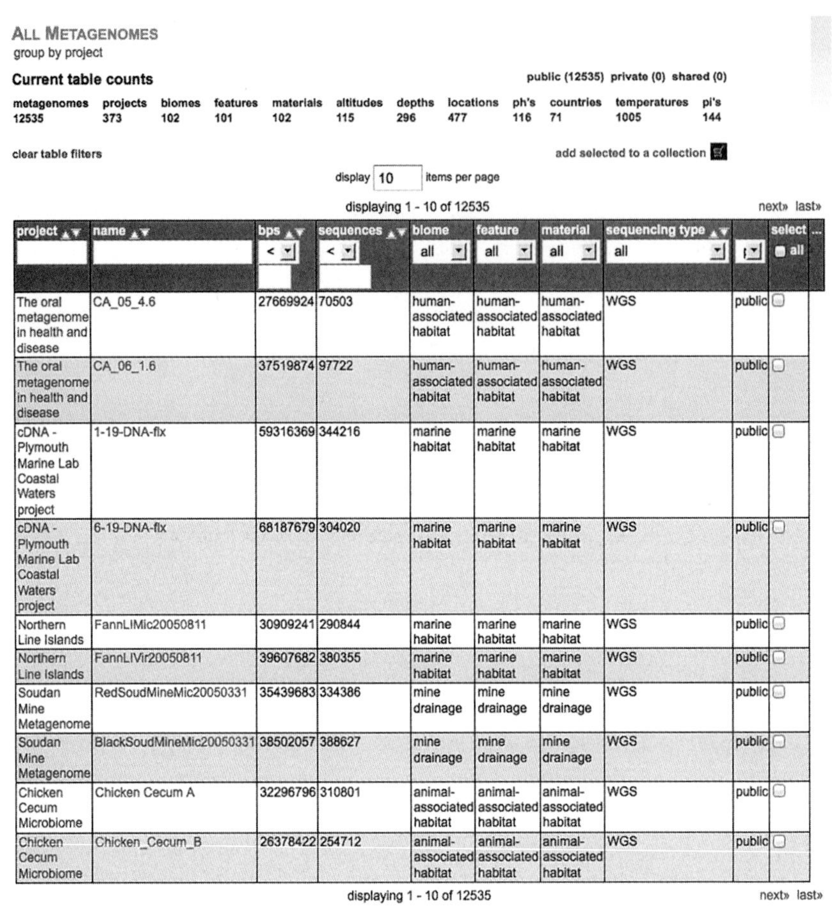

ALL METAGENOMES
group by project

Current table counts public (12535) private (0) shared (0)

metagenomes	projects	biomes	features	materials	altitudes	depths	locations	ph's	countries	temperatures	pi's
12535	373	102	101	102	115	296	477	116	71	1005	144

clear table filters add selected to a collection

display 10 items per page

displaying 1 - 10 of 12535 next» last»

project	name	bps	sequences	biome	feature	material	sequencing type		select ...
		<	<	all	all	all	all		all
The oral metagenome in health and disease	CA_05_4.6	27669924	70503	human-associated habitat	human-associated habitat	human-associated habitat	WGS	public	
The oral metagenome in health and disease	CA_06_1.6	37519874	97722	human-associated habitat	human-associated habitat	human-associated habitat	WGS	public	
cDNA - Plymouth Marine Lab Coastal Waters project	1-19-DNA-flx	59316369	344216	marine habitat	marine habitat	marine habitat	WGS	public	
cDNA - Plymouth Marine Lab Coastal Waters project	6-19-DNA-flx	68187679	304020	marine habitat	marine habitat	marine habitat	WGS	public	
Northern Line Islands	FannLIMic20050811	30909241	290844	marine habitat	marine habitat	marine habitat	WGS	public	
Northern Line Islands	FannLIVir20050811	39607682	380355	marine habitat	marine habitat	marine habitat	WGS	public	
Soudan Mine Metagenome	RedSoudMineMic20050331	35439683	334386	mine drainage	mine drainage	mine drainage	WGS	public	
Soudan Mine Metagenome	BlackSoudMineMic20050331	38502057	388627	mine drainage	mine drainage	mine drainage	WGS	public	
Chicken Cecum Microbiome	Chicken Cecum A	32296796	310801	animal-associated habitat	animal-associated habitat	animal-associated habitat	WGS	public	
Chicken Cecum Microbiome	Chicken_Cecum_B	26378422	254712	animal-associated habitat	animal-associated habitat	animal-associated habitat	WGS	public	

displaying 1 - 10 of 12535 next» last»

Figure 22.8 The metagenome browser page enables sorting and data search. Users can select the metadata they wish to view and search. Some of the metadata are hidden by default and can be viewed by clicking on the "..." header on the right side of the table and selecting the desired columns; this can also be used to hide unwanted columns. (For color version of this figure, the reader is referred to the online version of this chapter.)

3.3.1 Technical detail on the overview page

The overview page provides the MG-RAST ID for a dataset, a unique identifier that is usable as accession number for publications. Additional information like the name of the submitting PI and organization and a user-provided metagenome name are displayed at the top of the page as well. A static URL for linking to the system that will be stable across changes to the MG-RAST Web interface is provided as additional information (Fig. 22.9).

We provide an automatically generated paragraph of text describing the submitted data and the results computed by the pipeline. Via the project information, we display additional information provided by the data submitters at the time of submission or later.

One of the key diagrams in MG-RAST is the sequence breakdown pie chart (Fig. 22.10) classifying the submitted sequences submitted into several categories according to their annotation status. As detailed in the description of the MG-RAST v3 pipeline above, the features annotated in MG-RAST are protein-coding genes and ribosomal proteins.

Figure 22.9 Top of the metagenome overview page. (For color version of this figure, the reader is referred to the online version of this chapter.)

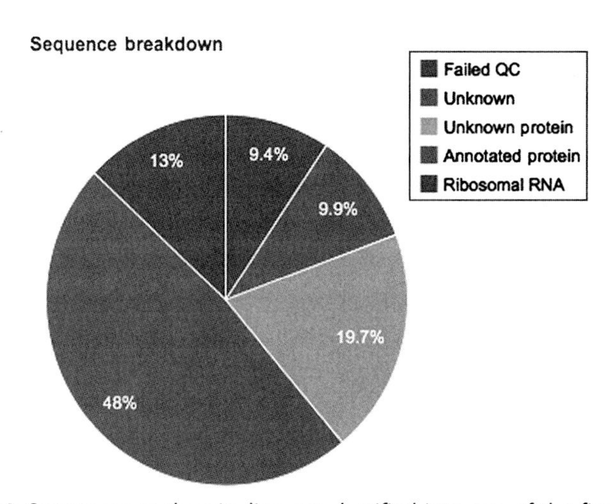

Figure 22.10 Sequences to the pipeline are classified into one of the five categories: gray = failed the QC, red = unknown sequences, yellow = unknown function but protein coding, green = protein coding with known function, and blue = ribosomal RNA. For this example, over 50% of sequences were either filtered by QC or failed to be recognized as either protein coding or ribosomal. (See color plate.)

It should be noted that for performance reasons no other sequence features are annotated by the default pipeline. Other feature types, for example, small RNAs or regulatory motifs (e.g., CRISPRS (Bolotin, Quinquis, Sorokin, & Ehrlich, 2005)) will not only require significantly higher computational resources but also are frequently not supported by the unassembled short reads that comprise the vast majority of today's metagenomic data in MG-RAST.

The overview page also provides metadata (data describing data) for each dataset to the extent that data have been made available. Metadata enables other researchers to discover datasets and compare annotations. MG-RAST requires standard metadata for data sharing and data publication. This is implemented using the standards developed by the Genomics Standards Consortium. Figure 22.11 shows the metadata summary for a dataset.

All metadata stored for a specific dataset are available in MG-RAST; we merely display a standardized subset in this table. A link at the bottom of the table (more metadata) provides access to a table with the complete metadata. This enables users to provide extended metadata going beyond the GSC minimal standards. A mechanism to provide community consensus extensions to the minimal checklists in the environmental packages are explicitly encouraged, but not required when using MG-RAST.

3.3.2 Metagenome QC

The analysis flowchart and analysis statistics provide an overview of the number of sequences at each stage in the pipeline. (Fig. 22.12). The text block next to the analysis flowchart presents the numbers next to their definitions.

3.3.3 Technical data

This component provides quick links to a general statistical overview of the different analysis steps performed (see analysis flowchart), a comprehensive list of all metadata for the dataset, sequence length, GC distributions, and a breakdown of blat hits per data source (e.g., hits to RefSeq (Pruitt, Tatusova, & Maglott, 2007), UniProt (Tatusov et al., 2013), or SEED (Overbeek et al., 2005)).

The analysis statistics and analysis flowchart provide sequence statistics for the main steps in the pipeline from raw data to annotation, describing the transformation of the data between steps.

GSC MIxS INFO

Investigation Type	**metagenome**
Project Name	The oral metagenome in health and disease
Latitude and Longitude	**39.481448, 0.353066**
Country and/or Sea, Location	**Spain Valencia**
Collection Date	**2010-03-01 10:00:00 UTC**
Environment (Biome)	**human-associated habitat**
Environment (Feature)	**human-associated habitat**
Environment (Material)	**human-associated habitat**
Environmental Package	**human-oral**
Sequencing Method	**454**
More Metadata	

Figure 22.11 The information from the GSC MIxS checklist providing minimal metadata on the sample. (For color version of this figure, the reader is referred to the online version of this chapter.)

Sequence length and GC histograms display the distribution before and after quality control steps.

Metadata is presented in a searchable table which contains contextual metadata describing sample location, acquisition, library construction, and sequencing using GSC compliant metadata. All metadata can be downloaded from the table.

3.3.4 Taxonomic and functional information on the overview page

3.3.4.1 Organism breakdown

The taxonomic hit distribution display breaks down taxonomic units into a series of pie charts of all the annotations grouped at various taxonomic ranks (Domain, Phylum, Class, Order, Family, Genus). The subsets are selectable for downstream analysis; this also enables downloads of subsets of reads, for example, those hitting a specific taxonomic unit.

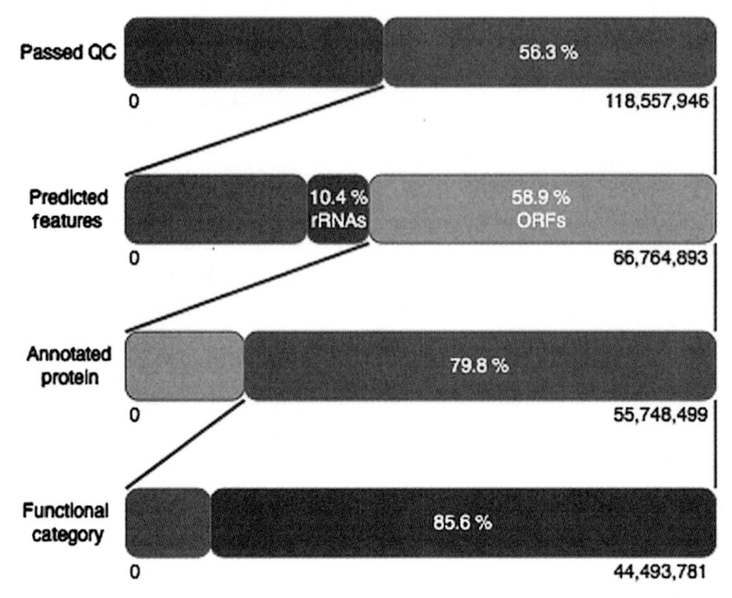

Figure 22.12 The analysis flowchart provides an overview of the fractions of sequences "surviving" the various steps of the automated analysis. In this case, about 20% of sequences were filtered during quality control. From the remaining 37,122,128 sequences, 53.5% were predicted to be protein coding, 5.5% hit ribosomal RNA. From the predicted proteins, 76.8% could be annotated with a putative protein function. Out of 32 million annotated proteins, 24 million have been assigned to a functional classification (SEED, COG, EggNOG, KEGG), representing 84% of the reads. (See color plate.)

The rank abundance plot (Fig. 22.13) provides a rank-ordered list of taxonomic units at a user-defined taxonomic level, ordered by their abundance in the annotations.

The *rarefaction* curve of annotated species richness is a plot of the total number of distinct species annotations as a function of the number of sequences sampled. The slope of the right-hand part of the curve is related to the fraction of sampled species that are rare. When the rarefaction curve is flat, more intensive sampling is likely to yield only few additional species. The rarefaction curve is derived from the protein taxonomic annotations and is subject to problems stemming from technical artifacts. These artifacts can be similar to the ones affecting amplicon sequencing (Reeder & Knight, 2009), but the process of inferring species from protein similarities may introduce additional uncertainty.

Finally, in this section, we display an estimate of the alpha diversity based on the taxonomic annotations for the predicted proteins. The alpha diversity is presented in context of other metagenomes in the same project.

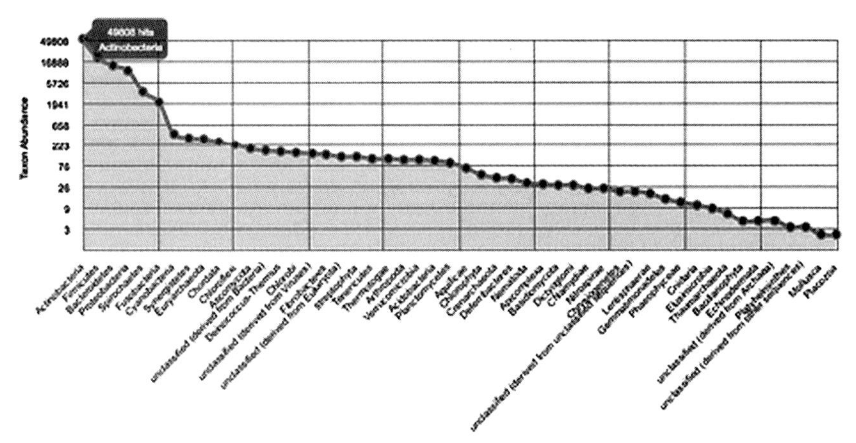

Figure 22.13 Organism breakdown: sample rank abundance plot by phylum. (For color version of this figure, the reader is referred to the online version of this chapter.)

3.3.4.2 Functional breakdown

This section contains four pie charts providing a breakdown of the functional categories for KEGG (Kanehisa et al., 2012), COG (Vasudevan, Wolf, Yin, & Natale, 2003), SEED subsystems (Overbeek et al., 2005), and EggNOGs (Jensen et al., 2008). Clicking on the individual pie chart slices will save the respective sequences to the workbench.

The relative abundance of sequences per functional category can be downloaded as a spreadsheet and users can browse the functional breakdowns via the Krona tool (Ondov, Bergman, & Phillippy, 2011) integrated in the page.

A more detailed functional analysis, allowing the user to manipulate parameters for sequence similarity matches, is available via the analysis page.

3.4. The analysis page

The MG-RAST annotation pipeline produces a set of annotations for each sample; these annotations can be interpreted as functional or taxonomic abundance profiles. The analysis page can be used to view these profiles for a single metagenome, or compare profiles from multiple metagenomes using various visualizations (e.g., heatmap) and statistics (e.g., PCoA, normalization).

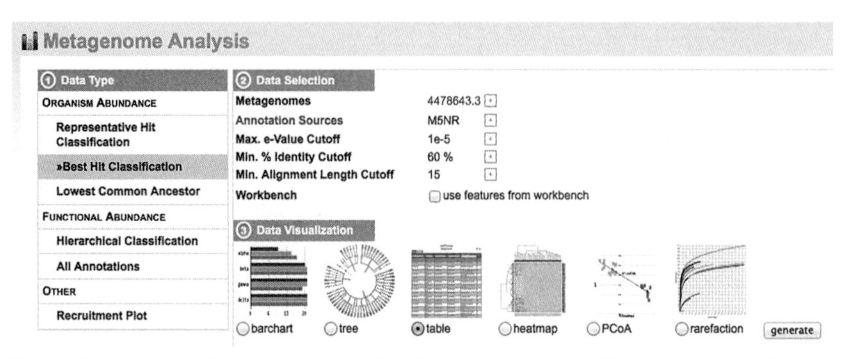

Figure 22.14 Using the analysis page is a three-step process. First, select a profile and hit (see below) type. Second, select a list of metagenomes and set annotation source and similarity parameters. Third, choose a comparison. (For color version of this figure, the reader is referred to the online version of this chapter.)

The page breaks down in three parts following a typical workflow (Fig. 22.14):

1. Selection of an MG-RAST analysis scheme, that is, selection of a particular taxonomic or functional abundance profile mapping. For taxonomic annotations, since there is not always a unique mapping from hit to annotation, we provide three interpretations: best hit, representative hit, and lowest common ancestor (LCA) as explained below. Functional annotations can be either grouped into mappings to functional hierarchies or displayed without a hierarchy. In addition, the recruitment plot displays the recruitment of protein sequences against a reference genome.
2. Selection of sample and parameters. This dialog allows the selection of multiple metagenomes which can be compared individually, or selected and compared as groups. Comparison is always relative to the annotation source, e-value, and percent identity cutoffs selectable in this section. In addition to the metagenomes available in MG-RAST, sets of sequences previously saved in the workbench can be selected for visualization.
3. Data visualization and comparison. Depending on the selected profile type, the profiles for the metagenomes can be visualized and compared using "bar charts," "trees," spreadsheet like "tables," "heatmaps," "PCoA," "rarefaction plots," "Circular recruitment plot," and KEGG maps.

3.4.1 Representative hit, best hit, and lowest common ancestor interpretation

MG-RAST searches the nonredundant M5NR and M5RNA databases in which each sequence is unique. These two databases are built from multiple sequence database sources, and the individual sequences may occur multiple times in different strains and species (and sometimes genera) with 100% identity. In these circumstances, choosing the "right" taxonomic information is not a straightforward process.

To optimally serve a number of different use cases, we have implemented three different ways of finding the "right" function or taxon information. This impacts the end-user experience as they have three different methods to choose the number of hits reported for a given sequence in their dataset. The details on the three different classification functions implemented are given in the following sections.

3.4.1.1 Best hit

The best hit classification reports the functional and taxonomic annotation of the best hit in the M5NR for each feature. In those cases where the similarity search yields multiple same-scoring hits for a feature, we do not choose any single "correct" label. For this reason, we have decided to "double count" all annotations with identical match properties and leave determination of truth to our users. While this approach aims to inform about the functional and taxonomic potential of a microbial community by preserving all information, subsequent analysis can be biased because of a single feature having multiple annotations, leading to inflated hit counts. If you are looking for a specific species or function in your results, the "best hit" function is likely what you are looking for.

3.4.1.2 Representative hit

The representative hit classification selects a single unambiguous annotation for each feature. The annotation is based on the first hit in the homology search and the first annotation for that hit in our database. This makes counts additive across functional and taxonomic levels and thus allows, for example, to compare functional and taxonomic profiles of different metagenomes.

3.4.1.3 Lowest common ancestor

To avoid the problem of multiple taxonomic annotations for a single feature, we provide taxonomic annotations based on the widely used LCA method introduced by MEGAN (Huson, Auch, Qi, & Schuster, 2007). In this method, all hits that have a bit score close to the bit score of the best hit are collected. The taxonomic annotation of the feature is then determined by computing the LCA of all species in this set. This replaces all taxonomic annotations from ambiguous hits with a single higher-level annotation in the NCBI taxonomy tree.

The number of hits (occurrences of the input sequence in the database) may be inflated if the "best hit" filter is used, or your favorite species might be missing despite a very similar sequence similarity result if using the "representative hit" classifier function (in fact, 100% identical match to your favorite species exists).

One way to consider both "representative" and "best" hit is that they overinterpret the available evidence; with the LCA classifier function, any input sequence is only classified down to a trustworthy taxonomic level. While naively this seems to be the best function to choose in all cases as it classifies sequences to varying depths, this causes problems for downstream analysis tools that might rely on everything being classified to the same level.

3.4.2 Normalization

Normalization refers to a transformation that attempts to reshape an underlying distribution. A large number of biological variables exhibit a log-normal distribution, meaning that when you transform the data with a log transformation, the values exhibit a normal distribution. Log transformation of the counts data makes a normalized data product that is more likely to satisfy the assumptions of additional downstream tests like ANOVA or *t*-tests.

Standardization is a transformation applied to each distribution in a group of distributions so that all distributions exhibit the same mean and the same standard deviation. This removes some aspects of intersample variability and can make data more comparable. This sort of procedure is analogous to commonly practiced scaling procedures but is more robust in that it controls for both scale and location.

The analysis page calculates the ordination visualizations with either raw or normalized counts, at the user's option. The normalization procedure is to take

$$\text{normalized\_value\_}i = \log 2(\text{raw\_counts\_}i + 1).$$

And then the standardized values are calculated from the normalized values by subtracting the mean of each sample's normalized values and dividing by the standard deviation of each sample's normalized values.

$$\text{standardized\_}i = (\text{normalized\_}i - \text{mean}(\{\text{normalized\_}i\}))/\text{stddev}(\{\text{normalized\_}i\})$$

You can read more about these procedures in a number of texts—we recommend Terry Speed's "Statistical Analysis of Gene Expression in Microarray Data" (ISBN1584883278).

When data exhibit a nonnormal, normal, or unknown distribution, nonparametric tests (e.g., Mann–Whitney or Kruskal–Wallis) should be used. Boxplots are an easy way to check—and the MG-RAST analysis page provides boxplots of the standardized abundance values for checking the comparability of samples (Fig. 22.15).

3.4.3 Heatmap/dendrogram

The heatmap/dendrogram (Fig. 22.16) is a tool that allows an enormous amount of information to be presented in a visual form that is amenable to human interpretation. Dendrograms are trees that indicate similarities between annotation vectors. The MG-RAST heatmap/dendrogram has two dendrograms, one indicating the similarity/dissimilarity among metagenomic samples (x-axis dendrogram) and another indicating the similarity/dissimilarity among annotation categories (e.g., functional roles; the y-axis dendrogram). A distance metric is evaluated between every possible pair of sample abundance profiles. A clustering algorithm (e.g., ward-based clustering) then produces the dendrogram trees. Each square in the heatmap dendrogram represents the abundance level of a single category in a single sample. The values used to generate the heatmap/dendrogram figure can be downloaded as a table by clicking on the "download" button.

3.4.4 Bar chart and tree

The bar chart and tree tools map raw or normalized abundances onto functional or taxonomic hierarchies. The bar chart tool presents mapping onto the highest category of a hierarchy (e.g., Domain) and allows a drill down into the hierarchy. In addition, reads from a specific level can be added into the workbench.

3.4.5 Ordination

MG-RAST uses principle coordinate analysis (PCoA) to reduce the dimensionality of comparisons of multiple samples that consider functional or taxonomic annotations.

PCoA is a well-known method for dimensionality reduction of large datasets. Dimensionality reduction is a process that allows the complex

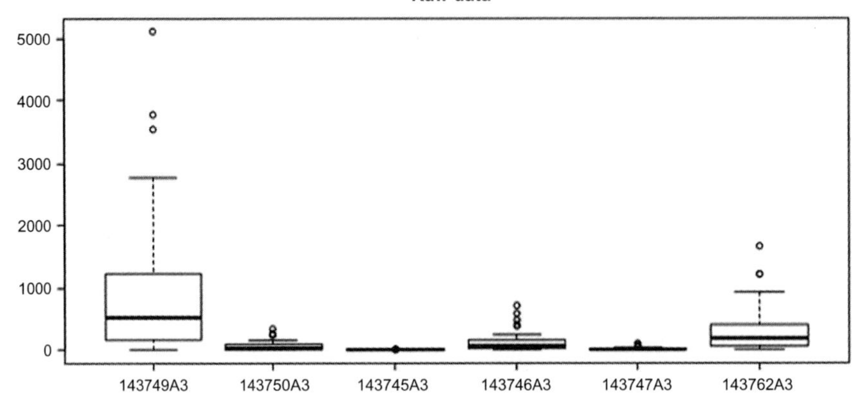

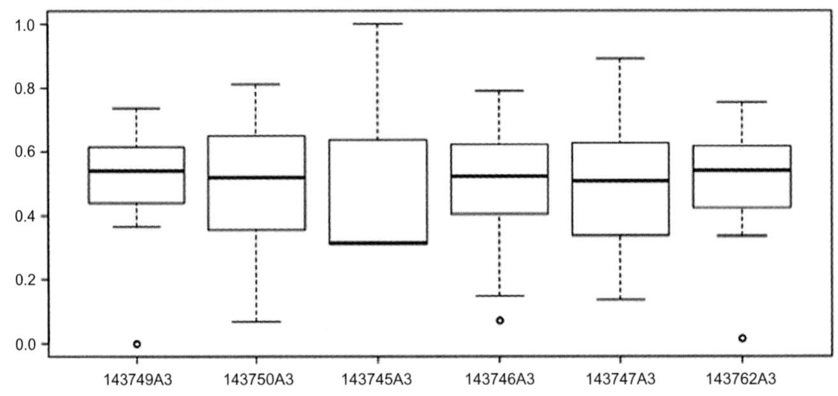

Figure 22.15 Boxplots of the abundance data for raw values (top) as well as values that have undergone the normalization and standardization procedure described above (bottom). It is clear that after normalization and standardization, samples exhibit value distributions that are much more comparable, and that exhibit a normal distribution; the normalized and standardized data are suitable for analysis with parametric tests, the raw data are not.

variation found in large datasets (e.g., the abundance values of thousands of functional roles or annotated species across dozens of metagenomic samples) to be reduced to a much smaller number of variables that can be visualized as simple 2- or 3-dimensional scatter plots. The plots enable interpretation of the multidimensional data in a human-friendly presentation. Samples that exhibit similar abundance profiles (taxonomic or functional) group together, whereas those that differ are found further apart. A key feature of PCoA-based analyses is that users can compare components not just to each other, but to metadata-recorded variables (e.g., sample pH, biome, DNA

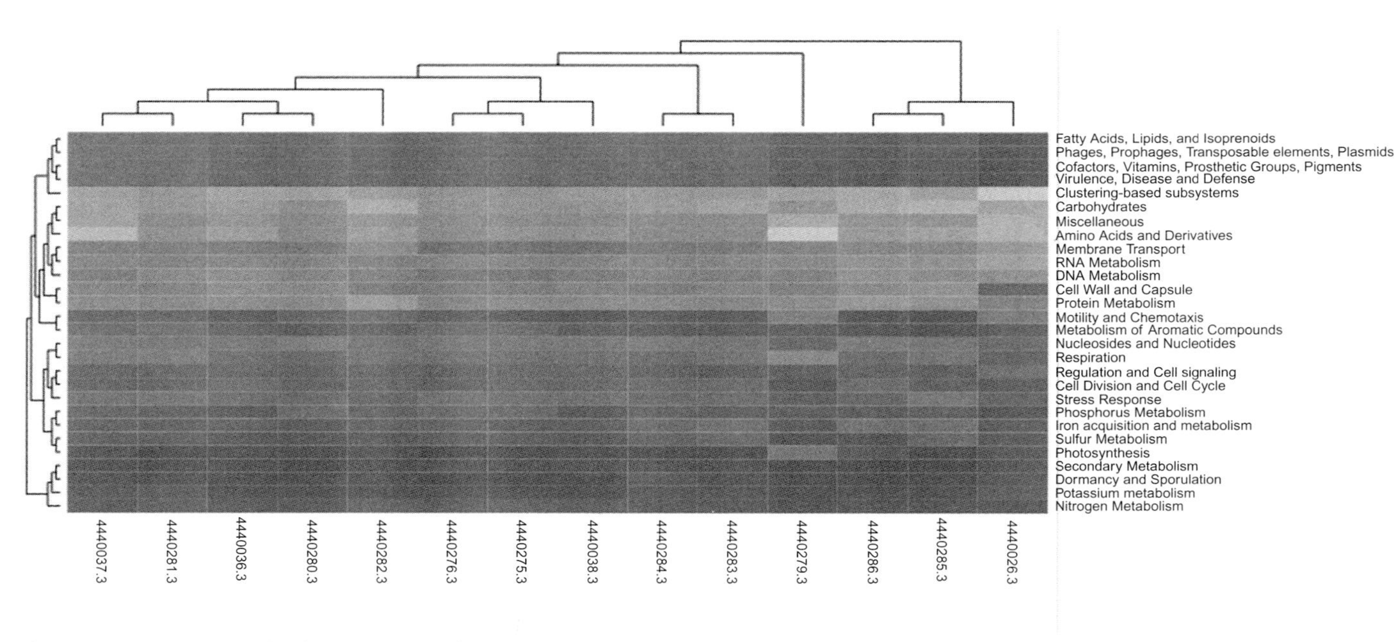

Figure 22.16 Heatmap/dendrogram example in MG-RAST. The MG-RAST heatmap/dendrogram has two dendrograms, one indicating the similarity/dissimilarity among metagenomic samples (*x*-axis dendrogram) and another to indicate the similarity/dissimilarity among annotation categories (e.g., functional roles; the *y*-axis dendrogram). (See color plate.)

extraction protocol, etc.) to reveal correlations between extracted variation and metadata-defined characteristics of the samples. It is also possible to couple PCoA with higher-resolution statistical methods to identify individual sample features (taxa or functions) that drive correlations observed in PCoA visualizations. This can be accomplished with permutation-based statistics applied directly to the data before calculation of distance measures used to produce PCoAs, or by applying conventional statistical approaches (e.g., ANOVA or Kruskal–Wallis test) to groups observed in PCoA-based visualizations.

3.4.6 Table

The table tool creates a spreadsheet-based abundance table that can be searched and restricted by the user. Tables can be generated at user-selected levels of phylogenetic or functional resolution. Table data can be visualized using Krona (Ondov et al., 2011) and can be exported in BIOM format to be used in other tools, for example, QIIME (Caporaso et al., 2010), or the tables can be exported as tab-separated text. Abundance tables serve as the basis for all comparative analysis tools in MG-RAST, from PCoA to heatmap–dendrograms.

3.4.7 KEGG maps

The KEGG map tool allows the visual comparison of predicted metabolic pathways in metagenomic samples. It maps the abundance of identified enzymes onto a KEGG (Kanehisa et al., 2012) map of functional pathways. Metagenomes can be assigned into one of the two groups and those groups can be visually compared.

4. HOW TO DRILL DOWN USING THE WORKBENCH

One of the new features of MG-RAST v3 is the workbench. It is the main mechanism for exchanging subsets of data between analysis views. It also allows you to download the FASTA files of a selection of proteins.

When you initially go to the analysis page, your workbench will be empty. It is displayed as the leftmost tab in the data tabular view. So, how do you get data into the workbench? There are two simple ways to select data subsets—from any generated table or from the drill down of a bar chart.

Try this example: Start by selecting the lean and the obese mouse cecum samples (MG-RAST IDs 4440463.3 and 4440464.3) (Turnbaugh et al.,

2006) in the data selection and creating a table. To do this, go to the analysis page and select the analysis view "Organism Classification." Expand the metagenome selection by clicking the plus symbol next to metagenomes. Select public from the drop-down box (to view only public datasets) and type "mouse" into the filter box. Select the two samples, click the button with the right arrow, and then click the ok button. The default data visualization is *table*, so you can click the "generate" button (Fig. 22.17).

After a short wait, a new tab will appear in the tabview below (see Fig. 22.18), showing the data table with organism classifications for the two samples. The last column of this table will have a button labeled "to workbench" as the column header. Each cell in that column will have a checkbox. Checking a checkbox and clicking the "to workbench"-button will send the proteins identified by that row to the workbench (Fig. 22.19). Note that you only have one workbench and putting a new set of proteins into it will replace the current content. So what if I want to select all

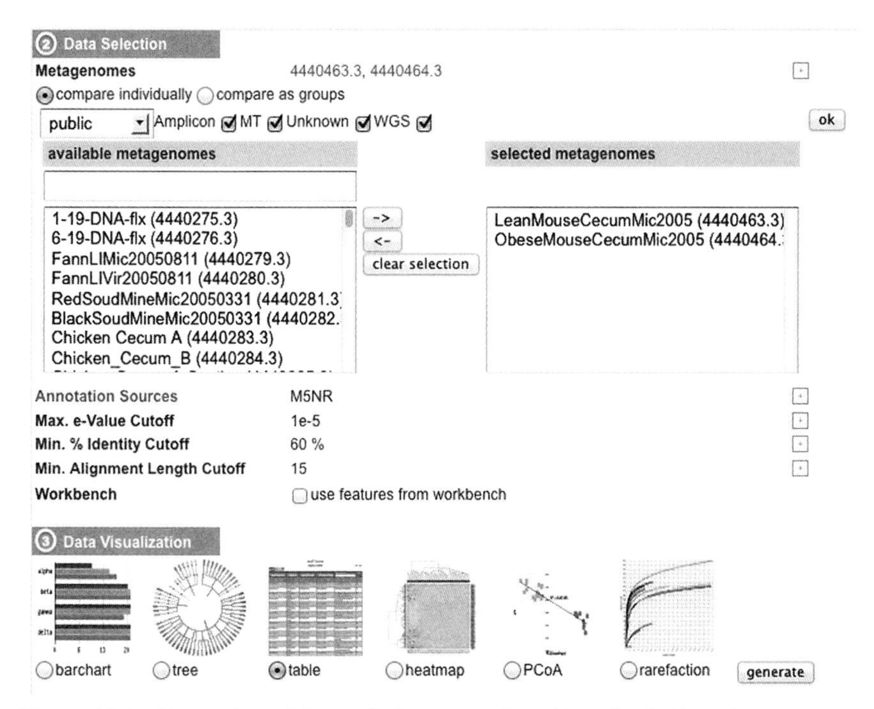

Figure 22.17 Screenshot of the analysis page and workbench tab. Note that users can search and select metagenomes to analyze, the annotation courses and parameters to set, along with the analysis and visualization they want to perform. (For color version of this figure, the reader is referred to the online version of this chapter.)

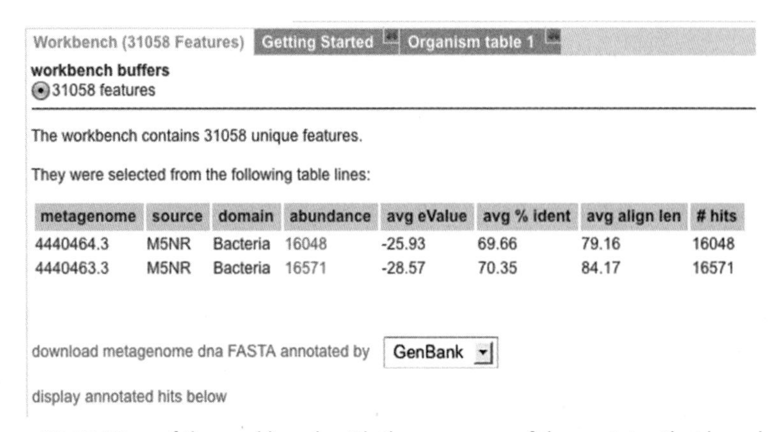

Figure 22.18 View of the workbench with the summary of the proteins that have been added. (For color version of this figure, the reader is referred to the online version of this chapter.)

metagenome	source	domain	abundance	avg eValue	avg % ident	avg align len	# hits	to workbench
all	M5NR	all	<	<	<	<	<	
4440463.3	M5NR	Archaea	52	-14.51	65.47	58.12	52	☐
4440464.3	M5NR	Archaea	108	-15.49	65.72	61.37	108	☐
4440464.3	M5NR	Bacteria	16048	-25.93	69.66	79.16	16048	☑
4440463.3	M5NR	Bacteria	16571	-28.57	70.35	84.17	16571	☑
4440463.3	M5NR	Eukaryota	138	-25.53	74.19	73.10	69	☐
4440464.3	M5NR	Eukaryota	160	-18.54	74.21	59.04	116	☐

Figure 22.19 Use the table to select results you want to add to your workbench for further analyses. (For color version of this figure, the reader is referred to the online version of this chapter.)

Bacteria? Do I really need to click through all those checkboxes? No—you can use the grouping feature of the table, so you only have to click one checkbox per metagenome.

Above the table, you will find a drop-down box labeled "group table by" (Fig. 22.20). Select "domain" and the table will be grouped, so there is only one row per metagenome and domain.

Now check the two boxes in the "Bacteria" rows and click the "to workbench" button.

A pop-up message will appear, telling you how many proteins have been sent to the workbench. If you take a look at the tabular view now, you will notice that the workbench tab shows the number of proteins it currently contains. If you click on that tab, you will get information about what the workbench contains. On this tab, you will also find a "download as FASTA" button.

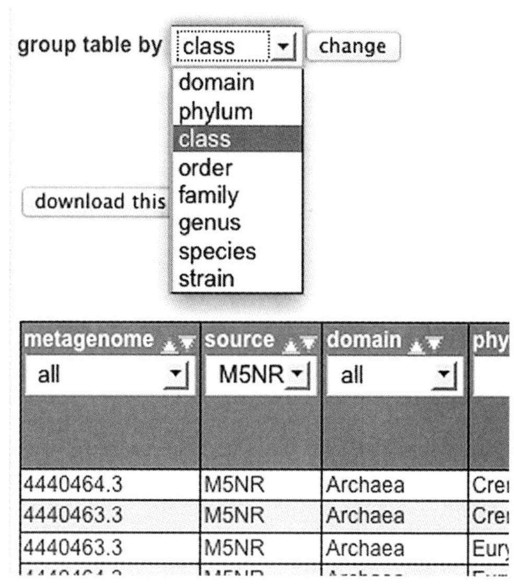

Figure 22.20 Use the tables to group results. (For color version of this figure, the reader is referred to the online version of this chapter.)

Aside from being able to download the sequences of your selected proteins, you can also use them to generate other visualizations. This includes switching from organism to functional classification. To do this, simply check the "use proteins from workbench" checkbox in the data selection when generating a new visualization, for example, a circular tree using the proteins we just buffered.

The table is not the only visualization that allows one to put a subselection into the workbench. You can also use the bar chart to do this (Fig. 22.21). Simply click on the "to workbench" button next to the headline of a drilldown. Note that you cannot put the topmost bar chart into the workbench, as it is not yet a subselection of proteins.

4.1. 2.6 Downloads

The workbench feature stores subselections of data and allows those to be used as input for further selection or displays, for example, select all *Escherichia coli* reads and then display the functional categories present just in *E. coli* reads across multiple datasets. In addition, the workbench allows downloading the annotated reads for the subselection stored in the workbench as fasta (Fig. 22.22).

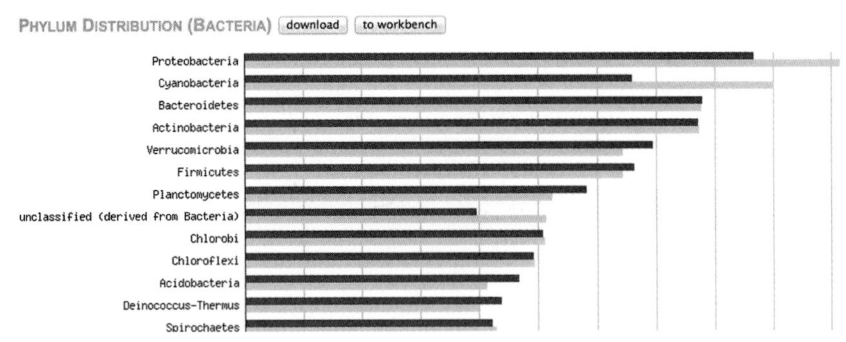

Figure 22.21 In addition to the results table, users can download results or add to their workbench from bar charts. (For color version of this figure, the reader is referred to the online version of this chapter.)

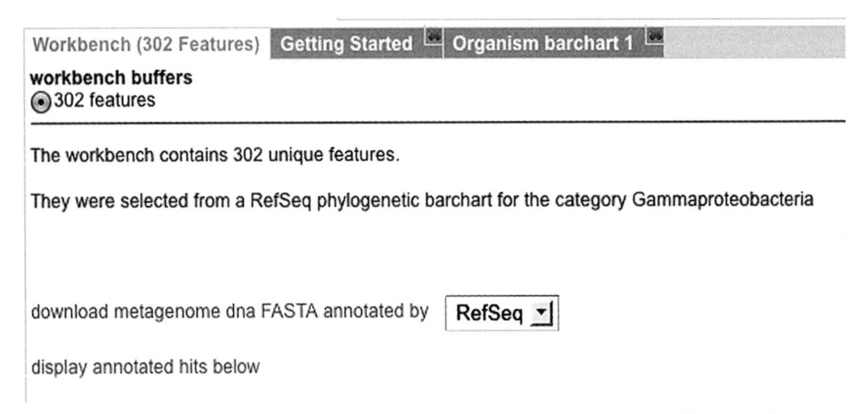

Figure 22.22 The workbench facilitates the download of selected reads using the name space of the selection. (For color version of this figure, the reader is referred to the online version of this chapter.)

Once processing datasets in MG-RAST is finished, a download page is created for the project. On this page, all data products created during the computation are made available as files. In addition, datasets which have been published in MG-RAST have links to an ftp site at the top of this page where you can download additional information.

4.2. Viewing evidence

For individual proteins, the MG-RAST page allows users to retrieve the sequence alignments underlying the annotation transfers (see Fig. 22.23). Using the M5NR (Wilke et al., 2011) technology, users can retrieve alignments against the database of interest with no additional overhead.

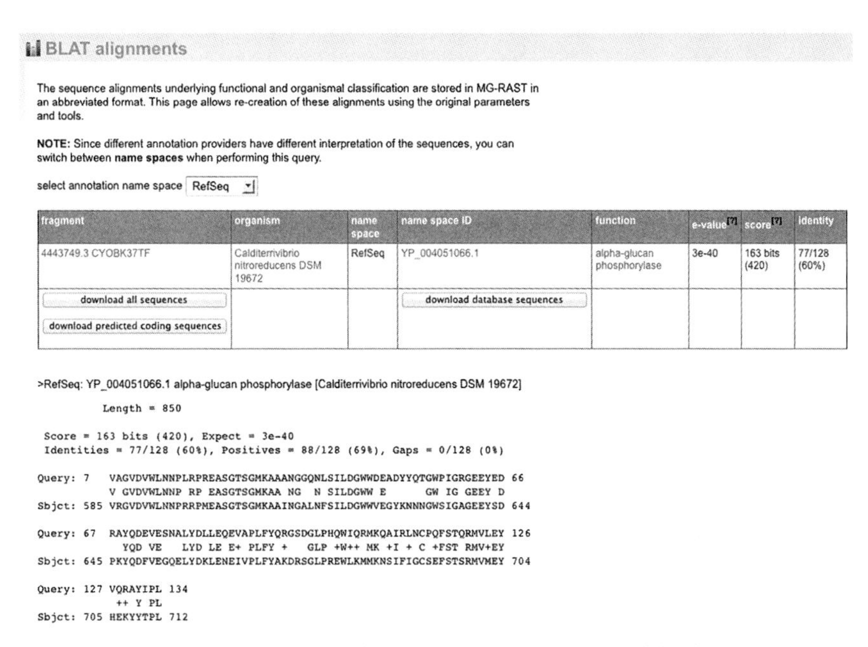

Figure 22.23 BLAT hit details with alignment. (For color version of this figure, the reader is referred to the online version of this chapter.)

5. MG-RAST DOWNLOADS

One of the critical insights when developing MG-RAST version 3 was the need to make a maximum number of data products available for download for downstream analysis. For this purpose, we have created the download page that contains all automatically created data products in a single location for each metagenome. In addition, a global download page provides access to all public datasets grouped by projects.

Below, we list the data products available on the download page for each metagenome using a specific example (MG-RAST ID: 4465825.3).

Uploaded File(s)

DNA (4465825.3.25422.fna)

Uploaded nucleotide sequence data in FASTA format.

Preprocessing

Depending on the options chosen, the preprocessing step filters sequences based on length, number of ambiguous bases, and quality values if available.

passed, DNA (4465825.3.100.preprocess.passed.fna)
A FASTA formatted file containing the sequences which were accepted and will be passed on to the next stage of the analysis pipeline.
removed, DNA (4465825.3.100.preprocess.removed.fna)
A FASTA formatted file containing the sequences which were rejected and will not be passed on to the next stage of the analysis pipeline.
Dereplication
The optional dereplication step removes redundant "technical replicate" sequences from the metagenomic sample. Technical replicates are identified by binning reads with identical first 50 base pairs. One copy of each 50-base-pair identical bin is retained.
passed, DNA (4465825.3.150.dereplication.passed.fna)
A FASTA formatted file containing one sequence from each bin which will be passed on to the next stage of the analysis pipeline.
removed, DNA (4465825.3.150.dereplication.removed.fna)
A FASTA formatted file containing the sequences which were identified as technical replicates and will not be passed on to the next stage of the analysis pipeline.

Screening

The optional screening step screens reads against model organisms using bow tie to remove reads which are similar to the genome of the selected species.
passed, DNA (4465825.3.299.screen.passed.fna)
A FASTA formatted file containing the reads which had no similarity to the selected genome and will be passed on to the next stage of the analysis pipeline.

Prediction of protein-coding sequences

Coding regions within the sequences are predicted using FragGeneScan, an *ab initio* prokaryotic gene calling algorithm. Using a hidden Markov model for coding regions and noncoding regions, this step identifies the most likely reading frame and translates nucleotide sequences into amino acid sequences. The predicted coding regions, possibly more than one per fragment, are called features.
coding, Protein (4465825.3.350.genecalling.coding.faa)
A amino acid sequence FASTA formatted file containing the translations of the predicted coding regions.
coding, DNA (4465825.3.350.genecalling.coding.fna)
A nucleotide sequence FASTA formatted file containing the predicted coding regions.

RNA Clustering

Sequences from step 2 (before dereplication) are prescreened for at least 60% identity to ribosomal sequences and then clustered at 97% identity using UCLUST. These clusters are checked for similarity against the ribosomal RNA databases (Greengenes, LSU, SSU, and RDP).

rna97, DNA (4465825.3.440.cluster.rna97.fna)

A FASTA formatted file containing sequences that have at least 60% identity to ribosomal sequences and are checked for RNA similarity.

rna97, Cluster (4465825.3.440.cluster.rna97.mapping)

A tab-delimited file that identifies the sequence clusters and the sequences that comprise them.

The columns making up each line in this file are:

1. Cluster ID, for example, rna97_998
2. Representative read ID, for example, 11909294
3. List of IDs for other reads in the cluster, for example, 11898451,11944918
4. List of percentage identities to the representative read sequence, for example, 97.5%, 100.0%

RNA similarities

The two files labeled "expand" are comma- and semicolon-delimited files that provide the mappings from md5s to function and md5s to taxonomy:

annotated, Sims (4465825.3.450.rna.expand.lca)

annotated, Sims (4465825.3.450.rna.expand.rna)

Packaged results of the blat search against all the DNA databases with md5 value of the database sequence hit followed by sequence or cluster ID, similarity information, annotation, organism, database name.

raw, Sims (4465825.3.450.rna.sims)

This is the similarity output from BLAT. This includes the identifier for the query which is either the FASTA ID or the cluster ID, and the internal identifier for the sequence that it hits.

The fields are in BLAST m8 format:

1. Query ID (either fasta ID or cluster ID), for example, 11847922
2. Hit ID, for example, lcl|501336051b4d5d412fb84afe8b7fdd87
3. percentage identity, for example, 100.00
4. alignment length, for example, 107
5. number of mismatches, for example, 0
6. number of gap openings, for example, 0
7. q.start, for example, 1
8. q.end, for example, 107
9. s.start, for example, 1262
10. s.end, for example, 1156

11. e-value, for example, 1.7e-54
12. score in bits, for example, 210.0

filtered, Sims (15:04 4465825.3.450.rna.sims.filter)

This is a filtered version of the raw Sims file above that removes all but the best hit for each data source.

Gene Clustering

Protein-coding sequences are clustered at 80% identity with UCLUST. This process does not remove any sequences but instead makes the similarity search step easier. Following the search, the original reads are loaded into MG-RAST for retrieval on-demand.

aa90, Protein (4465825.3.550.cluster.aa90.faa)

An amino acid sequence FASTA formatted file containing the translations of one sequence from each cluster (by cluster IDs starting with aa90_) and all the unclustered (singleton) sequences with the original sequence ID.

aa90, Cluster (4465825.3.550.cluster.aa90.mapping)

A tab-separated file in which each line describes a single cluster.
The fields are:

1. Cluster ID, for example, aa90_3270
2. protein-coding sequence ID including hit location and strand, for example, 11954908_1_121_+
3. additional sequenceIDs including hit location and strand, for example, 11898451_1_119_+,11944918_19_121_+
4. sequence % identities, for example, 94.9%,97.0%

Protein similarities

annotated, Sims (4465825.3.650.superblat.expand.lca)

The expand.lca file decodes the md5 to the taxonomic classification it is annotated with.
The format is:

1. md5(s), for example, cf036dfa9cdde3a8a4c09d7fabfd9ba5;1e538305b8319dab322b8f28da82e0a1
2. feature ID (for singletons) or cluster ID of hit including hit location and strand, for example, 11857921_1_101_-
3. alignment %, for example, 70.97;70.97
4. alignment length, for example, 31;31
5. e-value, for example, 7.5e-05;7.5e-05
6. Taxonomic string, for example, Bacteria; Actinobacteria; Actinobacteria (class); Coriobacteriales; Coriobacteriaceae; Slackia; Slackia exigua;-

annotated, Sims (4465825.3.650.superblat.expand.protein)

Packaged results of the blat search against all the protein databases with md5 value of the database sequence hit followed by sequence or cluster ID, similarity information, functional annotation, organism, database name.

Format is:

1. md5 (identifier for the database hit), for example, 88848aa7224ca2f3ac117e7953edd2d9
2. feature ID (for singletons) or cluster ID for the query, for example, aa90_22837
3. alignment % identity, for example, 76.47
4. alignment length, for example, 34
5. e-value, for example, 1.3e-06
6. protein functional label, for example, SsrA-binding protein
7. Species name associated with best protein hit, for example, Prevotella bergensis DSM 17361 RefSeq 585502

raw, Sims (4465825.3.650.superblat.sims)

Blat output with sequence or cluster ID, md5 value for the sequence in the database and similarity information.

filtered, Sims (4465825.3.650.superblat.sims.filter)

Blat output filtered to take only the best hit from each data source.

6. DISCUSSION

We have described MG-RAST, a community resource for the analysis of metagenomic sequence data. We have developed a new pipeline and environment for automated analysis of shotgun metagenomic data as well as a series of interactive tools for comparative analysis. The pipeline is also being used for the analysis of metatranscriptome data as well as amplicon data of various kinds. This service is being used by thousands of users worldwide, many contributing their data and analysis results to the community. We believe that community resources, such as MG-RAST, will fill a vital role in the bioinformatics ecosystem in the years to come.

MG-RAST has become a community clearinghouse for metagenomic data and analysis, with over 12,000 public datasets that can be freely used. Because analysis was performed in a uniform way, these datasets can be used as building blocks for new comparative analysis; so long as new datasets are analyzed similarly, results are robustly comparable between new and old dataset analysis. These datasets (and the resulting analysis data products) are made available for download and reuse as well.

Community resources like MG-RAST provide an interesting value proposition to the metagenomics community. First, it enables low-cost meta-analysis. Users utilize the data products in MG-RAST as a basis for comparison without the need to reanalyze every dataset used in their studies. The high computational cost of analysis (Wilkening, Wilke, Desai, & Meyer, 2009) makes precomputation a prerequisite for large-scale meta-analyses. In 2001, Angiuoli et al. determined the real currency cost of reanalysis for the over 12,000 datasets openly available on MG-RAST to be in excess of 30 million U.S. dollars if Amazon's EC2 platform is used (Angiuoli et al., 2011). This figure does not consider the 66,000 private datasets that have been analyzed with MG-RAST.

Second, it provides incentives to the community to adopt standards, in terms of both metadata and analysis approaches. Without this standardization, data products are not readily reusable, and computational costs quickly become unsustainable. We are not arguing that a single analysis is necessarily suitable for all users, rather we are pointing out that if one particular type of analysis is run for all datasets, the results can be efficiently reused, amortizing costs. Open access to data and analyses foster community interactions that make it easier for researchers' efforts to achieve consensus with respect to establishing best practices as well as identifying methods and analyses that could provide misleading results.

Third, community resources drive increased efficiency and computational performance. Community resources consolidate the demand for analysis resources sufficiently to drive innovation in algorithms and approaches. Due to this demand, the MG-RAST team has needed to scale the efficiency of their pipeline by a factor of nearly 1000 over the past 4 years. This drive has caused improvements in gene calling, clustering, sequence quality analysis, as well as many other areas. In less specialized groups with less extreme computational needs, this sort of efficiency gain would be difficult to achieve. Moreover, the large quantities of datasets that flow through the system have forced the hardening of the pipeline against a large variety of sequence pathology types that would not be readily observed in smaller systems.

We believe that our experiences in the design and operation of MG-RAST are representative of bioinformatics as a whole. The community resource model is critical if we are to benefit from the exponential growth in sequence data. These data have the potential to enable new insights into the world around us, but only if we can analyze them effectively. It is only due to this approach that we have been able to scale to the demands of our users effectively, analyzing over 200 billion sequences thus far.

We note that scaling to the required throughput by adding hardware to the system or simply renting time using an unoptimized pipeline on, for example, Amazon's EC2 machine would not be economically feasible. The real currency cost on EC2 for the data currently analyzed in MG-RAST (26 Terabasepairs) would be in excess of 100 million U.S. dollars using an unoptimized workflow like CLOVR (Angiuoli et al., 2011).

All of MG-RAST is open source and available on https://github.com/MG-RAST.

7. FUTURE WORK

While MG-RAST v3 is a substantial improvement over prior systems, much work remains to be done. Dataset sizes continue to increase at an exponential pace. Keeping up with this change remains a top priority, as metagenomics users continue to benefit from increased resolution of microbial communities. Upcoming versions of MG-RAST will include (1) mechanisms for speeding pipeline up using data reduction strategies that are biologically motivated; (2) opening up the data ecosystem via an API that will enable third-party development and enhancements; (3) providing distributed compute capabilities using user-provided resources; and (4) providing virtual integration of local datasets to allow comparison between local data and shared data without requiring full integration.

ACKNOWLEDGMENTS

This work used the Magellan machine (Office of Advanced Scientific Computing Research, Office of Science, U.S. Department of Energy, Contract Grant DE-AC02-06CH11357) at Argonne National Laboratory and the PADS resource (National Science Foundation Grant OCI-0821678) at the Argonne National Laboratory/University of Chicago Computation Institute. This work was supported in part by the U.S. Department of Energy under Contract DE-AC02-06CH11357, the Sloan Foundation (SLOAN #2010-12), NIH NIAID (HHSN272200900040C), and the NIH Roadmap HMP program (1UH2DK083993-01).

REFERENCES

Angiuoli, S. V., Matalka, M., Gussman, A., Galens, K., Vangala, M., Riley, D. R., et al. (2011). CloVR: A virtual machine for automated and portable sequence analysis from the desktop using cloud computing. *BMC Bioinformatics*, *12*, 356.

Benson, D. A., Cavanaugh, M., Clark, K., Karsch-Mizrachi, I., Lipman, D. J., Ostell, J., et al. (2012). GenBank. *Nucleic Acids Research*, *41*(Database issue), D36–D42.

Bolotin, A., Quinquis, B., Sorokin, A., & Ehrlich, S. D. (2005). Clustered regularly interspaced short palindrome repeats (CRISPRs) have spacers of extrachromosomal origin. *Microbiology*, *151*(Pt. 8), 2551–2561.

Caporaso, J. G., Kuczynski, J., Stombaugh, J., Bittinger, K., Bushman, F. D., Costello, E. K., et al. (2010). QIIME allows analysis of high-throughput community sequencing data. *Nature Methods*, 7(5), 335–336.

Cox, M. P., Peterson, D. A., Biggs, P. J., & Solexa, Q. A. (2010). At-a-glance quality assessment of Illumina second-generation sequencing data. *BMC Bioinformatics*, *11*, 485.

Edgar, R. C. (2010). Search and clustering orders of magnitude faster than BLAST. *Bioinformatics*, *26*, 2460–2461.

Field, D., Amaral-Zettler, L., Cochrane, G., Cole, J. R., Dawyndt, P., Garrity, G. M., et al. (2011). The Genomic Standards Consortium. *PLoS Biology*, *9*(6), e1001088.

Gabriel, E., Fagg, G. E., Bosilca, G., Angskun, T., Dongarra, J. J., Squyres, J. M., et al. (2004). In *Proceedings, 11th European PVM/MPI Users' Group Meeting*.

Gene Ontology Consortium (2013). Gene Ontology annotations and resources. *Nucleic Acids Research*, *41*(Database issue), D530–D535.

Gomez-Alvarez, V., Teal, T. K., & Schmidt, T. M. (2009). Systematic artifacts in metagenomes from complex microbial communities. *The ISME Journal*, *3*, 1314–1317.

Huse, S. M., Huber, J. A., Morrison, H. G., Sogin, M. L., & Welch, D. M. (2007). Accuracy and quality of massively parallel DNA pyrosequencing. *Genome Biology*, *8*(7), R143.

Huson, D. H., Auch, A. F., Qi, J., & Schuster, S. C. (2007). MEGAN analysis of metagenomic data. *Genome Research*, *17*, 377–386.

Jensen, L. J., Julien, P., Kuhn, M., von Mering, C., Muller, J., Doerks, T., et al. (2008). eggNOG: Automated construction and annotation of orthologous groups of genes. *Nucleic Acids Research*, *36*(Database issue), D250–D254.

Kanehisa, M., Goto, S., Sato, Y., Furumichi, M., & Tanabe, M. (2012). KEGG for integration and interpretation of large-scale molecular data sets. *Nucleic Acids Research*, *40*(Database issue), D109–D114.

Keegan, K. P., Trimble, W. L., Wilkening, J., Wilke, A., Harrison, T., D'Souza, M., et al. (2012). A platform-independent method for detecting errors in metagenomic sequencing data: DRISEE. *PLoS Computational Biology*, *8*(6), e1002541.

Kent, W. J. (2002). BLAT—The BLAST-like alignment tool. *Genome Research*, *12*, 656–664.

Langmead, B., Trapnell, C., Pop, M., & Salzberg, S. L. (2009). Ultrafast and memory-efficient alignment of short DNA sequences to the human genome. *Genome Biology*, *10*(3), R25.

Markowitz, V. M., Chen, I. M., Palaniappan, K., Chu, K., Szeto, E., Grechkin, Y., et al. (2012). IMG: The Integrated Microbial Genomes database and comparative analysis system. *Nucleic Acids Research*, *40*(Database issue), D115–D122.

Meyer, F., Paarmann, D., D'Souza, M., Olson, R., Glass, E. M., Kubal, M., et al. (2008). The metagenomics RAST server - a public resource for the automatic phylogenetic and functional analysis of metagenomes. *BMC Bioinformatics*, *19*(9), 386.

McDonald, D., Clemente, J. C., Kuczynski, J., Rideout, J., Stombaugh, J., Wendel, D., et al. (2012). The Biological Observation Matrix (BIOM) format or: How I learned to stop worrying and love the ome-ome. *Gigascience*, *1*(1), 7.

Ondov, B. D., Bergman, N. H., & Phillippy, A. M. (2011). Interactive metagenomic visualization in a Web browser. *BMC Bioinformatics*, *12*(1), 385.

Overbeek, R., Begley, T., Butler, R. M., Choudhuri, J. V., Chuang, H. Y., Cohoon, M., et al. (2005). The subsystems approach to genome annotation and its use in the project to annotate 1000 genomes. *Nucleic Acids Research*, *33*(17), 5691–5702.

Pruitt, K. D., Tatusova, T., & Maglott, D. R. (2007). NCBI reference sequences (RefSeq): A curated non-redundant sequence database of genomes, transcripts and proteins. *Nucleic Acids Research*, *35*(Database issue), D61–D65.

Reeder, J., & Knight, R. (2009). The 'rare biosphere': A reality check. *Nature Methods*, *6*, 636–637.

Rho, M., Tang, H., & Ye, Y. (2010). FragGeneScan: Predicting genes in short and error-prone reads. *Nucleic Acids Research*, *38*, e191.

Riesenfeld, C. S., Schloss, P. D., & Handelsman, J. (2004). Metagenomics: Genomic analysis of microbial communities. *Annual Review of Genetics*, *38*, 525–552.

Tatusov, R. L., Fedorova, N. D., Jackson, J. D., Jacobs, A. R., Kiryutin, B., Koonin, E. V., et al. (2003). The COG database: an updated version includes eukaryotes. *BMC Bioinformatics*, *11*(4), 41.

Tatusov, R. L., Fedorova, N. D., Jackson, J. D., Jacobs, A. R., Kiryutin, B., Koonin, E. V., et al. (2013). Update on activities at the Universal Protein Resource (UniProt) in 2013. *Nucleic Acids Research*, *41*(Database issue), D43–D47.

Trimble, W. L., Keegan, K. P., D'Souza, M., Wilke, A., Wilkening, J., Gilbert, J., et al. (2012). Short-read reading-frame predictors are not created equal: Sequence error causes loss of signal. *BMC Bioinformatics*, *13*, 183.

Turnbaugh, P. J., Ley, R. E., Mahowald, M. A., Magrini, V., Mardis, E. R., & Gordon, J. I. (2006). An obesity-associated gut microbiome with increased capacity for energy harvest. *Nature*, *444*(7122), 1027–1031.

Vasudevan, S., Wolf, Y. I., Yin, J. J., & Natale, D. A. (2003). The COG database: An updated version includes eukaryotes. *BMC Bioinformatics*, *4*, 41.

Wilke, A., Harrison, T., Wilkening, J., Field, D., Glass, E. M., Kyrpides, N., et al. (2011). The M5nr: A novel non-redundant database containing protein sequences and annotations from multiple sources and associated tools. *BMC Bioinformatics*, *13*, 141.

Wilke, A., Wilkening, J., Glass, E. M., Desai, N. L., & Meyer, F. (2011). An experience report: Porting the MG-RAST rapid metagenomics analysis pipeline to the cloud. *Concurrency and Computation: Practice and Experience*, *23*(17), 2250–2257.

Wilkening, J., Wilke, A., Desai, N., & Meyer, F. (2009). Using clouds for metagenomics: A case study. New Orleans, LA: IEEE Cluster 2009.

Yilmaz, P., Kottmann, R., Field, D., Knight, R., Cole, J. R., & Amaral-Zettler, L. (2011). Minimum information about a marker gene sequence (MIMARKS) and minimum information about any (x) sequence (MIxS) specifications. *Nature Biotechnology*, *29*(5), 415–420.

CHAPTER TWENTY-THREE

A User's Guide to Quantitative and Comparative Analysis of Metagenomic Datasets

Chengwei Luo[*,†,‡,1], Luis M. Rodriguez-R[*,†,1], Konstantinos T. Konstantinidis[*,†,‡,2]

[*]Center for Bioinformatics and Computational Genomics, Georgia Institute of Technology, Atlanta, Georgia, USA
[†]School of Biology, Georgia Institute of Technology, Atlanta, Georgia, USA
[‡]School of Civil and Environmental Engineering, Georgia Institute of Technology, Atlanta, Georgia, USA
[1]These authors contributed equally to the work.
[2]Corresponding author: e-mail address: kostas@ce.gatech.edu

Contents

1. Introduction 526
2. How to Assemble a Metagenomic Dataset 528
3. How to Determine the Fraction of the Community Captured in a Metagenome 533
   3.1 Single species analysis 533
   3.2 Whole community analysis based on single gene markers 533
   3.3 Whole community analyses based on whole genomes 534
4. How to Identify the Taxonomic Identity of a Metagenomic Sequence 535
   4.1 Composition-based methods 536
   4.2 Alignment-based methods 536
   4.3 The MeTaxa algorithm 537
   4.4 Combination and optimization 538
5. How to Determine Differentially Abundant Genes, Pathways, and Species 538
   5.1 Modifications for other scenarios 539
6. Limitations and Perspectives for the Future 540
Acknowledgments 542
References 542

Abstract

Metagenomics has revolutionized microbiological studies during the past decade and provided new insights into the diversity, dynamics, and metabolic potential of natural microbial communities. However, metagenomics still represents a field in development, and standardized tools and approaches to handle and compare metagenomes have not been established yet. An important reason accounting for the latter is the continuous changes in the type of sequencing data available, for example, long versus short sequencing reads. Here, we provide a guide to bioinformatic pipelines developed to

Methods in Enzymology, Volume 531
ISSN 0076-6879
http://dx.doi.org/10.1016/B978-0-12-407863-5.00023-X

accomplish the following tasks, focusing primarily on those developed by our team: (i) assemble a metagenomic dataset; (ii) determine the level of sequence coverage obtained and the amount of sequencing required to obtain complete coverage; (iii) identify the taxonomic affiliation of a metagenomic read or assembled contig; and (iv) determine differentially abundant genes, pathways, and species between different datasets. Most of these pipelines do not depend on the type of sequences available or can be easily adjusted to fit different types of sequences, and are freely available (for instance, through our lab Web site: http://www.enve-omics.gatech.edu/). The limitations of current approaches, as well as the computational aspects that can be further improved, will also be briefly discussed. The work presented here provides practical guidelines on how to perform metagenomic analysis of microbial communities characterized by varied levels of diversity and establishes approaches to handle the resulting data, independent of the sequencing platform employed.

1. INTRODUCTION

Culture-independent whole-genome shotgun (WGS) DNA sequencing has revolutionized the study of the diversity and ecology of microbial communities during the last decade (Handelsman et al., 2007). However, the tools to analyze metagenomic data are clearly lagging behind developments in sequencing technologies, and several important bioinformatic challenges remain (Hugenholtz & Tyson, 2008; Kunin, Copeland, Lapidus, Mavromatis, & Hugenholtz, 2008). For instance, metagenomic studies of environmental samples typically recover only short (e.g., <10 kb long) fragments of the genome, which only rarely contain rRNA genes, the backbone of bacterial identification and taxonomic classification (Brenner, Staley, & Krieg, 2001), either because of chance (<0.1% of the genome is represented by rRNA genes) or the high similarity among rRNA genes from distinct organisms that prevents their correct assembly from metagenomic data (Miller, Baker, Thomas, Singer, & Banfield, 2011). Accordingly, identifying and studying novel taxa based on metagenomic approaches remain challenging due to the lack of appropriate non-rRNA-based methods and reference genomes.

Most metagenomic surveys to date have sampled only a small fraction of the total diversity within the target community, especially in highly complex soil/sediment microbial communities (Delmont, Simonet, & Vogel, 2012; Tyson et al., 2004), and the amount of additional sequencing required to cover the whole diversity has typically remained speculative. Generally speaking, this fraction is termed coverage and depends on both the sequencing effort and the diversity of the microbial community in the sample. Incomplete coverage does not prevent researchers from reaching valuable conclusions about the communities under study, but it constitutes a source of uncertainty and limits several downstream analyses such as assessing the importance of low-abundance (rare) community members. Estimating the diversity of a sample in terms of the number of species or operational taxonomic units (OTUs) present is challenging. Current approaches mainly rely on the construction of rarefaction curves (or similar approaches) based on the identification of OTUs (e.g., Caporaso et al., 2010; Schloss & Handelsman, 2005). The application of these techniques in short-read datasets, however, requires either the use of a reference database (to assign/recruit reads to reference sequences and then cluster reference sequences in OTUs) or clustering of assembled sequences (reference-free approach). The former is biased by the limited number of reference genes available in the databases, with the probable exception of the 16S rRNA gene (Cole, Konstantinidis, Farris, & Tiedje, 2010). The latter is biased by the use of phylogenetic markers that are much more conserved than the average gene in the genome (in order to be sufficiently similar to allow clustering/alignments) such as the ribosomal rRNA genes. However, important levels of genomic and ecological differentiation frequently underlie identical 16S rRNA gene sequences (Acinas et al., 2004; Konstantinidis, Ramette, & Tiedje, 2006).

A related challenge for metagenomics is how to identify differentially present genes, pathways, and species between datasets. The issue is complicated not only by the low coverage achieved in typical metagenomic datasets but also by the difficulty in defining microbial species (hence, OTUs are typically preferred instead; reviewed in Caro-Quintero & Konstantinidis, 2012; Gevers et al., 2005; Rossello-Mora & Amann, 2001) and the short-read length of current next-generation sequencing (NGS) technologies, which limits identification and quantification of target phylogenetic markers. For instance, short-read (i.e., 50–200 bp) NGS technologies have become increasingly popular due to their high throughput and low cost per sequenced base, but it remains unclear whether these technologies can be used to routinely and robustly assemble complete gene and/or individual genome sequences from complex communities. The low coverage typically achieved in metagenomic studies also represents a major challenge for assembly, in addition to short-read length (Delmont, Prestat, et al., 2012). NGS technologies are also changing continuously and, thus, their sequencing errors and artifacts need to be examined, and the associated bioinformatic pipelines to be updated, on a regular basis (e.g., Luo, Tsementzi, Kyrpides, Read, & Konstantinidis, 2012).

Here, we describe the bioinformatic approaches others and we have developed to achieve several of the tasks mentioned earlier, focusing primarily on "how to" accomplish the tasks and the limits of each approach, depending on the coverage obtained, type of sequencing technology employed, and objective of the study. A fundamental concept underlying our own approaches is the sequence-discrete populations. Our recent review and synthesis of the major metagenomic studies performed to date on various populations and habitats revealed that natural microbial communities are predominantly composed of discrete populations, with the intrapopulation sequence diversity typically ranging between ~95% and ~100% genome-aggregate average nucleotide identity, or gANI, depending on the population considered (Caro-Quintero & Konstantinidis, 2012; Konstantinidis & DeLong, 2008). The 95% gANI level corresponds tightly to 70% DNA–DNA hybridization, which is commonly used to demarcate bacterial species (Goris et al., 2007). Whether or not these populations should be equated to species remains unclear (Caro-Quintero & Konstantinidis, 2012), but the 95% gANI level appears to represent robust means to define populations, and hence, OTUs. Accordingly, we employed 95% gANI as needed during our analyses, and our pipelines employ genomic relatedness measures such as gANI, which offers important advantages compared to traditional approaches based on rRNA genes for the same purposes.

2. HOW TO ASSEMBLE A METAGENOMIC DATASET

Due to the large difference between the desired sequence length for analysis (e.g., the average bacterial gene length is 950 bp and a typical *Escherichia coli* genome is around 4.5 Mbp) and the sequencing read length provided by NGS (e.g., frequently <200 bp long), assembling WGS reads is usually the first step of metagenomic studies and represents the foundation for various downstream analyses. It is a critical yet challenging step, largely due to short read length and the fact that metagenomes represent mixtures of different genotypes, some closely related to each other. The objective is to obtain assemblies with long average contig length (typically measured by N50, which is defined as the longest length for which the collection of all contigs of that length or longer contains at least half of the total of the lengths of the contigs) and high quality (e.g., low frequencies of chimeras and base call errors).

Depending on the number of taxa present in the target community lacking sequenced representatives, the assembly process may be reference guided or *de novo*, or a mixture of both. When appropriate representatives are available (we recommend >90% gANI between reference genome and target population), reference-guided assembly is usually optimal. For instance, more than 3000 reference genomes are currently, or will be soon, available as part of the Human Microbiome Project (HMP; www.hmpdacc.org). Therefore, to reconstruct bacterial genotypes from human microbiome datasets, it is common to first prepare a nonredundant set of reference genome sequences at a given clustering threshold (e.g., 95% gANI) and then map the metagenomic reads to these references using mapping tools such as BLAST (Altschul et al., 1997), BLAT (Kent, 2002), MAQ (Li, Ruan, & Durbin, 2008), or Burrows–Wheeler transformation-based algorithms, which are suitable for fast short-read mapping (e.g., Illumina or ABI SOLiD platforms), including BWA (Li & Durbin, 2009) and Bowtie (Langmead & Salzberg, 2012). Reads mapped to a reference can then be binned together for population assembly (see below), substantially reducing complexity and hence, improving assembly quality. A second round of assembly can be subsequently applied, as necessary, in which the resulting contigs are assembled together with all reads in an attempt to recover genomic islands present in the target population but absent from the reference genome (and thus, missed during the reference-guided assembly).

A more challenging scenario occurs when few or no available references exist and thus, *de novo* assembly is needed, as is often the case for metagenomes from most natural habitats. In general, *de novo* assemblers fall into two categories: overlap-based and graph algorithm-based (Miller, Koren, & Sutton, 2010). The former perform well with long sequences such as those generated by Roche 454 and Sanger sequencers (Luo, Tsementzi, Kyrpides, & Konstantinidis, 2012). Exemplary assemblers of this category include the widely used Newbler package (Margulies et al., 2005), Celera assembler (Myers et al., 2000), and Arachne (Batzoglou et al., 2002). Graph algorithm-based assemblers have recently gained popularity for metagenomic studies mainly due to the prevalence of short-read sequencing data. Early generation assemblers from this category employed greedy algorithms and include SSAKE (Warren, Sutton, Jones, & Holt, 2007), VCAKE (Jeck et al., 2007), and SHARCGS (Dohm, Lottaz, Borodina, & Himmelbauer, 2007). They were later outperformed by *de Bruijn* graph-based algorithms such as Velvet (Zerbino & Birney, 2008), Euler (Chaisson, Brinza, & Pevzner, 2009; Chaisson & Pevzner, 2008), SOAPdenovo (Li et al., 2010),

ABySS (Simpson et al., 2009), and AllPaths (Butler et al., 2008), each of which builds its core data structure using different variations of a *K*-mer graph. A *K*-mer graph is composed of nodes, which are short nucleotide sequences (*K*-mers), and edges, which connect the nodes. Transitional relationship is a commonly implemented approach to connect two *K*-mers; for example, the 4-mer ATGA can transition to the 4-mer TGAC by removing the 5′-end letter A and adding the 3′-end letter C.

The latter tools were originally designed for assembly of single genomes, not metagenomes; and several properties of metagenomes violate basic assumptions of the corresponding algorithms. For instance, Velvet assumes even coverage along the target genome—an assumption that does not hold true for metagenomes, where the relative abundance of each species is almost always different (uneven). Several methods were more recently developed to overcome these limitations. For example, Meta-IDBA (Peng, Leung, Yiu, & Chin, 2011) and MetaVelvet (Namiki, Hachiya, Tanaka, & Sakakibara, 2012) tackle the problem by first isolating graphs into components that likely belong to the same population (or coverage bin) and then performing a variant-tolerating assembly for each individual component. In our previous study, we developed a robust hybrid protocol that combines the power of *de Bruijn* graph algorithms (Velvet and SOAPdenovo) and overlap-based approaches (Newbler package) to provide higher-quality assemblies, with larger N50 values (Luo, Tsementzi, Kyrpides, & Konstantinidis, 2012). In short, this protocol first removes redundancy among preassembled contigs from several independent runs of Velvet for preferably its metagenomic variant MetaVelvet (Namiki et al., 2012), and SOAPdenovo using a wide range of *K*-mers from 21 to 63 (three runs per algorithm are recommended) and then combines and assembles the remaining contigs into final contigs using Newbler. This hybrid protocol showed a twofold increase in average contig length and returned about 50% more assembled reads, while maintaining similar assembled sequence quality when compared with assemblies solely constructed using Velvet or SOAPdenovo in various metagenomes, including freshwater planktonic (Oh et al., 2011) and ocean beach sand samples (Rodriguez-R, Konstantinidis, et al., unpublished).

The ultimate goal in metagenome assembly is to recover whole-genome sequences. Initially, these efforts were focused on relatively simple communities, such as the acid mine drainage system (Denef & Banfield, 2012; Simmons et al., 2008), archaeal symbionts of marine sponges (Hallam et al., 2006), and the gut microbiome of premature infants (Morowitz et al., 2011), and combined manual inspection with popular assembly software. It is important to note that in all these studies, a mosaic genome representing the average genome of the target population, rather than a single genotypic variant present in the sample, was recovered.

More recently, successful assembly of genomes from complex communities using more automated approaches has been reported. For instance, Iverson and colleagues were able to recover nearly completed genomes of marine *Euryarchaeota*, *Thaumarchaeota*, and *Flavobacteria*, each representing 4–10% of a surface water metagenome, by using paired-end read information of a jumping library (insert size 2–3 kb as opposed to the typical ~300 bp size) to link precontigs (Iverson et al., 2012). Wrighton and colleagues recovered 49 incomplete genomes from groundwater metagenomes (completeness level varied between 41% and 95%) spanning different phyla by integrating self-organizing, map-based sequence binning methods and iterative coassembling techniques (Wrighton et al., 2012). As metagenomic sequencing becomes more and more affordable, related metagenomes along spatial and temporal (i.e., time-series) gradients will be increasingly common. Hence, there is a need to develop robust genome assembly methods suitable for time-series metagenomes.

In our experience, it is nearly impossible to obtain assemblies that resolve the genomes of closely related, co-occurring genotypes (strains) of the same population for almost any sample or method employed as the intrapopulation divergence is usually too small for assemblers to differentiate, frequently at the same level as sequencing errors. However, in some special cases of low-diversity communities or deep-branching populations (i.e., no close relatives co-occurring in the community), it has been possible to resolve a small set of target gene sequences at the strain level with the aid of visualization tools such as Strainer (Eppley, Tyson, Getz, & Banfield, 2007) or computationally intensive expectation–maximization (EM) algorithm-based methods such as EMIRGE (Miller et al., 2011). We have also recently performed a comprehensive *in silico* evaluation of the strain level resolution of our hybrid assembly protocol for different scenarios of intrapopulation genetic structure, and the reader is directed to the original publication for further details (Luo, Tsementzi, Kyrpides, & Konstantinidis, 2012).

Metagenomic assemblies should be carefully evaluated before being used for further analysis. Factors that can affect assembly quality include the intrinsic characteristics of the target community (e.g., richness and evenness of species, G + C% content and size of genomes, and abundance of repeated sequences), the experimental design (e.g., sequencing throughput, library size, and the choice of sequencing platform), and the parameter settings of the assembler. To evaluate the effect of these factors, simulated systems have been extensively used. For instance, Charuvaka and Rangwala evaluated the relationship between assembly quality (e.g., chimera frequency, average contig length) and community complexity and choice of *K*-mer size using simulated reads (Charuvaka & Rangwala, 2011). In our previous study, we spiked-in real Illumina reads of isolate genomes into Illumina metagenomes, instead of using simulated reads (Luo, Tsementzi, Kyrpides, & Konstantinidis, 2012), and this approach provides reliable means to evaluate several parameters of the assembly (Fig. 23.1). The evaluation showed that, with about 20 × coverage of the target population, its genome can be recovered at high-draft status (Branscomb & Predki, 2002), while at lower coverage levels, chimeric contigs increase in frequency and probably cause, in part, community diversity to be (artificially) overestimated. The relationships among population coverage and different types of sequencing errors and artifacts were examined more thoroughly for 100-bp-long Illumina data, and the reader is referred to the original publications for further details (Luo, Tsementzi, Kyrpides, & Konstantinidis, 2012; Luo, Tsementzi, Kyrpides, Read, et al., 2012).

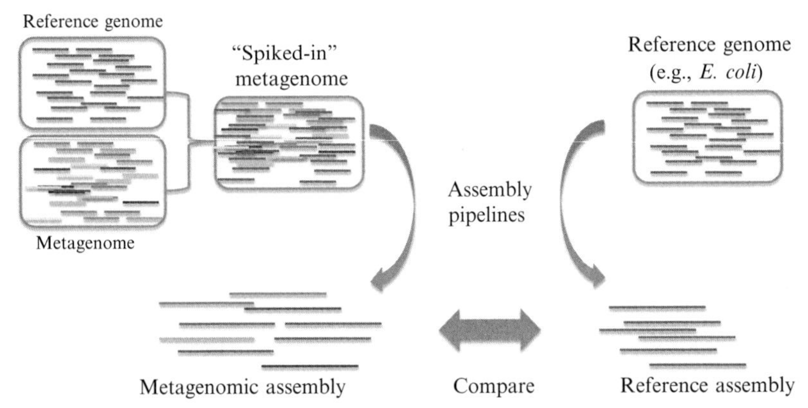

Figure 23.1 An approach to assess assembly parameters and output based on *in silico*-generated "spiked-in" metagenomes. To assess the impact of assembly parameters (e.g., *K*-mer, consensus cutoff, and minimum coverage), reads of reference genome(s), generated ideally from the same sequencing platform as the target metagenome, are spiked into the metagenome to form an *in silico* dataset. This *in silico* dataset is sequentially assembled, and the assembled contigs are compared against the reference genome sequence or the contigs assembled from the reference genome data alone. Note that a similar approach can be used for other purposes, for instance, to assess tools for functions other than assembly. (For color version of this figure, the reader is referred to the online version of this chapter.)

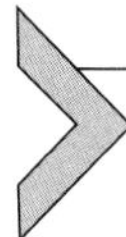

3. HOW TO DETERMINE THE FRACTION OF THE COMMUNITY CAPTURED IN A METAGENOME

Almost all metagenomic datasets published today lack completeness, that is, they do not capture all the DNA molecules or species within the target community. Several approaches have been devised to assess the level of community coverage, depending on the specific aim of the study and the additional information available about the community.

3.1. Single species analysis

In the simplest case, the objective is to determine the whole-genome coverage of one or a few target species or, more precisely, the coverage breadth, to differentiate from the average number of times each position is sequenced (called sequencing depth and sometimes referred to as coverage depth); the captured fraction of the remaining genomes of the community is not relevant. In 1988, Lander and Waterman (1988) provided a model aiming to solve this problem. In this model, the coverage breadth of a genome can be predicted from the sequencing depth. The latter can be easily estimated as the total size of the dataset multiplied by the abundance of the target species and divided by its genome size. For example, if a species constitutes about 4% of the community, with a genome of 4 Mbp, the expected sequencing depth in a metagenome of 300 Mbp would be 3×. That is, each position of the genome is sequenced three times, on average. Next, the following expression can be applied to predict the coverage breadth:

$$\mathrm{Cov_{breadth}} = 1 - \mathrm{e}^{-\mathrm{Cov_{depth}}}$$

For the given example above, the expected coverage breadth will be 95%. Although the Lander–Waterman model is still widely used, more refined models have been proposed (e.g., Wendl, 2006; Wendl, Kota, Weinstock, & Mitreva, 2012; Wendl et al., 2001), but no software implementations of the latter models are currently available.

3.2. Whole community analysis based on single gene markers

A more complex, but also more common, scenario is to target a specific gene marker, as opposed to a given taxon, to assess its total diversity and/or new variants in the sample. Typical examples include the analysis of rRNA genes, which can also provide taxonomic information, and thus, estimations of the total number of taxa in the community. All gene-based analyses have a common step, that is, to cluster sequences into OTUs. This allows the application of a well-known family of indexes derived from the nonparametric Good's coverage estimator (Good, 1953). Good's estimator has been shown to be statistically efficient (Esty, 1986), and yet it benefits from its simplicity:

$$\widetilde{C} = 1 - \frac{n_1}{N}$$

where n_1 is the number of clusters (or OTUs) found with only one observation and N is the total number of clusters or taxa, if clustering was based on rRNA gene sequences, found. Alternatively, one can estimate the total number of clusters or taxa (as opposed to coverage) using the Chao1 index (Chao, 1984), which is derived from Good's estimator. This is remarkable, because it means that the number of clusters in the unobserved part of the community can be estimated from the distribution of the observed clusters. Most practical applications, however, only marginally benefit from capturing extremely rare variants. In other words, the objective is usually to capture near-completeness as opposed to full completeness. This difference may seem subtle but, because of the long-tail distribution of clusters within many communities, it can translate to several orders of magnitude difference in sequencing effort. A more useful representation for this purpose are collector's curves (also referred to as rarefaction curves), which are derived from the capture–recapture methods of population ecology and essentially represent plots of the number of clusters observed as a function of sample size. Mothur (Schloss et al., 2009) is a popular package to generate collector's curves from sequencing data, but the same or similar methods have been implemented in a wide variety of packages and libraries. Essentially, these curves allow visual inspection of the level of sample saturation by sequencing because a more pronounced plateau (if any) indicates more saturated sampling (hence, near-completeness). This concept was recently generalized as "complexity curves" in the package preseq (Daley & Smith, 2013), providing a method to accurately project these curves to completeness, assuming sampling is unsaturated and the curve is close enough to saturation for the projection to be reliable.

3.3. Whole community analyses based on whole genomes

A more challenging scenario is to estimate the coverage at the whole-genome level in shotgun metagenomes, where the genome size and abundance of each member of the community are typically inaccessible and clustering is not feasible. There are three main solutions to this problem. The first is to approximate the distribution of abundances and genome sizes.

This technique is exemplified by the work of Stanhope (2010), and for the most part by the work of Hooper et al. (2010), and represents an intuitive solution but current implementations largely depend on the availability of optimal assemblies, making it inappropriate for very small datasets and/or very complex communities. A second solution is to identify gene markers sufficiently general to allow characterization of the captured portion, and subsequently apply techniques derived from targeting single gene markers (discussed earlier). For example, clusters can be formed using rRNA gene sequences extracted from the shotgun datasets. Alternatively, genomic-based taxonomic classifications can be used to define clusters. However, the latter approach depends on a comprehensive database of reference genomes, which is typically not available for most natural communities, uses only a small fraction of the datasets, and is subject to large influence of randomness. Nonetheless, in both cases described earlier, coverage and richness can be predicted and collector's curves can be constructed and projected.

Finally, intrinsic characteristics of the metagenomes can be used to estimate the captured fraction, sidestepping the biases introduced by suboptimal assemblies or incomplete reference databases. We recently presented Nonpareil (Rodriguez-R & Konstantinidis, unpublished), a method that examines the redundancy among all reads of a metagenome to estimate the average coverage of the genomes in the community. In addition, Nonpareil allows projecting the average coverage at increased sequencing efforts to predict the coverage that could be attained at any given size of dataset. This method enables fast calculation of coverage in entire metagenomic datasets, even those that are several Gbp (giga base pairs) in size, provides estimations of the amount of sequencing required to cover the complete or nearly complete diversity of the sample, and reflects the relative diversity of samples when compared with reference datasets or between samples. Our analyses of both *in silico*-constructed as well as real datasets from the HMP suggest that Nonpareil outperforms other tools for the same purposes and is applicable to microbial communities that show a wide range of diversity and complexity. Nonpareil is available for online querying through http://enve-omics.gatech.edu/ and as a stand-alone binary at https://www.github.com/lmrodriguezr/nonpareil.

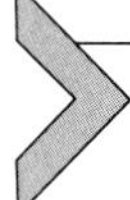

4. HOW TO IDENTIFY THE TAXONOMIC IDENTITY OF A METAGENOMIC SEQUENCE

Identifying the taxonomic affiliation of a sequence recovered in a metagenome remains challenging, primarily for the following reasons. First, the current collection of genome sequences is far from comprehensive and, thus, does not represent well the organisms in most natural environments (Cole et al., 2010). Second, no universally accepted definition of bacterial species exists, and hence, it is difficult to decide the degree of novelty of a new taxon (Konstantinidis & Tiedje, 2007). Finally, horizontal gene transfer, which is pronounced in the microbial world (Gogarten & Townsend, 2005), creates inconsistencies between sequence and organismal phylogeny, further complicating the issue. To tackle this issue, various algorithms have been developed, which can be classified into two categories: composition based and alignment based (Mande, Mohammed, & Ghosh, 2012). The former utilize sequence statistics with robust taxonomic signal; the latter are based on homology searches between query sequences and a database and employ sequence similarity as a proxy for taxonomic relatedness. Each approach has its own advantages and disadvantages, and the choice of which tool to implement often depends on the specific objective of the study.

4.1. Composition-based methods

The most important advantage of composition-based methods is that they are almost reference independent (most of them still need reference genomes to train the underlying algorithms) and, therefore, can assign taxonomy to sequences that do not match any reference sequences (unlike alignment-based methods). Also, they are, in general, faster and require less computational resources. Popular composition-based algorithms include, but are not limited to, PhyloPythia (McHardy, Martin, Tsirigos, Hugenholtz, & Rigoutsos, 2007; Patil et al., 2011), NBC (Rosen, Reichenberger, & Rosenfeld, 2011), Phymm (Brady & Salzberg, 2009), RAIPhy (Nalbantoglu, Way, Hinrichs, & Sayood, 2011), and TACOA (Diaz, Krause, Goesmann, Niehaus, & Nattkemper, 2009). However, these methods do not usually perform well on short sequences (e.g., <500 bp long), largely due to the insufficient information provided by such sequences.

4.2. Alignment-based methods

These methods classify query sequences according to their relatedness to available reference sequences, based on the corresponding alignments. The primary alignment engines are BLAST (Altschul, Gish, Miller, Myers, & Lipman, 1990); BLAST-like tools such as BLAT (Kent, 2002); hidden Markov model-based tools such as HMMer (Finn, Clements, & Eddy, 2011); or Burrows–Wheeler transform-based methods such as MAQ (Li et al., 2008),

BWA (Li & Durbin, 2009), and Bowtie (Langmead & Salzberg, 2012). Alignment-based methods are computationally more expensive. However, they are probably indispensible for every metagenomic study as their results are used for additional downstream analyses such as gene annotation and community profiling. Further, they would assign an increasingly larger number of query sequences as the available reference genome sequences from isolation or single-cell efforts (Stepanauskas, 2012) increase. Alignment-based classifiers include MG-RAST (Meyer et al., 2008), MEGAN (Huson, Auch, Qi, & Schuster, 2007), MARTA (Horton, Bodenhausen, & Bergelson, 2010), CARMA (Krause et al., 2008), AMPHORA (Wu & Eisen, 2008), TreePhyler (Schreiber, Gumrich, Daniel, & Meinicke, 2010), and others. We have recently presented MeTaxa, an algorithm that employs unique design elements to classify at least 5% more sequences than any existing alignment-based tool (Luo, Rodriguez-R & Konstantinidis, unpublished; and also available through http://enve-omics.gatech.edu/). MeTaxa is briefly described in the following section.

4.3. The MeTaxa algorithm

MeTaxa differs from other alignment-based methods in that it takes into account all genes encoded on a query sequence, weighting each gene based on its (predetermined) classifying power. The weights reflect: (i) how well the gene resolves the classification at a given taxonomic level based on its degree of sequence conservation (e.g., 16S rRNA resolves poorly the species level in contrast to the genus level or higher) and (ii) how consistent the gene phylogeny is with species phylogeny, the latter being approximated by the genome-aggregate average amino acid identity (gAAI). Parameterized weights and alignment-based matches against a reference database are subsequently integrated via a maximum likelihood algorithm. MeTaxa reports the probability for each possible taxonomic classification of the query sequence as well as the degree of novelty for sequences representing novel taxa (e.g., novel species, genus, or phylum) based on previously determined gAAI standards that correspond well to taxonomic ranks (Konstantinidis & Tiedje, 2005). The standardized approach to assess novelty represents another important improvement provided by MeTaxa compared to previous approaches. The gene weights are precalculated "offline" based on the publicly available completed and draft genomes and are included in the MeTaxa package. Users need only to provide, as input to MeTaxa, a BLAST tabular-like output from the search of each query sequence against their preferred reference database, for example, NR, KEGG, Swissprot, etc. MeTaxa can return high-precision predictions for thousands of input query sequences in a matter of a few minutes on a personal laptop computer.

4.4. Combination and optimization

As the two categories of methods have their own advantages and disadvantages, hybrid protocols have been more recently reported. For instance, PhymmBL combines BLAST output (alignment-based) and Phymm algorithm (composition-based) to achieve higher performance (Brady & Salzberg, 2009). In general, alignment-based methods are usually more accurate, while for query sequences without significant matches to the reference database, composition-based methods are probably the only options available. Among the composition-based methods, we have obtained good results with NBC, although the best method of choice would depend on the specific objective of the study and the type of data available.

5. HOW TO DETERMINE DIFFERENTIALLY ABUNDANT GENES, PATHWAYS, AND SPECIES

Any standardized annotation of metagenomic sequences, whether it involves genes, pathways, species, or any other functional or taxonomic categorization, can essentially be the subject of comparison across samples. To detect annotations that are differentially abundant between datasets with confidence, a statistical approach is necessary to account for under sampling of community diversity and the stochastic nature of WGS metagenomes. This task, generally referred to as profile comparison, can be divided into three main steps. First, metagenomic sequences must be annotated. Although annotation is feasible for short metagenomic reads, a more reliable annotation is often achieved based on assembled contigs. For example, entire contigs can be assigned a taxonomic affiliation, and predicted genes encoded on the contigs a putative function. Next, the abundance of annotations (features) is determined by mapping the original reads onto the features, generating a table of read counts. Finally, the statistical significance of the differences is evaluated. Several tools have been specifically designed to carry out statistical tests for metagenomic datasets such as the Statistical Analysis of Metagenomic Profiles, or STAMPS, package (Parks & Beiko, 2010). STAMPS can analyze any set of features across sets of metagenomes, in any of three modes: comparison of two samples, comparison of several samples in two groups (e.g., treatment vs. control), and comparison of multiple samples. It should also be noted that there is rich literature on the comparison of differentially abundant features from other types of studies, for example, transcriptomics (RNA-seq and CHiP-Seq), mostly varying on the assumptions about the underlying distribution of counts (e.g., binomial, Poisson, overdispersed Poisson, negative binomial). A guide and an evaluation of several methods were recently presented (Fang, Martin, & Wang, 2012; Schreiber et al., 2010). The type of data and most assumptions used in these fields are essentially the same as in metagenomics; hence, the methods are applicable to the problem discussed earlier.

We have developed a simple and robust statistical method to identify differentially abundant genes, pathways, or organisms between well-replicated control versus treatment metagenomes. In brief, the method combines resampling techniques (Jackknife), the DESeq package (Anders & Huber, 2010), and binomial hypothesis testing. Suppose we have m treatment and n control samples, a Jackknife method is used to generate all possible combinations of $\lfloor m/2 \rfloor$ treatment and $\lfloor n/2 \rfloor$ control samples ($\lfloor x \rfloor$ denotes the floor function of a real number, x, which maps it to the largest previous integer). For each combination, a normalized count table (see below for normalization) is generated by mapping sequences (e.g., reads) to different features (e.g., genes, pathways, population genomes); each row in the table represents a feature and each column represents a sample. DESeq is then applied to detect the difference between treatment and control samples for each feature. For a specific feature (row in the table), the log2 fold changes determined by the DESeq analysis of all combinations of samples follow a distribution; the mean represents the best estimate of fold change and the variance reflects the uncertainty of the estimate. A binomial test is then carried out to test the significance of the log2 fold change (1 for significantly different log2 fold change; 0 otherwise), and the P-value is adjusted for false discovery rate using the Benjamini–Hochberg method (Benjamini & Hochberg, 1995).

5.1. Modifications for other scenarios

Our method was originally developed to compare well-replicated (6 replicates per treatment at minimum; 10 replicates or more recommended) complex soil metagenomes generated by the Illumina HiSeq platform. To extend it to other types of sequence data or samples characterized by different complexities, modifications might be required, most often at the normalization and sparse counts (i.e., features with zero counts in certain samples) steps, and some limitations could emerge. Count table normalization is necessary in order to compare samples with different sequencing depths; the most popular approach is to present the number of sequences as a fraction of the total sequences of the corresponding sample. However, the latter approach undermines the statistical power derived from count data and thus is not recommended. Instead, we suggest normalizing counts using quantile-based methods, like the ones described previously (Bullard, Purdom, Hansen, & Dudoit, 2010; Fang et al., 2012). The DESeq algorithm in our abovementioned approach normalizes samples based on similar methods. Additional normalization steps may be required in some cases, however. For instance, when samples differ substantially in sequencing quality (e.g., different percentage of reads passing quality trimming) or when the sample datasets are too large to determine the number of reads for every feature, a resampling technique should be used to subsample the datasets at random to the same size. For features with low relative abundance, the sparse counts among samples will frequently lead to inaccurate testing results. A pragmatic way to account for this is to set a cutoff on the number of sequences mapped to a feature, and the features with lower counts are discarded from further analysis. To determine an appropriate cutoff, a Fisher's exact test-based method is proposed to simulate the impact of different cutoffs on the accuracy (White, Nagarajan, & Pop, 2009), while an alternative is to modify the hypothesis testing as discussed in Tusher, Tibshirani, and Chu (2001). The smaller the number of metagenomes compared (e.g., $n=2$), the more important is to account for the sparse count issue.

6. LIMITATIONS AND PERSPECTIVES FOR THE FUTURE

We presented here a practical guide to the analysis and interpretation of metagenomic data that should be useful in future studies across different habitats and microbial groups. It is important to realize, however, that the field of metagenomics is currently undergoing a major expansion, and new tools and approaches are being developed, including pipelines that integrate various tools to offer a comprehensive analysis of metagenomic datasets such as the Kbase project (http://kbase.science.energy.gov/) and MetAMOS (Treangen et al., 2013). It was not possible to mention all recent developments as part of the present document nor was that our intention. Our goal was instead to provide practical recommendations based on current knowledge and types of sequence data available, and a reference point for future developments (Fig. 23.2). We anticipate that several of the approaches and tools described earlier will require modification in the not-so-distant future, mostly due to new types of sequence data that will become available.

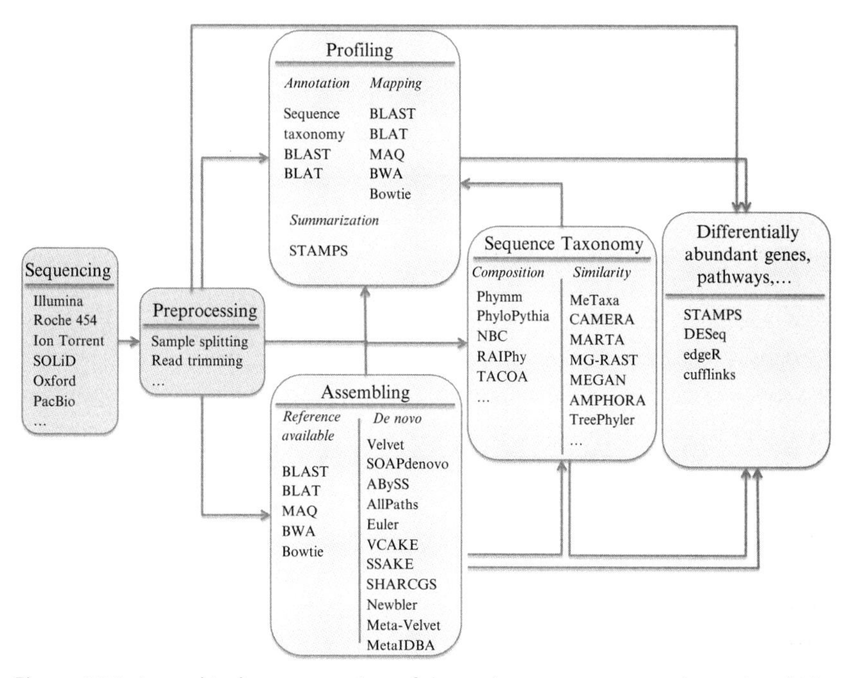

Figure 23.2 A graphical representation of the major components and associated bioinformatic tools of a typical metagenomics study. The bioinformatic analysis of metagenomic datasets starts with read quality processing (e.g., read trimming, multiplexing barcode removal), followed by community profiling, assembling, sequence taxonomic assignment (individual read or assembled contig level), and finally identification of differentially abundant genes, pathways, and organisms. The graph shows the relevant tools that can be used at each step and the arrays connecting different analysis represent the possible workflow. See text for more details. (For color version of this figure, the reader is referred to the online version of this chapter.)

For instance, it is foreseeable that metagenome assembly will become less challenging when single molecule sequencing, which can provide long sequence fragments on the order of tens of kilobases (Eid et al., 2009; Stoddart, Heron, Mikhailova, Maglia, & Bayley, 2009), becomes more routine. Related to the latter, single-cell technologies, which can provide the draft genome sequence of individual cells in a sample, are becoming increasingly more throughput, reliable (e.g., no DNA contamination), and affordable (Stepanauskas, 2012), and can greatly assist metagenomic studies by providing reference points in the analysis. Integrated approaches that combine multiple omics techniques, including transcriptomics and proteomics, have clear advantages, depending also on the specific objective(s) of the study. For instance, combining shotgun metagenomics with single-cell genomics is advantageous for population genetic studies but probably does not offer as much when the goal is to compare the gene content of different communities or the same community after a perturbation.

Even with the availability of longer sequences or single-cell genomes, however, a major computational challenge remains in how to handle and analyze the increasing volume of data that is produced by such approaches. Clearly, the tools and algorithms available do not scale up with the amount of data produced by new sequencers and single-cell technologies. Innovating solutions in terms of hardware implementations, algorithm optimizations (e.g., Pell et al., 2012), and (redundant) data reduction are highly needed to cope with the data available. Until then, obtaining the complete picture of complex microbial communities that are composed of hundreds to thousands of distinct species will remain somewhat a utopia.

ACKNOWLEDGMENTS

We thank Mike Weigand for helpful suggestions regarding the chapter. Our work is supported in part by the U.S. DOE Office of Science, Biological and Environmental Research Division (BER), Genomic Science Program, Award No. DE-SC0006662 and DE-SC0004601, and by U.S. National Science Foundation under Award No. 1241046.

REFERENCES

Acinas, S. G., Klepac-Ceraj, V., Hunt, D. E., Pharino, C., Ceraj, I., Distel, D. L., et al. (2004). Fine-scale phylogenetic architecture of a complex bacterial community. *Nature, 430*(6999), 551–554.

Altschul, S. F., Gish, W., Miller, W., Myers, E. W., & Lipman, D. J. (1990). Basic local alignment search tool. *Journal of Molecular Biology, 215*(3), 403–410. http://dx.doi.org/10.1016/S0022-2836(05)80360-2.

Altschul, S. F., Madden, T. L., Schaffer, A. A., Zhang, J., Zhang, Z., Miller, W., et al. (1997). Gapped BLAST and PSI-BLAST: A new generation of protein database search programs. Research Support, U.S. Gov't, P.H.S. Review, *Nucleic Acids Research, 25*(17), 3389–3402.

Anders, S., & Huber, W. (2010). Differential expression analysis for sequence count data. *Genome Biology, 11*(10), R106. http://dx.doi.org/10.1186/gb-2010-11-10-r106.

Batzoglou, S., Jaffe, D. B., Stanley, K., Butler, J., Gnerre, S., Mauceli, E., et al. (2002). ARACHNE: A whole-genome shotgun assembler. *Genome Research, 12*(1), 177–189. http://dx.doi.org/10.1101/gr.208902.

Benjamini, Y., & Hochberg, Y. (1995). Controlling the false discovery rate: A practical and powerful approach to multiple testing. *Journal of the Royal Society B, 57*(1), 289–300.

Brady, A., & Salzberg, S. L. (2009). Phymm and PhymmBL: Metagenomic phylogenetic classification with interpolated Markov models. Research Support, N.I.H., Extramural, *Nature Methods, 6*(9), 673–676. http://dx.doi.org/10.1038/nmeth.1358.

Branscomb, E., & Predki, P. (2002). On the high value of low standards. Research Support, Non-U.S. Gov't Research Support, U.S. Gov't, Non-P.H.S. Review, *Journal of Bacteriology, 184*(23), 6406–6409, discussion 6409.

Brenner, D., Staley, J., & Krieg, N. (2001). In *Bergey's manual of systematic bacteriology*. (2nd ed.). Vol. 1. New York: Springer-Verlag.

Bullard, J. H., Purdom, E., Hansen, K. D., & Dudoit, S. (2010). Evaluation of statistical methods for normalization and differential expression in mRNA-Seq experiments. Evaluation Studies Research Support, N.I.H., Extramural Research Support, Non-U.S. Gov't Research Support, U.S. Gov't, Non-P.H.S. *BMC Bioinformatics*, *11*, 94. http://dx.doi.org/10.1186/1471-2105-11-94.

Butler, J., MacCallum, I., Kleber, M., Shlyakhter, I. A., Belmonte, M. K., Lander, E. S., et al. (2008). ALLPATHS: De novo assembly of whole-genome shotgun microreads. *Genome Research*, *18*(5), 810–820. http://dx.doi.org/10.1101/gr.7337908.

Caporaso, J. G., Kuczynski, J., Stombaugh, J., Bittinger, K., Bushman, F. D., Costello, E. K., et al. (2010). QIIME allows analysis of high-throughput community sequencing data. *Nature Methods*, 7(5), 335–336. http://dx.doi.org/10.1038/nmeth.f.303 nmeth.f.303 [pii].

Caro-Quintero, A., & Konstantinidis, K. T. (2012). Bacterial species may exist, metagenomics reveal. *Environmental Microbiology*, *14*(2), 347–355. http://dx.doi.org/10.1111/j.1462-2920.2011.02668.x.

Chaisson, M. J., Brinza, D., & Pevzner, P. A. (2009). De novo fragment assembly with short mate-paired reads: Does the read length matter? *Genome Research*, *19*(2), 336–346. http://dx.doi.org/10.1101/gr.079053.108.

Chaisson, M. J., & Pevzner, P. A. (2008). Short read fragment assembly of bacterial genomes. *Genome Research*, *18*(2), 324–330. http://dx.doi.org/10.1101/gr.7088808.

Chao, A. (1984). Non-parametric estimation of the number of classes in a population. *Scandinavian Journal of Statistics*, *11*, 265–270.

Charuvaka, A., & Rangwala, H. (2011). Evaluation of short read metagenomic assembly. *BMC Genomics*, *12*(Suppl. 2), S8. http://dx.doi.org/10.1186/1471-2164-12-S2-S8.

Cole, J., Konstantinidis, K. T., Farris, R. J., & Tiedje, J. M. (2010). Microbial diversity and phylogeny: Extending from rRNAs to genomes. In W.-T. Liu & J. Jansson (Eds.), *Environmental molecular biology* (pp. 1–20). Norwich, UK: Horizon Scientific Press.

Daley, T., & Smith, A. D. (2013). Predicting the molecular complexity of sequencing libraries. *Nature Methods*, *10*(4), 325–327. http://dx.doi.org/10.1038/nmeth.2375.

Delmont, T. O., Prestat, E., Keegan, K. P., Faubladier, M., Robe, P., Clark, I. M., et al. (2012). Structure, fluctuation and magnitude of a natural grassland soil metagenome. *ISME Journal*, *6*(9), 1677–1687. http://dx.doi.org/10.1038/ismej.2011.197.

Delmont, T. O., Simonet, P., & Vogel, T. M. (2012). Describing microbial communities and performing global comparisons in the 'omic era. *ISME Journal*, *6*(9), 1625–1628. http://dx.doi.org/10.1038/ismej.2012.55.

Denef, V. J., & Banfield, J. F. (2012). In situ evolutionary rate measurements show ecological success of recently emerged bacterial hybrids. *Science*, *336*(6080), 462–466. http://dx.doi.org/10.1126/science.1218389.

Diaz, N. N., Krause, L., Goesmann, A., Niehaus, K., & Nattkemper, T. W. (2009). TACOA: Taxonomic classification of environmental genomic fragments using a kernelized nearest neighbor approach. Research Support, Non-U.S. Gov't, *BMC Bioinformatics*, *10*, 56. http://dx.doi.org/10.1186/1471-2105-10-56.

Dohm, J. C., Lottaz, C., Borodina, T., & Himmelbauer, H. (2007). SHARCGS, a fast and highly accurate short-read assembly algorithm for de novo genomic sequencing. *Genome Research*, *17*(11), 1697–1706. http://dx.doi.org/10.1101/gr.6435207.

Eid, J., Fehr, A., Gray, J., Luong, K., Lyle, J., Otto, G., et al. (2009). Real-time DNA sequencing from single polymerase molecules. *Science*, *323*(5910), 133–138. http://dx.doi.org/10.1126/science.1162986, 1162986 [pii].

Eppley, J. M., Tyson, G. W., Getz, W. M., & Banfield, J. F. (2007). Strainer: Software for analysis of population variation in community genomic datasets. *BMC Bioinformatics*, *8*, 398. http://dx.doi.org/10.1186/1471-2105-8-398.

Esty, W. W. (1986). The efficiency of Good's nonparametric coverage estimator. *The Annals of Statistics*, *14*, 1257–1260.

Fang, Z., Martin, J., & Wang, Z. (2012). Statistical methods for identifying differentially expressed genes in RNA-Seq experiments. *Cell & Bioscience*, *2*(1), 26. http://dx.doi.org/10.1186/2045-3701-2-26.

Finn, R. D., Clements, J., & Eddy, S. R. (2011). HMMER web server: Interactive sequence similarity searching. *Nucleic Acids Research*, *39*, W29–W37. http://dx.doi.org/10.1093/nar/gkr367, Web Server issue.

Gevers, D., Cohan, F. M., Lawrence, J. G., Spratt, B. G., Coenye, T., Feil, E. J., et al. (2005). Opinion: Re-evaluating prokaryotic species. *Nature Reviews. Microbiology*, *3*(9), 733–739. http://dx.doi.org/10.1038/nrmicro1236, nrmicro1236 [pii].

Gogarten, J. P., & Townsend, J. P. (2005). Horizontal gene transfer, genome innovation and evolution. Research Support, U.S. Gov't, Non-P.H.S. Review, *Nature Reviews. Microbiology*, *3*(9), 679–687. http://dx.doi.org/10.1038/nrmicro1204.

Good, I. J. (1953). The population frequencies of species and the estimation of population, parameters. *Biometrika*, *40*, 237–264.

Goris, J., Konstantinidis, K. T., Klappenbach, J. A., Coenye, T., Vandamme, P., & Tiedje, J. M. (2007). DNA-DNA hybridization values and their relationship to whole-genome sequence similarities. *International Journal of Systematic and Evolutionary Microbiology*, *57*(Pt. 1), 81–91. http://dx.doi.org/10.1099/ijs.0.64483-0, 57/1/81 [pii].

Hallam, S. J., Konstantinidis, K. T., Putnam, N., Schleper, C., Watanabe, Y., Sugahara, J., et al. (2006). Genomic analysis of the uncultivated marine crenarchaeote Cenarchaeum symbiosum. *Proceedings of the National Academy of Sciences of the United States of America*, *103*(48), 18296–18301. http://dx.doi.org/10.1073/pnas.0608549103, 0608549103 [pii].

Handelsman, J., Tiedje, J., Alvarez-Cohen, L., Ashburner, M., Cann, I., Delong, E., et al. (2007). *The new science of metagenomics: Revealing the secrets of our microbial planet.* Washington, DC: The National Academies Press.

Hooper, S. D., Dalevi, D., Pati, A., Mavromatis, K., Ivanova, N. N., & Kyrpides, N. C. (2010). Estimating DNA coverage and abundance in metagenomes using a gamma approximation. Research Support, U.S. Gov't, Non-P.H.S.*Bioinformatics*, *26*(3), 295–301. http://dx.doi.org/10.1093/bioinformatics/btp687.

Horton, M., Bodenhausen, N., & Bergelson, J. (2010). MARTA: A suite of Java-based tools for assigning taxonomic status to DNA sequences. *Bioinformatics*, *26*(4), 568–569. http://dx.doi.org/10.1093/bioinformatics/btp682.

Hugenholtz, P., & Tyson, G. W. (2008). Microbiology: Metagenomics [News]. *Nature*, *455*(7212), 481–483. http://dx.doi.org/10.1038/455481a.

Huson, D. H., Auch, A. F., Qi, J., & Schuster, S. C. (2007). MEGAN analysis of metagenomic data. *Genome Research*, *17*(3), 377–386. http://dx.doi.org/10.1101/gr.5969107.

Iverson, V., Morris, R. M., Frazar, C. D., Berthiaume, C. T., Morales, R. L., & Armbrust, E. V. (2012). Untangling genomes from metagenomes: Revealing an uncultured class of marine Euryarchaeota. Research Support, Non-U.S. Gov't Research Support, U.S. Gov't, Non-P.H.S. *Science*, *335*(6068), 587–590. http://dx.doi.org/10.1126/science.1212665.

Jeck, W. R., Reinhardt, J. A., Baltrus, D. A., Hickenbotham, M. T., Magrini, V., Mardis, E. R., et al. (2007). Extending assembly of short DNA sequences to handle error. *Bioinformatics*, *23*(21), 2942–2944. http://dx.doi.org/10.1093/bioinformatics/btm451.

Kent, W. J. (2002). BLAT—The BLAST-like alignment tool. Research Support, U.S. Gov't, P.H.S. *Genome Research*, *12*(4), 656–664. http://dx.doi.org/10.1101/gr.229202, Article published online before March 2002.

Konstantinidis, K. T., & DeLong, E. F. (2008). Genomic patterns of recombination, clonal divergence and environment in marine microbial populations. *ISME Journal*, *2*(10), 1052–1065. http://dx.doi.org/10.1038/ismej.2008.62, ismej200862 [pii].

Konstantinidis, K. T., Ramette, A., & Tiedje, J. M. (2006). The bacterial species definition in the genomic era. *Philosophical Transactions of the Royal Society of London. Series B, Biological Sciences*, *361*(1475), 1929–1940.

Konstantinidis, K. T., & Tiedje, J. M. (2005). Towards a genome-based taxonomy for prokaryotes. Research Support, Non-U.S. Gov't Research Support, U.S. Gov't, Non-P.H.S. *Journal of Bacteriology*, *187*(18), 6258–6264. http://dx.doi.org/10.1128/JB.187.18.6258-6264.2005.

Konstantinidis, K. T., & Tiedje, J. M. (2007). Prokaryotic taxonomy and phylogeny in the genomic era: Advancements and challenges ahead. *Current Opinion in Microbiology*, *10*(5), 504–509. http://dx.doi.org/10.1016/j.mib.2007.08.006, S1369-5274(07)00120-8 [pii].

Krause, L., Diaz, N. N., Goesmann, A., Kelley, S., Nattkemper, T. W., Rohwer, F., et al. (2008). Phylogenetic classification of short environmental DNA fragments. *Nucleic Acids Research*, *36*(7), 2230–2239. http://dx.doi.org/10.1093/nar/gkn038.

Kunin, V., Copeland, A., Lapidus, A., Mavromatis, K., & Hugenholtz, P. (2008). A bioinformatician's guide to metagenomics. Research Support, U.S. Gov't, Non-P.H.S. Review, *Microbiology and Molecular Biology Reviews*, *72*(4), 557–578. http://dx.doi.org/10.1128/MMBR.00009-08, Table of Contents.

Lander, E. S., & Waterman, M. S. (1988). Genomic mapping by fingerprinting random clones: A mathematical analysis. Research Support, Non-U.S. Gov't Research Support, U.S. Gov't, Non-P.H.S. Research Support, U.S. Gov't, P.H.S. *Genomics*, *2*(3), 231–239.

Langmead, B., & Salzberg, S. L. (2012). Fast gapped-read alignment with Bowtie 2. *Nature Methods*, *9*(4), 357–359. http://dx.doi.org/10.1038/nmeth.1923.

Li, H., & Durbin, R. (2009). Fast and accurate short read alignment with Burrows–Wheeler transform. *Bioinformatics*, *25*(14), 1754–1760. http://dx.doi.org/10.1093/bioinformatics/btp324.

Li, H., Ruan, J., & Durbin, R. (2008). Mapping short DNA sequencing reads and calling variants using mapping quality scores. *Genome Research*, *18*(11), 1851–1858. http://dx.doi.org/10.1101/gr.078212.108.

Li, R., Zhu, H., Ruan, J., Qian, W., Fang, X., Shi, Z., et al. (2010). De novo assembly of human genomes with massively parallel short read sequencing. *Genome Research*, *20*(2), 265–272. http://dx.doi.org/10.1101/gr.097261.109.

Luo, C., Tsementzi, D., Kyrpides, N. C., & Konstantinidis, K. T. (2012). Individual genome assembly from complex community short-read metagenomic datasets. *ISME Journal*, *6*(4), 898–901. http://dx.doi.org/10.1038/ismej.2011.147.

Luo, C., Tsementzi, D., Kyrpides, N. C., Read, T., & Konstantinidis, K. T. (2012). Direct comparisons of Illumina vs. Roche 454 sequencing technologies on the same microbial community DNA sample. *PLoS One*, 7(2), e30087. http://dx.doi.org/10.1038/ismej.2011.147.

Mande, S. S., Mohammed, M. H., & Ghosh, T. S. (2012). Classification of metagenomic sequences: Methods and challenges [Review]. *Briefings in Bioinformatics*, *13*(6), 669–681. http://dx.doi.org/10.1093/bib/bbs054.

Margulies, M., Egholm, M., Altman, W. E., Attiya, S., Bader, J. S., Bemben, L. A., et al. (2005). Genome sequencing in microfabricated high-density picolitre reactors. *Nature*, *437*(7057), 376–380. http://dx.doi.org/10.1038/nature03959.

McHardy, A. C., Martin, H. G., Tsirigos, A., Hugenholtz, P., & Rigoutsos, I. (2007). Accurate phylogenetic classification of variable-length DNA fragments. *Nature Methods*, *4*(1), 63–72. http://dx.doi.org/10.1038/nmeth976.

Meyer, F., Paarmann, D., D'Souza, M., Olson, R., Glass, E. M., Kubal, M., et al. (2008). The metagenomics RAST server—A public resource for the automatic phylogenetic and functional analysis of metagenomes. Research Support, N.I.H., Extramural Research Support, U.S. Gov't, Non-P.H.S. Research Support, U.S. Gov't, P.H.S. *BMC Bioinformatics*, *9*, 386. http://dx.doi.org/10.1186/1471-2105-9-386.

Miller, C. S., Baker, B. J., Thomas, B. C., Singer, S. W., & Banfield, J. F. (2011). EMIRGE: Reconstruction of full-length ribosomal genes from microbial community short read sequencing data. Research Support, U.S. Gov't, Non-P.H.S. *Genome Biology*, *12*(5), R44. http://dx.doi.org/10.1186/gb-2011-12-5-r44.

Miller, J. R., Koren, S., & Sutton, G. (2010). Assembly algorithms for next-generation sequencing data. *Genomics*, *95*(6), 315–327. http://dx.doi.org/10.1016/j.ygeno.2010.03.001.

Morowitz, M. J., Denef, V. J., Costello, E. K., Thomas, B. C., Poroyko, V., Relman, D. A., et al. (2011). Strain-resolved community genomic analysis of gut microbial colonization in a premature infant. *Proceedings of the National Academy of Sciences of the United States of America*, *108*(3), 1128–1133. http://dx.doi.org/10.1073/pnas.1010992108.

Myers, E. W., Sutton, G. G., Delcher, A. L., Dew, I. M., Fasulo, D. P., Flanigan, M. J., et al. (2000). A whole-genome assembly of Drosophila. *Science*, *287*(5461), 2196–2204.

Nalbantoglu, O. U., Way, S. F., Hinrichs, S. H., & Sayood, K. (2011). RAIphy: Phylogenetic classification of metagenomics samples using iterative refinement of relative abundance index profiles. *BMC Bioinformatics*, *12*, 41. http://dx.doi.org/10.1186/1471-2105-12-41.

Namiki, T., Hachiya, T., Tanaka, H., & Sakakibara, Y. (2012). MetaVelvet: An extension of Velvet assembler to de novo metagenome assembly from short sequence reads. *Nucleic Acids Research*, *40*(20), e155. http://dx.doi.org/10.1093/nar/gks678.

Oh, S., Caro-Quintero, A., Tsementzi, D., DeLeon-Rodriguez, N., Luo, C., Poretsky, R., et al. (2011). Metagenomic insights into the evolution, function, and complexity of the planktonic microbial community of Lake Lanier, a temperate freshwater ecosystem. Research Support, U.S. Gov't, Non-P.H.S. *Applied and Environmental Microbiology*, 77(17), 6000–6011. http://dx.doi.org/10.1128/AEM.00107-11.

Parks, D. H., & Beiko, R. G. (2010). Identifying biologically relevant differences between metagenomic communities. Research Support, Non-U.S. Gov't, *Bioinformatics*, *26*(6), 715–721. http://dx.doi.org/10.1093/bioinformatics/btq041.

Patil, K. R., Haider, P., Pope, P. B., Turnbaugh, P. J., Morrison, M., Scheffer, T., et al. (2011). Taxonomic metagenome sequence assignment with structured output models. Letter Research Support, N.I.H., Extramural Research Support, Non-U.S. Gov't, *Nature Methods*, *8*(3), 191–192. http://dx.doi.org/10.1038/nmeth0311-191.

Pell, J., Hintze, A., Canino-Koning, R., Howe, A., Tiedje, J. M., & Brown, C. T. (2012). Scaling metagenome sequence assembly with probabilistic de Bruijn graphs. *Proceedings of the National Academy of Sciences of the United States of America*, *109*(33), 13272–13277. http://dx.doi.org/10.1073/pnas.1121464109.

Peng, Y., Leung, H. C., Yiu, S. M., & Chin, F. Y. (2011). Meta-IDBA: A de Novo assembler for metagenomic data. *Bioinformatics*, *27*(13), i94–i101. http://dx.doi.org/10.1093/bioinformatics/btr216.

Rosen, G. L., Reichenberger, E. R., & Rosenfeld, A. M. (2011). NBC: The Naive Bayes Classification tool webserver for taxonomic classification of metagenomic reads. Research Support, U.S. Gov't, Non-P.H.S. *Bioinformatics*, *27*(1), 127–129. http://dx.doi.org/10.1093/bioinformatics/btq619.

Rossello-Mora, R., & Amann, R. (2001). The species concept for prokaryotes. *FEMS Microbiology Reviews*, *25*(1), 39–67.

Schloss, P. D., & Handelsman, J. (2005). Introducing DOTUR, a computer program for defining operational taxonomic units and estimating species richness. Research Support, Non-U.S. Gov't Research Support, U.S. Gov't, Non-P.H.S. *Applied and Environmental Microbiology*, *71*(3), 1501–1506. http://dx.doi.org/10.1128/AEM.71.3.1501-1506.2005.

Schloss, P. D., Westcott, S. L., Ryabin, T., Hall, J. R., Hartmann, M., Hollister, E. B., et al. (2009). Introducing mothur: Open-source, platform-independent, community-supported software for describing and comparing microbial communities. Research Support, Non-U.S. Gov't, *Applied and Environmental Microbiology*, *75*(23), 7537–7541. http://dx.doi.org/10.1128/AEM.01541-09.

Schreiber, F., Gumrich, P., Daniel, R., & Meinicke, P. (2010). Treephyler: Fast taxonomic profiling of metagenomes. *Bioinformatics*, *26*(7), 960–961. http://dx.doi.org/10.1093/bioinformatics/btq070.

Simmons, S. L., Dibartolo, G., Denef, V. J., Goltsman, D. S., Thelen, M. P., & Banfield, J. F. (2008). Population genomic analysis of strain variation in Leptospirillum group II bacteria involved in acid mine drainage formation. *PLoS Biology*, *6*(7), e177. http://dx.doi.org/10.1371/journal.pbio.0060177.

Simpson, J. T., Wong, K., Jackman, S. D., Schein, J. E., Jones, S. J., & Birol, I. (2009). ABySS: A parallel assembler for short read sequence data. *Genome Research*, *19*(6), 1117–1123. http://dx.doi.org/10.1101/gr.089532.108.

Stanhope, S. A. (2010). Occupancy modeling, maximum contig size probabilities and designing metagenomics experiments. *PLoS One*, *5*(7), e11652. http://dx.doi.org/10.1371/journal.pone.0011652.

Stepanauskas, R. (2012). Single cell genomics: An individual look at microbes. Research Support, U.S. Gov't, Non-P.H.S. Review, *Current Opinion in Microbiology*, *15*(5), 613–620. http://dx.doi.org/10.1016/j.mib.2012.09.001.

Stoddart, D., Heron, A. J., Mikhailova, E., Maglia, G., & Bayley, H. (2009). Single-nucleotide discrimination in immobilized DNA oligonucleotides with a biological nanopore. *Proceedings of the National Academy of Sciences of the United States of America*, *106*(19), 7702–7707. http://dx.doi.org/10.1073/pnas.0901054106.

Treangen, T. J., Koren, S., Sommer, D. D., Liu, B., Astrovskaya, I., Ondov, B., et al. (2013). MetAMOS: A modular and open source metagenomic assembly and analysis pipeline. *Genome Biology*, *14*(1), R2. http://dx.doi.org/10.1186/gb-2013-14-1-r2.

Tusher, V. G., Tibshirani, R., & Chu, G. (2001). Significance analysis of microarrays applied to the ionizing radiation response. *Proceedings of the National Academy of Sciences of the United States of America*, *98*(9), 5116–5121. http://dx.doi.org/10.1073/pnas.091062498.

Tyson, G. W., Chapman, J., Hugenholtz, P., Allen, E. E., Ram, R. J., Richardson, P. M., et al. (2004). Community structure and metabolism through reconstruction of microbial genomes from the environment. *Nature*, *428*(6978), 37–43.

Warren, R. L., Sutton, G. G., Jones, S. J., & Holt, R. A. (2007). Assembling millions of short DNA sequences using SSAKE. *Bioinformatics*, *23*(4), 500–501. http://dx.doi.org/10.1093/bioinformatics/btl629.

Wendl, M. C. (2006). A general coverage theory for shotgun DNA sequencing. Research Support, N.I.H., Extramural, *Journal of Computational Biology: A Journal of Computational Molecular Cell Biology*, *13*(6), 1177–1196. http://dx.doi.org/10.1089/cmb.2006.13.1177.

Wendl, M. C., Kota, K., Weinstock, G. M., & Mitreva, M. (2012). Coverage theories for metagenomic DNA sequencing based on a generalization of Stevens' theorem. *Journal of Mathematical Biology*. http://dx.doi.org/10.1007/s00285-012-0586-x.

Wendl, M. C., Marra, M. A., Hillier, L. W., Chinwalla, A. T., Wilson, R. K., & Waterston, R. H. (2001). Theories and applications for sequencing randomly selected clones. Research Support, U.S. Gov't, P.H.S. *Genome Research*, *11*(2), 274–280. http://dx.doi.org/10.1101/gr.133901.

White, J. R., Nagarajan, N., & Pop, M. (2009). Statistical methods for detecting differentially abundant features in clinical metagenomic samples. *PLoS Computational Biology*, *5*(4), e1000352. http://dx.doi.org/10.1371/journal.pcbi.1000352.

Wrighton, K. C., Thomas, B. C., Sharon, I., Miller, C. S., Castelle, C. J., VerBerkmoes, N. C., et al. (2012). Fermentation, hydrogen, and sulfur metabolism in multiple uncultivated bacterial phyla. *Science*, *337*(6102), 1661–1665. http://dx.doi.org/10.1126/science.1224041.

Wu, M., & Eisen, J. A. (2008). A simple, fast, and accurate method of phylogenomic inference. *Genome Biology*, *9*(10), R151. http://dx.doi.org/10.1186/gb-2008-9-10-r151.

Zerbino, D. R., & Birney, E. (2008). Velvet: Algorithms for de novo short read assembly using de Bruijn graphs. *Genome Research*, *18*(5), 821–829. http://dx.doi.org/10.1101/gr.074492.107.

AUTHOR INDEX

Note: Page numbers followed by "*f*" indicate figures, and "*t*" indicate tables and "*np*" indicate footnotes.

A

Abarenkov, K., 337
Abe, T., 135, 137
Abellan, J. J., 261–262
Abraham, P. E., 272–273, 314–316
Abram, F., 290–291
Abubucker, S., 341*f*
Abulencia, C. B., 5
Acinas, S. G., 526–527
Ackerley, D. F., 125
Adamczyk, J., 29*t*, 33
Adams, R. M., 294–295
Adey, A., 149–153, 157–158
Adkins, J. N., 290–291
Advani, A., 155–156, 158–159
Aho, T., 194–195
Ahring, B. K., 46–47, 62
Aird, D., 450
Aitchison, J. D., 283–285, 284*f*
Aitken, J. D., 372
Akkermans, A. D. L., 225–226
Alberti, A., 146–147, 153, 156–157, 159–160
Albrecht, M., 36, 37–39
Alexa, A., 201–202, 254–255
Alexander, D. H., 449
Allaire, J., 431–432
Allen, A. E., 48–49, 170–171, 188–191, 261–262
Allen, E. E., 62–64, 145, 336, 526–527
Allen, H. K., 137
Allen, L. Z., 159–160
Allers, E., 159–160, 307–308
Allewell, N. M., 195
Allgaier, M., 147–148, 151*t*, 156–158, 207–208, 337–338, 344, 346, 348
Allison, S. D., 260
Alm, E. J., 154, 357, 360, 367
Altinatas, I., 184
Altman, W. E., 334–335, 449, 529–530
Altschul, S. F., 310, 383*t*, 388–389, 466, 467, 529, 536–537
Alvarez-Cohen, L., 95, 102, 526
Amann, R. I., 4–5, 6–8, 11*f*, 16, 22–23, 29*t*, 32, 33, 35–36, 40, 93–94, 252–253, 293–294, 307–308, 354, 527–528
Amaral-Zettler, L., 307–308, 494–495
Amman, R. A., 33
Amorese, D., 31, 38–39
Amstislavskiy, V., 211–212
Anantharaman, K., 189*t*, 191–192
Anders, S., 539
Andersen, G. L., 335–336, 375, 388–389
Andersen, R. A., 52
Anderson, N. G., 145
Anderson, N. L., 145
Andersson, A. F., 272, 307–308
Andresen, B., 145, 151*t*
Andrews, P., 46–47
Andrews-Pfannkoch, C., 148–149
Angenent, L. T., 357
Angiuoli, S. V., 519–521
Angly, F., 238
Angskun, T., 493–494
Antipov, D., 452*t*, 454
Antonopoulos, D., 334–335, 375
Anuar, F., 372
Aoi, Y., 5
Ara, T., 133, 134
Aravind, L., 111–112, 115–117
Arbeli, Z., 255–256
Arciniega, M., 52, 53
Arkin, A. P., 360–361
Armbrust, E. V., 53–54, 154–155, 459–460, 531
Armour, C. D., 201–202
Arnheim, N., 47
Aro, E. M., 145–146
Artz, R. R., 422–424
Arumugam, M., 92, 401–402, 484
Asan, C., 154–155

Asan, X., 149–153, 157–158
Ascher, J., 252–253
Ashburner, M., 526
Ashkin, A., 22–23, 64–65, 85–86
Assefa, S. A., 452*t*, 456–457
Astor, Y. M., 307–308
Astrovskaya, I., 540–542
Atlas, R. M., 396–399
Attiya, S., 334–335, 449, 529–530
Aubert, M., 252
Auch, A. F., 320–325, 467, 505–506, 536–537
August, P. R., 128–129, 134, 135, 136
Avery, L., 422–424
Ayres, D. L., 32–33
Azam, F., 38–39
Aziz, R., 54–55

B

Baba, M., 133, 134
Baba, T., 133, 134
Babak, T., 201–202
Babu, M., 192–194
Bachoon, D. S., 195
Backhed, F., 396
Backhed, H. K., 372
Bader, J. S., 334–335, 449, 529–530
Badger, J. H., 341*f*
Bae, J. W., 148, 207–208
Bagby, S. C., 144–145
Bailey, M. J., 255–256
Bailey, S., 144–145
Bailly, J., 254–256, 262
Baker, B. J., 272, 290–291, 293, 314–316, 337–338, 340, 343, 344, 346, 347–348, 526, 531
Baker, D. R., 5, 8
Baker, H. E., 194–195
Bakker, P. A. H. M., 252
Balaji, P., 334–335, 375
Balasubramanian, S., 334–335, 449
Baldwin, M. A., 291–293
Baliga, N. S., 411
Ball, M., 124–125, 127, 138
Ballarini, A., 336–337, 453–454
Baltrus, D. A., 529–530
Banaru, M., 211–212
Banfield, J. F., 220–222, 290–291, 336, 337–338, 340, 343, 344, 345, 346–348, 526, 530–531
Bankevich, A., 452*t*, 454
Bano, N., 189*t*, 201–202, 238
Bantscheff, M., 294–295
Baranyi, C., 111–112
Barekati, Z., 222
Barerro-Canosa, J., 159–160
Barnett, T. C., 192–194
Barns, S. M., 56–57, 64–65
Barofsky, D. F., 272–273, 278, 279*f*, 314–316
Barot, S., 252
Barrett, T., 466
Barrios, E., 252
Barron, A. E., 448–449
Barros, A., 252
Barry, K. W., 5, 46–47, 62, 124, 147–148, 156–157, 459
Bartels, D., 54–55
Bartfai, R., 151*t*, 156–158
Bartha, R., 396–399
Bartram, A. K., 357
Bass, D., 307–308
Battchikova, N., 145–146
Batzoglou, S., 338–339, 529–530
Bauer, M., 452*t*, 459
Bayer, C., 337
Bayley, H., 540–542
Beal, E. J., 15–16
Bebb, C. E., 47
Bebout, B. M., 189*t*
Becher, D., 272–273
Becker, J. W., 124–125, 210
Bedard-Haughn, A., 252
Begley, T., 472, 494, 500, 503
Behnke, A., 307–308
Behr, B., 62
Behrens, S., 10
Beiko, R. G., 336–337, 538–539
Beinart, R. A., 191–192, 211, 238
Beisker, W., 5, 8
Béjà, O., 111–112, 115–117
Belcaid, M., 148–149
Belmar, L., 306–307
Belmonte, M. K., 529–530
Belnap, C. P., 272, 290–291, 294–295
Belnap, J., 252

Belogubova, E. V., 32
Bemben, L. A., 334–335, 449, 529–530
Bench, S. R., 47, 48–49, 50, 51, 52–55, 211, 238
Beneze, E., 159–160, 307–308
Bengtsson, J., 337
Benjamini, Y., 539
Benndorf, D., 314–316
Bennke, C. M., 272–273
Benson, D. A., 58, 466, 494
Bent, S. J., 364
Bent, Z. W., 201–202
Bentley, D. R., 334–335, 449
Benyon, R. J., 294–295
Bercik, P., 372
Berendsen, R. L., 252
Berg, H. C., 290–291
Bergelson, J., 536–537
Berg-Lyons, D., 98, 220–222, 334–336, 345, 356, 375, 376–377, 427, 428*f*
Bergman, K., 69–70
Bergman, N. H., 503, 510
Bergonzelli, G. E., 372
Berlin, A. M., 23–24, 49–50, 52, 153
Berriman, M., 157–158, 452*t*, 456–457
Berthiaume, C. T., 53–54, 154–155, 459–460, 531
Bertilsson, S., 307–308
Best, A., 54–55
Bettermann, A. D., 128–129, 134, 135–136
Bhavsar, J., 159
Bhupathiraju, V. K., 95, 102
Bibbs, L., 272
Bibby, J., 402
Bielawski, J. P., 144–145
Biggs, P. J., 452*t*, 453, 490
Bik, E. M., 5, 14–15, 46–47, 62, 67–68, 86
Binder, B. J., 4–5
Bircher, J. S., 386–388, 411, 412
Bird, D. F., 48–49
Birney, E., 154, 452*t*, 453, 455*f*, 456, 529–530
Birol, I., 154, 157, 529–530
Birren, B. W., 23–24, 49–50, 52, 111–115, 125, 132, 209, 425
Bittinger, K., 99, 356, 372–373, 374, 375, 389, 407, 438–439, 467, 493–494, 495, 510, 526–527
Bjorkholm, B., 372
Bjorkholm, J. E., 85–86
Blackall, L. L., 4–5, 6–8
Blackburn, M. C., 154, 357
Blackshields, G., 389, 390
Blagoev, B., 294–295
Blainey, P. C., 46–47, 62, 63*f*, 64–68, 81–82, 85–86
Blake, R. C. II., 272, 293, 314–316
Blazewicz, S. J., 467–468, 471*f*, 475*f*, 478*f*, 483*f*
Block, S. M., 69–70, 71–74
Blouin, M., 252
Bodenhausen, N., 536–537
Bodrossy, L., 252–253
Boersema, P. J., 291–293
Boetius, A., 15–16, 23, 29*t*, 33, 36, 37–39, 40, 64
Bohannan, B. J. M., 252
Boisvert, S., 452*t*, 454
Bokulich, N. A., 357, 380, 380*t*, 391, 438
Bolger, A. M., 452*t*, 453
Bollet, C., 382
Bolotin, A., 500
Bond, P. L., 127, 133, 136, 290–291, 314–316
Boo, S. M., 52
Boon, N., 93–94
Booth, M., 188–191
Borgogni, F., 252–253
Borgstrom, E., 157
Bork, P., 320–325
Borlee, B. R., 137
Borodina, T., 211–212, 529–530
Bosilca, G., 493–494
Botha, A., 255–256
Botting, C. H., 290–291
Boucher, C., 452*t*, 456
Bouvier, T., 93–94
Bouwman, L. A., 252–253
Bowen, B., 290–291
Bowerman, S., 5, 8
Bowien, S., 137–138
Boysen, C., 113–115
Bradford, M. A., 252–253
Bradford, M. M., 281
Bradley, A. S., 129, 132, 134, 135–136
Brady, A., 336–337, 452*t*, 453–454, 459, 536, 538
Brady, S. F., 128–129, 134, 135–136, 465–466

Braff, J. C., 125
Bragg, J. G., 144–145
Branscomb, E., 531–532
Bratbak, G., 155
Brehm-Stecher, B. F., 46–47
Breier, J. A., 191–192
Breiman, L., 420
Breiner, H. W., 307–308
Breitbart, M., 145–146, 148, 151*t*, 238, 252
Brenner, D., 526
Brenner, S. E., 211, 335–336, 361–362, 388–389
Brewer, H., 35–36
Brewer, P. G., 15–16
Brian, P., 127
Brinza, D., 529–530
Brisco, M. J., 62
Brodie, E. L., 334, 347–348, 361–362, 374, 375, 388–389, 390
Bronner, G., 335
Bronstein, P. A., 148–149
Brooks, E. M., 211
Brost, R., 127
Brown, C. G., 334–335, 449
Brown, C. T., 23–24, 30–31, 33, 35–37, 39–40, 334–335, 375, 542
Brown, D. G., 357
Brown, K. A., 127, 354
Brown, M. V., 272
Brown, N. P., 389, 390
Brüchert, V., 306, 307–308
Bruckner, C. G., 306
Brühl, A., 11*f*
Brulc, J. M., 238
Brum, J. R., 159–160
Bruns, M. A., 255–256
Brunsell, N., 252
Brussaard, L., 252–253
Bryant, J. A., 195–196, 263, 307–308
Bryant, S. H., 466
Buchan, A., 201–202, 238
Buck, D., 345
Buee, M., 252–253
Bugrysheva, J. V., 192–194
Bullard, J. H., 539–540
Bulow, S. E., 308
Bunce, M., 31
Bunge, J., 307–308
Bureau, F., 252
Burgdorf, K. S., 92, 484
Burggraf, S., 64–65
Burke, C., 136
Burke, D. J., 291–293
Burke, G. R., 447
Burkill, P. H., 4, 5, 93–94, 293–294, 307–308
Burlingame, A. L., 291–293
Burnhum, K. E., 290–291
Burns, M. J., 155
Burow, L. C., 189*t*
Burton, J. N., 154, 452*t*, 455–456
Buscot, F., 258–259
Bushman, F. D., 99, 356, 372–373, 374, 375, 389, 407, 438–439, 467, 493–494, 495, 510, 526–527
Buslov, K. G., 32
Bussema, C., 335
Butler, J., 529–530
Butler, R. M., 472, 494, 500, 503
Button, D., 38–39

C

Cable, A., 85–86
Cai, Q., 154–155
Cai, Y., 360
Callister, S. J., 290–291, 294, 348
Campbell, B. J., 238
Canese, K., 466
Canfield, D. E., 197, 198–199, 306, 307–308
Canino-Koning, R., 542
Cann, I., 526
Cantarel, B. L., 92, 372
Canty, A. J., 203*t*
Cao, X., 211
Caporaso, J. G., 98, 99, 220–222, 334–336, 345, 356, 357, 361–362, 372–373, 374, 375, 376–377, 389, 402–405, 407, 425, 427, 428*f*, 438–439, 467, 493–494, 495, 510, 526–527
Cappellano, C. M., 124–125, 127, 138
Cardenas, E., 388–389
Carlioz, A., 382
Carmack, E., 291
Caro-Quintero, A., 527–528, 530
Carozza, M. J., 294–295

Carter, B. J., 47, 48–49, 50, 52–53, 54–56, 58
Carter, N. P., 47
Carvalhais, L. C., 212, 252–253
Carvalho, F. A., 372
Carver, R., 294–295
Casamayor, E. O., 238
Casella, G., 252, 382–385
Castelle, C. J., 62–64, 337–338, 343–344, 348, 531
Castle, J. C., 201–202
Cavalieri, D., 372
Cavanaugh, C. M., 335–336
Cavanaugh, M., 58, 494
Cebron, A., 252–253
Ceccherini, M. T., 252–253
Cepeda, A., 226–227
Ceraj, I., 526–527
Cermak, L., 258–259
Chadd, E. H., 69–70
Chaffron, S., 290–291, 314–316
Chai, B., 334, 388–389
Chaisson, M. J., 529–530
Chakerian, J., 402–405
Chami, M., 93–94
Chan, C.-K. K., 452*t*, 459
Chan, L. K., 188–191, 238–239
Chang, B. X., 308
Chang, H. W., 148
Chao, A., 363, 364, 533–534
Chao, C. J., 135–136
Chapman, A. R., 62
Chapman, J., 62–64, 145, 336, 526–527
Chappell, N. A., 252
Charkowski, A. O., 233
Charlson, E. S., 379, 421
Charuvaka, A., 531–532
Chase, J. H., 425
Chavarria, K. L., 296, 467–468, 471*f*, 475*f*, 478*f*, 483*f*
Chavez, F. P., 189*t*, 196, 238
Chazdon, R. L., 364
Chen, F., 195
Chen, H., 192–194, 240–241
Chen, I.-M. A., 54–55, 494
Chen, J., 54–55, 272
Chen, R., 201–202
Chen, T., 335–336, 337
Chen, V. B., 402
Chen, W. S., 450
Chen, X., 12
Chen, Y. C., 154–155, 291–293
Chen, Z., 201–202, 334–335
Cheng, C. T., 201–202, 254–255
Cheng, J., 372
Chenna, R., 389, 390
Chetvernin, V., 466
Cheung, K. J., 192–194
Cheung, V. G., 47
Chevreux, B., 452*t*, 453
Chiang, G. T., 159
Chin, C.-S., 449
Chin, F. Y., 154, 530
Chin, M., 209
Chinwalla, A. T., 341*f*, 533
Chisholm, S. W., 4–5, 23–24, 46–47, 49–50, 52, 62, 124, 144–145, 147–148, 156–157, 170–171, 192–194, 193*f*, 261–262
Chistoserdov, A. Y., 307–308
Chistoserdova, L., 5, 8, 22–23
Chitsaz, H., 452*t*, 456
Cho, J. C., 272
Cho, Y. J., 201–202
Choe, L. H., 290–291
Choi, P. J., 192–194
Chokhawala, H., 336, 459–460
Chomczynski, P., 222
Choudhuri, J. V., 472, 494, 500, 503
Chourey, K., 294–295, 296
Christen, R., 307–308
Christianson, L., 111–112
Chu, G., 539–540
Chu, K., 54–55, 494
Chu, S., 85–86
Chuang, H.-Y., 472, 494, 500, 503
Chung, D., 255–256
Church, G. M., 144–145, 192–194, 290–291
Church, M. J., 238–239, 308, 311*t*
Cianciotto, N. P., 135–136
Ciulla, D. M., 170–171, 178–181, 199–201
Civiti, T. D., 222
Claesson, M. J., 361–362
Clapham, P., 159
Clardy, J., 135–136, 465–466

Clark, I. M., 252–253, 527–528
Clark, K., 58, 494
Clark, T., 211
Claude, J., 402–405
Claverie, J.-M., 64
Clemente, J. C., 361, 372–373, 377, 379, 388, 390, 422–424, 429–430, 495
Clements, J., 536–537
Clifton, V. L., 222
Clingenpeel, S., 14–15, 23–24, 31, 86–87
Clipson, N., 252–253
Clokie, M. R. J., 144–145
Clum, A., 5, 14–15, 447–448
Coates, G., 159
Cochrane, G., 494–495
Codreanu, S. G., 297
Coe, M., 252
Coeffet-Legal, M. F., 127
Coenye, T., 354, 527–528
Cohan, F. M., 354, 527–528
Cohoon, M., 472, 494, 500, 503
Cole, J. R., 58, 334, 361–362, 374, 375, 388–389, 467, 468, 494–495, 526–527, 535–536
Coleman, D. C., 115–116, 354
Coleman, M. K., 170–171, 294–295, 320–325
Coleman, M. L., 193*f*, 261–262
Collins, G., 290–291
Collins, S. M., 372
Collman, R. G., 379, 421
Colwell, R. K., 363, 364
Colwell, R. R., 144
Condon, J., 62
Connor, T. R., 208–209
Consolandi, C., 207–208
Constan, L., 35–36
Contreras, M., 92, 372
Copeland, A., 5, 14–15, 62, 447–448, 457, 526
Corbeil, J., 452*t*, 454
Cordes, T., 195–196, 221*f*
Corrier, K., 145
Corthier, G., 137
Costello, E. K., 92, 99, 336, 356, 372–373, 374, 376, 403*f*, 407, 420, 427, 428*f*, 438–439, 467, 493–494, 495, 510, 526–527, 530–531
Cottrell, M. T., 136
Coupland, P., 208–209
Courtois, S., 124–125, 127, 128–129, 138
Courty, P. E., 252–253
Cox, M. P., 452*t*, 453, 490
Coyne, K. J., 201
Cozzarelli, N., 201–202
Crapoulet, N., 64
Creamer, R., 252
Crouse, J., 31, 38–39
Crowell, J. E., 291–293
Cruaud, C., 146–147, 153, 156–157, 159–160
Cucci, T. L., 95
Cui, X., 47
Cullender, T. C., 372
Culley, A. I., 148–149
Cummings, D., 189*t*
Curtis, D. J., 201–202
Curtis, T. P., 252, 381
Cuvelier, M. L., 48–49

D

Daims, H., 4–5, 11*f*
Dalevi, D., 534–535
Daley, T., 533–534
Dalin, E., 336, 466
Dalsgaard, T., 197, 198–199, 306, 307–308
Dammeyer, T., 144–145
Damon, C., 238, 252–253, 262, 263
Daniel, R., 136, 137–138, 536–537, 538–539
Danielsson, M., 450
Darfeuille, F., 201–202
Darling, A. E., 32–33, 148–149, 336
Datsenko, K. A., 133
Daubin, V., 460
Davenport, E. R., 361–362
Davenport, K. W., 457
Davenport, R. J., 381, 430
David, M. M., 467–468, 471*f*, 475*f*, 478*f*, 483*f*
Davis, I. W., 402
Dawyndt, P., 494–495
de Beer, D., 33
De Brabandere, L., 197, 306, 307–308
De Filippo, C., 372
De Hollander, M., 335

de Jong, P. J., 125
De la Iglesia, R., 48–49
de la Torre, J. R., 111–112
de Menezes, A., 252–253, 263
De Preter, K., 233
De Vloed, F., 233
de Vos, W. M., 225–226, 290–291
de Winter, A., 52
Dean, F. B., 47, 62
Dean, R. A., 119
DeAngelis, K. M., 467–468, 471*f*, 475*f*, 478*f*, 483*f*
Debaud, J. C., 254–255
Debroas, D., 145–146, 148–149, 335
Decaens, T., 252
Dehal, P. S., 360–361
DeJongh, M., 54–55
Dekas, A. E., 23–24, 30–31, 33, 35–37, 39–40
del Giorgio, P. A., 48–49, 92, 93–94, 95
Delcher, A. L., 529–530
DeLeon-Rodriguez, N., 530
Delmont, T. O., 252–253, 526–528
Delmotte, N., 290–291, 314–316
DeLong, E. F., 4–5, 15–16, 23, 31, 33, 35–36, 46, 111–112, 115–117, 124–125, 128–129, 132, 133, 134–136, 138, 139, 188–192, 189*t*, 193*f*, 195–196, 197, 259–260, 261–262, 263, 306, 307–308, 311*t*, 526, 528
DeMaere, M. Z., 272–273, 290–291, 314–316
Demir, E., 201
Dempster, A. P., 338–339
Denef, V. J., 220–222, 272, 290–291, 336, 530–531
Deng, L., 146–148, 151*t*, 153, 156–157, 159–160
Deng, Y., 252–253
Denman, S. E., 459
Dennis, P. G., 212, 252–253
Depret, G., 128–129
Dequin, S., 238, 252–253
Derveaux, S., 233
Desai, C., 258–259
Desai, N. L., 159–160, 489, 519–520
DeSantis, T. Z., 334, 335–336, 347–348, 361–362, 374, 375, 388–389, 390
Desany, B. A., 47, 48–49, 51, 53–55
Desiderio, R., 272
Dethlefsen, L., 372
Deutscher, M. P., 192–194, 254–255
Devereux, R., 4–5
Devol, A. H., 308
Dew, I. M., 529–530
Dey, N., 335–336, 361–362, 388–389
Di Palma, S., 291–293
Di Paola, M., 372
Diaz Heijtz, R., 372
Diaz, N. N., 536–537
Diaz, R. J., 306
DiBartolo, G., 272, 530–531
Dick, G. J., 189*t*, 191–192
Dill, B. D., 272, 290–291
Dinsdale, E. A., 238
Distel, D. L., 526–527
Disz, T., 54–55
Do, C. B., 338–339
Docktor, M., 367
Dodsworth, J. A., 65–67, 85–86
Doerks, T., 494, 503
Doherty, M. K., 294–295
Dohm, J. C., 449, 529–530
Dominati, E., 252
Dominguez-Bello, M. G., 92, 372
Donaghey, J., 211
Donahoo, R. S., 290–291
Dongarra, J. J., 493–494
Dorninger, C., 4–5
Doucette, G., 196
Doyle, E., 252–253
Drake, J., 449
Drancourt, M., 382
Dreyfus, M., 205, 253–254
Dröge, J., 457
D'Souza, M., 484, 490–491, 493–494, 536–537
Duan, X., 201–202
Dudoit, S., 211, 539–540
Duffy, S., 145–146
Duhaime, M., 146–148, 146*f*, 151*t*, 153, 156–158, 159–160
Duller, S., 111–112
Dumont, M. G., 138
Dunbar, J., 252
Duncan, A., 92, 372

Dupont, C. L., 64, 459–460
Durban, A., 261–262
Durbin, R., 149–153, 157, 529, 536–537
Durham, I., 456
Durso, L., 337
Dusane, D. H., 12
Dushoff, J., 363
Dutton, R. J., 92
Dvorkin, M., 452*t*, 454
Dyer, T. A., 334
Dyhrman, S. T., 189*t*
Dziedzic, J. M., 64–65, 85–86

E

Earl, A. M., 335–336, 346, 386–388
Eberl, L., 296, 314–316
Eckburg, P. B., 372
Eddy, S. R., 389, 536–537
Edgar, R. C., 347–348, 361, 373–374, 383*t*, 385–386, 388, 389, 390, 431, 488, 493–494
Edgcomb, V., 307–308
Edwards, C., 32
Edwards, R. A., 54–55, 170–171, 207–208, 238, 290–291
Egan, R., 336, 459–460
Egan, S., 136
Egholm, M., 334–335, 447–448, 529–530
Egli, T., 93–94
Ehrlich, S. D., 500
Eid, J., 540–542
Eilam, R., 62
Eilers, H., 33, 293–294
Eisen, J. A., 62–64, 145, 148–149, 335, 336, 536–537
Elifantz, H., 294, 348
Ellersdorfer, G., 290–291, 314–316
Ellison, S. L., 155
Elshahed, M. S., 62, 335–336
Elvert, M., 33, 36, 37–39
Elwood, H. J., 56–57
Embaye, T., 23–24, 30–31, 33, 35–37, 39–40
Enault, F., 145–146, 148–149, 335
Endres, D. M., 365
Eng, J., 291
Engel, P., 458–459
Enger, J., 69–70
Engholm, M., 449
English, A. C., 449
English, C. A., 155
Engman, A., 55–56
Enright, A.-M., 290–291
Eppley, J. M., 124–125, 189*t*, 195–196, 238, 263, 307–308, 531
Epstein, S. S., 307–308
Erickson, B. E., 290–291
Erickson, B. K., 290–291
Ericsson, M., 69–70
Ernst, J., 29*t*, 33
Esty, W. W., 533–534
Ettema, C. H., 252
Ettwig, K. F., 16
Evans, F., 272–273, 290–291, 314–316
Evans, J., 390

F

Fadrosh, D. W., 148–149
Fagg, G. E., 493–494
Faith, D. P., 399
Faith, J. J., 357, 380, 380*t*, 391, 438
Faligault, H. E., 457
Fan, H. C., 62
Fan, L., 272, 290–291, 335–336, 337
Fang, X., 529–530
Fang, Z., 538–540
Farrell, M. L., 360
Farrelly, V., 335
Farris, R. J., 334, 388–389, 526–527, 535–536
Fasulo, D. P., 529–530
Faubladier, M., 252–253, 527–528
Fedorova, N. D., 494
Fehr, A., 540–542
Feike, J., 189*t*, 191–195, 197–198
Feil, E. J., 354, 527–528
Feingersch, R., 111–112
Feldgarden, M., 335–336, 346, 386–388
Fenn, J. B., 290
Fennell, T., 450, 467
Fente, C., 226–227
Ferdelman, T. G., 15–16
Ferneyhough, B., 345
Fernie, A. R., 452*t*, 453
Ferrer, M., 252–253
Ficarro, S. B., 291–293

Fiehn, O., 222
Field, D., 159–160, 170–171, 189*t*, 207–208, 420, 479, 494–495
Fierer, N., 92, 98, 159–160, 252–253, 334–336, 372, 375, 376–377, 396–399, 401–402, 403*f*, 425
Fike, D. A., 29
Filiatrault, M. J., 148–149, 212
Findeiss, S., 201–202
Finkelstein, D., 194–195
Finn, M. G., 291–293
Finn, R. D., 536–537
Fischer, C., 290–291
Fischer, L. M., 372
Fish, J., 388–389
Fitzsimons, N. A., 225–226
Flanigan, M. J., 529–530
Fleige, S., 233, 256–257
Florens, L., 294–295, 320–325
Fodor, A. A., 334–335, 372
Fonseca, Á., 10
Fontana, L., 422–424
Forney, L. J., 354, 364
Forstova, J., 258–259
Foster, J., 390
Foster, R. A., 47, 48–49, 51, 52–56, 58
Fournier, M., 294–295
Fox, G. E., 334
Fraissinet-Tachet, L., 252–253, 254–255, 262
Francis, C. A., 46–47, 62, 64–65, 67–68, 81–82, 86
Franco, C., 226–227
Francoijs, K. J., 151*t*, 156–158
Francou, F. X., 124–125, 127, 138
Frank, D. N., 203*t*, 430
Frank, I., 52, 53
Frankenberg-Dinkel, N., 144–145
Franzke, D., 15–16
Fraser-Liggett, C. M., 375
Frazar, C. D., 53–54, 154–155, 272–273, 278*np*, 279*f*, 283–285, 290–291, 314–316, 459–460, 531
Frazer, K., 345
Freedman, D. L., 272
Freeland, H., 306
Freeman, K. H., 307–308
Frias-Lopez, J., 170–171, 189*t*, 191–192, 193*f*, 195–196, 201–202, 203*t*, 210–211, 261–262
Fricker, A. D., 372
Friedman, J., 367
Friendly, M., 413
Frigaard, N.-U., 128–129
Frischkorn, K. R., 337–338, 343, 345, 346–347, 348
Fritsch, E. F., 132
Frostegard, A., 128–129
Froula, J. L., 202–205, 260–261
Frumkin, D., 62
Fu, L., 401–402
Fuchs, B. M., 4, 5, 33, 93–94, 272–273, 293–294, 307–308, 310, 334, 347–348, 468
Fuentes, C. L., 255–256
Fuhrman, J. A., 144, 145–146, 159–160
Fukatsu, T., 194–195
Fukuda, R., 293–294
Fulthorpe, R. R., 252, 382–385
Fung, J. M., 272
Fureder, S., 252–253
Furlan, M., 178–181, 207–208
Furlong, E. E., 153
Furumichi, M., 494, 503, 510
Futschik, M., 192–194

G

Gaboyer, F., 48–49
Gabriel, E., 493–494
Gail, F., 56–57
Gajer, P., 401–402
Galens, K., 519–521
Galeote, V., 238, 252–253
Gallagher, S. R., 256–257
Galperin, M. Y., 472–473
Gammon, C. L., 29
Ganapathy, G., 449
Gans, J., 252
Garcia, J. A., 5
Gardebrecht, A., 272–273
Garrett, W. S., 99
Garrity, G. M., 58, 361–362, 374, 375, 388–389, 467, 468, 494–495
Gaskell, S. J., 294–295
Gasol, J. M., 92, 93–94

Gassmann, M., 174–175, 257
Gayral, J. P., 382
Geng, C., 154–155
Gerber, S. A., 294–295
Gerin, J. L., 145
Gerken, J., 32–33, 58, 388–389, 431
Gerstein, M., 52
Getz, W. M., 531
Gevers, D., 99, 335–336, 341*f*, 346, 354, 357, 367, 372–373, 377, 379, 380, 380*t*, 386–388, 391, 425, 429–430, 438, 527–528
Gharib, S. A., 272
Ghosh, T. S., 452*t*, 457, 459, 535–536
Giambelluca, T., 252
Giannoukos, G., 170–171, 178–181, 199–201, 203*t*, 207–208, 209–210, 211–212, 335–336, 346, 386–388
Gibson, J. F., 53–54, 252–253, 334
Gieseke, A., 15–16, 23, 29*t*, 33, 36, 40, 64
Giesler, T. L., 47, 62
Gifford, S. M., 188–191, 189*t*, 201, 209, 238–240, 242
Giglione, C., 145–146
Gil, R., 460
Gilad, Y., 361–362
Gilbert, J. A., 170–171, 184, 189*t*, 203*t*, 207–208, 334–335, 375, 479, 493–494
Gilbert, M. T., 31
Gill, J., 201–202
Giller, P. S., 252
Gillespie, D. E., 135–136
Gilna, P., 170–171, 207–208
Giovannoni, S. J., 62–64, 272, 334
Girguis, P. R., 35–36, 191–192
Gish, W., 310, 383*t*, 388–389, 466, 467, 536–537
Glaab, E., 235
Gladden, J. M., 337–338, 344, 346, 348
Glaser, F., 145–146
Glass, E. M., 489, 494, 536–537
Glaubitz, S., 307–308
Glavas, S., 155–156, 158–159
Glenn, T. C., 159
Glöckner, F. O., 58, 452*t*, 459
Gloux, K., 137
Glover, L. A., 252–253
Gnerre, S., 154, 452*t*, 455–456, 529–530
Godiska, R., 127
Godzaridis, E., 452*t*, 454
Godzik, A., 283–285, 290–291, 314–316, 383*t*
Goesmann, A., 536–537
Goffredi, S. K., 23–24, 30–31, 33, 35–37, 39–40
Gogarten, J. P., 535–536
Gold, N. J., 376, 430
Goltsman, D., 272
Goltsman, D. S., 530–531
Gomez-Alvarez, V., 157–158, 490–491
Gonzalez, A., 372–373, 402, 422–424, 425, 427, 428*f*
González, J. M., 62
Good, I. J., 533–534
Goodlett, D. R., 272–273, 278*np*, 279*f*, 281, 283–285, 284*f*, 290–291, 314–316
Goodman, R. M., 127, 128–129, 136, 465–466
Goodrich, J. K., 372, 388–389
Goransson, P., 128–129
Gordon, J. I., 92, 357, 375, 380, 380*t*, 391, 396, 438, 510–511
Gordon, J. P., 64–65
Gorfer, M., 468
Goris, J., 528
Gorokhova, E., 194–195
Gorski, S., 170–171
Gosalbes, M. J., 261–262
Gotelli, N. J., 363
Goto, S., 472, 494, 503, 510
Gottschalk, G., 136
Gower, J. C., 402, 422–424
Gray, J., 540–542
Grayston, S. J., 252–253
Grechkin, Y., 54–55, 494
Green, J. L., 335, 337
Greene-Diniz, R., 361–362
Gregory, A., 159–160
Gregory, L., 345
Griffin, N. M., 294–295
Griffiths, A. D., 153
Griffiths, R. I., 255–256
Grills, G., 148–149
Grimsley, G. R., 281
Grindhaug, S. H., 468

Grossart, H. P., 197
Grossman, T. H., 128–129, 134, 135, 136
Grubert, F., 52
Grussenmeyer, T., 222
Gu, W., 457
Guan, C., 137
Guazzaroni, M. E., 314–316
Guerri, G., 252–253
Gumrich, P., 536–537, 538–539
Gunasekaran, P., 258–259
Gundersen, K., 48–49
Guo, Z. B., 259–260
Gupta, A., 360
Gupta, N., 290–291, 319
Gurevich, A. A., 452*t*, 454
Gussman, A., 519–521
Gygi, S. P., 294–295

H

Haange, S.-B., 252–253, 290–291
Haas, B. J., 170–171, 178–181, 199–201, 209, 335–336, 346, 361, 386–388
Haase, C., 192–194
Hachiya, T., 154, 530
Hadd, A., 111–112, 115–117
Hadwin, A. K. M., 252, 382–385
Haegeman, B., 363
Hagberg, P., 69–70
Haider, M. Z., 238, 252–253
Haider, P., 536
Haiser, H. J., 92, 93–94, 103, 238
Hajibabaei, M., 53–54, 252–253
Hales, B. A., 32
Halfvarson, J., 283–285, 290–291, 314–316
Halgamuge, S. K., 452*t*, 459
Hall, B. G., 32–33
Hall, D., 238
Hall, G., 32
Hall, J. R., 310, 356, 360, 361, 373, 382–385, 383*t*, 388–389, 533–534
Hall, N., 381
Hallam, S. J., 35–36, 128–129, 145–147, 153, 156–157, 159–160, 306–310, 311*t*, 530–531
Hallen, L., 211–212
Halpern, A. L., 5, 14–15, 62–64, 67–68, 83–84, 139, 145, 459–460
Ham, A. J., 297
Hamada, T., 111–112
Hamady, M., 92, 372, 375, 376, 386–388, 401–402, 411, 412, 430
Hamelin, J., 363
Hamer, U., 252
Hammes, F., 93–94
Hamp, T. J., 372
Han, C., 62
Han, Y., 449
Hanada, S., 262
Handelsman, J., 128–129, 135–136, 137, 382–385, 401–402, 465–466, 488, 526–527
Handley, K. M., 337–338, 343, 345, 346–347, 348
Handsaker, B., 467
Handy, S. M., 201
Hannah, M. A., 201, 238–239
Hansen, K. D., 211, 539–540
Hanstorp, D., 69–70
Hao, X., 335–336, 337
Harayama, S., 124–125
Harismendy, O., 345
Harmelin, A., 62
Harms, H., 290–291
Haroon, M. F., 4–5, 6, 8–9, 15–16, 22–23, 68–69
Harrington, E. D., 62, 64, 65–67
Harris, J. K., 376, 430
Harris, S. R., 208–209
Harrison, T., 490–491, 494
Hartmann, M., 310, 337, 356, 360, 361, 373, 382–385, 383*t*, 388–389, 533–534
Harvey, H. R., 281
Hasegawa, M., 133, 134
Haselkorn, P. T., 31
Hassink, J., 252–253
Hatfull, G., 153
Haugland, R. P., 95
Hayat, M. A., 6
Hayes, J. M., 15–16
Haynes, M., 148, 178–181, 207–208
Hayward, B., 307–308
Hazen, T. C., 337–338, 344, 346, 348
Hazuki, M., 457
He, S. M., 202–205, 259*t*, 260–261
He, Y., 208–209
He, Z., 252–253

Head, I. M., 381
Hearn, J., 192–194
Heath, L. S., 452*t*, 459
Heck, A. J. R., 291–293
Heidelberg, J. F., 62–64, 111–112, 145, 238
Heiner, C., 449
Heldal, M., 155
Helder, J., 252
Heller, P., 52, 53
Hellmann, J. J., 252
Hellweger, F. L., 144–145
Helynck, G., 124–125, 127, 138
Henn, M. R., 23–24, 49–50, 52, 153
Henne, A., 136, 137–138
Hennings, U., 197
Herbert, B. N., 95
Herbst, F. A., 314–316
Herlemann, D. P., 307–308
Hernandez, M., 95, 102
Heron, A. J., 540–542
Hespell, R. B., 334
Hess, M., 336, 459–460
Hesselsoe, M., 252–253
Hettich, R. L., 290–291, 294–295
Hewitt, K. M., 425
Hewson, I., 48–49, 50, 52–53, 54–55, 170–171, 188–191, 189*t*, 211, 238, 261–262
Hickenbotham, M. T., 529–530
Hidalgo, G., 372
Hill, M. O., 363
Hillenkamp, F., 290
Hiller, K., 195–196, 221*f*
Hillier, L. W., 533
Himmelbauer, H., 449, 529–530
Hinrichs, K.-U., 15–16, 23, 33, 36, 46
Hinrichs, S. H., 536
Hintze, A., 542
Hirano, K., 85–86
Ho, E. X. P., 335–336, 337–338, 344, 345, 348
Hochberg, Y., 539
Hodson, R. E., 195, 255–256
Hoeijmakers, W. A., 151*t*, 156–158
Hoffmann, S., 201–202
Höhna, S., 32–33
Holler, T., 29*t*, 33, 35–36
Hollibaugh, J. T., 189*t*, 191–192
Hollister, E. B., 310, 356, 360, 361, 373, 382–385, 383*t*, 388–389, 533–534
Holmes, S., 402–405, 415
Holmfeldt, K., 145, 148
Holstege, F. C., 238–239
Holt, R. A., 529–530
Holzgreve, W., 222
Homer, N., 467
Hongoh, Y., 46–47
Hooper, S. D., 534–535
Hopke, J., 127, 135, 138
Hopmans, E. C., 307–308
Horikoshi, K., 308
Horn, M., 4, 6–8, 310
Horner, J., 431–432
Hörth, P., 293
Horton, M., 536–537
House, C. H., 15–16, 23, 33, 36, 37–38, 46
Hove-Jensen, B., 124–125, 138
Howard, J. T., 449
Howe, A., 542
Howes, C. G., 306, 307–310, 459–460
Hradecna, Z., 129, 131–132, 134
Hsu, A. L., 452*t*, 459
Hu, N., 208–209
Hu, S., 15–16
Hu, X., 240–241
Hu, Y., 52
Huang, H. S., 354
Huang, K. H., 62, 144–145, 170–171, 178–181, 199–201, 425
Huang, W. E., 46–47, 154, 452*t*, 456
Huang, Y., 170–171, 207–208
Huber, H., 10
Huber, J. A., 334–335, 345, 490
Huber, R., 64–65
Huber, W., 539
Hubert, R., 47
Hugenholtz, P., 4–5, 6–9, 15–16, 22–23, 62–64, 68–69, 145, 147–148, 151*t*, 156–158, 159–160, 207–208, 334, 335–336, 337–338, 344, 346, 347–348, 361–362, 374, 375, 388–389, 390, 452*t*, 457, 459, 466, 526–527, 536
Hugenholz, F., 290–291
Hugenholz, P., 290–291
Hughes, E., 62
Hughes, J. B., 252
Hung, S. S., 203*t*

Hunt, D. E., 526–527
Hunt, R. H., 93, 95
Huntley, J., 98, 334–335, 375, 376–377
Hurst, G. B., 294–295
Hurt, R. A., 255–256
Hurtado-Gonzales, O., 290–291
Hurwitz, B. H., 145–148
Huse, S. M., 334–335, 345, 354, 373, 490
Huson, D. H., 124, 170–171, 208–209, 320–325, 336–337, 452*t*, 458–459, 466, 467, 468, 479, 505–506, 536–537
Hussong, R., 221*f*
Hutchins, D. A., 201
Hutchison, D., 5
Hutchison, S. K., 192–194
Huttenhower, C., 336–337, 341*f*, 354–355, 453–454
Hwang, M., 290–291
Hyatt, D., 290–291

I

Ibrahim, S. F., 47, 49–50
Ignacio-Espinoza, J. C., 144–145, 146–147, 153, 156–157, 159–160
Ikemura, T., 135, 137
Iliozer, H., 137
Imelfort, M., 15–16
Imyanitov, E. N., 32
Inácio, J., 10
Ingraham, J. L., 201–202
Innerebner, G., 290–291, 314–316
Ioannidis, P., 252–253
Isaacson, T., 111–112
Ishiguro, K., 95
Ishoey, T., 5, 14–15, 67–68, 83–84, 159–160
Itzkovitz, S., 62
Ivanov, A. R., 293
Ivanova, N. N., 5, 14–15, 46–47, 62, 67–68, 86, 336, 452*t*, 459, 466, 534–535
Iversen, N., 252–253
Iverson, V., 53–54, 154–155, 459–460, 531
Izard, J., 99, 170–171, 178–181, 199–201

J

Jackman, S. D., 154, 157, 529–530
Jackson, J. D., 494
Jackson, S., 201–202
Jacobs, A. R., 494
Jafari, M., 293
Jaffe, D. B., 529–530
Jaffe, J. D., 144–145, 290–291
Jahn, C. E., 233
Jain, S., 189*t*
Jaitly, N., 290–291
Janouskovec, J., 55–56
Jansson, J., 159–160, 296, 375
Jaroszewski, L., 383*t*
Jayakumar, A., 306
Jeannin, P., 128–129
Jeck, W. R., 529–530
Jenkins, B. D., 238–239
Jensen, L. J., 494, 503
Jensen, M. M., 197, 306, 308
Jensen, S., 196
Jeon, S., 307–308
Jewell, L. D., 31
Ji, Q. C., 282
Jia, Y., 144–145
Jiang, X. T., 354
Jimenez-Hernandez, N., 261–262
Jin Yu, H., 201–202
Johansson, M. E., 372
John, S. G., 146–148
Johnson, E. A., 46–47
Johnson, Z. I., 144–145
Johnston, A. W., 127, 133, 136
Jones, M. D., 307–308
Jones, M. J., 459
Jones, R., 252
Jones, S. J., 154, 157, 529–530
Jorgensen, B. B., 307–308
Jørgensen, S. L., 468
Jost, G., 191–192, 306
Jost, L., 365, 396–399
Jousson, O., 336–337, 453–454
Joye, S. B., 239*f*
Julien, P., 494, 503
Jung, K. H., 111–112
Jung, S., 283–285, 284*f*
Jürgens, K., 191–192, 306, 307–308
Justice, N. B., 290–291

K

Kainejais, L. H., 290–291
Kakirde, K. S., 127, 252–253
Kalnejais, L. H., 290–291
Kalyuzhnaya, M. G., 5, 8, 22–23

Kamagata, Y., 262
Kandler, O., 354
Kanehisa, M., 472, 494, 503, 510
Kang, H. L., 113–115
Kanno, M., 262
Karas, M., 290
Karl, D. M., 111–112, 238–239, 308, 311*t*
Karsch-Mizrachi, I., 58, 494
Kartal, B., 307–308
Kassen, R., 252
Katoh, K., 389
Katsura, S., 85–86
Kauffman, A. K. M., 146–148
Keegan, K. P., 252–253, 490–491, 493–494, 527–528
Keer, J. T., 155
Keiblinger, K. M., 290–291, 296, 314–316
Keller, J., 15–16
Keller, K., 334, 347–348, 361–362, 374, 375, 388–389, 390
Kellermann, M. Y., 33
Kelley, S. T., 401–402, 536–537
Kelly, L., 62, 144–145, 153
Kelso, J., 159
Kembel, S. W., 335, 337
Kemmeren, P., 238–239
Kemp, P. F., 201
Kendall, J., 46–47
Kennedy, R., 402–405
Kent, A. D., 252, 382–385
Kent, J. T., 402
Kent, W. J., 431, 488, 490, 493–494, 529, 536–537
Kerby, M. B., 448–449
Kerkhof, L. J., 307–308
Kern, S. E., 62, 146–148
Khodursky, A., 201–202
Kiene, R. P., 191–192
Kienle, S., 294–295
Kiesslich, K., 307–308
Kim, J., 252–253
Kim, K. H., 148, 207–208
Kim, M. S., 148
Kim, S., 201–202, 319
Kim, T.-W., 336, 459–460
Kim, U. J., 111–112, 113–115, 125
Kim, Y., 258–259
Kimel, S., 70
Kinzler, K. W., 62
Kircher, M., 159, 450–451
Kirchman, D. L., 136, 238
Kirschner, M. W., 294–295
Kiryutin, B., 494
Kiss, H., 62
Kitano, H., 220–222
Kitzman, J. O., 149–153, 157–158
Kjelleberg, S., 136, 272, 290–291
Klaassens, E. S., 290–291
Klammer, A. A., 449
Klappenbach, J. A., 528
Kleber, M., 529–530
Kleikemper, J., 201–202, 238
Kleinsteuber, S., 290–291
Klepac-Ceraj, V., 526–527
Klindworth, A., 310
Klockow, C., 272–273
Knief, C., 290–291, 314–316
Knietsch, A., 137–138
Knight, R. D., 92, 159–160, 252–253, 335–336, 341*f*, 356, 357, 361–362, 372–373, 374, 375, 376, 380, 380*t*, 381, 388–389, 391, 396–399, 401–405, 403*f*, 420, 425, 430, 438, 494–495, 502
Knights, D., 376, 379, 402, 420, 421, 422–424
Knittel, K., 29*t*, 33, 35–36, 40, 310, 334, 347–348, 468
Koenig, J. E., 372
Koenig, S. S., 401–402
Kohler, C., 222
Koike, I., 293–294
Kolbe, D. L., 389
Kolvek, S. J., 127, 135
Kong, Y. H., 367–368
Konstantinidis, K. T., 53–54, 146–147, 153, 156–157, 159–160, 208–209, 526–528, 529–532, 535–536, 537–538
Konwar, K. M., 306–308
Koonin, E. V., 111–112, 115–117, 472–473, 494
Kopecky, J., 258–259
Kora, G., 294–295
Koren, O., 361–362, 372, 420
Koren, S., 449, 529–530, 540–542
Kort, R., 29*t*, 33, 40
Kota, K., 533

Kottmann, R., 494–495
Kowalchuk, G. A., 335
Kozarewa, I., 149–153, 157–158, 450
Kratchmarova, I., 294–295
Krause, L., 536–537
Krieg, N., 526
Krisch, H. M., 144–145
Krishnan, P., 70
Kristensen, D. B., 294–295
Kristoffersen, S. M., 192–194
Krobitsch, S., 211–212
Krüger, S., 191–192, 197
Krumholz, L. R., 4–5, 335–336
Krupke, A., 47, 52–53, 55–56, 58
Krupovic, M., 145–146, 148–149
Kubal, M., 536–537
Kuczynski, J., 99, 356, 372–373, 374, 376, 377, 379, 390, 396–399, 401–402, 407, 420, 421, 422–424, 429–430, 438–439, 467, 493–494, 495, 510, 526–527
Kudo, T., 46–47
Kuhn, K., 294–295
Kuhn, M., 472–473, 494, 503
Kuhn, R., 314–316
Kukkillaya, V. U., 335–336, 337–338, 344, 345, 348
Kukutla, P., 206–207
Kulam-Syed-Mohideen, A. S., 334
Kulasekara, H. D., 283–285
Kuligina, E. Sh., 32
Kulikov, A. S., 452*t*, 454
Kuma, K., 389
Kumar, A. R., 12
Kunin, V., 457, 526
Kuo, W. P., 293
Kurhela, E., 194–195
Kuster, B., 294–295
Kustu, S., 201–202
Kuypers, M. M., 307–308
Kvist, T., 46–47, 62
Kyrpides, N. C., 53–54, 208–209, 452*t*, 459, 494, 527–528, 529–530, 531–532, 534–535

L

La Scola, B., 64
Labrenz, M., 191–192, 306, 307–308
Lacerda, C. M. R., 290–291
Laczny, C., 221*f*
Ladau, J., 337
Lader, E., 195
Laiko, V. V., 291–293
Laird, N. M., 338–339
Laitinen, K., 372
Lakay, F. M., 255–256
Lam, P., 197, 306, 307–308
Lamour, K. H., 290–291
Lamy, D., 62, 75
Lander, E. S., 529–530, 533
Landfear, D., 95, 102
Landry, Z. C., 145, 146–148, 156–157
Lane, D. J., 32, 56–57, 62–64, 334, 335–336
Lane, P. D., 201–202
Lang, A. S., 148–149
Langan, S. J., 422–424
Langevin, S. A., 201–202
Langmead, B., 343–344, 347–348, 492, 529, 536–537
Lanyon, C., 175–177
Lanza, F., 252–253
Lanzén, A., 124, 170–171, 208–209, 337, 381, 430, 468
Lapidus, A., 457, 526
Lappin-Scott, H. M., 95
Larkin, M. A., 389, 390
Larner, S., 222
Larsen, N., 334, 347–348, 361–362, 373, 374, 375, 388–389, 390
Lasken, R. S., 5, 14–15, 46–47, 62, 67–68, 83–84, 159–160, 447–448
Lauber, C. L., 92, 98, 220–222, 252–253, 334–336, 345, 356, 372–373, 375, 376–377, 379, 402, 403*f*, 425, 427, 428*f*, 429–430
Lauro, F. M., 272–273, 290–291, 314–316
Lavelle, P., 252
Lavik, G., 197, 306, 307–308
Laviolette, F., 452*t*, 454
Lawrence, J. G., 354, 527–528
Lay, C., 335–336, 337–338, 344, 345, 348
Le Paslier, D., 401–402
Leadbetter, J. R., 159–160
Lebrun, L., 221*f*
LeCleir, G., 201–202, 238
Leclerc, M., 137

Lee, J. A., 14–15, 23–24, 31, 86–87, 144–145
Lee, J. E., 201–202
Lee, N., 29*t*, 33
Lee, S., 201, 211
Lee, Z. M.-P., 335
Lefever, S., 233
Lefsrud, M. G., 272
Lehembre, F., 252–253, 262
Lehner, A., 29*t*, 33
Leiber, M., 174–175, 257
Leisch, F., 431–432
Leisse, A., 201, 238–239
Lemaire, M., 254–255
Lemeer, S., 294–295
Lemke, A., 33
Lemmer, H., 32, 35
Lengelle, J., 252–253
Leo, L., 201–202, 254–255
Lepère, C., 48–49
Lerat, E., 460
Leslin, C., 307–308
Lesniewski, R. A., 189*t*
Letelier, R. M., 306, 307–308
Letunic, I., 320–325
Leung, H. C., 154, 530
Levin, J. Z., 170–171, 178–181, 199–201
Ley, R. E., 92, 372, 375, 386–388, 396, 411, 412, 420, 510–511
L'Haridon, R., 137
Li, D. F., 354
Li, F., 282
Li, G. W., 192–194
Li, H., 467, 529, 536–537
Li, R., 92, 484, 529–530
Li, S., 208–209
Li, W., 170–171, 207–208, 383*t*
Li, Y., 127, 133, 136, 208–209, 294–295
Li, Z., 154, 294–295, 452*t*, 456
Liang, H., 70
Liang, Y., 252–253
Liaw, A., 420
Lidstrom, M. E., 5, 8, 22–23
Liebler, D. C., 297
Liesack, W., 258–259
Liles, M. R., 128–129, 134, 135–136, 252–253
Lim, Y. W., 178–181, 207–208
Lin, A., 184
Lin, C. W., 363
Lin, S., 201–202, 203*t*
Lin, X. J., 307–308
Lindeberg, M., 148–149
Lindell, D., 144–145, 192–194
Lindquist, E. A., 189*t*
Link, A. J., 291
Linnarsson, S., 149–153
Liou, G. F., 69–70
Lipman, D. J., 58, 310, 383*t*, 388–389, 466, 467, 494, 536–537
Lipton, M. S., 272, 290–291
Liu, B., 154, 452*t*, 456, 540–542
Liu, C., 194–195
Liu, J., 336–337
Liu, L., 208–209, 360
Liu, M., 272, 290–291
Liu, Z., 335–336, 376, 388–389, 396–399, 401–402
Livny, J., 209
Lladser, M. E., 402
Llewellyn, M., 252
Lloyd, D., 95
Lo, I., 272
Lochte, K., 33
Loerch, P., 201–202
Lohse, M., 452*t*, 453
Loiacono, K. A., 134, 136
Lolo, M., 226–227
Lombardo, M.-J., 64
Lombardot, T., 452*t*, 459
Lomonosov, A. M., 153
Long, E., 272–273, 290–291, 314–316
Long, F., 294–295
Lopez-Cortes, N., 314–316
López-García, P., 56–57
Lores, I., 314–316
Lösekann, T., 5, 14–15, 29*t*, 33, 40, 46–47, 62, 67–68, 86
Lottaz, C., 529–530
Loua, T., 413
Low, L., 335–336, 337–338, 344, 345, 348
Loy, A., 6–8, 29*t*, 33, 111–112
Lozupone, C. A., 98, 220–222, 252, 334–335, 345, 356, 374, 375, 376, 386–388, 389–390, 396–399, 401–402, 411, 412

Lu, S., 62
Lucas, S. B., 31
Ludwig, T., 390
Ludwig, W. G., 32–33, 35–36, 252–253, 310, 334, 347–348, 354, 373, 468
Luis, P., 252–253, 262
Lundeberg, J., 157
Lundin, S., 157
Lunn, J. E., 452*t*, 453
Luo, C., 53–54, 208–209, 527–528, 529–530, 531–532, 536–537
Luo, H., 188–191, 238–239
Luo, R., 154, 452*t*, 456
Luo, Z. W., 240–241
Luong, K., 540–542
Luton, P. E., 35–36
Lyle, J., 540–542

M

Maaløe, O., 201–202
Maccallum, I., 154, 452*t*, 455–456, 529–530
MacDonald, N. J., 336–337
Maciukenas, M. A., 373
Mackay, A., 252
Mackelprang, R., 467–468, 471*f*, 475*f*, 478*f*, 483*f*
MacNeil, I. A., 127, 134, 135, 136, 138
Madamwar, D., 258–259
Madden, T. L., 529
Maglia, G., 540–542
Maglott, D. R., 500
Magrini, V., 510–511, 529–530
Magris, M., 372
Mahaffy, J. M., 145, 151*t*
Mahowald, M. A., 510–511
Maixner, F., 6–8
Malek, J., 5, 46–47, 62, 124, 147–148, 156–157
Malfatti, S. A., 55–56, 459
Malmstrom, R. R., 23–24, 49–50, 52, 62, 124–125, 154, 210, 357
Mammitzsch, K., 306
Manary, M. J., 92, 372
Manchester, K. L., 256–257
Mancino, V., 111–115, 125, 132
Mande, S. S., 452*t*, 457, 459, 535–536
Maniatis, T., 132
Manichanh, C., 92, 137, 484
Maniloff, J., 334
Mann, M., 290, 297, 319
Mannino, F. L., 425
Manza, L. L., 297
Marcelino, L. A., 32
Marcy, Y., 5, 14–15, 46–47, 62, 67–68, 83–84, 86
Mardia, K. V., 402
Mardis, E. R., 510–511, 529–530
Maresca, J. A., 125
Margulies, E. H., 154
Margulies, M., 334–335, 449, 529–530
Marin, R. I., 124
Marin, R. 3rd., 195, 196
Marine, R., 153
Markewitz, D., 465–466
Markiel, A., 411
Markle, J., 203*t*
Markowitz, V. M., 54–55, 494
Marks, P., 449
Marra, M. A., 533
Marrs, E., 175–177
Marsh, T. L., 115–117, 373
Marshall, I. P., 189*t*
Martelin, R., 382
Martens-Habbena, W., 93
Marti, V., 431–432
Martin, H. G., 5, 14–15, 46–47, 62, 67–68, 86, 452*t*, 459, 466, 536
Martin, J., 239*f*, 449, 538–540
Martinez, A., 124–125, 127, 129, 132, 133, 134–136, 138, 139
Martinez-Garcia, M., 62, 75
Martini, A. M., 31
Martin-Platero, A. M., 360
Martinson, V. G., 458–459
Martiny, A. C., 5, 46–47, 62, 124, 147–148, 156–157, 459–460
Masella, A. P., 357
Maslau, S., 345
Massart, S., 372
Matalka, M., 519–521
Materna, A. C., 154, 357
Matsuzawa, Y., 85–86
Mattes, T. E., 278*np*, 278
Mauceli, E., 529–530
Mauck, C. K., 401–402
Maurice, C. F., 92, 93–94, 95, 103, 238

Mavromatis, K., 5, 14–15, 336, 447–448, 457, 466, 526, 534–535
May, P., 235
Mayer, T., 64–65
McCaig, A. E., 252–253
McCarren, J., 124–125, 189*t*, 210
McCaughey, M. J., 373
McCleland, M. L., 291–293
McCord, B. R., 255–256
McCrow, J. P., 48–49
McDonald, D., 376, 388–389, 390, 396–399, 401–402, 403*f*, 495
McDonald, I. R., 138
McDonald, W. H., 294–295, 297
McDonel, P. E., 354
McElroy, K., 335–336, 337
McEwen, G. K., 154
McFeters, G. A., 95
McGarrell, D. M., 334
McGettigan, P. A., 389, 390
McGinnis, D. F., 197
McGrath, K. C., 201–202, 203*t*, 254–255, 258, 259*t*, 261
McGuire, A. L., 425
McHardy, A. C., 452*t*, 457, 459, 466, 536
McIndoo, J., 46–47
McKeegan, K. D., 23, 33, 46
McKendree, W., 360
McLean, J. S., 159–160
McLean, P. G., 372
McMurdie, P. J., 415, 432
McRose, D., 55–56
McSorley, F. R., 124–125, 138
McWilliam, H., 389, 390
Mead, D. A., 127, 145, 151*t*
Medlin, L., 56–57
Meeske, C., 307–308
Meier, H., 29*t*, 32–33, 35, 373, 390
Meinicke, P., 536–537, 538–539
Meinnel, T., 145–146
Melefors, O., 155–156, 158–159
Melin, J., 22–23
Mende, D. R., 401–402
Mendez, C. B., 146–148
Meng, C. K., 290
Merchant, M., 291–293
Merrimen, B., 449
Merten, C. A., 153
Messié, M., 48–49
Mettel, C., 258–259, 259*t*, 260
Metzker, M. L., 157–158, 159, 252–253, 449
Meyer, F., 184, 334–335, 375, 488, 489, 519–520, 536–537
Meyer, M., 450–451
Meyerdierks, A., 29*t*, 33, 35–36
Miao, V., 127
Middelboe, M., 148
Miguez, C. B., 138
Mikhailova, E., 540–542
Milanova, D., 448–449
Miller, C. A., 293
Miller, C. S., 62–64, 337–338, 340, 343–344, 345, 346–348, 526, 531
Miller, J. R., 529–530
Miller, S. I., 283–285
Miller, W., 310, 383*t*, 388–389, 466, 467, 529, 536–537
Miltion, J., 449
Milton, J., 334–335
Milucka, J., 15–16
Mincer, T. J., 128–129, 308, 311*t*
Minoche, A. E., 449
Minor, C., 134, 136
Mironov, A., 253–254
Miropolsky, L., 99
Misawa, K., 389
Mitchison, T. J., 297
Mitra, S., 336–337, 452*t*, 458–459, 467, 479
Mitreva, M., 533
Miura, T., 262
Miyamoto, D. T., 297
Miyata, T., 389
Miyauchi, R., 5
Miyazaki, K., 135, 136
Mize, G. J., 291
Mizrahi-Man, O., 361–362
Mizuno, A., 85–86
Mkakosya, R., 372
Mohammed, M. H., 452*t*, 457, 459, 535–536
Mohammed, S., 291–293
Mohn, W. W., 307–310, 311*t*
Mohrholz, V., 306, 307–308
Moll, S., 148–149
Monier, A., 48–49

Montoya, J. P., 189*t*
Moore, E. K., 281
Moore, J. A., 136
Morales, R. L., 53–54, 154–155, 459–460, 531
Moran, M. A., 170–171, 188–194, 189*t*, 201, 212, 238–240, 239*f*, 242, 255–256, 261–262
Moran, N. A., 447, 448, 450, 455*f*, 458–459, 460
Moraru, C., 159–160
Moreira, D., 56–57
Morgan, J. L., 148–149, 336
Moriarty, J., 363
Morin, E., 252–253
Morley, A. A., 62
Morowitz, M. J., 336, 530–531
Morris, D. R., 291
Morris, R. M., 53–54, 154–155, 272–273, 278, 278*np*, 279*f*, 283–285, 290–291, 314–316, 459–460, 531
Morrison, H. G., 149–153, 157–158, 334–335, 345, 354, 490
Morrison, M., 536
Morse, S., 222
Mosier, A. C., 46–47, 62, 64–65, 67–68, 81–82, 86
Moskovets, E. V., 293
Mosley, A. L., 320–325
Mou, X., 189*t*
Mouser, P. J., 294, 348
Moustafa, A., 52
Moya, J., 460
Mozer, M. C., 379, 421
Muegge, B. D., 422–424
Mueller, J. A., 148–149
Mueller, O., 174–175, 257
Mueller, R. S., 272, 290–291
Müller, B., 197
Muller, E. E. L., 195–196, 220–222, 221*f*, 235
Muller, J., 472–473, 494, 503
Muller, R., 126
Muller, S., 95
Munk, A. C., 457
Murrell, J. C., 138
Murugapiran, S. K., 65–67, 85–86
Musat, N., 47, 52–53, 55–56, 58
Mussmann, M., 111–112
Mutter, G. L., 194–195
Myers, E. W., 310, 383*t*, 388–389, 466, 467, 529–530, 536–537

N

Nadalig, T., 33
Nagaraj, N., 297
Nagarajan, N., 539–540
Nagata, T., 293–294
Nagel, A., 452*t*, 453
Naik, H., 308
Najar, F. Z., 335–336
Nakorchevsky, A., 291
Nalbantoglu, I., 372
Nalbantoglu, O. U., 536
Nam, D. K., 211
Nam, Y. D., 148
Namiki, T., 154, 530
Nancharaiah, Y. V., 12
Nannipieri, P., 252–253
Naqvi, S. W. A., 306, 307–308
Narasimhan, V., 336–337, 453–454
Narayanasamy, S., 221*f*
Natale, D. A., 472–473, 503
Nattkemper, T. W., 536–537
Nauhaus, K., 36, 37–39
Navidi, W., 47
Navin, N., 46–47
Nawrocki, E. P., 388–389
Nazar, R. N., 255–256
Neal, P. R., 334–335, 345, 363
Nebel, M., 307–308
Neef, A., 32, 35
Neidhardt, F. C., 201–202
Nelson, A., 175–177
Nelson, J. R., 47, 62
Nelson, S. F., 47
Nelson, W. C., 111–112
Nemergut, D. R., 376
Neoh, S. H., 62
Neuberg, D., 194–195
Neufeld, J. D., 357
Neuman, K. C., 69–70, 71–74
Newbold, C., 452*t*, 456–457
Newington, J. E., 252
Ng, C., 272
Nguyen, L. P., 111–112, 115–117

Niazi, F., 47, 48–49, 50, 51, 52–55, 192–194
Nicinski, J., 95
Nicora, C. D., 272, 290–291
Niedringhaus, T. P., 448–449
Niehaus, K., 536
Nielsen, P. H., 367–368
Niemann, H., 33
Ning, Z., 157–158
Noda, S., 46–47
Norbeck, A. D., 35–36, 272
Nordenskjöd, M., 47
Nordstrom, H., 155–156, 158–159
Normand, P., 124–125, 127, 138
Novo, D. J., 93, 95
Novotna, J., 258–259
Novotny, M. A., 5, 14–15, 159–160
Nudler, E., 253–254
Nulton, J., 252
Nunn, B. L., 272–273, 278*np*, 279*f*, 281, 283–285, 290–291, 314–316
Nurk, S., 452*t*, 454
Nusbaum, C., 209
Nüsslein, K., 31
Nutter, G., 449
Nuytens, J., 233
Nystrom, T., 69–70

O

Oberbach, A., 252–253, 290–291
Ochman, H., 211–212
Odic, D., 148
O'Donnell, A. G., 255–256
O'Dwyer, J. P., 337
O'Farrell, P. H., 290
O'Flaherty, V., 290–291
Ogata, H., 64
Ogawa, H., 293–294
Oger-Desfeux, C., 252–253, 262
Oh, P., 294–295
Oh, S., 530
Ohi, R., 297
Ohnishi, K., 124–125
Ohnuki, T., 135, 136
Oittinen, T., 194–195
Oki, K., 5
Okumura, Y., 133, 134
Okuta, A., 124–125
Olsen, G. J., 62–64, 334, 335–336, 373
Olson, R. J., 4–5, 536–537
O'Mullan, G. D., 306
Ondov, B. D., 503, 510, 540–542
Ong, S. H., 335–336, 337–338, 344, 345, 348
Ong, S.-E., 294–295
Oono, Y., 415
Opel, K. L., 255–256
Orcutt, K. M., 48–49
O'Reilly, J., 290–291
Orphan, V. J., 15–16, 23–24, 29, 30–31, 33, 35–38, 39–40, 46
Orsi, W., 307–308
Orsini, M., 255–256
Osburne, M. S., 135, 138, 153
Ostell, J., 58, 494
O'Sullivan, O., 361–362
Ott, J., 111–112
Ottesen, E. A., 124, 159–160, 189*t*, 193*f*, 195, 196, 238, 259–260
Otto, G., 540–542
Otto, T. D., 208–209, 452*t*, 456–457
Ouverney, C., 5, 14–15, 46–47, 62, 67–68, 86
Overbeek, R., 373, 472, 494, 500, 503
Owen, J. G., 125
Ozier, O., 411

P

Paarmann, D., 536–537
Pace, B., 281, 334, 335–336
Pace, N. R., 4–5, 62–64, 334, 335–336, 354, 375
Pachter, L., 211
Page, A. P., 35–36
Palaniappan, K., 54–55, 494
Palejev, D., 52
Palese, L. L., 314–316
Palumbo, A. V., 255–256, 354
Pamp, S. J., 62, 64, 65–67
Pan, C., 272, 290–291, 294–295
Pan, X., 52
Panchaud, A., 283–285, 284*f*
Pandey, A., 294–295
Pandey, V., 449
Pang, X., 259–260
Papa, E., 367
Parachin, N. S., 125

Paradis, E., 402–405
Pare, M. C., 252
Parfrey, L. W., 372–373, 377, 379, 429–430
Pargett, D., 196
Parker, S. S., 252
Parkhomchuk, D., 211–212
Parkinson, N. J., 345
Parks, D. H., 336–337, 538–539
Parsley, L. C., 252–253
Partensky, F., 111–112
Partha, R., 111–112
Passalacqua, K. D., 192–194
Paszkiewicz, K., 54–55
Pati, A., 452*t*, 459, 534–535
Patil, K. R., 536
Patterson, M., 145, 252
Paul, J. H., 144
Paull, C. K., 36
Paulmier, A., 306
Peano, C., 207–208
Pedreira, S., 226–227
Pei, D. C., 259–260
Pelaez, A. I., 314–316
Pell, J., 542
Pelletier, D. A., 56–57
Pelletier, E., 401–402
Pelosi, C., 252
Peltier, S., 184
Peng, Y., 154, 530
Penn, J., 127
Peplies, J., 58, 310, 334, 347–348, 468
Peres, G., 252
Peretó, J., 460
Perez-Cobas, A. E., 261–262
Perlmutter, N. G., 93, 95
Pernthaler, A., 4–5, 23–24, 30–31, 33, 35–37, 39–40
Pernthaler, J., 4–5, 23
Perrotta, A. R., 360
Perry, J., 175–177
Persoh, D., 258–259
Peter, B., 201–202
Peterson, D., 452*t*, 453
Peterson, D. A., 490
Petiti, L., 207–208
Petrosino, J. F., 425
Petsch, S. T., 31
Pevzner, P. A., 319, 452*t*, 456, 529–530
Pfaffl, M. W., 233, 256–257
Pfannkuche, O., 33
Pham, V. H. T., 252–253
Pharino, C., 526–527
Philippe, H., 56–57
Phillippy, A. M., 456, 503, 510
Phillips, R., 159–160
Phipps, D., 95
Pickering, M., 201–202, 238
Pickup, R. W., 32
Pieterse, C. M. J., 252
Pietramellara, G., 252–253
Pietrelli, A., 207–208
Pignatelli, M., 261–262
Pijl, A. S., 335
Pinard, R., 52
Pinel, N., 221*f*
Pirrung, M., 402–405
Planas, D., 48–49
Plant, R., 52
Plhackova, K., 258–259
Podar, M., 5
Pohl, C., 197
Pointing, S. B., 252
Poisson, G., 148–149
Pol, A., 16
Polerecky, L., 15–16
Pollice, A., 314–316
Polson, S. W., 153
Polz, M. F., 32, 335–336
Ponder, B. A. J., 47
Pong, R., 208–209
Pop, M., 343–344, 347–348, 456, 492, 539–540
Pope, P. B., 459, 536
Poretsky, R. S., 170–171, 188–191, 189*t*, 201–202, 211, 238, 261–262, 530
Poroyko, V., 530–531
Porte, N., 431–432
Portune, K. J., 201
Potanina, A., 46–47, 62, 64–65, 67–68, 81–82, 86
Potts, J. M., 422–424
Poulain, J., 48–49
Poulet, A., 145–146, 148–149
Poullet, J. B., 372

Poulos, B. T., 146–148, 151*t*, 153, 156–157, 159–160
Poulton, N. J., 48–49
Powell, S., 472–473
Power, M. E., 294–295
Prairie, Y. T., 48–49
Prakash, T., 46–47
Pratt, J. M., 294–295
Preckel, T., 293
Prediger, E., 449
Predki, P., 531–532
Preheim, S. P., 360, 367
Prestat, E., 252–253, 527–528
Preston, C. M., 35–36, 62, 75, 111–112, 124, 128–129, 195, 196, 308, 311*t*
Price, D. C., 22–23, 52
Price, M. N., 360–361, 388–389
Primo, V., 293
Prior, B. A., 255–256
Probst, A., 388–389
Proshkin, S., 253–254, 261
Prosser, J. I., 252–253
Pruesse, E., 32–33, 58, 310, 334, 347–348, 388–389, 431, 468
Prufert-Bebout, L., 189*t*
Pruitt, K. D., 500
Przybylski, D., 154, 452*t*, 455–456
Puhler, A., 157–158
Pukall, R., 336, 466
Pulleman, M., 252
Purdom, E., 539–540
Purvine, S. O., 290–291
Putnam, I. F., 307–308
Putnam, N., 530–531

Q

Qi, G., 159
Qi, J., 124, 208–209, 320–325, 336–337, 467, 484, 505–506, 536–537
Qian, W., 529–530
Qian, Y., 372
Qin, J., 92, 484
Qiu, X. Y., 255–256, 354
Qu, J., 449
Quail, M. A., 149–153, 157–158, 208–209
Quake, S. R., 22–23, 46–47, 62, 64–68, 81–82, 86
Quast, C., 32–33, 58, 310, 334, 347–348, 388–389, 431, 468
Quince, C., 335, 361, 381, 388, 430
Quinquis, B., 500

R

Rabkin, B. A., 4–5, 6, 8–9, 22–23, 68–69
Radajewski, S. M., 138
Radax, R., 337
Radonjic, M., 238–239
Radpour, R., 222
Raes, J., 92, 401–402, 484
Raferty, M. J., 290–291
Raftery, M. J., 272–273, 314–316
Raghavan, R., 211–212
Raghoebarsing, A. A., 16
Rahm, B. G., 272
Rahmouni, A. R., 253–254
Rainey, F. A., 335
Rainey, P. B., 252
Rajah, V. D., 22–23, 52
Rajaram, S., 415
Rajendhran, J., 258–259
Rajput, J. K., 12
Ram, R. J., 62–64, 145, 272, 293, 314–316, 336, 526–527
Ramage, D., 411
Ramazzotti, M., 372
Rambold, G., 258–259
Ramette, A., 526–527
Ramey, R. R., 386–388, 411, 412
Ramirez, K. S., 252–253
Ranger, J., 252–253, 262
Rangwala, H., 531–532
Raoult, D., 382
Rapp, E., 314–316
Rapp, H. T., 337
Rappé, M. S., 62–64
Rassoulzadegan, F., 144
Rattei, T., 337
Ravel, J., 153, 159, 401–402
Ravenschlag, K., 15–16, 23, 29*t*, 33, 36, 40, 64
Raymond, F., 452*t*, 454
Read, T., 53–54, 208–209, 527–528, 531–532
Reardon, K. F., 290–291
Rector, T., 192–194

Redestig, H., 201, 238–239
Reeder, J., 381, 430, 502
Régnier, P., 205, 253–254
Reich, D., 155–156, 157
Reichenberger, E. R., 467, 536
Reichl, U., 314–316
Reid, R. W., 372
Reignier, J., 201–202
Reinhardt, J. A., 529–530
Relman, D. A., 62, 64, 65–67, 336, 372, 530–531
Remington, K., 62–64, 139, 145
Renaut, J., 195–196, 221*f*
Renesto, P., 64
Repeta, D. J., 124–125, 210
Reppas, N. B., 5, 46–47, 62, 124, 147–148, 156–157
Revsbech, N. P., 198–199
Rey, F. E., 92
Reyes-Prieto, A., 52
Rho, M., 488, 492, 493–494
Ribeiro, F. J., 154, 452*t*, 455–456
Rich, J. J., 308
Rich, V., 128–129
Richards, S. R., 449
Richardson, D. C., 402
Richardson, D. J., 127, 133, 136
Richardson, P. M., 35–36, 62–64, 145, 336, 526–527
Richardson, R., 278, 279*f*
Richter, D. Jr., 465–466
Richter, L., 32–33, 373
Richter, R. A., 64
Rick, J., 294–295
Rickert, D., 15–16, 23, 29*t*, 33, 36, 40, 64
Ricketts, T. H., 252
Rideout, J. R., 390, 495
Ridgway, H. F., 95
Riemann, L., 145, 148
Riesenfeld, C. S., 136, 488
Riesenfeld, S. J., 337
Rigoutsos, I., 452*t*, 459, 466, 536
Rijpstra, W. I. C., 16
Riley, D. R., 519–521
Riley, P. W., 35–36
Rineau, F., 252–253
Rinke, C., 14–15, 23–24, 31, 86–87
Rinn, J. L., 211
Rinta-Kanto, J. M., 189*t*, 191–192, 201, 238
Rissanen, A. J., 194–195
Ritchie, D. A., 32
Riva, A., 252, 382–385
Rivers, A., 188–191, 238–240, 239*f*
Robb, E. J., 255–256
Robe, P., 252–253, 527–528
Robert, M., 306
Roberts, A., 211
Robertson, B. R., 38–39
Robertson, C. E., 203*t*
Robins, K. J., 125
Rocap, G., 272–273, 278*np*, 279*f*, 283–285, 290–291, 314–316
Rocha, M. S., 69–70
Rodbumrer, J., 128–129
Roderick, M. L., 252
Rodgers, L., 46–47
Rodrigue, S., 23–24, 49–50, 52, 62, 154, 357
Rodriguez, G. G., 95
Rodriguez-Mora, M. J., 307–308
Roe, B. A., 335–336
Roesch, L. F., 252, 382–385
Roh, S. W., 148
Roh, Y., 255–256
Rohland, N., 155–156, 157
Rohwer, F., 144–145, 148, 536–537
Rojas, M., 334, 347–348, 361–362, 374, 375, 388–389, 390
Roman, B., 196
Romano-Spica, V., 255–256
Rondon, M. R., 128–129, 134, 135–136, 465–466
Ronen, R., 452*t*, 456
Ronquist, F., 32–33
Rosario, K., 145–146
Roschitzki, B., 290–291, 296, 314–316
Rosen, G. L., 467, 536
Rosenberg, R., 306
Rosenfeld, A. M., 467, 536
Rosenow, C., 192–194
Rosenquist, M., 283–285, 290–291, 314–316
Roslev, P., 252–253
Ross, C. A., 65–67, 85–86
Ross, M. G., 450
Ross, M. M., 291–293

Ross, R. P., 361–362
Rossello-Mora, R., 527–528
Rossi, E., 207–208
Rossnagel, P., 64–65
Roth, A., 472–473
Rothberg, J. M., 449
Roume, H., 195–196, 220–222, 221*f*, 226, 227, 230–231, 232, 233, 234*f*
Roux, S., 145–146, 148–149, 335
Ruan, J., 467, 529–530, 536–537
Rubin, D. B., 338–339
Ruckert, C., 157–158
Rühland, C., 10
Ruiz, A., 252–253
Ruiz-Pino, D., 306
Rusch, D. B., 62–64, 139, 145, 459–460
Ruscheweyh, H.-J., 336–337, 452*t*, 458–459, 467
Ruse, C. I., 291
Rush, J., 294–295
Russ, C., 450
Russell, A. L., 283–285, 290–291, 314–316
Russell, D. W., 31, 153
Ryabin, T., 310, 356, 360, 361, 373, 382–385, 383*t*, 388–389, 533–534
Ryan, J. P., 124, 189*t*, 195, 196, 238
Ryu, S., 272

S

Sabehi, G., 111–112
Sacchi, N., 222
Sagova-Mareckova, M., 258–259
Sahm, K., 4, 5
Sakakibara, Y., 154, 530
Sakamoto, J., 401–402
Salamon, P., 145, 151*t*, 252
Sale, K., 159
Salka, I., 197
Salowsky, R., 174–175, 257
Salzberg, S. L., 336–337, 343–344, 347–348, 452*t*, 453–454, 459, 492, 529, 536–537, 538
Sama, A., 195
Samatova, N. F., 294–295
Sambrook, J., 31, 132
Samuelsson, A., 372
Sanchez, J. J., 31
Sanders, J. G., 189*t*, 191–192
Sanders, M. J., 157–158
Sardiu, M. E., 320–325
Sarkar, N., 205
Sarkis, G., 52
Sass, H., 93
Satinsky, B., 188–191, 238–239, 239*f*
Sato, Y., 494, 503, 510
Saunders, J. R., 32
Savage, D. C., 372
Savidor, A., 290–291
Savitski, M. M., 294–295
Sawyer, S., 450–451
Saxena, R. M., 192–194
Sayood, K., 536
Scalfone, N., 372
Scally, A., 149–153, 157
Scannell, J. W., 252
Schackwitz, W., 5, 14–15, 447–448
Schaeffer, S. M., 252–253
Schäfer, J., 294–295
Schaffer, A. A., 529
Schattenhofer, M., 39–40
Schatz, M. C., 449, 456
Scheffer, T., 536
Schein, J. E., 154, 157, 529–530
Schenck, R. O., 159–160
Schenk, P. M., 212, 252–253
Scherl, A., 283–285
Schieltz, D. M., 291
Schimel, J. P., 252–253
Schindelin, J. E., 365
Schirle, M., 294–295
Schjenken, J. E., 222
Schlapbach, R., 290–291, 314–316
Schleifer, K. H., 11*f*, 32, 35–36, 252–253, 354
Schleper, C., 124, 170–171, 208–209, 337, 530–531
Schlichting, N., 290–291
Schloss, P. D., 137, 189*t*, 310, 335–336, 346, 356, 360, 361, 373, 382–385, 383*t*, 388–389, 401–402, 488, 526–527, 533–534
Schloter, M., 252–253
Schmerberg, C. M., 282
Schmid, E., 290–291, 296, 314–316
Schmid, M., 15–16, 306, 308
Schmidt, G., 294–295

Schmidt, S., 201–202, 254–255
Schmidt, T. M., 157–158, 335, 490–491
Schmieder, R., 178–181, 207–208
Schmitt, K., 47
Schmitt, S., 194–195
Schmitz, R. A., 136
Schnapp, L. M., 272
Schneider, T., 290–291, 296, 314–316
Schoenfeld, T., 145
Scholin, C. A., 124, 189*t*, 195, 196, 238
Schreiber, F., 536–537, 538–539
Schreiber, L., 29*t*, 33, 35–36
Schroeder, A., 174–175, 257
Schroth, G., 336, 459–460
Schubert, C. J., 15–16, 23, 29*t*, 33, 36, 40, 64, 197
Schulz, V., 52
Schuster, S. C., 124, 170–171, 193*f*, 208–209, 261–262, 320–325, 336–337, 452*t*, 458–459, 467, 505–506, 536–537
Schwacke, R., 263
Schwalbach, M. S., 145, 146–148, 156–157
Schwarz, J., 294–295
Schweer, T., 32–33, 58, 310, 388–389, 431
Schwientek, P., 157–158
Scopes, R. K., 281
Scott, J. R., 192–194
Scranton, M. I., 307–308
Scribbins, D., 175–177
Sczyrba, A., 14–15, 23–24, 31, 62, 75, 86–87, 336, 459–460
Segall, A. M., 145, 151*t*
Segata, N., 99, 336–337, 453–454
Selinger, D. W., 192–194
Shabanowitz, J., 291–293
Shaffer, S. A., 283–285, 284*f*
Shah, M. B., 145, 272–273, 283–285, 290–291, 296, 314–316
Shah, V., 290–291
Shan, J., 144–145
Shang, J., 54–55
Shannon, P., 411
Shapiro, H. H., 92
Shapiro, H. M., 93, 95
Sharma, A. K., 195–196, 263
Sharma, C. M., 201–202, 203*t*
Sharma, S., 188–192, 189*t*, 201, 238, 239*f*
Sharma, V. K., 46–47
Sharon, I., 62–64, 111–112, 145–146, 336, 337–338, 343–344, 346, 348, 531
Sharp, R. J., 35–36
Sharpton, T. J., 337
Shaw, C. A., 157–158
Shaw, G., 222
Sheffield, J., 252
Sheflin, L. G., 211
Sheik, C. S., 191–192, 335–336
Shen, B., 54–55
Shen, T. J., 364
Shen, Z., 291–293
Sheneman, L., 390
Sheng, H. F., 354
Sheng, Y., 112–113, 132
Sherr, B. F., 95
Sherr, E. B., 95
Shi, T., 48–49, 50, 52–53, 54–55, 211, 238
Shi, Y., 15–16, 124–125, 170–171, 188–191, 189*t*, 193*f*, 195–196, 198–199, 210–211, 261–262
Shi, Z., 529–530
Shiea, J., 291–293
Shilova, I. N., 52, 53
Shin, D., 70
Shizuya, H., 111–112, 115–117, 125
Shlyakhter, I. A., 529–530
Shmoish, M., 111–112
Shokralla, S., 53–54, 252–253
Shore, S., 294–295
Short, C. M., 238–239
Shrestha, P. M., 258–259
Sieracki, M. E., 22–23, 47, 48–49, 52, 62, 95
Sikora, M., 222
Silva, F. J., 460
Simister, R. L., 194–195
Simmons, S. L., 220–222, 530–531
Simon, M. I., 38–39, 125
Simonet, P., 128–129, 526–527
Simpson, J. T., 154, 157, 529–530
Singer, A. U., 144–145
Singer, S. W., 290–291, 337–338, 340, 343, 344, 346, 347–348, 526, 531
Singh, K., 202–205, 260–261
Singh, N. K., 452*t*, 459
Sittka, A., 201–202
Siuzdak, G., 291–293
Slepak, T., 111–112, 113–115, 125

Sloan, D. B., 211–212, 448, 450, 455*f*
Sloan, W. T., 252
Smejkal, G. B., 293
Smidt, H., 290–291
Smillie, C., 367
Smith, A. D., 533–534
Smith, D. P., 272–273, 314–316
Smith, F., 149–153, 157
Smith, G. J., 48–49
Smith, G. P., 334–335, 449
Smith, J. J., 95
Smith, M. I., 208–209, 372
Smith, R., 222
Smith, W., 459
Smolders, A. J. P., 16
Sneath, P. H. A., 382
Snyder, M. P., 448–449
Soergel, D. A. W., 335–336, 361–362, 388–389
Sogin, M. L., 56–57, 334–336, 345, 354, 373, 490
Sokal, R. R., 382
Solberg, O. D., 201–202
Solexa, Q. A., 490
Solonenko, N., 145, 159–160
Solonenko, S., 146–147, 153, 156–157, 159–160
Sommer, D. D., 540–542
Song, Y. C., 35–36, 306, 307–310, 459–460
Song, Y. J., 259–260
Sordillo, L. M., 259–260
Sorokin, A., 500
Sowell, S. M., 272–273, 278, 279*f*, 314–316
Spall, J. L., 53–54, 252–253
Spaulding, S. E., 290–291
Spaulding, S. W., 211
Spencer, M. D., 372
Spencer-Martins, I., 10
Spor, A., 372
Spratt, B. G., 354, 527–528
Spring, S., 5
Spudich, J. L., 111–112
Squyres, J. M., 493–494
Stackebrandt, E., 334, 335
Stahl, D. A., 4–5, 6–8, 16, 29*t*, 33, 62–64, 334, 335–336
Staley, J., 526
Stamatakis, A., 390
Stamer, S. L., 297
Stange-Thomann, N., 153
Stanhope, S. A., 534–535
Stanley, K., 529–530
Stapels, M. D., 272
Steefel, C. I., 337–338, 346, 348
Steen, H., 294–295
Steen, R., 192–194
Steglich, C., 192–194
Stein, J. L., 115–117
Stemman, O., 294–295
Stepanauskas, R., 5, 14–15, 22–23, 47, 48–49, 52, 62, 536–537, 540–542
Stephens, P. J., 149–153, 157
Sterflinger-Gleixner, K., 290–291, 314–316
Steritz, M., 206–207
Stern, T., 62
Stetter, K. O., 64–65
Stevens, H., 307–308
Stevens, R., 334–335
Steward, G. F., 148–149
Stewart, C., 175–177
Stewart, F. J., 124–125, 189*t*, 191–192, 193*f*, 195–196, 197, 198–199, 200*f*, 201–202, 203*t*, 205–207, 209, 210–211, 259–260, 259*t*, 261–262, 263, 306, 307–308
Stewart, P. S., 12
Stickel, S., 56–57
Stilwell, C. P., 307–310, 311*t*
Stitt, M., 452*t*, 453
Stockdale, H., 95
Stocker, S., 174–175, 257
Stockwell, T. B., 5, 14–15, 67–68, 83–84
Stoddart, D., 540–542
Stodghill, P. V., 148–149
Stoeck, T., 307–308
Stoecker, K., 4–5
Stombaugh, J., 99, 356, 372–373, 374, 390, 402–405, 407, 425, 427, 428*f*, 438–439, 467, 493–494, 495, 510, 526–527
Stoye, J., 157–158
Strady, E., 197
Strauber, H., 95
Strimmer, K., 402–405
Strous, M., 307–308
Strunk, O., 32–33, 373
Stubbe, J., 144–145
Studholme, D. J., 54–55

Stuhrmann, T., 307–308
Stührmann, T., 306
Stukenber, P. T., 291–293
Stunnenberg, H. G., 151*t*, 156–158
Su, C. L., 259–260
Subramanian, S., 357, 380, 380*t*, 391, 438
Suenaga, H., 135, 136
Sugahara, J., 530–531
Suhai, S., 452*t*, 453
Suller, M. T. E., 95
Sullivan, M. B., 144–148, 146*f*, 151*t*, 153, 156–158, 159–160
Summons, R. E., 129, 132, 134, 135–136
Sun, C. L., 290–291
Sun, S. L., 188–192, 189*t*, 261–262
Sun, Y., 360
Sung, Y., 148
Sunner, J., 291–293
Suspitsin, E. N., 32
Suttle, B., 290–291
Suttle, C. A., 144, 148–149
Sutton, G. G., 139, 529–530
Suzuki, M. T., 111–112, 115–117, 308
Svenning, M. M., 263
Swain, M. T., 452*t*, 456–457
Swan, B. K., 62, 75
Swanson, S. K., 294–295
Sweetman, G., 294–295
Swerdlow, H. P., 334–335, 449
Swift, P., 189*t*
Swingley, W. D., 65–67, 85–86
Sykes, P. J., 62
Sylva, S. P., 15–16, 36
Szczepanowski, R., 157–158
Szeto, E., 54–55, 494
Szklarczyk, D., 472–473
Szybalski, W., 129, 131–132, 134

T

Tabb, D. L., 294–295
Tachiiri, Y., 111–112, 125
Tadmor, A. D., 159–160
Tagliabue, L., 207–208
Takai, K., 308
Takai, Y., 133, 134
Takasaki, K., 262
Tam, N. F. Y., 354
Tamaki, H., 262
Tamames, J., 314–316
Tanabe, M., 494, 503, 510
Tanaka, H., 154, 530
Tang, H., 456, 466, 467, 488, 492, 493–494
Tang, S.-L., 452*t*, 459
Tang, Y., 54–55
Taniguchi, Y., 192–194
Tanner, S., 290–291
Tarkka, M. T., 290–291
Tartaro, K., 52
Tatusov, R. L., 472–473, 494
Tatusova, T., 500
Taubert, M., 290–291
Taylor, G. T., 307–308
Taylor, L. T., 36, 308, 311*t*
Taylor, M. W., 194–195
Taylor, T. D., 46–47
Teal, T. K., 157–158, 490–491
Teeling, H., 272–273, 452*t*, 459
Telenius, H., 47
Temperton, B., 145, 146–148, 156–157, 189*t*
Teslenko, M., 32–33
Thamdrup, B., 197, 198–199, 306, 307–308
Thanh, N. C., 70
Thelen, M. P., 272, 290–291, 293, 314–316, 530–531
Theuerl, S., 258–259
Thingstad, T. F., 144
Thomas, B. C., 62–64, 272, 290–291, 336, 337–338, 340, 343–344, 345, 346–348, 526, 530–531
Thomas, J. J., 291–293
Thomas, N., 422–424
Thomas, S., 189*t*
Thomas, T., 136, 184, 272, 290–291, 335–336, 337
Thomas-Hall, S. R., 201–202, 254–255
Thompson, A. W., 47, 52–53, 55–56, 58, 62, 294–295
Thompson, J. R., 32
Thompson, L. R., 144–145
Thomsen, T. R., 367–368
Thorpe, J., 148–149
Thrash, J. C., 145, 146–148, 156–157
Thurber, R. V., 148
Tibshirani, R., 539–540

Tiedje, J. M., 58, 255–256, 354, 361–362, 374, 375, 388–389, 467, 468, 526–527, 528, 535–536, 537–538, 542
Tighe, D., 5, 14–15, 23–24, 31, 86–87, 447–448
Tiirola, M., 194–195
Till, H., 252–253
Timberlake, S. C., 154, 357
Ting, Y. S., 272–273, 278*np*, 279*f*, 283–285, 290–291, 314–316
Tiong, C. L., 134, 136
Togo, A. V., 32
Tolonen, A. C., 144–145
Tolosa, J. M., 222
Tom, L. M., 296
Torsvik, T., 255–256
Torsvik, V. L., 128–129, 255–256
Tortell, P. D., 307–310, 311*t*
Toulza, E., 48–49
Townsend, J. P., 535–536
Trachana, K., 472–473
Tran, B., 201–202
Tran, H. V., 307–308
Trang, T. C., 70
Trapnell, C., 211, 343–344, 347–348, 492
Treangen, T. J., 540–542
Trehan, I., 92, 372
Trimble, W. L., 490–491, 493–494
Tringe, S. G., 48–49, 65–67, 85–86, 202–205, 239*f*, 260–261, 306, 307–310, 334, 459–460
Tripp, H. J., 47, 48–49, 51, 53–55, 189*t*
Troge, J., 46–47
Truszkowski, J. M., 357
Tryon, R. C., 402
Tsai, I. J., 452*t*, 456–457
Tsementzi, D., 53–54, 208–209, 527–528, 529–530, 531–532
Tseng, Q., 153
Tsirigos, A., 452*t*, 459, 466, 536
Tsuneda, S., 5
Tunnacliffe, A., 47
Turk, K. A., 47, 48–49, 51, 53–55
Turnbaugh, P. J., 92, 93–94, 95, 98, 103, 220–222, 238, 334–335, 345, 372, 375, 381, 386–388, 396, 411, 412, 430, 510–511, 536
Turner, D. J., 157–158, 450
Turner, E. H., 149–153, 157–158
Tusher, V. G., 539–540
Tveit, A., 263
Tyson, G. W., 4–5, 6–9, 22–23, 62–64, 68–69, 124–125, 129, 133, 134–135, 136, 138, 139, 145, 170–171, 193*f*, 195–196, 212, 252–253, 261–262, 272, 293, 314–316, 336, 526–527, 531

U

Uchiyama, T., 135, 137
Ulloa, O., 124–125, 191–192, 198–199, 261–262, 306–308
Umbarger, H. E., 201–202
Unger, M. A., 80
Unterseher, M., 337
Urban, A. E., 52
Urich, T., 124, 170–171, 208–209, 263, 337
Uroz, S., 252–253
Ussler, W., 36

V

Vaishampayan, P., 337
Valas, R., 64
Vallon, L. S., 238, 252–253
van Bakel, H., 238–239
van de Pas-Schoonen, K. T., 16
van de Peppel, J., 238–239
van de Vossenberg, J., 306, 308
van den Engh, G., 47, 49–50
van der Mark, P., 32–33
Van der Plas, A., 306, 307–308
van Leenen, D., 238–239
van Nostrand, J. D., 252–253
Vandamme, P., 528
VanderGheynst, J. S., 337–338, 344, 346, 348
Vandernoot, V. A., 201–202, 203*t*
Vangala, M., 519–521
Varela, R., 307–308
Vasudevan, S., 503
Vaughan, E. E., 225–226, 290–291
Vaulot, D., 47, 48–49, 52–53, 55–56, 58
Vázquez, B., 226–227
Vee, V., 449
Venter, J. C., 62–64, 145, 459–460
Venugopalan, V. P., 12

VerBerkmoes, N. C., 62–64, 145, 272–273, 283–285, 290–291, 293, 294–295, 296, 314–316, 337–338, 343–344, 346, 348, 531
Vergin, K. L., 145, 146–148, 156–157, 272
Vermeulen, J., 233
Verner, M. C., 254–255
Vestris, G., 64
Vetriani, C., 307–308
Vila-Costa, M., 189*t*, 238
Vinten, A. J., 422–424
Vlassis, N., 235
Vogel, J., 170–171
Vogel, T. M., 526–527
Vogelstein, B., 62
Vogt, C., 290–291
Volossiouk, T., 255–256
von Bergen, M., 290–291
von Haller, P. D., 283–285
von Mering, C., 494, 503
Vos, M., 335
Vu, K. T., 70
Vyawahare, S., 153

W

Wagner, M., 4–5, 6–8, 11*f*
Wakeham, S. G., 307–308
Walcher, M., 5
Waldbauer, J. R., 129, 132, 134, 135–136
Waldmann, J., 452*t*, 459
Waldron, L., 99, 336–337, 453–454
Waldron, P. J., 31
Waldrop, M. P., 467–468, 471*f*, 475*f*, 478*f*, 483*f*
Walenz, B. P., 5, 14–15, 67–68, 83–84, 449
Walker, B. J., 154, 452*t*, 455–456
Walker, J. J., 376, 430
Wallner, G., 5, 8
Walsh, D. A., 306, 307–310, 311*t*, 459–460
Walsh, S., 95
Walters, W. A., 98, 220–222, 334–336, 345, 356, 361–362, 372–373, 375, 376–377, 379, 429–430
Walther, T. C., 319
Wang, C., 211
Wang, H., 336–337
Wang, J. T., 62, 259–260, 336–337, 411
Wang, M., 449
Wang, Q. O., 58, 334, 361–362, 374, 375, 388–389, 467, 468
Wang, S., 372
Wang, Y., 93–94, 154–155
Wang, Z., 538–540
Waniek, J. J., 307–308
Wanner, B. L., 133
Wanunu, M., 157–158
Ward, A. D., 46–47
Ward, B. B., 306, 308
Ward, D. V., 335–336, 346, 386–388
Wardle, D. A., 252
Warnecke, F., 290–291
Warren, R. L., 529–530
Waschkowitz, T., 137–138
Washburn, M. P., 291, 320–325
Wasserstrom, A., 62
Watanabe, K., 135, 137
Watanabe, Y., 530–531
Waterbury, J. B., 144–145
Waterman, M. S., 456, 533
Waterston, R. H., 533
Way, S. F., 536
Wayne, J. M., 35–36
Weber, N., 336–337, 452*t*, 458–459, 467
Weber, S., 367
Weckwerth, W., 222
Wegener, G. S., 15–16
Wegley, L., 148
Weil, M. R., 192–194
Weinbauer, M. G., 144
Weinberger, S., 291–293
Weinstock, G. M., 533
Weisburg, W. G., 56–57
Weitz, J. S., 363
Welch, D. M., 334–335, 345, 354, 373, 490
Wells, M. L., 48–49
Welsh, R. M., 55–56
Wendeberg, A., 39–40
Wendel, D., 390, 402–405, 495
Wendisch, V. F., 201–202
Wendl, M. C., 533
Wendt, J., 306
Wenz, C., 293
Wenzel, K., 222
Wenzel, S. C., 126
Werner, J. J., 357, 361–362
Wesolowski-Louvel, M., 254–255

Westcott, S. L., 310, 335–336, 346, 356, 360, 361, 373, 382–385, 383*t*, 388–389, 533–534
Westermann, A., 170–171
Westermann, P., 46–47, 62
Westram, R., 32–33, 373
Wetter, T., 452*t*, 453
Wexler, M., 127, 133, 136, 272, 290–291, 314–316
Wheeler, D. L., 466
Wheelis, M. L., 354
Whereat, E. B., 201
White, A. E., 189*t*, 211, 238
White, J. R., 539–540
White, R. E., 252
Whitehouse, C. M., 290
Whiteley, A. S., 46–47, 255–256
Whiting, A., 127
Whitman, W. B., 115–116, 354
Whitney, F., 306
Whittaker, R. H., 396–399
Wickham, G. S., 4–5
Wickham, H., 432
Widdel, F., 15–16, 23, 29*t*, 33, 36, 37–39, 40, 64
Wiebe, W. J., 115–116, 354
Wiener, M., 420
Wiggins, A. G., 127
Wild, J., 127, 129, 131–132, 134
Wilhartitz, I. C., 296, 314–316
Wilhelm, L. J., 272
Wilhelm, S. W., 144
Wilke, A., 489, 490–491, 493–494, 514, 519–520
Wilkening, J., 489, 490–491, 493–494, 519–520
Wilkins, D., 272
Wilkins, M. J., 294, 348
Wilkinson, L., 413
Williams, K. H., 294, 337–338, 346, 348
Williams, T. J., 272–273, 290–291, 314–316
Williamson, L. L., 128–129, 137
Williamson, S. J., 139, 148–149, 159–160
Willis, D. K., 233
Willmitzer, L., 201, 238–239
Willner, D., 178–181, 207–208
Wilm, A., 335–336, 337–338, 344, 345, 348
Wilmes, P., 195–196, 220–222, 221*f*, 235, 272, 290–291, 314–316
Wilson, R. K., 533
Wilson, W. H., 22–23, 52, 155
Winstanley, C., 259–260
Wiśnewski, J. R., 291–293, 297
Woebken, D., 189*t*, 306, 307–308
Woese, C. R., 354, 373
Wohl, E., 252
Wolf, Y. I., 503
Wolinsky, M., 252
Wolters, D. A., 291–293
Wommack, K. E., 144, 145, 159
Won, S., 201–202
Wong, K., 154, 157, 529–530
Wong, S. F., 290
Wood, E. F., 252
Wood-Charlson, E. M., 148–149
Worden, A. Z., 55–56
Workman, J. L., 294–295
Woyke, T., 5, 14–15, 23–24, 31, 62, 75, 86–87, 447–448
Wright, J. J., 306–310, 459–460
Wrighton, K. C., 62–64, 337–338, 343–344, 345, 346–347, 348, 531
Wu, D., 336, 466
Wu, K. Y., 115–117, 154–155
Wu, L. Y., 252–253, 255–256, 354
Wu, M., 335, 536–537
Wubet, T., 290–291
Wurtzel, O., 202–205, 260–261
Wurzbacher, C., 197
Wyatt, P. B., 124–125, 138
Wysoker, A., 467

X

Xia, Y., 367–368
Xie, C., 466
Xie, G., 62
Xie, X. S., 62
Xie, Y., 154, 431–432, 452*t*, 456
Xiong, X., 203*t*
Xu, J., 206–207
Xu, L., 240–241

Y

Yadhukumar, A., 373
Yadhukumar Buchner, A., 32–33

Yakushev, E. V., 197
Yamada, T., 401–402
Yamane, T., 64–65, 85–86
Yang, E. C., 52
Yang, H., 336–337
Yang, Y., 54–55
Yarza, P., 32–33, 58, 388–389, 431
Yates, J. R. III., 291–293, 297
Yatsunenko, T., 92, 372
Yau, S., 272
Ye, W., 155–156, 158–159
Ye, Y., 334–335, 466, 467, 488, 492, 493–494
Yi, H., 201–202, 203*t*
Yilmaz, P., 22–23, 32–33, 58, 388–389, 431, 494–496
Yilmaz, S., 4–5, 6, 8–9, 68–69, 147–148, 151*t*, 156–158, 159–160, 202–205, 207–208, 260–261
Yin, J. J., 503
Yip, C. L., 127, 135
Yiu, S. M., 154, 530
Yoon, H. S., 22–23, 52
Yooseph, S., 64, 139
Yoshikawa, K., 85–86
Youle, M., 178–181, 207–208
Young, C. R., 124–125, 189*t*, 195, 196, 210, 238
Young, M., 145
Youssef, N. H., 62, 335–336
Yu, F., 360
Yu, J., 294–295
Yu, L., 238
Yuan, J., 154, 452*t*, 456
Yuri, G., 184

Z

Zabinsky, R., 5, 8
Zaglauer, A., 32, 35
Zahrieh, D., 194–195
Zaikova, E., 306, 307–310, 311*t*, 459–460
Zakharov, S. D., 70
Zaneveld, J., 252–253, 376, 379, 421
Zechel, D. L., 124–125, 138
Zehr, J. P., 48–49, 50, 52–53, 54–55, 170–171, 188–191, 189*t*, 238–239, 261–262
Zeisel, S. H., 372
Zeng, Q., 144–145
Zengler, K., 5
Zerbino, D. R., 154, 452*t*, 453, 455*f*, 456, 529–530
Zhang, B., 272
Zhang, G., 345
Zhang, H., 201–202, 354
Zhang, J., 529
Zhang, K., 5, 46–47, 62, 124, 147–148, 156–157
Zhang, L., 47
Zhang, R., 240–241
Zhang, S., 272
Zhang, W., 54–55
Zhang, X., 252–253, 457
Zhang, Y., 336–337
Zhang, Z., 240–241, 529
Zhao, F., 336–337
Zhao, W., 189*t*
Zhao, Y., 145, 146–148, 156–157, 466, 467
Zheng, Z., 155–156, 158–159
Zhong, L., 272, 290–291
Zhou, D. S., 259–260, 354, 357
Zhou, G., 211
Zhou, H. W., 354
Zhou, J. Z., 255–256
Zhou, N., 403*f*
Zhou, Y., 334–335
Zhu, H., 119, 529–530
Zhuang, Y., 201–202
Ziebis, W., 29
Zielinski, B., 189*t*
Zimmer, D. P., 201–202
Zimmermann, L., 252–253
Zimmermann, S., 238
Zinder, S. H., 272, 278, 279*f*
Zinjarde, S. S., 12
Zong, C., 62
Zougman, A., 297
Zubkov, M. V., 4, 5, 93–94, 293–294, 307–308
Zwart, K. B., 252–253
Zybailov, B., 320–325

SUBJECT INDEX

Note: Page numbers followed by "*f*" indicate figures, and "*t*" indicate tables.

A

AFIS. *See* Automatic flow injection sampler (AFIS)
Alkaline lysis, 82–83
Alpha-diversity analysis
  and beta, statistical significance (*see* Statistical tests, alpha-and beta-diversity)
  microbial communities, 401
  phylogenetic and nonphylogenetic indices, 399
  rarefaction depths, 399, 400*f*
  Shannon entropy, 399
Anaerobic methanotrophic archaea (ANME)
  aggregates, *M. nitroreducens* cells, 16, 16*f*
  AOM, 15–16
  FITC *vs.* Cy3, 16–17, 17*f*
  genome sequencing, 15–16
Anaerobic oxidation of methane (AOM), 15–16
ANME. *See* Anaerobic methanotrophic archaea (ANME)
Automatic flow injection sampler (AFIS), 197

B

Bacterial artificial chromosome (BAC) libraries
  E. coli DH10B competent cell preparation, 112–113
  HMW DNA
    extraction from GELase, 117–119
    ligation reaction, 119
    partial digestion, 116–117
    seawater sample, 115–116
  size range inserts, cloned DNA
    clone purification, 120
    HindIII cloning, 120
    NotI digested BAC clones, 120, 121*f*
  steps, construction of, 112
  as successful tool, 111–112
  transformation, ligation products, 119–120
  vector preparation for cloning
    commercial ready-to-use, 115
    digested and dephosphorylated, 113–115
    HindIII restriction enzyme digestion, 113–115
    pIndigo536 vector, 113–115
    steps, 113–115
Beta-diversity analysis
  and alpha, statistical significance (*see* Statistical tests, alpha-and beta-diversity)
  compositional structures, 401–402
  default visualizations, 402–405
  distance matrix, 402
  hierarchical clustering, 402–405
  Jackknifing, 402
  microbial communities, 401
  PCoA, 402, 403*f*
  taxonomic information, 402, 404*f*
  TopiaryExplorer, 402–405
  Unifrac measures, 401–402
  UPGMA clustering, 402–405, 405*f*
Biological Observation Matrix (BIOM), 495
Biomacromolecular isolation
  chromatographic spin column-based methods, 222
  sequential isolation (*see* Sequential biomacromolecular isolation)
Biomolecular isolation
  concomitant biomolecules, 222
  quality control, 230–235
  sample fixation, 222–223
  sample preprocessing, 222–223
  sequential biomacromolecular isolation, 228–230
  systematic measurements
    definition, 220–222
    ecosystems biology, 220–222
    heterogeneity, within-and between-sample, 220–222, 221*f*

Biomolecular isolation (*Continued*)
high-throughput "omic" technologies, 220
Bovine serum albumin (BSA), 81–82

C

Catalyzed reporter deposition (CARD)
and FISH in community composition, 307–308
SYBR® Green-based qPCR assays, 308
Catalyzed reporter deposition–fluorescence *in situ* hybridization (CARD–FISH)
amplification reaction, 29–30
CARD hybridization buffer, 29
and DNA recovery, 29, 29*t*
cDNA synthesis. *See* Complementary deoxyribonucleic acid (cDNA) synthesis
Cell damage
membrane polarity, 95
and metabolic activity, 93, 96–98
Cell lysis, biomolecular fractions
biological wastewater treatment, 230–231
cell integrity, 230–231, 231*f*
disruption, 228
efficiency of, 228
red and green fluorescence micrographs, 231
Cell sorting. *See* Fluorescence-activated cell sorting (FACS)
Chimeric sequences identification, 386–388
Clusters of Orthologous Groups (COGs)
Alignment Viewer, 474–477
analyzer, 473, 476*f*
classifications, 473
functional classification schemes, 472–473
KEGG Viewer, 472
mappings, 473
Community analysis
correlation analysis, 367–368
distance metrics
applications, 364
incidence-based distances, 364
JSD, 365
diversity estimates, 363
OTUs, metadata
multivariate statistical techniques, 365
Random Forest classification, 367
SLIME, 367
statistical learning technique, 367
Comparative analyses, microbial communities
bioinformatics approaches, 355
community analysis, 362–368
Human Microbiome project, 354–355
library preparation, 354
llumina sequencing platform, 355
metagenomics approaches, 354
microbial DNA, amplification, 354
OTU, 354
prokaryotic cells, 354
sequence processing, 355–356
sequencing terminology, 356–362
Complementary deoxyribonucleic acid (cDNA) synthesis
antisense total RNA, 210–211
double-stranded cDNA, 210–211
fragmentation, 181–183
in vitro transcription, 209–210
invitrogen, 210–211
libraries, metatranscriptomic, 181–183
master mix, 181, 182
metatranscriptome analyses, 211
preparation, 181
and soil sample processing, 262
strand orientation, 211–212
terminal tagging master mix, 182
thermocycler, 181, 182
transcript abundance patterns, 211
Conventional filtration methods
advantage and disadvantage, 273
sample collection flowchart, 273, 274*f*
Cryomilling, 226–227

D

Data analysis and genome assembly
assembly algorithms, 454
bacterial endosymbionts, 454
de Bruijn graph approach, 456
de novo assembly program, 456
endosymbiont genomes, 454
high-fidelity polymerase, 457
intermediate levels of sequencing coverage, 454, 455*f*
presorting reads, genome, 453–454
procedures, in silico, 457

quality assessment of read data, 452–453
SEQuel method, 456–457
sequence data, 454
sequence diversity, 454–456
in silico reconstruction, 451
software, disentangling, assembling, and refining genomes, 451, 452*t*
Data-dependent acquisition (DDA) analysis
cell lysate, *Thalassiosira pseudonana*, 285–286
exploratory proteomic profiling, 283–285
protein identifications, 283–285
DBC. *See* Distribution-based clustering (DBC)
Demultiplexing and quality-filtering
barcodes and primer sequences, 380
flowgram clustering, 381
high-throughput sequencing, 379
Illumina quality-filtering approach, 381
machine-learning benefit, 380
mapping file, 381
multiple lanes, illumina fastq data, 381
QIIME quality-filtering process, 381
quality-filtering parameters, guidelines, 380, 380*t*
sequence matching, 380
Denaturing gradient gel electrophoresis (DGGE), 307–308
Bis-(1,3-Dibutylbarbituric acid)trimethine oxonol (DiBAC4), 95
Disentangling associated genomes
biological and genomic characteristics, 460–461
data analysis and genome assembly, 451–457
endosymbiont genomes, 446, 447*f*
genomic characterization, 446
natural microbial communities, 446
postassembly analysis, 457–460
sequencing, 446–451
Distribution-based clustering (DBC), 356
DNA extraction method, 467
DNA internal standards
addition, 247–248
description, 242
preparation
genomic standard stock, 247
required materials, 247
DNA processing
extraction and concentration, 31
lysis and reversing PFA cross-links, 31
DNA/RNA purification
microbial ribosomal RNA, 172–173
mRNA, 172–173
DNase treatment, 177
Downstream analysis, QIIME
alpha-diversity analysis, 399–401
beta-diversity, 401–405
categorical analyses, 395
common diversity analyses, 391
de novo and closed-reference OTU table, 393
diversity analysis, 396–399
HTML file, 396, 397*f*
machine learning, 420–422
methodological artifact, 391
optimal sampling depth, 391
OTU networks and heatmaps, 411–413
phylogenetic tree, 391
procrustes analysis, 422–424, 424*f*
quality-filtering, 391
rarefaction level, 391–392
significance, OTU category, 417–420
SitePainter, 425–427
taxa summaries, dataset, 396, 398*f*
Duplicate read inferred sequencing error estimation (DRISEE)
assessment, 491
detected errors, 491
implementation, 491
whole genome shotgun metagenomic sequence data, 491

E

EMIRGE. *See* Expectation-maximization reconstruction of genes from the environment (EMIRGE)
Environmental DNA (eDNA) extraction
Quant-iTT PicoGreenc dsDNA reagent, 309
sample collection and filtration protocols, 308–310
Sterivex filter, 309

Environmental microbial consortia, Magneto-FISH method
CARD, 23
DNA processing, 31
FISH probes, 33–37
flow sorting approaches, 23–24
high-throughput sequencing technology, 22
in situ techniques, 22–23
library preparation, 23–24
Magneto-FISH capture experiments, 24–25
metagenomics, optimization, 37–40
microbial systems, 41
molecular analyses, 41
optimization, environmental systems, 40–41
PCR and cloning, 31–32
phylogenetic analysis, 32–33
quantification, 31
single-cell analysis, 22–23
Environmental sample processor (ESP), 196, 197
Ethidium bromide (EtBr)-stained denaturing agarose gel electrophoresis, 257
Eukaryotic mRNA depletion
bacterial species, 178
Invitrogen's Dynabead Oligo(dT)$_{25}$ kit, 177–178
poly-A and T tail, 177–178
and rRNA, 178
total RNA extraction, 177–178
Expectation-maximization (EM) algorithm, 338–339
Expectation-maximization reconstruction of genes from the environment (EMIRGE)
algorithm convergence, 340–341, 341*f*
assembly algorithm, 337–338
candidate database/rRNA genes, 347–348
EM algorithm, 338–339
iterations, 340–341, 340*f*
reconstructed sequences with abundances information, 342–343
running EMIRGE
input reads, quality control and filtering, 343–344
on metagenome shotgun sequencing data, 344–345
reconstructed sequencesquality control and filtering, 346–347
on SSU rRNA gene amplicons/ transcriptome data, 345–346
shotgun metagenomics, 337–338
stabilization, 340–341
uncertainty in template-guided SSU rRNA assembly, 338, 339*f*

F

FACS. *See* Fluorescence-activated cell sorting (FACS)
False discovery rate (FDR), 419
FCM. *See* Flow cytometry (FCM)
FISH probes
archaeal diversity, 35–36
archaeal probes, 33
clone libraries, 35–36
diagnostic metabolic genes, 36–37
differential distribution, 36
DNA recoveries, 33–35
gel electrophoresis, 32
Magneto-FISH-captured archaeal diversity, 33
PCR amplification, 33–35
phylogenetic relationships, 33, 34*f*
Fixation-free sample preparation methods, 6
Flow cytometry (FCM)
acquisition, 97
active and damaged subsets, human gut microbiota, 93, 94*f*
analysis, 97
benefits of, 93
cell growth and transcriptional activity, 93
and cell sorting, 92
5-cyano-2,3-ditolyltetrazolium chloride (CTC), 95
DiBAC4, 95
dyes, 93
and FACS, 48–49
fluorescent dyes, 93
low (LNA) and high nucleic acid (HNA) content, 93–94
membrane integrity, 95
polarity, membrane, 95
Fluorescence-activated cell sorting (FACS)

characterization, uncultivated symbionts, 47–48, 48*f*
coastal seawater sample, 50, 51*f*
contaminating DNA, 47
description, 46–47
experimental procedures, 98–99
and FISH, 5–6
flow cytometer features, 48–49
fluidic system, 8
freeze–thaw cycle, 52–53
functional screens, 135, 137
labeling validation and preparation, 12
laminar flow fluidics, 47
and MDA, 47
methods, 46–47
qPCR assay, 50
sequencing (FACS-Seq)
avoid contamination, 105
experimental procedures, 98–99
on human gut microbiota, 102–103
overall workflow, 93, 94*f*
single-cell sorting experiments, 13
sort gates, 50
sorting
environmental samples, 49–50
Influx™, 49
nozzle diameter, 49
sample collection tubes, 49
sheath fluid, 49
unfixed *B. megaterium* cells, 13, 14*f*
Fluorescence *in situ* hybridization (FISH)
and CARD in community composition, 307–308
E. coli
and *Bacillus megaterium*, 9, 9*t*
labeling efficiency, 11, 11*f*
epifluorescence microscopy, 4–5
ethanol dehydration, 8–9
and FACS, 5–6
formamide and milli-Q water volumes, 10, 10*t*
hybridization, 10–11
probe selection, 6–8
qPCR, 308
uncultivated symbioses, 46
washing conditions, 11–12
Fosmid environmental library construction
DNA isolation
extraction steps, 128–129
HMW DNA by CsCl, 129
from marine samples, 128
microbial biomass, harvesting, 128
spin dialysis, 129
pCC1FOS™ vector, 129–131
storage
arrays, 131
colony pools, 131
DNA pools, 131
Fosmid libraries
community DNA and RNA sequencing, 124
construction (*see* Fosmid environmental library construction)
functional screens (*see* Functional screens)
genomic library vector
cosmid and fosmid libraries, 125
DNA inserts, 125
heterologous screening host
alternative heterologous host, 126
E. coli, 126
screening environmental libraries, 127
shuttle BAC vectors, 127
preparation, fosmid DNA
L-arabinose, 131–132
electroporation, 132
standard alkaline lysis methods, 132
retransforming fosmid pools, 132
sequencing fosmids, 132–133
Functional screens
antibacterial and antifungal screens, 136
antibiotic resistance screens, 136
characterization of, 138
chitinase screens, 136
description, 124–125
enzyme-activity-based screens, 135
expression enhancement
copy-up conditions, 134
growth conditions/temperatures, 134
pCC1FOS™, 134
extraction and cloning, environmental DNA, 124–125, 126*f*
format
on agar plates, 134–135
FACS, 135
labor intensive, 135

Functional screens (*Continued*)
genes for 4-hydroxybutyrate utilization, 136
heterologous hosts, 133
high-throughput, library extracts, 136
increasing the odds, 137–138
intracellular biosensor screens, 137
modulators of eukaryotic cell growth, 137
pigmentation screens, 135–136
transcriptomic approaches, 124–125

H

High molecular weight (HMW) DNA
extraction from GELase, 117–119
ligation reaction, 119
partial digestion
for cloning, 116–117
PFGE, 116–117, 118*f*
protocol, 116–117
seawater sample
collection, 115–116
enzymatic reactions, 116
protein digestion, 115–116
purification, 115–116
High-pressure liquid chromatography (HPLC), 318–320
High-throughput assay
AutoGenprep 960, 132
of library extracts, 136
HMP. *See* Human Microbiome Project (HMP)
HMW DNA. *See* High molecular weight (HMW) DNA
Homology-based assignment, 458–459
Human microbiome
Crohn's disease, 372
DNA sequencing technologies, 372
healthy human physiology, 372
Illumina's MiSeq and HiSeq DNA sequencing instruments, 372–373
PCR and Sequencing on Illumina MiSeq, 375–377
phenotypes, mice, 372
processing 454 data
Illumina processing, 430
quality-filtering, 430
variable-length barcodes, 430
workflow, 430
QIIME (*see* Quantitative insights into microbial ecology (QIIME))
shotgun metagenomics, 431
18S rRNA gene sequencing, 430–431
testing linear gradients, time series analysis, 427–429
Human Microbiome Project (HMP), 529
Hybridization, rRNA depletion
biotinylated antisense probes, 205–206
environmental samples, 205–206
mRNA abundance, 206–207
protocols, 205–206
RT-PCR, 206–207

I

In situ preservation, RNA
AFIS, 197
ESP, 196, 197
filtration and fixation sampler (IFFS), 197
marine metatranscriptome studies, 196
microbial community, 193*f*, 196
rRNA depletion, 189*t*, 196
Internal standards
addition, 247–248
DNA (*see* DNA internal standards)
metagenome normalization, 249
metatranscriptome normalization, 248–249
mRNA (*see* mRNA internal standards)
and natural nucleic acid, 240–241
quantitative meta-omics datasets
application of, 238
benefits of, 239–240
improved quantification, 238, 239*f*
internal control sequences, 240
recoveries and quantification, 248
sample processing, 238–239
Intracellular biosensor screens, 137
In vitro transcription, amplification
bacterioplankton RNA, 209–210
MessageAmp techniques, 210
rRNA, 210
total RNA, 209–210

J

Jensen-Shannon divergence (JSD)
biogeochemical zones, stratified lake, 365, 366*f*

definition, 364–365
extraction techniques, 365
theoretic interpretation, 365
JSD. *See* Jensen-Shannon divergence (JSD)

L

LabView graphical programming language, 83–84
LCA. *See* Lowest common ancestor (LCA)
LC-MS/MS. *See* Liquid chromatograph-tandem mass spectrometry (LC-MS/MS)
Library preparation methods, viral metagenomics
adaptor ligation
addition of, 155–156
and end repair, 155
Illumina and LADS, A-tailing, 155
sequencing preparations, 156–157, 156*f*
amplification protocols, 157–158
common steps, 149, 150*f*
fragmentation
acoustic shearing/tagmentation, 153
enzymatic digestion, 149–153
goals, methods, 149
hydrodynamic shearing, 149–153
nebulization, 149–153
protocols, 149, 151*t*
gel sizing, 157
paired-end sequencing
assembly algorithms, 154
insert sizes, 154–155
mate-pair creation, 154–155
small-and large-insert, 154
Pippin Prep, 157
quantification, 158–159
sequencing reaction and technologies, 159
Linker amplification
adaptor ligation, 155–156, 156*f*
library preparation protocols, 151*t*
sample-to-sequence, 147–148
shotgun library method, 145
Liquid chromatograph-tandem mass spectrometry (LC-MS/MS)
equipment and reagents, 295–296
sample fractionation, 293
shotgun proteomics, 290
two-dimensional LC system, 297, 298*f*
typical 24-h, 293
Lowest common ancestor (LCA)
horizontal gene transfer, 468–469
page breaks, 504
reference sequences, 468–469
sequencing technique, 470

M

Machine learning
algorithms, 420
artificial diversity, 421
Baseline error, 421
BIOM tables, 422
contamination quantification problem, 421
learning algorithm, 421
OTU category significance, 420
QIIME, 420
SourceTracker algorithm, 422, 423*f*
supervised learning, 420
whole mouse homogenization, 421–422
Magneto-FISH
liquid CARD-FISH, 29–30
magnetic capture, 30–31
permeabilization and inhibition, endogenous peroxidases, 25–28
Mass spectrometry (MS)
bio-mass, 283–285
data-dependent and-independent approaches, 283–285, 284*f*
DDA analysis, 283–286
exploratory proteomic profiling, 283–285
PAcIFIC, 283–285
Q-Exactive, 285–286
sensitivity, 283–285
MDA. *See* Multiple displacement amplification (MDA)
MEGAN. *See* MEtaGenome ANalyzer (MEGAN)
Messenger RNA (mRNA), 172–173, 206–207
Metabolite fractions, biomolecular analyses, validation of, 232–233
sampling and sample preprocessing, 232

MEtaGenome ANalyzer (MEGAN)
command-line mode, 468
comparison file, 479, 482*f*
CSV file, 477–479
distance matrix, 479
ecological indices, 479
functional analysis
classification schemes, 470, 472–473
COG analyzer, 473, 476*f*
KEGG Viewer, 472, 473, 475*f*
RefSeq mappings, 473
SEED analysis, 473, 474*f*
high-level comparison, 477, 478*f*
metagenomic sequence data, 467
microbial community, 477–479
PCoA analysis, 479–483, 483*f*
relative abundance, taxonomic content, 479, 480*f*
Sample Viewer window, 479
sequence alignment
gene-specific multiple alignments, 477
protein-guided assemblies, 477
reference-guided multiple sequence alignment, 474–477
taxonomy
environmental shotgun sequencing, 468
LCA algorithm, 469
NCBI, 470, 471*f*
nodes, 470
Percent Identity Filter, 469
read pairing information, 469
reference database, 468–469
sequencing protocols, 469
taxonomic rank, genus, 470, 473*f*
types, charts, 470
word cloud, genus rank analysis, 470, 472*f*
Metagenome overview page
accession number, publications, 498
analysis page, 496
automated pipeline, 496
computational resources, 500
environmental packages, 500
functional breakdown, 503
GSC MIxS checklist, 500, 501*f*
metagenome QC, 500, 502*f*
organism breakdown, 501–502
project information, 499
sequences, classifiication, 499, 499*f*
technical data, 500–501
Metagenomic datasets
assembly process, 529
automated approaches, 531
binomial test, 539
bioinformatic approaches, 528
components and associated bioinformatic tools, 540–542, 541*f*
de Bruijn graph algorithms, 530
experimental design, 531–532
genome assembly methods, 531
gut microbiome, premature infants, 530–531
habitats and microbial groups, 540–542
HMP, 529
intrinsic characteristics, target community, 531–532
iterative coassembling techniques, 531
K-mer graph, 529–530
map-based sequence binning methods, 531
modifications, scenarios, 539–540
NGS, 527–528
non-rRNA-based methods, 526
paired-end read information, 531
reference-guided assembly, 529
short-read sequencing data, 529–530
in silico-generated "spiked-in" metagenomes, 531–532, 532*f*
single gene markers, whole community analysis, 533–534
soil/sediment microbial communities, 526–527
STAMPS, 538–539
statistical approach, 538–539
taxonomic identity, metagenomic sequence, 535–538
tools and algorithms, 542
transcriptomics and proteomics, 540–542
types, sequencing errors and artifacts, 531–532
visualization tools, 531
WGS, 526
whole community analyses, genomes, 534–535
Metagenomics

archaeal diversity, 39–40
clone diversity, 39
DNA extraction efficiency, 38–39
DNA standards, quantification of (*see* DNA internal standards)
high-throughput sequencing protocols, 37
library preparation DNA concentration, 39–40
limitations of, 238–239
meta-omics data, 238
methodological variation, 40
microbial assemblage, 37–38
mRNA standards, quantification of (*see* mRNA internal standards)
post-DNA amplification, 37
sample retention efficiency, DNA recovery, 37–38, 38*t*
sequencing, uncultured symbiosis
bioinformatic tools, 54–55
contigs assembly, 54–55
de novo genome assembly, 53–54
enzyme Phi29 and MDA reactions, 52
freeze–thaw cycle, 52–53
genome sizes, 51
MDA on sorted cells, 53
nucleic acid sequencing technology, 53–54
pyrosequencing technology, 53–54
symbiotic partners, separation of, 52
viruses (*see* Viral metagenomics)
Metaproteomics
classical proteomics approaches, 290
experimental options
biomass, 293–294
24-h LC-MS/MS experiment, 293
isoelectric focusing and gel electrophoresis, 293
labeling experiments, 294–295
peptide quantification, 294–295
shotgun proteomic experiments, 294
techniques, 291–293
workflow, 291–293, 292*f*
LC-MS/MS (*see* Liquid chromatograph-tandem mass spectrometry (LC-MS/MS))
OMZs, detection protocol, 314–316
primary goals, 290–291
protein extraction method, 296
proteolysis and cleanup, 296–297
tandem mass spectrometry analysis, 298–299
two-dimensional liquid chromatography separation, 297–298
Metatranscriptomics
amplification, RNA, 209–212
cDNA synthesis, 209–212
collection and RNA preservation
in situ preservation, 196–197
pump-cast collection, 197–198
shipboard filtration, 192–196
description, 188–191
diverse marine habitats, 191–192
DNA standards, quantification of (*see* DNA internal standards)
host-microbe systems
cDNA synthesis, 181–183
description, 170–171
DNA/RNA purification, 172–173
DNase treatment, 177
eukaryotic mRNA depletion, 177–178
high-quality RNA, 170–171
index barcodes, 183–184
materials, 171–172
metagenomics, 170–171
microbial communities, 170–171
mRNA, 170–171
quantification, 173–174
ribosomal depletion, 178–181
RNA integrity and size distribution, 174–175
sequencing and data analysis, 184
total RNA extraction, 175–177
limitations of, 238–239
marine microbial, 191–192
meta-omics data, 238
microbial community cDNA, 191–192, 193*f*
mRNA standards (*see* mRNA internal standards)
planktonic micro-organisms, 191–192
RNA extraction and standardization
ambion, 198–199
bioanalyzer electropherogram, 199–201, 200*f*
column purifications, 199

Metatranscriptomics (*Continued*)
invitrogen, 199–201
marine microorganisms, 198
mirVanaT protocol, 198–199
quantitation, 199–201
RNase decontamination, 198
rRNA depletion (*see* Ribosomal ribonucleic acid (rRNA) depletion)
sample preparation, 188–191
standardization protocol, 201
Methyl-coenzyme reductase alpha subunit (mcrA), 32–33
MG-RAST
analytical and computational approaches, 488, 489*f*
bioinformatics, 520
BLAT hit details, alignment, 514, 515*f*
community resources, 519–520
comparative analysis, 519
data tabular view, 510
downloads
data products, 514
gene clustering, 518
prediction, protein-coding sequences, 516
preprocessing, 515–516
protein similarities, 518–519
RNA clustering and similarities, 517–518
screening, 516
uploaded file(s), 515
FASTA files, 510
FragGenescan, 488
functional classification, 513
metadata and analysis approaches, 520
new version
dereplication, 490–491
DRISEE (*see* Duplicate read inferred sequencing error estimation (DRISEE))
k-mer profiles, 491–492
nucleotide histograms, 492–494
preprocessing, 490
quality assessment, 491
RNA detection, clustering, and identification, 490
user-generated data, 490
organism classifications, 511–512, 512*f*
pairwise comparisons, 488
pipeline and technology platform, 489–495
proteins identification, 511–512, 512*f*
scaling, workload, 488
screenshot, analysis page and workbench tab, 510–511, 511*f*
shotgun metagenomic data, 519
tables, group results, 512, 513*f*
unoptimized pipeline, 520–521
user-interface metaphors, 488
web interface, 495–510
workbench, bar charts, 513, 514*f*
Microbial communities
experimental procedures
FACS-Seq, 98–99
microbial activity and cell damage, 96–98
sample preparation, 95–96
experimental validation
of cell sorting, 103, 104*f*
CTC staining, 102
DiBAC and Pi concentrations, 102
FACS-Seq protocol, human gut microbiota, 102–103
microbial physiology in human fecal samples, 99, 100*f*
of staining with gut isolates, 99–102, 101*f*
strains from phyla, 99–102
FCM (*see* Flow cytometry (FCM))
freshwater planktonic, 225
human body, 92
MEGAN
BLAST, 468
DNA extraction method, 467
DNA sequencing, 466
environmental metagenomics, 465–466
functional analysis, 470–473
machine-learning techniques, 466
microbial diversity, 466
pairwise alignments, 466
permafrost soil metagenome dataset, 467–468
reference-based methods, 466
sample comparison, 477–483
sampling and sequencing, 465–466

sequence comparison software, 467
taxonomic and functional levels, 467
QIIME
artificial diversity, 438
downstream analysis (*see* Downstream analysis, QIIME)
instrument output, 377
OTU quality-filtering, 438
upstream analysis steps, 379–390
workflow overview, 377, 378*f*
troubleshooting
avoid contamination, 105
cell counts, 105
cell proportions, 105
no PCR amplification, 105
storage and oxygen exposure, 103–105
within-and between-sample heterogeneity, 220–222, 221*f*
Microbial community DNA
next-generation sequencing, 124
nucleotide sequence-based screening, 124–125
transcriptomic approaches, 124–125
Microbial metatranscriptomics, soil sample. *See* Soil sample processing
Microdevice design
cell selection and lysis, 79
48-chamber microfluidic chip, 78–79, 78*f*
electronic solenoid, 79–80
nominal control channel, 80
PDMS device, 80
pressurized control channels, 79
single-genome amplification, 78–79
whole-genome amplification, 79
mRNA amplification, 261–262
mRNA internal standards
data normalization, 242
DNA template and standard synthesis
commercially available plasmids, 242
custom DNA fragments, 242
in vitro transcription, genetic construct, 242–243, 243*f*
template size, 243–244
preparation
linearization, plasmid template, 246
plasmid into Top10 *E. coli* cells, transformation, 245
required materials, 244–245
resuspension of plasmid DNA, 245
synthesis and purification, 246–247
recovery of, 240–241, 241*f*
Multidimensional scaling (MDS), 425
Multiple displacement amplification (MDA)
assembly and annotation, genomic data, 54–55
description, 14–15, 47
enzyme Phi29 and kits, 52
rRNA depletion methods, 207–208
on sorted cells, 53
template-independent products (TIPs) production, 52
uncultivated microbial symbionts, genomic analysis, 47, 48*f*

N

Nested PCR
application of, 56–57, 57*f*
FACS coupled to MDA, 47–48, 48*f*
16S and 18S rRNA genes, 56
sequencing rRNA genes, 58
with universal primers, 56
Next-generation sequencing (NGS)
technologies, 527–528
viral metagenomics (*see* Library preparation methods, viral metagenomics)
Norgen Biotek All-in-One Purification Kit-based method, 229–230
Normalization, web interface
abundance data for raw values, 507, 508*f*
ANOVA/t-tests, 506
non-parametric tests, 507
raw/normalized counts, 506
standardization, 506
Nucleic acids
deoxyribonucleic acid (DNA), 233–235
proteins, 235
ribonucleic acid (RNA), 233
Nucleotide histograms, MG-RAST pipeline
amplicon datasets, 492
biased distributions, 492, 492*f*
contamination, artificial sequences, 492, 493*f*
database sequences, 494
gene prediction and AA clustering, 493–494

Nucleotide histograms, MG-RAST pipeline (*Continued*)
histidine biosynthesis, 494
ideal distributions, 492, 493*f*
pipeline stages, 494
protein identification and annotation mapping, 494
screening, 492
shotgun datasets, 492
untrimmed barcodes, 492, 493*f*

O

OMZs. *See* Oxygen minimum zones (OMZs)
Operational taxonomic units (OTUs)
heatmaps
applications in molecular biology, 413
exploratory analysis, microbiomes, 414
finer-scale taxonomic labels, 417
interactive, table, 414, 416*f*
NMDS, 417
open-reference dataset, 417, 418*f*
ordination plots/network analyses, 417
proper filtering, data, 417
protein expression studies, 417
QIIME, 414
taxonomy assignment, 414
transgenic and wild-type mice, 414, 415*f*
visualizations, 415
networks
bacterial community analysis, 412, 413*f*
clustering, samples, 411–412
descriptive visualizations, 412
G-test, 412
network analysis, 413
phylogenetic classification, 412
tree-based PCoA graphs, 413
types, nodes, 412
user metadata mapping file, 411
picking
approaches comparison, 386, 387*t*
Average-neighbor, 371–444, 384*f*
closed-reference approach, 382, 384*f*
clustering algorithms, 382–385, 385*f*
construction, table, 390
de novo approach, 382, 384*f*
description, algorithm, 382, 383*t*
DNA sequence data, 382
Furthest-neighbor, 382–385
greedy algorithms, 385–386
hierarchical clustering, 382–385
Nearest-neighbor, 382–385
novel diversity discovery, 386
open-reference approach, 382, 384*f*
representative sequence, 386
traditional taxonomy, 382
Optical hardware
alignment procedure and safety
dielectric broadband mirrors, 76
Galilean beam expander, 75–76
imaging plane, microscope, 77
laser goggles, 75
objective alignment tool, 77
optomechanical hardware, 77
phosphorescent cards, 76–77
reflective surfaces, 75
laser and microscope
automated/semiautomated microscopy approach, 71
beam attenuation, 71
competitive pricing, 71–74
Gaussian transverse mode, 70–71
idealized light path, 71, 72*f*
microfluidic controller, 70
selection, light source, 70
solid-state, continuous-wave lasers, 70–71
trapping systems, 70
objective lens, 74
Optical tweezing, cell isolation
coccoid cells, 65–67
flow cytometry and micromanipulation, 67
fluorescence-activated flow cytometry, 65
methods, 65
microfluidic approaches, 69
microscopy and cell sorting, 68–69
production, SAGs, 64–65
reagent-borne contamination, 68
sample-borne contaminants, 68
SC-WGS, 67
single-cell whole-genome amplification optofluidics, 64–65, 66*f*
sorting, optical trap, 69–70
spherical objects, 65–67
wavelength, light, 65

OTUs. *See* Operational taxonomic units (OTUs)
Oxygen minimum zones (OMZs)
community structure and function
amplicon-based methods, 307–308
CARD and FISH, 307–308
eDNA extraction, 308–310
environmental parameter data, 314, 315*f*
qPCR primer sequences, 310, 311*t*
quality controls, 312–314
sample collection and filtration protocols, 308–310
SAR324 qPCR standards and environmental samples, 312–314, 313*f*
SSU rRNA genes sequences, 310
SYBR® Green-based qPCR assays, 308
T-RFLP and DGGE, 307–308
definition, 306
detection protocol, proteins
digestion and sample cleanup, 317–318
extraction and peptide detection protocol, 316–317
HPLC coupled tandem mass spectrometry, 318–320
metaproteomics, 314–316
peptide matching, 318–320
taxonomic distributions in Saanich Inlet, 320–325, 321*f*
trypsin digest, 317
in microbial energy metabolism, 306
taxonomic and plurality sequencing studies, 306–307

P

PFGE. *See* Pulse field gel electrophoresis (PFGE)
Phylogeny construction, 389–390
pIndigo536 BAC vector, 113–115, 119
Pipeline and technology platform
analysis, MG-RAST version 3, 489, 490*f*
BIOM compliant tools, 495
cloud computing, 489
data discovery, 495
GSC, 494–495
hardware resources, 489
large-scale sequence analysis, 489
new MG-RAST pipeline, 490–494
purpose-built software platform, 489
Planktonic tandem MS-based community proteomics. *See* Tandem mass spectrometry (tandem MS)-based community proteomics
Polymerase chain reaction (PCR)
and cloning, 31–32
on Illumina MiSeq
Earth Microbiome Project Web site, 376
enzymatic treatments, 16S amplicon, 375
HMP, 375
library preparation, 375
methods for DNA extraction, 375
MiSeq, 376
reagent cartridges, 376–377
sequencing adaptors, 375
Postassembly analysis
binning metagenomes
complex communities, 458–459
simple communities, 458
bins, genomes, 459–460
methods, group assembled fragments, 457
taxonomic bins/draft genomes analyzsis, 460
Potassium hydroxide treatment, 82
Precursor acquisition independent from ion count (PAcIFIC), 283–285
Probe hybridization, 6
Pulse field gel electrophoresis (PFGE)
concentrations, *HindIII*, 117, 118*f*
HindIII digested DNA, 116–117, 118*f*
lambda Ladder DNA, 117
NotI digested BAC clones, 120, 121*f*

Q

Qiagen AllPrep DNA/RNA/Protein Mini Kit-based method, 228–229
qPCR. *See* Quantitative real-time PCR (qPCR)
Quality control
cell lysis, 230–231
metabolite fractions
sampling and sample preprocessing, 232

Quality control (*Continued*)
validation of metabolomic analyses, 232–233
nucleic acids
deoxyribonucleic acid, 233–235
proteins, 235
ribonucleic acid, 233
Quantitative insights into microbial ecology (QIIME)
biom-format project, 432
CRAN, 432
data graphics, 431–432
data transformations, 433–434
diversity metrics, 373
downstream and upstream processes, 439
facet dimension, 434–435, 436*f*
fidelity, 438–439
formal validation/testing, 437–438
graphical instructions, 433
interactive data analysis, 431–432
MDS ordination, 435–436, 437*f*
microbial community analysis (*see* Microbial communities)
open-reference dataset, 433–434, 435*f*
OTU richness estimates, 433, 434*f*
PCoA, 374
phyloseq, 432
sequence analysis, 373–374
software dependencies, 373–374
software packages, 373–374
statistical methods, 373
subsets and transformations, 433
taxonomy databases and reference databases, 374
visualization tools, 373
Quantitative real-time PCR (qPCR)
negative controls, 58
row and column sorts, screening of, 50, 51*f*
16S and 18S rRNA genes, nested PCR, 56–57, 57*f*
sorting single cells and screening, 56

R

Reference-based chimera detection, 388
Reverse transcription polymerase chain reaction (RT-PCR), 206–207
Ribonucleic acid (RNA)
amplification
antisense total RNA, 210–211
double-stranded cDNA, 210–211
in vitro transcription, 209–210
invitrogen, 210–211
metatranscriptome analyses, 211
strand orientation, 211–212
transcript abundance patterns, 211
downstream analysis, 233
electropherograms, 233
extraction method, 233, 234*f*
integrity number (RIN), 233
Qiagen AllPrep DNA/RNA/Protein Mini Kit-based method, 228–229
rRNA depletion (*see* Ribosomal ribonucleic acid (rRNA) depletion)
Ribosomal depletion
AMPure RNAClean XP Beads, 180
host and bacterial communities, 178–181
incubation, 180
magnetic beads, 179
Ribo-Zero Magnetic Gold Kit, 178–181
total RNA, 178–181
Ribosomal ribonucleic acid (rRNA) depletion
epicentre kit, 202–205
eukaryotic RNA removal, 205
hybridization, 202–205
and MDA, 207–208
metatranscriptomic studies, 201–205
MICROBExpressT and MICROBEnrichT, 202–205
microbial communities, 201–202
monophosphate, 202–205
mRNA enrichment, 201–205
oligonucleotides, 205
removal efficiencies, 207–208
Ribo-ZeroT, 207–208
sequencing costs, 208–209
subtractive hybridization, 205–207
total RNA samples, 201–205, 203*t*
transcripts, 209
RNA integrity
bioanalyzer, 175
gel matrix and dye, concentrations, 174
microfluidics platform, 174–175
nano and pico protocol, 174
number (RIN), 174–175

rRNA depletion. *See* Ribosomal ribonucleic acid (rRNA) depletion
rRNA subtraction methods, soil sample
  currently available, 259, 259*t*
  exonuclease treatment, 260–261
  mRNA sequencing coverage, 259
  size separation by gel electrophoresis, 261
  subtractive hybridization, 259–260

S

SAGs. *See* Single amplified genomes (SAGs)
Sample fixation, 222–223
Sample heterogeneity, 220–222, 221*f*
Sample preparation, tandem MS-based community proteomics
  approaches, 278
  flowchart, 278, 279*f*
  soluble and membrane cell fractionations
    concentrated samples, combining, 279–280
    French press, lysing cells, 280
    membrane-enriched, generation, 280–281
    sonication, lysing cells, 280
Sample preprocessing
  biological wastewater treatment plant biomass, 223–225
  freshwater planktonic microbial communities, 225
  human fecal sample, 225–226
  methodological workflow, 223, 224*f*
  and small molecule extraction
    from cell pellet, 227
    from cell supernatant, 226
    cryomilling, 226–227
Scalable vector graphics (SVG)
  mouse and gastrointestinal tract, 426, 426*f*
  Open Image, 426*f*, 427
  text files, 425
Sequencing, microbial genome
  biological characteristics, 448–449
  endosymbiont genomes, 446
  genome analysis, 446
  genomic DNA
    insect-associated endosymbiont, 447
    microbial species and assembly efforts, 448
    polymorphism, 447
    qPCR, 448
    small-scale test sequencing run, 448
    WGA, 447–448
  Illumina, 449
  multiplexed, 450–451
  nucleotide composition bias, library preparation, 450
  paired-end libraries, 449–450
  technological modifications, 449–450
  third-generation sequencing methods, 449
Sequencing terminology
  artifact removal and classification, OTUs, 361–362
  creating OTUs, 360–361
  DBC, 356
  dereplication, 357–360
  fastq quality filtering, 356
  methodological artifacts, 356
  optimal filtering criteria, 360
  overlapping paired-end sequencing, 357
  per base sequence quality, 357, 358*f*
  quality-filtering performance, 357, 359*f*
  sequencing data, 356–357
  16S-rRNA amplicon data, 357
  trimming of sequences, 357–360
Sequential biomacromolecular isolation
  cell disruption, 228
  and cell lysis, 230–231
  description, 228
  kit-based methods, 228
  Norgen Biotek All-in-One Purification, 229–230
  Qiagen AllPrep DNA/RNA/Protein Mini, 228–229
Shipboard filtration, RNA
  biomass samples, 195
  DNA analysis, 195–196
  fixatives, 194–195
  flash freezing, 194–195
  marine metatranscriptome studies, 194–195
  microbial community cDNA, 193*f*, 195–196
  mRNA degradation, 192–194
  nontoxic, 195
  planktonic microorganisms, 192–194
  protocols, 195

Shipboard filtration, RNA (*Continued*)
RNALater®, 195
SterivexT filters, 194
Shotgun metagenomics, 431
Single amplified genomes (SAGs), 62
Single cell and population genome recovery
bioreactor community, 17
FACS, 13–14
FISH (*see* Fluorescence *in situ* hybridization (FISH))
genome sequencing, 4
in-solution FISH, 8–12
isolation, anaerobic methanotrophic archaea, 15–17
metagenomics, 4
post-sorting, 14–15
sample preparation, 6
Single-genome analysis
application-specific imaging modalities, 85
cell isolation, optical tweezing, 64–70
cell types, 85–86
computer-based control, laser system, 85
culture-based isolation, 62
dual-screen graphical user interface, 83–84, 84*f*
FACs, 64
gene sequencing, 62–64
genomic approaches, 62, 63*f*
high-speed video imaging, 85
metagenomics, 62–64
microbial genomes, 62
microdevice design, 78–80
microfluidic technique, 86
molecular approaches, 87–88
optical hardware setup, 70–77
optofluidic technology, 86–87
reactor density, 86–87
single-cell genomics, 85–86
sorting and amplifying single cells, 80–83
Stanford valve manifold controller design, 84–85
user-independent sorting, 83–84
valve control, 84–85
Small subunit ribosomal RNA (SSU rRNA)
amplified SSU gene fragments, 334–335
chimeric amplification products, 335–336
"clade-specific" genes, 336–337
EMIRGE (*see* Expectation-maximization reconstruction of genes from the environment (EMIRGE))
fragments, 337
gene clone library sequencing, 307–308
molecular sequencing, 334
PCR primers, 335–336
phylogenetic marker, key qualities, 334
quality controls, qPCR, 312
quantification, taxon, 335
SAR324 primer design, 310
sequences, 310
T-RFLP and DGGE, 307–308
Soil sample processing
cDNA synthesis, 262
collection, 254–255
combined approaches, 263
culture-independent methods, 252–253
current state and future prospects, 263
data analysis, 263
ecosystem services, 252
extraction, RNA, 254–255
functional redundancy, 252–253
inhibitor removal, 258–259
with low microbial biomass, 258
microbes, 252
mRNA amplification and cDNA, 261–262
preservation, 254–255
quality and quantity, RNA
assessing, 256–257
EtBr-stained denaturing agarose gel electrophoresis, 257
microfluidics-based technologies, 257
RNA integrity number (RIN), 257
ultraviolet–visible spectrophotometry, 256–257
RNA extraction, 255–256
RNase and alien DNA, 253–254
sampling, 254–255
size and diversity, 252
steps, cDNA preparation, 254, 254*f*
subtraction methods, rRNA, 259–261
SSU rRNA. *See* Small subunit ribosomal RNA (SSU rRNA)
Statistical Analysis of Metagenomic Profiles (STAMPS), 538–539
Statistical tests, alpha-and beta-diversity

ANOVA, 410–411
comparisons, distance matrices, 407
distribution of distances, 408–410, 409*f*
Mantel correlation statistics, 407, 408*f*
multivariate analyses, 410–411
partial Mantel test, 407–408
QIIME, 406
rarefaction, 406
sequence dataset, 407
Type I and Type II error, 406
unweighted UniFrac distances, 408–410, 410*f*
visual and statistical tests, 408–410
Subtraction hybridization protocol, rRNA, 259–260
Supervised composition-based methods, 459
SVG. *See* Scalable vector graphics (SVG)
Symbiosis
cellular structures, 46
FACS (*see* Fluorescence-activated cell sorting (FACS))
FISH, 46
identity and metabolism
analysis of individual associations, 55
nested PCR (*see* Nested PCR)
qPCR (*see* Quantitative real-time PCR (qPCR))
sample handling, 55–56
metagenomic sequencing, 51–55
morphological features, 46

T

Tandem mass spectrometry (tandem MS)-based community proteomics
description, 272
digesting procedure
concentrations, protein and peptide, 281
desalting, 282–283
disulphide bonds, 281
reduction and alkylation, 282
trypsin digestion, 282
and MS (*see* Mass spectrometry (MS))
phosphate transporters, 272–273
proteomics, 272
sample collection
assembling, TFF system, 275
concentrating cells, 275–276, 277*t*
conventional filtration, 273
and filtration methods, 273
flowchart, 273, 274*f*
pre-filtration, 275
quantification, concentration efficiency, 276–278
TFF (*see* Tangential flow filtration (TFF) methods)
sample preparation (*see* Sample preparation, tandem MS-based community proteomics)
targets, 272–273
Tangential flow filtration (TFF) methods
advantage and disadvantage, 273–274
assembling, 275
conventional filtration, 273
Pellicon 2 TFF unit, 276
Taxonomic identity, metagenomic sequence
alignment-based methods, 536–537
combination and optimization, 538
composition-based methods, 536
MeTaxa algorithm, 537–538
Taxonomy assignment, 388–389
Terminal restriction length polymorphism (T-RFLP), 307–308
Time series analysis
ANOVA, 429
beta-diversity plots, 427–428, 428*f*
data, gradient, 428, 429*f*
metadata mapping file, 428
microbiome surveys, 427
PCoA, 427–428
Total RNA extraction
aqueous layers, 176
downstream applications, 175, 177
MoBio Laboratory's PowerMicrobiome, 175–177
RNase-free water, 177
trizol extractions, 175

V

Viral metagenomics
environmental
culture-independent methods, 145
cyanophages, 144–145
impacts, ecosystems, 144

Viral metagenomics (*Continued*)
Pacific Ocean virome (POV), 146–147
"photosynthetic virus" paradigm, 145–146
protein clusters, 146–147
quantitative data, 146–147
sample-to-sequence workflow, 146–147, 146*f*
library preparation (*see* Library preparation methods, viral metagenomics)
sample-to-sequence pipeline
DNA extraction, 147–148
nucleic acid extraction step, 148–149
purification, 147–148
SYBR Gold particle counts, 148

W

Web interface
analysis page
bar chart and tree tools, 507
best hit, 505
data visualization and comparison, 504
heatmap/dendrogram, 507, 509*f*
KEGG maps, 510
LCA, 504
lowest common ancestor, 505–506
normalization, 506–507
ordination, 507–510
representative hit, 505
selection, sample and parameters, 504
table tool, 510
taxonomic/functional abundance profile mapping, 504
three-step process, 504–505, 504*f*
Biome of interest, 495–496
dynamic filtering, 495
metadata-enabled data discovery, 496
metagenome analysis, 495
metagenome overview page, 496–503
MG-RAST system, 495
search of metagenomes, 495–496, 497*f*
upload and metadata pages, 496
WGS. *See* Whole-genome shotgun (WGS)
Whole-genome shotgun (WGS), 526

Y

Y-adaptor, 155–156

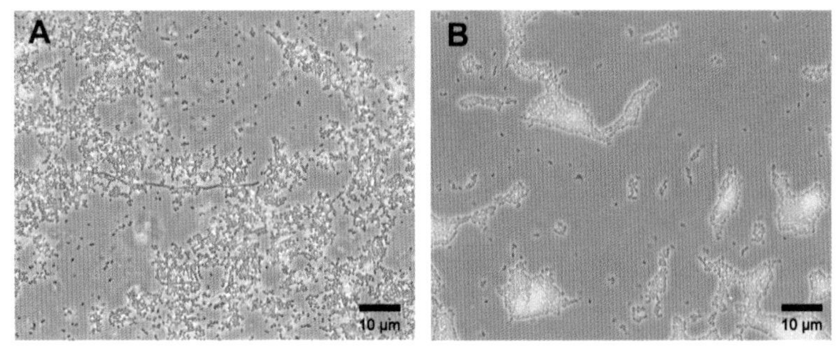

Mohamed F. Haroon *et al.*, Figure 1.2 In-solution FISH performed on unfixed *E. coli* cells using a (A) 1.5 h and (B) 2.5 h incubation times. Cells were hybridized using EUB338mix (Daims, Brühl, Amann, Schleifer, & Wagner, 1999) with a FITC fluorophore before a subsample was dried onto a glass slide and visualized by epifluorescence microscopy (630× magnification). The shorter hybridization time resulted in only ~20% of cells fluorescencing, as opposed to the 2.5 h incubation where nearly all cells were detectably fluorescent.

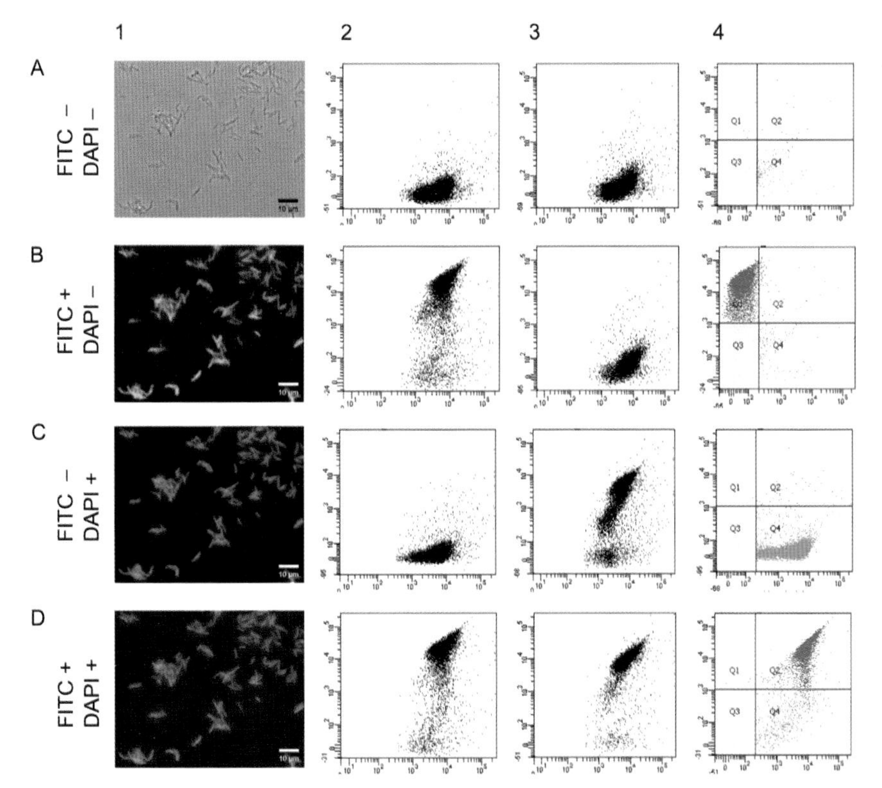

Mohamed F. Haroon *et al.*, Figure 1.3 FISH–FACS outputs for unfixed *B. megaterium* cells (A) unlabeled, (B) labeled with EUBmix338-FITC, (C) stained with DAPI, and (D) labeled with EUBmix338-FITC and stained with DAPI. From left to right, (1) cells imaged using epifluorescence microscopy, flow cytometry scattergrams of (2) SSC versus FITC, (3) SSC versus DAPI, and (4) DAPI versus FITC. In column 4, scattergrams are partitioned into four quadrants. The Q1 quadrant gated cells (in red) were labeled with EUBmix388-FITC only, Q2 quadrant gated cells (in purple) were labeled with both EUBmix338-FITC and DAPI, Q3 quadrant (in yellow) is background, while Q4 quadrant gated cells (in green) are stained with DAPI only.

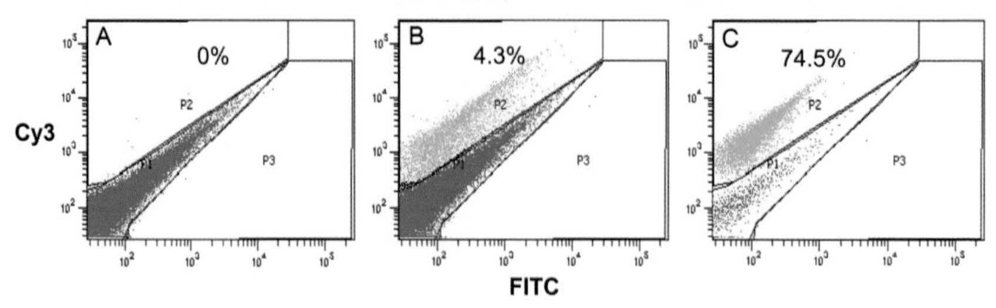

Mohamed F. Haroon *et al.*, Figure 1.5 FACS scattergrams showing FITC versus Cy3 for bioreactor samples. (A) Unlabeled sample, (B) first rapid sort with specific Darch872-Cy3 probe which targets *M. nitroreducens*, and (C) second sort to ensure high purity of target population single cells. Each plot contains 50,000 events. The P2 region of the plot indicates the events gated for sorting.

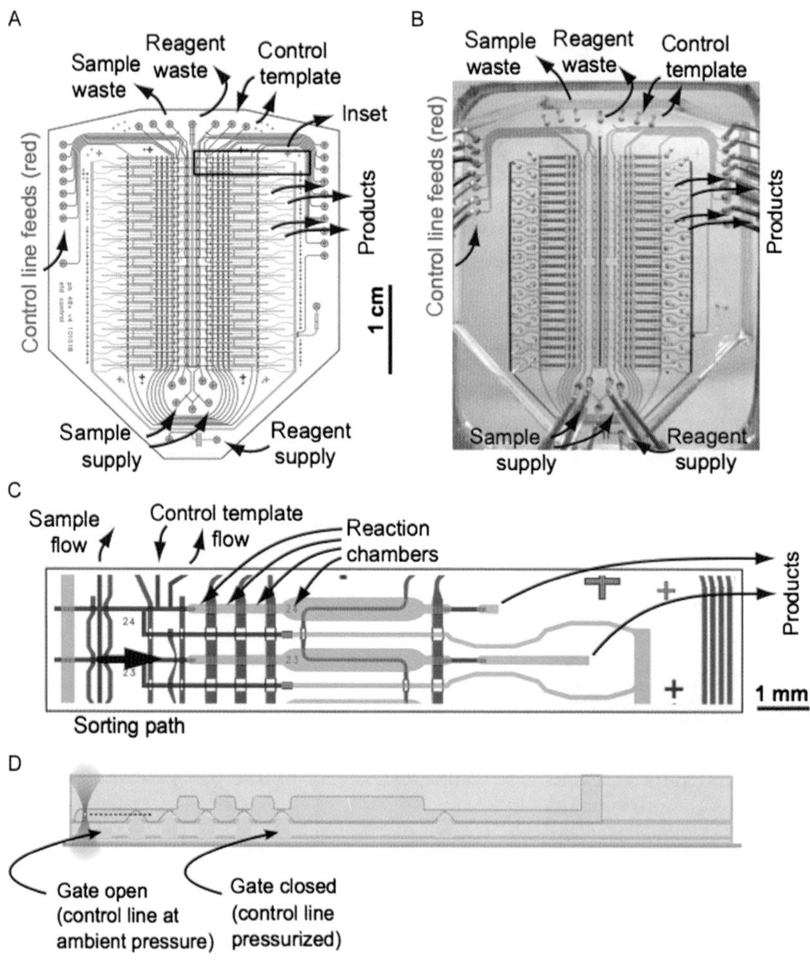

Zachary C. Landry *et al.*, Figure 4.4 48-chamber microfluidic chip for single cell sorting and whole-genome amplification. (A) Design of 48x_v4 device with key ports marked. (B) Photograph of 48-chamber device with corresponding ports marked. (C) Inset showing detail of two reaction chambers in the 48x_v4 device. (D) Cross-sectional schematic corresponding to part (C) indicating the trapping of a cell in the (upper) flow layer and the actuation of valves by pressurization of channels in the (lower) control layer.

A

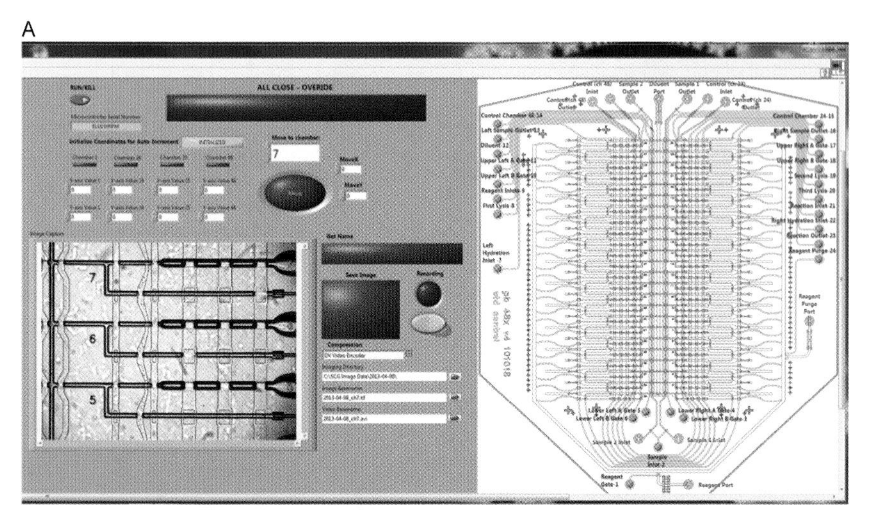

B C

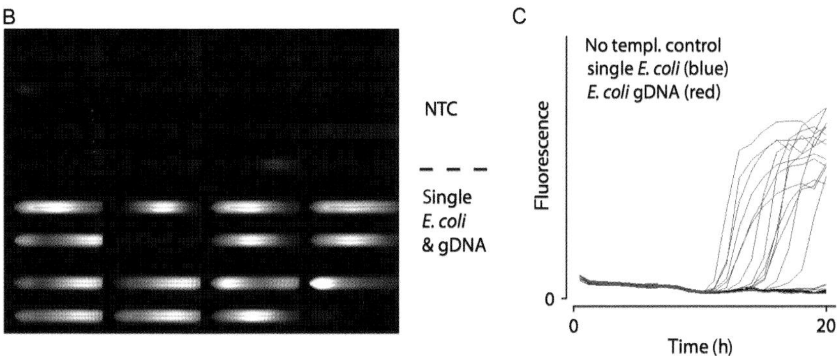

Zachary C. Landry *et al.*, Figure 4.5 Example of dual-screen graphical user interface with integrated microfluidic controls, stage automation, and video and image recording (A). Automated fluorescence data acquisition (B) and real-time kinetics curve of double-stranded DNA formation in individual microfluidic reaction chambers (C).

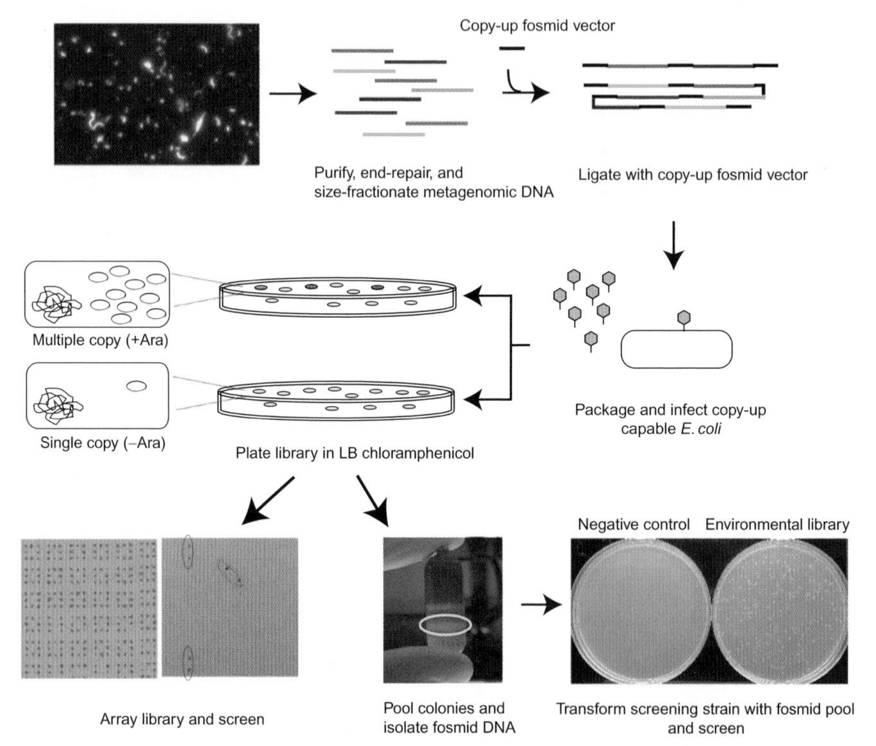

Asunción Martínez and Marcia S. Osburne, Figure 7.1 Process for extraction and cloning of environmental DNA into fosmid vector and screening of metagenomic libraries. Large genomic fragments obtained from uncultured microbial microorganisms are end-repaired, ligated with fosmid vector, and packaged *in vitro*. After infection of a copy-up *E. coli* strain, the metagenomic library can be screened directly, either in the absence of arabinose (single copy) or in the presence of arabinose which increases the copy number of the fosmid and may result in increased expression (represented here as colored colonies). Alternatively, the clones can be arrayed and frozen for screening (bottom left), or pooled and used to prepare a fosmid DNA pool to transform additional hosts (bottom right).

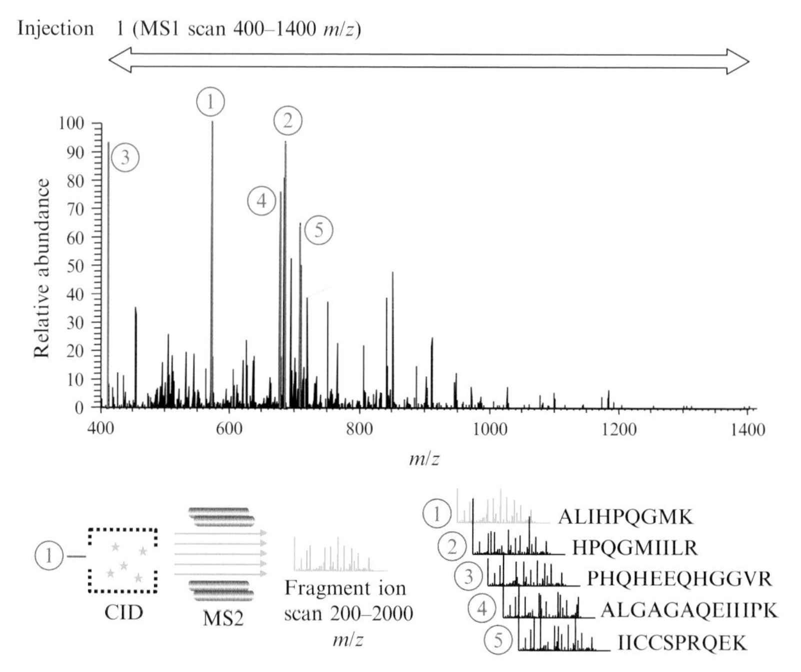
A
Injection 1 (MS1 scan 400–1400 m/z)
Relative abundance
m/z
CID
MS2
Fragment ion scan 200–2000 m/z
ALIHPQGMK
HPQGMIILR
PHQHEEQHGGVR
ALGAGAQEIIIPK
IICCSPRQEK

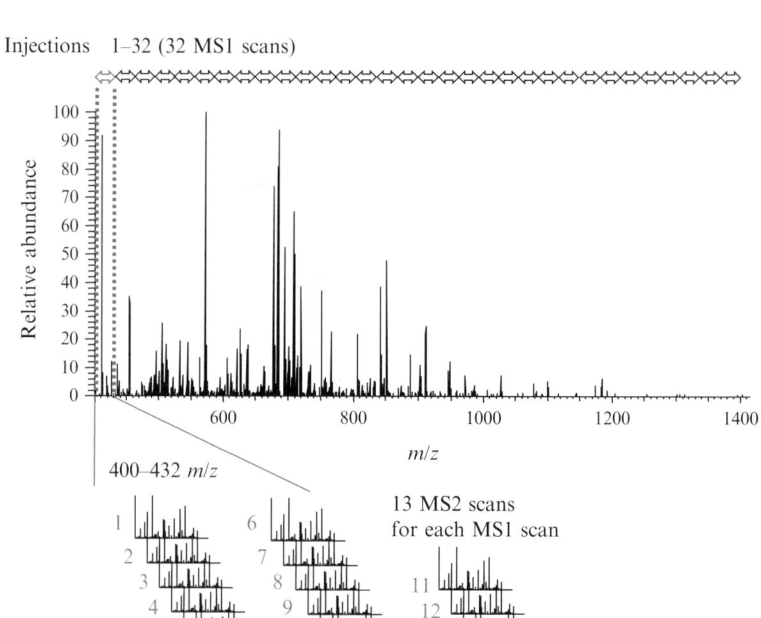
B
Injections 1–32 (32 MS1 scans)
Relative abundance
m/z
400–432 m/z
13 MS2 scans for each MS1 scan

Robert M. Morris and Brook Nunn, Figure 14.3 Data-dependent and data-independent tandem MS-based proteomic approaches. (A) During data-dependent analyses, the MS is programmed to collect MS2 data on the five most intense peaks observed in the MS1 precursor scan (1–5 in red). Those ions are then isolated in the ion trap or quadrapole and undergo collision-induced dissociation (CID), creating peptide fragment ions. The resulting fragment ions are measured in the second mass spectrometer and MS2 fragmentation spectra are collected. MS2 spectra collected represent putative peptides that are identified by searching a peptide database. (B) During data-independent acquisition MS2 data are systematically collected across the 400–1400 *m/z* scan range for the purpose of identifying low-abundance peptides in samples with variable or wide protein concentration ranges (Panchaud, Jung, Shaffer, Aitchison, & Goodlett, 2011). This method, termed Precursor acquisition independent of ion count (PAcIFIC), requires that each sample be injected into the mass spectrometer a total of 32 times to cover the full *m/z* range. Each injection experiences a 60–90 min HPLC gradient and 13 MS2 scans are performed per duty cycle, each collecting fragmentation spectra in a 2.5 *m/z* channel.

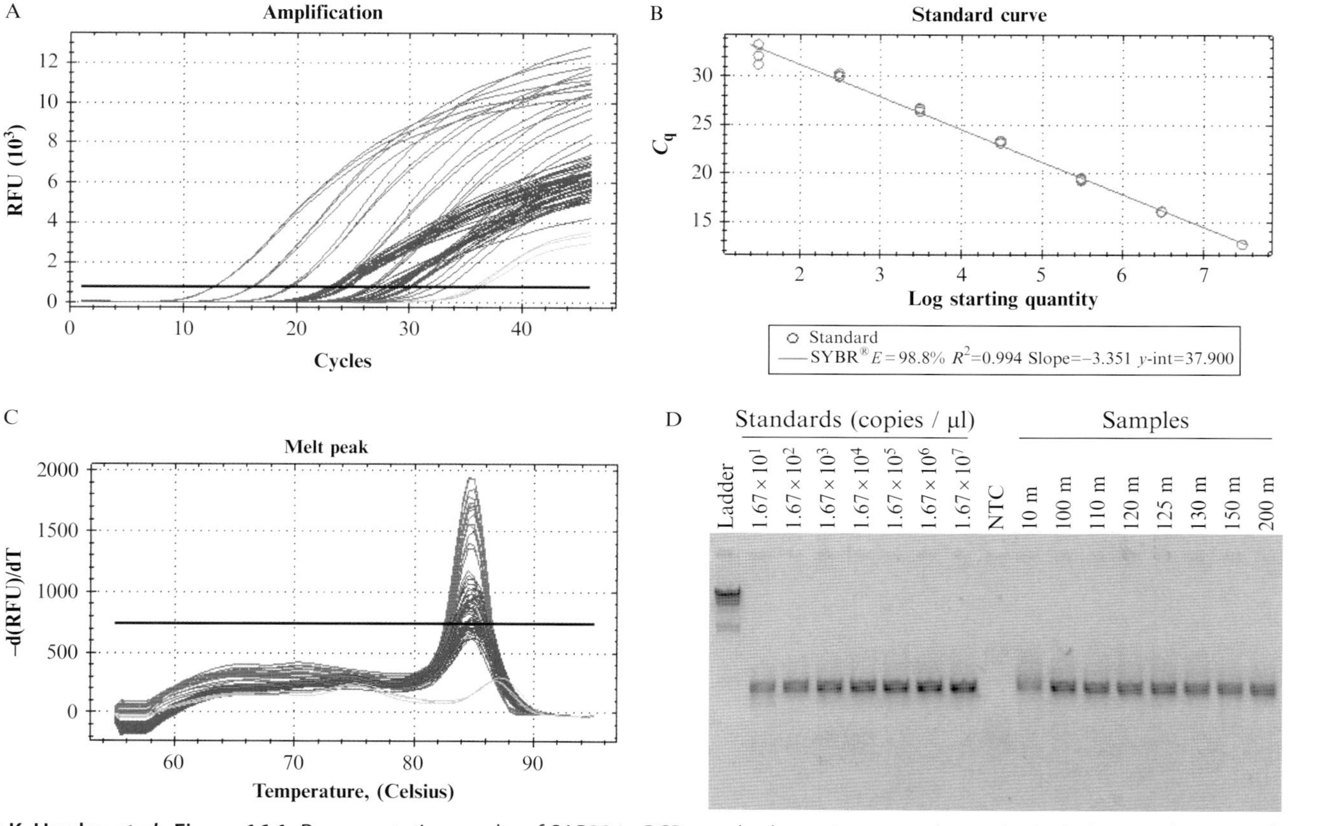

Alyse K. Hawley *et al.*, Figure 16.1 Representative results of SAR324 qPCR standards, environmental samples including quality control metrics. (A) Real-time run curves for standards (green), samples (red), and no template control (gray). (B) Standard curve of quantification cycle (C_q) versus Log of standard concentration. (C) Melting curve for standards (green), samples (red), and no template control (gray), showing single melting temperature for each sample and standard, indicating a single amplification product. (D) Representative gel image for qPCR product showing single band to confirm single amplification product observed in melting curve profiles.

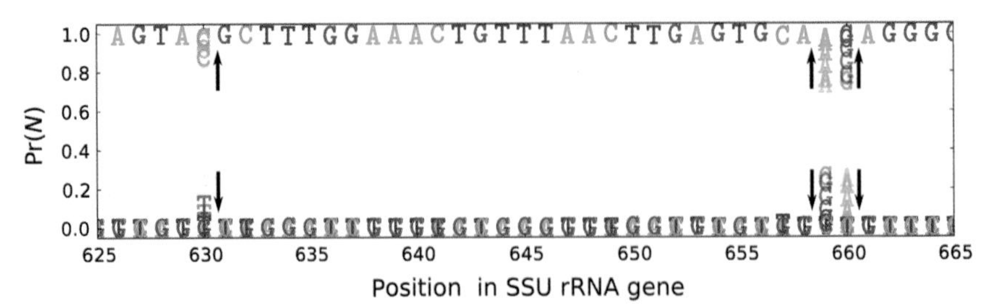

Christopher S. Miller, Figure 17.2 Iterative adjustment of individual base probabilities in the consensus sequences of candidate SSU rRNA genes. Shown are the probabilities (Pr(*N*); *y*-axis) for each of four possible bases at each position (*x*-axis) for selected nucleotides of a representative candidate gene. Probabilities for iterations 1–8 are overlaid with increasing opacity. Most positions in the gene are unambiguous: a single base has a probability near one, while the remaining three bases have probabilities near zero, and this does not change from iteration to iteration. For other positions (highlighted with arrows indicating increasing iteration number), EMIRGE gradually adjusts originally less certain bases to reflect increased evidence from confidently mapped reads in later iterations, ultimately revealing a highly probable consensus across all positions.

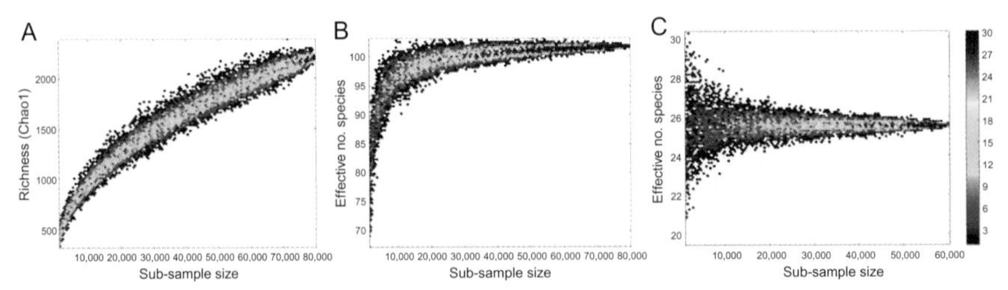

Sarah P. Preheim *et al.*, Figure 18.3 Richness is a function of sequencing depth, whereas Shannon and Simpson diversity indices stabilize with sequencing depth. (A) Chao-1 estimator (Chao1) of richness for the same library subsampled to different depths. (B) Effective number of species calculated with the Shannon diversity index (SDI) for the same library subsampled to different depths. Because SDI is computed on a log scale, we transform the SDI to the effective number of species present, which is on linear scale, by computing e^{SDI}. (C) Effective number of species calculated as the inverse of the Simpson diversity index (1/Simpsons diversity index) for the same library subsampled to different depths.

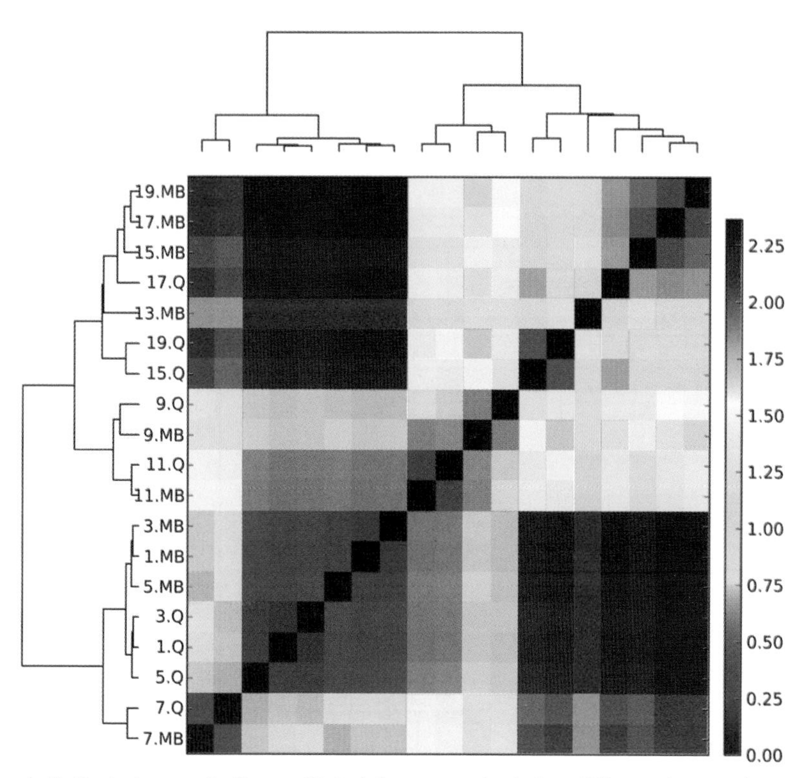

Sarah P. Preheim *et al.*, Figure 18.4 A heatmap depicting different biogeochemical zones of a stratified lake. The bacterial community for each depth was compared to all other libraries using the square root of the Jensen–Shannon divergence ($JSD^{1/2}$) and plotted as a heatmap. Bacterial communities cluster into two major zones (upper, oxic and lower, anoxic zone) as shown by the largest branch on the dendrogram (i.e., tree) and four subzones. Libraries are labeled with the depth (in meters) and the extraction technique. MB, MoBio Power Water DNA extraction kit; Q, Qiagen Blood and Tissue Kit. For example, 1.MB refers to 1 m depth in the water extracted with the MoBio kit and 19.Q refers to 19 m depth with the Qiagen kit. The dendrogram above and to the left of the heatmap are identical.

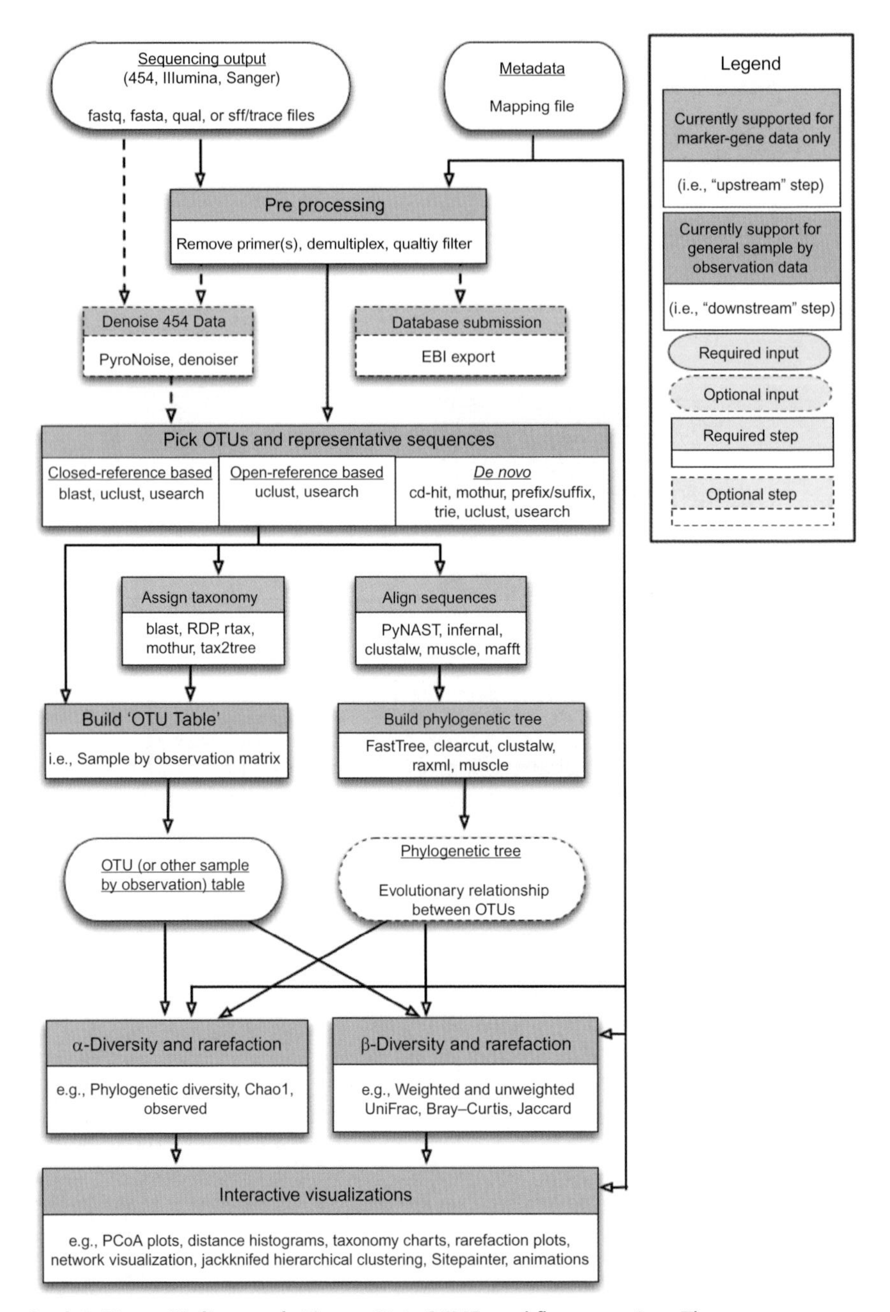

José A. Navas-Molina *et al.*, Figure 19.1 QIIME workflow overview. The upstream process (brown boxes) includes all the steps that generate the OTU table and the phylogenetic tree. This step starts by preprocessing the sequence reads and ends by building the OTU table and the phylogenetic tree. The downstream process (blue boxes) includes steps involved in analysis and interpretation of the results, starting with the OTU table and the phylogenetic tree and ending with alpha- and beta-diversity analyses, visualizations, and statistics.

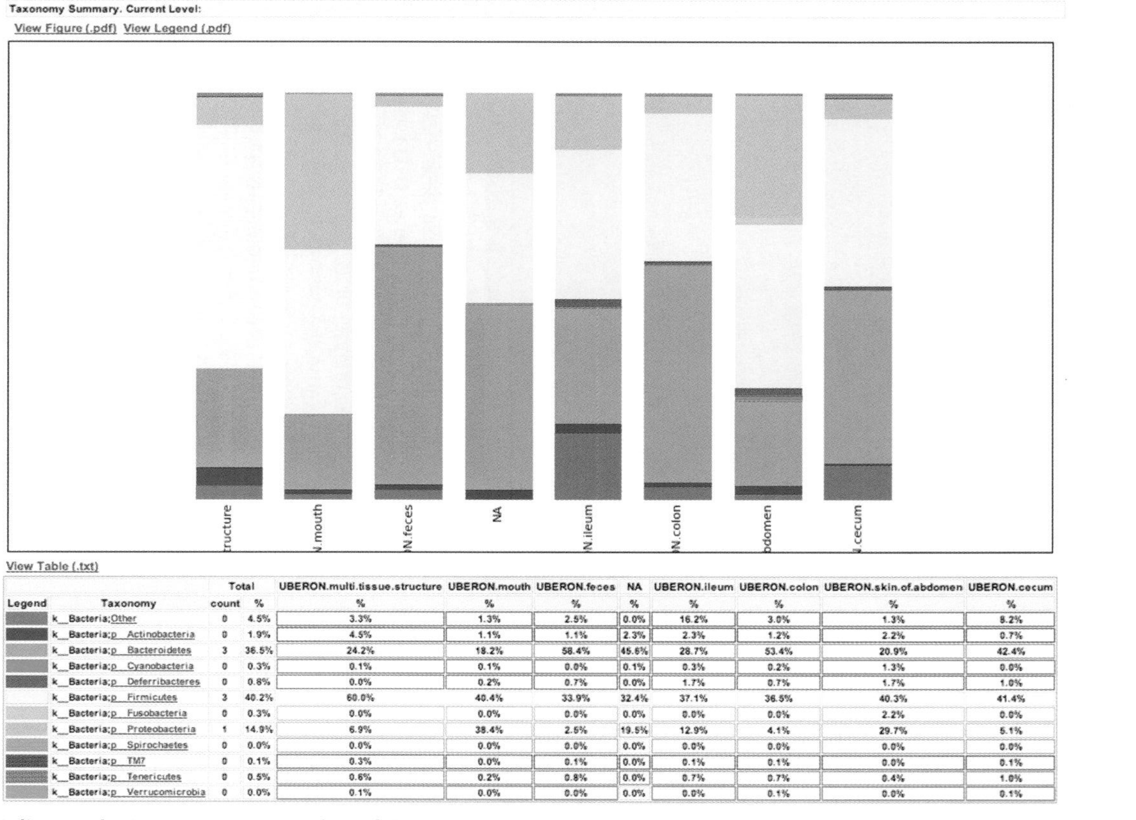

Legend	Taxonomy	Total count	Total %	UBERON.multi.tissue.structure %	UBERON.mouth %	UBERON.feces %	NA %	UBERON.ileum %	UBERON.colon %	UBERON.skin.of.abdomen %	UBERON.cecum %
	k__Bacteria;Other	0	4.5%	3.3%	1.3%	2.5%	0.0%	16.2%	3.0%	1.3%	8.2%
	k__Bacteria;p__Actinobacteria	0	1.9%	4.5%	1.1%	1.1%	2.3%	2.3%	1.2%	2.2%	0.7%
	k__Bacteria;p__Bacteroidetes	3	36.5%	24.2%	18.2%	58.4%	45.6%	28.7%	53.4%	20.9%	42.4%
	k__Bacteria;p__Cyanobacteria	0	0.3%	0.1%	0.1%	0.0%	0.1%	0.3%	0.2%	1.3%	0.0%
	k__Bacteria;p__Deferribacteres	0	0.8%	0.0%	0.2%	0.7%	0.0%	1.7%	0.7%	1.7%	1.0%
	k__Bacteria;p__Firmicutes	3	40.2%	60.0%	40.4%	33.9%	32.4%	37.1%	36.5%	40.3%	41.4%
	k__Bacteria;p__Fusobacteria	0	0.3%	0.0%	0.0%	0.0%	0.0%	0.0%	0.0%	2.2%	0.0%
	k__Bacteria;p__Proteobacteria	1	14.9%	6.9%	38.4%	2.5%	19.5%	12.9%	4.1%	29.7%	5.1%
	k__Bacteria;p__Spirochaetes	0	0.0%	0.0%	0.0%	0.0%	0.0%	0.0%	0.0%	0.0%	0.0%
	k__Bacteria;p__TM7	0	0.1%	0.3%	0.0%	0.1%	0.0%	0.1%	0.1%	0.0%	0.1%
	k__Bacteria;p__Tenericutes	0	0.5%	0.6%	0.2%	0.8%	0.0%	0.7%	0.7%	0.4%	1.0%
	k__Bacteria;p__Verrucomicrobia	0	0.0%	0.1%	0.0%	0.0%	0.0%	0.0%	0.1%	0.0%	0.1%

José A. Navas-Molina *et al.*, Figure 19.5 A snapshot of the taxa summary of the example dataset using the web interface. Samples have been grouped and averaged by body site, and taxonomic composition is shown on the phylum level. Each column in the plot represents a body site, and each color in the column represents the percentage of the total sample contributed by each taxon group at phylum level. The taxa summaries plot help us to see which taxon groups are more prevalent in a sample. For example, the fecal samples are dominated by Bacteroidetes, while mouth and skin samples are dominated by Proteobacteria. We can also see that Fusobacteria is only present at appreciable levels in the skin samples.

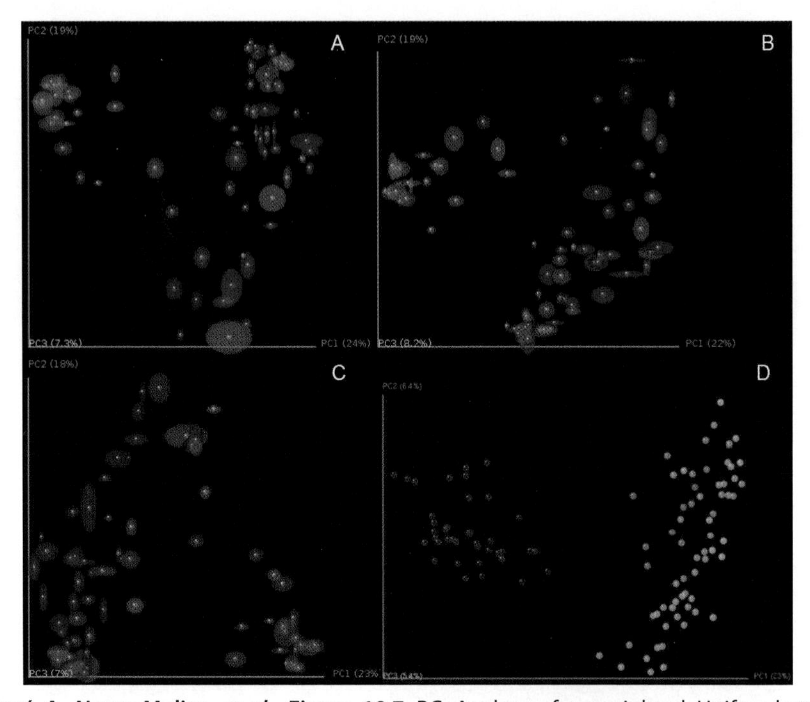

José A. Navas-Molina *et al.*, Figure 19.7 PCoA plots of unweighted Unifrac beta-diversity. Panels A–C show jackknifed replicate results for the example data set using *de novo* OTU picking, closed-reference OTU picking, and open-reference OTU picking, illustrating different results from the three OTU picking approaches (Table 19.3). Each dot represents a sample, either from a WT mouse (orange) or TG mouse (blue). The two groups are not clearly separated, probably because the data set is contaminated (recall that this is a class project and different participants varied in their dissection skills). The size of the ellipsoids shows the variation for each sample calculated from jackknife analysis. These plots are generated by the command `jackknifed_beta_diversity.py -i $PWD/denovo_otus/otu_table_filtered.biom -t $PWD/denovo_otus/rep_set.tre -m $PWD/IQ_Bio_16sV4_L001_map.txt -o $PWD/diversity_analysis/jk_denovo -e 7205 -a -O 64` (the input parameters should be adapted for using the OTU tables from different OTU picking approaches). Panel D shows the beta-diversity PCoA plot of a data set from the "keyboard" data set (Fierer et al., 2010) which links individuals to their computer keyboard through microbial community similarity. Each dot represents a microbial community sampled from either fingertips or keyboard keys from three individuals, annotated by the three colors shown in the plot. In contrast to panels A–C, panel D shows the microbial communities well separated by individual in the PCoA plot.

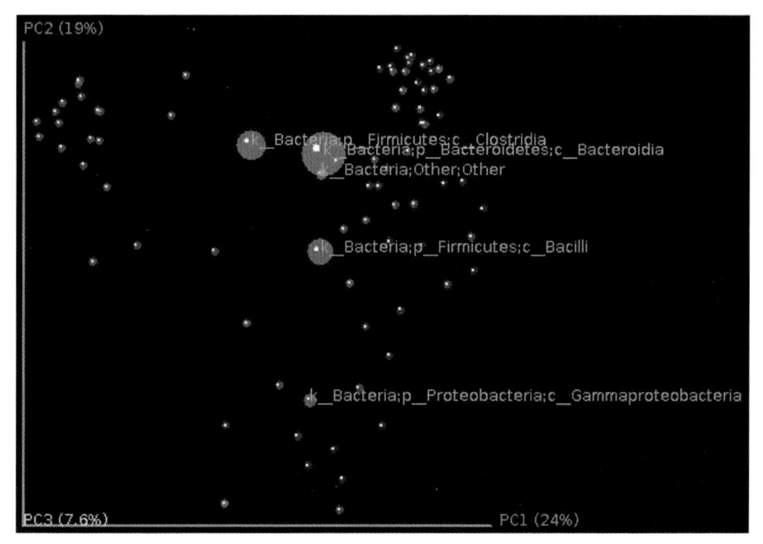

José A. Navas-Molina *et al.*, Figure 19.8 Biplot of the example data set. This is the unweighted Unifrac beta-diversity plot, similar to Fig. 19.7, with labels for the most five abundant phylum-level taxa added. The size of the sphere for each taxon is proportional to the mean relative abundance of that taxon across all samples. This plot is created by the command `make_3d_plots.py -i $PWD/diversity_analysis/open_ref/bdiv_even7205/unweighted_unifrac_pc.txt -m $PWD/IQ_Bio_16sV4_L001_map.txt -t $PWD/diversity_analysis/open_ref/taxa_plots/table_mc7205_sorted_L3.txt -n_taxa_keep 5 -o $PWD/diversity_analysis/3d_biplot`.

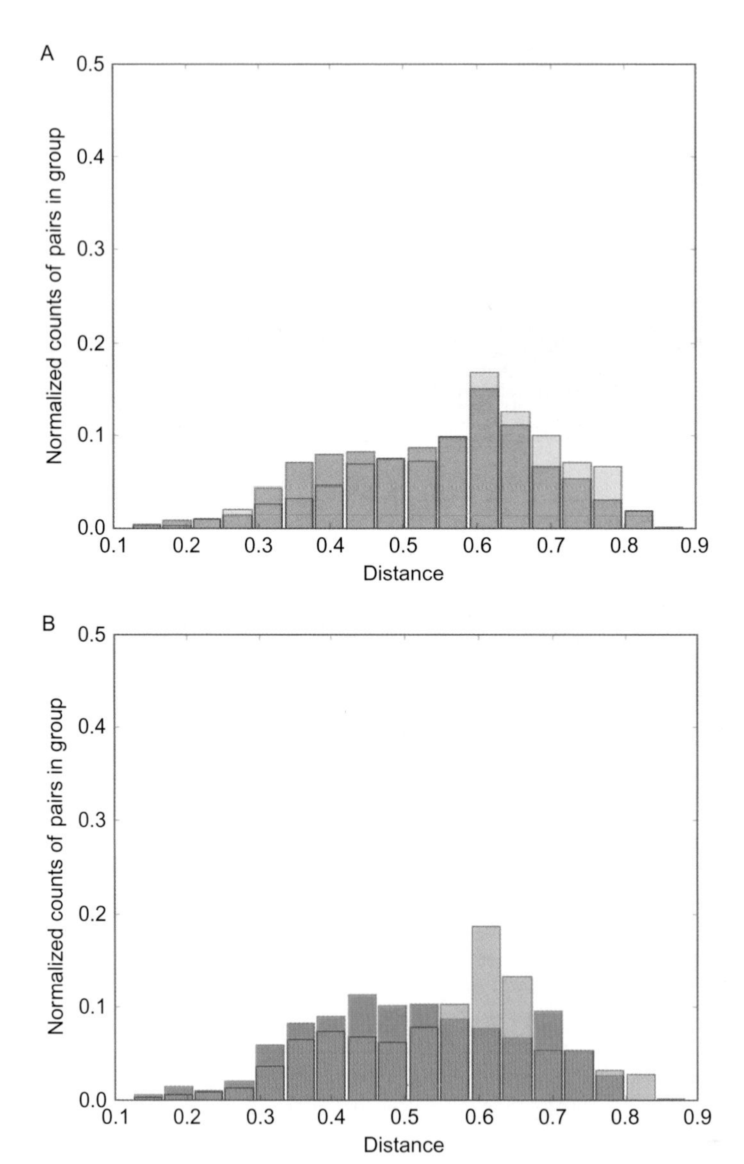

José A. Navas-Molina *et al.*, Figure 19.11 (A) Histogram showing distribution of distances between (light brown) and within (dark brown) mice gut microbiota taking into account both wild-type and transgenic mouse groups. (B) Distribution of within distances in gut bacterial community of wild-type mice (light orange) and transgenic ones (blue).

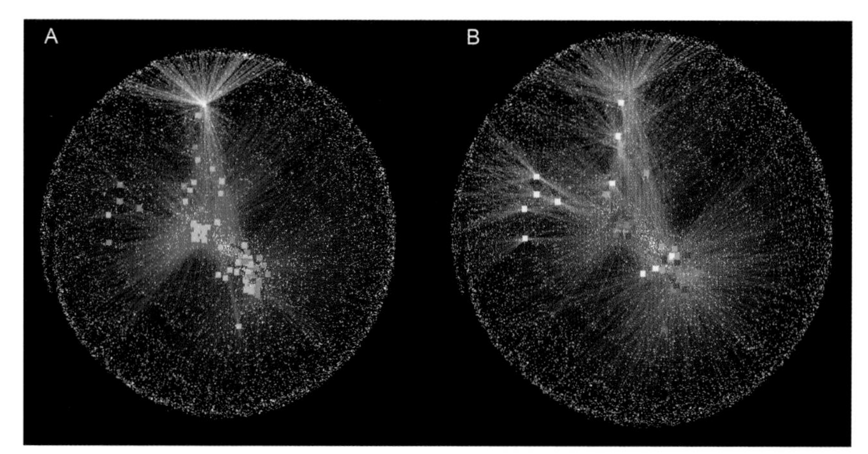

José A. Navas-Molina *et al.*, Figure 19.13 OTU-network bacterial community analysis applied in wild-type and transgenic mice. (A) Network colored by genotype (wild type: blue; transgenic: red). Control sample (yellow dot) is external in the network and several OTUs are not shared with mice. Although we can see some degree of clustering, discrimination by genotypes is difficult to assess. (B) Network colored by body site (mouth: yellow; skin: red; ileum: blue; colon: pink; cecum: orange; feces: brown; and multitissue samples: green). A control sample is colored in gray.

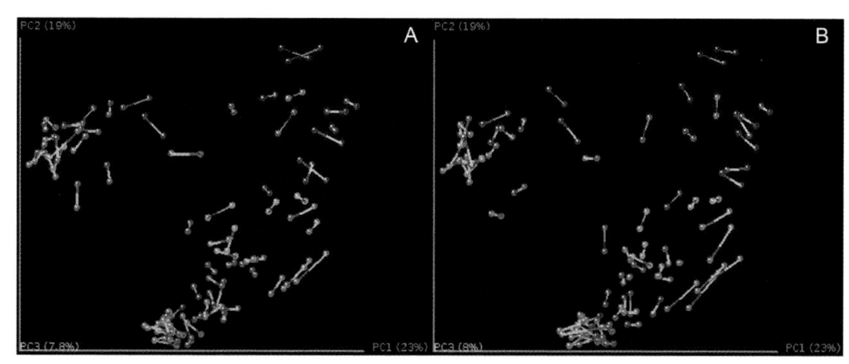

José A. Navas-Molina *et al.*, Figure 19.18 Procrustes analysis of different picking algorithms, where we can see that different OTU-clustering methods yield similar PCoA distributions. PCoA plots are colored by BODY_HABITAT. (A) Comparing samples with clusters picked using the *de novo* picking protocol against the closed-reference. (B) Comparing samples with clusters picked using the open-reference picking protocol against the closed-reference.

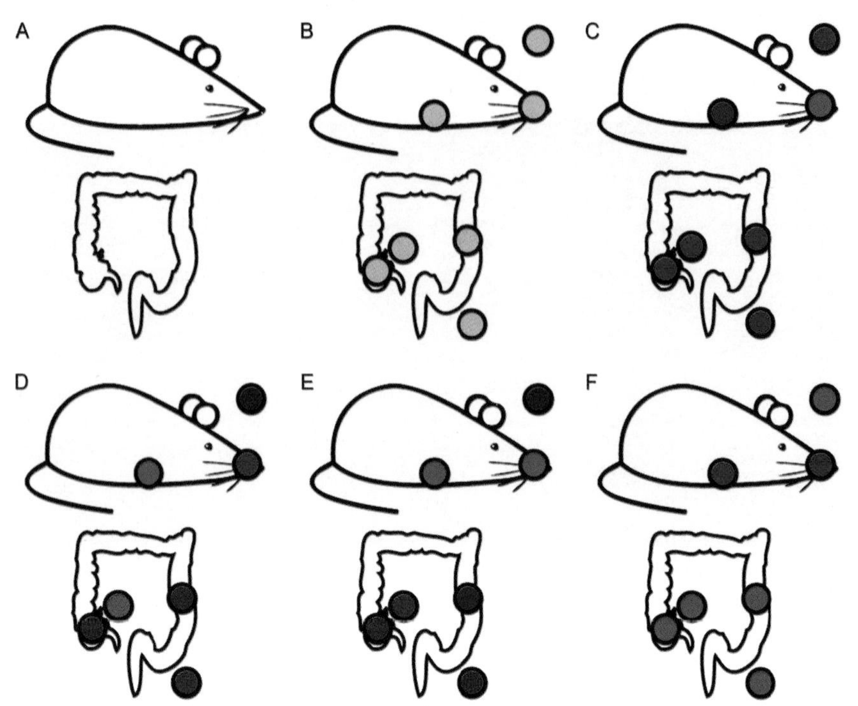

José A. Navas-Molina *et al.*, Figure 19.19 Image representing the mouse and its gastrointestinal tract. (A) Raw image without samples. (B) Image in SitePainter with samples. (C) and (D) PCoA axis 1 and 2, in red high values, in blue low values, similar colors represent similar communities. (E) and (F) Taxonomic distributions of (E) Betaproteobacteria and (F) Gammaproteobacteria, in red high abundance, in blue low abundance.

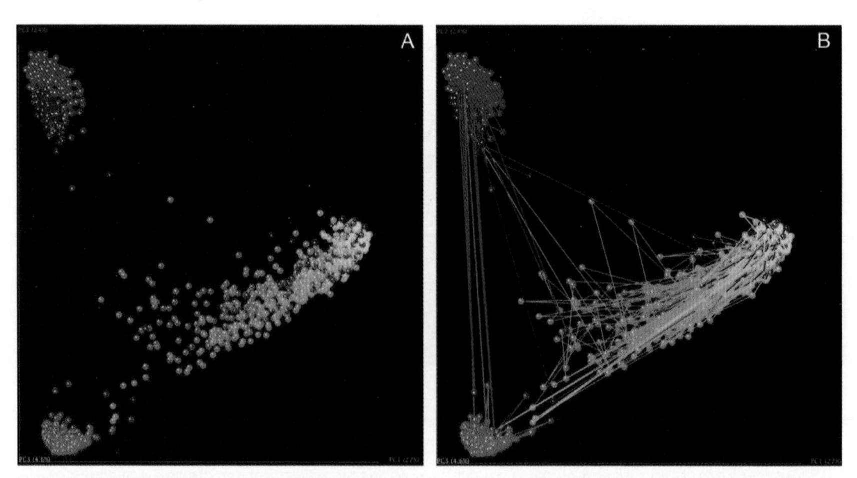

José A. Navas-Molina *et al.*, Figure 19.20 Beta-diversity plots for the moving pictures dataset using unweighted UniFrac as the dissimilarity metric (Caporaso et al., 2011). (A) PCoA plot colored by the body site and subject. (B) PCoA plot colored by the body site and subject with connecting lines between samples. Note in (B) that these lines allow us to track the individual body sites with a different approach.

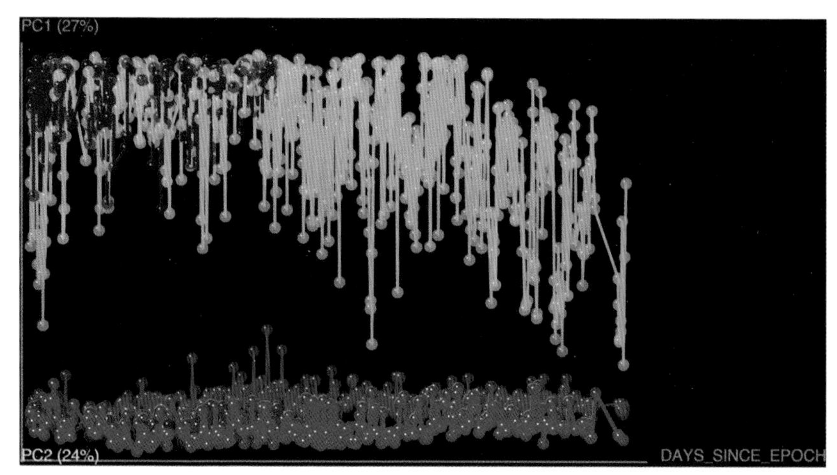

José A. Navas-Molina *et al.*, Figure 19.21 Three-dimensional plots in which two of the axes are PC1 and PC2 and the other is the day when that sample was collected in reference to the epoch time. Although this is not explicitly a beta-diversity plot, this representation allows differentiation of the individual trajectories over time.

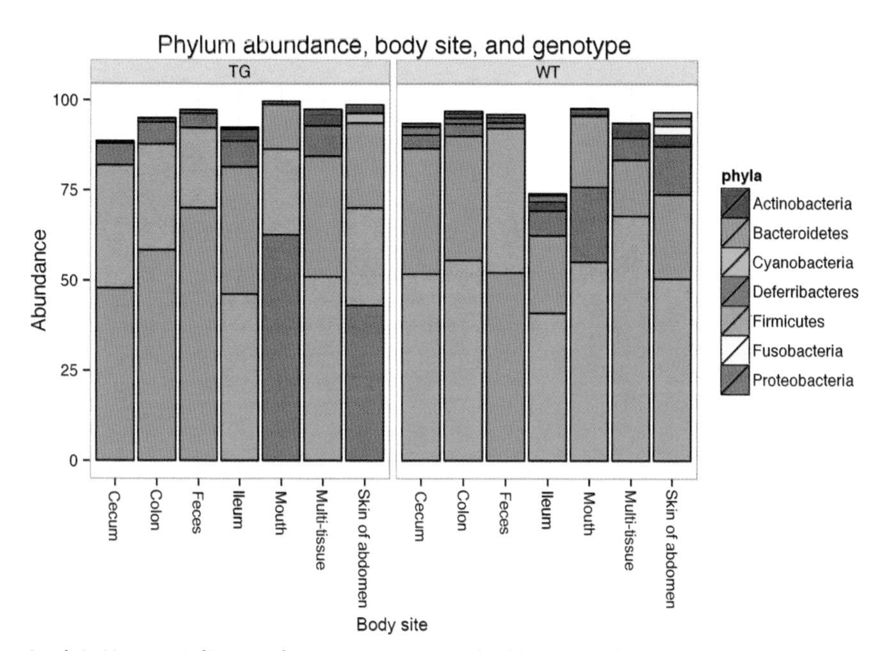

José A. Navas-Molina *et al.*, Figure 19.23 Stacked bar plot of the abundance values in the open-reference dataset. The bars are shaded according to phyla with each rectangle representing the relative abundance of a phylum in a particular sample group. The OTU rectangle in each stack is ordered according to abundance. The horizontal and vertical axes indicate the body site of the samples and the average fractional abundance of the OTU within the sample group, respectively. The separate panels "TG" and "WT" indicate the mouse genotype, achieved automatically by the `facet_grid(~GENOTYPE)` layer in the command.

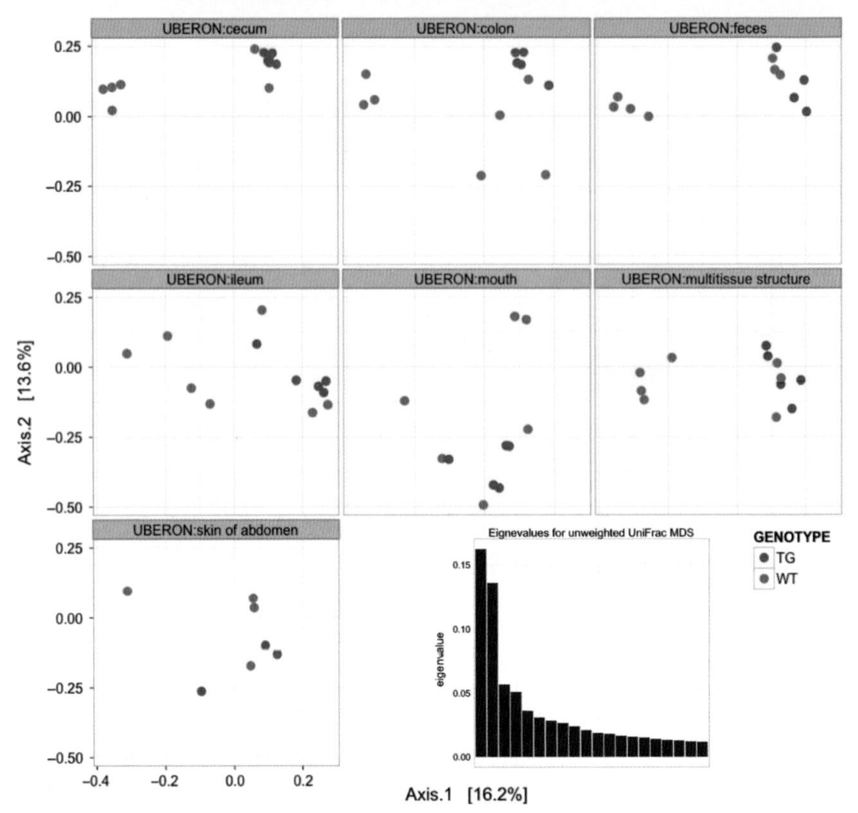

José A. Navas-Molina *et al.*, Figure 19.25 MDS ordination results on the unweighted UniFrac distances of the open-reference dataset. The samples are separated into different panels according to body site and shaded red or blue if they were from transgenic or wild-type mice, respectively. The horizontal and vertical axis of each panel represents the first and second axis of the ordination, respectively, with the relative fraction of variability indicated in brackets. (Inset) A scree plot showing the distribution of eigenvalues associated with each ordination axis.

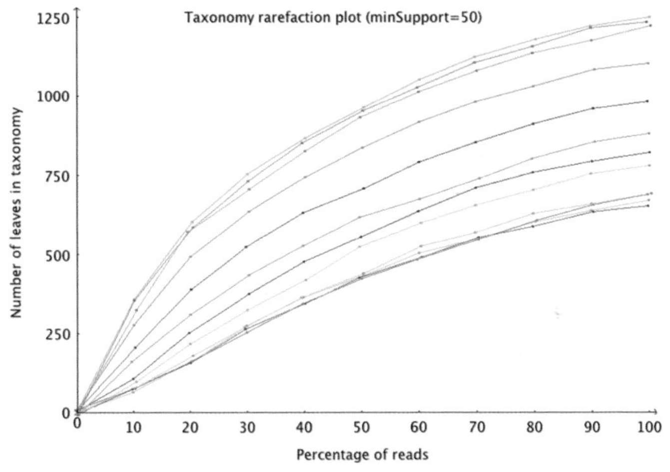

Daniel H. Huson and Nico Weber, Figure 21.3 Rarefaction curve for taxonomic assignment of all 12 samples at the taxonomic rank of genus. MEGAN generates this plot by randomly and repeatedly subsampling 10%, 20%, ... 100% of the reads and counting the number of leaves in the resulting graph. Such curves are used to estimate how many additional taxa may be discovered if one increases the amount of sequencing by a certain amount.

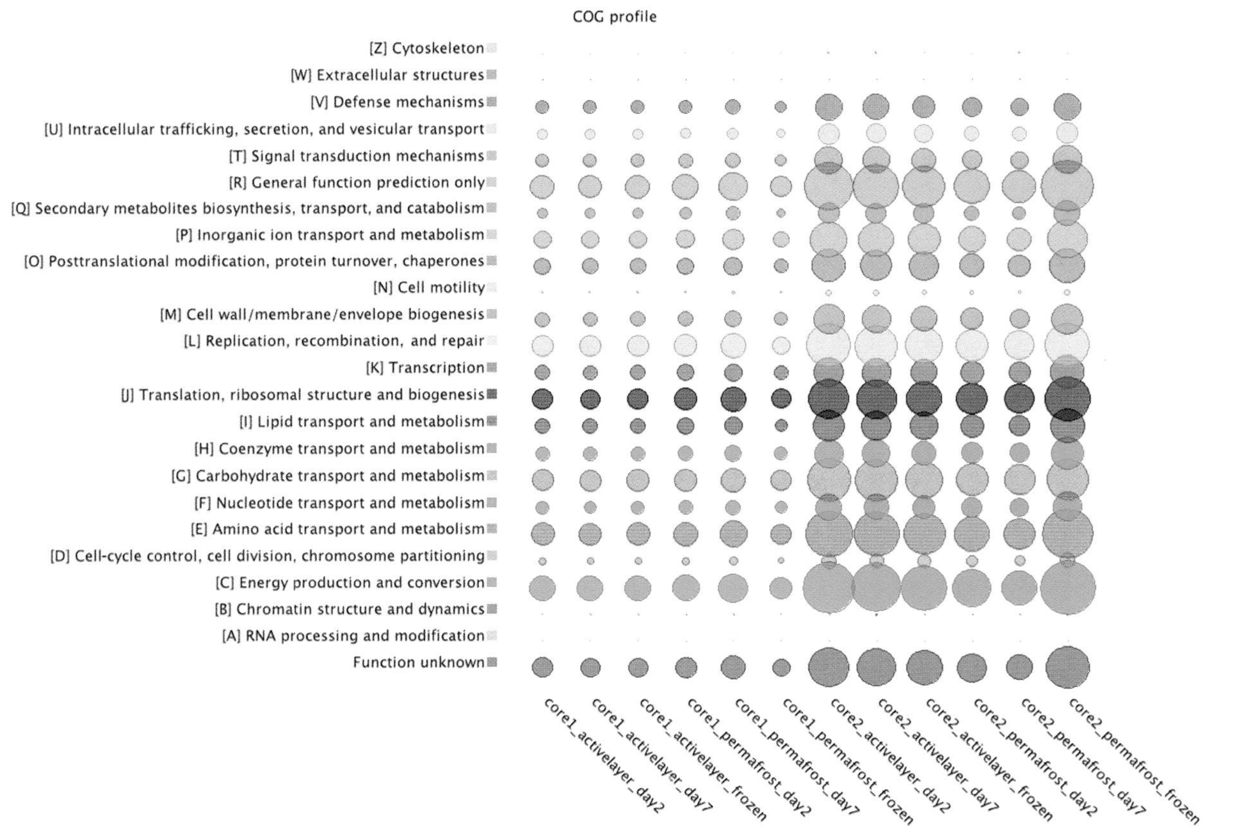

Daniel H. Huson and Nico Weber, Figure 21.6 Visual output of the COG analyzer. The information is charted here as a bubble chart. Samples are displayed along the *X*-axis. The *Y*-axis represents different COG categories. Nodes represent the number of assigned reads and can be scaled linearly by square root or logarithmically.

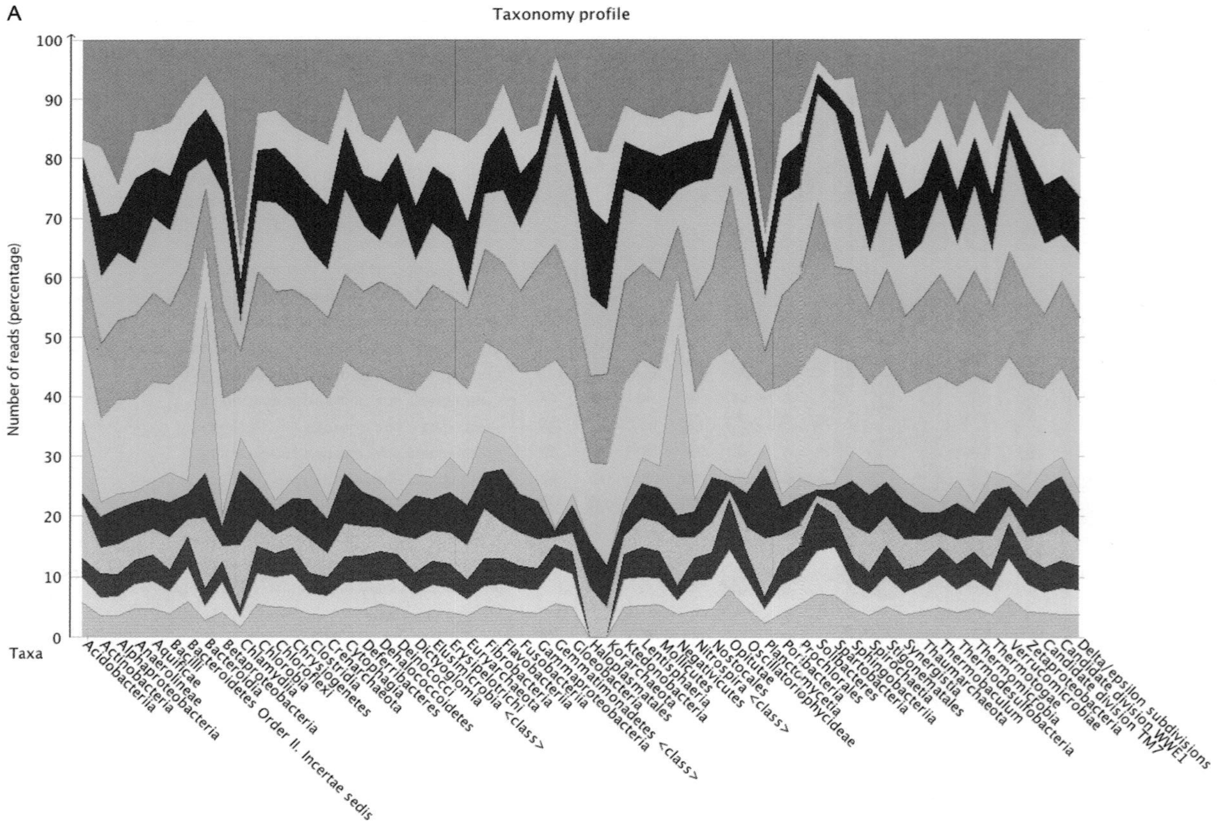

A

Taxonomy profile

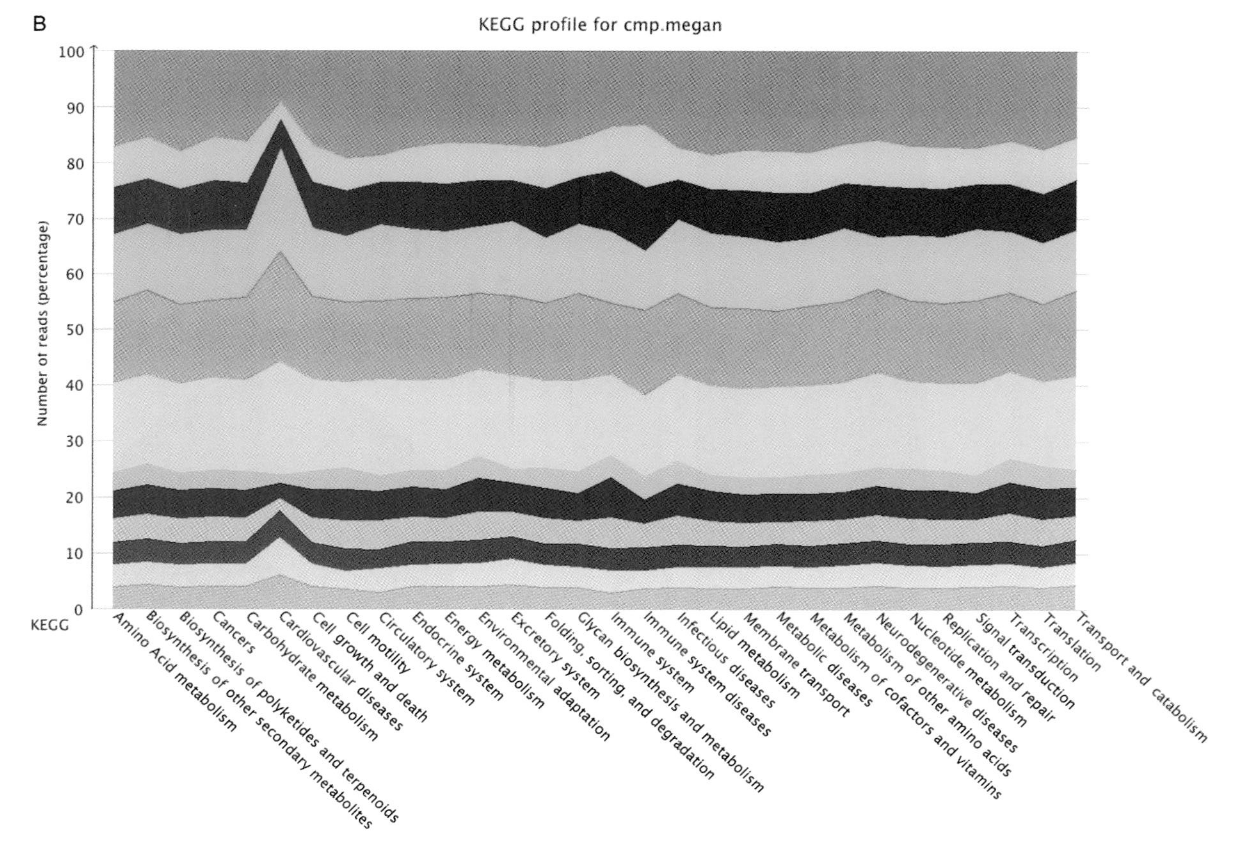

Daniel H. Huson and Nico Weber, Figure 21.8 Relative abundance of the taxonomic content (A) and the KEGG functional content (B) of multiple samples. The taxonomic content is much more unstable and therefore more difficult to interpret. Function on the other hand is more conserved over multiple samples. Significant changes may therefore be easier to detect, because overall noise is less present.

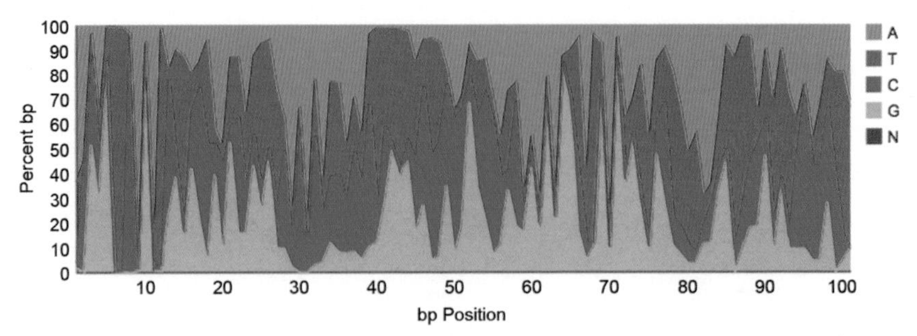

Andreas Wilke *et al.*, Figure 22.3 Nucleotide histogram with biased distributions.

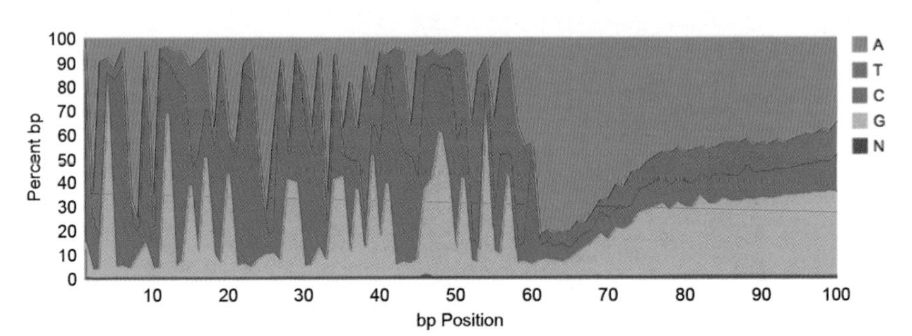

Andreas Wilke *et al.*, Figure 22.6 Nucleotide histogram with contamination.

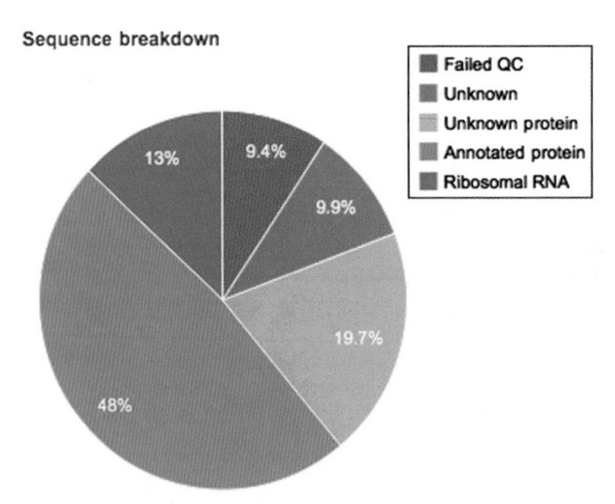

Andreas Wilke *et al.*, Figure 22.10 Sequences to the pipeline are classified into one of the five categories: gray = failed the QC, red = unknown sequences, yellow = unknown function but protein coding, green = protein coding with known function, and blue = ribosomal RNA. For this example, over 50% of sequences were either filtered by QC or failed to be recognized as either protein coding or ribosomal.

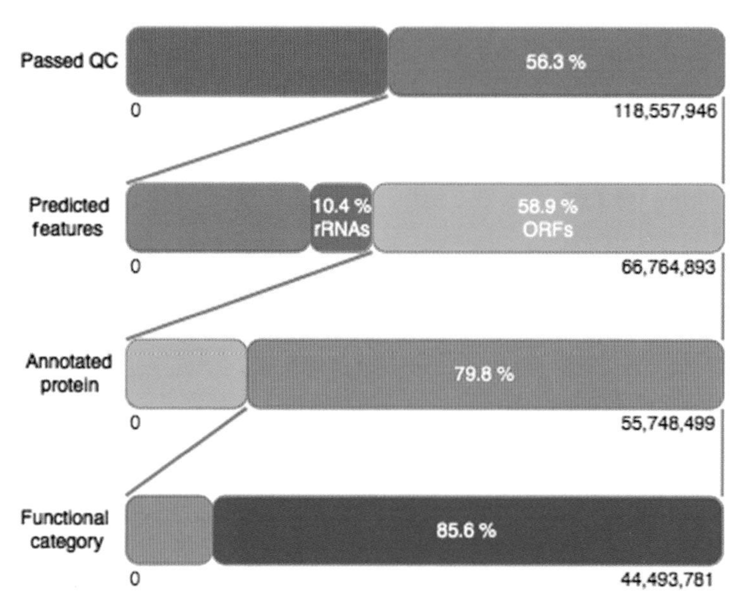

Andreas Wilke *et al.*, Figure 22.12 The analysis flowchart provides an overview of the fractions of sequences "surviving" the various steps of the automated analysis. In this case, about 20% of sequences were filtered during quality control. From the remaining 37,122,128 sequences, 53.5% were predicted to be protein coding, 5.5% hit ribosomal RNA. From the predicted proteins, 76.8% could be annotated with a putative protein function. Out of 32 million annotated proteins, 24 million have been assigned to a functional classification (SEED, COG, EggNOG, KEGG), representing 84% of the reads.

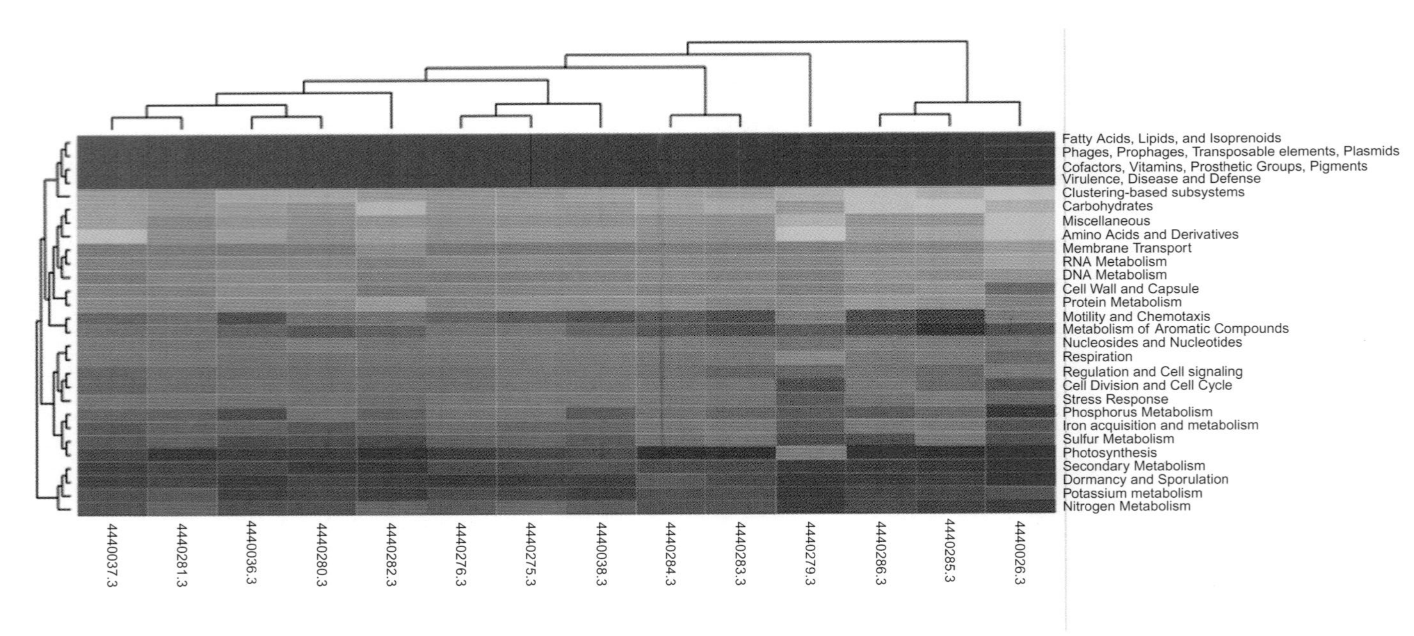

Andreas Wilke *et al.*, Figure 22.16 Heatmap/dendrogram example in MG-RAST. The MG-RAST heatmap/dendrogram has two dendrograms, one indicating the similarity/dissimilarity among metagenomic samples (*x*-axis dendrogram) and another to indicate the similarity/dissimilarity among annotation categories (e.g., functional roles; the *y*-axis dendrogram).